VOLUME TWO

DEVELOPMENTS IN MARINE GEOLOGY

ARCTIC OCEAN SEDIMENTS: PROCESSES, PROXIES, AND PALEOENVIRONMENT

VOLUME TWO

DEVELOPMENTS IN MARINE GEOLOGY

ARCTIC OCEAN SEDIMENTS: PROCESSES, PROXIES, AND PALEOENVIRONMENT

By

RUEDIGER STEIN

Alfred Wegener Institute for Polar and Marine Research, Bremerhaven, Germany

ELSEVIER

Amsterdam • Boston • Heidelberg • London • New York • Oxford

Paris • San Diego • San Francisco • Singapore • Sydney • Tokyo

Elsevier
Radarweg 29, PO Box 211, 1000 AE Amsterdam, The Netherlands
The Boulevard, Langford Lane, Kidlington, Oxford OX5 1GB, UK

First edition 2008

British Library Cataloguing in Publication Data
A catalogue record for this book is available from the British Library

Library of Congress Cataloging-in-Publication Data
A catalog record for this book is available from the Library of Congress

ISBN: 978-0-444-52018-0
ISSN: 1572-5480

Printed and bound by CPI Group (UK) Ltd, Croydon, CR0 4YY

Transferred to Digital Print 2011

For

Annika, Jan, Hauke, Nils, and Anna-Lena,
and Kirsten-ule

Contents

Preface

Although it is generally accepted that the Arctic Ocean is a very sensitive and important region for changes in the global climate, this region is one of the last major physiographic provinces of the earth whose short- and long-term geological history is not very well known. Since the first recovery and description of deep-sea sediments during the famous 1893–1896 *Fram*-Expedition of Fridtjof Nansen (Nansen, 1897; Böggild, 1906), the progress in getting a better understanding of the Arctic Ocean system and its relationship to global change has been slow in comparison to studies in other ocean regions. This lack of knowledge is mainly caused by the major technological/logistic problems in reaching this permanently ice-covered region with normal research vessels and in retrieving long and undisturbed sediment cores. Prior to 1990, the available samples and geological data from the central Arctic Basins are derived mainly from drifting ice islands such as T-3 (e.g., Clark et al., 1980) and CESAR (Jackson et al., 1985), and a few ships expeditions such as *Ymer*-80 (Boström & Thiede, 1984), *Polarstern* ARK-IV/3 (Thiede, 1988), and *Polar Star*-1988 (Phillips et al., 1992). Comprehensive summaries about the knowledge on Arctic Ocean geology based on data available prior to 1990 were published in Herman (1989), Bleil and Thiede (1990), and Grantz et al. (1990).

In the following years, several international and multidisciplinary expeditions were carried out, for example, the *Polarstern/Oden* Expedition in 1991 (Fütterer, 1992), the *Polar Sea* Expedition in 1993 (Grantz et al., 1998), the *Louis St. Laurent/Polar Sea* Expedition in 1994 (Wheeler, 1997), the *Polarstern* expeditions in 1998, 2001, 2004, and 2007 (Jokat, 1999; Thiede, 2002; Stein, 2005; Schauer, 2008), and the *Akademik Federov* Expedition in 2000 (Kaban'kov et al., 2004), the *Oden* expeditions in 2001 and 2007 (Grönlund, 2001; Jakobsson, Polyak, & Darby, 2007b), and the *Healy/Oden* Expedition in 2005 (Darby et al., 2005). Furthermore, major multidisciplinary circum-Arctic research programmes and initiatives were developed, for example, SEARCH (The Study of Environmental Change; Morison et al., 2001), SBI (Arctic Shelf-Basin Interactions; Grebmeier & Harvey, 2005), APARD (Arctic Paleo-River Discharge; Stein, 1998), and ACD (Arctic Coastal Dynamics; Rachold, Are, Atkinson, Cherkashov, & Solomon, 2004b).

Prior to 2004, the geological sampling in the Arctic Ocean was restricted to obtaining near-surface sediments, that is, only the upper 15 m could be sampled by means of gravity and piston coring. Thus, all studies were restricted to the Late Pliocene-Quaternary time interval, with one exception. In four short sediment cores from Alpha Ridge where older strata are cropping-out, upper Cretaceous and lower Tertiary sediments could be sampled by gravity coring from an ice flow (see Chapter 7). That means, the old pre-Pliocene palaeoenvironmental history of the central Arctic Ocean was almost unknown. This situation changed with the first scientific drilling on Lomonosov Ridge, which was carried out in August/September 2004 within the framework of the Integrated Ocean Drilling Program (IODP). During this IODP–ACEX (Arctic Coring Expedition) expedition, which is a break-through for palaeoenvironmental research, a more than 400 m thick sedimentary sequence of Neogene, Palaeogene, and Campanian could be drilled successfully (Backman et al., 2006; Moran et al., 2006). This record will allow for the first time detailed multidisciplinary studies of the early Arctic Ocean history and its change from greenhouse to icehouse conditions.

During the past about two decades, numerous multidisciplinary studies of the data and sediment material collected during these ship expeditions and carried out in international cooperation have greatly advanced (and will further advance) our knowledge on Arctic Ocean palaeoenvironment and its variability through Cenozoic times. A comprehensive compilation of data on Artic Ocean palaeoenvironment and its short- and long-term variability based on the huge amount of new data including the ACEX drilling data, however, has not been available yet. Thus, the main scope of this book is to (partly) fill this gap in knowledge.

The book is divided into three parts: Part I (Chapters 1 and 2) gives (i) a short introduction into the Arctic Ocean system including the alarming story of recent Arctic climate change, the history of Arctic research, the tectonic evolution, and the glacial history of the High Latitudes, and (ii) a description of the characteristics of the modern Arctic Ocean. Part II (Chapters 3 and 4) is dealing with (i) glacio-marine sedimentary processes and (ii) marine-geological proxies and methods used for (palaeo-) environmental reconstructions in the Arctic Ocean. Some more general background information as well as examples of using these proxies in Arctic Ocean environmental studies are given. It should be mentioned that some of the described proxies have their main strength outside the polar ice-covered regions. Nevertheless, they are included here because they yield important information about (palaeo-) environmental conditions in the marginal ice-covered (Subarctic) zones or are used for palaeoenvironmental reconstructions of the old pre-glacial Arctic Ocean. A more detailed discussion of selected proxies is then included in Part III of this book (Chapters 5–7) where results of case studies dealing with reconstructions of modern and ancient Arctic Ocean environment are presented. Concerning Part III, I concentrate on themes such as, for example, the modern and ancient terrigenous input, the Quaternary glacial history as reflected in marine sediment cores, the organic-carbon record and its palaeoenvironmental significance, and the long-term climate history as reflected in the ACEX record, that is, themes in which I am personally more involved. Thus, neither the spatial coverage of records nor the major themes discussed in this book can be regarded as complete.

Although major progress in Arctic Ocean research has been made during the past decades, the knowledge of its short- and long-term palaeoceanographic and palaeoclimatic history as well as its plate-tectonic evolution is still behind that from the other world's oceans and coordinated multidisciplinary research projects are needed. In the final chapter of this book (Chapter 8), some key objectives of future Arctic geoscientific research are shortly presented and discussed.

ACKNOWLEDGEMENTS

I am very grateful to Jan Backman, Hans Brumsack, Hajo Eicken, Kirsten Fahl, Hannes Grobe, Robie Macdonald, Jens Hefter, Jens Matthiessen, Frank Niessen, Rainer Sieger, Robert Spielhagen, Kristen St. John, and Jutta Wollenburg for discussions and material, and/or for their efforts in offering many suggestions for improvement of the manuscript. Many of the data and interpretations included in this book were obtained by my former PhD students Marion Behrends, Daniel Birgel, Bettina Boucsein, Klaus Dittmers, Christoph Kierdorf, Jochen Knies, Matthias Kraus, Uwe Langrock, Claudia Müller, Seung-il Nam, Frank Schoster, Carsten Schubert, Christoph Vogt, Petra Weller, and Daniel Winkelmann, and published in numerous joint papers. Many thanks to all of them. Part of these studies was funded by the German Research Foundation (DFG) and the German Ministry of Education, Science, Research and Technology, which is gratefully acknowledged. I would also like to thank Dieter K. Fütterer, former head of the AWI Geology Department, who gave me 99% freedom to do my Arctic research over the years. Special thanks to Hervé Chamley, editor of the Elsevier Series devoted to *Developments in Marine Geology*, for having invited me writing this book and for his support during all stages of this "project". Finally, I am most thankful to Anna-Lena, Nils, and Kirsten for their ongoing support and love at home — despite my almost 24-hours-per-day mental stay in the Arctic during the final stage of writing this book.

Ruediger Stein
Bremerhaven
April 2008

List of Abbreviations

ACEX	Arctic Coring Expedition
AMS	Acceleration Mass Spectrometry
AO	Arctic Oscillation
BG	Beaufort Gyre
BIT	Branched and Isoprenoid Tetraether index
Cal. yr. BP	Calendar years Before Present
CCD	Carbonate compensation depth
DOC	Dissolved organic carbon
DSDP	Deep Sea Drilling Project
GDGT	Glycerol dialkyl glycerol tetraether
GIN Sea	Greenland–Iceland–Norwegian Sea
GLAMAP	Glacial Atlantic Ocean Mapping
HBI	Highly branched isoprenoid
HC	Hydrocarbons
IBCAO	International Bathymetric Chart of the Arctic Ocean
ICAM	International Conference on Arctic Margins
IODP	Integrated Ocean Drilling Program
IPY	International Polar Year
IRD	Ice-rafted debris
LGM	Last Glacial Maximum
LIS	Laurentide Ice Sheet
MIS	Marine Isotope Stage
MIZ	Marginal Ice Zone
MWP	Meltwater pulse
NADW	North Atlantic Deep Water
NAO	North Atlantic Oscillation
NGS	Norwegian-Greenland Seaway
NHG	Northern Hemisphere Glaciation
OAE	Oceanic Anoxic Event
OC	Organic carbon
ODP	Ocean Drilling Program
PETM	Palaeocene–Eocene Thermal Maximum
POC	Particulate organic carbon
PONAM	Polar North Atlantic Margin
QUAX	Quantitative phase analysis by X-ray diffraction
QUEEN	Quaternary Environment in the Eurasian North
SBIS	Svalbard-Barents Sea Ice Sheet
SCICEX	SCience ICe EXercise
SEM	Scanning electronic microscopy

SIC	Sea-ice cover
SIS	Scandinavian Ice Sheet
SSS	Sea-surface salinity
SST	Sea-surface temperature
TEX	Tetraether index
TOC	Total organic carbon
TPD	Transpolar Drift
TSM	Total suspended matter
UCM	Unresolved complex mixture of hydrocarbons
WSC	West Spitsbergen Current
XRD	X-ray diffraction
YD	Younger Dryas

Part 1: Introduction and Background

CHAPTER ONE

Introduction to the Arctic: Significance and History

1.1. The Arctic Ocean and Its Significance for the Earth's Climate System

For several reasons, the Arctic Ocean (Figure 1.1) is unique in comparison to the other world oceans:

- it is surrounded by continents and the world's largest shelf seas, with limited connections to the Pacific and Atlantic Oceans via Bering Strait and Fram Strait, respectively, making the Arctic Ocean a "mediterranean" sea (Jakobsson, 2002);
- it is seasonally to permanently covered by sea ice;
- it is strongly influenced by huge river discharge which is equivalent to 10% of the global runoff (Aagaard & Carmack, 1989; Holmes et al., 2002);
- it has a strong seasonal forcing (runoff, ice formation, sunlight); and
- most of the terrestrial surface around the Arctic Ocean is occupied by permafrost.

These characteristics, which are outlined in more details in Chapter 2, have a large influence on the environment of the Arctic Ocean itself, the global Earth system, and climate change. The freshwater supply (see Chapter 2.5), for example, is essential for the maintenance of the low-salinity layer of the central Arctic Ocean and, thus, contributes significantly to the strong stratification of the near-surface water masses, encouraging sea-ice formation. Changes in the freshwater balance would influence the extend of sea-ice cover. The melting and freezing of sea ice result in distinct changes in the surface albedo, the energy balance, the temperature and salinity structure of the upper water masses, and the biological processes. Freshwater and sea ice are exported from the Arctic Ocean through Fram Strait into the North Atlantic. Changes in these export rates of freshwater would result in changes of North Atlantic as well as global oceanic circulation patterns. The interplay of the cold Arctic freshwater-rich surface-water layer and its ice cover with the relatively warm and saline Atlantic water is important for the renewal of deep waters driving the global thermohaline circulation (e.g., Broecker, 1997; Clark, Pisias, Stocker, & Weaver, 2002). Because factors such as the global thermohaline circulation, sea-ice cover and earth albedo have a strong influence on the earth's climate system, climate change in the Arctic could cause major perturbations in the global environment.

During the 1990s, it became widely recognized that the Arctic was undergoing dramatic change (Macdonald, 1996; Dickson et al., 2000; Morison,

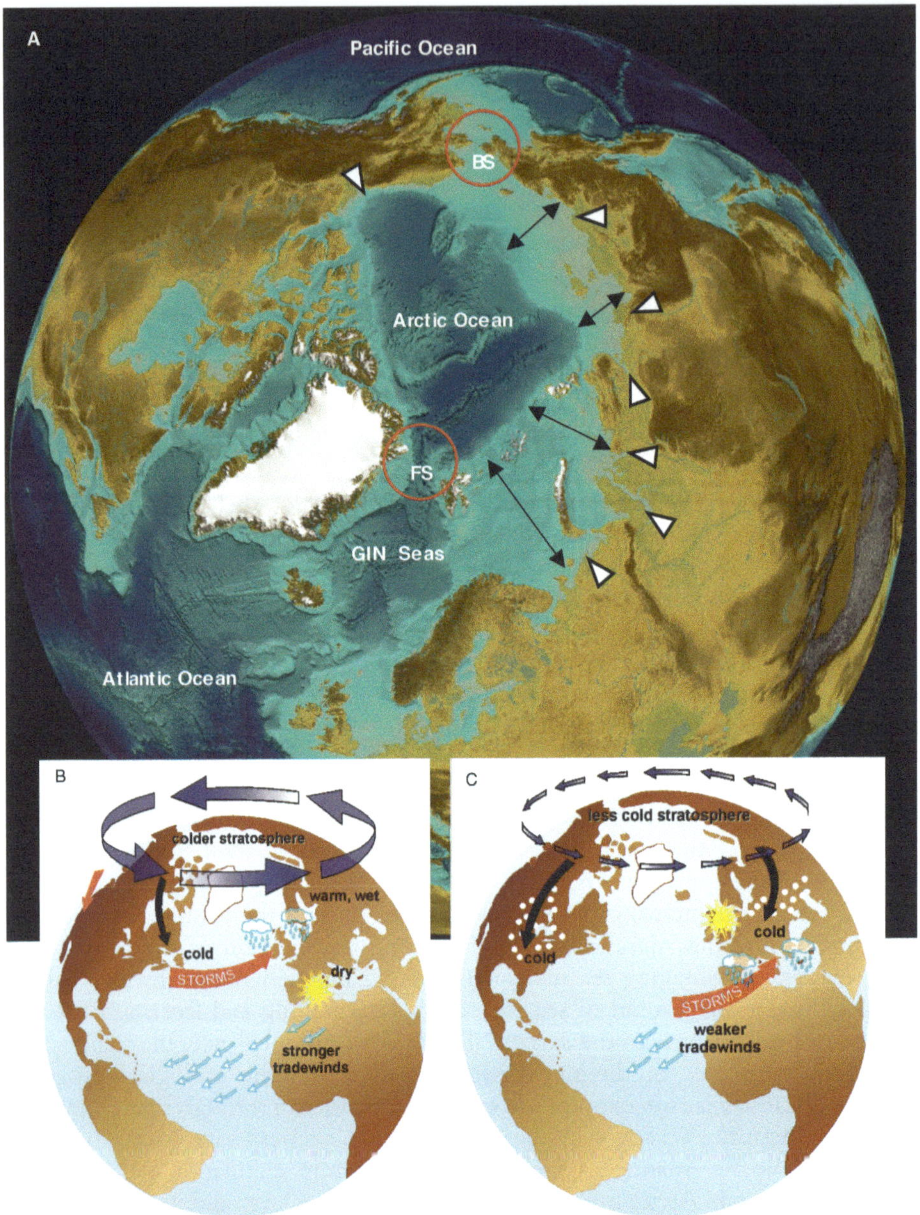

Figure 1.1 (A) Overview map of the Arctic Ocean (from Wille, 2005, supplemented). The connection to the Pacific and Atlantic Oceans, Bering Strait (BS), and Fram Strait (FS), respectively, loci of major riverine input (open triangles), and broad Eurasian shelf seas (black arrows) are highlighted. Schematic diagram of the polar vortex and North Atlantic storm tracks showing (B) the effect of the positive phase of the Arctic Oscillation and (C) the effect of the negative phase of the Arctic Oscillation (from Macdonald, Sakshaug, & Stein, 2004a; data source http://www-nsidc.colorado.edu/arcticmet/patterns/arctic.oscillation.html).

Aagaard, & Steele, 2000; Serreze et al., 2000; Moritz, Bitz, & Steig, 2002). Over the past decades, a significant increase in Siberian river discharge, associated with a warmer climate and enhanced precipitation in the river basins, has been observed (Semiletov et al., 2000; Serreze et al., 2000). At the same time, an increase in the amount and temperature of Atlantic water inflow into the Arctic, a reduced sea-ice cover, a thawing of permafrost, and a retreat of small Arctic glaciers are obvious (Dickson et al., 2000; Serreze et al., 2000). Reducing the sea-ice cover (see Chapter 2.3) causes a reduced albedo effect, and thawing of permafrost (see Chapter 2.6) may release greenhouse gases into the atmosphere, both positive feedbacks to further warming. Furthermore, increased glacial melt and river runoff add more freshwater to the ocean, raising global sea level and possibly slowing the global thermohaline circulation.

The Arctic sea ice, a key indicator and agent of climate change, affecting surface reflectivity, cloudiness, humidity, exchanges of heat and moisture at the ocean surface, and ocean currents, had been undergoing retreat over the past three decades (Figure 1.2), as recognized by the science community with some alarm (e.g., Johannessen et al., 2004; ACIA, 2004, 2005; Francis, Hunter, Key, & Wang, 2005; Serreze, Holland, & Stroeve, 2007; Stroeve, Holland, Meier, Scambos, & Serreze, 2007). Observed changes not only included a reduction in total area covered by sea ice (Maslanik, Serreze, & Barry, 1996; Johannessen, Shalina, & Miles, 1999; Johannessen et al., 2004; Parkinson, Cavalieri, Gloersen, Zwally, & Comiso, 1999; Vinnikov et al., 1999; Levi, 2000), but also an increase in the length of the ice melt season (Smith, 1998; Stabeno & Overland, 2001; Rigor, Wallace, & Colony, 2002), a loss of multiyear ice (Nghiem et al., 2007), and a general decrease in the thickness of ice over the central Arctic Ocean (Rothrock, Yu, & Maykut, 1999) (see Chapter 2.3 for more details).

On the basis of global coupled atmosphere–ice–ocean climate model simulations, Johannessen et al. (2004) predicted that the summer ice cover may be reduced by ~80% at the end of this century, i.e., the Arctic Ocean may nearly become ice-free during summer (Figure 1.3, lower records). The decrease in the summer ice cover is much greater than in the winter (Figure 1.3, upper records), or annual means modelled previously (Vinnikov et al., 1999; Johannessen, Shalina, Kuzmina, Miles, & Nagurnyi, 2001). The reduction of future sea ice may be even more rapid: an alarming record-low in minimum sea-ice cover was observed in September 2007, which is ~40% less than that of 1979, the start of sea-ice observation by satellites (Figure 1.2; Kerr, 2007). Such a minimum cover was forecasted by modelling to occur in the middle of this century (Figure 1.3; Johannessen et al., 2004).

All these environmental changes described above, seem to be related to a cyclic variation of the Northern Hemisphere/Arctic atmospheric circulation pattern, i.e., the "Arctic Oscillation (AO)" and the "North Atlantic Oscillation (NAO)" (Hurrell, 1995; Thompson & Wallace, 1998; Dickson et al., 2000; Peterson et al., 2002). With respect to changes in Eurasian arctic river discharge, Peterson et al. (2002) showed strong correlations with both global surface air temperature and the NAO index (Figure 1.4). During a positive NAO/AO Index phase, warm wet air is brought to Northern Europe and the Russian Arctic (Figure 1.1), causing increased precipitation in the drainage area of the Siberian rivers, which result in increased

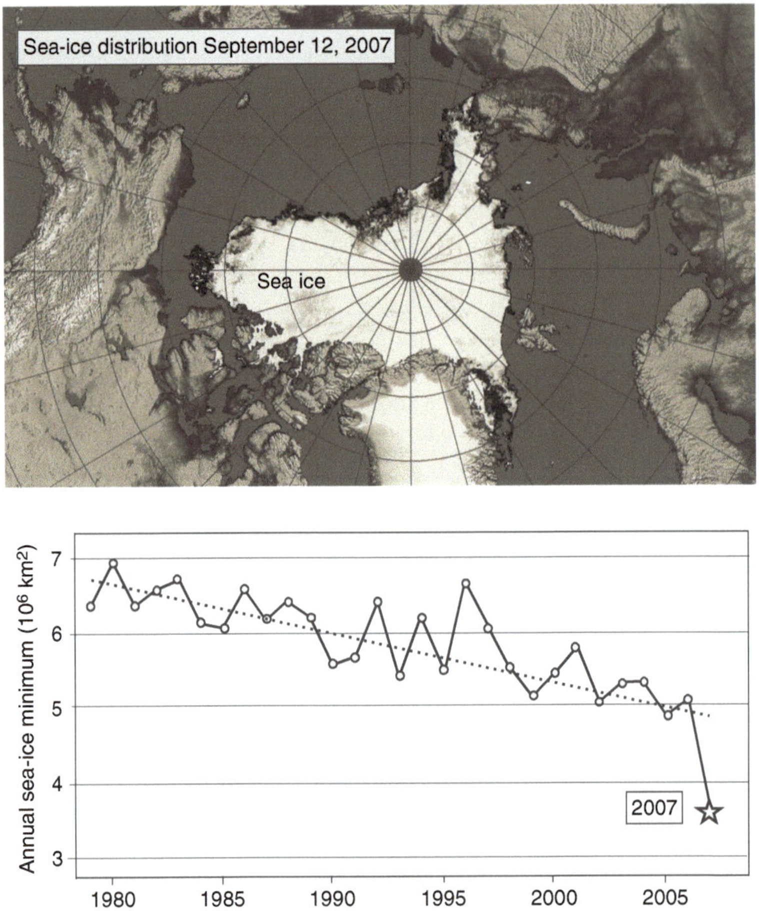

Figure 1.2 Distribution of sea-ice cover in the Arctic Ocean on 12 September 2007 (data source http://iup.physik.uni-bremen.de:8084/amsr/amsre.html), and a 1979–2007 record of annual Arctic sea-ice minimum Arctic sea ice using NASA's measurement techniques (redrawn from Kerr, 2007). The new record-low of September 2007 is marked by an open star.

river discharge to the Arctic Ocean. Details about the interaction between NAO/AO and climate change, however, are still under discussion.

Concerns about global warming have stimulated a wide range of polar research. This research is motivated in part because climate change models predict greater future temperature changes in polar regions. Based on the report of the Intergovernmental Panel on Climate Change (IPCC, 2001, 2007), the winter warming of the High Northern Latitudes by the end of the century will be at least 40% greater than the global mean, based on a number of models and emissions scenarios, while the warming predicted for the central Arctic is ~3–4°C during the next 50 years, or more than twice the global mean (Johannessen et al., 2004). Furthermore, models suggest that the Arctic will respond very sensitively to and amplify the effects of global climate change (e.g., Walsh, 1991; Serreze et al., 2000).

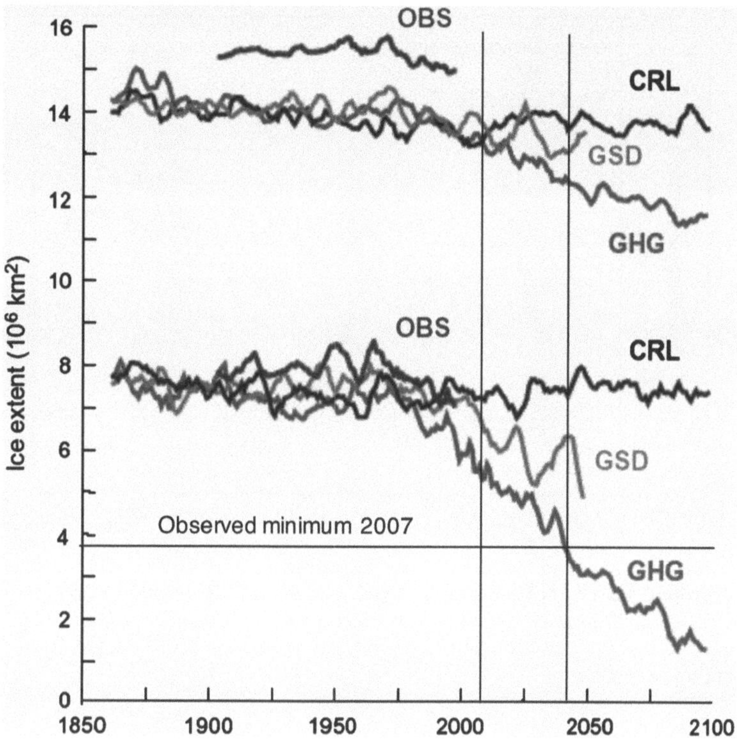

Figure 1.3 Observed (OBS) and modelled (using the coupled atmosphere–ice–ocean climate model ECHAM4 of the Max Planck Institute for Meteorology; Roeckner, Bengtsson, Feichter, Lelieveld, & Rodhe, 1999) Northern Hemisphere sea-ice extent in late winter (March) and late summer (September) for the period 1860–2100. The modelled scenarios include a control run (CRL) and changes in greenhouse gases (GHGs) and GHGs plus sulfate aerosols (GSD) from the IPCC IS92 emissions scenarios (Scenario B2) [Emission scenario B2 (IPCC, 2001) assumes "a world concerned with environmental protection and social equity, with solutions focused at the local and regional levels. It is a world in which global population grows to reach 10.4 billion by 2100, there is an intermediate level of economic development, and there is diverse technological change around the world. In a B2 world, by the year 2100, coal supplies 22% of the primary energy, and 49% of the world's energy is derived from sources that emit no carbon dioxide" (ACIA, 2004, 2005).] (from Johannessen et al., 2004, supplemented). In winter (upper records), the difference between the observed and modelled ice extent is due to seasonal ice in the peripheral seas and bays outside the model domain. The observed record-low of September 2007 is indicated (see Figure 1.2).

Although there is a general consensus that the polar regions — and especially the Arctic Ocean and surrounding areas — are (in real time) and have been (over historic and geologic time scales) subject to rapid and dramatic change, the causes of the recent changes are a subject of intense scientific and environmental debate. As outlined by Johannessen et al. (2004), it remains open to debate whether the warming in recent decades is an enhanced greenhouse-warming signal or (at least partly) natural decadal and multidecadal climate variability (Polyakov & Johnson, 2000;

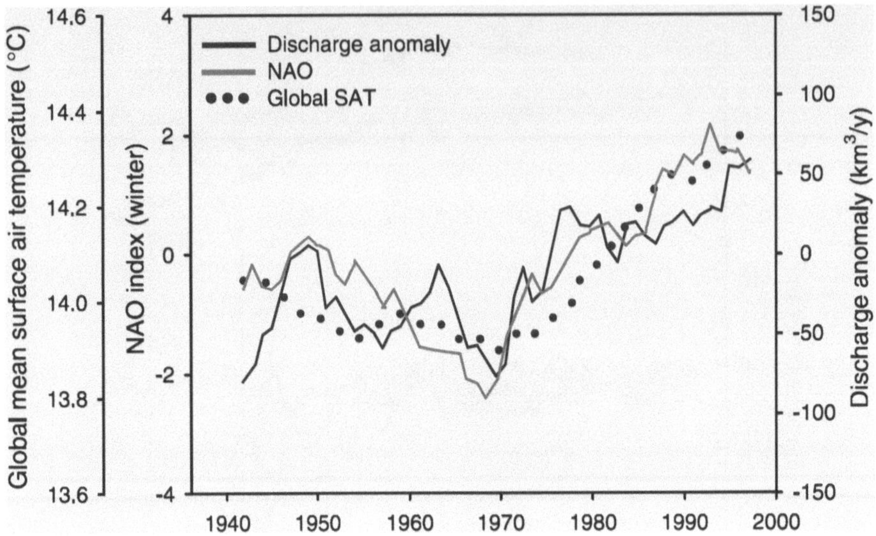

Figure 1.4 Ten-year running averages of the Eurasian arctic river discharge anomaly, winter (December through March) NAO index, and global mean SAT from 1936 to 1999 (redrawn from Peterson et al., 2002). "Anomaly" refers to variation from the long-term mean. NAO data were taken from www.cgd.ucar.edu/jhurrell/nao.html, and temperature data are from www.giss.nasa.gov/data/update/gistemp/

Polyakov, Johnson, Colony, Bhatt, & Alekseev, 2002). The latter is possibly expressed by the Arctic warming observed in the 1920s and 1930s followed by cooling until the 1960s (e.g., Kelly, Jones, Sear, Cherry, & Tavakol, 1982). Based on observational and modelling analysis, however, Johannessen et al. (2004) concluded that neither the warming trend nor the decrease of sea-ice extent and volume over the past two decades can be explained by natural processes alone. The uncertainties in statements related to causes of recent climate change and its extrapolation into the future are in part related to the lack of homogeneous, century-scale instrumental data sets needed to resolve the inherent time scales of variability in the Arctic (Venegas & Mysak, 2000), a region characterized by high variability. Here, high-resolution palaeoclimatic records going back beyond the time scale of direct measurements may help to solve some of the uncertainties in the debate of recent climate change.

In general, palaeoclimate records document the natural climate, rates of change, and variability prior to anthropogenic influence. The instrumental records of temperature, salinity, precipitation and other environmental observations span only a very short interval (<150 years) of Earth's climate history and provide an inadequate perspective of natural climate variability, as they are biased by an unknown amplitude of anthropogenic forcing. Palaeoclimate reconstructions, on the other hand, can be used to assess the sensitivity of the Earth's climate system to changes of different forcing parameters (e.g., CO_2) and to test the reliability of climate models by evaluating their simulations for conditions very different from the modern climate. A precise knowledge of past rates and scales of climate change are the only mode to separate natural and anthropogenic forcings and will enable us to further increase the reliability of prediction of future climate change. Thus, understanding the mechanisms of natural climate change is one of the major challenges for mankind in the coming years. In this context, the polar regions certainly play a key role, and detailed climate records from the Arctic Ocean spanning time intervals from the Palaeogene Greenhouse world to the Neogene-Quaternary Icehouse world will give new insight into the functioning of the Arctic Ocean within the global climate system. Not until we better understand the history and the role of the Arctic Ocean for global climate we can attempt to predict the future development of the Earth System (see also Chapter 8 for further discussion).

1.2. History of Arctic Ocean Research

A compilation of historical background data of the geographic exploration of the Arctic regions on land and in coastal zones, the exploration and mapping of the Arctic seafloor, and the geological and geophysical exploration of the Arctic Basin until the end of the 1980s was published by Weber and Roots (1990). In this chapter, some milestones of the Arctic research activities are shortly presented, concentrating on the geoscientific exploration of the Arctic Ocean and including the post-1990 activities.

1.2.1. Arctic Marginal Seas

Not counting the geographic exploration of the Arctic coastlines by fishermen and commercial traders, the first purely scientific expedition to the Arctic Ocean was the two-ship voyage led by the Hon. Captain C. J. Phipps in 1773, planned to reach the North Pole. Due to heavy ice conditions directly north of Spitsbergen, however, the ships were hindered to steam further to the north, and they explored the ice edge from near Greenland to White Island (Weber & Roots, 1990). As outlined in Weber and Roots (1990), main results of this expedition were the first scientific description of the polar bear (*Ursus maritimus* Phipps), the first measurements of the West Spitsbergen and East Greenland currents, the first temperature and salinity profiles and depth soundings of Arctic Ocean waters, and the measurement of the gravity difference between northernmost Spitsbergen and Greenwich.

Systematic investigations of the natural properties of the Arctic Ocean began in the 19th century. Investigations in the Arctic pack ice and under a hostile climate were extremely difficult given the existing ships at that time. The first detailed description of surface sediments of the Barents Sea was done by the Russian F. P. Litke in 1821–1824 during a voyage to the shores of Novaya Zemlya (Kulikov, Lapina, & Ivanov, 1999). It was the German Carl Koldewey who sailed to Fram Strait in 1868 to study the nature of the ice margin (Thiede, 1996). The Swedish–Russian *Vega* Expedition under the leadership of A. E. Nordenskjold

(1878–1879) sailed from Tromsø along the Siberian coast to Bering Strait (Nordenskjold, 1882). During this expedition, soundings were made along the entire track, establishing a shallow-draft shipping route from the White Sea to the Bering Sea, the later Northeast Passage (Weber & Roots, 1990). The Dutch *Varka* Expedition (1882–1883) recovered first information about the distribution of surface sediments and iron–manganese nodules in the southern Kara Sea (Kulikov et al., 1999).

A milestone in international scientific cooperation in the Arctic was the First International Polar Year (IPY) (1882–1883), initiated by the Austrian explorer and naval officer Lt. Karl Weyprecht in 1775 (Weber & Roots, 1990). The key concept of the First IPY was that geophysical phenomena could not be surveyed by one nation alone; rather, an undertaking of this magnitude would require a coordinated international effort. Twelve countries participated, and 15 expeditions to the poles were completed (13 to the Arctic and 2 to the Antarctic). Beyond the advances to science and geographical exploration, a principal legacy of the First IPY was setting a precedent for international science cooperation (http://www.ipy.org). This strategy has been later continued through the Second IPY (1932–1933) and the International Geophysical Year (GPY; 1957–1958) (Bartels et al., 1959; Weber & Roots, 1990).

Extensive research activities continued in the Barents, White, and Kara seas during the first half of the last century prior to the Second World War. Here, a systematic exploration of bottom sediments of the Russian Arctic began in the 1920s and large data sets on lithology, grain size, mineralogical and chemical compositions of the sediments were collected and published in the Russian literature (Samoilov & Gorshkova, 1924; Samoilov & Klenova, 1927; Gorshkova, 1931; see Kulikov et al., 1999 and references therein for summary). After the Second World War, the research in the Russian Arctic was again intensified, and first comprehensive monographs dealing with the surface sediment characteristics were published in the following decades (e.g., Saks, 1952; Kordikov, 1953; Lapina, 1959, 1965; Klenova, 1960; Belov & Lapina, 1961; Kulikov, 1961; Lapina, Belov, Kulikov, Semenov, & Spiridonov, 1970). Since the 1960s, also long sediment cores could be recovered by means of gravity and piston coring, allowing to study not only the Holocene, but also late Pleistocene sediments (Kulikov et al., 1999; Tarasov, Matishov, Samoilovich, & Kukina, 1999). Between 1962 and 1964, a large number of sediment cores were also collected in the Chukchi, East Siberian, and Laptev seas by the US Coast Guard research vessels *Northwind* and *Burton Island* (Naugler, Silverberg, & Creager, 1974). In the 1970s, geophysical investigations were introduced in the Russian shelf studies (Kulikov et al., 1999). In 1981, the "Atlas of grounds of the Arctic Ocean" was issued, including numerous maps of mineralogical and geochemical provinces as well as physiochemical properties of bottom sediments (Korotkevich, Egizarov, & Faleev, 1981).

With the development of oil and gas fields in the Russian Arctic shelf seas multidisciplinary studies of seafloor environments and geoecology were carried out in the 1990s. Between 1991 and 1994, for example, "VNIIOkeangeologia" St. Petersburg, together with branches of the research–industrial association "Sevmorgeologia" conducted cruises into the Kara Sea with the research vessels *Academik Alexander Karpinsky* (1991), *Geolog Fersman* (1992, 1993), and *Professor Logachev*

(1994) (Kulikov et al., 1999). During the multidisciplinary Kara Sea Expedition of RV *Dmitry Mendeleev* in 1993 (Lisitzin & Vinogradov, 1995), numerous new data and concepts on sedimentary processes and the ecosystems of the Kara Sea as well as the Ob and Yenisei estuaries were obtained and published (e.g., Lisitzin, 1995 and references therein).

Over the decades, a huge amount of sedimentological, geochemical, and palaeontological data were collected by Russian scientists. Most of the data, however, were not open for the western literature. Because of the political reorganization of the former Soviet Union at the end of the 1980s, Russian institutions were getting more and more involved in the international exploration of the Arctic and making their experience, technical facilities, and data available to the international scientific community (Figure 1.5). Numerous joint bilateral/international expeditions into the Eurasian continental margin and neighbouring deep-sea basins have been carried out over the following years. As an example, the extensive German–Russian research programmes "Laptev Sea System" (Kassens et al., 1999; Bauch & Kassens, 2005 and references therein) and "Siberian River Run-off — SIRRO" (Stein, Fahl, Fütterer, & Galimov, 2003b and references therein) with a large number of multidisciplinary expeditions in the Laptev Sea (1994–recent) and in the Kara Sea (1997–2003), respectively, are mentioned here. Overall objectives of these programmes include(d) studies of high-resolution variability of river discharge and its relationship to biological, (geo-) chemical, and geological processes, glacial to Holocene climate variability and permafrost dynamics. Results of the international multidisciplinary research in the Eurasian Arctic are discussed in Parts II and III of this book.

In the Canadian and Alaskan Arctic (Beaufort Sea, Chukchi Sea), numerous expeditions were also carried out since 1970, and a large number of oceanographic, geophysical, and geological data were collected (see Weber & Roots, 1990 for more detailed information on expeditions prior to 1990). More recently, major international multidisciplinary research programmes were initiated such as, for example, the Arctic Shelf-Basin Interactions (SBI) Program (Grebmeier & Harvey, 2005).

1.2.2. Central Arctic Ocean

In 1893, the door to central Arctic research was opened with the Norwegian Fridtjof Nansen's famous expedition (1893–1896) on his newly built and dedicated polar research vessel *Fram* into the eastern Arctic ice (Nansen, 1897, 1902, 1904). The drift of the *Fram* in the pack ice (see Chapter 3.2, Figure 3.9) was both the first scientific expedition to the central Arctic Ocean and a technical break-through. For the first time, the drift allowed the gathering of reliable scientific data from the interior ice-covered Arctic Ocean. Böggild (1906) described the first deep-sea sediment samples from the "North Polar Sea", which had been taken by Nansen's *Fram* expedition. Furthermore, Nansen's bathymetric map of the Arctic Ocean revealed a deep ocean basin surrounded by the large land masses of Eurasia, North America, and Greenland and connected to the North Atlantic by a deep-water passage.

The progress in central Arctic research since the *Fram* expedition (1893–1896) has been slow in comparison to studies in the Arctic marginal seas as well as other

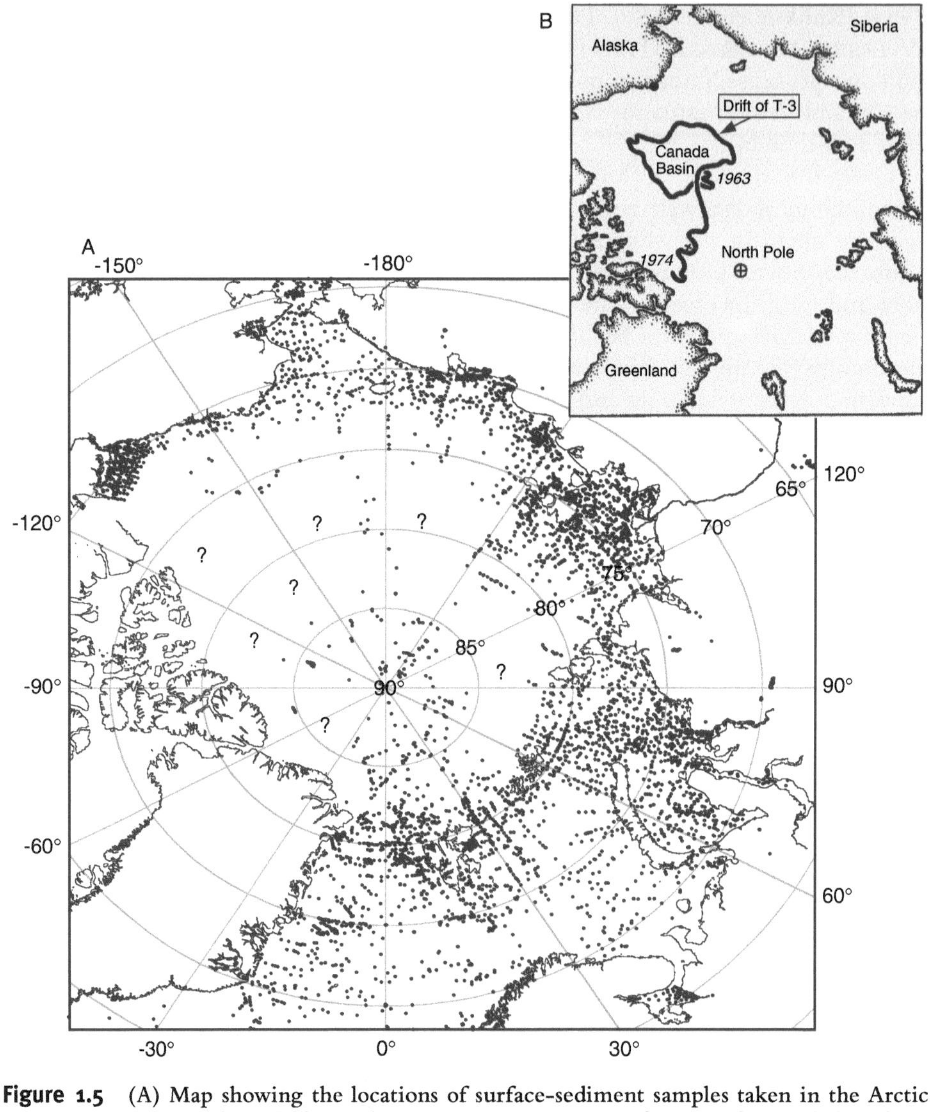

Figure 1.5 (A) Map showing the locations of surface-sediment samples taken in the Arctic Ocean until 2006 and available in the PANGAEA repository for georeferenced data from earth system research (http://www.pangaea.de/). (B) Track of Fletscher's Ice Island (T-3) drift 1963–1974 (Clark, Whitman, Morgan, & Mackey, 1980). All FL cores studied by Clark et al. (1980), Darby, Naidu, Mowatt, & Jones (1989) and numerous others were taken along the track of this drift.

world ocean regions. This lack of knowledge is mainly caused by the major technological/logistic problems in reaching this permanently ice-covered region with normal research vessels. Until the 1950s, only Russian scientists equipped ice islands, large thick ice floes that have calved off the ice shelves in the Arctic and drift with the pack ice across the Arctic basins, on a regular basis to gather basic

Table 1.1 List of Research Activities on Ice Islands (from Jokat, 2004).

Year of expedition	Expedition
Since 1937	Numerous Soviet drifting stations
1952–1974	Fletcher's Ice Island (T-3)
1957–1958	Alpha
1959–1960	Charlie
1961–1965	Arlis
1971, 1972, 1975–1976	Aidjex
1979	Lorex
1979–1982	Fram
1983	Cesar
1984–??	Hobson's Choice

oceanographic and meteorological information as well as geophysical data (Kiselev, 1986; Kristoffersen, 1990). In the following decades, US and Canadian institutions mounted projects on several ice islands to collect new geoscientific data (Table 1.1; Jokat, 2004). Under the Arctic Sciences Program of the Office of Naval Research (ONR), a series of large international field programmes were initiated (Table 1.2, Figure 1.6; Conlon & Curtin, 2004), including the Arctic Ice Dynamics Joint Experiment (AIDJEX, 1970–1975), FRAM I–IV (1979–1981; Johnson, 1983), the Marginal Ice Zone Experiment (MIZEX, 1983–1984, 1987), the Coordinated Eastern Arctic Experiment (CEAREX, 1987–1988), the Leads Experiment (LEADEX, 1992), the Sea Ice Mechanics Initiative (SIMI, 1993–1994), and the Surface Heat Budget of the Arctic Experiment (SHEBA, 1997–1998). Within SHEBA, a programme focusing on ice–albedo and cloud–albedo feedback processes, the Canadian icebreaker *Des Grosselliers* was frozen into the pack ice and allowed to drift for 14 months (Conlon & Curtin, 2004).

Geophysical investigations were part of a number of these research programmes on ice islands and provided the first information on the general structure of the seafloor and the sediment distribution. Coinciding with measurements from ice island, the interior Arctic Ocean which was not accessible for surface ships until the 1980s, was surveyed by long-range aircraft of Russian and US institutions to acquire aeromagnetic data (Karasik, 1968, 1973; Vogt, Taylor, Kovacs, & Johnson, 1979, 1982). The magnetic data together with sparse bathymetric data obtained from drift stations and the few unclassified submarine data, allowed the identification of the large-scale structures in the Arctic and a first-order tectonic interpretation of the ridges and basins (Jokat, 2004).

From the ice islands ~700 short sediment cores were taken over the years, with ~580 alone from the ice island T-3 (Weber & Roots, 1990; for drift path of T-3 see Figure 1.5). A large number of these sediment cores were intensely studied for grain-size, mineralogical, micropaleontological, and geochemical characteristics and stratigraphy (e.g., Clark, Whitman, Morgan, & Mackey, 1980; Jackson, Mudie, & Blasco, 1985; Darby, et al., 1989). Prior to 1987, most of the available information

Table 1.2 Research Activities on Ice Islands between 1950 and 1998 and Major Insights by Discipline Achieved Over the Years of the Arctic Program at the Office of Naval Research (Conlon & Curtin, 2004).

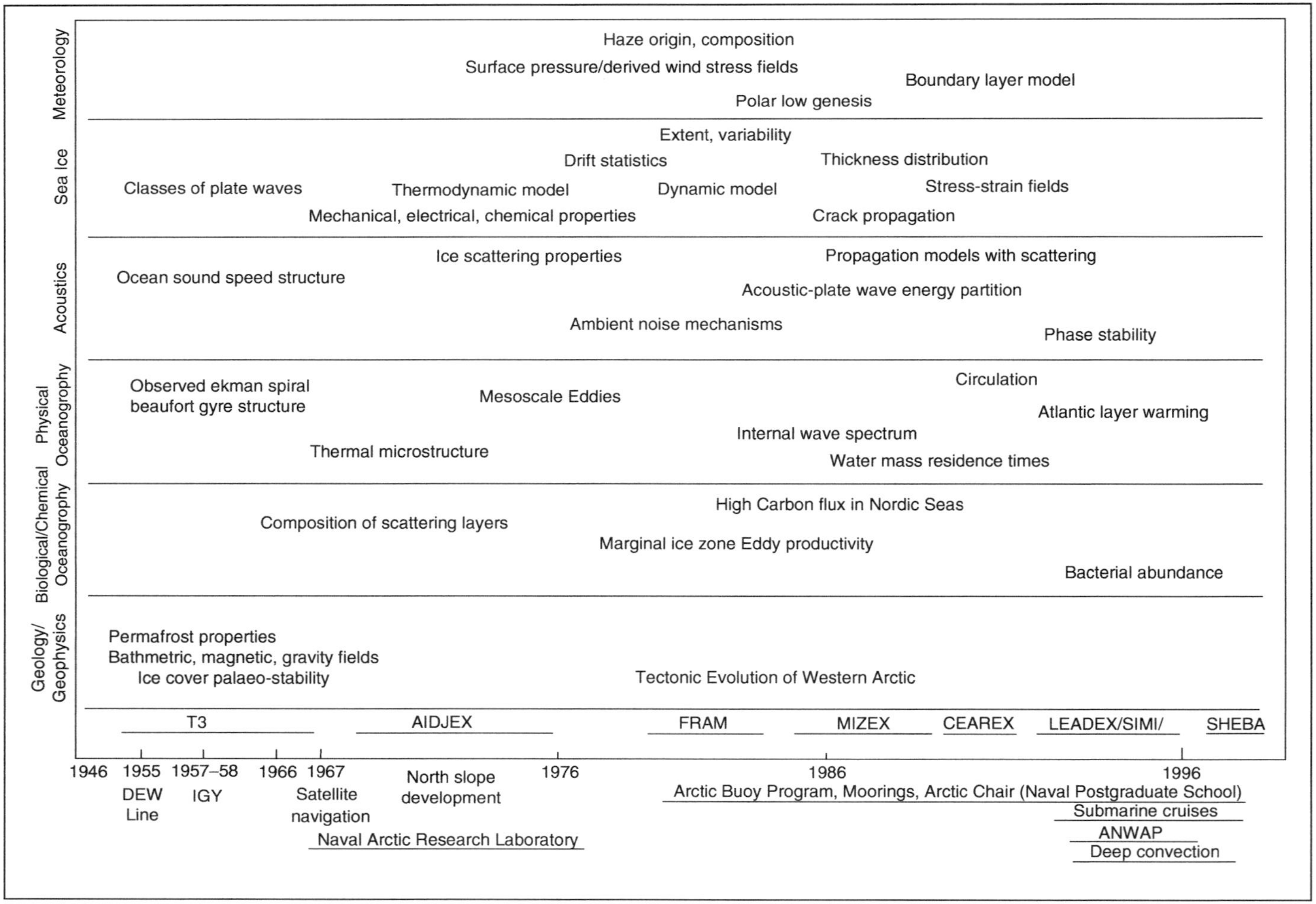

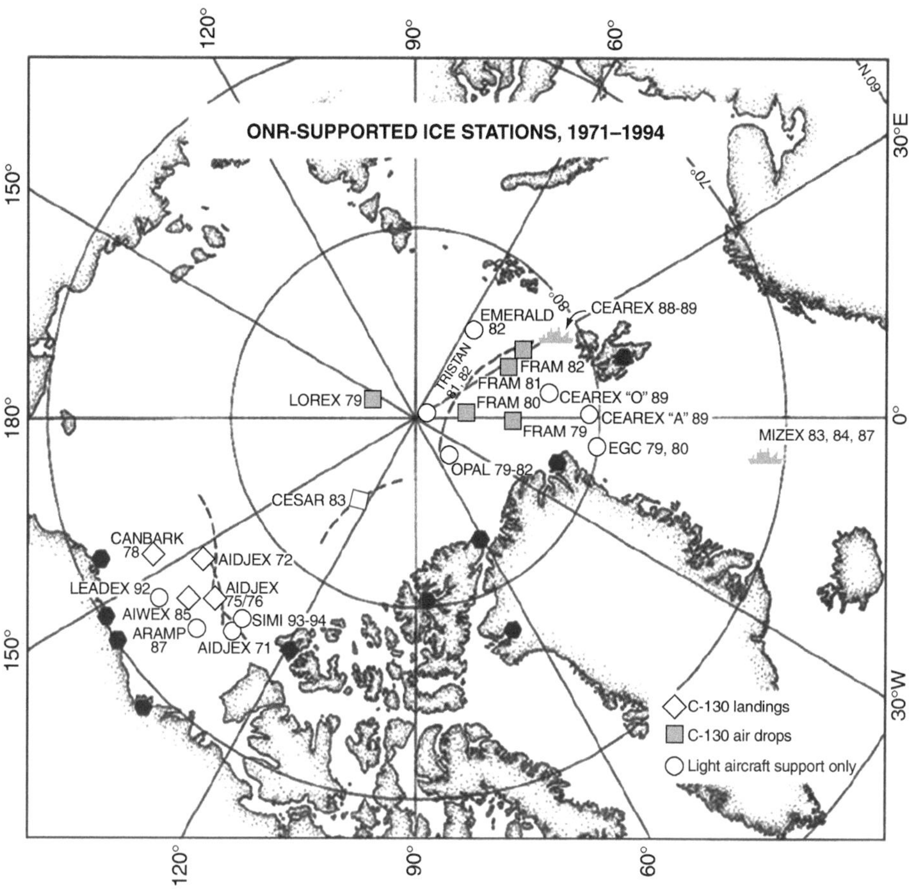

Figure 1.6 Ice stations supported by the US Office of Naval Research from 1971 through 1994. Modified from Conlon and Curtin (2004).

on sediment characteristics and palaeoenvironmental history of the central Arctic Ocean was obtained from these sediment cores (see Chapter 6.3 for results).

Comprehensive summaries about the knowledge on Arctic Ocean geology and palaeoceanography at the end of the 1980s are published in Herman (1989), Bleil and Thiede (1990), and Grantz, Johnson, and Sweeney (1990).

Within the SCience ICe EXercise (SCICEX), an unprecedented collaboration between the US Navy and the marine research community, eight submarine cruises were carried out between 1993 and 1999 with USS *Pargo* (SCICEX-93), USS *Cavalla* (SCICEX-95), USS *Pogy* (SCICEX-96), USS *Archerfish* (SCICEX-97), and USS *Hawkbill* (SCICEX-98 and SCICEX-99). The programme was designed to use nuclear-powered submarines to map and sample the ice canopy; the physical, chemical, and biological water properties; the seafloor topography; and the shallow subsurface of the Arctic Ocean (Edwards & Coakley, 2003, 2004). SCICEX publications have contributed to most every field of science, providing novel observations, testable hypotheses for future work, and an increased understanding of both Arctic and global processes. Some examples related to results from SCICEX cruises are the discovery of young volcanoes on Gakkel Ridge (Edwards et al., 2001), evidence for thick ice shelves possibly extending into the Arctic Ocean (Polyak, Edwards, Coakley, & Jakobsson, 2001; see Chapter 6.3.2 for discussion), and evidence for thinning of the present-day sea ice (Rothrock, Zhang, & Yu, 2003; see Chapter 2.3 for more details) and changes in temperature and circulation patterns of intermediate waters (e.g., Smith, Ellis, & Boyd, 1999; Smethie, Schlosser, Bonisch, & Hopkins, 2000; Gunn & Muench, 2001). Furthermore, the SCICEX bathymetry data were major contributions for the compilation of the International Bathymetric Chart of the Arctic Ocean (IBCAO) bathymetry map (Jakobsson, Grantz, Kristoffersen, & Macnab, 2003a; see Chapter 2.1).

Multidisciplinary cruises of scientific research ships started to be scheduled in the ice-covered high Arctic in the 1980s. After the prelude on the Swedish YMER 80-expedition into the Fram Strait, Yermak Plateau, and Nansen Basin (Schytt, Boström, & Hjort, 1981; Boström & Thiede, 1984), at least 15 major international expeditions with oceanographic, biological, geological, and geophysical objectives were carried out in the Arctic Ocean Basin (Table 1.3; for *Polarstern* expeditions see Figure 1.7). During the 1991 expedition, *Oden* and *Polarstern* as the first conventionally driven research ice breakers jointly succeeded to reach the North Pole (Anderson & Carlsson, 1991; Fütterer, 1992). Onboard *Polarstern*, unique undisturbed surface sediments and long sediment cores were recovered from the Nansen, Amundsen, and Makarov Basins, the Gakkel and Lomonosov Ridges, and the Morris-Jesup Rise and the Yermak Plateau (Fütterer, 1992). The most recent expeditions in 2007/2008 (see Table 1.3) are coordinated research activities under the umbrella of the IPY 2007/2008. Multidisciplinary studies of the data and sediment material collected during these ship expeditions have greatly advanced (and will further advance) our knowledge on Arctic Ocean palaeoenvironment and its variability through Quaternary times. Some of the results are discussed in Chapter 6.3.

1.2.3. Drilling in the High Northern Latitudes

Because gravity and piston coring is restricted to the uppermost metres of the sediment cover mainly representing the Quaternary, the knowledge about the early Arctic Ocean history remained fragmentary over the past decades (see Chapter 7 for more details). For deciphering the long-term Arctic palaeoenvironmental and tectonic history, long sedimentary sections are needed to be obtained by drilling.

Concerning the recovery of long pre-Quaternary Arctic sedimentary sections, over decades activities were restricted to (still ongoing) petroleum exploration drilling in the Arctic marginal seas (e.g., Mørk & Bjorøy, 1984; Dixon, Dietrich, Snowdon, Morrell, & McNeil, 1992; Leith et al., 1992; Montgomery, 2005; see Chapter 7). In this context, also Russian drilling programmes in the western Eurasian shelf seas have to be mentioned. During the 1980s, for example, a large number of deep wells were drilled in the Barents Sea by the Arctic Murmansk

Table 1.3 List of Selected Geoscientific Icebreaker Cruises to the Central Arctic Ocean. In Addition, There were Numerous National and International Expeditions Carried-Out in the Circum-Arctic Marginal Seas as well as the Norwegian-Greenland Sea Including Fram Strait, Not Listed Here (from Jokat, 2004, Updated).

Year of expedition and area	Ship and country	Reference (Cruise Report or related paper)
1980, Nansen Basin, Yermak Plateau	*Ymer* (Sweden)	Schytt et al. (1981); Boström and Thiede (1984)
1987, Nansen Basin, Gakkel Ridge	*Polarstern* (Germany)	Thiede (1988)
1988, Chukchi Plateau	*Polar Star* (USA)	Phillips et al. (1992)
1991, Nansen/Amundsen Basin/Lomonosov Ridge	*Oden/Polarstern* (Sweden/ Germany)	Anderson and Carlsson (1991) and Fütterer (1992)
1993, Canada Basin/Chukchi Plateau	*Polar Sea* (USA)	Grantz et al. (1998)
1993, Nansen Basin	*Polarstern* (Germany)	Fütterer (1994)
1994, Transarctic (Canada Basin/Lomonosov Ridge)	*Polar Sea/Louis St. Laurent* (USA/Canada)	Aagaard, Barrie, and Carmack (1996) and Wheeler (1997)
1995, Amundsen/Makarov Basin/Lomonosov Ridge	*Polarstern* (Germany)	Rachor (1995)
1996, Lomonosov Ridge/ Makarov Basin	*Oden* (Sweden)	Backman et al. (1997)
1998, Alpha Ridge/ Lomonosov Ridge	*Polarstern/Arktika* (Germany/Russia)	Jokat (1999) and Jokat et al. (1999)
2000, Mendeleev Ridge	*Akademic Fedorov* (Russia)	Kaban'kov, Andreeva, Ivanov, and Petrova (2004)
2001, Nansen Basin/Gakkel Ridge (AMORE)	*Healy/Polarstern* (USA/ Germany)	Thiede (2002)
2001, Nansen Basin/ Lomonosov Ridge	*Oden* (Sweden)	Grönlund (2001)
2004, Yermak Plateau	*Polarstern* (Germany)	Stein (2005)
2004, Lomonosov Ridge (ACEX)	*Vidar Viking/Oden/ Sovetskij Soyuz* (IODP)	Backman et al. (2006)
2005, Transarctic (HOTRAX)	*Healy/Oden* (USA/ Sweden)	Darby, Jakobsson, and Polyak (2005)
2007, Eur. Cont. Marg/ Nansen Basin/Lomo-, Alpha Ridge	*Polarstern* (Germany)	Schauer (2008)
2007, Lomonosov Ridge off Greenland (LOMROG)	*Oden/50 Let Pobedy* (Sweden/Russia)	Jakobsson et al. (2007b)
2008, East Siberian Sea/ Alpha-Mendeleev Ridge (scheduled)	*Polarstern* (Germany)	Chief scientist: W. Jokat/AWI

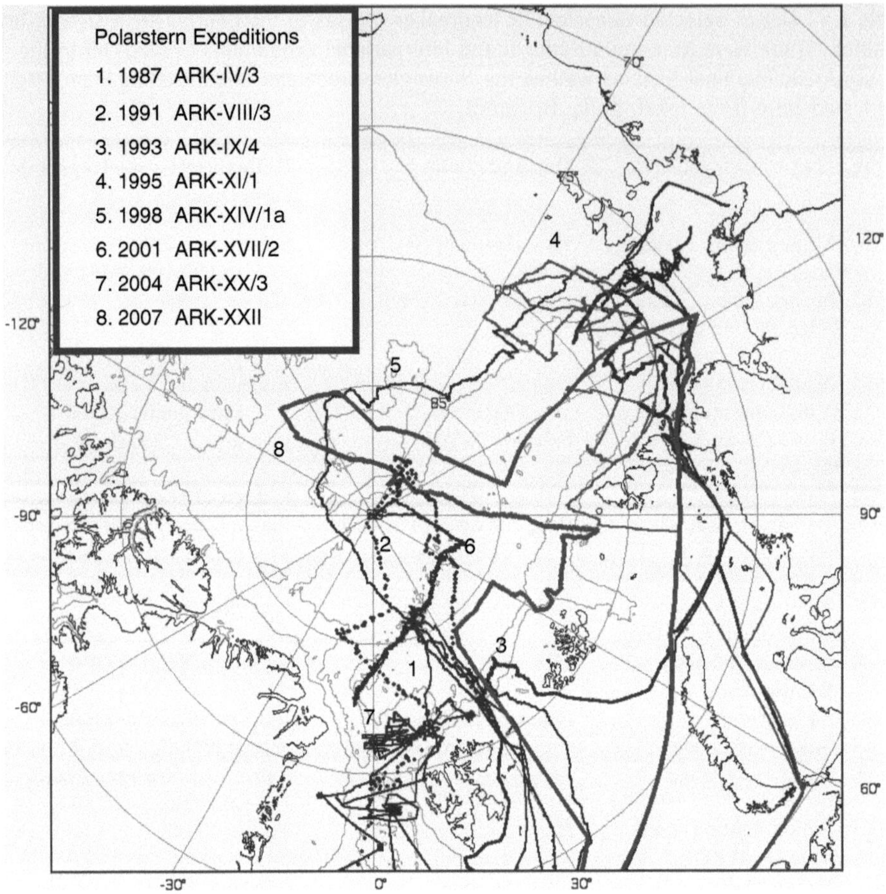

Figure 1.7 Cruise plots of expeditions carried out by RV *Polarstern* in central Arctic Ocean between 1987 and 2008 (see Table 1.2 for references).

Government of Exploration Drilling (AMURB) and the Scientific Production Association "Arcticmorneftegazrazvedka" (Kulikov et al., 1999; Tarasov et al., 1999). At the same time, the Arctic Marine Engineer-Geological Expedition (AMIGE) carried out hundreds of drillings in the Barents and Kara seas from RV "Bavenit" and RV "Kimberlit", and new data on thicknesses, composition, and age of these deposits were obtained (Tarasov et al., 1999). Recently, large amount of data obtained from drill holes in the Eurasian Arctic became available in the PANGAEA repository for georeferenced data from earth system research (http://www.pangaea.de/), such as grain size, heavy mineral, clay mineral, and organic carbon data from Tertiary, Cretaceous, Jurassic, and Triassic sequences recovered in drill holes in the Kara and Barents seas (e.g., Bondarev & Kosheleva, 2007; Bro, 2007; Pchelina & Komarnitsky, 2007; Shkola, 2007a, 2007b; Zaripov, 2007).

A major step forward in scientific drilling in the marginal Arctic was the voyage of the *Joides Resolution* of the Ocean Drilling Program (ODP) during Leg 151 (July–September 1993) to the Norwegian–Greenland–Iceland Sea and the Fram Strait/Yermak Plateau area which brought a scientific drilling vessel higher into the High Northern Latitudes with seasonally ice-covered waters than ever before (Myhre et al., 1995; Thiede, Myhre, Firth, Johnson, & Ruddiman, 1996). The drillship was escorted by the Finnish icebreaker *Fennica*. ODP Leg 151 opened completely new scientific perspectives and represents henceforth an important historic step in the scientific exploration of the Arctic (see Chapter 7.2.3 for some more details).

The most recent milestone in scientific drilling in the High Northern Latitudes was the "Arctic Coring Expedition" — ACEX (Integrated Ocean Drilling Program — IODP Expedition 302, the first Mission Specific Platform (MSP) expedition within the IODP) in 2004 (Backman et al., 2005, 2006). With the successful completion of ACEX, a new era in Arctic research has begun. For the first time, a scientific drilling in the permanently ice-covered Arctic Ocean was carried out, penetrating 428 m of Quaternary, Neogene, Palaeogene, and Campanian sediment on the crest of Lomonosov Ridge close to the North Pole between 87°N and 88°N (Figure 1.8; Backman et al., 2006; Moran et al., 2006; Backman & Moran, 2008). Cores were recovered in five holes across three sites (holes MSP0002A, MSP0003A, MSP0004A, MSP0004B, and MSP0004C) situated within 15.5 km of each other on seismic Line AWI-91090 (Figures 1.8 and 1.9; Jokat, Uenzelmann-Neben, Kristoffersen, & Rasmussen, 1992) and interpreted as a single site (composite section) because of the internally consistent seismic stratigraphy across that distance.

IODP Expedition 302 was an outstanding success for two reasons (Backman et al., 2006). First, the biggest technical challenge was to maintain the drillship's location while drilling and coring in heavy sea ice over the Lomonosov Ridge. ACEX has proven that with an intensive ice-management strategy, i.e., a three-ship approach with two powerful icebreakers (*Sovetskiy Soyuz* and *Oden*) protecting the drillship (*Vidar Viking*) by breaking upstream ice floes into small pieces (Figure 1.8), successful scientific drilling in the permanently ice-covered central Arctic Ocean is possible. The icebreakers kept the drillship on location in 90% cover of multiyear ice for up to nine consecutive days, a benchmark feat for future drilling in this harsh environment. Second, the first scientific results comprise a milestone in Arctic Ocean research, and future results of ongoing studies on ACEX material will certainly bring new insights into the Arctic Ocean climate history and its global significance (see Chapter 7.2 for details).

Although major progress in Arctic Ocean research has been made over the past decades, the knowledge of its short- and long-term palaeoceanographic and palaeoclimatic history as well as its plate-tectonic evolution is much behind that from the other world's oceans. Furthermore, large regions of the Arctic Ocean are still untouched at all (Figure 1.5). This lack of knowledge will hopefully be filled during international, multidisciplanary research programmes to be carried out in the future. The rationale of future Arctic geoscientific research is outlined in Chapter 8.

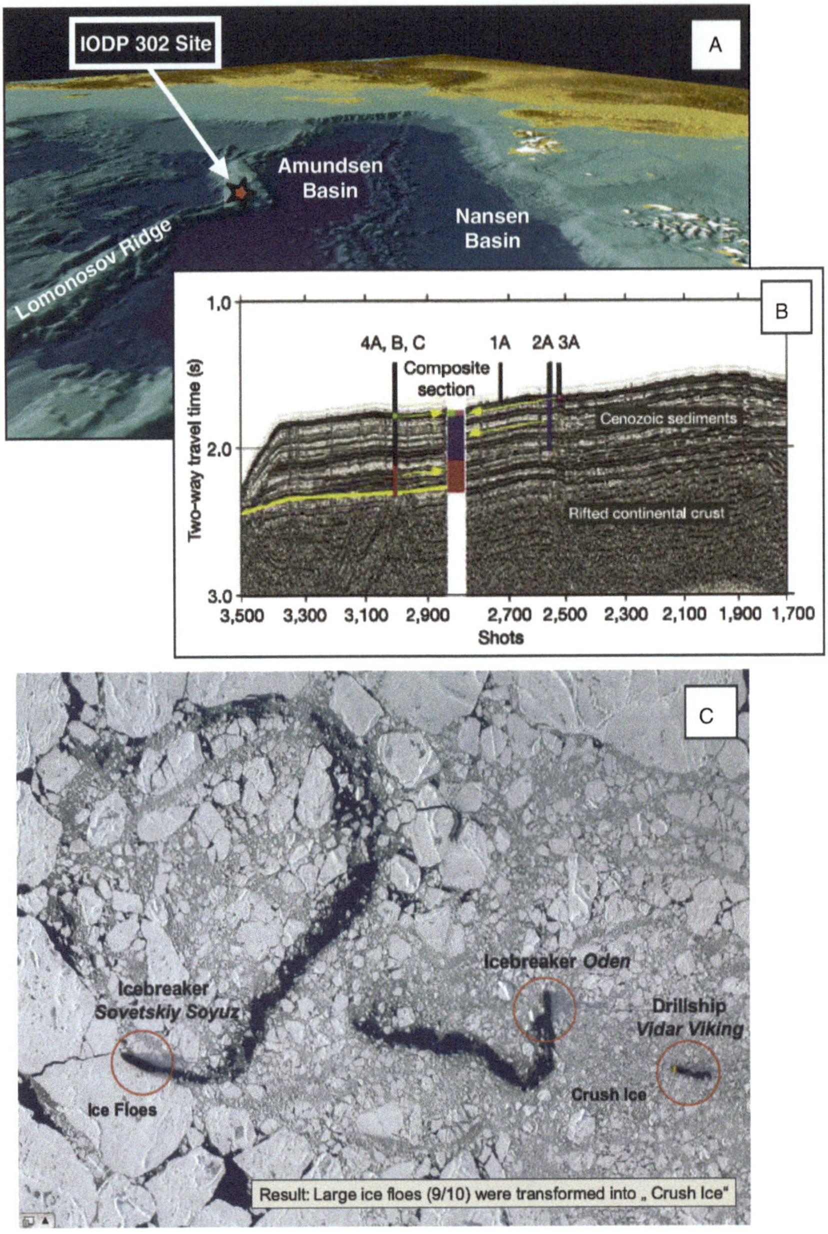

Figure 1.8 Location of IODP Expedition 302 (ACEX) drill site, seismic profile, and sea-ice operations on Lomonosov Ridge (Backman et al., 2006; Moran et al., 2006, supplemented). (A) 3D bathymetry of the central Arctic Ocean (based on the IBCAO Chart; Jakobsson, Cherkis, Woodward, Coakley, & Macnab, 2000b; Jakobsson et al., 2003a) with location of the ACEX drill site. (B) Seismic profile AWI 91090 across the Lomonosov Ridge, interpreted as continental crust truncated by a regional unconformity overlain by a continuous sediment sequence (Jokat, Uenzelmann-Neben, Kristoffersen, & Rasmussen, 1992). The four ACEX sites were positioned on this profile, shown as solid vertical lines. At each of the four sites (1–4), multiple holes (A, B, C) were drilled and sampled to the depth of the solid vertical line. The palaeoceanographic record overlies acoustic basement, a seismic unconformity (highlighted in yellow), that was confirmed to be Cretaceous continental crust (Moran et al., 2006). (C) Photograph showing the sea-ice operations during the ACEX expedition (Backman et al., 2006). The nuclear icebreaker *Sovetskiy Soyuz* is crashing the large ice floes, to be further cut-down then by the icebreaker *Oden*. This leaves the drillship *Vidar Viking* working under crushed-ice conditions and staying for some days on station to carry out the drilling.

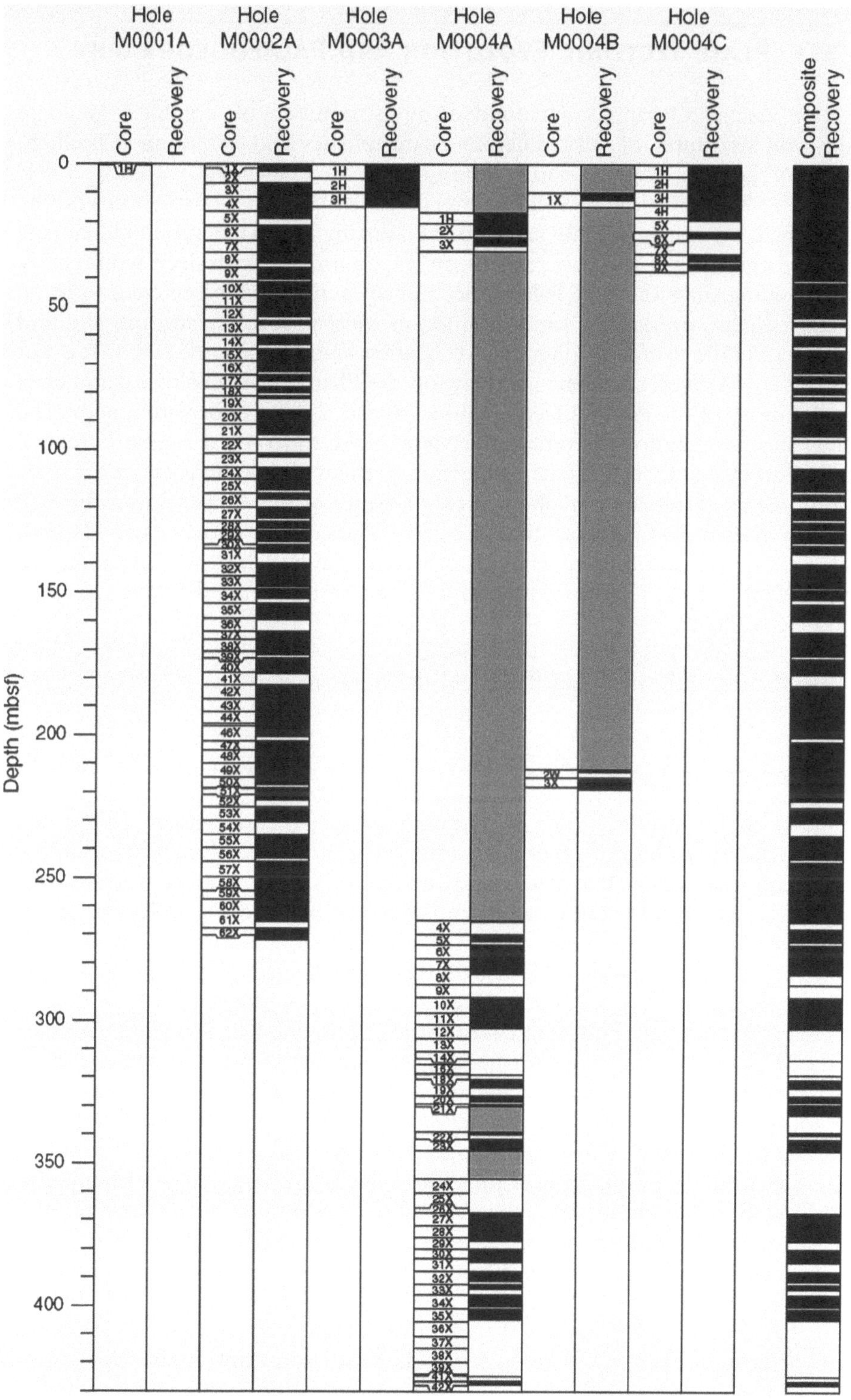

Figure 1.9 Core recovery summary diagram for all Expedition 302 (ACEX) holes (Backman et al., 2006). Black, recovered core; white, no recovery; shaded, washed intervals. Total recovery was 68.4%.

1.3. Plate Tectonic Evolution and Palaeogeography

The Arctic Ocean is surrounded by a number of complex geologic terrains and structures of very different characteristics and evolution (Okulitch, Lopatin, & Jackson, 1989; Grantz, Johnson, & Sweeney, 1990; Trettin, 1991; Dolginow & Kropatschjow, 1994). In general, three main types of structures can be distinguished: (1) young oceanic basins formed during Late Meozoic and Cenozoic times; (2) large shelved and low terrains on the continents with deep sedimentary basins subsiding since the Late Palaeozoic; and (3) continental basement consisting of ancient shields, overlain by late Precambrian and Palaeozoic platforms, and fold belts separating the platforms (Figure 1.10; Zonenshain & Natapov, 1989). It is not intended here to give a complete overview on the Phanerozoic geological and plate tectonic history of the Arctic Ocean region. Instead, a brief information about the general tectonic evolution relevant to the Mesozoic–Cenozoic palaeoenvironmental history of the Arctic Ocean and its connection to the world ocean is presented here. For more detailed information about plate tectonics of the Arctic Ocean through Palaeozoic–Mesozoic–Cenozoic times refer to Zonenshain and Natapov (1989), Grantz et al. (1990 and further references therein), Lawver, Müller, Srivastava, and Roest (1990), Grantz, Clark, Phillips, and Srivastava (1998), and Lawver, Grantz, and Gahagan (2002).

Depending on its age and tectonic evolution the Arctic Ocean is divided into two main basins, the Amerasian Basin and the Eurasian Basin.

1.3.1. Amerasian Basin

In contrast to the evolution of the Eurasian Basin (see below), the origin and evolution of the Amerasian Basin is more speculative and poorly understood. Detailed information on the nature and the age of the Amerasian Basin and its subbasins (Canada and Makarov basins) and ridges (Alpha and Mendeleev ridges) is not available due to the sparse and insufficient geophysical and geological database. As far as it is known, the oldest Arctic deep-sea basin (Canada Basin) was formed in the Cretaceous by seafloor spreading (Vogt et al., 1982; Grantz et al., 1998). During or after the opening of the Canada Basin, the Alpha–Mendeleev Ridge complex and the Makarov Basin were formed. The complex horst and graben ridge topography of volcanic rocks is covered by 0.5–2 km of sediments (Hall, 1973, 1979; Jokat, 2003). The magnetic anomaly pattern over the ridge is partly irregular and generally correlated with ridge topography (Taylor, Kovacs, Vogt, & Johnson, 1981). A tectonic reconstruction of the Amerasian Basin based on the stratigraphy of Northwind Ridge and sea-floor magnetic anomalies in the Canada Basin is given by Grantz et al. (1998) (Figure 1.11).

A number of hypotheses have been forwarded to explain the origin of the Alpha–Mendeleev Ridge, such as this ridge may be: (1) of continental origin; (2) a former spreading centre; or (3) a result of 'hotspot' activity (Jokat, 2003 and further references therein). The existing database from the Alpha–Mendeleev Ridge is mainly from the pioneering seismic reflection survey and sediment

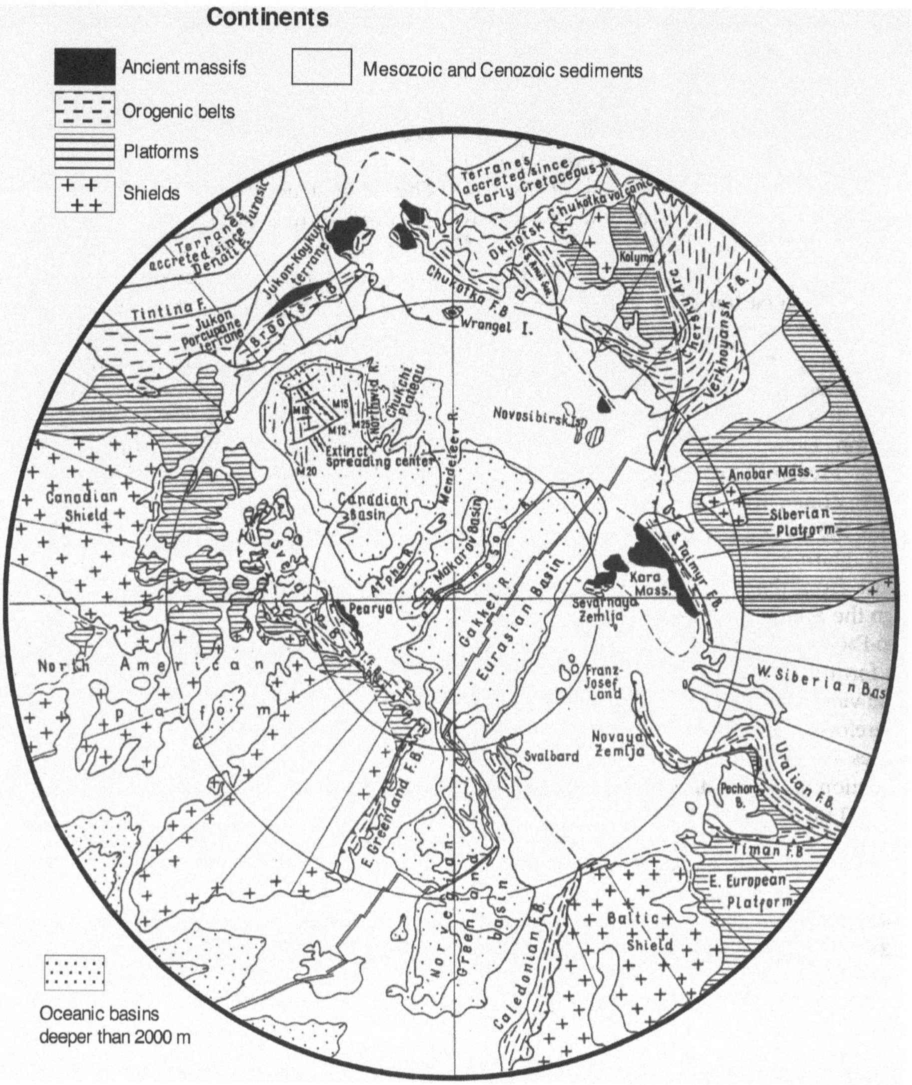

Figure 1.10 Main geological structures of the Arctic Ocean and surrounding continents. Dotted areas in the central part of the figure indicate oceanic basins deeper than 2,000 m. The active spreading centre is shown as double line. On the continents, shields, platforms, and orogenic belts are indicated as crosses, horizontally hatched, and dashed lines, respectively. Ancient massifs, remnants of the Arctida continent are black, Mesozoic and Cenozoic sedimentary basins blank (white). Modified from Zonenshain and Natapov (1989).

sampling effort from US ice drift station T-3 (Hunkins, 1961; Hall, 1973) and the Canadian CESAR expedition (Jackson et al., 1985). More recently (1998), a two-ship experiment with the Russian nuclear icebreaker *Arktika* and RV *Polarstern* probed the central part of Alpha Ridge (Jokat, 1999, 2003; Jokat, Stein,

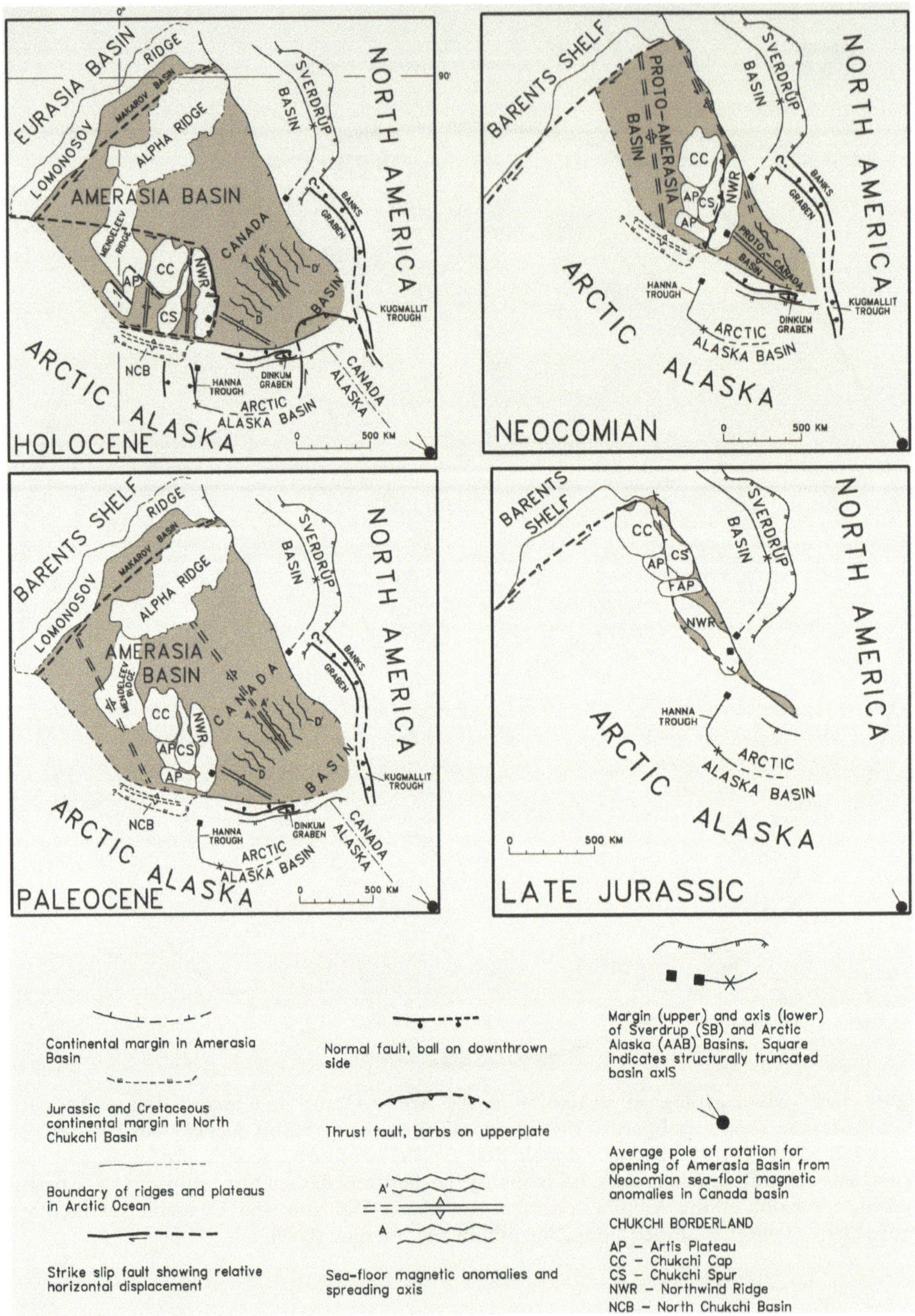

Figure 1.11 Stepwise reconstruction of the Amerasia basin from Late Jurassic to Holocene time, using the closing Tertiary extension in the Chukchi borderland (Grantz et al., 1993) and Late Jurassic to Neocomian extension in the Amerasia Basin. Taken from Grantz et al. (1998) where details of reconstruction and database are described.

Rachor, & Schewe, 1999). In total 320 km of multichannel seismic data were acquired along three profiles supplemented by four sonobuoys. The sediment velocities range from 1.6 to 2.7 $km\,s^{-1}$ and the sediment thicknesses vary between 500 and 1,200 m (Figure 1.12). The units lie conformably on the basement. In general, the sediments can be divided into two units, the ages of which are quite hypothetical: the upper unit is most likely of Cenozoic, the lower of Cretaceous age. The interpretation of the seismic velocities ranging from 4.3 to 6.7 $km\,s^{-1}$ suggests oceanic basement. In combination with a basalt sample recovered by gravity coring (Figure 1.12), there is little doubt of the oceanic origin of Alpha Ridge, at least in its western sector (Jokat et al., 1999). Recent high-precision whole rock $^{40}Ar/^{39}Ar$ incremental heating dating of this basalt gave a plateau age of 82 ± 1 Ma (Jokat, 2003). This age strongly supports geophysical models that suggest a Late Cretaceous age for Alpha Ridge (Vogt et al., 1982; Lawver & Müller, 1994).

1.3.2. Eurasian Basin

The origin and evolution of the Eurasian Basin is quite well understood. Since Heezen and Ewing (1961) recognized, that the mid-ocean rift system extended from the North Atlantic to the Arctic Ocean, it has been assumed that the more than 1,800 km long Lomonosov Ridge was originally a continental fragment broken off of the Eurasian continental margin and became separated by seafloor spreading. Regional aeromagnetic data indicated the presence of seafloor spreading anomalies in the basins north and south of the Gakkel Ridge, the active spreading centre located in the middle of the Eurasian Basin (Figure 1.13; Kristoffersen (1990) and further references therein). The interpretation of the magnetic anomalies in terms of seafloor spreading and their correlation with the geomagnetic time scale allowed to link the evolution of the Eurasian Basin to the opening of the Norwegian–Greenland Sea. According to this correlation, seafloor spreading was probably initiated in the Eurasian Basin between chron 24 and 25 in the late Palaeocene (Vogt et al., 1979; Kristoffersen, 1990). The average spreading rate was very slow, ranging between 1 and 2 $cm\,yr^{-1}$.

On the basis of multichannel seismic profiles crossing the entire Nansen and Amundsen basins, thick sedimentary sequences were deposited since that time (Figure 1.14; Jokat & Micksch, 2004). In the Nansen Basin, maximum sediment thicknesses of 4.5 km are reached close to the Barents Sea continental margin, vanishing towards the Gakkel Ridge. In the Amundsen Basin, on the other hand, sediments are only 1.7–2.0 km thick. The distinctly higher sediment thicknesses in the Nansen Basin are caused by the huge sediment input from the Eurasian continental margin (see also Chapter 3.3.1, Figure 3.18).

1.3.3. Arctic Ocean Gateways and Connection with the World Oceans

The modern polar cryosphere is an expression of an extreme climate situation, which developed during (global) stepwise Cenozoic cooling (e.g., Zachos, Pagani,

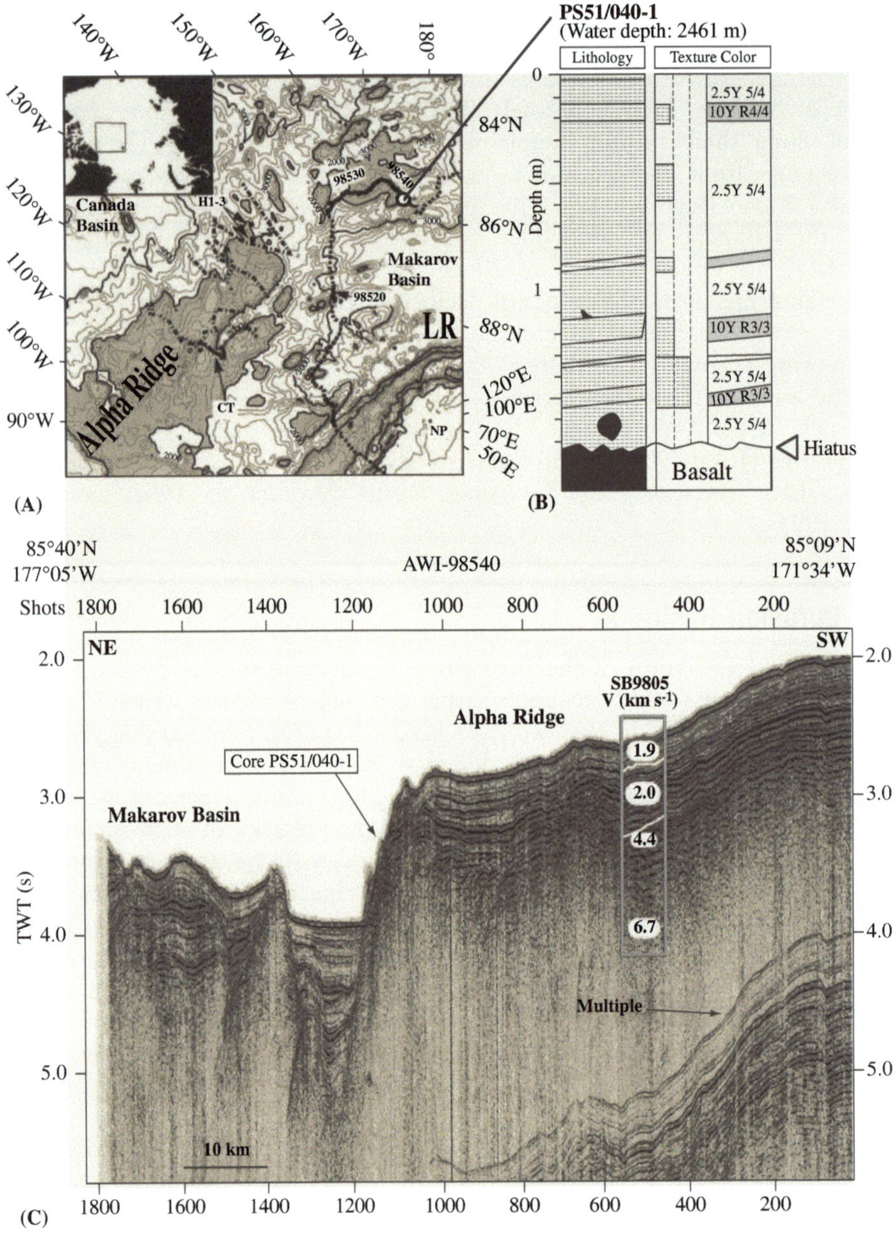

Figure 1.12 (A) Location of seismic profiles 98520–98540 (bold lines). The seafloor topography (contoured in 200 m) is taken from the new Arctic bathymetry grid (Jakobsson et al., 2000b). The grey-shaded areas indicate water depths shallower than 2,000 m. The dotted line indicates the cruise track of the Arctic-98 cruise (1998 June 27–July 27; Jokat et al., 1999); the dashed-dot-dashed line shows the drift paths of the T3-ice island across Alpha Ridge; CT, drift pattern of the CESAR experiment; H1–3 soundings performed by Hunkins (1961); NP, North Pole; LR, Lomonosov Ridge (from Jokat, 2003). (B) Profile 98540 running perpendicular to Alpha Ridge. The basement is rougher close to the escarpment near shot point 1200. Weak reflections indicate that older sediments might outcrop at the escarpment. Note that the prominent reflector at 2.4 s TWT (shot point 200) seems to be also present in the graben northeast of the escarpment. At least the seismic signature is similar. The arrow indicates the location where basalt samples were recovered by gravity coring (Jokat et al., 1999) (from Jokat, 2003). (C) Lithology of Core PS51/040-1 (from Stein, Fahl, Niessen, & Siebold, 1999b) which is dominated by alternation of beige- and brownish-coloured silty clays typical for Quaternary sediments from the central Arctic ridges (e.g., Jakobsson et al., 2000a; Stein et al., 2001). The lowermost part of the cores is composed of basaltic material which has been used for absolute dating (see text).

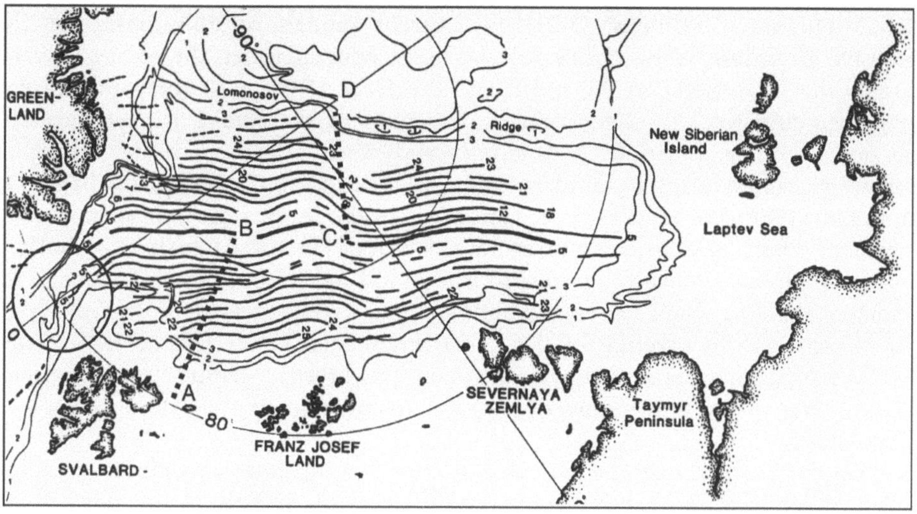

Figure 1.13 Magnetic lineations in the Eurasian Basin (from Kristoffersen, 1990, based on Kovacs et al., 1983). Locations of seismic profiles A–B and C–D (see Figure 1.14) are shown.

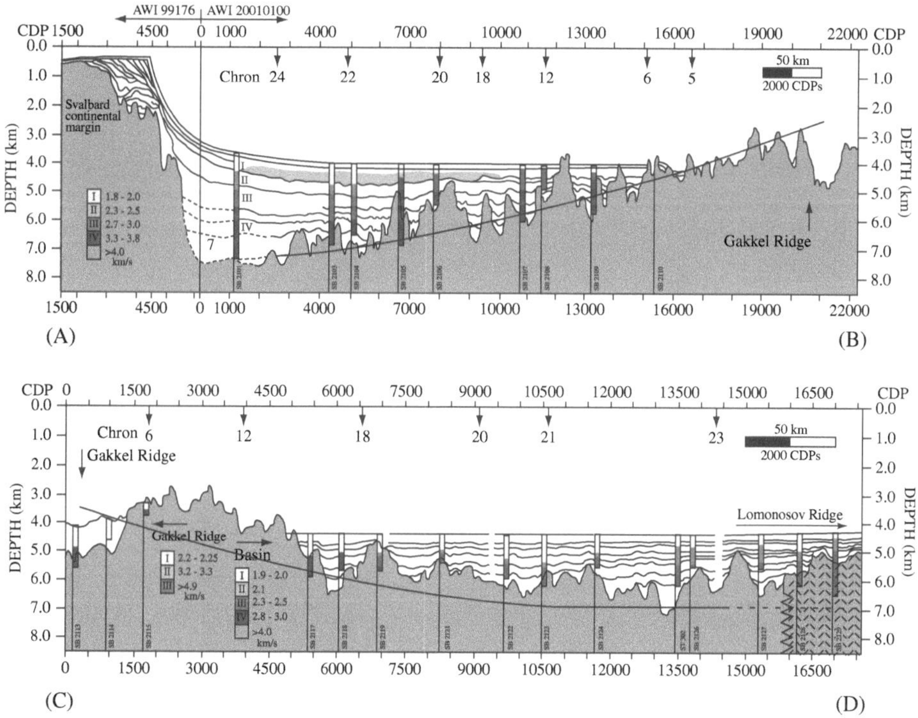

Figure 1.14 (A) Line drawing of seismic lines 99176 and 20010100 across the Nansen Basin north of Kvitoya; and (B) line drawing of profile 20010300 across the Amundsen Basin at 70° E (from Jokat & Micksch, 2004). The deployment positions of the sonobuoys are marked and their generalized velocity depth functions are displayed. The hatched area close to Lomonosov Ridge marks a zone, which is interpreted as the continent–ocean transition into the Amundsen Basin. Magnetic anomalies are shown on top of both profiles. For location of profiles see Figure 1.13, for details and database see Jokat and Micksch (2004).

Sloan, Thomas, & Billups, 2001). In current models on the causes for the intensive glaciations at the earth's poles during Cenozoic times, the polar gateways, namely the Fram Strait in the north and the Drake Passage in the south, played an outstanding role. In the south, the opening of the Drake Passage finalized the oceanographic isolation of Antarctica and possibly lead to the beginning of massive glaciation of this continent some 40 Ma ago (e.g., Miller, Fairbanks, & Mountain, 1987; Miller, Wright, & Fairbanks, 1991). In the north, no such isolated continent exists. Continents surround the Arctic Ocean, which was isolated from any global circulation system during most of its geological history. This gradually changed when Greenland and Svalbard started to move apart from each other initiating the opening of the Fram Strait (e.g., Kristoffersen, 1990; Eide, 2002), the major oceanic gateway between the North Atlantic and the Arctic Ocean. Fram Strait is the only deep-water connection to the world ocean (see Chapter 2.1).

The Fram Strait opening was due to the propagation of seafloor spreading from the northern Greenland Sea into the Eurasia Basin. In both basins chron 24/25 (53–56 Ma) marks the beginning of oceanic crust formation (Figure 1.13), but according to published tectonic models, it took ~30 million years until the gateway was fully opened. A deep-water exchange between the Arctic and Atlantic oceans probably occurred since the middle Miocene (Figure 1.15; Eide, 2002; Jakobsson et al., 2007a; Jokat, Leinweber, Ehlers, Boebel, & Schenke, 2008; see also Chapter 7.2.2, Figure 7.25). Shallow marine sediments in the Barents Sea show that a connection between the North Atlantic and the Arctic was established long before the opening of this gateway. A complex system of fracture zones and mid-ocean ridges provided a link to the global current circulation system, and consequently enabled an exchange of warm Atlantic and cold Arctic water masses. The opening of the Fram Strait gateway might have allowed for the formation of deep-water currents. Therefore, this event might have had a more dramatic effect on the regional to global palaeoceanographic and palaeoclimatic evolution, probably similar to the opening of the Drake Passage in the Southern Hemisphere. It is believed that this exchange of cold and warm water masses was one of the main triggers for the increased frequency of Northern Hemisphere Glaciations (NHGs). Furthermore, the dense, cold bottom waters of most of the world's oceans originating in the Nordic Seas and influenced by the Arctic/Atlantic water mass exchange, influence global thermohaline circulation, which drives world climate (see Chapter 6).

On the basis of plate tectonic reconstructions, the Cretaceous and early Tertiary Arctic Ocean was probably isolated from the world ocean in terms of deep-water connection (e.g., Ziegler, 1988; Mutterlose et al., 2003; Backman et al., 2006; Jakobsson et al., 2007a; Figures 1.15 and 1.16). The isolation of the Arctic Ocean at those times was an important prerequisite for the development of widespread anoxic/euxinic conditions during the late Jurassic/early Creatacous (e.g., Leith et al., 1992; Langrock, Stein, Lipinski, & Brumsack, 2003a; Langrock, Stein, Lipinski, & Brumsack, 2003b; Mutterlose et al., 2003) as well as the Palaeocene/Eocene (Backman et al., 2006; Moran et al., 2006; Sluijs et al., 2006;

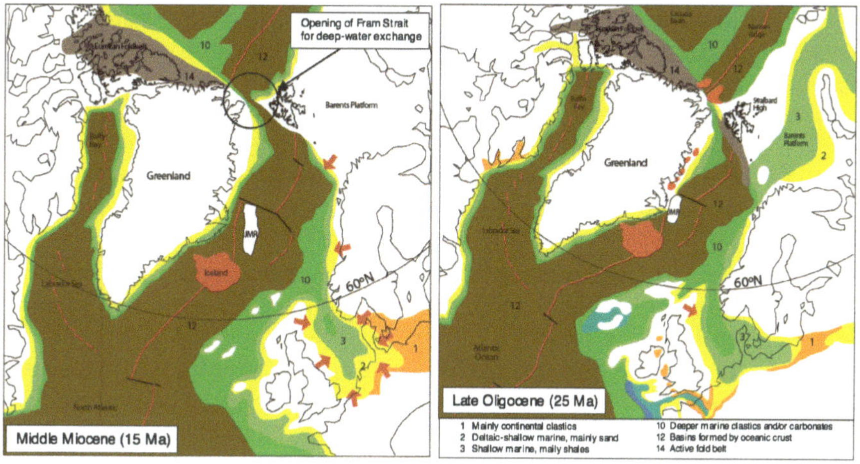

Figure 1.15 Reconstruction of the North Atlantic/Arctic gateway palaeogeography for the time slices (A) Late Oligocene (~25 Ma) and Middle Miocene (15 Ma) (from Eide, 2002, supplemented). 1, Mainly continental clastics; 2, deltaic-shallow marine, mainly sand; 3, shallow marine, mainly shales; 10, deeper marine clastics and/or carbonates; and 12, basins formed by oceanic crust.

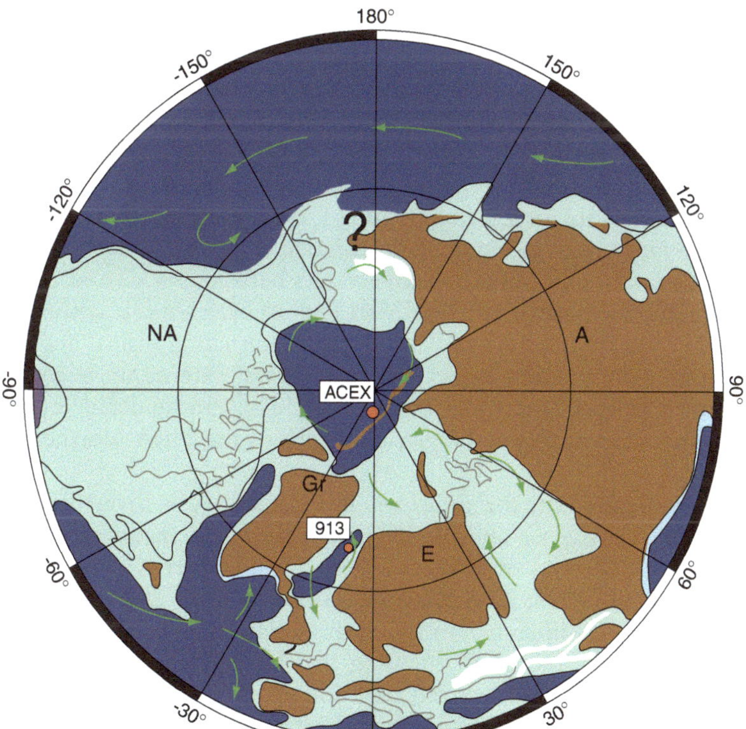

Figure 1.16 The map showing a palaeogeographic reconstruction of the high-northern latitudes around 50 Ma and location of ACEX Site and ODP Site 913 (from Backman et al., 2006, supplemented).

Stein et al., 2006a see Chapter 7 for more details). Shallow-water exchange, however, might have existed towards the Atlantic and Tethys as well as towards the Pacific (Figure 1.16).

1.4. Glaciations in Earth's History

The geologic record indicates that the Earth has experienced the development of long-lasting cold climates several times, which led to extensive glaciations, i.e., during the Precambrian (~2500–2000 Ma and 650–750 Ma), the Carboniferous-Permian (~350–250 Ma), and the Cenozoic (since ~40 Ma) (e.g., Schwarzbach, 1974; Frakes, Francis, & Syktus, 1992; Ruddiman, 2002). Whereas the pre-Cenozoic glaciations were probably unipolar, a bipolar glaciation with ice sheets in the southern and northern polar regions was developed during Cenozoic times (e.g., Bleil & Thiede, 1990). For a long time, however, it has been postulated and agreed to that the onset of Cenozoic glaciations was not contemporaneous. Whereas glaciation of Antarctica began close to the Eocene/Oligocene boundary (e.g., Kennett & Shackleton, 1976; Miller et al., 1987; Zachos et al., 2001), but probably as early as ~43 Ma (e.g., Lear, Elderfield, & Wilson, 2000), it was suggested that the NHG began no earlier than ~14 Ma (e.g., Thiede et al., 1998; Winkler, Wolf-Welling, Stattegger, & Thiede, 2002). Very recently new records from ODP/IODP cores from the Greenland Sea and the central Arctic Ocean, however, push back the date of Northern Hemisphere cooling and onset of glaciation to about 45 Ma, and suggest that the Earth's transition from the Greenhouse to the Icehouse world was bipolar (Backman et al., 2006; Moran et al., 2006; St. John & Willard, 2006; Eldrett, Harding, Wilson, Butler, & Roberts, 2007; St. John, 2008). During the Neogene, glaciations became more expended near 14 Ma and, especially, after ~3.2 Ma (Figure 1.17; e.g., Tiedemann, Sarnthein, & Shackleton, 1994; Driscoll & Haug, 1998; Thiede et al., 1998; Zachos et al., 2001). The long-term increase in oxygen isotope values between ~3.2 and 2.5 Ma marks the development of a permanent Northern Hemisphere ice cap with varying size. The last ~2.5 Ma are characterized by distinct alternating glacial and interglacial climate stages, with glacial stages (and ice sheets extension) becoming more extensive during the last ~0.9 Ma (Figure 1.17). For a more detailed discussion of the long-term record of Cenozoic glacial history of the High Northern Latitudes see Chapter 7.2.3.

During the Quaternary, the mid to high latitudes were covered repeatedly by thick large-sized ice sheets, one of the most prominent features of the ice ages (Figure 1.18; e.g., Denton & Hughes, 1981; Dyke, Morris, Green, & England, 1992; Svendsen et al., 1999, 2004; Ehlers & Gibbard, 2003, 2007; Astakhov, 2004). The glacial/interglacial difference in ice-sheet extension and volume is much more pronounced in the continents surrounding the Arctic Ocean than on Antarctica. Traces for glaciation are widespread and found at numerous locations on the Northern Hemisphere, as compiled in a comprehensive overview work carried out by the INQUA Work Group on "Extent and Chronology of Glaciations"

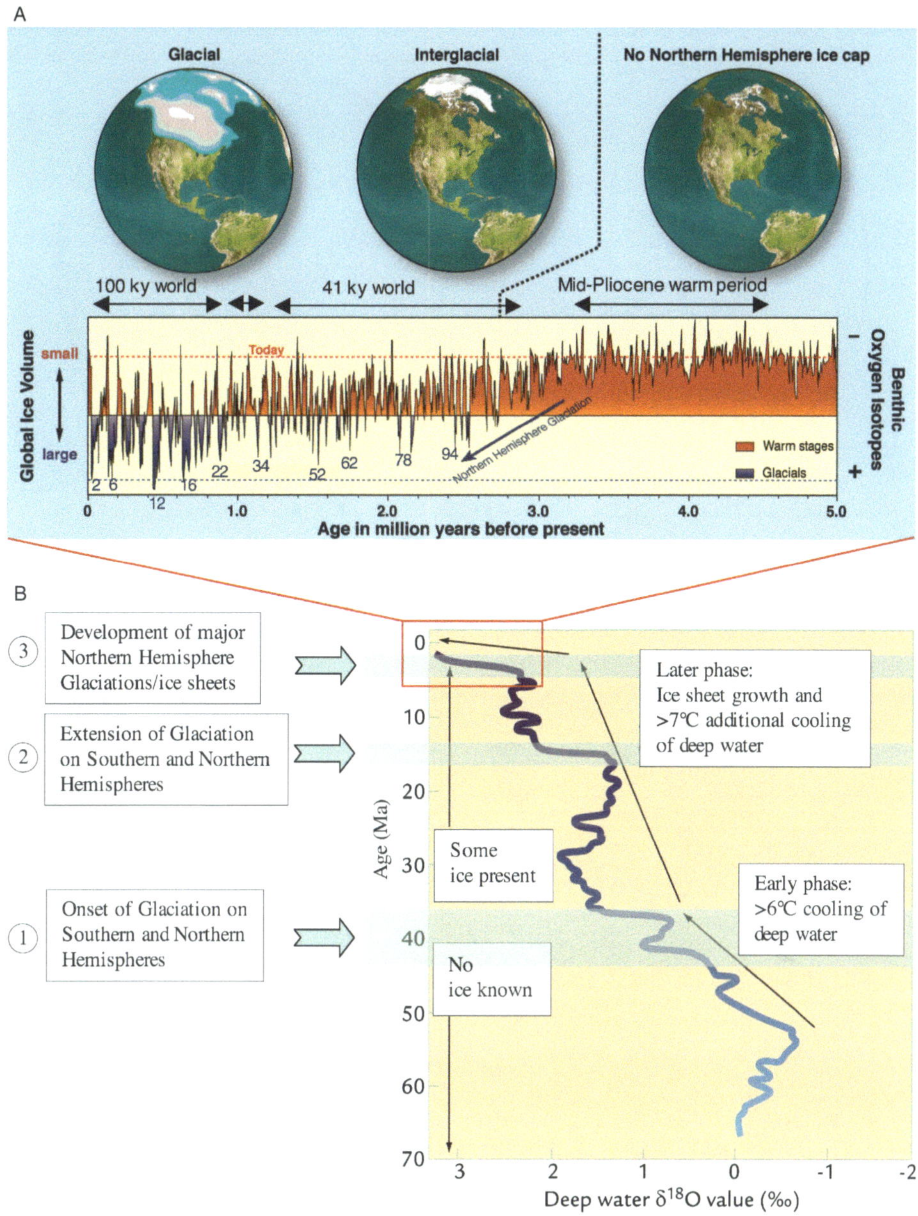

Figure 1.17 (A) Generalized benthic oxygen isotope curve reflecting the global climate evolution of the last 5 Ma (Tiedemann, Sarnthein, & Shackleton, 1994). The long-term increase in oxygen isotope values from ~3.2 to 2.5 Ma marks the development of a permanent Northern Hemisphere ice cap with varying size. The last ~2.5 Ma are characterized by distinct alternating glacial and interglacial climate stages, with glacial stages (and ice sheets extension) becoming more extensive during the last ~0.9 Ma. Figure (A) taken from http://www.awi.de/de/forschung/fachbereiche/geowissenschaften/marine_geology_and_paleontology/. (B) Generalized benthic oxygen stable isotope record throughout the last 70 Ma and palaeoenvironmental interpretation (based on Zachos et al., 2001, modified from Ruddiman, 2002).

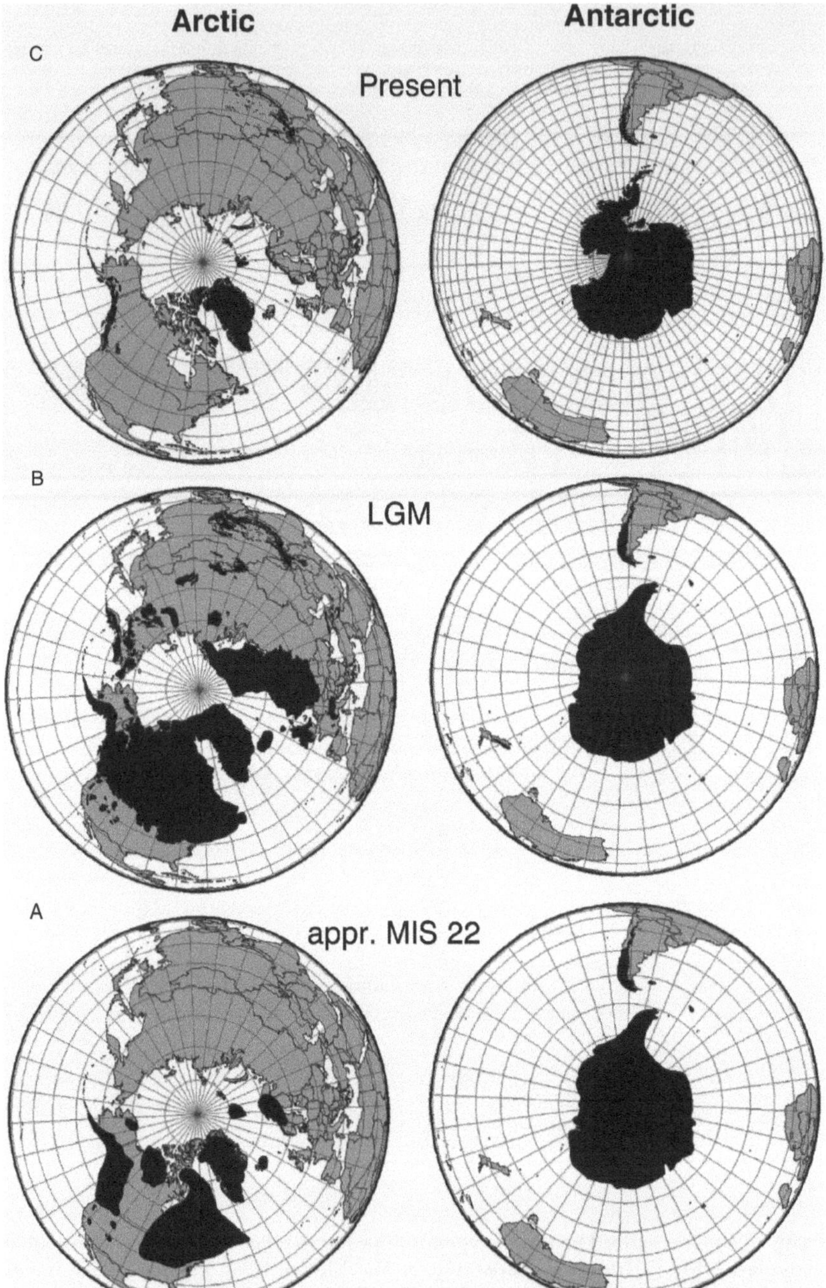

Figure 1.18 Extent of (A) late Matuyama Chron (appr. MIS 22), (B) LGM, and (C) modern glaciation in the Northern and Southern Hemispheres (based on Ehlers & Gibbard, 2004a, 2004b, 2004c, taken from Ehlers & Gibbard, 2007). The extent of MIS 22 ice sheet gives minimum extension, larger extension probable in Eurasian.

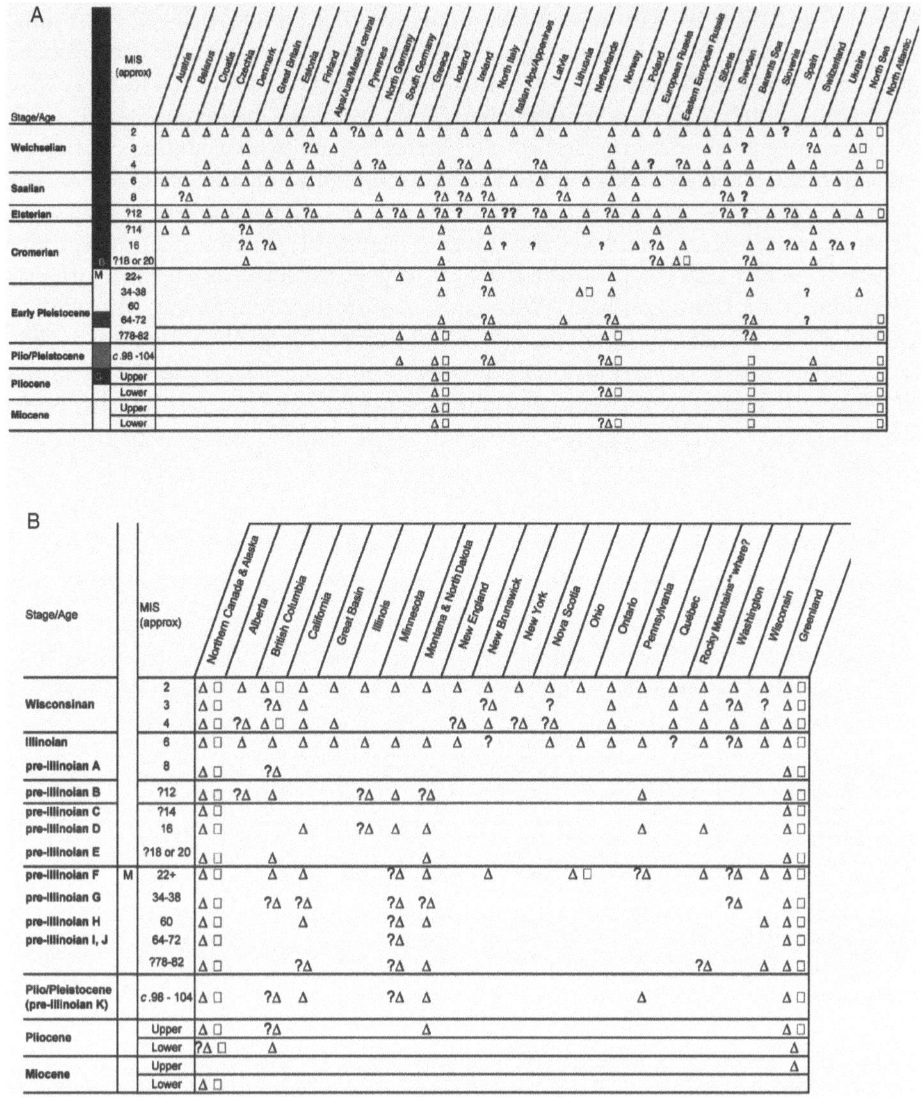

Figure 1.19 Occurrence of glaciation in (A) Europe and (B) North America through the Cenozoic based on numbers of observations presented in contributions to the INQUA project "Extent and Chronology of Quaternary Glaciations" (from Ehlers & Gibbard, 2007, based on Ehlers & Gibbard, 2004a, 2004b).

(Figure 1.19; Ehlers & Gibbard, 2004a, 2004b, 2004c, 2007). Based on these data, the first major glaciation with substantial ice volumes comparable to the middle/late Pleistocene glaciations probably occurred during the upper Matuyama Chron (Marine Isotope Stage (MIS) 22) (Figure 1.18A), most widespread glaciations occurred during MIS 16, 12, 6, and 2 (Figures 1.17B and 1.18B). Today, large ice sheets on Earth are restricted to Greenland and Antarctica (Figure 1.18C), reaching

$\sim 3 \times 10^6$ and 29×10^6 km^3 in volume, respectively, i.e., a total of $\sim 32 \times 10^6$ km^3. During the Last Glacial Maximum (LGM), the global ice volume was probably about three times higher, reaching 92×10^6 km^3 in total (Denton & Hughes, 2002; Huybrechts, 2002; Zweck & Huybrechts, 2005; see also Chapter 6.2.1).

These ice sheets had a huge impact on Earth's climate by rearranging continental drainage systems and by changing the Earth's topography and albedo (Clark & Mix, 2002). The decay of the ice sheets resulted in a release of huge amount of melt water and a major post-glacial sea-level rise. During the LGM, for example, the sea level was between 115 and 140 m lower than today (e.g., Shackleton, 1987; Fairbanks, 1989; Bard et al., 1996; Lambeck, Yokoyama, & Purcell, 2002; Milne, Mitrovica, & Schrag, 2002; Peltier, 2002; Zweck & Huybrechts, 2005). For a more detailed (partly still controversial) discussion of the evolution of the Quaternary ice sheets in the Northern Hemisphere, with special emphasis on the Eurasian ice sheets, see Chapter 6.2.

CHAPTER TWO

Modern Physiography, Hydrology, Climate, and Sediment Input

2.1. Bathymetry and Physiography

Information on bathymetry and physiography of the Arctic Ocean presented in this chapter, are obtained from the IBCAO, the most recent compilation of bathymetric data (Figure 2.1; Jakobsson et al., 2000b; Jakobsson et al., 2003a; Jakobsson, Grantz, Kristoffersen, & Macnab, 2004; Jakobsson, 2002; for latest update see Jakobsson et al., 2008). The IBCAO map is a digital 2.5 × 2.5 km grid-model on a polar stereographic projection that provides a detailed and coherent description of the morphology of the Arctic Ocean seafloor north of 64°N. As stated by Jakobsson et al. (2000b, 2003a, 2004), these data included echo soundings from United States and British navy nuclear submarine cruises between 1958 and 1988, echo soundings from nuclear submarine cruises from the SCICEX program 1993–1999 (Edwards & Coakley, 2003, 2004; see also Chapter 1.2), and echo soundings from icebreaker and research vessel cruises carried out by Canada, Germany, Norway, Russia, Sweden, and the United States, as well as published bathymetric contour maps such as the Russian 1:5,000,000 scale bathymetric contour map "Bottom Relief of the Arctic Ocean". Major part of the data was not available for the compilation of earlier maps of the Arctic Ocean.

The central Arctic Ocean with its deep basins is surrounded by continental massifs and the shallow continental shelves. The only break in this ring of continental crust is the Fram Strait between northeastern Greenland and northwestern Svalbard, the only deep-water connection with the world ocean via the Atlantic Ocean (Figure 2.1). Towards the Pacific Ocean, a shallow-water connection exists through Bering Strait. With an area of $\sim 9.5 \times 10^6\,km^2$, the entire Arctic Ocean comprises 2.6% of the total area of the world's oceans, but <1% of the volume ($\sim 13 \times 10^6\,km^3$) (Tables 2.1 and 2.2; Jakobsson, 2002; Jakobsson et al., 2003a, 2004). The relatively small total volume is due to the large shelf areas (see later description), resulting in a shallow mean depth of the entire Arctic Ocean of 1,361 m.

The Arctic Ocean proper is divided into two major subbasins, the Eurasian Basin bounded by the Lomonosov Ridge and the shallow shelves of the Barents, Kara, and Laptev seas and northern Greenland, and the Amerasian Basin bounded by the Lomonosov Ridge and the shelves of the East Siberian, Chukchi, and Beaufort seas and the Canadian Arctic Archipelago. The Eurasian Basin is subdivided by the Gakkel Ridge into the Amundsen Basin and the Nansen Basin,

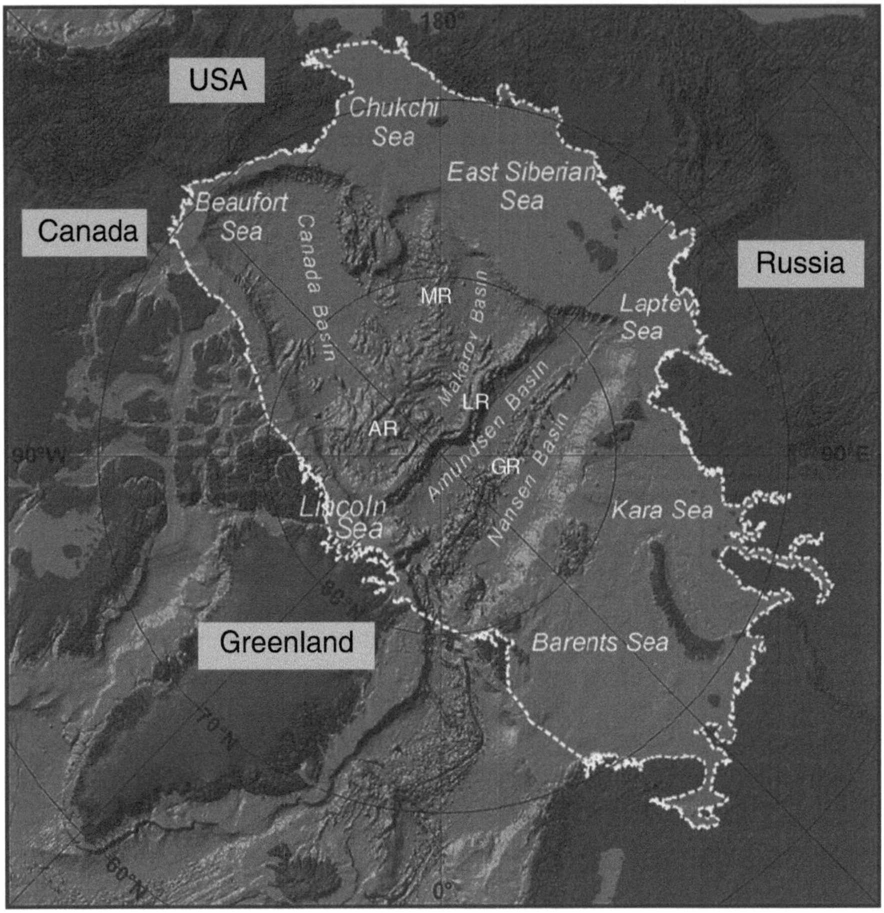

Figure 2.1 The Arctic Ocean, its shallow marginal seas, and major basins (from Jakobsson et al., 2003a, supplemented). Abbreviations used in this figure: GR, Gakkel Ridge; LR, Lomonosov Ridge; AR, Alpha Ridge; MR, Mendeleev Ridge. The dashed white line circumscribes the area of the Arctic Ocean that was adopted for Tables 2.1 and 2.2.

whereas the Alpha-Mendeleev Ridge complex subdivides the Amerasian Basin into the Canada Basin and the Makarov Basin (Figure 2.1).

The shallow continental shelves of the Arctic Ocean, defined as the area from the coastline out to the shelf break, make up as much as 52.7% of the total area of the Arctic Ocean (Figure 2.2, Table 2.1; Jakobsson et al., 2003a, 2004). These large shelf areas make the Arctic Ocean unique in comparison to the other world oceans, where the combined area of continental shelves and slopes have been estimated to range from only 9.1% and 17.7% (Menard & Smith, 1966). The Arctic shelves, however, display large regional differences in terms of size and depth (Table 2.2). Whereas the Beaufort Sea shelf is very narrow and has a mean water depth of 124 m, the East Siberian and Laptev seas are characterized by very broad shelf areas and average water depths of only 58 and 48 m, respectively (Figure 2.1). In the Kara Sea,

Table 2.1 Physiographic Provinces of the Arctic Ocean and Their Calculated Areas.

Physiographic province	Area (1,000 km^2)	Relative area (%)
1. Continental shelves	5,025	52.7
2. Continental slopes	541	5.7
3. Continental rises	733	7.7
4. Perched continental rises	362	3.8
5. Abyssal plain	1,122	11.8
6. Perched abyssal basins	222	2.3
7. Isolated basins	23	0.2
8. Ridges	1,506	15.8
9. Submarine highlands	(65)	(0.7)
Total (rounded from sum of non rounded original calculations)	9,534	100

Source: From Jakobsson et al. (2003a, 2004).
Note: The area presented for the province "Submarine Highlands" (written within brackets) has been included into the physiographic provinces that these features are located within. Furthermore, the total Area of continental shelves are given to $\sim$5,025 × 10^3 km^2, which is $\sim$27 × 10^3 km^2 smaller than what one gets summarizing the areas of all shelf seas in Table 2.2. The reason for this is that the Chukchi Spur was classified as a Ridge in Jakobsson et al. (2003a), whereas it was included in the Chukchi Sea in Jakobsson (2002).

Table 2.2 Area, Volume and Mean Depth in the Arctic Ocean Calculated with Redefined Limits for the Constituent Seas.

Seas	Area (10^3 km^2)	Volume (10^3 km^3)	Mean depth (m)
Barents Sea	1,512	302	200
White Sea	85	5	56
Kara Sea	926	121	131
Laptev Sea	498	24	48
East Siberian Sea	987	57	58
Chukchi Sea	620	50	80
Beaufort Sea	178	22	124
Lincoln Sea	64	16	257
Central Arctic Ocean Basin	4,489	12,339	2,748
SNCAA	146	49	338
SNEG	30	4	119
SNWS	6	1	93
Totals and mean depth	9,541	12,990	1,361

Source: From Jakobsson et al.(2004), modified from Jakobsson (2002).
Notes: The presented total area of the Arctic Ocean in Table 2.1 is $\sim$0.07% smaller than the presented area in this table. This is due to the different calculation techniques used in Jakobsson (2002) versus Jakobsson et al. (2003a). Abbreviations used in this table is SNCAA, Shelf of Northern Canadian Arctic Archipelago; SNEG, Shelf of Northeastern Greenland; SNWS, Shelf of Northwestern Svalbard.

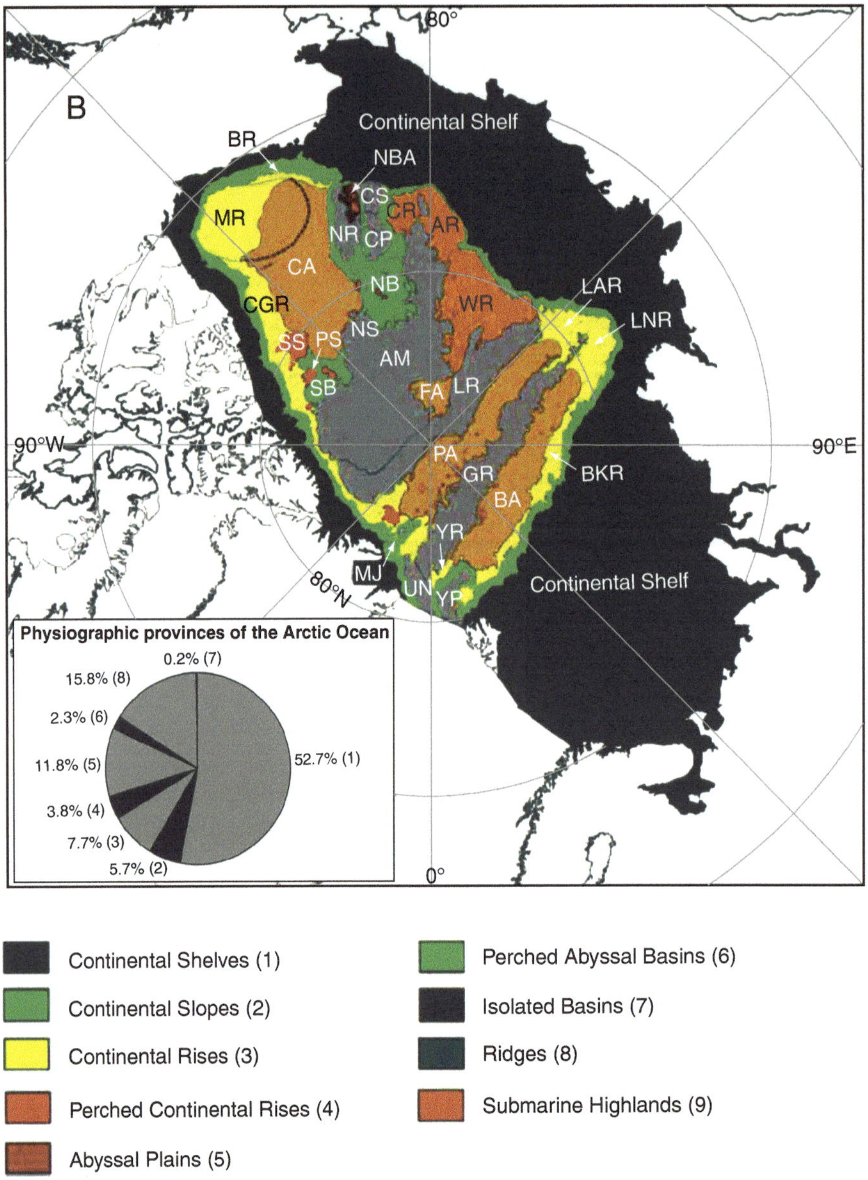

Figure 2.2 First-order physiographic provinces defined according to their bottom slope or other criteria (from Jakobsson et al., 2003a, supplemented). Abbreviations used in this figure: AM, Alpha-Mendeleev Ridge system; AR, Arlis Perched Rise; BA, Barents Abyssal Plain; BR, Beaufort Rise; BKR, Barents/Kara Rise; CA, Canada Abyssal Plain; CGR, Canada-Greenland Rise; CP, Chukchi Plateau; CR, Chukchi Perched Rise; CS, Chukchi Spur; FA, Fletcher Abyssal Plain; GR, Gakkel Ridge; LAR, Lena/Amundsen Rise; LNR, Lena/Nansen Rise; LR, Lomonosov Ridge; MJ, Morris Jesup Rise; MR, Mackenzie Rise; NB, Nautilus Basin; NBA, Northwind Basin; NR, Northwind Ridge; NS, Nautilus Spur; PA, Pole Abyssal Plain; PS, Pearya Spur; SB, Stefansson Basin; SS, Sever Spur; UN, unnamed mid-ocean ridge segment; WR, Wrangel Perched Rise; YP, Yermak Plateau; YR, Yermak Rise.

In the inlay figure, relative areas of classified first-order physiographic provinces underlying the Arctic Ocean and its shallow marginal seas, are shown. Numbers within parentheses refer to the physiographic provinces.

the third-largest Arctic marginal sea with a size of $927 \times 10^3\,km^2$ and a mean water depth of 131 m, large submarine canyons, that is, the Voronin Trough (400 m water depth) and St. Anna Trough (600 m water depth), indent the northern part of the shelf. These troughs are important for the water exchange between the shelf and the open ocean as well as the sediment export. The Barents Sea, with $1{,}512 \times 10^3\,km^2$ the largest Arctic marginal sea, is characterized by major deep depressions and troughs.

Beyond the shelf breaks are the relatively steep continental slopes, followed by the more gently sloping continental rises most of which extend from the foot of the slopes to the flat abyssal plains. The continental slopes and continental rises (excluding the perched rises; see later description) make up 5.7% and 7.7% of the Arctic Ocean, respectively (Table 2.1; Jakobsson, 2002; Jakobsson et al., 2003a). Some of the rises, however, begin at the continental slope and extend only as far as sediment dams that are located from some hundreds to more than 1,000 m above the abyssal plain that lies downslope. These are the continental-rise sedimentary prisms that formed off the western Chukchi and East Siberian shelves and separated from the Canada and Fletcher Abyssal Plains by the Northwind Ridge, the Chukchi Spur, the Chukchi Plateau, the Mendeleev Ridge, and the Lomonosov Ridge (Figure 2.2; Jakobsson et al., 2003a). These rises that are perched from some 100 to more than 1,000 m above the abyssal plain are classified as "perched continental rises" and underlie 3.8% of the Arctic Ocean (Jakobsson et al., 2003a).

Jakobsson et al. (2004) divided the deep-water area of low relief into four main abyssal plains (Figure 2.2), the Barents and Pole Abyssal Plains in the Eurasian Basin (separated by the Gakkel Ridge), and the Canadian Abyssal Plain and the Fletcher Abyssal Plain in the Amerasian Basin (separated by the Alpha-Mendeleev Ridge). In these abyssal plains, sediment thickness reaches several kilometres (Grantz et al., 1990; Jackson & Oakey, 1990; Jokat et al., 1992; Jokat, Weigelt, Kristoffersen, & Rasmussen, 1995). Beneath the Canada Abyssal Plain, the sediments are 6–14 km thick, the thickest sedimentary deposits in the Arctic Basin (Grantz et al., 1990). The Pole Abyssal Plain is the deepest abyssal plain in the Arctic Ocean with depths exceeding 4,000 m. The Barents Abyssal Plain is significantly shallower due to the large input of terrestrial sediment from the large rivers that drain into the adjacent Kara and Laptev seas (Johnson, 1969) and from glacial erosion on the Barents-Kara shelf during the Quaternary (Elverhøi et al., 1998b). In total, the Arctic Ocean abyssal plains make up ~11.8% of the Arctic Ocean (Table 2.1; Jakobsson, 2002; Jakobsson et al., 2003a, 2004).

There are two small basins that are perched above the northern part of the Canada Abyssal Plain and classified by Jakobsson et al. (2003a, 2004) as "perched abyssal basins". These are the Nautilus Basin consisting of irregularly shaped plains lying at depths of 3,200–3,800 m through which rise submarine highlands as shallow as 2,300 m, and the Stefansson Basin which lies at depths of 3,000–3,500 m and contains isolated submarine highlands with peaks that reach to 2,000 m below sea level. The perched abyssal basins, including the area of their submarine highlands, make up 2.3% of the Arctic Ocean (Figure 2.2, Table 2.1).

The submarine ridges make up 15.8% of the Arctic Ocean (Table 2.1). The main ridges are the Gakkel Ridge, the Lomonosov Ridge, and the Alpha-Mendeleev Ridge. The Gakkel Ridge, an active spreading ridge, is the continuation of the North Atlantic Mid-Ocean Ridge system that enters the Arctic Ocean through the Fram Strait (see Chapter 1.3). The deepest area of the Arctic Ocean is found in the axial valley of the Gakkel Ridge near 81°20′ N 120° 45′ W close to the Laptev Sea margin where the IBCAO grid indicates 5,243 m (Jakobsson, 2002). The Lomonosov Ridge dividing the Arctic Ocean into the Eurasian and Amerasian basins and separated as continental sliver from the Eurasian continental margin by seafloor spreading near 56 Ma (Wilson, 1963; Kristoffersen, 1990; see Chapter 1.3), spans across the Arctic Ocean from the northern Greenland margin to the shelf off the New Siberian Islands. This submarine ridge reaches depths shallower than 700 m and is important for the Arctic Ocean oceanographic circulation (Anderson et al., 1994; Rudels, Jones, Anderson, & Kattner, 1994; see Chapter 2.2). The shallow areas of the ridge crest $< \sim$1,000 m of water depth have probably been affected by extensive erosion (Jakobsson, 1999) that has been attributed to ice-grounding (Polyak et al., 2001) or possible ice-grounding in combination with strong bottom currents during the Late Pleistocene (Jakobsson et al., 2001). The largest submarine ridge system in Arctic Ocean is the Alpha-Mendeleev Ridge in the Amerasian Basin, a complex physiographic feature with numerous valleys and elevations that range from more than 2,000–740 m below sea level. (Jakobsson et al., 2003a; see Chapter 1.3 for more details concerning its tectonic evolution.) In addition to these large ridge systems, the Arctic Ocean contains a number of smaller systems which are all connected to their adjacent continental shelves: the Northwind Ridge and the Chukchi Spur–Chukchi Plateau composite ridge of the Chukchi Continental Borderland in the Amerasian Basin, and the Yermak Plateau and the Morris Jessup Rise in the Eurasian Basin (Jakobsson et al., 2003a, 2004; Figure 2.2).

2.2. Oceanic Circulation Pattern and Water-Mass Characteristics

A schematic view of the surface, intermediate, and deep-water circulation patterns within the Arctic Ocean and the Norwegian Sea is summarized in Figure 2.3, indicating the main oceanographic elements: (1) the Bering Sea inflow from the Pacific, (2) the Beaufort Gyre, (3) the Transpolar Drift (TPD), (4) the inflow of warm Atlantic waters via the Norwegian/West Spitsbergen Current (WSC) system, (5) the outflow of cold waters via the East Greenland Current, (6) the Greenland Gyre, and the (7) the Island Gyre, as well as the main Arctic Ocean water masses (Aagaard, Swift, & Carmack, 1985; Aagaard & Carmack, 1994).

Based on the distribution of potential temperature, salinity, and density, the Arctic Ocean's hydrographic structure consists of three main water masses: the upper waters, the intermediate waters (Atlantic Layer), and the deep waters (Figures 2.4 and 2.5; Aagaard et al., 1985; Schlosser, Swift, Lewis, & Pfirman, 1995; Macdonald & Bewers, 1996).

The upper waters can be further divided into the Polar Mixed Layer (PML) with depths typical between 30 and 50 m, and the Arctic halocline ranging 30–50 m and $\sim$200 m water depth (Figure 2.4). The PML is characterized by temperatures close

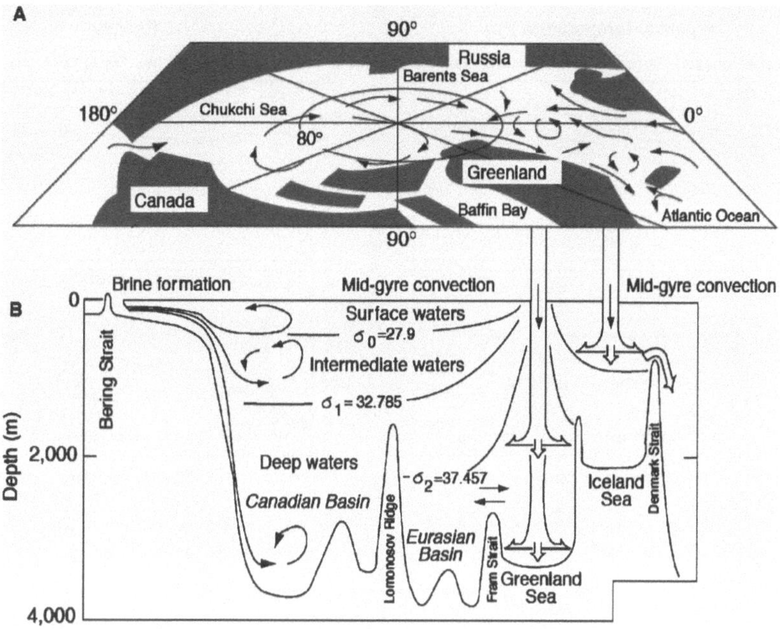

Figure 2.3 Schematic drawing of the circulation and water mass structure in the Arctic Ocean and the GIN Sea (from Myhre et al., 1995, after Aagaard, Swift, & Carmack, 1985). (A) The upper map shows the major elements of the Arctic Ocean and the GIN Sea. (B) The lower panel describes the water budget of northern polar deep-sea basin intermediate and deep-water movements. The selected isopycnals ($\sigma_0\sigma_1\sigma_2$) separate surface, intermediate and deep waters.

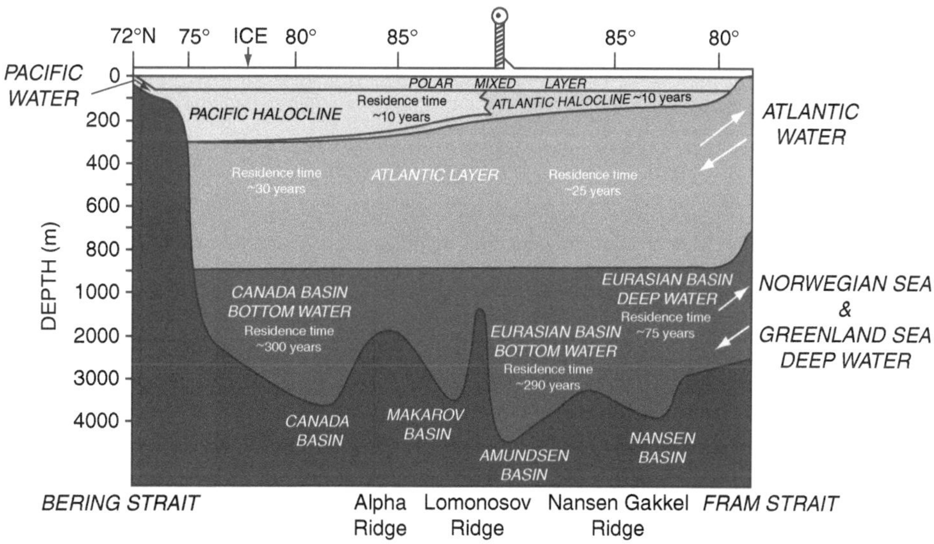

Figure 2.4 A schematic diagram showing the stratification of the Arctic Ocean (from Macdonald et al., 2004a, according to Macdonald & Bewers, 1996).

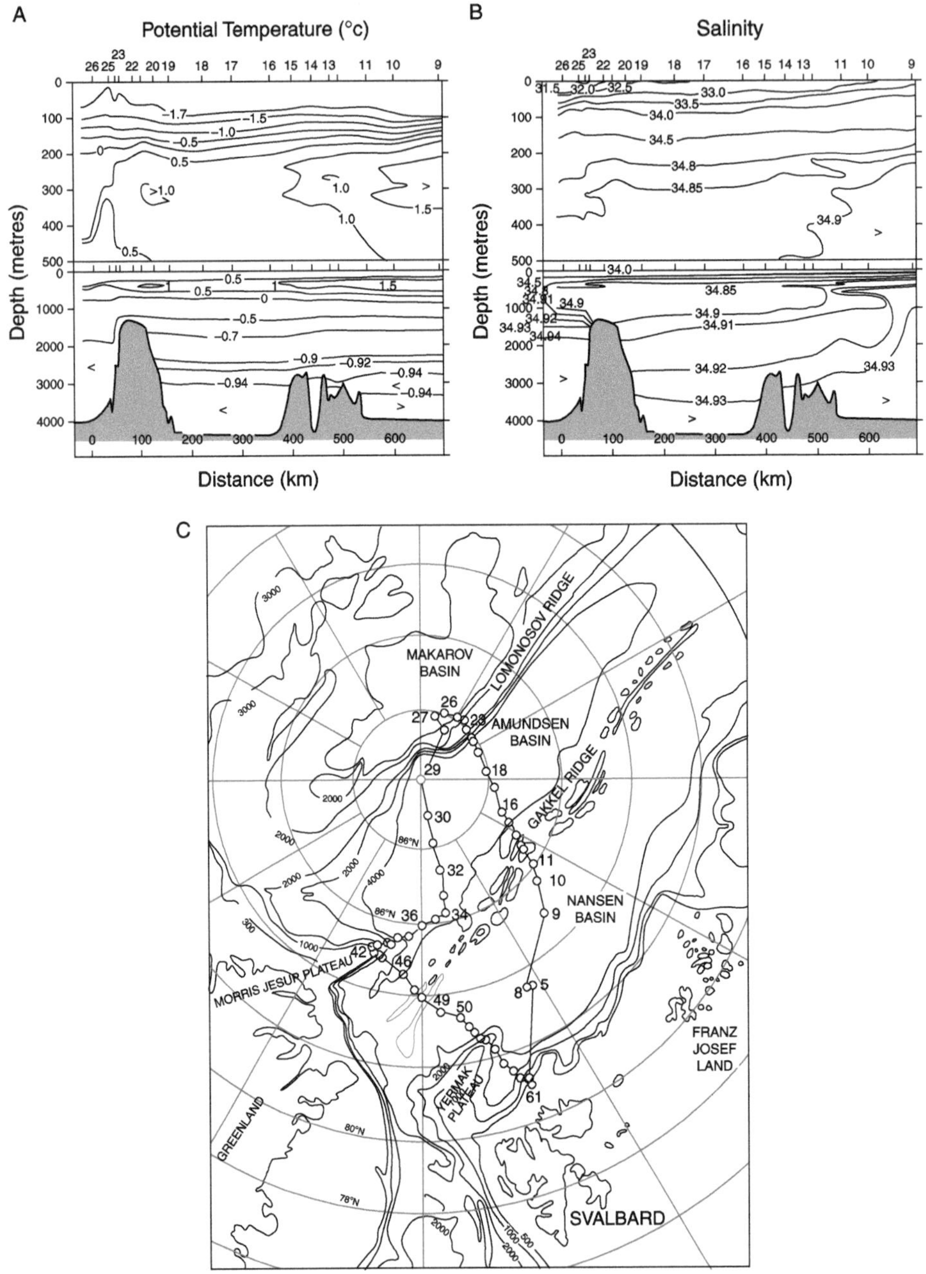

Figure 2.5 (A) Profile of potential temperature and (B) salinity determined at hydrographic stations during the Oden 91 expedition (from Rudels et al., 1994). See (C) for location of profile.

to the freezing point and very low salinities caused by river runoff (see also Chapter 2.5), and derives from local mixing of waters below the sea-ice cover. Brine produced by ice formation in winter tends to destabilize the water column allowing it to mix while in summer, melting of sea ice and freshwater runoff produces stratification with a fresher surface layer.

The Arctic halocline, a complex of cold, salt-stratified layers, is permanent over the deep basins and limits the exchange between the deep ocean and the surface ocean. The various components of the halocline can be ascribed to (i) ice melt mixed with water of Atlantic origin, (ii) Pacific origin in flow waters, and (iii) brine-enriched shelf drainage.

Below the surface waters, a more warm and salty intermediate water mass exists, known as the Atlantic Layer (Figure 2.4). The Atlantic Layer is ubiquitous over all Arctic basins and typically occurs in water depths between ~200 and 800 m. Traditionally, it is defined as the layer between the 0°C isotherms (e.g., Schlosser et al., 1995). Water within this layer enters the Arctic Ocean via the WSC Current, an extension of the North Atlantic-Norwegian Current, through Fram Strait where it cools and extends at intermediate water depths along the Eurasian continental margin (Aagaard et al., 1985; Carmack, 1990; Rudels et al., 1994; Fahrbach et al., 2001). In addition, Atlantic water is flowing across the Barents Sea and through the St. Anna Trough, reaching water depths of ~1,200 m in the Arctic Ocean (Rudels et al., 1994; Schauer, Muench, Rudels, & Timokhov, 1997a; Schauer et al., 1997b). The Barents Sea branch meets with the branch of Atlantic water coming from Fram Strait, and both branches flow at intermediate water depths in cyclonic loops following the continental slope and ridges of the deep basins (Figure 2.6; Rudels et al., 1994; Schauer, Loeng, Rudels, Ozhigin, & Dieck, 2002). The two contributions of Atlantic water appear to be of equal strength (Rudels et al., 1994). Although they have the same source (i.e., the Norwegian Current), the Barents Sea branch is considerably colder and less saline than the Fram Strait branch. Because the former is colder than 0°C, it will not be addressed as Atlantic water in the Arctic Ocean according to the traditional definition (Schauer et al., 2002).

The deep waters below the Atlantic Layer, representing ~60% of the total volume, are characterized by relatively high salinities. All salinities measured in the central Arctic Ocean in water depths below 1,500 m are in excess of ~34.92 (Schlosser et al., 1995). Deep-water characteristics are strongly influenced by bathymetry, especially the Lomonosov Ridge. The fairly isolated deep water in the Canada Basin is the warmest (−0.5°C) and most saline (>34.95), whereas the deep water in the Eurasian Basin is characterized by temperatures and salinities of −0.7°C and 34.94, respectively (Carmack, 1990). These salinities are too high to be explained solely by lateral deep-water advection from the Greenland Sea. An additional source for salt is needed. In this context, the formation of brines on polar shelves, that is, cold, saline, and well-oxygenated water masses which sink over the continental margin into the deep basins, is probably important (Aagaard et al., 1985). This hypothesis is supported by the comparison of the $^{18}O/^{16}O$ ratios of shelf water, Atlantic water, and the deep waters of the Arctic Ocean, indicating that the sources of the deep and bottom waters of the Eurasian Basin are located in the Barents and Kara seas (Bauch, Schlosser, & Fairbanks, 1995).

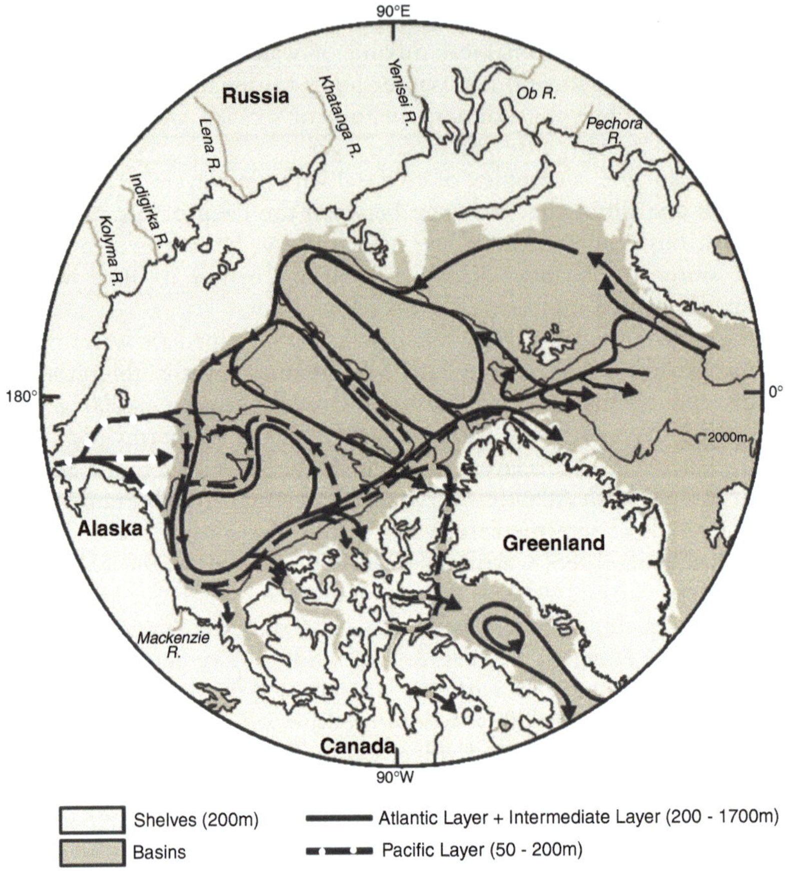

Figure 2.6 A schematic diagram showing the locations of boundary currents in the Arctic Ocean (from Macdonald et al., 2004a, according to Rudels et al., 1994; McLaughlin, Carmack, Macdonald, & Bishop, 1996; Jones, Anderson, & Swift, 1998; Jones, 2001).

2.2.1. Surface-water Characteristics: Temperature, Salinity, and Currents

Whereas the central Arctic Ocean mean sea-surface temperatures (SST) are $<-1°C$ throughout the year, a distinct seasonal variability is obvious in the marginal seas (Figure 2.7A and B; EWG, 1998). In the latter, mean SST reaches values of 0–5°C during summer at times of increased river discharge and Atlantic water inflow. During winter times, low mean SST of $<-1°C$ similar to those of the central Arctic Ocean were also measured in the marginal seas. An exception from this general picture is the Barents Sea, strongly influenced by Atlantic water inflow. Here mean summer SST may increase to 8°C and even during winter time, values of up to 4°C were determined in the southwestern part (Figure 2.7A and B).

The distribution of the sea-surface salinity (SSS) is strongly controlled by the river discharge and the Atlantic water inflow (Figure 2.7C and D). The high river discharge results in a very low salinity in the Arctic Ocean proper as well as the

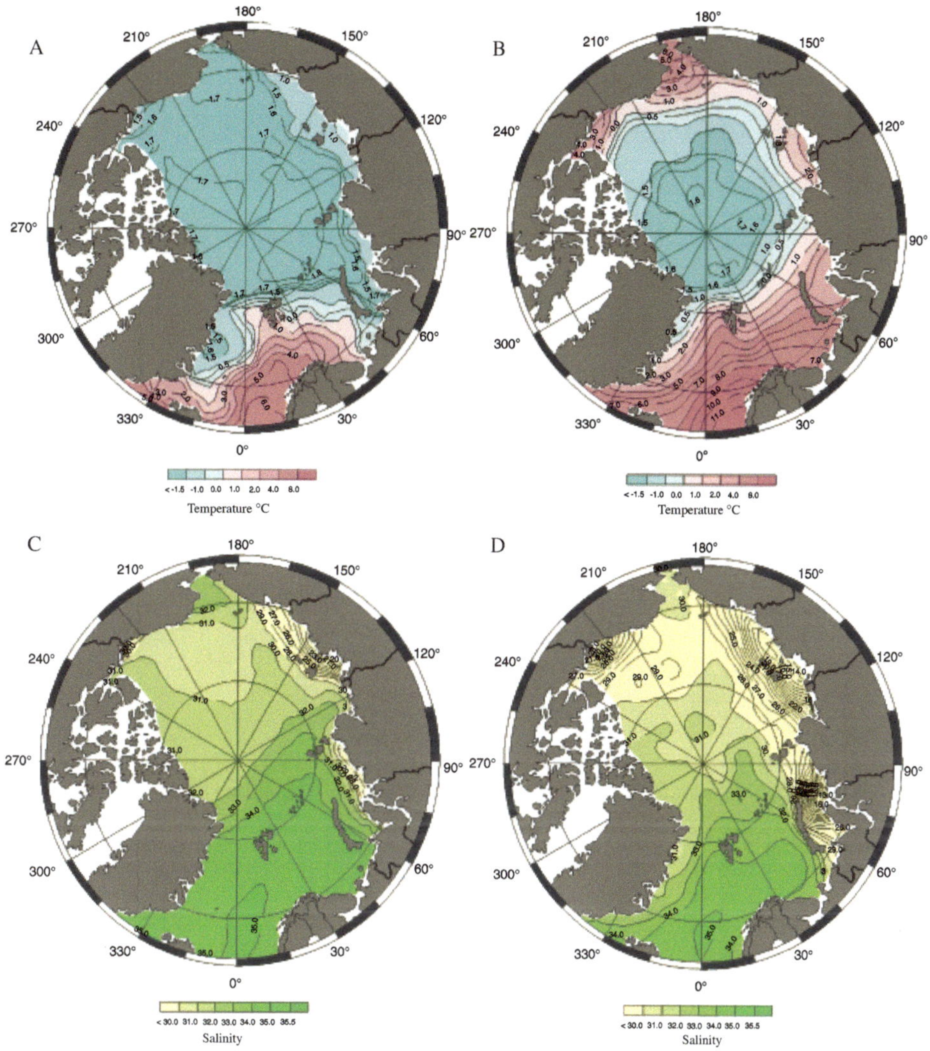

Figure 2.7 Surface-water temperature ((A) winter; (B) summer) and surface-water salinity ((C) winter; (D) summer) (from EWG, 1998).

marginal seas, except for the Barents Sea, influenced by more saline Atlantic water inflow. The Beaufort, East Siberian, Laptev, and Kara seas are characterized by mean SSS <29 per mil during the summer, strongly decreasing towards the river mouths, in most part of the central Arctic, mean SSS is below 32 per mil. In the Barents Sea and adjacent continental slope, mean SSS increases to >34 per mil. Owing to seasonal variability in river discharge, the SSS also displays a strong variability, especially in the marginal seas (Figure 2.7C and D).

The circulation of the surface waters is quite well understood from studies of sea-ice drift. Today, two major current systems dominate the surface-water circulation in the Arctic Ocean: the anticyclonic Beaufort Gyre occupying most of

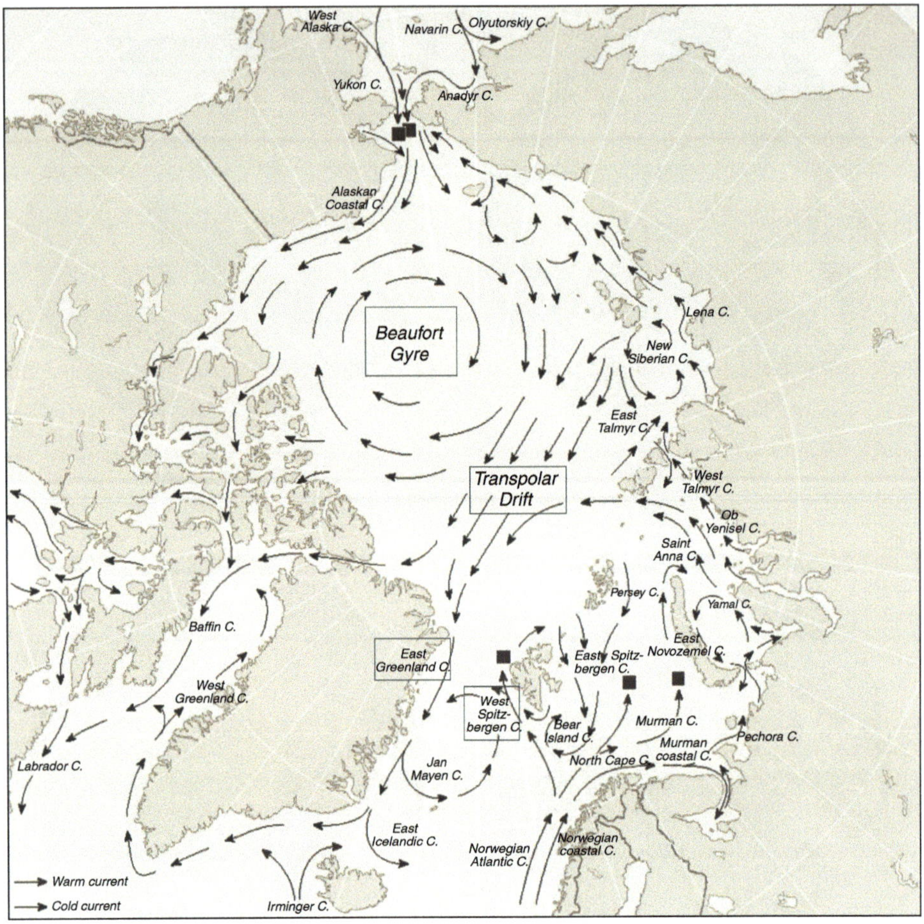

Figure 2.8 Surface-water circulation in the Arctic Ocean (from Macdonald, Harner, Fyfe, Loeng, & Weingartner, 2003, supplemented).

the Amerasian Basin, and the TPD in the Eurasian Basin (Figure 2.8). The latter system transports sea ice from the Siberian shelves through the central Arctic Ocean to the Fram Strait (see Chapter 2.3). Through Fram Strait, two major currents, with southward flowing waters on the west and northward flowing waters on the west, exchange water between the Arctic and the North Atlantic: The cold, ice-transporting East Greenland Current is the main current out of the Arctic Ocean. In contrast, the eastern WSC, an extension of the North Atlantic-Norwegian Current, carries warm, relatively saline water into the Arctic Ocean (see earlier).

Using a two-dimensional, wind-forced, barotropic model that includes frictional coupling between the ocean and ice, and was run over the time interval from 1946 to 1993, Proshutinsky and Johnson (1997) showed that the wind-driven motion of water in the central Arctic Ocean within the Beaufort Gyre and the TPD alternates between anticyclonic and cyclonic circulation patterns (Figure 2.9). Each circulation regime seems to persist for a period of 5–7 years, with anticyclonic

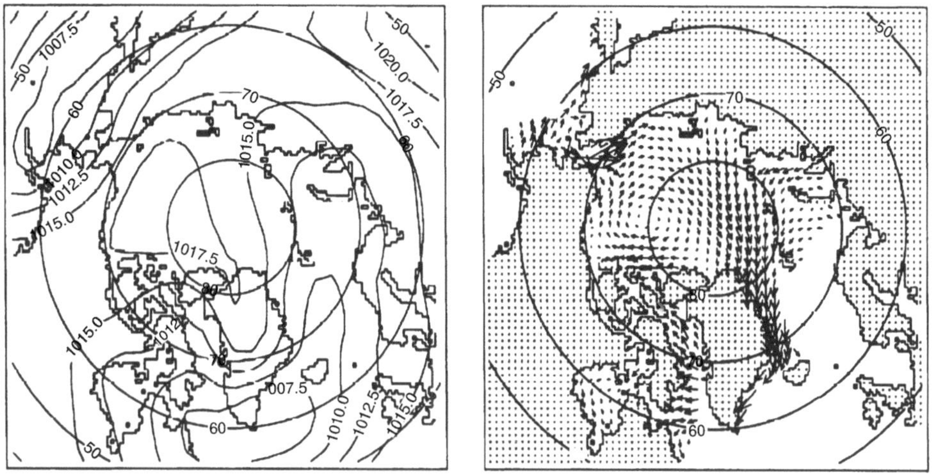

(a) Anticyclonic circulation pattern (1958-1963)

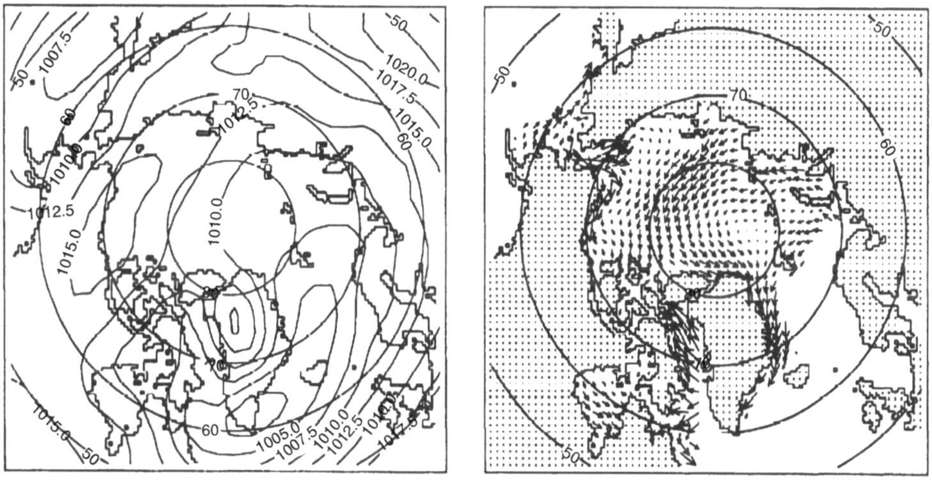

(b) Cyclonic circulation pattern (1989-1993)

Figure 2.9 (A) Sea-surface atmospheric pressure (left) and wind-driven ice motion in the Arctic averaged for the anticyclonic period 1958–1963. (B) Sea-surface atmospheric pressure (left) and wind-driven ice motion in the Arctic averaged for the cyclonic period 1989–1993 (from Proshutinsky & Johnson, 1997).

wind-driven motion occurring during 1946–1952, 1958–1963, 1972–1979, and 1984–1988, and cyclonic motion occurring during 1953–1957, 1964–1971, 1980–1993, and 1989–1993. The shifts from one regime to another are forced by changes in the location and intensity of the Islandic Low and the Siberian High (Proshutinsky & Johnson, 1997). The model's ice motion agrees quite well with observed ice motion from buoys (Thorndike & Colony, 1982). During an anticyclonic phase, the Beaufort Gyre system as well as the TPD system are strongly intensified (Figure 2.9). The former is transporting sea ice (and entrained sediments) from the Canadian Arctic into the central Arctic, the latter shows a direct transport of sea ice (and entrained sediments) across the central Arctic towards Fram Strait. During a strong cyclonic phase, on the other hand, sea ice (and entrained sediments) from the Laptev and Kara seas may be transported into the central Arctic including the Canada Basin (Figure 2.9). These changes in circulation patterns are also reflected in the sedimentary records (see Chapter 6.3.6).

In terms of transport rates, the exchange with the North Atlantic is most important. About 1–1.5 Sv of waters enters the Arctic through Fram Strait (even 3–5 Sv according to Woodgate, Schauer, & Fahrbach, 1998), and ~2 Sv flow into the Arctic via Barents Sea. On the west side of Fram Strait, 3–3.5 Sv are exported into the North Atlantic (9±2 Sv according to Woodgate, Fahrbach, & Rohard, 1999). The export of sea ice through Fram Strait and its melt in the Greenland-Norwegian Sea together with the export of copious amounts of freshwater from rivers play an important role in controlling the deep-water formation in the northern North Atlantic (Figure 2.3) and, thus, is of global significance (e.g., Aagaard & Carmack, 1989; Untersteiner & Carmack, 1990). Transport through other passages are smaller. From the Pacific, ~1 Sv flows into the Arctic via the shallow Bering Strait (e.g., Roach et al., 1995). This nutrient- and silicate-rich, less saline water dominates the surface waters for much of the Canadian Basin. Through the Canadian Archipelago, ~1–2.7 Sv are exported (Loeng, Ozhigin, Adlandsvik, & Sagen, 1993; Melling, 2000).

2.3. Sea-Ice Cover: Extent, Thickness, and Variability

An important phenomenon of the Arctic Ocean is the permanent sea-ice cover with its strong seasonal variation in the marginal (shelf) areas (Figure 2.10; Johannessen et al., 2004, and further references therein). Sea ice is a very critical component of the Arctic system that responds sensitively to changes in atmospheric

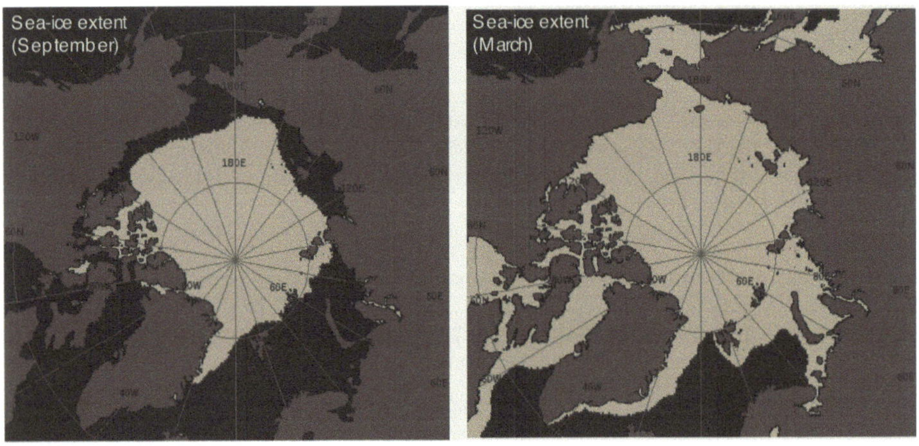

Figure 2.10 Map showing the average distribution of sea ice in the Arctic Ocean in September (1979–2004) and March (1979–2005); according to Maurer (2007; http://nsidc.org/data/atlas/).

circulation, incoming radiation, atmospheric, and oceanic heat fluxes, as well as the hydrological cycle. Ice significantly reduces the heat flux between ocean and atmosphere; through its high albedo it has a strong influence on the radiation budget of the entire Arctic. The albedo of open water is as low as 0.10, whereas the ice albedo ranges between 0.6 and 0.8 (Barry, 1996). Thus, over ice surfaces as compared to open water up to eight times as much of the incoming shortwave radiation is reflected, resulting in lower surface temperatures. Furthermore, the sea-ice cover strongly affects the biological productivity, as a more closed sea-ice cover restricts primary production due to low light influx in the surface waters (see Chapter 2.4). Sea ice is also an important agent for sediment transport from the shelves into, across, and out of the Arctic Ocean (see Chapter 3.1). As the sea ice is sensitive to atmospheric and oceanic variability, and because sea ice is involved in several key climate feedbacks (ice-albedo feedback, cloud-radiation feedback, etc.), sea ice probably plays a substantial role in climate system variability.

There are two primary forms of sea ice: seasonal (or first-year) ice and perennial (or multiyear) ice. The thickness of first-year ice ranges from a few tenths of a metre near the southern margin of the marine cryosphere to 2.5 m in the high Arctic at the end of winter. Some first-year ice survives the summer and becomes multiyear ice. In the present climate, old multiyear ice floes without ridges are ~3 m thick at the end of winter. In addition, land-fast ice (or fast ice) occurs in the coastal circum-Arctic area, which is immobilized for up to 10 months each year by coastal geometry or by grounded ice ridges (stamukhi) (ACIA, 2004). Under recent climatic conditions, ice ridges ground to form stamukhi in depths of up to 30 m, as the pack ice is repeatedly crushed against the fast ice by storm winds. Within the Canadian Archipelago in late winter, land-fast ice bridges channels up to 200 km wide and covers an area of 1 million km^2. Some of this ice is trapped for decades as multiyear land-fast ice (Reimnitz, Eicken, & Martin, 1995).

2.3.1. Sea-Ice Cover and Its Variability

Sea-ice concentration (percent ice area per unit area) and derived parameters such as ice extent (the area within the ice-ocean margin defined as 15% ice concentration) can be reliably retrieved through satellite-borne passive-microwave sensors, which are available continuously since 1978 and operating independently of cloud cover and light conditions (Gloersen et al., 1992; Johannessen et al., 2004). Satellite data have shown that the area of sea ice decreases from roughly 14–15 million km^2 in March to 6–7 million km^2 in September, as much of the first-year ice melts during the summer (Figure 2.11) (Gloersen et al., 1992; Cavalieri, Gloersen, Parkinson, Comiso, & Zwally, 1997; Johannessen et al., 2004). The area of multiyear sea ice, mostly over the Arctic Ocean basins, the East Siberian Sea, and the Canadian polar shelf, is ~4–5 million km^2 (e.g., Johannessen et al., 1999; Nghiem et al., 2007).

In the central Arctic Ocean, the seasonal change in ice coverage only amounts to a small fraction (14%). Most of it, 86% of the seasonal variation, takes place in the marginal seas, over the shallow shelf regions. In winter these areas are usually completely covered with ice, whereas in summer they are mostly ice-free (Figure 2.10) (Kolatschek, Eicken, Alexandrov, & Kreyscher, 1996). An exception

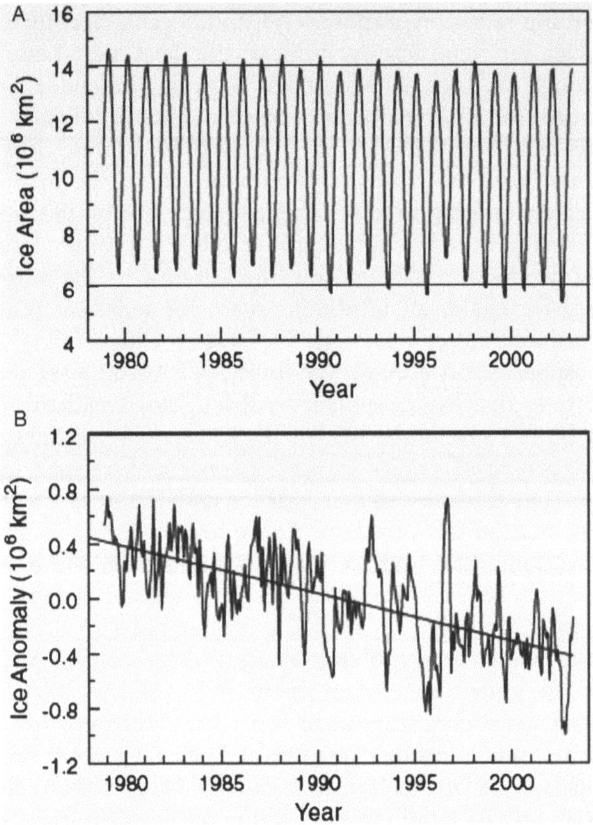

Figure 2.11 Total sea-ice area (the area within the ice-ocean margin minus open-water area) for the Northern Hemisphere, 1978–2003, as retrieved from satellite passive-microwave remote-sensing data (from Johannessen et al., 2004). (A) Monthly ice area, indicating that the predominant variability is the seasonal cycle. (B) Monthly anomalies or departures from the mean and seasonal cycle. The linear trend was $-0.34 \times 10^6\,\text{km}^2$ per decade during the period. The largest negative anomaly is found in the most recent data, with record-low ice cover in September 2002.

is the Barents Sea and the area around Svalbard, strongly influenced by the warm Norwegian/WSC system. There, large parts are ice-free throughout the year (Figure 2.10).

Over the last decades, Arctic sea ice has undergone a dramatic change. Satellite data indicate a decrease of $\sim 0.8 \times 10^6\,\text{km}^2$ ($\sim$7.4%) in the Northern Hemisphere annual sea-ice area in the period between 1978 and 2003 (Figure 2.11). During this period, the decrease was larger in summer. A decrease of $0.94 \times 10^6\,\text{km}^2$ (14%) was calculated for September versus a decrease of $0.68 \times 10^6\,\text{km}^2$ (5%) for March (Johannessen et al., 2004; see also Rigor & Wallace, 2004; Francis et al., 2005; Stroeve et al., 2007). An alarming record-low in minimum sea-ice cover was observed in September 2007, which is $\sim$40% less than that of 1979, the start of sea-ice observation by satellites (see Chapter 1.1, Figure 1.2; Kerr, 2007). Such a minimum was forcasted by modelling to occur in the middle of this century (see Chapter 1.1, Figure 1.3; Johannessen et al., 2004). Concerning the perennial (multiyear) sea-ice extent, also a major decrease has been observed during the last 30 years, with a very distinct and rapid decrease from $4.69 \times 10^6\,km^2$ to $3.61 \times 10^6\,km^2$ (i.e., a loss by $1.08 \times 10^6\,km^2$ or 23%) between March 2005 and March 2007 (Nghiem et al., 2007).

2.3.2. Sea-Ice Thickness

The knowledge of the sea-ice thickness is not as thorough and only estimates of low spatial and temporal coverage are available. These data are mainly based on sonar recordings from military submarines taken from underneath the ice (Wadhams, 1997; Rothrock et al., 1999). Similar data at fixed locations are provided by upward looking sonars mounted on oceanographic moorings, and some time series exist over fixed locations, but there is no large geographical coverage. Drilling is by far the most accurate means to estimate ice thickness (Eicken et al., 1995), but obviously supplies only very limited data sets.

The average thickness of sea ice in the Arctic Ocean is ~2.6–3 m (Figure 2.12; Bourke & Garrett, 1987; Eicken et al., 1995; Johannessen et al., 2004). The lowest values of ice thickness are found in the Eurasian sector of the Arctic, with values between 1 and 2 m, corresponding to the growth of ice during one season. In the direction to the North Pole and to the northern coast of Greenland and the Canadian Arctic Archipelago, ice thickness is continuously increasing (Figure 2.13). The maximum thickness is reached near the Canadian Archipelago with some 7–8 m (Bourke & Garrett, 1987). There is little information about the thickness of the seasonal sea ice that covers more than half of the marine Arctic.

As the sea-ice cover, sea-ice thickness appears also to decrease during the last decades. Based on sonar data, the mean ice thickness decreased from 3.1 to 1.8 m, corresponding to an ~40% reduction over the last three to four decades (Rothrock et al., 1999). However, analyses of sonar data alone from different transects, years, and seasons yield a range of estimates and the question is, how representative these measurements are (Johannessen et al., 2004). An analysis of submarine (Rothrock et al., 1999; Rothrock et al., 2003) and modelled (Holloway & Sou, 2001) ice thickness over the same time period has demonstrated that ice motion and high interannual variability could mislead inference of trends from sonar transect data, for example 12% (Holloway & Sou, 2001) versus the 40% decrease reported by Rothrock et al. (1999). Based on 20-year time series of monthly, area-averaged ice thickness derived from field-based measurements of surface elastic-gravity waves from Russian North Polar drifting stations between 1970 and 1991, a decrease of only ~10 cm (<4%) over the 20 years was determined (Johannessen et al., 2004).

2.3.3. Sea-Ice Drift

The drift of sea ice in the Arctic today is a direct response to surface atmospheric pressure gradients and resulting wind patterns, mainly related to the AO (Thompson & Wallace, 1998; Rigor et al., 2002). Any change in these pressure

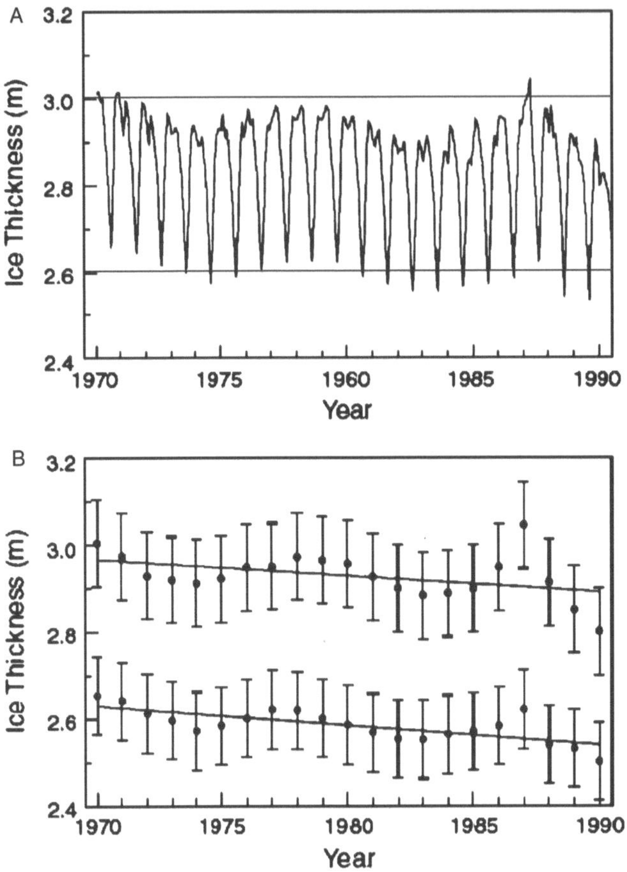

Figure 2.12 Arctic sea-ice thickness variability from 1970 to 1990 (from Johannessen et al., 2004, supplemented). (A) Monthly area-averaged thickness estimates as derived from surface-based measurements of ice-surface vibrations made from Russian North Pole drifting stations in the perennial ice pack of the Arctic Ocean (Nagurnyi, Korostelev, & Abaza, 1994; Nagurnyi, Korostelev, & Ivanov, 1999). (B) Interannual variability and linear trends for winter (April) and summer (August), with errors bars denoting the 95% confidence interval of the ice thickness estimates.

gradients, such as the AO, has a profound effect on both the pattern and velocity of ice motion in the Arctic Ocean. Variations in wind associated with the AO modulate the location of the TPD (Mysak, 2001; Mysak & Venegas, 1998; Kwok, 2000; see Chapter 6.3.1, Figure 6.59), the major ice drift system controlling sea-ice export from the Arctic through Fram Strait. During a positive AO situation, the AO induces the TPD to intensify, shifting farther eastward from Siberia towards North America. This change in the TPD also causes an increased export through Fram Strait of the thicker ice in the Beaufort Gyre (Hilmer & Jung, 2000; Rigor et al., 2002; Kwok & Rothrock, 1999), reducing the surface salinity in the Greenland Sea, which subsequently diminishes the thermohaline circulation by slowing the production of NADW (Aagaard & Carmack, 1989; Raymo, Rind, &

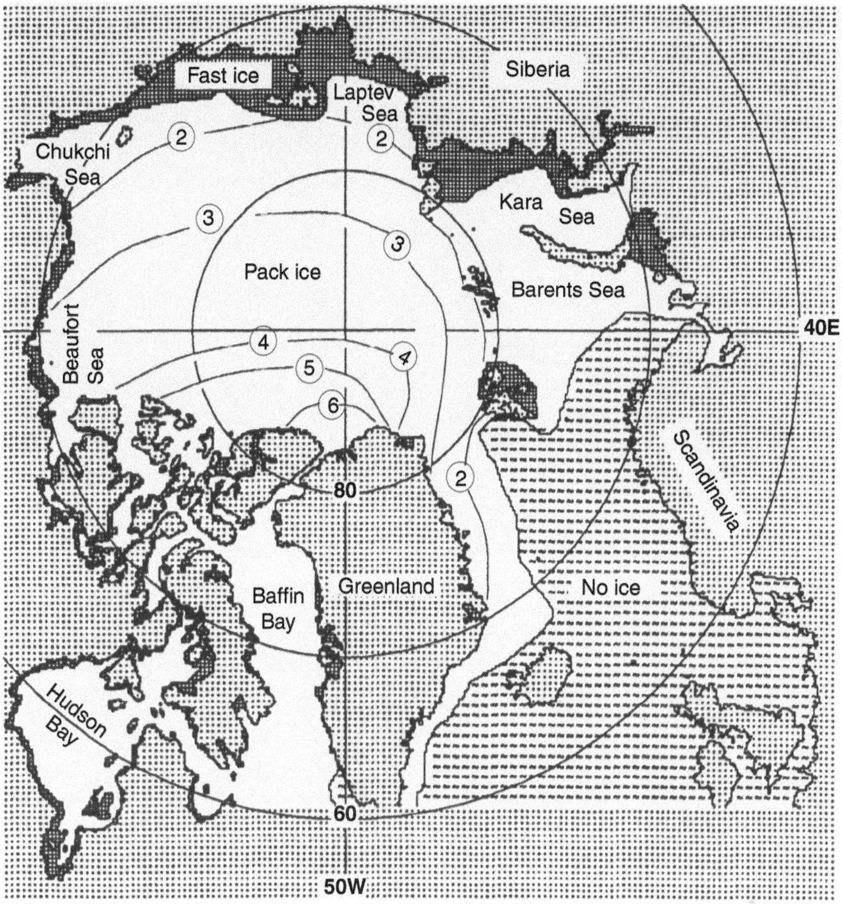

Figure 2.13 Contours of ice thickness in metres in the Arctic for average winter conditions (from Proshutinsky & Johnson, 1997, supplemented). The model ice thickness is explicitly prescribed and does not change in the simulations presented here. Data combined from Bourke and Garret (1980), Hibler (1989), and Romanov (1993).

Ruddiman, 1990). During a negative AO situation, the TPD is more restricted to the Russian Arctic, and the clockwise Beaufort Gyre is dominant in the Arctic Ocean near North America. The causes of these changes that generate the fluctuations in the AO and that correlate to fluctuations in the NAO are yet to be established (e.g., Dickson et al., 2000).

The mean sea-ice drift pattern is quite well known from drift buoy and geostrophic wind field data (Thorndike & Colony, 1982; Colony & Thorndike, 1985; Hibler, 1989; Pfirman, Colony, Nürnberg, Eicken, & Rigor, 1997). North of the Alaskan coast the ice circulates in a big anticyclonic structure, the Beaufort Gyre, corresponding to the high-pressure system over the Beaufort Sea. Speed varies from almost zero in the centre to 3 cm s^{-1} in the marginal zones. During part of the year, reversals in the drift can lead to divergence and the generation of open

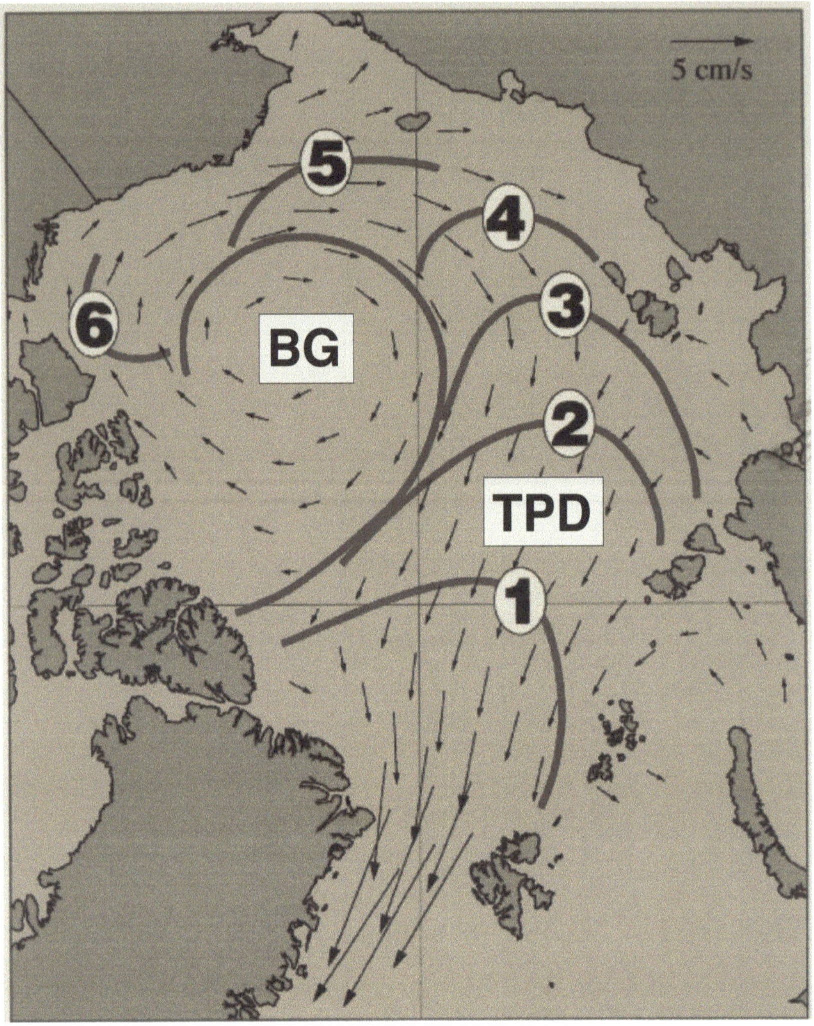

Figure 2.14 Mean field of ice drift in the Arctic Ocean derived from buoy drift from 1979 to 1994 (from Hovland, 2001, supplemented; based on Colony & Thorndike, 1985; Thorndike, 1986; Pfirman et al., 1997). Velocities are indicated by arrows. Numbered lines indicate the average number of years required for ice in this location to exit the Arctic through Fram Strait. BG, Beaufort Gyre; TPD, Transpolar Drift.

water in the gyre (Barry, Serreze, & Maslanik, 1993). Towards the Siberian Arctic this gyre opens up to a linear structure, the TPD. Its origin is in the area of the Chukchi-East Siberian Sea and it leads to the West. Ice drifts with mean velocities of $\sim 2\,\mathrm{cm\,s^{-1}}$ past Franz Josef Land and the North Pole, accelerating to mean values of up to $10\,\mathrm{cm\,s^{-1}}$ through Fram Strait between Svalbard and Greenland (Kolatschek et al., 1996). It takes 3–4 years for the ice to move from East Siberia to Fram Strait (Figure 2.14) (Pfirman et al., 1997; Hovland, 2001). Ice which gets caught in the Beaufort Gyre remains longer in the Arctic Basin, typically it takes

5–10 years for the ice to complete one cycle (Thorndike, 1986). Ice Island T-3 (see Figure 1.5) made two rounds in the gyre and exited the Fram Strait ~30 years after its discovery north of Canada.

The movement of the ice near Alaska and the Chukchi Sea within the Beaufort Gyre is parallel to the coast, whereas the direction of the TPD in the Laptev Sea and western East Siberian Sea is in general away from the coast. Thus, with the Beaufort Gyre dominating circulation in the North American Arctic as well as the Chukchi and eastern East Siberian Seas, these regions contribute on average only small amounts of exported first-year ice to the ice budget of the Arctic Ocean. In contrast, the TPD results in a large export of first-year ice from the Laptev, western and central East Siberian and Kara Seas (Timokhov, 1994; Eicken, 2004). Especially the Laptev Sea is the most important "ice factory" of the Arctic. Based on direct measurements, remote sensing and modelling, the net export of sea ice from the marginal seas into the Arctic Ocean and the export through Fram Strait into the Greenland Sea were estimated (Eicken, 2004, and further references therein). Estimates of ice export from the Laptev Sea are as high as $670\,km^3\,yr^{-1}$ which is more than the export from Barents Sea ($35\,km^3\,yr^{-1}$), Kara Sea ($240\,km^3\,yr^{-1}$), East Siberian Sea ($150\,km^3\,yr^{-1}$), and Chukchi Sea ($10\,km^3\,yr^{-1}$) together. Through Fram Strait, about $2850\,km^3\,yr^{-1}$ of ice are exported.

2.3.4. Sea-Ice Sediments

A large proportion of the sea ice is "dirty ice" containing high amounts of sediment (e.g., Pfirman, Gascard, Wollenburg, Mudie, & Abelmann, 1989; Wollenburg, 1993; Nürnberg et al., 1994; Pfirman et al., 1997; Eicken et al., 2000). These sediments are transported with the sea-ice drift throughout the entire Arctic Ocean. Details on sea-ice sediment characteristics, processes related to sediment entrainment into sea ice and the significance of sediment transport by sea ice are described in Chapter 3.1.

2.4. Primary Production and Vertical Carbon Fluxes in the Arctic Ocean

2.4.1. Primary Production in the Arctic Ocean and Surrounding Marginal Seas

A comprehensive review of the primary production in the permanently ice-covered central Arctic Ocean and its marginal seas is given by Sakshaug (2004). Here, only a short summary is presented, and for more detailed information it is referred to Sakshaug's review paper and further references therein.

Marine bio-production in the Arctic and Subarctic is based primarily on microscopic unicellular algae (phytoplankton) in the water column and microalgae associated with ice. The latter may even become dominant in the central Arctic Ocean where algae in multiyear ice can contribute over half of the total primary production of the whole water column, compared to 3% in the first-year ice of the surrounding regions (Gosselin, Levasseur, Wheeler, Horner, & Booth, 1997; Sakshaug, 2004). Owing to its light limitation and sea-ice cover, the central Arctic Ocean is the least productive region of the world's oceans, even when the estimated productivity values were adjusted upward by an order of magnitude during the last ~10 years (see later description) (Sakshaug, 2004). In Table 2.3, average annual primary production (in $gC\,m^{-2}y^{-1}$) as well as new production (in $gC\,m^{-2}y^{-1}$) and total areal primary production (total input of algae-derived organic carbon in 10^6t y^{-1}) in the different Arctic areas are summarized.

Table 2.3 Approximate Annual Particulate Primary Production (New and Total; DOC not included) in $gC\,m^{-2}y^{-1}$ and Total Areal Primary Production in 10^6 t C in Different Arctic and Subarctic Seas.

	Area $10^3\,km^2$	Primary production ($gC\,m^{-2}y^{-1}$)		Total areal PP ($10^6\,tC\,y^{-1}$)
		Total	New	
Central Arctic Ocean	4.489	>11	<1	>50
Arctic shelves	5,052	32	8	279
Barents Sea	1,512	<20–200	<8–100	136
Barents north slope	—	35	16	—
White Sea	90	25	6	2
Kara Sea	926	30–50	7–12	37
Laptev Sea	498	25–40	6–10	16
East Siberian Sea	987	25–40	6–10	30
Chukchi Sea	620	20–>400	5–>160	42
Beaufort Sea	178	30–70	7–17	8
Lincoln Sea	64	20–40	5–10	3
Other (Canadian Arctic)	182	20–40	5–10	5
NE Water Polynya	<50	20–50	13–32	—
North Water Polynya	<31	150	90	—
Total Arctic Ocean	9,541	>26	<5	>329
Subarctic region				
Atlantic sector	5,000	97	50	483
Baffin Bay	690	60–120	25–50	62
Hudson Bay	820	50–70	25–35	49
Greenland Sea	600	70	40	42
Labrador Sea	1,090	100	45	110
Norwegian Sea	1,400	80–150	35–65	160
Icelandic Sea	400	100–200	45–90	60
Bering Shelf	1,300	>230	—	>300
Okhotsk Sea	1,600	100–200	—	240

Source: From Sakshaug (2004).

Notes: For data sources and methods of evaluation see Sakshaug (2004).

2.4.1.1. Central Arctic Ocean

The earliest estimates of the primary production of the permanently ice-covered central Arctic Ocean suggested $<1\,\text{gC}\,\text{m}^{-2}\,\text{y}^{-1}$ (English, 1961), the lowest for any sea. However, production in the multiyear ice was ignored in these estimates, like it also was in a later estimate (Subba Rao & Platt, 1984). Based on measurements carried out during the 1994 central Arctic Ocean transect, Gosselin et al. (1997) indicate that because of significant primary production in the multiyear ice and leads, the annual primary production of the deep Arctic Ocean may reach $>11\,\text{gC}\,\text{m}^{-2}\,\text{y}^{-1}$ (Figure 2.15), assuming a 120-day growth season. About 60% of the primary production value may be related to algae in multiyear ice. Using the primary production value of Gosselin et al. (1997), the annual supply of algae-derived organic carbon is more than $50 \times 10^6\,\text{tC}$ (Table 2.3).

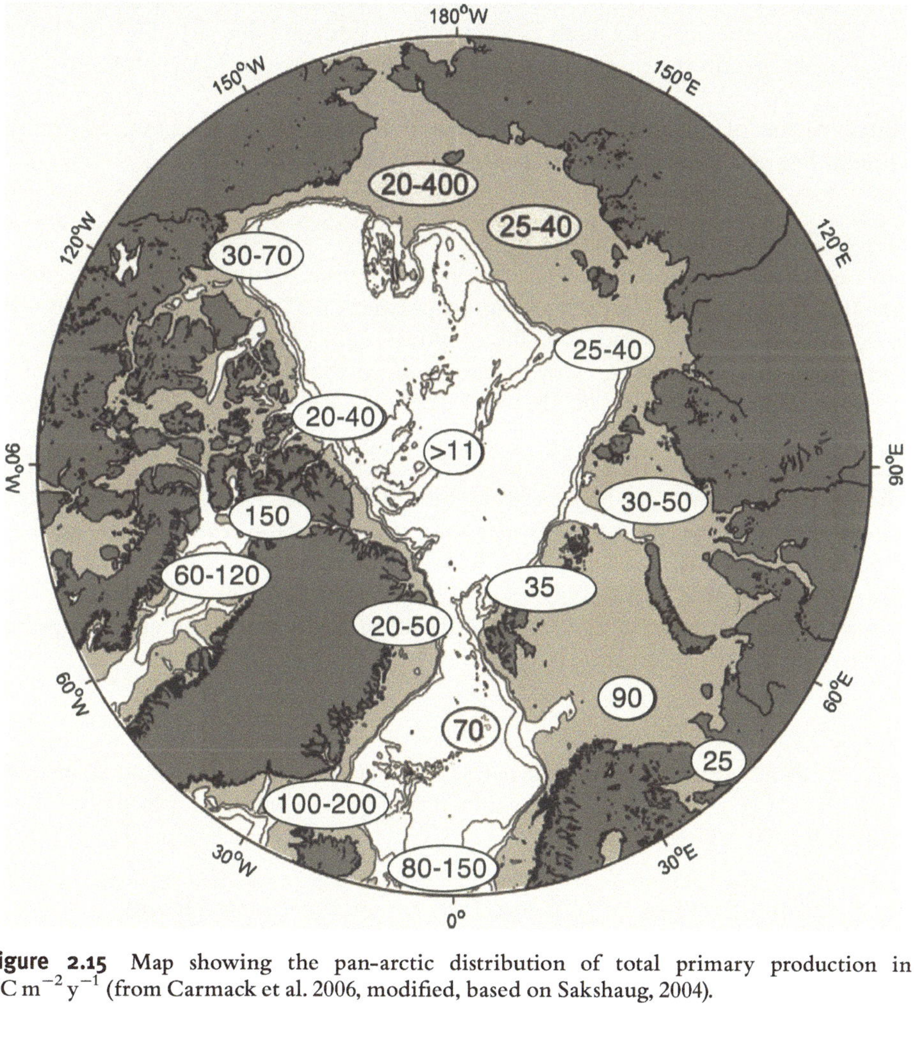

Figure 2.15 Map showing the pan-arctic distribution of total primary production in $\text{gC}\,\text{m}^{-2}\,\text{y}^{-1}$ (from Carmack et al. 2006, modified, based on Sakshaug, 2004).

2.4.1.2. Arctic marginal seas

In the southwestern part of the Barents Sea, quite high annual primary production values of up to 80–150 (200) gC m^{-2} are given (Sakshaug, 2004). This is due to the fact that the Atlantic water inflow makes the southwest part of the Barents Sea permanently ice-free and local currents and topography favour continual upwelling. The other Eurasian shelves, including the northern Barents Sea, are characterized by a stratified layer of Arctic water, which is poor in nutrients because the ice-cover effectively prevents upwelling of the deep water at the shelf edge. Thus, total annual primary production is much lower. In his review, Sakshaug (2004) estimate the total primary production of the whole Barents Sea (including the northern more ice-covered part and the southwestern ice-free part), Kara Sea, Laptev Sea, and East Siberian Sea to reach values of 90, 30–50, 25–40, and 25–40 gC m^{-2} y^{-1}, respectively (Figure 2.15, Table 2.3). These estimates result in a total annual input of algae-derived organic carbon of 136, 37, 16, and 30 × 10^6 tC (Table 2.3). High primary production in the Chukchi Sea of up to 400 gC m^{-2} y^{-1} in part reflects the very high nutrient fluxes through Bering Strait (Walsh & Dieterle, 1994; Sakshaug, 2004). In the Beaufort Sea, total primary production may reach 30–70 gC m^{-2} y^{-1}, averaging 50 gC m^{-2} y^{-1} (Macdonald, Wong, & Erickson, 1987). In terms of total annual marine organic carbon input, 42 and 8 × 10^6 tC were calculated for the Chukchi Sea and Beaufort Sea, respectively (Table 2.3).

2.4.1.3. Polynyas

Increased primary production may occur in polynyas. Polynyas are open waters surrounded by sea ice. They are common between fast and drifting ice along the coasts of the Arctic Ocean (coastal polynyas or flaw leads), in the Canadian Archipelago, the northern Greenland Sea, and on the Bering shelf (AMAP, 1998; Niebauer, Bond, Yakunin, & Plotnikov, 1999). Well-studied polynyas are the Northeast Water (NEW) Polynya on the Northeast Greenland coast and the North Water (NOW) Polynya in the Canadian Archipelago (Figure 2.15). In the NEW Polynya, daily primary production averages ~0.56 gC m^{-2} from late May to mid-August. Because the growth season is at most 90 days, annual total primary production is presumably 20–50 gC m^{-2} (Table 2.3; Smith et al., 1997). Data for the northern part of the NOW Polynya, 31,000 km^2 in size, show that annual total primary production is very high (Table 2.3; 150 gC m^{-2}), due to efficient upwelling (Klein et al., 2002; Tremblay, Gratton, Fauchot, & Price, 2002).

2.4.1.4. Total Arctic Ocean

For the whole Arctic Ocean, Sakshaug (2004) gives an average annual primary production of 26 gC m^{-2} and a total annual flux of algae-derived organic carbon of about 330 × 10^6 tC (Table 2.3).

2.4.2. Vertical Fluxes of Organic Carbon in the Arctic Ocean

In order to estimate the significance of the Arctic Ocean for the global organic carbon budget, also an understanding and quantification of the vertical organic carbon export, that is data on the particulate organic carbon (POC) flux to the seafloor obtained in sediment traps, are required. Here, the data base for the Arctic Ocean is very limited and restricted to some short-term deployments along its fringes (north of Spitsbergen and the Nansen, Amundsen, and Makarov basins) and in the Kara Sea (Wassmann et al., 2004 and further references therein), and four long-term deployments from the Lomonosov Ridge (1995/1996; Fahl & Nöthig, 2007), the Kara Sea (2000–2003; Gaye et al., 2007), and the Beaufort Sea (1987/1988; O'Brien, Macdonald, Melling, & Iseki, 2006) as well as the Canadian Ice Island, a major ice floe north of the Canadian Archipelago (1986–1987 and 1989–1990, when the drift of the ice island was negligible; Hargrave et al., 1989; Hargrave, von Bodungen, Stoffyn-Egli, & Mudi, 1994; Hargrave, 2004) (Figure 2.16). In addition, some long-term sediment traps were deployed in some subarctic areas, that is, in the Norwegian-Greenland Sea including the NEW Polynya and Fram Strait (Honjo, Manganini, Karowe, & Wooward, 1987; Honjo, Manganini, & Wefer, 1988; Hebbeln & Wefer, 1991; Wassmann et al., 2004 and references therein), the northern Baffin Bay (NOW Polynya; Hargrave et al., 2002), and the Bering Sea/Subarctic Pacific (Takahashi, Fujitani, Yanada, & Maita, 2000; Takahashi, Fujitani, & Yanada, 2002; Sasaki & Fukuchi, 2004) (Figure 2.16; Table 2.4). Based on all these long-term sediment-trap data, it is obvious that the vertical POC fluxes in the Arctic Ocean and its surounding areas is highly variable, ranging between 1.5 and 14 gC $m^{-2} y^{-1}$ at roughly 200 m water depths (Table 2.4). The minimum values are from Lomonosov Ridge, the maximum values from the NOW Polynya.

When interpreting the carbon fluxes it has to be considered that resuspension and lateral transport might influence the organic carbon (and total sediment) fluxes as shown at some locations where data from shallow and deep traps deployed at the same station are available (e.g., in the NOW Polynya and on Lomonosov Ridge; Table 2.4, Hargrave, 2004; Fahl & Nöthig, 2007). For a more detailed discussion of POC fluxes to the Arctic seafloor, see the review paper by Wassmann et al. (2004) and further references therein.

Over the past decade, ^{234}Th (half-life = 24.1 days) has been increasingly used to estimate the flux of POC exported from the euphotic zone in the global ocean (e.g., Moran et al., 2003). This naturally occurring, particle-reactive, radionuclide is produced continuously in seawater by alpha decay of soluble ^{238}U (half-life = 4.47×10^9 years) and has proven to be a useful tracer of upper ocean particle export on a time scale of days to months (S.B. Moran in Wassmann et al., 2004). Numerous studies on ^{234}Th-derived POC fluxes were also carried out for the Arctic Ocean and its surrounding subpolar regions (Cochran, Barnes, Achman, & Hirschberg, 1995; Moran, Ellis, & Smith, 1997; Moran et al., 2003; Moran & Smith, 2000; Amiel, Cochran, & Hirschberg, 2002; Coppola, Roy-Barman, Wassmann, Mulsow, & Jeandel, 2002; Baskaran, Swarzenski, & Porcelli, 2003; Trimble & Baskaran, 2005). From these studies, a wide spatial variability in POC export flux is obvious (see overview by S.B. Moran in Wassmann et al., 2004). In general, shelf waters exhibit much higher fluxes compared to the central basins, which is consistent with the elevated rates of productivity in Arctic shelf-slope waters. In general, the ^{234}Th-derived POC fluxes are also consistent with sediment trap observations that

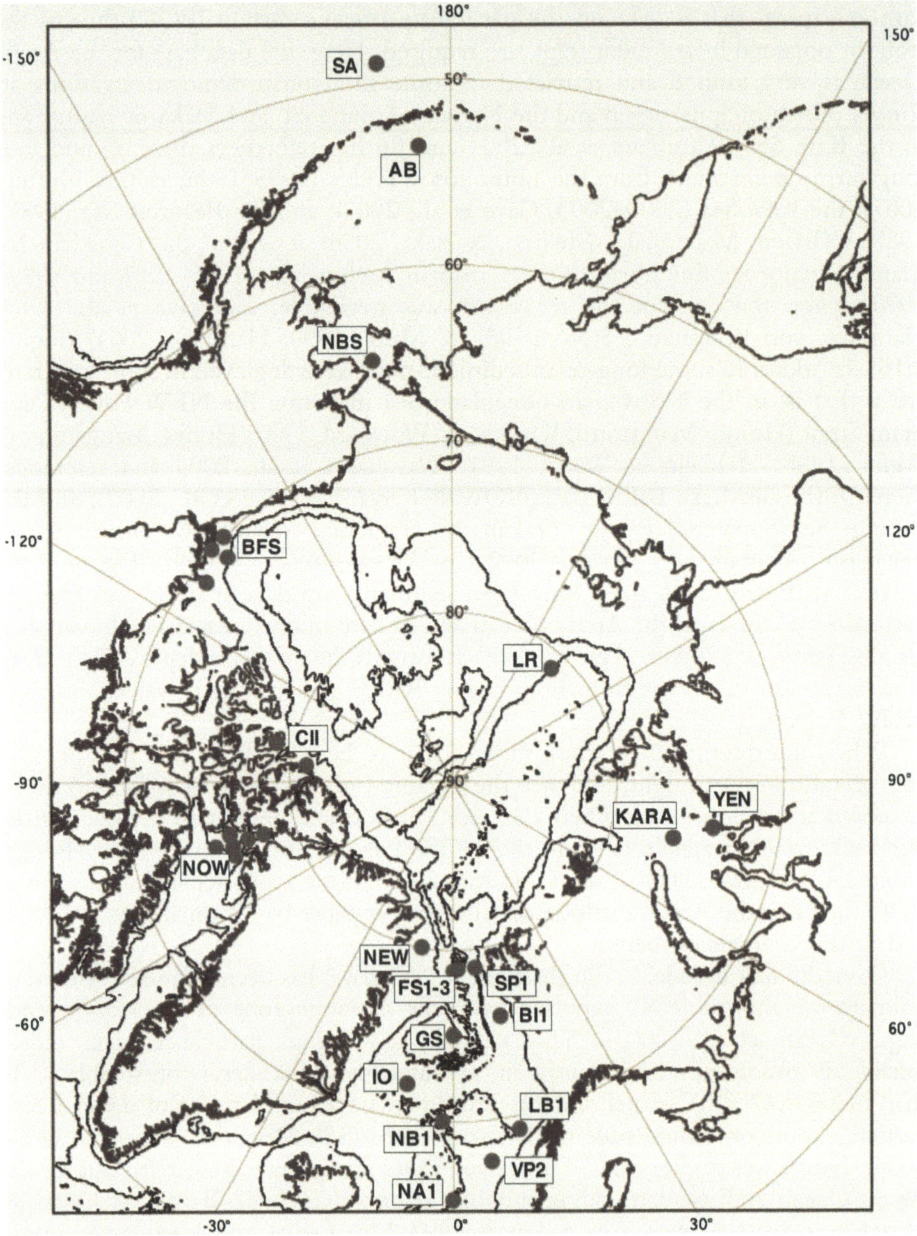

Figure 2.16 Location map of long-term deployments in the subarctic and arctic region. For explanation of site acronyms and references see Table 2.4.

indicate the export of POC can be quite high in ice-free regions, particularly over the broad continental shelves. When going into details, however, significant differences in absolute flux numbers obtained by the two approaches, may occur (for comparison and more detailed discussion, see Wassmann et al., 2004).

Table 2.4 Estimates of Vertical POC Fluxes from Long-term Deployments in the Arctic Ocean and Adjacent Regions.

Site	Position	Water depth	Trap depth (m)	Time interval	OCF (gC m^{-2} y^{-1})		References
LR	81°N 138°E	1,700	150	1995–1996	1.5		Fahl and Nöthig (2007)
		1,700	1,550	1995–1996	1.05		Fahl and Nöthig (2007)
Yen-2	74°N 80°E	31	20	2000–2001	10.6		Gaye et al. (2007)
Kara-1	76°12′N 75°45′E	73	54	2001–2002	8.2		Gaye et al. (2007)
CII	81°N 96°W	var.[1]	100	1986–1987	0.07		Hargrave (2004)
	79°N 102°W	var.[1]	100	1989–1990	0.09		Hargrave (2004)
	79°N 102°W	var.[1]	100	1989–1990	0.18		Hargrave (2004)
BFS[1]							
SS-1	70°45′N 127°20′W	214	159	1987–1988	3.7		O'Brien et al. (2006)
SS-2	71°30′N 128°40′W	173	118	1987–1988	3.9		O'Brien et al. (2006)
SS-3	70°55′N 134°25′W	180	125	1987–1988	1.6		O'Brien et al. (2006)
SS-4	69°55′N 138°35′W	268	213	1987–1988	5.8		O'Brien et al. (2006)
NOW					(97–98)	(98–99)	
(S5)	76°N 77°W	365	258	1997–1999	8.0	11.7	Hargrave, Walsh, and Murray (2002)
(S5)	76°N 77°W	365	315	1997–1999	8.4	nd	Hargrave et al. (2002)
(S4)	76°N 74°W		205	1997–1999	1.3	2.5	Hargrave et al. (2002)
(S2)	76°N 72°W	560	198	1997–1999	3.8	6.7	Hargrave et al. (2002)
(S2)	76°N 72°W	560	507	1997–1999	8.1	13.8	Hargrave et al. (2002)
NEW	80°N 11°W	330	130	1992–1993	1.0 (2.7)		Bauerfeind (2004)
FS-1	78°52′N 01°22′E	2,527	2,146	1984–1985			Honjo et al. (1987)
FS-2	79°00′N 04°55′E	2,430	1,925	1985–1986			Honjo et al. (1987)
FS-3	78°46′N 00°11′E	2,487	1,488	1987–1988			Hebbeln and Wefer (1991)

Table 2.4 *(Continued)*

Site	Position	Water depth	Trap depth (m)	Time interval	OCF ($gC\,m^{-2}\,y^{-1}$)	References
SP-1	78°53′N 06°45′E	1,618	1,087	1987–1988	9.7	Hebbeln (2000)
BI-1	75°51′N 11°30′E	2,123	1,700	1984–1985		Honjo et al. (1988)
GS	75°N 00°W		200–300	1993–1995	1.4	Peinert and Noji (2004)
IO	72°N 7-10°W	2,500	500	1988–1992	2.3 (1.4–3.8)	Peinert et al. (2001)
	72°N 7-10°W	2,500	1,000	1988–1992	1.3 (1.1–1.8)	Peinert et al. (2001)
	72°N 7-10°W	2,500	2,000	1988–1992	0.9 (0.5–0.9)	Peinert et al. (2001)
LB-1	69°12′N 10°59′E	3,000	2,600	1983–1984		Honjo et al. (1987)
VP-2	67°37′N 05°50′E	1,429	700	1986–1987		Bathmann, Peinert, Noji, and Bodungen (1990)
NB-1	70°01′N 01°58′W	3,284	2,764	1985–1986		Honjo et al. (1987)
NA-1	65°31′N 00°05′W	3,058	2,630	1985–1986		Honjo et al. (1987)
NBS	64°58′N 169°10′W	62	49	1988 (June–September)	44 $gC\,m^{-2}$ 96 d^{-1}	Sasaki and Fukuchi (2004)
AB	53°30′N 177°W	3,788	3,188	1990–1995	2.5	Takahashi et al. (2000)
SA	49°N 174°W	5,406	4,806	1990–1995	1.4	Takahashi et al. (2000)

Notes: LR, Lomonosov Ridge; Kara Kara Sea; Yen Yenisei; BFS, Beaufort Sea; CII, Canadian Ice Island; NOW, North Water Polynya; NEW, Northeast Water Polynya; FS, Fram Strait; GS, Greenland Sea; IO, IS Odden; SP, Spitsbergen; BI, Bear Island; NBS, Northern Bering Sea; AB, Aleutian Basin (Bering Sea); Subarctic Bering Sea. References are listed. NOW, left column: 1997–1998; right column: 1998–1999.

NEW: in parenthesis POC flux corrected for DOC leakage. IO: in parenthesis: range of vertical annual export between 1988 and 1992. In addition, sediment traps studied by Berner (1991) for clay mineralogy are listed (see Chapter 5.1.2, Figure 5.10 for some results).

[1]Water depths vary between 120 and 305 m due to the drift if the Canadian Ice Island.

2.5. River Discharge

One of the main characteristics of the Arctic Ocean is its huge river discharge (Figure 2.17). Comprehensive compilations of river discharge data of the Arctic Ocean are published by, for example, Aagaard and Carmack (1989), Gordeev, Martin, Sidorov, and Sidorova (1996), Lammers, Shiklomanov, Vörösmarty, and Peterson (2001), Holmes et al. (2002), Bobrovitskaya, Kokorev, and Lemeshko (2003), and Rachold et al. (2004a). Data can also be downloaded from several web sites. Eurasian data are available from R-ArcticNet (http://www.r-arcticnet.sr.unh.edu/) and Arctic RIMS (http://rims/unh.edu), and North American data from the Water Survey of Canada's Hydrometric Database (HYDAT) (http://www.wsc.ec.gc.ca/). Detailed information on the freshwater budget of the Arctic Ocean is published in Lewis (2000). The discharge data presented and discussed in this chapter concentrate on

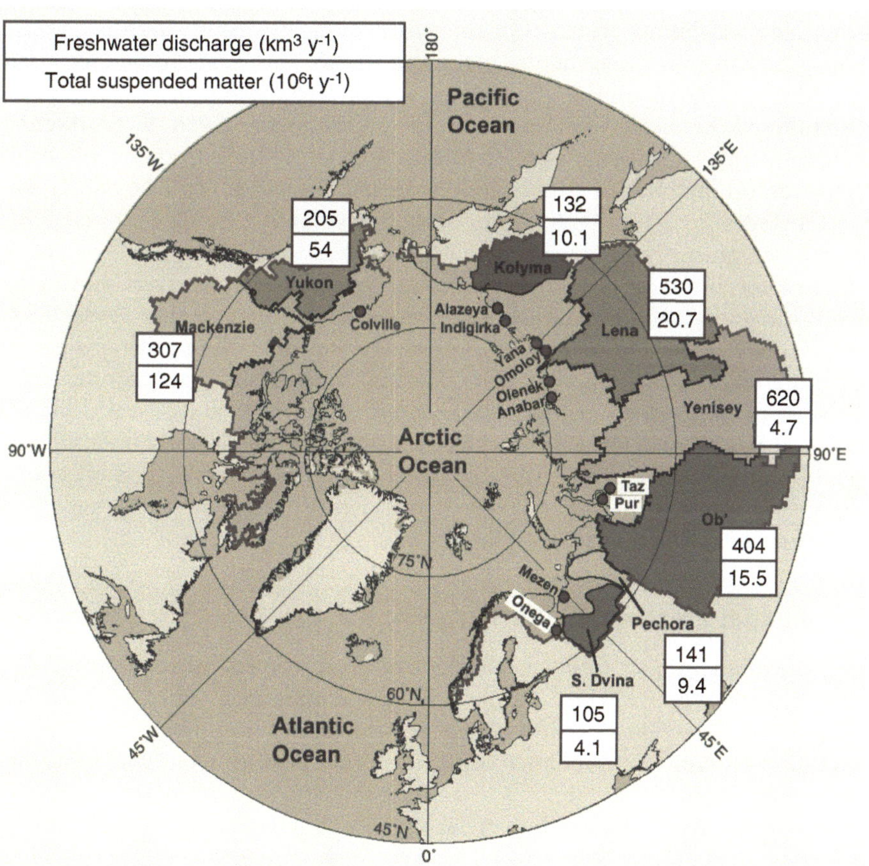

Figure 2.17 Map of pan-arctic watershed showing drainage basins of the eight largest rivers and the mouth locations of 11 other arctic rivers (from Holmes et al., 2002, supplemented). Average annual freshwater discharge and total suspended-matter supply of these eight rivers are shown (data from Holmes et al., 2002).

freshwater discharge and fluxes of total suspended matter (TSM) and organic carbon, and are mainly based on those published by Gordeev et al. (1996) and Holmes et al. (2002) (Table 2.5; Figure 2.17). For data on nutrient supply, it is referred to papers by Gordeev et al. (1996) and Holmes et al. (2000, 2001).

Holmes et al. (2002) and Bobrovitskaya et al. (2003) give information about the history of Arctic Ocean river discharge monitoring. Water discharge monitoring on downstream reaches of Arctic rivers began much earlier in the Former Soviet Union than in North America. The duration of water discharge measurements at 22 hydrological stations in the Eurasian Arctic varies from 20 to 120 years; measurements were made from 1881 to 2000 (Bobrovitskaya et al., 2003). For the major Eurasian rivers Lena, Yenisei, and Ob, for example, monitoring began in the 1930s (Figure 2.18A), for the Severnaya Dvina even a 120 years long record from 1881 to 2000 is available (Bobrovitskaya et al., 2003). Records of consistent measurements of sediment flux at the downstream monitoring stations of the same rivers are much shorter, starting in the 1940s for the Ob river and in the late 1960s for the Lena and Yenisei rivers. On the other hand, for the rivers Mackenzie and Yukon, discharge measurements including freshwater and suspended matter, began in the early to mid-1970s (Figure 2.18A; Holmes et al., 2002). It should be mentioned here, that sediment input values for a given river often vary substantially in the different publications. A review and synthesis of available information on methods of sample collection and approaches to flux calculations as well as a discussion of the reasons for sometimes conflicting flux estimates in the literature, are given by Holmes et al. (2002).

Freshwater discharge as well as sediment fluxes of the Arctic rivers show a strong seasonal variability, with highest in late spring/early summer (Figure 2.18B; Holmes et al., 2002). For the Ob and Yenisei rivers, for example, the maximum discharge rate is observed in June, when ~45–65% of the annual freshwater runoff and 80% of the annual suspended matter are released (Shiklomanov & Skakalsky, 1994; Gordeev et al., 1996). Increases of both freshwater discharge and sediment fluxes are generally steeper during the spring than declines during the fall. One exception to this general pattern is the Yenisei, where sediment flux drops off very rapidly after the summer peak (Figure 2.18B; Holmes et al., 2002).

2.5.1. Annual Freshwater Discharge

Today, the annual direct freshwater inflow by major rivers into the Arctic Ocean reaches a total of ~3,300 km^3 (Table 2.5). For comparison, the amount of water discharged by the world rivers to the present-day ocean is estimated to be between 32 and 37 $\times 10^3\,km^3\,y^{-1}$ (Milliman & Meade, 1983; McLennan, 1993; Chakrapani, 2005). That means, the Arctic Ocean discharge is equivalent to 10% of the global runoff. Major contributors are the Yenisei (620 $km^3\,y^{-1}$), the Ob (429 $km^3\,y^{-1}$), the Lena (525 $km^3\,y^{-1}$), and the MacKenzie (249 $km^3\,y^{-1}$) (Table 2.5, Figure 2.17; Gordeev et al., 1996; Holmes et al., 2002; Rachold et al., 2004a). The freshwater discharge into the specific marginal Arctic seas is very different. The Kara Sea is by far the most freshwater influenced marginal sea with an annual input of 1,480 km^3, followed by the Laptev Sea, Barents Sea (including the White Sea),

Table 2.5 Average Riverine Water and Suspended Matter Discharges and Fluxes of DOC, POC, and TOC to the Arctic Ocean Note that sometimes POC+DOC is not equal to TOC because of different sources of data.

River	Water discharge	Total suspended matter		Content			Flux		
				DOC	POC	TOC	DOC	POC	TOC
	($km^3 y^{-1}$)	(mgl^{-1})	($10^6 t y^{-1}$)	(mgl^{-1})			($10^6 t y^{-1}$)		
White and Barents Seas									
Onega	15.9	18	0.30	—	—	20.7	—	—	0.33
N. Dvina	110	37	4.1	11.6	2.6	15.3	1.28	0.28	1.68
Mezen	27.2	33	0.6	12.1	1.8	7.0	0.25	0.04	0.19
Pechora	141	72	9.4	12.7	0.3	13.0	1.66	0.04	1.70
Other area	179	19	3.5	—	—	13.7	—	—	2.45
Total	474	39	17.9	—	—	13.7	4.18	0.45	6.35
Kara Sea									
Ob	404	37	15.5	9.1	0.9	7.1	3.68	0.36	2.87
Nadym	18	22	0.4	—	—	5.0	—	—	0.09
Pyr	34.3	18	0.7	—	—	6.7	—	—	0.23
Taz	44.3	21	0.7	—	—	—	—	—	—
Yenisei	620	8	4.7 (14.4)	8.5	0.3	7.4	4.86	0.17 (0.5)	4.59
Pyasina	86	39	3.4	—	—	—	—	—	—
Other area	275	20	5.5	—	—	7.2	—	—	1.98
Total	1,480	21	30.9 (40.6)	—	—	7.2	—	—	10.6

Table 2.5 *(Continued)*

River	Water discharge	Total suspended matter		Content			Flux		
				DOC	POC	TOC	DOC	POC	TOC
	($km^3 y^{-1}$)	(mgl^{-1})	($10^6 t y^{-1}$)	(mgl^{-1})			($10^6 t y^{-1}$)		
Laptev Sea									
Khatanga	85.3	20	1.7	—	—	6.3	—	0.04	0.54
Anabar	17.3	24	0.4	—	—	5.1	—	—	0.09
Olenjok	32.8	38	1.1	10.2	0.83	7.2	0.32	0.026	0.24
Lena	523	39	20.7	6.6	1.1	7.7	3.6	1.2	4.8
Omoloy	7	18	0.04	2.8	0.3	—	0.003	0.001	—
Yana	31.9	130	4.0	2.8	1.6	6.7	0.085	0.05	0.21
Other area	40.3	16	0.65	—	—	9.2	—	—	0.37
Total	738	39	28.6	—	—	9.2	—	—	6.8
East Siberian Sea									
Indigirka	54.2	207	11.1	4.8	3.5	7.7	0.24	0.17	0.42
Alazeya	8.8	80	0.1	—	—	—	—	—	—
Kolyma	122	83	10.1	4.6	3.1	8.1	0.46	0.31	0.99
Other area	48.2	80	3.85	—	—	8.0	—	—	0.38
Total	233	110	25.15	—	—	8.0	—	—	1.86
Chukchi Sea (excluding Alaska)									
Amguema	9.2	6	0.05	—	—	6.7	—	—	0.06
Other area	11.2	58	0.65	—	—	6.7	—	—	0.07
Total	20.4	34	0.7	—	—	6.7	—	—	0.13

Total Eurasian Arctic									
	2,944	36	102.2 (111.9)	6.6	1.3	8.8	19.4	3.81 (4.14)	25.7
Chukchi Sea (Alaska) and Beaufort Sea									
Kobuk	—	—	—	—	—	—	—	—	0.04
Kuparuk	—	—	—	—	—	—	—	—	0.014
Mackenzie	330	168	124	5.2	7.2	12.5	1.3	2.1	4.1
Other area	37	—	1.1	—	—	—	0.19	0.055	0.24
Total Canadian Arctic									
	367	—	125.1	5.1	5.8	11.6	1.9	2.15	4.3
Total Arctic									
	3,311	63	227.3 (237)	—	—	9.1	—	5.96 (6.3)	30.0

Source: Extracted from Rachold et al. (2004a) and mainly based on Gordeev et al. (1996) and Holmes et al. (2002).

Notes: For details on data sources and complete references see Rachold et al. (2004a). Numbers in brackets for Yenisei give measurements for times preceding the dam constructed near Krasnoyarsk in 1967 (Telang et al., 1991; Holmes et al., 2002).

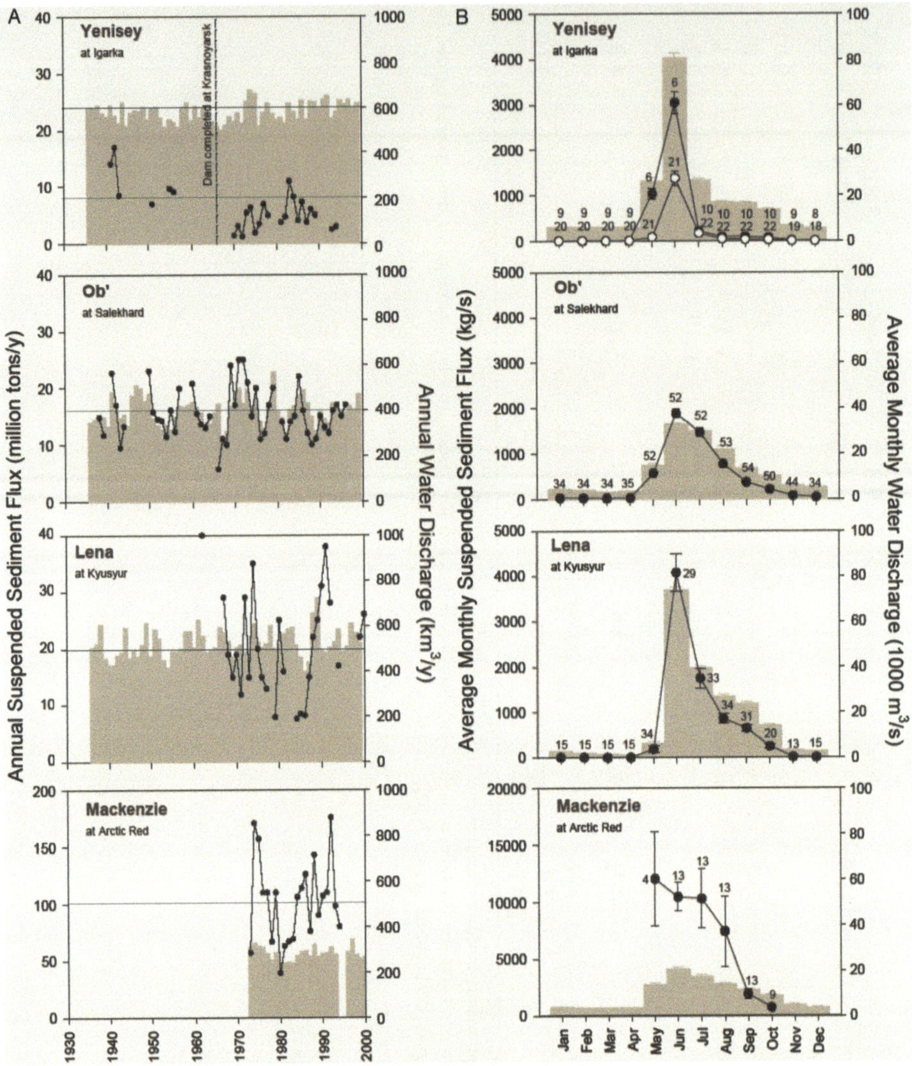

Figure 2.18 (A) Time series of annual sediment flux and water discharge estimates and (B) average monthly sediment flux and water discharge estimates for the four largest rivers (by water discharge) in the pan-arctic watershed (from Holmes et al., 2002, supplemented). Numbers above sediment flux data points in (B) indicate number of months of data in averages. For the Yenisey River, closed symbols are pre-Krasnoyarsk dam averages whereas open symbols are post-dam. Note that the scales for sediment flux in the Mackenzie River are five times (A) and (B) four times larger than those of the other rivers. For data sources and further references see Holmes et al. (2002).

and the East Siberian Sea, where a discharge of 738, 474, and 233 $km^3 y^{-1}$, respectively, was determined (Table 2.5). In the Chukchi Sea (excluding the Alaskan part), only ~20 $km^3 y^{-1}$ were measured. This gives a total annual freshwater discharge of 2,944 $km^3 y^{-1}$ for the Eurasian Arctic. The total direct discharge into the Canadian Arctic shelf seas, that is, the Beaufort Sea and the Alaskan part of the Chukchi Sea, is 367 $km^3 y^{-1}$ (Table 2.5). Although the Yukon River (205 $km^3 y^{-1}$; Figure 2.17) does not discharge directly into the Arctic Ocean, the discharge might (partly) be added to the value of 367 $km^3 y^{-1}$ because it makes major contributions of freshwater to the Arctic Ocean via prevailing ocean currents (Guay & Falkner, 1997). This may increase the total influx of freshwater into the Canadian Arctic to a value of ~570 $km^3 y^{-1}$. Including the Yukon River discharge, the total annual discharge of freshwater into the Arctic Ocean would be as high as ~3,500 $km^3 y^{-1}$.

2.5.2. Annual Total Suspended Matter Fluxes

When looking at the total suspended matter (TSM) fluxes, the order of importance of the Arctic rivers is quite different than the freshwater discharge (Table 2.5, Figure 2.17; Gordeev et al., 1996; Holmes et al., 2002; Rachold et al., 2004a). The two extremes are the Yenisei and Mackenzie rivers. The Yenisei, with 620 $km^3 y^{-1}$ clearly number one in term of freshwater discharge, is characterized by very low TSM flux values of $4.7 \times 10^6 t y^{-1}$ ($14.4 \times 10^6 t y^{-1}$ prior to the 1967 dam construction in Krasnoyarsk; Telang et al., 1991; Holmes et al., 2002) (Figure 2.18A). The Mackenzie with a freshwater discharge of 243 $km^3 y^{-1}$, on the other hand, has a TSM flux value of as high as $124 \times 10^6 t y^{-1}$, which is more than 50% of the total TSM flux of the entire Arctic Ocean, reaching $227 \times 10^6 t y^{-1}$ ($237 \times 10^6 t y^{-1}$ using the pre-dam Yenisei value) (Table 2.5). The TSM fluxes in the Eurasian Arctic and Canadian Arctic reach values of $102 \times 10^6 t y^{-1}$ ($112 \times 10^6 t y^{-1}$ using the pre-dam Yenisei value) and $125 \times 10^6 t y^{-1}$, respectively. The value for the Canadian Arctic may become even higher when part of the TSM flux of the Yukon River ($54 \times 10^6 t y^{-1}$, Figure 2.17) transported by ocean currents into the Arctic Ocean (see later description), would be added.

The differences in annual TSM fluxes are at least partly linked to differences in TSM concentration among the rivers, as indicated by plots of sediment concentration relative to water discharge (Figure 2.19; Holmes et al., 2002). The Mackenzie, Yukon, and Kolmya rivers characterized by high TSM concentrations, differ distinctly from the Yenisei, Lena, Ob, and Severnaya Dvina rivers characterized by low TSM concentrations. The latter are generally in line with what has been observed in other lowland rivers (Milliman & Syvitski, 1992). The relatively narrow ranges of discharge in the Kolyma, Pechora, and Severnaya Dvina rivers make these rivers more difficult to categorize in this way. Variations in sediment concentration as a function of water discharge among the largest Arctic rivers also reflect geographical patterns, as described by Holmes et al. (2002) (Figure 2.19). The drainage basins of the Mackenzie, Yukon, and Kolyma rivers share features of geology and climate that set them apart from the drainage basins of the Yenisei, Lena, Ob, and Severnaya Dvina rivers (Gordeev et al., 1996; Semiletov et al., 2000).

When using the TSM values of the Arctic rivers described above for estimating the riverine sediment input into the Arctic Ocean, one should have in mind that these numbers were determined at the northernmost gauging stations of the rivers (Figure 2.17), which are still quite south of the Arctic Ocean itself. Much of this TSM may be retained in the so-called marginal filter (Lisitzin, 1995), the zone

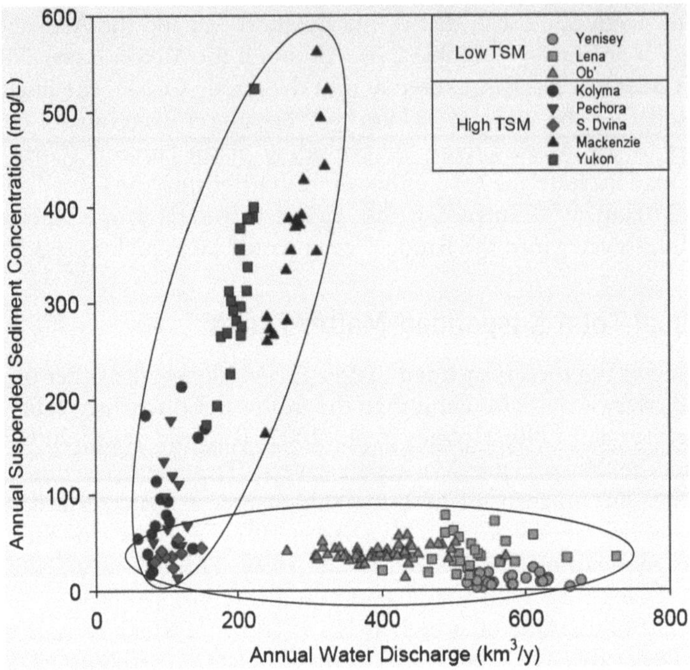

Figure 2.19 Annual sediment flux as a function of annual water discharge for each year of record in the eight largest arctic rivers (from Holmes et al., 2002, supplemented).

where freshwater and salt water mix (salinities of ~2–10) and rapid accumulation (precipitation) of fine-grained suspension occurs due to coagulation processes (Figure 2.20). Especially in the estuaries and deltas major amounts of sediments may be trapped and not be further transported into the Arctic Ocean.

How much of riverine sediment is really reaching the Arctic Ocean is still not clear, and numbers differ in different publications. For example, Alabyan, Chalov, Korotaev, Sidorchuk, and Zaitsev (1995) state that only 10–17% of the sediment in the Lena at Kyusyer Station makes it through the Lena Delta, whereas Rachold et al. (2000, 2004a) persuasively argue that essentially all of the Lenas suspended sediment reaches the Laptev Sea. For the Mackenzie, it has been estimated that about half of the riverine suspended sediment is transported through the extensive Mackenzie Delta (Macdonald et al., 1998; Macdonald et al., 2004a; Macdonald, Naidu, Yunker, & Gobail, 2004b). As stated by Holmes et al. (2002), it seems unlikely that a significant portion of the suspended sediment in the Yenisei and Ob rivers is transported through their lengthy estuaries on annual time scales. There is, however, considerable disagreement about more quantitative estimates of the proportion of Yenisei and Ob TSM that reaches the Kara Sea. Based on measurements of recent suspended matter concentration, Lisitzin (1995) estimates that 90% of the suspended matter is accumulating in the Yenisei and Ob estuaries (see also Figure 2.20). A recent calculation of TSM fluxes at the river mouths based on TSM sampling carried out in 2000/2001, has resulted in a Yenisei TSM discharge reaching the Kara Sea of $5 \times 10^6\,t\,y^{-1}$, comparable to the estimate from

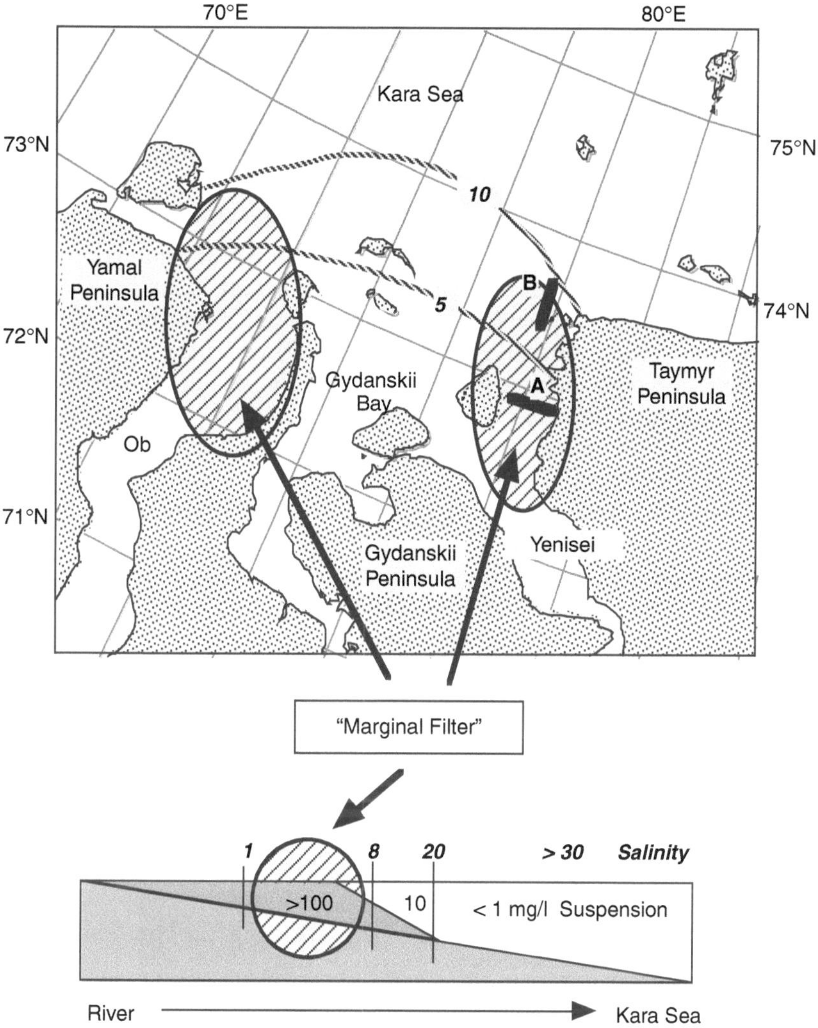

Figure 2.20 Map of the southern Kara Sea with the Ob and Yenisei. Lines with italic numbers (5, 10) indicate salinities, hatched area the "marginal filter", and black bars (A and B) the location of sediment echograph profiles presented in Figure 2.21. In addition, surface-water salinity and suspended matter data in the marginal filter system from Lisitzin (1995) are shown (see Stein, Fahl, Dittmers, Niessen, & Stepanets, 2003a; Stein et al., 2004a for more details).

Igarka Station, which would imply that the lower Yenisei is mainly a by-pass system (Gebhardt, Gaye-Haake, Unger, Lahajnar, & Ittekkot, 2004). The Ob TSM discharge reaching the Kara Sea is estimated to be about $3.8 \times 10^6\,t\,y^{-1}$ which is about one-fourth of the amount discharged at Salekhard Station, that is, about 75% of sediments are trapped in the Ob Estuary (Gebhardt et al., 2004). The estimate for the Yenisei by Gebhardt et al. (2004) neither agree with Lisitzin's estimates nor with estimates based on sediment data. Based on late Holocene (0–6 Cal. kyr BP) sediment budget calculations, ~70% of the TSM input is trapped in the estuaries (Dittmers, Niessen, & Stein, 2003; Stein & Fahl, 2004a). Accumulation rate records from the estuaries indicate a further increase during the last 2 Cal. kyr BP which may suggest an even higher accumulation within the marginal filter during more recent times (Stein, Fahl, Dittmers, Niessen, & Stepanets, 2003a, Stein & Fahl, 2004a), supporting the estimate of Lisitzin (1995). The thick pile of young (Holocene) soft sediments in the Yenisei Estuary reaching a maximum thickness in the central part of the estuary (Figure 2.21; Stein, 2001; Dittmers et al., 2003; Stein et al., 2003a), also support the effectiveness of the marginal filter as sediment trap and contradict a "by-pass system" on a longer (Holocene) time scale. The disagreement between Gebhardt et al. (2004) and the Holocene sediment budget values might be explained by the higher "pre-dam" Yenisei discharge values (see earlier) which are more appropriate to use for interpreting the sediment values (Stein & Fahl, 2004a) or simply by the fact that the 2001 August TSM sampling which was the base for the calculation of Gebhardt et al. (2004), was not representative of the discharge system of the entire summer (as also mentioned by the authors themselves).

In summary, it can be stated that the flux estimates provided in published synthesis papers allow assessment of sediment flux from a large percentage of the pan-arctic watershed; however, further research is needed to determine how much of this sediment actually reaches the sea (Holmes et al., 2002).

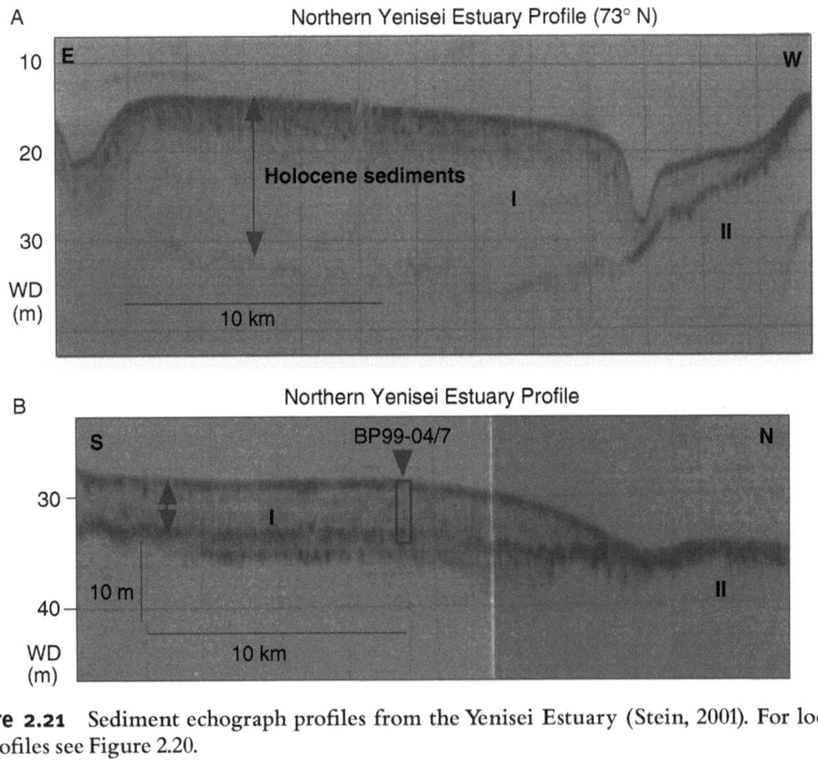

Figure 2.21 Sediment echograph profiles from the Yenisei Estuary (Stein, 2001). For location of profiles see Figure 2.20.

2.5.3. Annual Riverine Input of Organic Matter

Concerning the riverine organic carbon input, only a short summary taken from the latest synthesis published by Rachold et al. (2004a) is presented here. For more detailed information on organic carbon supply and accumulation in the Arctic Ocean, including a comprehensive list of further references, it is referred to Rachold et al. (2004a) and Stein and Macdonald (2004a).

The first assessments of total organic carbon (TOC) discharge for the Russian Arctic rivers were obtained by Skopintsev and Krylova (1955) and Alekin and Brazhnikova (1964). More recent estimates of organic carbon discharge from the Russian territory to the Arctic Ocean appeared in Artemiev (1997), Gordeev et al. (1996), Gordeev and Tsirkunov (1998), and Gordeev (2000). The most recent work by Romankevich and Vetrov (2001) is an important summary of all available information on the carbon cycle in the Russian Arctic Seas including the riverine discharge of organic carbon. The transport of organic carbon to the oceans by the North American rivers was summarized in the work of Mulholland and Watts (1982). During the 1990s several studies on the topics of discharge and origin of riverine organic matter and its distribution in the Arctic Ocean have been published by Telang et al. (1991), Cauwet and Sidorov (1996), Macdonald et al. (1998), Lara et al. (1998), Opsahl, Benner, and Amon (1999), Rachold and Hubberten (1999), and Lobbes, Fitznar, and Kattner (2000).

Based on a review of the existing information (Rachold et al., 2004a), best estimates of TOC, DOC (dissolved organic carbon), and POC (particulate organic carbon) fluxes from rivers of the Eurasian and Canadian Arctic are listed in Table 2.5. At present, reliable TOC data are available for most rivers, whereas only major rivers have been studied separately for DOC and POC concentrations. Rachold et al. (2004a) estimate that total TOC discharge to the Eurasian Arctic and Canadian Arctic is $25.7 \times 10^6\,\mathrm{tC\,y^{-1}}$ and $4.3 \times 10^6\,\mathrm{tC\,y^{-1}}$, respectively, giving a total of $30 \times 10^6\,\mathrm{tC\,y^{-1}}$ for the whole Arctic Ocean. The annual flux of POC is estimated to be $\sim 6 \times 10^6\,\mathrm{tC}$ in total, with $\sim 4 \times 10^6\,\mathrm{tC}$ determined for the Eurasian Arctic and $2.1 \times 10^6\,\mathrm{tC}$ for the Canadian Arctic (Table 2.5). That means, the DOC/POC ratio is much higher for the Eurasian rivers (DOC/POC = 5) than for the Canadian rivers (DOC/POC = 1).

Riverine organic matter is generally derived from allochthonous sources (eroded soil and plant material) with a much smaller component from autochthonous sources (freshwater aquatic production) (e.g., Meybeck, 1982; Ittekkot, 1988). For the Arctic rivers, the predominance of allochthonous organic carbon has been approved in numerous studies based on C/N ratios, $\delta^{13}C$ values of the organic matter, and specific biomarkers (e.g., Yunker & Macdonald, 1995; Lara et al., 1998; Rachold & Hubberten, 1999; Fernandes & Sicre, 2000; Goñi, Yunker, Macdonald, & Eglinton, 2000; Naidu et al., 2000; Lobbes et al., 2000; Fahl et al., 2003; Stein & Macdonald, 2004a and further references therein).

As for the TSM supply, it has to be mentioned that the organic carbon flux numbers listed in Table 2.5 are determined at the northernmost gauging stations of the rivers and not directly at the river mouths. That means, significant amount of organic carbon is certainly trapped in the marginal filter and nor further transported into the Arctic Ocean itself (e.g., Macdonald et al., 1998; Stein et al., 2003a; Stein & Macdonald, 2004a).

2.5.4. The Significance of Arctic Riverine Input

The present state of the Arctic Ocean and its influence on the global climate system strongly depend on the large river discharge. Furthermore, climate models predict significant global future warming with greatest temperature changes in polar regions (Houghton et al., 1996). Contemporaneously, precipitation will increase significantly (ACIA, 2005), resulting in increased river discharge (see later description). Thus, examination of inputs from arctic rivers to the ocean has been proposed as a means for tracking the effects of climate change because fluxes from rivers provide an integrative signal of processes occurring in their watersheds (Holmes et al., 2002).

The overall importance of Arctic river discharge for the Arctic Ocean and its significance for the global ocean system can be summarized as follows:

(1) The fluvial freshwater supply is essential for the maintenance of the ca. 200 m thick low-salinity layer of the central Arctic Ocean and, thus, contributes significantly to the strong stratification of the surface-near water masses, encouraging sea-ice formation. Changes in the freshwater balance would influence the extend of sea-ice cover. The melting and freezing of sea ice result in distinct changes in the surface albedo, the energy balance, the temperature and salinity structure of the upper water masses, and the biological processes.

(2) The freshwater exported from the Arctic Ocean through Fram Strait influences the global thermohaline circulation. Today, the annual liquid freshwater export with the East Greenland Current is $\sim$1,160 km^3. Estimates based on sea-ice export even reach values of 1,680 km^3 y^{-1} (Aagaard & Carmack, 1989) to 2,850 km^3 y^{-1} (Eicken, 2004). Changes in these export rates of freshwater would result in changes of deep-water formation and, thus, in global thermohaline circulation ("global conveyor belt"; Broecker, 1997) and ventilation. Because factors such as the global thermohaline circulation as well as sea-ice cover and earth albedo have a strong influence on the earth's climate system, the freshwater input to the Arctic and its change through time may have triggered global climate changes in the past (e.g., the initiation of the Northern Hemisphere Glaciation during mid-Pliocene times; Driscoll & Haug, 1998).

(3) Major quantities of nutrients controlling surface-water productivity, are supplied by rivers (e.g., Martin, Guan, Elbaz-Poulichet, Thomas, & Gordeev, 1993; Gordeev et al., 1996). Relatively high contents of chlorophycean algae, phaeopigments, and specific marine biomarkers in the surface sediments near the river mouths and on the shelf, indicating increased primary productivity, document this strong fluvial influence (Heiskanen & Keck, 1996; Fahl & Stein, 1997; Kunz-Pirrung, 1999; Boucsein & Stein, 2000).

(4) The Arctic rivers transport large amounts of dissolved and particulate material (i.e., chemical elements, siliciclastic and organic matter, etc.) onto the shelves (see earlier) where it is accumulated or further transported by different

mechanisms (sea ice, icebergs, turbidity currents, etc.) towards the open ocean (e.g., Stein & Korolev, 1994; Dethleff, Rachold, Tintelnot, & Antonow, 2000). Thus, river-derived material contributes in major proportions to the entire Arctic Ocean sedimentary and chemical budgets. Furthermore, the different rivers carry suspension loads characterized by different mineralogical and geochemical tracers, dependant on the geology of the hinterland. These tracers (such as clay minerals and heavy minerals as well as major, minor, and rare earth elements) can be used as indicator for specific source areas and reconstruction of pathways of terrigenous matter (and thus ocean current patterns) in the ocean (see Chapter 5.1.2; e.g., Behrends, Hoops, & Peregovich, 1999; Wahsner et al., 1999; Phillips & Grantz, 2001; Schoster, Behrends, Müller, Stein, & Wahsner, 2000).

(5) The Arctic rivers may also transport major amounts of anthropogenic pollutants (radioactive elements, heavy metals, etc.) which may be trapped in coastal-near sediments and/or transported towards the open ocean (Shiklomanov & Skakalsky, 1994; Landa et al., 1998; Nies et al., 1998). Anthropogenic pollutants are not further discussed in this book. Here, it is referred to a synthesis paper by Crane et al. (2001).

2.5.5. Recent Changes in Riverine Input and Climate Change

A significant increase in Siberian river discharge, associated with a warmer climate and enhanced precipitation in the river basins, has been observed during the past decades (Semiletov et al., 2000; Serreze et al., 2000; Peterson et al., 2002). As the Arctic and Subarctic freshwater system responds to global warming, increasing freshwater fluxes into the Nordic Seas and Atlantic subpolar basins may in turn act as a feedback on climate (Rahmstorf, 2002; Clark et al., 2002). More specifically, increasing freshwater inputs may slow-down North Atlantic Deep Water (NADW) formation (McClelland, De'ry, Peterson, Holmes, & Wood, 2006).

The trends in Arctic river discharge observed over the last decades, however, are not evenly distributed nor necessarily of the same sign from region to region. While river discharge to the Arctic Ocean from Eurasia has been increasing (Semiletov et al., 2000; Peterson et al., 2002), discharge to the Arctic Ocean from North America has not changed significantly or even slightly decreased (De'ry & Wood, 2005). But in total, discharge to the Arctic Ocean increased by $5.6\,km^3\,y^{-1}\,y^{-1}$ during 1964–2000, the net result of a large increase from Eurasia moderated by a small decrease from North America (Figure 2.22; McClelland, De'ry, Peterson, Holmes, & Wood, 2006).

Contemporaneously with the increased river discharge over the last decades, an increase in the amount and temperature of Atlantic water inflow into the Arctic, a reduced sea-ice cover, a thawing of permafrost and a retreat of small Arctic glaciers are obvious (Dickson et al., 2000; Serreze et al., 2000), and all these environmental changes seem to be related to a cyclic variation of the Northern Hemisphere/Arctic atmospheric circulation pattern, that is, the "Arctic Oscillation (AO)" and the

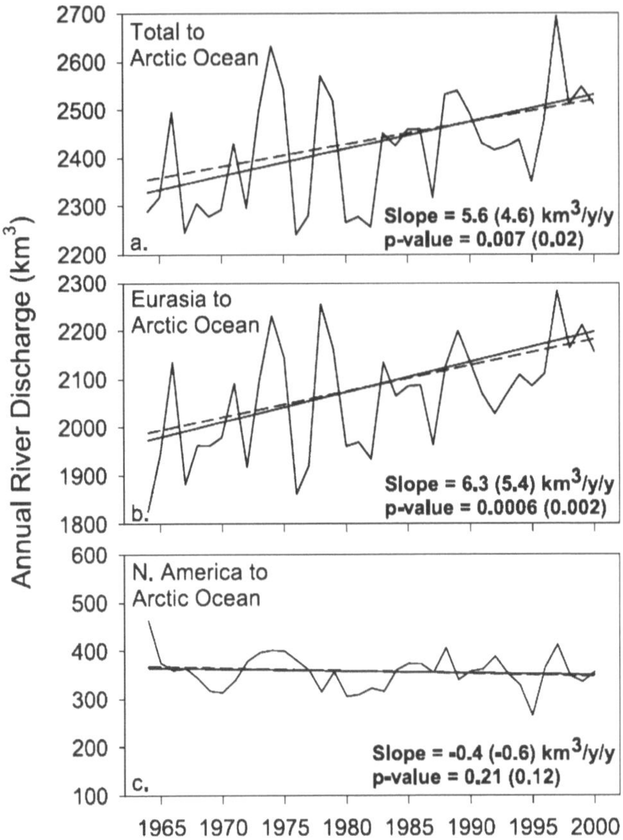

Figure 2.22 River discharge to the Arctic Ocean from 1964 through 2000 (from McClelland et al., 2006). Linear trends are represented by Kendall-Theil Robust Lines with p-values determined using the Mann-Kendall test. Dashed lines and values shown in parentheses reflect adjustments made to remove discharge deficits from years of reservoir filling. For background and data base see McClelland et al. (2006).

"North Atlantic Oscillation (NAO)" (see Chapter 1.1, Figure 1.1; Hurrell, 1995; Thompson & Wallace, 1998; Dickson et al., 2000; Peterson et al., 2002).

2.6. Permafrost

Regions occupied by permafrost, a critical component of the cryosphere and the Arctic system, are widespread around the Arctic Ocean (Figure 2.23). In general, two classes of frozen ground are distinguished: perennially frozen ground (permafrost), defined as any subsurface material that remains at or below 0°C continuously for at least two consecutive years, and seasonally frozen ground, the "active layer", which freezes and thaws on an annual basis (Zhang, Barry, Knowles, Heginbottom, & Brown, 1999; Zhang, Barry, Knowles, Ling, & Armstrong, 2003;

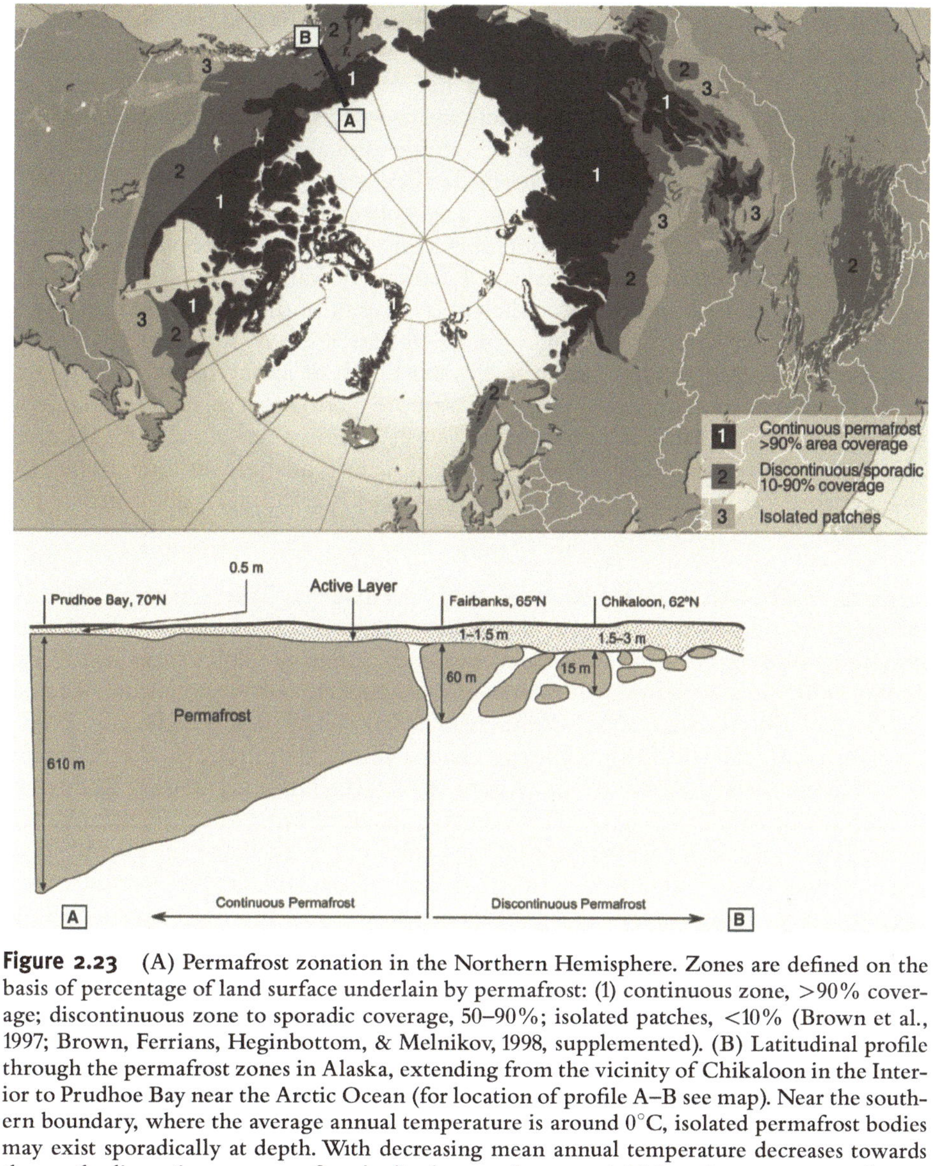

Figure 2.23 (A) Permafrost zonation in the Northern Hemisphere. Zones are defined on the basis of percentage of land surface underlain by permafrost: (1) continuous zone, >90% coverage; discontinuous zone to sporadic coverage, 50–90%; isolated patches, <10% (Brown et al., 1997; Brown, Ferrians, Heginbottom, & Melnikov, 1998, supplemented). (B) Latitudinal profile through the permafrost zones in Alaska, extending from the vicinity of Chikaloon in the Interior to Prudhoe Bay near the Arctic Ocean (for location of profile A–B see map). Near the southern boundary, where the average annual temperature is around 0°C, isolated permafrost bodies may exist sporadically at depth. With decreasing mean annual temperature decreases towards the north, discontinuous permafrost bodies become larger and thicker. At average annual temperatures around −5°C, permafrost in Alaska is essentially continuous, and in northern Alaska it extends to depths of over 400 m (Nelson et al., 1999; U.S. Arctic Research Commission Permafrost Task Force, 2003).

Romanovskii et al., 2000; U.S. Arctic Research Commission Permafrost Task Force, 2003). The thickness of permafrost ranges from very thin layers only a few centimeters thick to several hundreds of metres (Figure 2.23). In unglaciated areas of Siberia, even thicknesses of about 1,500 m may be reached (Washburn, 1980). In general, the thickness of lowland permafrost increases steadily with increasing latitude. In the Northern Hemisphere permafrost regions occupy ~24% of the terrestrial surface (Brown, Ferrians, Heginbottom, & Melnikov, 1997; Zhang et al., 1999, 2003). Along the Siberian shelf, a major part of the coastline consists of ice-rich deposits ("Ice Complex") which are strongly influenced by wave erosion and thermal denudation, resulting in an average annual retreat of the coastline by several metres (see Chapter 2.7 for more details).

Furthermore, thermal modelling (e.g., Lachenbruch & Sass, 1982; Hubberten & Romanovskii, 2003; Romanovskii, Hubberten, Gavrilov, Eliseeva, & Tipenko, 2005) and geophysical data (e.g., Rekant, Cherkashev, Vanstein, & Krinitsky, 2005) indicate that large areas of the Arctic shelves, as a result of their exposure during the Last Glacial Maximum, are thought to be almost entirely underlain by subsea permafrost from the coastline down to a water depth of about 100 m (Rachold et al., 2007). However, the distribution of offshore permafrost is not well known, mainly due to the lack of direct observations (Brown et al., 1997; Danilov, Komarov, & Vlasenko, 1998). There are only a few offshore drilling results on subsea permafrost available (e.g., Dallimore, 1991; Osterkamp, Baker, Harrison, & Matava, 1989; Rachold et al., 2007).

Many of the potential environmental and socioeconomic impacts of global warming in the Arctic are associated with permafrost. The effects of climatic warming on permafrost and the overlying "active layer" can severely disrupt ecosystems and human infrastructure, but may intensify global warming (e.g., Nelson et al., 1993; Nelson, Shiklomanov, & Mueller, 1999; Jorgenson, Racine, Walters, & Osterkamp, 2001; Frey, McClelland, Holmes, & Smith, 2007). Permafrost can facilitate further climate change through the release of greenhouse gases (Rivkin, 1998; Robinson & Moore, 1999; Robinson, Turetsky, Kettles, & Wieder, 2003; Anisimov, 2007). Considerable quantities of carbon are sequestered in the upper layers of permafrost, and vast amounts of methane, in a solid form as gas hydrates, are trapped in permafrost and at shallow depths in cold ocean sediments (e.g., Romanovskii et al., 2005). If the temperature of the permafrost or water at the seabed rises a few degree, a widespread increase in the thickness of the thawed layer and the decomposition of hydrates could lead to the release of large quantities of CH_4 (and CO_2) to the atmosphere (Michaelson, Ping, & Kimble, 1996; Anisimov, Shiklomanov, & Nelson, 1997; Goulden et al., 1998; ACIA, 2004) The release of these greenhouse gases in turn would create a positive feedback mechanism that can amplify regional and global warming (U.S. Arctic Research Commission Permafrost Task Force, 2003 and references therein).

2.7. Coastal Erosion

During the last two decades, several studies have underlined the importance of coastal erosion for the sediment budget of the Arctic Seas (e.g., Reimnitz, Graves, & Barnes, 1988; Macdonald et al., 1998; Are, 1999; Rachold et al., 2000; Brown, Jorgenson, Smith, & Lee, 2003; Grigoriev, Rachold, Hubberten, & Schirrmeister, 2004 and references therein). There are, however, pronounced regional differences when comparing the different Arctic marginal seas. According to Reimnitz et al. (1988), for example, coastal erosion in the Colville River area of the Alaskan coast supplied seven times more sediments to the Alaskan Beaufort Sea than did rivers. Rachold et al. (2000) concluded that the sediment flux to the Laptev Sea through coastal erosion is two times larger than the river input. In the Canadian Beaufort Sea, on the other hand, the Mackenzie River input is the dominant source of sediments and coastal erosion is much less important (Macdonald et al., 1998).

The quality of coastal erosion data is quite variable for the different marginal seas. Furthermore, information about the background of the data is not always available, that is, a quality check of the data is not possible in those cases. Whereas fair to good data are published for the Canadian Arctic (e.g., Reimnitz et al., 1988; Hill, Blasco, Harper, & Fissel, 1991; Dallimore, Wolfe, & Solomon, 1996; Macdonald et al., 1998) and – especially – the Laptev and East Siberian seas (Are, 1999; Rachold et al., 2000; Grigoriev et al., 2004), information on coastal erosion in the Barents, Kara, and Chukchi seas are still limited (Romankevich & Vetrov, 2001; Vasiliev, Kanevskiy, Cherkashov, & Vanshtein, 2005; see Grigoriev et al., 2004 for further references most of which are in Russian language).

For areas where coastal erosion are quite well studied, the estimates of sediment and TOC input via coastal erosion are based on the following steps as outlined in Grigoriev et al. (2004):

- The quantification of coastal retreat rates at representative key sections, which can be carried out by long-term field measurements, comparison of different-time topographic maps, satellite images, and aerial photographs.
- The determination of the coastal morphology (mainly cliff height and slope) to quantify the volume of material supplied ($m^3\,y^{-1}$) based on linear coastal retreat rates ($m\,y^{-1}$).
- The analysis of the composition of the coastal sediments, which includes the determination of ice content and specific density of the sediments. These data are essential to quantify the sediment mass input through coastal erosion.
- The determination of the TOC concentration of the coastal sediments at the key sites to estimate the TOC flux.
- Finally, the data obtained for key sections must be extrapolated to the entire coastline. This can be done by segmenting the coast into homogeneous segments depending on the locations of substantial changes in sediment texture, ice content, or TOC concentration.

As an example, data on coastal erosion from the Laptev Sea are shown in Table 2.6 and Figure 2.24. These data were mainly obtained within the German-Russian Laptev Sea 2000 Project (Kassens et al., 1999) in which several key locations were studied in detail for coastal retreat rates and composition of the coastal sediments between 1998 and 2000 (Grigoriev et al., 2004). In the Laptev Sea (as well as the East Siberian Sea) a major part of the coastline consists of ice-rich deposits ("Ice Complex"; Figure 2.24). The Ice Complex may have an ice content of up to 80% and is strongly eroded by waves and thermal denudation, resulting in an average annual retreat of the coastline by 2.5 m (Figure 2.24). Erosion of the Ice Complex is of major importance for the sediment budget of the Laptev Sea, contributing ~76% of the coastal sediment flux ($44.4 \times 10^6\,t\,y^{-1}$). In total, the sediment and

Table 2.6 Coastal Types of the Laptev Sea and Their Characteristics Associated with Coastal Erosion, Including Total Sediment and Total Organic Carbon (TOC) Fluxes.

	Ice complex and thermokarst deposits	Ice-poor Pleistocene-Holocene coasts	Rocky and other types of non-icy coasts
Total length of the coast (km)	2,400	1,600	3,200
Average retreat rate ($m\,y^{-1}$)	2.5	1	0.05
Average ice content (%)	50	20	2
Average cliff height (m)	10	5	20
Average TOC concentration (%)	4	1	0.3
Coastal sediment flux ($10^6\,t\,y^{-1}$)	44.4	9.5	4.5
Coastal TOC flux ($10^6\,t\,y^{-1}$)	1.78	0.1	0.015

Source: From Grigoriev et al. (2004).

Figure 2.24 Coastal retreat rates and photographs documenting coastal erosion in the Laptev Sea (taken from the Arctic Coastal Dynamics (ACD) web site (http://www.awi-potsdam.de/acd). (1) Bolshoy Lyahkhovsky Island (73.33°N, 141.35°E). Ice complex with large wedges, forming an ice wall near the coast (Photo AWI); (2) Muostakh Island (71.62°N, 129.94°E). Erosion of ice complex deposits; cliff height is ca. 20 m (Photo by V. Rachold).

organic carbon input into the Laptev Sea by coastal erosion reach values of 58.4×10^6 and $1.8 \times 10^6\,t\,y^{-1}$, respectively (Tables 2.6 and 2.7; Grigoriev et al., 2004). For the East Siberian Sea, similar coastal erosion values of $66.5 \times 10^6\,t\,y^{-1}$ for total sediment and $2.2 \times 10^6\,t\,y^{-1}$ for TOC were determined (Table 2.7).

The other Russian Arctic Seas have unfortunately not been studied in such detail, and publications on coastal erosion do not follow the methodology described

Table 2.7 Total Sediment and Total Organic Carbon (TOC) Flux to the Arctic Ocean Through Coastal Erosion.

	Sediment flux ($10^6\,t\,y^{-1}$)	TOC flux ($10^6\,t\,y^{-1}$)
White Sea[1]	60	0.3
Barents Sea[1]	59	0.5
Kara Sea[1]	109	1
Kara Sea[2]	40	0.35
Laptev Sea[3]	58.4	1.8
East Siberian Sea[3]	66.5	2.2
Chukchi Sea[4]	70	0.8
Beaufort Sea[5]	7.9	0.09
Total	430.8 (361.8)	6.69 (6.04)

Source: From Grigoriev et al. (2004, supplemented).
[1]Romankevich and Vetrov (2001).
[2]Vasiliev et al. (2005).
[3]Grigoriev et al. (2004).
[4]Refers to the entire Chukchi Sea (twice the value of Romankevich and Vetrov (2001) given for the Asian sector of the Chukchi Sea only.
[5]Sum of Canadian Beaufort Sea and Alaskan Beaufort Sea (based on Grigoriev et al., 2004 and references therein).

earlier. Estimates on coastal erosion in the Barents Sea, White Sea, and Kara Sea are summarized by Romankevich and Vetrov (2001), giving 59, 60, and $109 \times 10^6\,t\,y^{-1}$ for total sediment supply, and 0.5, 0.3, and $1 \times 10^6\,t\,y^{-1}$ for TOC input, respectively (Table 2.7). The same authors estimate coastal erosion in the Asian sector of the Chukchi Sea to reach 35×10^6 and $0.4 \times 10^6\,t\,y^{-1}$ for total sediment and TOC, respectively. Grigoriev et al. (2004) just multiplied these numbers by 2 to get an estimate for the entire Chukchi Sea (70 and $0.8 \times 10^6\,t\,y^{-1}$). Very recently, Vasiliev et al. (2005) published new coastal erosion data for the Kara Sea, which are about three times lower than Romankevich and Vetrov's values (Table 2.7; 40×10^6 and $0.35 \times 10^6\,t\,y^{-1}$ for total sediment and TOC, respectively). It should be mentioned, however, that the new Kara Sea estimates are based on only one key site studied in detail (Marre-Sale at the western coast of Taymyr Peninsula) and extrapolation of the single-site data to the entire Kara Sea coast. Although the Romankevich and Vetrov (2001) data are mentioned by Vasiliev et al. (2005), there is no discussion about these differences. Thus, it is difficult to judge which of the data sets is the more realistic one.

In the Beaufort Sea, sediment input by coastal erosion is less important, reaching a total of 7.9×10^6 and $\sim 0.1 \times 10^6\,t\,y^{-1}$ for total sediment and TOC, respectively (Table 2.7).

For the entire Arctic Ocean, Grigoriev et al. (2004) give a total sediment and TOC input by coastal erosion of $\sim 430 \times 10^6$ and $6.7 \times 10^6\,t\,y^{-1}$. Using the lower Kara Sea values published by Vasiliev et al. (2005), these values become reduced to $\sim 362 \times 10^6$ and $6 \times 10^6\,t\,y^{-1}$, respectively (Table 2.7).

2.8. Aeolian Input

Fine-grained particles (as well as organic matter and contaminants) may be transported by winds into the Arctic Ocean (Crane et al., 2001; Lisitzin, 2002; Shevchenko, 2003; Shevchenko, Lisitzin, Vinogradova, & Stein, 2003). There are, however, only a few regional estimates of the importance of aeolian material fluxes onto the Arctic surface (Mullen, Darby, & Clark, 1972; Macdonald et al., 1998; Zdanowicz, Zielinski, & Wake, 1998). Macdonald et al. (1998), for example, estimated the aeolian flux in the Mackenzie shelf area, reaching $1.4 \times 10^4\,t\,y^{-1}$. Based on studies of snow samples from the western central Arctic Ocean an aeolian flux of $3.3–14.0\,\mu g\,cm^{-2}$ and year was calculated (Darby, Burckle, & Clark, 1974; Darby et al., 1989). Even if all this material would be released during summer melting and would reach the seafloor (which is not very realistic), this would result in an average sedimentation rate of $0.02–0.09\,mm\,ky^{-1}$. That means, aeolian input is only of very minor importance for the sedimentary budget of the Arctic Ocean.

The most detailed study on aeolian input into the Arctic Ocean based on direct measurements of mass concentration and grain sizes of the insoluble coarse fraction in the Eurasian marginal seas and the central Arctic during expeditions between 1991 and 1999, has been carried out by V. Shevchenko (Shevchenko et al., 1998, 1999, 2000, 2001, 2003, 2004), summarized in Shevchenko (2003) and Shevchenko and Lisitzin (2004). During the measured time interval (August–September), the mass concentration of the insoluble coarse fraction varies from 0.02 to $1.38\,\mu g\,m^{-3}$ ($0.23\,\mu g\,m^{-3}$ on average, $n = 55$ samples). Using these numbers, the authors estimate the dry deposition fluxes of aeolian material to the sea surface, based on the assumption that the sedimentation rate of insoluble coarse ($>1\,\mu m$) particles, which form the major portion of aerosols by mass (Shevchenko et al., 2000), is equal to $2\,cm\,s^{-1}$ (Duce, Liss, & Merrill, 1991). As a result, this gives an average flux of $142\,mg\,m^{-2}\,y^{-1}$. To quantify the total flux of insoluble aerosols per year, it is assumed that the dry deposition is 25% of the total deposition, and that the aerosol concentration in summer is several times lower than in winter and spring, but the amount of aerosol deposition is three to four times larger in summer than in spring. Thus, the total flux of aeolian input calculated under these assumptions is equal to $570\,mg\,m^{-2}\,y^{-1}$ (Shevchenko & Lisitzin, 2004).

As outlined in Shevchenko and Lisitzin (2004), the flux of the aeolian material in the Arctic can be determined independently based on the mean content of suspended particulates in fresh snow, which averages $2.19\,mg\,l^{-1}$ (the mean value calculated from data presented by Mullen et al., 1972; Darby et al., 1974; Pfirman et al., 1989; Dibb, 1996; Dethleff et al., 1998; Shevchenko et al., 1999), and the mean precipitation over the Arctic Ocean which averages $285\,mm\,y^{-1}$ (Voskresenskii & Petrov, 1985). This direct method gives a value of $624\,mg\,m^{-2}\,y^{-1}$ for the vertical flux of aerosols.

The two independent methods described above give an average value of vertical aerosol flux of $\sim 597\,mg\,m^{-2}\,y^{-1}$, which has been used by Shevchenko and Lisitzin (2004) to estimate the aeolian input for the Arctic Ocean and its marginal seas. Based on these authors, $\sim 3 \times 10^6$ and $2.7 \times 10^6\,t\,y^{-1}$ of total (fine-grained) material are transported by winds into the marginal seas and the central Arctic Ocean, respectively (Table 2.8). This results in a total aeolian input of $\sim 5.7 \times 10^6\,t\,y^{-1}$. For the aeolian organic carbon input, a total of $1.7 \times 10^6\,t\,y^{-1}$ has been calculated (Table 2.8).

Table 2.8 Aeolian Input of Total Sediment and Particulate Carbon in the Arctic Ocean. These Calculations are Based on a Total (Dry and Wet) Aeolian Flux of Insoluble Particles of 597 mg $m^{-2}\,y^{-1}$ and an Average Particulate Carbon Content of the Aeolian Material of 30.2%.

	Area ($10^6\,km^2$)	Total sediment flux ($10^6\,t\,y^{-1}$)	Total TOC flux ($10^6\,t\,y^{-1}$)
Barents Sea	1.512	0.904	0.273
Laptev Sea	0.498	0.298	0.090
Kara Sea	0.926	0.553	0.167
East Siberian Sea	0.987	0.589	0.178
Chukchi Sea	0.62	0.371	0.112
Beaufort Sea	0.178	0.106	0.032
Central Arctic	4.489	2.68	0.809
Total Arctic	9.541	5.695	1.720

Source: From Shevchenko and Lisitzin (2004).

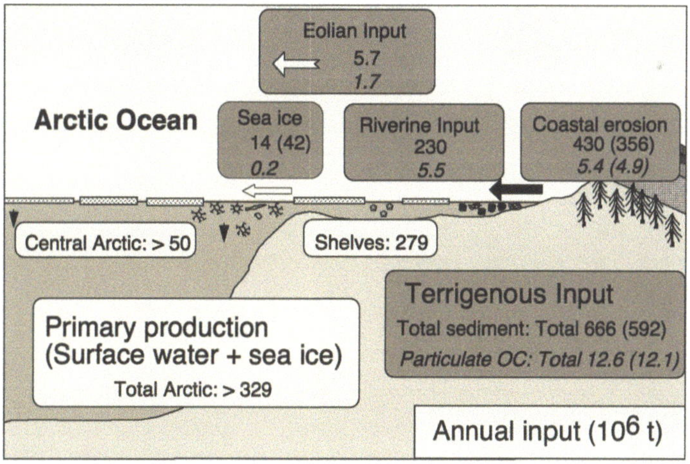

Figure 2.25 Modern total sediment and terrigenous and marine organic carbon input in the Arctic Ocean ($10^6\,t\,y^{-1}$) (Stein & Macdonald, 2004b, based on data from Rachold et al., 2004a; Sakshaug, 2004; Shevchenko & Lisitzin, 2004). Lower numbers given in brackets are obtained if lower input values for coastal erosion in the Kara Sea, as proposed for by Vasiliev et al. (2005), are used.

2.9. Modern Sediment Input: A Summary

In the previous Sections 2.4–2.7, the modern input of marine OC via primary production and the input of total sediments and terrigenous POC by rivers, coastal erosion, and winds are discussed. Figure 2.25 summarizes the total input into the Arctic Ocean marginal seas (Stein & Macdonald, 2004b and further references therein). Altogether, $\sim 666 \times 10^6\,t\,y^{-1}$ of total sediment are supplied whereas the terrigenous particulate OC input is estimated at $\sim 12.6 \times 10^6\,t\,y^{-1}$. By far the most important processes controlling the terrigenous sediment and OC in put are river discharge and coastal erosion. As discussed in Section 2.6, the coastal erosion estimates for the Kara Sea might be too high. If the lower numbers are used, the total input of sediment and terrigenous particulate OC would decrease to $592 \times 10^6\,t\,y^{-1}$ and $12.1 \times 10^6\,t\,y^{-1}$, respectively (Figure 2.25). The OC input from marine primary production accounts for $279 \times 10^6\,t\,y^{-1}$ over Arctic shelves and $>50 \times 10^6\,t\,y^{-1}$ in the central Arctic (Figure 2.25).

Part 2: Processes and Proxies

CHAPTER THREE

Glacio-Marine Sedimentary Processes

This chapter deals with a short overview on modern and ancient glacio-marine sedimentary processes in the High Northern Latitudes, concentrating on processes related to sea ice, icebergs, and sediment mass-wasting (Figure 3.1). The processes as well as their significance for palaeoenvironmental reconstructions and sedimentary budgets in the High Northern Latitudes will be described. Since most important information on understanding sediment mass-wasting processes in the High Northern Latitudes comes from glacial Subarctic regions (i.e., the Norwegian-Greenland Sea), these regions are included here too. For more detailed synthesis papers on glacio-marine processes, the reader is referred to Dowdeswell and Scourse (1990), Elverhøi, Dowdeswell, Funder, Mangerud, and Stein (1998a), Dowdeswell and Ó Cofaigh (2002), Mienert and Weaver (2003), and references therein. The most comprehensive synthesis of studies dealing with glacio-marine sedimentation on a global scale, that is, including data from the Arctic, Subarctic, Antarctic, and periglacial zones of the oceans, was published by Lisitzin (2002). Information on terrestrial records related to glacial processes such as glacial erosion, extent and variability of ice sheets etc., as well as results from ice-sheet modelling and correlations between terrestrial and marine records can be obtained from synthesis compilations edited by Elverhøi et al. (1998a), Thiede, Bauch, Hjort, and Mangerud (2001), and Thiede (2004). These compilations resulted from major European research projects bringing together marine and terrestrial geoscientists and carried out in the polar North Atlantic and the Eurasian continental margin and

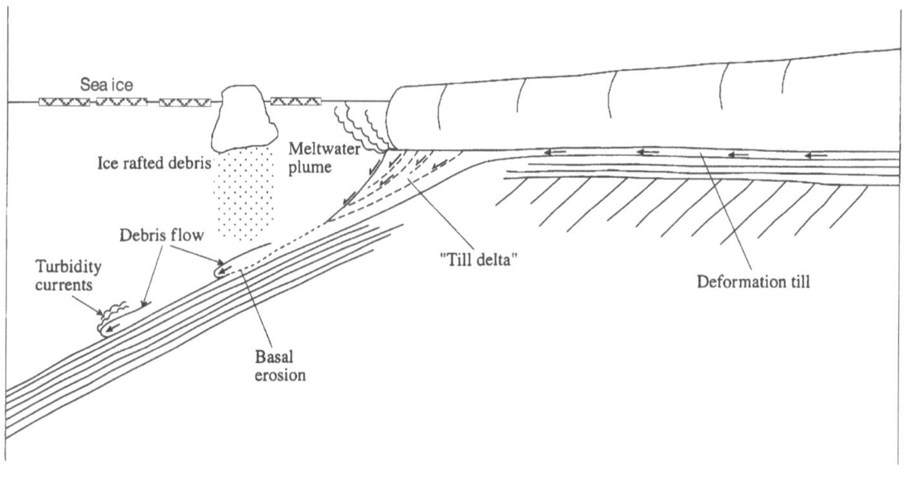

Figure 3.1 Schematic model showing the main sedimentary processes at the shelf break and upper continental slope during the presence of an ice sheet at the shelf break (from Laberg & Vorren, 1995).

adjacent deep-sea area ("Polar North Atlantic Margin — PONAM" and "Quaternary Environment in the Eurasian North — QUEEN").

3.1. Sea-Ice Processes: Sediment Entrainment and Transport

Information about modern sea-ice distribution and variability as well as its influence on primary production has been described in Chapters 2.3 and 2.4, respectively. Here, it will be concentrated on sea-ice sediments and the processes related to sediment entrainment and transport. Research over the past two decades has established that sediment-laden or "dirty" sea ice is a common phenomenon in the Arctic Ocean and its marginal seas (e.g., Larssen, Elverhøi, & Aagaard, 1987; Pfirman et al., 1989, 1997; Reimnitz, Barnes, & Weber, 1993a; Reimnitz, McCormick, McDougall, & Brouwers, 1993b; Wollenburg, 1993; Nürnberg et al., 1994; Eicken et al., 1997, 2000; Eicken, Gradinger, Graves, Mahoney, & Rigor, 2005; Dethleff, 2005). The data on these sea-ice-related processes are based on field observations (e.g., Reimnitz et al., 1993b; Dethleff, Nürnberg, Reimnitz, Saarso, & Savchenko, 1993, 2000; Nürnberg et al., 1994; Eicken et al., 1997, 2000; Darby, 2003), remote sensing (e.g., Reimnitz et al., 1993b; Eicken et al., 2000; 2005), and analysis of sea-ice trajectories and modelling (see Chapter 5.1.8; Pfirman et al., 1997; Eicken et al., 2005). Furthermore, detailed analysis of sediment cores obtained from Arctic and Subarctic areas revealed that sediment transport by sea ice probably was an important process influencing the sedimentation regime in the Arctic Ocean and the Greenland Sea in the recent geological past (e.g., Clark & Hanson, 1983; Bischof & Darby, 1997; Nørgaard-Pedersen, Spielhagen, Thiede, & Kassens, 1998; Behrends, 1999; see also Chapter 6.3).

These sea-ice sediments mainly consist of terrigenous material with clay minerals, quartz, and feldspars as main components (Figure 3.2) (Nürnberg et al., 1994). The mineralogy of sea-ice sediments may be very variable in time and space. Thus, studies of the mineralogical composition may allow to identify source areas of the sea-ice sediments and, based on these data, reconstruct transport pathways (see Chapter 5.1). The biogenic components indicate a highly diverse ice flora, with diatoms reaching the highest proportion (Abelmann, 1992; Nürnberg et al., 1994). Furthermore, silt- and clay-sized material is by far the most dominant grain-size class in the sea-ice sediments (Figure 3.3; Wollenburg, 1993; Nürnberg et al., 1994; Dethleff, 2005).

Dethleff (2005), for example, showed in his detailed study of sea-ice sediments from the Laptev Sea that most of the sea-ice sediments samples (95%) contain $<10\%$ coarse material and lie close to or even on the silt/clay line (Figure 3.4). In contrast, 64% of the shelf sediments contain more than 10% coarse material reaching a maximum of almost 90% sand at one station. From Figure 3.4 it is obvious that the shelf sediment has a wider range of grain sizes, while the compositions of individual sea-ice sediment samples are more similar to each other and clearly emphasize the fine fraction $<63\,\mu m$, which is mostly dominated by silt. The average silt distribution of Laptev Sea shelf sediments and sea-ice sediment samples, on the other hand, are very similar as shown in paired silt-grain-size distribution at three shallow

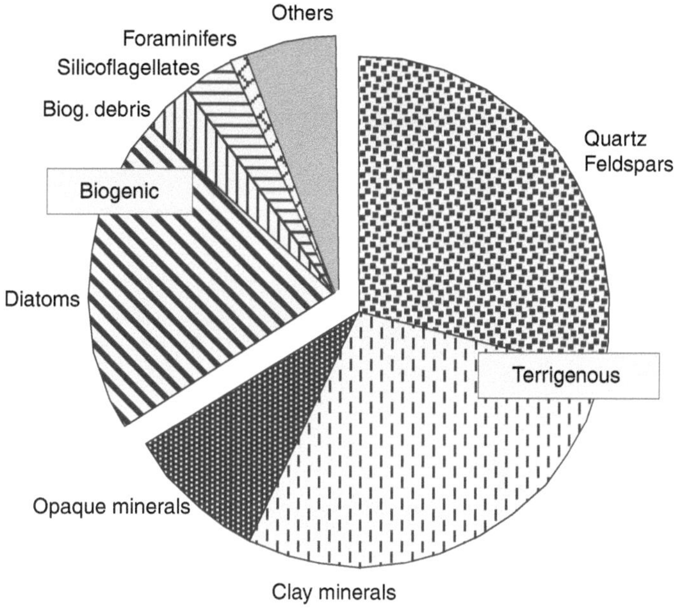

Figure 3.2 Composition of sea-ice sediments in the central Arctic Ocean (from Nürnberg et al., 1994, redrawn).

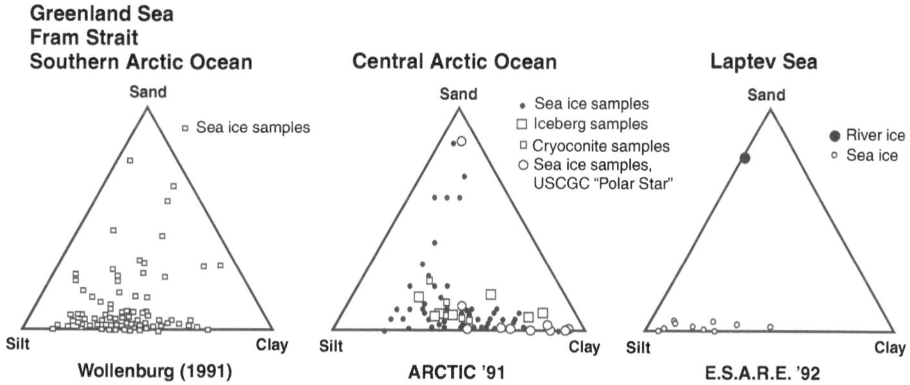

Figure 3.3 Grain-size distribution of Arctic Ocean sea-ice sediments (from Nürnberg et al., 1994).

sites with water depths between 4 and 17.5 m, where sea-ice sediments and shelf sediments were obtained exactly from the same locations (Figure 3.4; Dethleff, 2005). This indicates that silt is entrained into the ice as present on the shelf surface. Furthermore, the silt and clay distribution of Laptev sea-ice sediment samples is very similar to the fine material texture found in Beaufort Gyre sea-ice sediments (Reimnitz, McCormick, Bischof, & Darby, 1998). The similarity of the fine fraction texture in Laptev and Beaufort sea-ice sediments (see Section 4.2.2, Figure 4.9) points to a physically uniform hydrodynamic entrainment process active on shallow circum-Arctic shelves (Reimnitz et al., 1993a, 1993b, 1998; Dethleff, 2005).

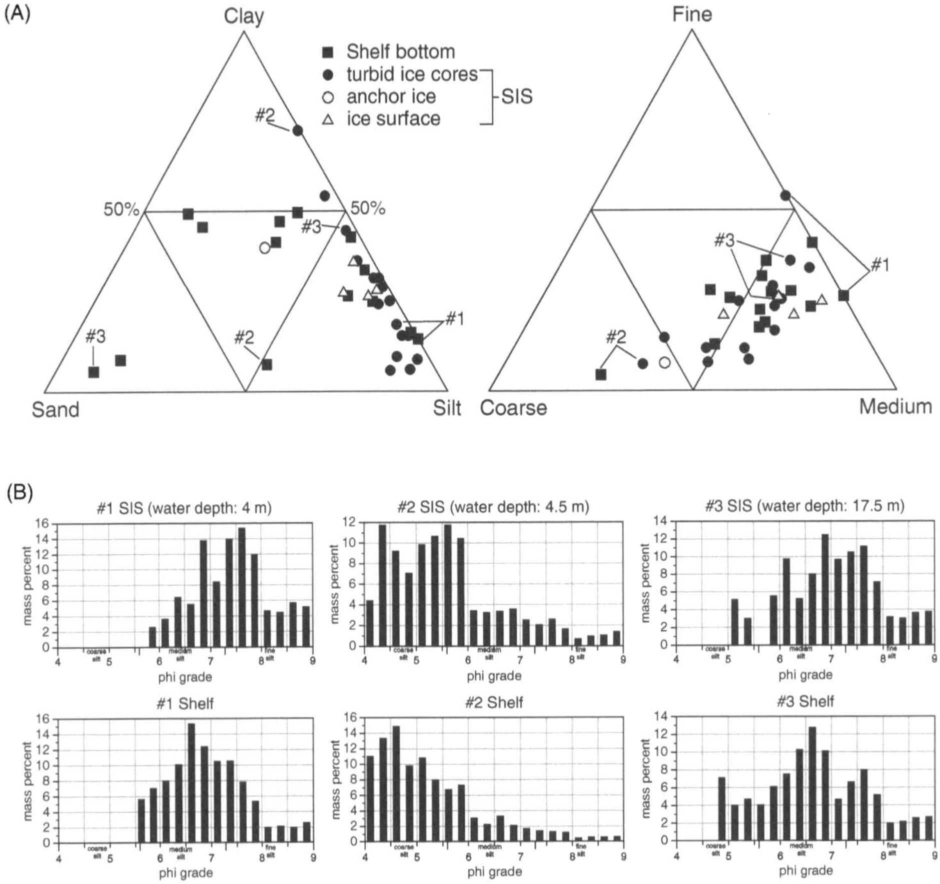

Figure 3.4 Surface sediment and sea-ice sediment data from the Laptev Sea (from Dethleff, 2005). (A) Ternary plot diagrams showing sand/silt/clay concentration and silt fractionation. Solid circles indicate material collected from upper ice core sections, and open circles denote anchor ice (both ESARE '92), while triangles indicate material collected from the ice surface (*Polarstern* Expedition ARK-IX/4). Numbers #1, 2, and 3 indicate examples shown in (B). (B) Exemplary comparison of silt fractionation in shelf sediments (bottom row) and sea-ice sediments (SIS) (top row).

What are the main processes controlling the entrainment of sediment into the sea ice? Sea-ice formation mainly occurs during fall and the early stages of winter over the broad Siberian shelves in comparatively shallow water of mean depths ~50 m in the Laptev and East Siberian Seas (Timokhov, 1994; Eicken, 2004). Tidal and wind mixing during fall freeze-up as well as thermohaline mixing associated with rapid ice growth and brine rejection promote the resuspension of sediments (Sherwood, 2000) from the seafloor and furthermore enhance the rate of formation of frazil ice in the water column (Smith, Muench, & Pease, 1990). The combination of these processes results in the entrainment of substantial amounts of resuspended particulate matter into the newly formed ice (Figure 3.5; Pfirman, Lange, Wollenburg, & Schlosser, 1990; Reimnitz, Clayton, Kempema, Payne, & Weber, 1993a, 1993b,

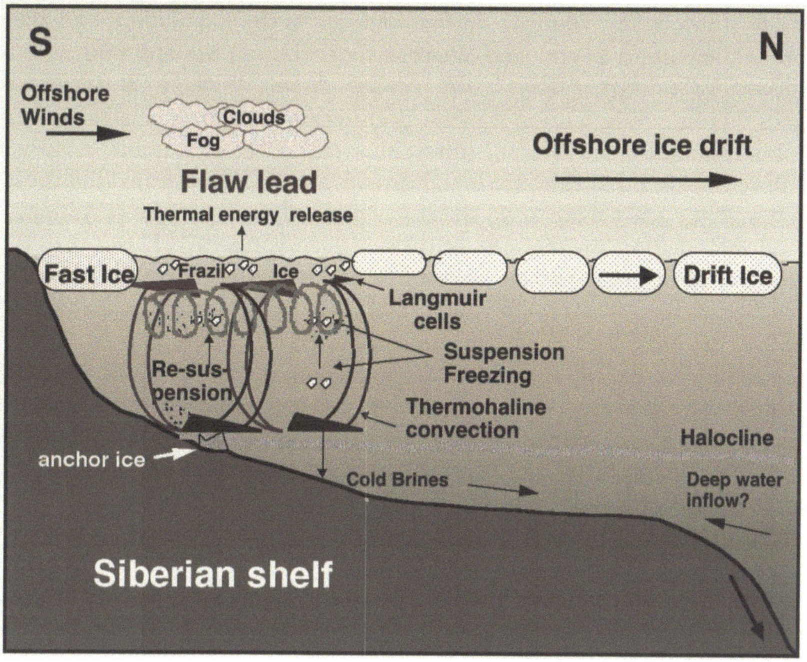

Figure 3.5 Processes controlling sediment entrainment into sea ice (based on Dethleff, 2005).

1993c; Dethleff et al., 1998; Eicken et al., 2000; Dethleff, 2005 and references therein). In addition, other processes such as anchor ice formation, deposition of river sediments onto a flooded fast ice cover, and aeolian deposition may also contribute to the particulate load. Because fine-grained shallow marine sediment inclusions are much more important than all other types of inclusions (Figure 3.3), the process of suspension freezing seems to be the most significant contributor to the phenomenon of sediment-laden sea ice in the North American and Siberian Arctic (Osterkamp & Gosink, 1984; Reimnitz, Marincovich, McCormick, & Briggs, 1992, 1993a, 1993b, 1993c; Nürnberg et al., 1994; Eicken et al., 2000; Eicken, 2004).

The quantification of sediment transport by sea ice presents a considerable challenge due to the patchy distribution of sediments, our lack of understanding of entrainment processes in detail, and the overall difficulty in obtaining data on the areal distribution and suspended particulate material (SPM) concentrations characteristic of sediment-laden ice (Pfirman et al., 1990; Reimnitz et al., 1993a, 1993b; Nürnberg et al., 1994; Dethleff et al., 2000; Eicken et al., 2000; Eicken, 2004; Dethleff, 2005). Thus, published quantitative estimates of sea-ice sediments differ significantly (see below). Data on sediment concentrations in sea ice based on direct sampling of the ice, have shown concentrations to range between <10 and $>10{,}000\,\mathrm{mg\,l^{-1}}$. In the central Arctic Ocean, highest sediment concentrations were found on surfaces of multiyear ice floes, ranging between 10 and $56{,}000\,\mathrm{mg\,l^{-1}}$ (Nürnberg et al., 1994). Average sediment concentrations in visibly discoloured, sediment-laden ice sampled in different regions range between $\sim 40\,\mathrm{mg\,l^{-1}}$ in the Beaufort Gyre, $190\,\mathrm{mg\,l^{-1}}$ in the coastal Beaufort Sea, 70 mg l^{-1} in first-year ice in the Laptev Sea, and 380 mg l^{-1} in ice from the eastern Laptev and Western East Siberian Sea and 660 mg l^{-1} in the upper layers of multiyear ice in the central Arctic (Eicken, 2004 and further references therein). When using these concentration values, especially the very high values, for estimating the overall importance of sea-ice sediments, however, one should have in mind that they are based on samples collected from (the most) dirty ice patches. Large sea-ice areas, on the other hand, do not contain sediments. Thus, additional information on the area of dirty sea ice is needed. Here, very different values are found in the literature, as summarized by Lisitzin (2002): Eurasian Basin north of Spitsbergen 1–10% (Pfirman et al., 1989); Barents Sea 20–30% (Vinje, 1987); Beaufort Sea 30–90% (Reimnitz et al., 1993a); central part of Arctic basins ~50% (Dethleff et al., 1993) or 50–90% (Eicken et al., 1997). Based on these data, Lisitzin postulates that (1) ~10–50% of the total Arctic sea-ice area is covered by dirty sea ice, and (2) the average sea-ice sediment content of the total Arctic sea-ice area (i.e., considering dirty and clean ice-covered areas) is ~20–30 mg l^{-1}.

Because absolute numbers of sea-ice sediment input are based on the concentration values, also these numbers differ significantly from paper to paper. Eicken (2004), for example, estimated an annual sea-ice sediment input from the Arctic shelves into the open ocean of ~ 14×10^6 tons, with the Laptev Sea and Kara Sea being the most important source regions for sea-ice sediments (10×10^6 and 2.4×10^6 tons, respectively; Figure 3.6).

In a very recent detailed study of sea-ice sediments in the Laptev Sea area, Dethleff (2005) came to significantly higher sea-ice sediment input values than

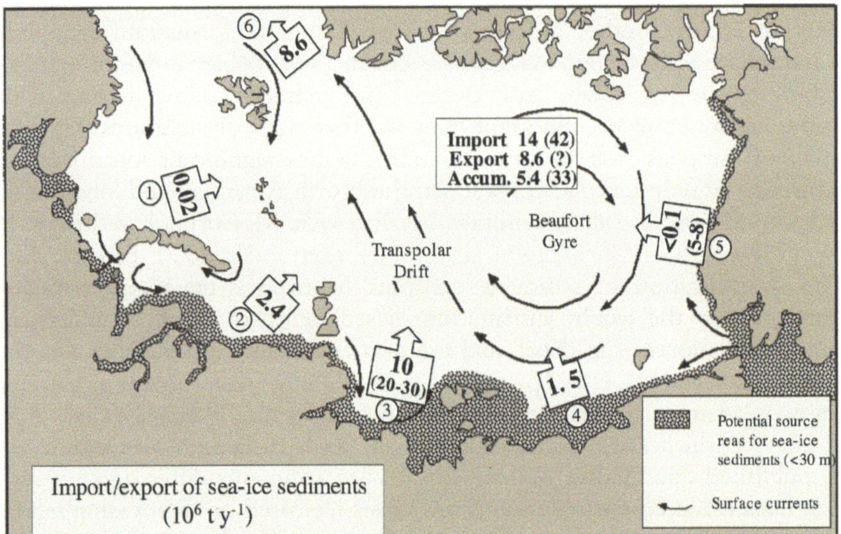

Figure 3.6 Import of sea-ice sediments from the marginal seas into the open Arctic Ocean and export from the Arctic Ocean into the North Atlantic Ocean (based on Eicken, 2004; Dethleff, 2005; Eicken et al., 2005; potential source areas according to Reimnitz et al., 1993a, 1993b). 1, Barents Sea; 2, Kara Sea; 3, Laptev Sea; 4, East Siberian Sea; 5, Beaufort Sea; 6, Fram Strait.

those published by Eicken (2004). Considering ice-formation rates and concentrations of sea-ice sediments in 10 individual Laptev Sea flaw lead sections, annually 26.6×10^6 tons particulate matter may be entrained into lead ice. This is a value very similar to the annual riverine sediment input into the Laptev Sea ($28.6 \times 10^6\,t\,y^{-1}$; Rachold et al., 2004a; see also Chapter 2.5). According to Dethleff (2005) roughly $20–30 \times 10^6\,t\,y^{-1}$ of sea-ice material may be exported from the Laptev Sea towards the central Arctic Ocean and Fram Strait.

On the basis of observations over the Chukchi and Beaufort shelves in 2001/2002 indicating widespread occurrence of sediment-laden ice over an area of more than 100,000 km^2 between 68°N and 74°N and 155°W and 170°W, and analysis of Synthetic Aperture Radarsat (SAR) imagery in conjunction with bathymetric data, Eicken et al. (2005) revised the values published in Eicken (2004) ($<0.1 \times 10^6\,t\,y^{-1}$) to significantly higher values, reaching $5–8 \times 10^6\,t\,y^{-1}$ of sea-ice sediment over the entire Chukchi and Beaufort shelves (Figure 3.6).

Unfortunately, such profound estimates of sea-ice sediment entrainment and export rates are not available from the Barents, Kara, and East Siberian seas yet. The number estimated for the East Siberian Sea, for example, might probably be a minimum value when thinking about the broad shallow shelf areas available for sediment entrainment into the sea ice (Figure 3.6). Thus, reliable estimates of total annual Arctic Ocean sea-ice sediment fluxes are difficult to give. Taken the revised numbers for the Laptev, Chukchi, and Beaufort seas described above, the total annual sea-ice sediment flux from the shelves into the Arctic Ocean proper of 14×10^6 t (Eicken, 2004) increases to at least to 42×10^6 t (Figure 3.6). Using these higher values, sedimentation from sea ice may become much more important for an Arctic Ocean sedimentary budget calculation (see Stein & Macdonald, 2004b).

Another interesting approach for estimating the importance of sea-ice sediments in Arctic Ocean sedimentation was done by Lisitzin (2002). Starting from a maximum ice volume in the Arctic Ocean during spring of $\sim 34.55 \times 10^3\,km^3$ and a minimum ice volume during autumn of $25.3 \times 10^3\,km^3$, $\sim 9.25 \times 10^3\,km^3$ of sea ice is melting and releasing entrained sediments during summer times. Taking an average concentration of sea-ice sediments of $30\,mg\,l^{-1}$ (see above), this results in very high number of 277.5×10^6 t of sediments per year. This is higher than the annual fluvial input of suspended matter ($227 \times 10^6\,t\,y^{-1}$; see Table 2.5). The difference becomes even much more drastic when considering that major amounts of the suspended riverine matter is trapped in the estuaries and deltas (see Chapter 2.5). As stated by Lisitzin (2002) most part of the ice formed during autumn and winter melts during spring and summer already on the shelf, that is, within the sea of origin, where most part of the entrained sediment is released. According to an estimate by Lisitzin (2002) $\sim 29.4 \times 10^6\,t\,y^{-1}$ of sea-ice sediments are leaving the shelf and transported into central Arctic Ocean, a number which is similar to those shown in Figure 3.6.

A major part of the sediment incorporated into the sea ice in the Arctic shelf areas and exported from the shelf is transported through the central Arctic Ocean via the Transpolar Drift towards Fram Strait. In areas of extensive melting, such as Fram Strait and the Norwegian-Greenland Sea, sediment particles are released and deposited at the sea floor (Wollenburg, 1993). In these areas, ice-rafted debris (IRD) may contribute significantly to the supply and accumulation of terrigenous material.

According to Eicken (2004), $\sim 8.6 \times 10^6\,t\,y^{-1}$ of the sea-ice sediments supplied from the shelf seas are exported through Fram Strait (Figure 3.6). If this estimate is correct, $\sim 5.4–33 \times 10^6\,t\,y^{-1}$ of sea-ice sediments (depending on the used input value, see above) will already be released within the Arctic Ocean and contribute to the sedimentation there. Here, it should be mentioned that Larssen et al. (1987) gave a much higher export rate of sea-ice sediments through Fram Strait, reaching $\sim 150 \times 10^6\,t\,y^{-1}$, a value which seems to be unrealistically high when looking at the input numbers of sea-ice sediments (Figure 3.6) as well as the Arctic Ocean sediment budget (see Chapter 6.4.4).

In order to get a first-order estimate about the importance of sea-ice sediment input for the sediment accumulation in the Arctic Ocean proper, that is, the continental slope, continental rise, abyssal plains, and ridges, total accumulation rates from these areas (Stein & Macdonald, 2004b) were compared with the sea-ice sediment input (Table 3.1), based on the following assumptions. First, for sea-ice sediment input (release) the higher estimate of $33 \times 10^6\,t\,y^{-1}$ is used. This number seems to be more realistic than the lower ones because for large areas such as the East Siberian Sea the published numbers for sediment input by sea ice are probably underestimating this process. Second, it is assumed that the sediment is released from sea ice uniformly over the whole area. Third, based on detailed records from the Laptev and Kara seas, the late Holocene sediment accumulation is, in average, only 60% of the mean Holocene accumulation (Stein & Macdonald, 2004b). Thus, the comparison of mean Holocene accumulation rates with modern input values (including sea-ice input) is biased, and a comparison with late Holocene values is probably more realistic (see Stein & Macdonald, 2004b for details). As result, about 55% and 85% of the modern sediments accumulated in the abyssal plains and on the ridges, respectively, are related to sea-ice sediment input, whereas at the continental rise no more than 17% are related to sea-ice sediment input (Table 3.1). For the whole central Arctic Ocean, on an average $\sim 23\%$ of the sediments are caused by sea-ice sediment input.

Table 3.1 Estimate of Sea-Ice Sedimentation in Percentage of Modern (late Holocene) Total Sediment Accumulation in the Arctic Ocean (Excluding the Shelf Areas). For Size Data and References see Table 2.1. Total Sediment Accumulation Data from Stein and Macdonald (2004b); in Brackets Average Holocene Accumulation Rates are Given. For Comparison with Modern Sea-Ice Sediment Input (i.e., Sediment Release from Sea Ice) it Seems to be More Realistic to Use the Late Holocene Total Accumulation Rates (see Stein & Macdonald, 2004b for Details). In Brackets Percentage of Sea-Ice Sediments, if Average Holocene Accumulation Rates would be used. See Text for Further Explanation.

Area	Size ($\times 10^3$ km^2)	Size (%)	Modern sediment accumulation ($10^6\,t\,y^{-1}$)	Sea-ice sediments ($10^6\,t\,y^{-1}$)	Sea-ice sediments (%)
Continental slope	541	12	64 (107)	4	6 (4)
Continental rise	1095	24	47 (79)	8	17 (10)
Abyssal plains	1367	30	18 (30)	10	55 (33)
Ridges	1506	33	13 (21)	11	85 (52)
Total	4509	100	142 (237)	33	23 (14)

3.2. Ice Sheet- and Iceberg-Related Processes

During the Quaternary (and Neogene) ice sheets have advanced and retreated across the high-latitude margins in a series of climate related glacial–interglacial or stadial–interstadial cycles (e.g., Svendsen et al., 2004 and further references therein; Dowdeswell et al., 2002). In the present interglacial, there are no areas in the Arctic where glacier ice is grounded at the continental shelf edge, whereas, in contrast, ice sheets advanced across the continental shelf during full-glacial periods (Dowdeswell et al., 2002; see Chapter 6.2.1 for more details and discussion). On many shelves, ice-sheet advance produced erosion, reworking, and hiatuses. Here, most impressive evidence are the major troughs cut into the shelves by glacial erosion (Figure 3.7;

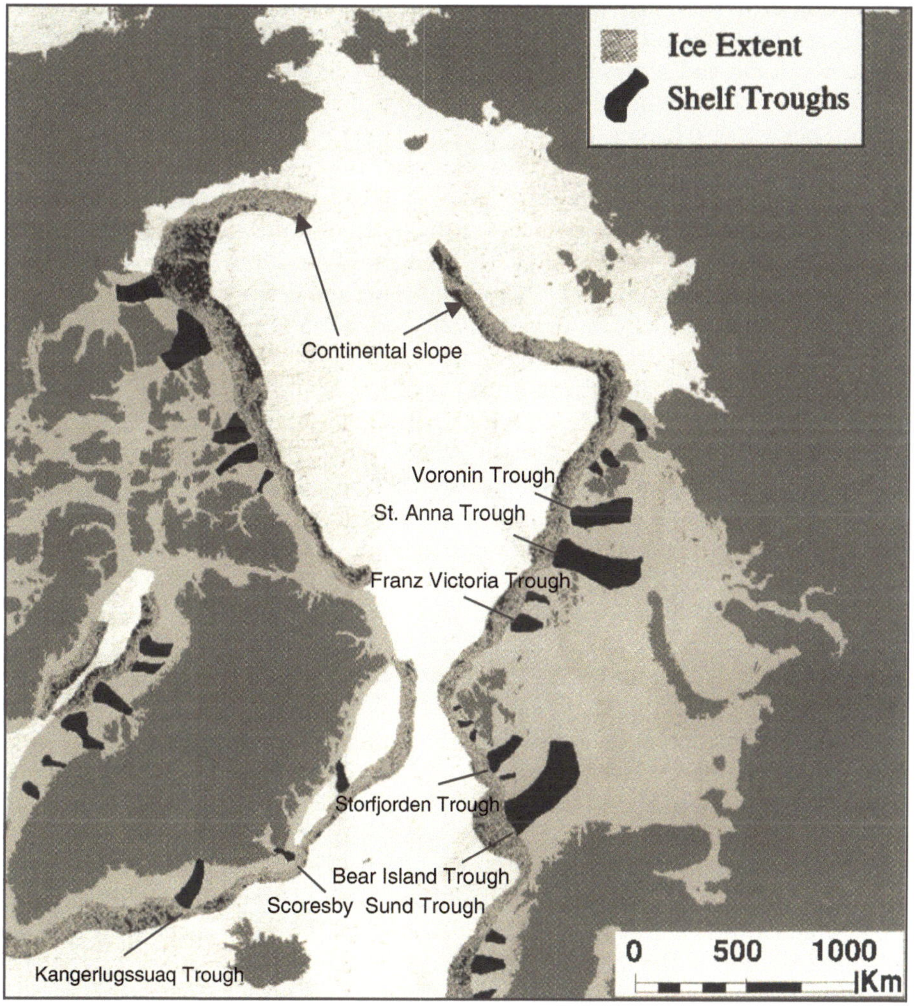

Figure 3.7 Distribution of shelf troughs in the circum-Arctic area (O'Grady & Syvitski, 2002, supplemented).

O'Grady & Syvitski, 2002). Furthermore, the ice advance affected the nature of the sedimentation on the shelves and greatly increased the rate of sediment delivery to the continental slope during glacial periods (e.g., Elverhøi et al., 1995a, 1998a; Dowdeswell et al., 2002) (see Section 3.3). Even during interglacials, sediment reworking may continue by ploughing action of iceberg keels (iceberg scouring) impinging on the sea floor (Figure 3.8; Dowdeswell, Whittington, & Hodgkins, 1992; Dowdeswell, Villinger, Whittington, & Marienfeld, 1993; Dowdeswell et al., 2002; Ó Cofaigh et al., 2002).

Sediments included in glaciers may consist of basal, englacial, and supraglacial debris (e.g., Anderson & Molnia, 1989; Figure 3.8). When reaching the shelf, typical grounded-ice-sheet-related sediments on the shelf itself are lodgement tills. Seawards beyond the grounding line, calving of icebergs occurs, and sediments entrained in the icebergs are transported towards the open ocean within the existing ocean current systems as IRD. When icebergs melt, entrained debris is released into the water column and settles down to the sea floor. Resulting ice-rafted sediments are poorly sorted, that is, they consist of fine-grained mud to oversized clasts ("dropstones") (see Chapter 4.2.2).

Sediment transport by icebergs is of secondary importance in the modern Arctic Ocean, because icebergs are very rare today due to the absence of large ice shelves. Arctic glaciers today typically terminate at the heads of fjords, with the consequence that sediments are delivered mainly to inner-fjord locations and relatively little glacier-derived debris reaches the open shelf, slope, and deep-sea environment (e.g., Marienfeld, 1991, 1992; Evans et al., 2002). Today, calving of icebergs and related iceberg sediment transport into the Arctic Ocean are restricted to Ellesmere Island, North Greenland, Svalbard, Franz-Josef Land, and Severnaya Zemlya (Figure 3.9; Sudgen, 1982; Darby et al., 1989). During glacial intervals and — especially — late glacial to interglacial transitions, on the other hand, iceberg transport may have become a major sediment-transport process in the Arctic Ocean (see Chapter 6.2).

Where the keels of icebergs contact the sea floor, scouring or ploughing of the substrate take place (Figure 3.8). These iceberg scour marks on the sea floor are most readily identified from side-scan sonar and swath bathymetry records (Figure 3.8B), which demonstrate their morphology and lateral continuity over distances of hundreds and sometimes thousands of metres (e.g., Josenhans, Zevenhuizen, & Klassen, 1986; Praeg, MacLean, Piper, & Shor, 1987; Solheim, Milliman, & Elverhøi, 1988; Dowdeswell et al., 1992; Shipp, Wellner, & Anderson, 2002). The intensity of scouring is dependent on the relationship between iceberg keel dimensions and water depth, on the rate of production and drift track of the icebergs, and on the nature of the substrate (e.g., Dowdeswell et al., 1993). Both modern and relict scours produced by iceberg keels have been reported from a number of higher latitude continental shelf areas in the Northern Hemisphere (e.g., Brett & Zarudzki, 1979; Woodworth-Lynas, Simms, & Rendell, 1985; Josenhans et al., 1986; Praeg et al., 1987; Solheim et al., 1988; Dowdeswell et al., 1992; Polyak et al., 2001) and Antarctica (e.g., Lien, Solheim, Elverhøi, & Rokoengen, 1989; Shipp et al., 2002). At present, icebergs in the Arctic Ocean have at most 50 m draughts (Robe, 1980), whereas icebergs off Antarctica and Greenland reach depths of 500–550 m (Barnes & Lien, 1988; Dowdeswell et al., 1992).

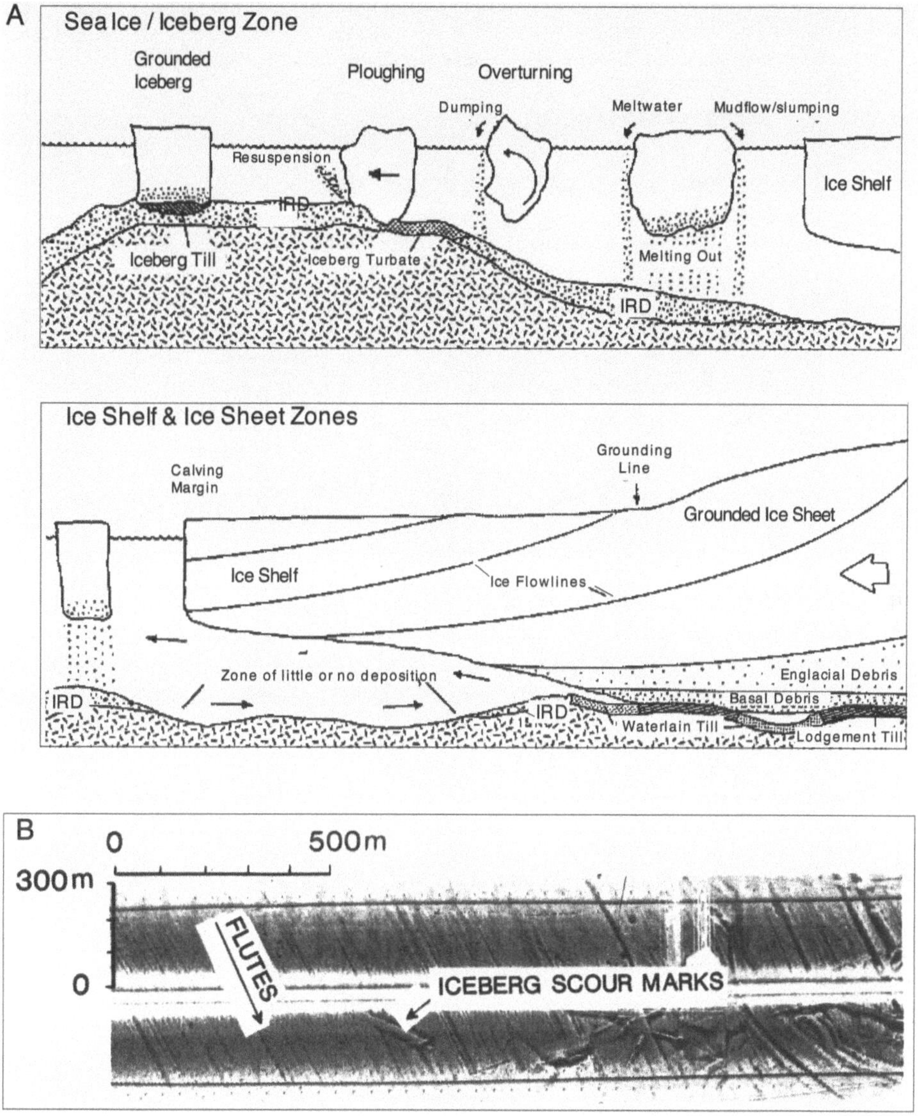

Figure 3.8 (A) Model of glacio-marine sedimentation in a proximal ice shelf setting and an open-shelf setting, developed from Antarctic shelf sediments (from Anderson & Molnia, 1989, according to Kellogg & Kellogg, 1988). (B) Sea-floor morphological features in the NW Barents Sea related to the presence of ice. The side-scan sonar image shows flutes interpreted to be of subglacial origin, and scour marks formed by iceberg keels at modern water depth of ~350 m (from Solheim et al., 1990; Dowdeswell et al., 1998).

During two RV *Polarstern* expeditions in 1988 and 1990, Dowdeswell et al. (1992, 1993) carried out an intensive study of iceberg scouring in the Scoresby Sund (Figure 3.10) and adjacent East Greenland continental shelf area. Although side-scan equipment was not deployed from *Polarstern*, the very irregular nature of

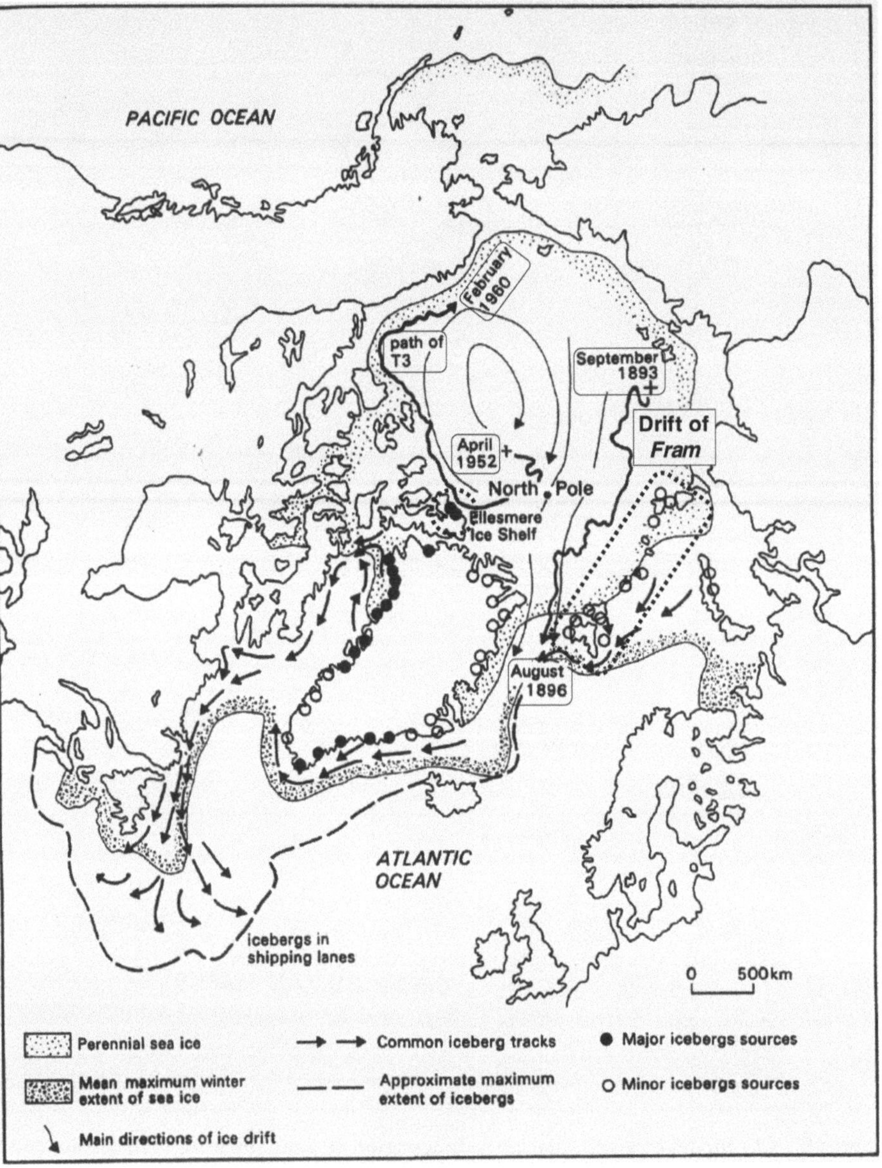

Figure 3.9 Sources of modern icebergs in the High Northern Latitudes (Sudgen, 1982, supplemented). Areas of iceberg input in the Arctic Ocean are surrounded by dotted lines. Furthermore, drift paths of the ice island T-3 (1952–1960; for time interval 1963–1974 see Figure 1.5) and the Fram (Nansen, 1897, 1904) are shown.

the sea floor in the study area related to iceberg scouring was clearly observed on both Parasound and 3.5 kHz records at water depths of <550 m. During September 1990, the sizes of 1,900 icebergs were measured in the Scoresby Sund fjord system. As a result, 31% of icebergs had a maximum width in excess of 200 m, and 5 were

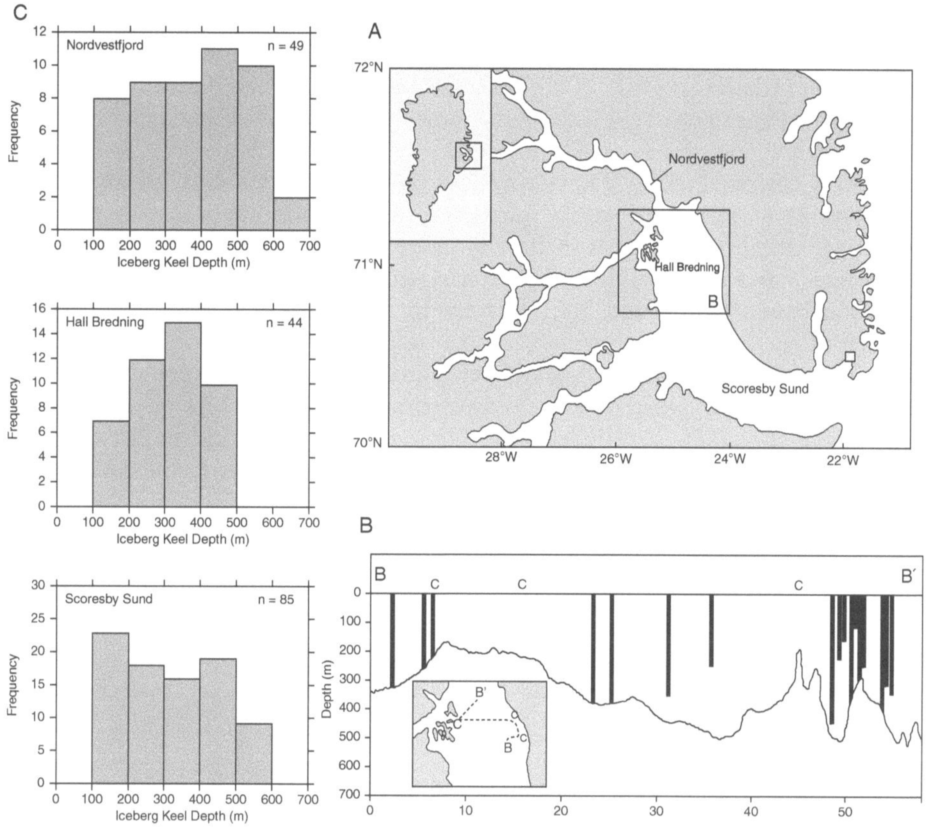

Figure 3.10 (A) Scoresby Sund area, East Greenland; (B) transect across Hall Bredning, showing bathymetry and iceberg keel depths, and (C) frequency distribution of iceberg-keel depth (from Dowdeswell et al., 1993, supplemented).

larger than 1 km in width. The modal iceberg keel depth in Nordvestfjord was 400–500 m, declining to 300–400 m for Hall Bredning, and 100–200 m for Scoresby Sund (Figure 3.10). In summary, Dowdeswell et al. (1992, 1993) observed iceberg scours in an area of ~25,000 km^2 of the East Greenland continental shelf, between 69°N and 72°N and at 75°N. Based on the rate of iceberg production from outlet glaciers of the Greenland Ice Sheet and the nature of iceberg drift on the Greenland continental shelf, these authors suggest that scouring by iceberg keels is likely to be an important process over a significant portion of the 500,000 km^2 area above the approximately 500 m deep shelf break.

Echo-soundings of the Chukchi Plateau registered an abnormally rough seafloor at depths to 300–400 m, interpreted as iceberg scouring (Hunkins, Herron, Kutschale, & Peter, 1962; Phillips & Grantz, 1997). Recent seafloor surveys with a 12-kHz swath bathymetry and side-scan and penetration sonar confirmed the abundance of randomly oriented scours (Polyak et al., 2001; Polyak, Darby, Bischoff, & Jakobsson, 2001, 2007; Darby, Jakobsson, & Polyak, 2005). Below ~400 m, however, side-scan and swath bathymetry images show that the seafloor exhibits coherent sets of evenly spaced, parallel, low-relief lineations (flutes) reaching more then 15 km in length and extending in some areas to almost 1,000 m water depth (Jakobsson et al., 2005; Polyak et al., 2007). These flutes at the surface on Chukchi Plateau are very similar to numerous examples from formerly glaciated continental margins (e.g., Davies et al., 1997; Shipp, Anderson, & Domack, 1999; Ó Cofaigh, Pudsey, Dowdeswell, & Morris, 2003a; Ó Cofaigh, Taylor, Dowdeswell, & Pudsey, 2003b). Furthermore, the authors could identify two separate generations of flutes, a shallow (younger) set and a deep (older) set (Figure 3.11, Flutes 1 and 2), probably formed during the LGM and an earlier glacial event, possibly between MIS 4 and 5.4, respectively. According to Polyak et al. (2007), both erosional events were presumably caused by the grounding of ice shelves originating from the Laurentide Ice Sheet. Similar sets of flutes were also recorded in northern Barents Sea (Solheim, Russwurm, Elverhøi, & Nyland-Berg, 1990; Dowdeswell, Elverhøi, & Spielhagen, 1998; see Figure 3.8) and in the St. Anna Trough up to at least 82°N (Polyak, Forman, Herlihy, Ivanov, & Krinitsky, 1997) in

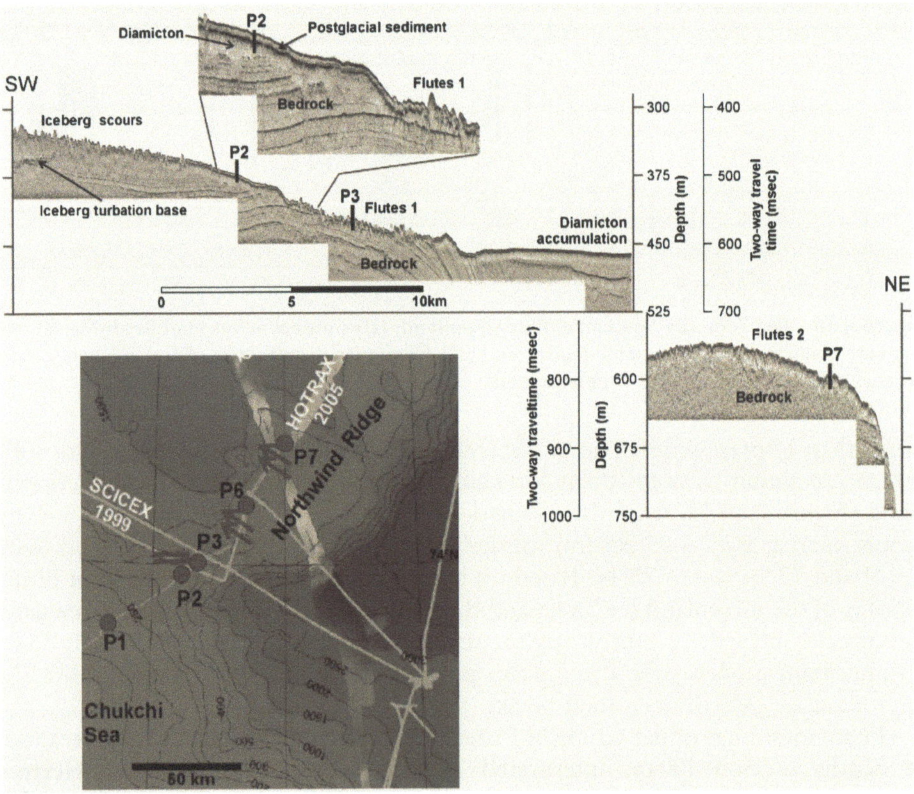

Figure 3.11 Subbottom chirp sonar records on Northwind Ridge (HOTRAX-2005 data: Darby et al., 2005) across core sites P2/P3 and P7 (see map for location). Blow-up fragment near P2 shows details of the upper strata. Data were collected with the USCGC *Healy* hull-mounted Knudsen 320 B/R chirp sonar system. Character of erosion (e.g., iceberg scouring versus fluting) was verified using side-scan and swath bathymetry images (modified from Polyak et al., 2007).

water depths of $\sim$150–350 m and >600 m, respectively, interpreted to be of subglacial origin, that is, caused by grounding ice of the LGM Barents-Kara-Sea Ice Sheet. For further discussion on history of late Quaternary ice sheets and its reconstruction from marine records refer to Chapter 6.

During periods of ice-sheet decay and iceberg melting, large amount of meltwater are released into the ocean surface waters. This may occur during glacial/interglacial transitions but also on much shorter time scales such as described in the North Atlantic as Heinrich Events (e.g., Heinrich, 1988; Broecker, Bond, Klass, Clark, & McManus, 1992; Bond et al., 1993; Broecker, 1994; Sarnthein et al., 2001; Hemming, 2004 and further references therein). These phases of strong meltwater input are reflected clearly in the light $\delta^{18}O$ record of the planktonic foraminifer *N. pachyderma* sin. in the North Atlantic during times of Heinrich Events (Bond et al., 1992; Hillaire-Marcel, de Vernal, Bilodeau, & Wu, 1994; Cortijo et al., 1997; Labeyrie et al., 1999; see Hemming, 2004 for overview) as well as in the central Arctic Ocean climate records (e.g., Stein et al., 1994a; Nørgaard-Pedersen et al., 2003; Spielhagen et al., 2004; see Chapter 6.3). Such meltwater events have a strong influence on the water-mass stratification and formation of deep water. Model studies, for example, have shown a dramatic decrease of NADW turnover when a sudden flux of fresh water is injected into the North Atlantic (e.g., Manabe & Stouffer, 1995; Alley, Anandakrishnan, & Jung, 2001; Clark et al., 2002; Ganopolski & Rahmstorf, 2002; Rahmstorf, 2002).

3.3. Sediment Mass-Wasting Processes

Sediment mass-wasting, that is, debris flows, slides, as well as turbidity currents, represents an important process for transporting large volumes of sediments in the submarine environment (e.g., Kuvaas & Kristoffersen, 1996; Dowdeswell et al., 1998, 2002; Elverhøi, Hooke, & Solheim, 1998b; Vorren et al., 1998; Ó Cofaigh et al., 2003a, 2003b; Andreassen, Nilssen, Rafaelsen, & Kuilman, 2004). In Figure 3.12, the role of cryospheric processes, in the form of both ice sheets and sea ice, on the nature and rate of sediment build-up is summarized in a conceptual model for glacier-influenced sedimentation on the Polar North Atlantic continental margin, separating three stages strongly depending on the sediment discharge onto the shelf (Dowdeswell et al., 1996, 1998, 2002; Vorren et al., 1998):

(1) During full-glacial periods, the ice sheet is often characterized by a series of fast-flowing ice streams that may have advanced onto the continental shelf and formed major troughs on the shelf by glacial erosion (Figure 3.7). By this process, large amounts of sediments were transported close to the shelf break, resulting in upper slope instabilities and downslope movement of sediments, and accumulation in trough mouth fans. Due to the location in front of glacial troughs, Vorren, Hald, and Lebesbye (1988); Vorren, Lebesbye, Andreassen, and Larsen (1989) proposed naming these features "trough mouth fans".
(2) Between ice streams the rate of ice-sheet flow is one to two orders of magnitude slower and sediment delivery to the continental slope is lower than at ice-stream margins. Here, large sediment slides are most characteristic.

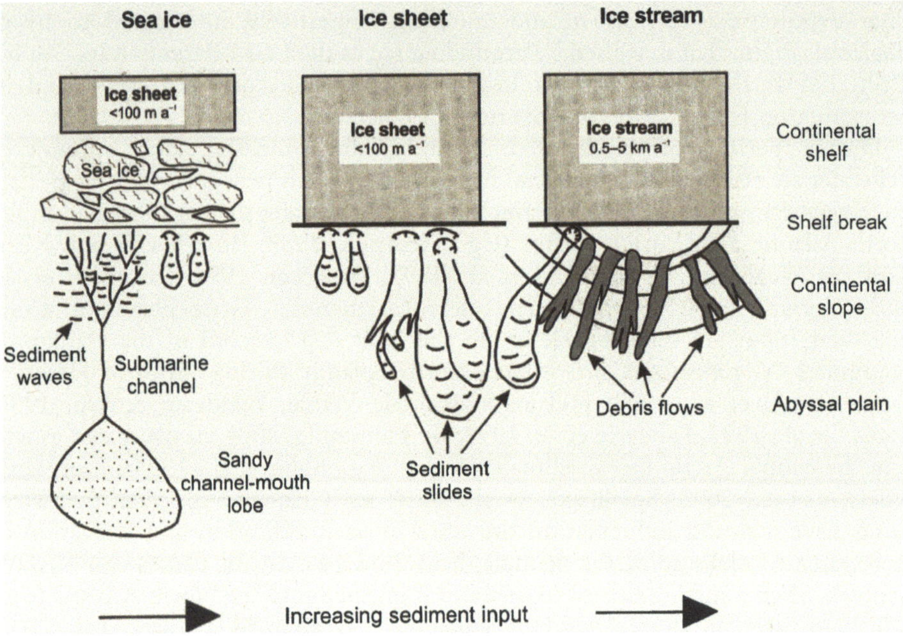

Figure 3.12 Conceptional model of ice dynamics and sedimentary products on a glacier-influenced continental margin, with increasing sediment delivery from left to right (from Dowdeswell et al., 1998, 2002, supplemented).

(3) In areas where the ice sheets did not reach the continental shelf break even during full glacials, sediment input remains relatively low throughout glacial–interglacial cycles. At these margins, large submarine channel systems are typical instead (Mienert, Kenyon, Thiede, & Hollender, 1993; Dowdeswell et al., 2002). By these authors, the large-scale channel systems are related to intermittent downslope flow of dense bottom water (produced by salt rejection during sea-ice formation) and associated turbidity currents. Ó Cofaigh et al. (2004), on the other hand, argued that channel formation is most likely the product of turbidity-current activity during successive glaciations of the Northeast Greenland continental shelf (see below).

Not all troughs ending at the shelf break have fans at their mouths (Vorren et al., 1998). The absence of large sediment fans on continental slopes in front of cross-shelf troughs, however, does not necessarily indicate the former absence of palaeo-ice streams in the geological record. As outlined by Ó Cofaigh et al. (2003b) and summarized in Figure 3.13, the ideal criteria for the formation of a well-developed trough mouth fan are: (1) a favourable depositional setting along a passive continental margin in front of a cross-shelf trough containing a large, fast-flowing ice stream; (2) abundant and readily erodible sediments on a wide continental shelf, which the ice stream can cannibalize into a subglacial deforming layer; and (3) a low-gradient ($<1^{\circ}$) continental slope on which mass movement is dominated by debris flows (Figure 3.13A). In settings with a much steeper continental slope

($>10°$), large volumes of subglacial debris may be delivered to the shelf edge, however, the steep slope results in the rapid development of turbidity currents and associated rapid transfer of this debris to the deep sea, preventing fan development (Figure 3.13B). Finally, Ó Cofaigh et al. (2003b) stated that in areas where an ice stream overlies a predominantly hard bed rock on the continental shelf, subglacial till production may be depressed and, thus, sedimentation rates at the ice-stream terminus may be low. Resulting sedimentary processes would be limited to occasional debris flows, turbidity currents, and suspension settling from meltwater plumes, as well as iceberg rafting (Figure 3.13C).

3.3.1. Trough Mouth Fans

In the Norwegian-Greenland Sea, several large submarine fans are present (Figure 3.14). These fans have been a focus of extensive research, and much is known about their extent, morphology, architecture, and sedimentology (e.g., Vorren et al., 1988, 1989, 1998; Laberg & Vorren, 1995, 2000; Faleide et al., 1996; Dowdeswell, Kenyon, & Laberg, 1997, 2002; Elverhøi et al., 2002). The largest is the Bear Island Fan with an area of $\sim 215{,}000\,km^2$, located at the mouth of the Bear Island Trough. Along the Barents Sea-Norwegian continental margins, others are the Storfjorden Fan ($35{,}000\,km^2$), the Bellsund Fan ($6{,}000\,km^2$), the Isfjorden Fan ($3{,}700\,km^2$), and the Kongsfjorden Fan ($2{,}700\,km^2$). At the East Greenland continental margin, the Scoresby Sund Fan ($19{,}000\,km^2$; Dowdeswell et al., 1997; Solheim et al., 1998) and a prograding fan offshore the Kangerlugssuaq Trough at 67°N (Figure 3.7; Stein & Syvitski, 1997) have to be mentioned. Along the Eurasian Arctic Ocean continental margin between Svalbard, Franz Josef Land, and Severnaya Zemlya, there is also evidence of large fans off the Franz Victoria, St. Anna, and Voronin troughs (Figure 3.7) (Vagnes, 1996; Polyak et al., 1997; Kleiber, Knies, & Niessen, 2000; see below).

In general, the fans mainly consist of a series of stacked debris flows which are clearly visible on side-scan sonar imagery and sub-bottom profiler records, forming a prograding sediment wedge at the mouth of a cross-shelf trough or large fjord system (e.g., Vorren, Richardsen, Knutsen, & Henriksen, 1991; Vorren et al., 1998; Faleide et al., 1996; Solheim et al., 1998). In the Bear Island Fan area, for example, the flows are composed of well-defined lobes with sharp terminations and dimensions ranging from 20 to 200 km in length, 2 to 10 km in width, and 10 to 50 m in thickness (Figure 3.15; Dowdeswell et al., 1997; Ó Cofaigh et al., 2002, 2003b). More recently, it has been recognized that these debris flow packages may be separated by, and pass laterally into, acoustically stratified sediments. Cores from these acoustically stratified units recovered massive to weakly laminated muds that may record suspension settling from turbid meltwater plumes and contour current activity (Taylor, Dowdeswell, Kenyon, & Ó Cofaigh, 2002; Ó Cofaigh et al., 2003b). This differs from previous depositional models of trough mouth fans (e.g., Vorren & Laberg, 1997) due to the increased contribution of meltwater sedimentation to fan formation.

From the temporal evolution of a fan system, important information about the glacial history of a continental margin can be obtained. In numerous seismic profiles

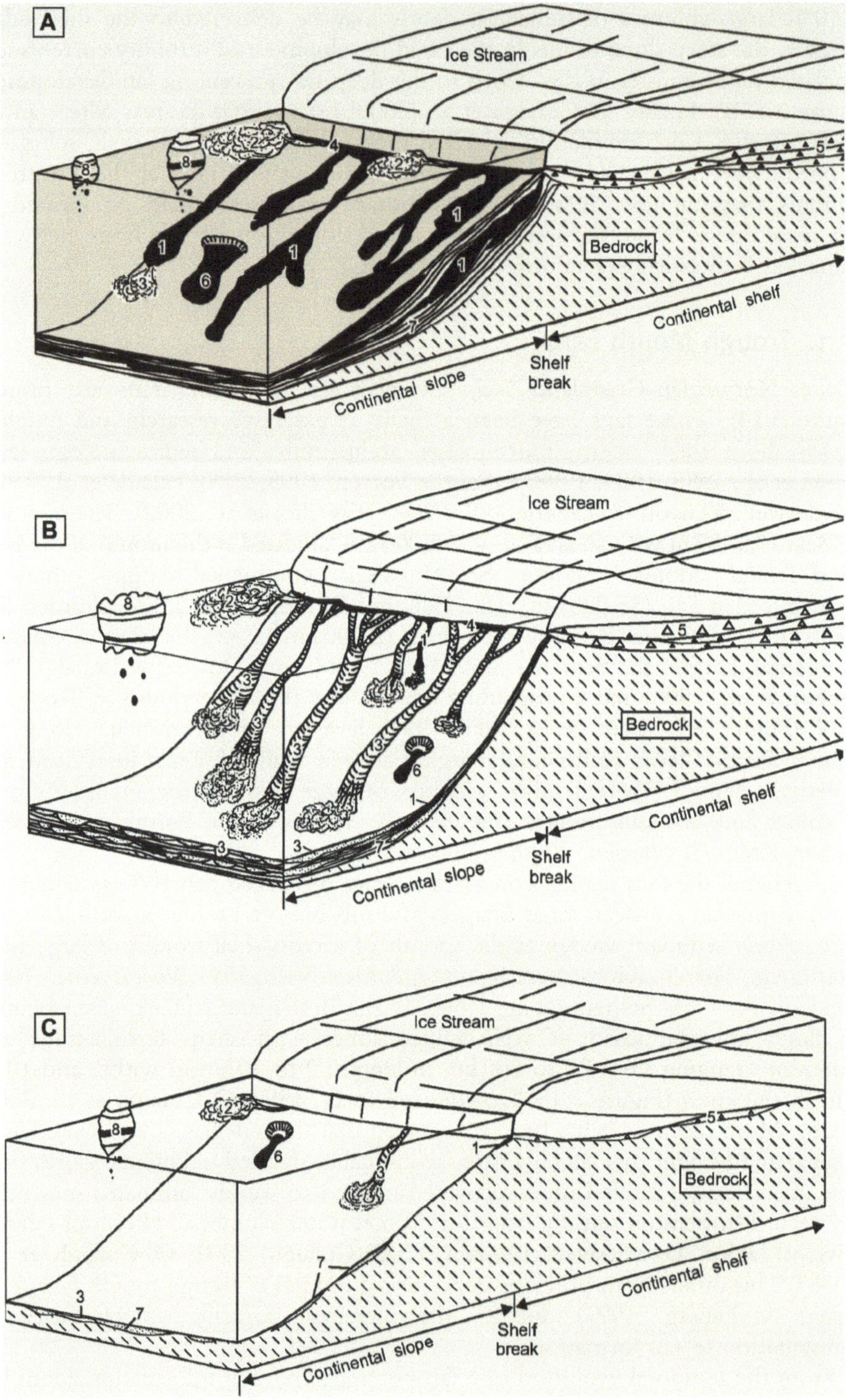
A
Ice Stream
Bedrock
Continental shelf
Shelf break
Continental slope
B
Ice Stream
Bedrock
Continental shelf
Shelf break
Continental slope
C
Ice Stream
Bedrock
Continental shelf
Shelf break
Continental slope

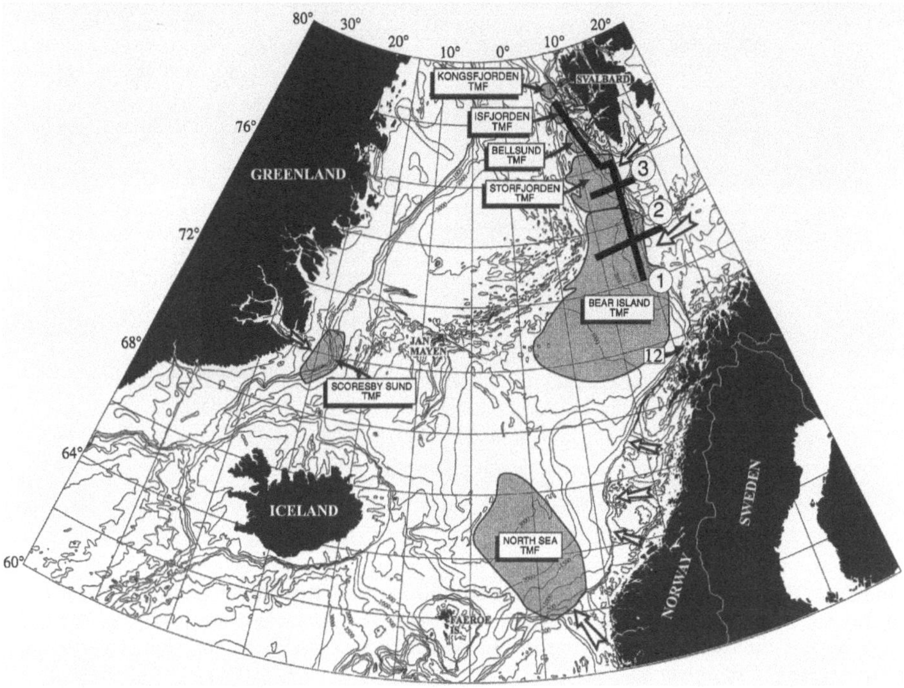

Figure 3.14 Bathymetry map of the Norwegian-Greenland Sea with location and extent of trough mouth fans. Arrows indicate large shelf troughs (from Vorren et al., 1998, supplemented). Location of seismic lines shown in Figure 3.16 are marked as black bars 1–3.

Figure 3.13 Conceptual model of continental slope sedimentation in front of an ice stream reaching the shelf edge during glacial maxima (from Ó Cofaigh et al., 2003b). (A) Classic trough mouth fan (e.g., Polar North Atlantic). Abundant sediment supply, wide continental shelf, low slope gradient ($<1°$). Debris-flow activity and suspension settling from turbid meltwater plumes dominate sedimentation: (1) Debris flows sourced from glacigenic sediment at the shelf edge; (2) buoyant turbid meltwater plume; (3) turbidity current formed by downslope evolution from a debris flow; (4) subglacial till extruded along the ice-stream front/grounding line; (5) subglacial (deformation) till; (6) slump generated debris flow; (7) stratified sediments deposited by suspension settling from meltwater plumes, contour current, and minor turbidity-current activity; (8) iceberg rafting. (B) Steep slope ($>10°$), abundant sediment supply, wide continental shelf (e.g., west Antarctic Peninsula continental margin). Large volumes of subglacial sediment are advected to the ice-stream terminus and upper continental slope and turbidity currents quickly develop due to the steep slope. Channels and gullies are eroded by the turbidity currents on the slope and debris is transferred rapidly downslope. Numbers as in (A). (C) Hypothetical case of low sediment supply and wide continental shelf (sediment-starved setting). The ice stream traverses a hard bed (e.g., crystalline bedrock) on the continental shelf, with fast-flow occurring by enhanced basal sliding. Limited advection of subglacial sediment to the ice-stream terminus due to the hard bedrock substrate. Numbers as in (A).

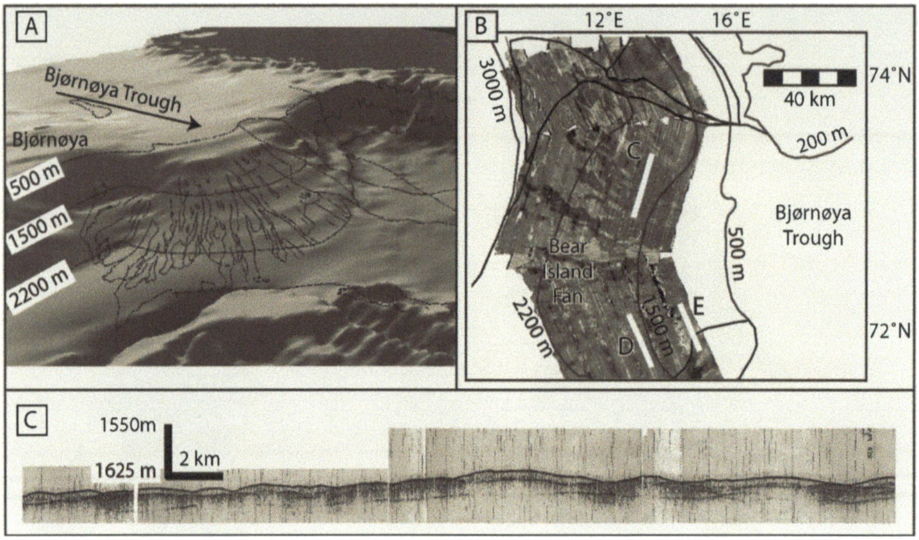

Figure 3.15 (A) Relationship between fan morphology and debris-flow location on the Bear Island Fan. (B) GLORIA 6.5 kHz side-scan sonar image of debris-flow lobes on the Bear Island Fan. Lower backscatter is shown as darker tones. (C) 3.5 kHz subbottom profile across debris flows on the Bear Island Fan showing the stacked nature of the flows and their convex upper surface (from Ó Cofaigh et al., 2003b).

across the fan-influenced area, seven regionally significant reflectors (R1–R7) have been identified between oceanic basement and the sea floor which can be correlated along the western Barents Sea/Svalbard continental margin (Figure 3.16; Faleide et al., 1996). The lowermost unconformity R7 is tied seismically to several exploration wells at the western Barents Sea continental margin (Eidvin & Riis, 1989; Eidvin, Jansen, & Riis, 1993) and ODP Sites on Vøring Plateau (Jansen, Bleil, Henrich, & Slettemark, 1988) and off the Isfjorden (Figure 3.16, Site 986; Solheim et al., 1998), where it corresponds to the base of the glacial sediments dated to slightly older than 2.3 Ma to ~2.6 Ma. The reflector R7 has also been tied to well 7316/5-1 (see Figure 3.16, profile 2, for location), where glacially derived upper Pliocene and Pleistocene sediments rest unconformably on a lower Miocene sequence (Eidvin, Goll, Grogan, Smelror, & Ulleberg, 1994; Faleide et al., 1996). A late Pliocene–Pleistocene age for the entire late Cenozoic sedimentary wedge of the Bear Island Fan is further supported by two volcanic clasts found in a drill core southwest of Bjørnøya just below the base of the wedge and dated by the ^{40}Ar-^{39}Ar method to 2.35 ± 0.12 Ma and 2.20 ± 0.12 Ma, respectively (Mørk & Duncan, 1993; Saettem et al., 1994; Faleide et al., 1996). This age model revises an earlier age model by Vorren et al. (1991) who suggested an age of ~15.5 Ma for the base of the glacial sediments.

On the basis of these seismic and geological records, glacially dominated deposition along the western Barents Sea continental margin started with reflector R7. Thick glacial sediments have been deposited above this reflector, reaching total thicknesses of ~4.5 and 3.5 km in the Storfjorden and Bear Island Fans

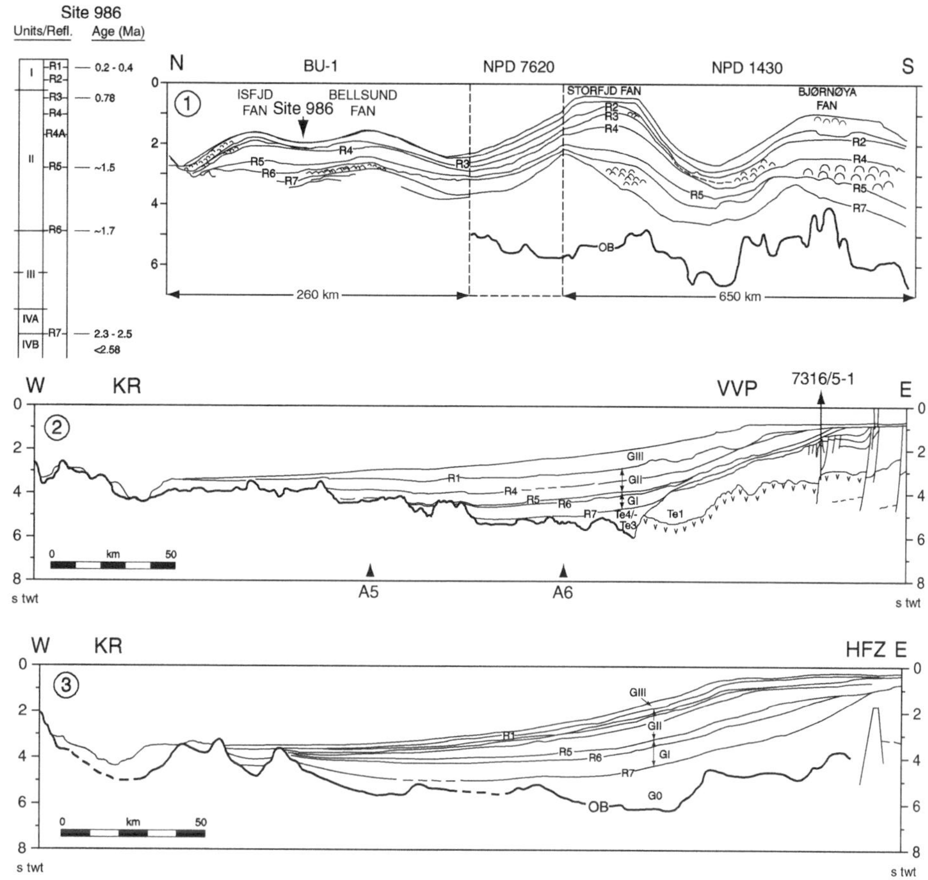

Figure 3.16 Composite regional seismic (N-S) strike-line covering the Isfjorden, Bellsund, Storfjorden, and Bear Island (Bjørnøya) Fans (1) with projected location of ODP Site 986 and age model for seismic reflectors R1–R7 (Solheim et al., 1998), and regional seismic (W-E) lines across the Bear Island Fan (2) and the Storfjorden Fan (3); for location see Figure 3.14. R1–R7, regionally significant reflectors identified between oceanic basement and the sea floor along the western Barents Sea-Svalbard continental margin; GI–GIII, sedimentary mega-sequences; OB (thick line), top oceanic basement; HFZ, Hornsund Fault Zone; KR, Knipovich Ridge; VVP, Vestbakken Volcanic Province; A5 and A6, magnetic anomalies. Location of well 7316/5-1 situated southwest of Bjørnøya is shown in line 2 (from Faleide et al., 1996).

(Fiedler & Faleide, 1996; Hjelstuen, Elverhøi, & Faleide, 1996). From average sediment thicknesses average sedimentation rates can be estimated. These values indicate that with the onset of glaciation, the sedimentation rates as well as the total sediment supply increased by one to two order of magnitudes (Fiedler & Faleide, 1996). Based on seismic stratigraphy and isopach mapping, the Bear Island Fan represent a total volume of $\sim 340,000\,km^3$, deposited since the onset of major NHG near 2.3 Ma (Fiedler & Faleide, 1996).

According to Faleide et al. (1996) and Fiedler and Faleide (1996), these sediments can be subdivided into three mega-sequences GI–GIII (Figure 3.16), interpreted to represent three main phases of glaciations and glacial erosion on the shelf: (1) the base of sequence GI (R7) is interpreted to represent the onset of extensive continental shelf glaciations at ~2.3–2.5 Ma; (2) the base of sequence GII (R5) marks the onset of large-scale mass movements in the Bear Island Fan (indicated by maximum sedimentation and shelf erosion values) and is tentatively related to an intensification of the glaciations at ~1 Ma; and (3) the sequence GIII corresponds to the sediments above the upper regional unconformity (URU) on the shelf and its base (R1) has been dated to ~0.44 Ma.

3.3.1.1. Eurasian continental margin

Detailed data on fan dimension and evolution along the Eurasian Arctic Ocean continental margin are much more limited in comparison to the Norwegian-Greenland Sea. A first evidence for the existence of fans derives from the bathymetry chart suggesting thick sediment wedges in front of major cross-shelf trough systems of similar dimensions as the Bear Island and Storfjorden troughs, that is, the Franz Victoria, St. Anna, and Voronin troughs (Figure 3.17).

First maps of the sediment thicknesses in the Eurasia Basin mainly based on limited number of seismic profiles indicate that in significant parts of the Eurasia Basin oceanic basement is overlain by more than 3 km of sediments (e.g., Jackson & Oakey, 1990; Kristoffersen, 1990). A maximum sediment thickness exceeding 6 km has been suggested between Franz Josefs Land and Severynaya Zemlya (Pogrebitskii, 1983). To obtain a first order estimate of the sediment isopachs in the Eurasia Basin, Vagnes (1996) used modern bathymetry and the standard age-depth curve for oceanic crust in the North Atlantic as a reference to define a bathymetric anomaly which was then inverted, assuming local isostatic equilibrium. As a result, the isopachs reveal two huge fans (deltas) outside the St. Anna and Voronin troughs. This "twin delta" (Vagnes, 1996) covers an area of ~75,000 km^2 and has a maximum sediment thickness of ~7 km (Figure 3.18). Similar, but smaller, deltas are found both east and west of this delta. By analogy to the western Barents Sea, Vagnes (1996) suggests that these deltas contain huge amounts of glacigenic sediments, related to glacial erosion by fast-flowing in the troughs. This supports earlier results by Vagnes, Faleide, and Gudlaugsson (1992) indicating that the entire Barents and Kara shelves underwent late Cenozoic glacial erosion on the same scale as the Western Barents Sea, that is, 0.5–1.5 km. The last glacial advance and erosional event occurred during the LGM when the entire St. Anna Trough was filled with a grounded ice sheet extending to the shelf edge to at least 82°N and a water depth of 630 m (Polyak et al., 1997). According to these authors, radiocarbon ages of 13.3 ka from deglacial sediments provide a minimum age for retreat of a grounded-ice margin from the deep, axial part of the St. Anna Trough (see Chapter 6.2 for more details).

New seismic, wide-angle sonobuoy and gravity data presented by Geissler and Jokat (2004) provide the first more detailed insight into the sedimentary and crustal structure of the northern Svalbard margin between 15°E and 30°E. Based on these records, the sedimentary cover below the continental shelf and slope, except for the Sophia Basin, is highly variable in thickness, ranging between 0 and 3.5 km. In the Nansen Basin, a maximum thickness of 4.5 km for the sedimentary rocks occurs at

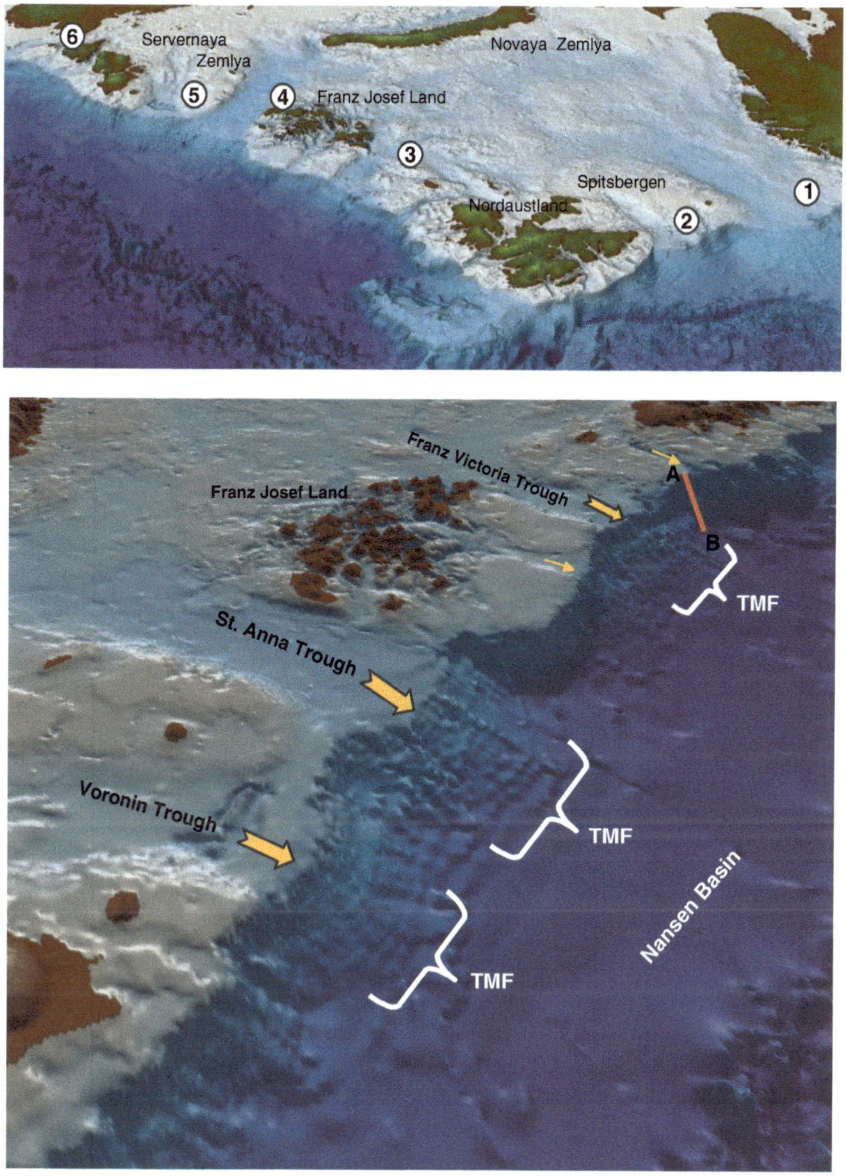

Figure 3.17 3D bathymetry map of the Barents-Kara-Sea continental margin, based on the IBCAO Chart (Jakobsson et al., 2000b, 2003a) and produced by using the Fledermaus software. Major trough systems are indicated (1, Bear Island Trough; 2, Storfjorden Trough; 3, Franz Victoria Trough; 4, St. Anna Trough; 5, Voronin Trough; 6, Vilkitsky Strait Channel). TMF, Trough mouth fan. The location of seismic profile (A–B) shown in Figure 3.19 is marked.

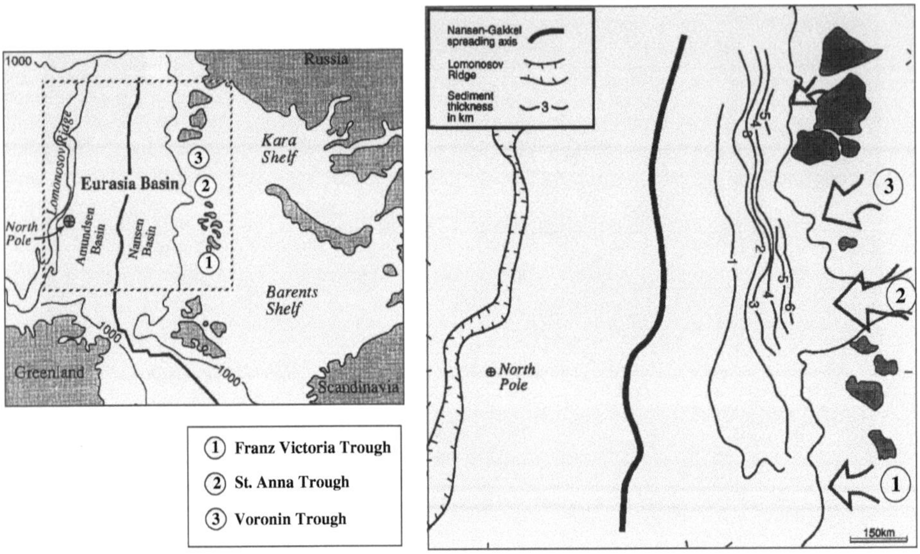

Figure 3.18 Map of the Eurasian Basin with adjacent shelf seas and major trough systems and sediment isopachs (in km) in the Nansen Basin estimated from "relative bathymetry anomalies" (from Vagnes, 1996, supplemented).

82°N. Thick glacial accumulations of late Cenozoic age do not appear to be present along the northern Svalbard-Barents Sea margin west of 30°E. Shelf progradation, however, is evident east of 30°E in the vicinity of troughs crossing the shelf. Based on these data, Geissler and Jokat (2004) suggest that the ice sheet covering the areas south of the investigated margin (Nordaustlandet and northern Spitsbergen) was mainly drained at a few locations through ice streams and that the ice sheet between these troughs may not have been thick enough to produce significant horizontal mass transport. That means, the main sedimentary input into the Nansen Basin during the late Cenozoic occurred along the glacial troughs of the ice streams. In between the troughs, a much steeper slope is typical (Figure 3.17), indicating that local sediment transport processes connected with a minor amount of sediments dominate the shelf.

At ~30°E, a thick sedimentary prism can be identified on the outer shelf, situated at the mouth of a glacial trough east of the Svalbard Archipelago between Nordaustlandet and Kvitøya (Figure 3.19, line AWI-99176; Geissler & Jokat, 2004). This sedimentary prism can be compared with similar structures in front of Woodfjorden and Hinlopen Strait and along the western Svalbard-Barents Sea continental margins. Here the uppermost seismic unit KV-3 (KV — Kvitøya) can be distinguished from unit KV-2 by a different reflection pattern. These prograding sequences within unit KV-3 are typical for glaciated continental margins at high latitudes (Bart & Anderson, 1995).

As shown in the bathymetry chart, a large glacial trough mouth fan also occurs east of 35°E in the prolongation of the Franz Victoria Trough (Figure 3.17). This is supported by a thick sediment wedge recorded in seismic line 98500 (Jokat et al., 1999; see Figure 3.19 for location of profile). Gravity data from this area suggest that

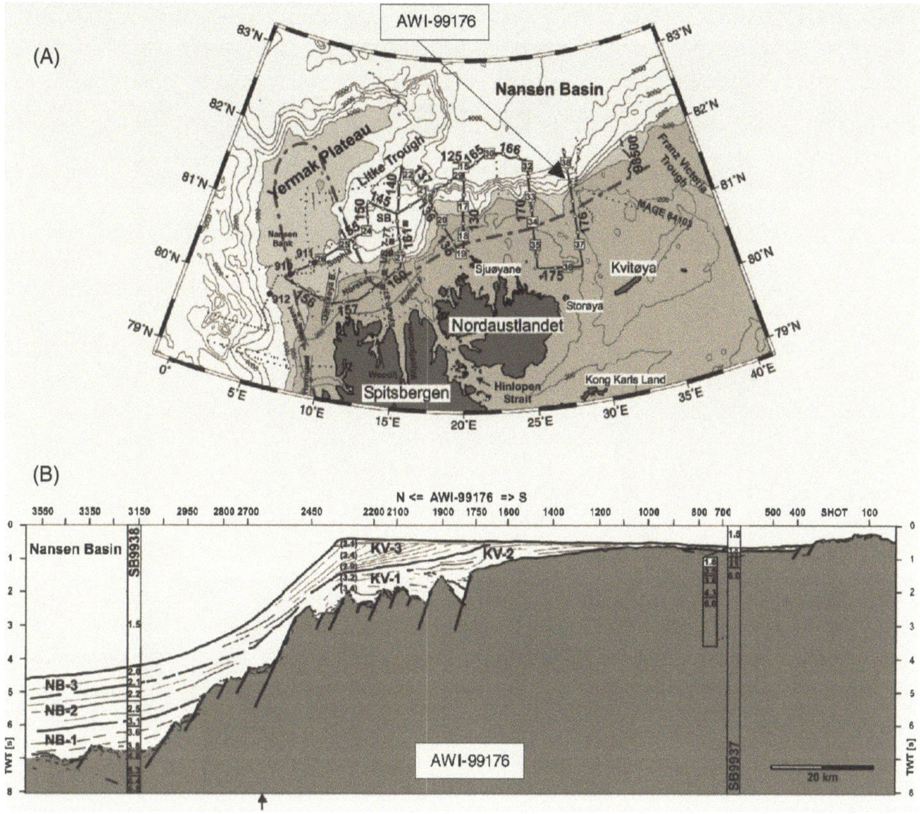

Figure 3.19 (A) Map of the area along the northern Svalbard continental margin. Bold lines display seismic profiles acquired during expedition ARK-XV/2 of RV *Polarstern* 1999. Some of the existing, older profiles (Jackson, Johnson, Sundvor, & Myhre, 1984; Eiken, 1994; Weigelt, 1998) are drawn as dotted lines. The names of the profiles were abbreviated to make the figure more readable (e.g., 166 denotes AWI-99166). The same was done for the sonobuoys (e.g., 19 denotes SB9919). Black circles mark the ODP sites from Leg 151 (Myhre et al., 1995). IBCAO bathymetry is taken from Jakobsson et al. (2000b). The dashed line represents the assumed continent–ocean boundary according to Sundvor and Austegard (1990). (B) Line drawing of seismic profile AWI-99176 at 30°E north of Nordaustlandet. Shelf progradation occurs in front of a glacial trough between Nordaustlandet and Kvitøya (from Geissler & Jokat, 2004).

the glacial fan extends well north of the 500 m isobath (Geissler & Jokat, 2004). Further direct evidence for mass-wasting processes and fan formation comes from PARASOUND profiles and sediment core records.

In PARASOUND profiles from the continental slope in front of the Franz Victoria Trough a sequence of stacked, transparent lenses and layers of ~10 m in thickness and several kilometers in width was recorded (Figure 3.20; Kleiber et al., 2000), very similar to the debris flow deposits described for the Bear Island Fan (e.g., Vorren et al., 1998). The top of this seismic sequence corresponds with diamictons found in sediment cores obtained from the Franz Victoria Trough area and interpreted as debris flows (Figure 3.21). Based on

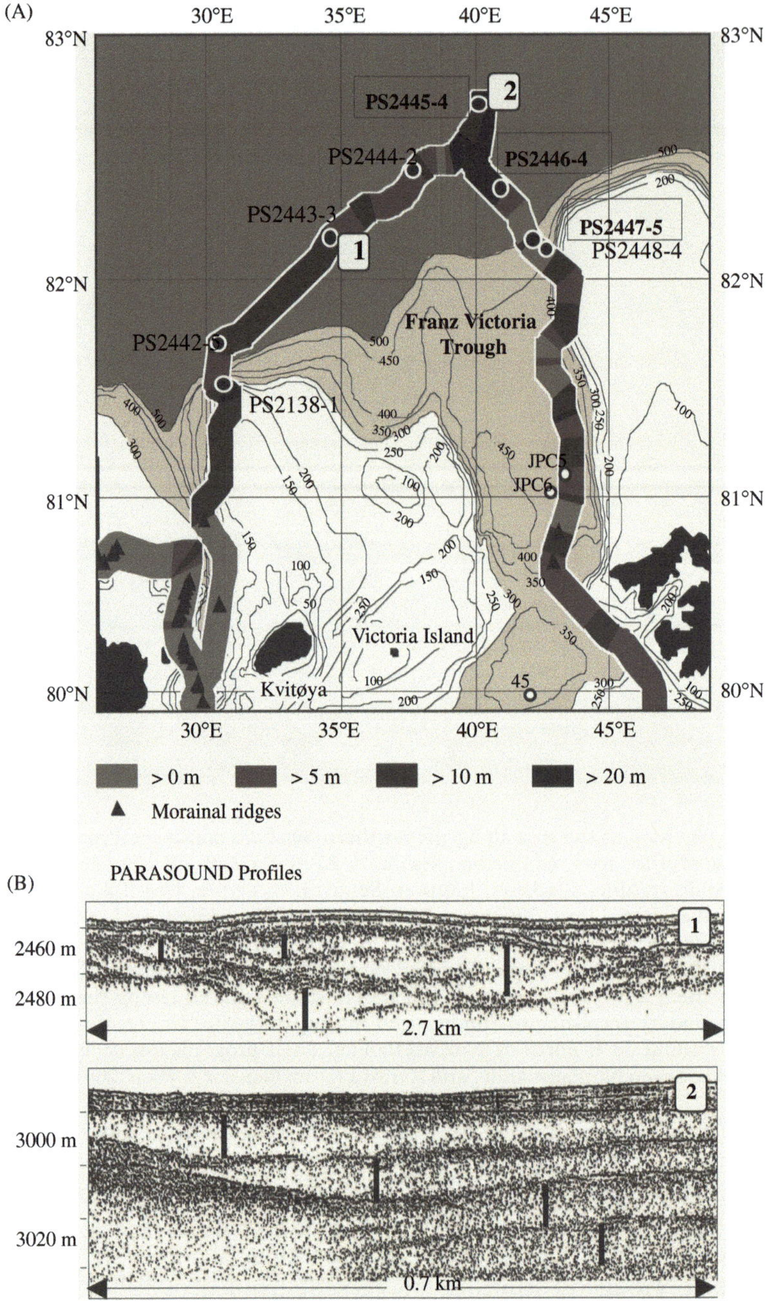

Figure 3.20 (A) Bathymetric map of the northern Barents Sea continental margin with the Franz Victoria Trough, showing the minimum thickness of the Quaternary sediments (indicated by the maximum penetration of the PARASOUND system along the ship tracks) and locations of morainal ridges and *Polarstern* cores. (B) PARASOUND profiles showing truncating, stacked transparent lenses and layers (for location see A). The profiles section were recorded while the ship remained on station during coring operation (from Kleiber et al., 2000, supplemented).

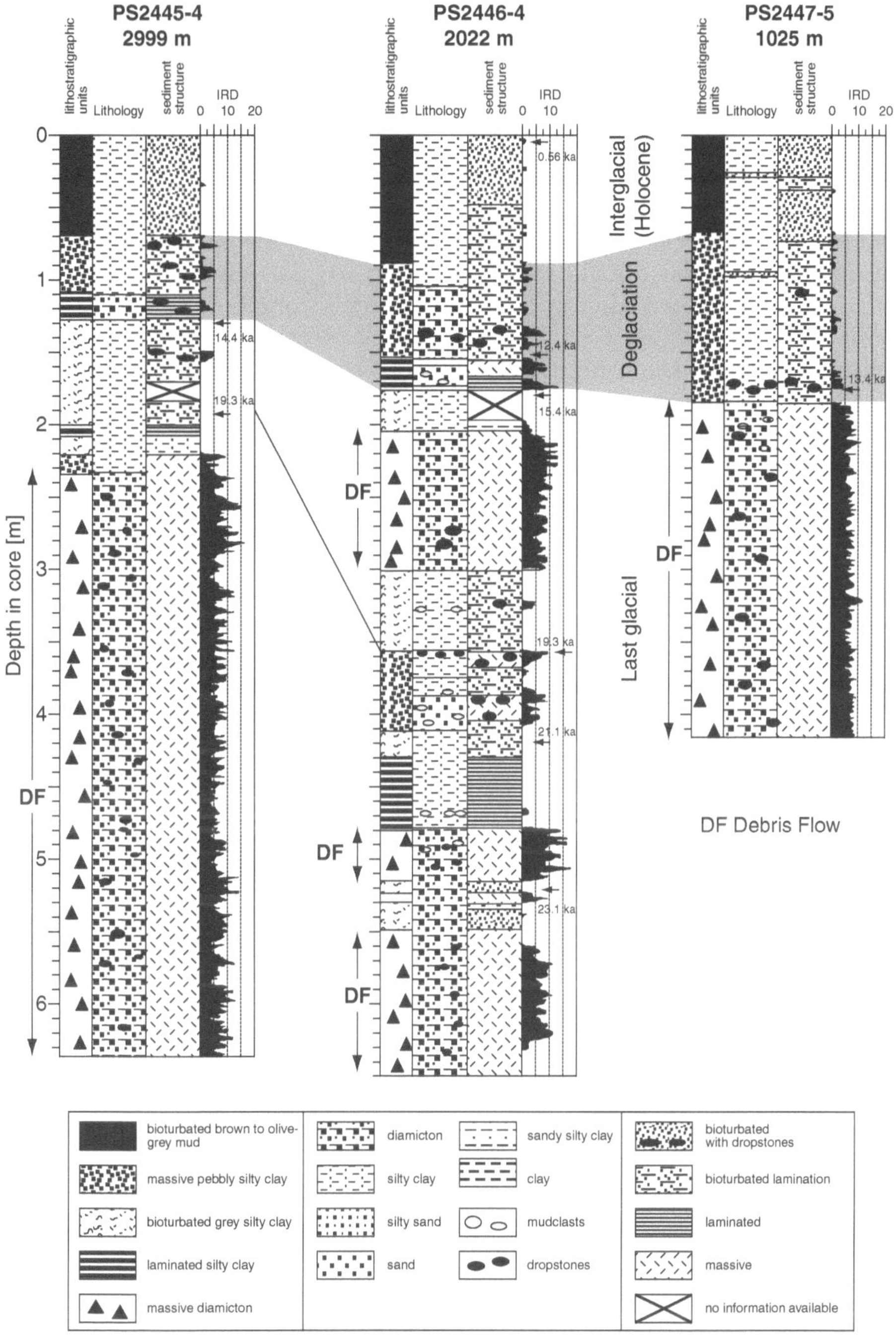

Figure 3.21 Lithostratigraphic units, lithology, sedimentary structure, and IRD content (numbers of detritus >2 mm in centimetre core intervals) of sediment cores PS2445-4, PS2446-4, and PS2447-5 (for location of cores see Figure 3.20). AMS^{14}C ages shown have been reservoir corrected using an age of 440 years (from Kleiber et al., 2000, supplemented).

accelerator mass spectrometry (AMS)^{14}C datings, these debris flows are of late Weichselian age. Kleiber et al. (2000) suggest that the thick sequence of debris flow lobes and diamicton layers were deposited when the grounding line of a Svalbard-Barents-Sea Ice Sheet reached the shelf break of the Franz Victoria Trough near 23.1 ka. The lack of morainal ridges in the PARASOUND profile in the northern Franz Victoria Trough is a further indication that the grounding line of the ice sheet reached the shelf edge (Kleiber et al., 2000; see Laberg & Vorren, 1996a, 1996b). The restricted distribution of the debris flows in front of the Franz Victoria Trough may indicate that the trough acted as conduit for the fast-flowing ice stream, responsible for the formation of fan deposits, as also proposed by Elverhøi et al. (1995b) for the Isfjorden Fan. The general lack of glacial fans off the northern Svalbard margin (Solheim, Andersen, Elverhoi, & Fiedler, 1996) may indicate that the continental margin west of the Franz Victoria Trough — except for the Hinlopen Strait area (see Section 3.3.2) — is not directly influenced by a fast-flowing ice stream reaching the shelf edge.

The grounding line of the northern Svalbard-Barents-Sea Ice Sheet retreat from the shelf edge of the Franz Victoria Trough at ~13.4 ka, indicated by a drastic change in sedimentary pattern on the upper continental slope. This occurs more or less contemporaneously to the retreat of grounded ice in the St. Anna Trough (see above; Polyak et al., 1997). The stepwise deglaciation of the northern Svalbard-Barents-Sea Ice Sheet is documented in the sedimentary records of the sediment cores from the Franz Victoria Trough by distinct IRD-pulses which appear to be contemporaneous with the onset of distinct pulses of IRD and glacio-marine sedimentation in the adjacent areas between 13 and 9.4 ka (Kleiber et al., 2000; see also Chapter 6.2.2).

Debris flows related to glacio-marine processes were also recorded further to the east. Based on a detailed acoustic (PARASOUND) survey, Kleiber, Niessen, and Weiel (2001) could identify four seismic units in the western Laptev Sea and Vilkitsky Strait area, the lowermost one (Unit IV) is dominated by transparent lenses interpreted as debris flows (Figure 3.22). These debris flows extend continuously down to the continental rise (3,200-m water depth), where a large debris-flow dominated submarine fan of ~28,000 km^2 is developed. The distribution of the debris-flow deposits suggests that the high sediment input to the upper continental slope in the eastern Vilkitsky Strait probably derives from advances of the eastern Kara Sea Ice Sheet to the shelf edge off Severnaya Zemlya and Taymyr Peninsula (i.e., to the mouth of the Vilkitsky-Khatanga Channel; see No. 6 in Figure 3.17). In contrast to the northern Barents Sea continental margin, the last maximum extent of an ice sheet on the Kara Sea, Severnaya Zemlya and Taymyr Peninsula, responsible for the debris-flow deposition, was probably older than MIS 2 and occurred during Middle Weichselian time (MIS 4) as suggested by Svendsen et al. (1999, 2004) (see Chapter 6.2).

In general, the debris flows along the Eurasian continental margin certainly contribute significantly to the sedimentary budget. For the late Weichselian (glacial) time interval represented in the studied sediment cores from the Franz Victoria Trough, for example, accumulation rates are more than three times higher than those of the "normal" Holocene not influenced by debris flows. That means, that along the Barents Sea continental margin, which was strongly influenced by glacigenic input of total sediment due to the huge LGM ice sheet, significantly

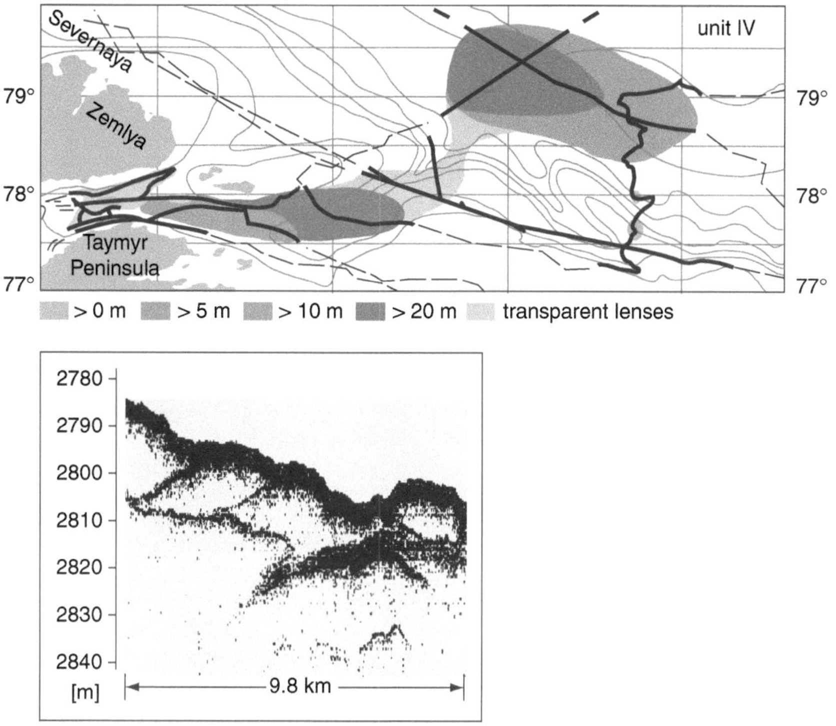

Figure 3.22 Bathymetric map of the area north of Severnaya Semlya showing the distribution and the related thicknesses of seismic unit IV representing debris flow deposits. Grey, stippled lines show the cruise tracks of the RV *Polarstern*. Typical PARASOUND profile of transparent lenses (Unit IV) (from Kleiber et al., 2001).

higher accumulation rates were typical for the last glacial interval, whereas during the following Holocene interglacial, accumulation decreased (e.g., Kleiber et al., 2000; Knies, Müller, Nowaczyk, Vogt, & Stein, 2000; see Chapter 6.2.2 for more detailed discussion).

3.3.2. Sediment Slides

Along the western Norwegian-Barents Sea continental margin, several large sediment slides have been identified by GLORIA long-range side-scan sonar mapping (Figure 3.23; Bugge, Belderson, & Kenyon, 1987, 1988; Jansen et al., 1987; Mienert et al., 1993, 2005; Laberg & Vorren, 1993, 2000; Dowdeswell et al., 1996; Vorren et al., 1998; Lindberg, Laberg, & Vorren, 2004; Mienert, 2004; Hjelstuen, Eldholm, & Faleide, 2007). The largest and intensely studied slide event is the Storegga Slide complex off mid-Norway, reaching an area of $\sim > 110{,}000\,km^2$ and a sediment volume of $\sim 6{,}000\,km^3$ (e.g., Bugge et al., 1988; Haflidason et al., 2004; Haflidason, Lien, Sejrup, Forsberg & Bryn, 2005; Mienert et al., 2005; Paull et al., 2008). North of the Storegga Slide, between 67°N and 73°N, several larger and smaller slides were observed, noticeably the

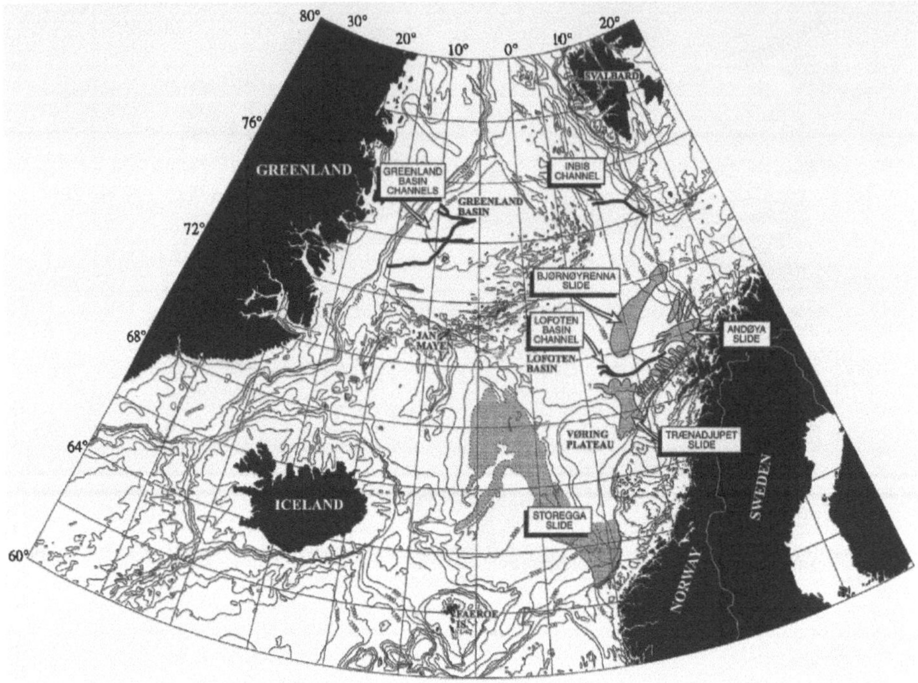

Figure 3.23 Bathymetry map of the Norwegian-Greenland Sea with location and extent of slides and large channels on the continental margin (from Vorren et al., 1998).

Traenadjupet, Andøya, and Bjørnøyrenna Slides (Dowdeswell et al., 1996; Vorren et al., 1998). At the East Greenland continental margin, detailed mapping by GLORIA long-range side-scan sonar and PARASOUND/3.5 kHz survey is restricted to areas between 70–80°N (Mienert, Hollender, & Kenyon, 1993, 1995; Pfirman & Kassens, 1995) and 68–70°N (Dowdeswell et al., 1997). These data do not show any large-scale sliding event.

A "mega-slide" event first described by Cherkis et al. (1999), also occurred at the northern Svalbard continental margin directly north of the Hinlopen Strait (Figure 3.24). This so-called Hinlopen-Yermak Slide has recently been studied under the umbrella of an EUROMARGIN project (e.g., Mienert, 2004 and references therein). Based on new high-resolution acoustic data (Hydrosweep swath-bathymetry and PARASOUND) and seismic data, the extent was revised and the true geometry of the mega-slide characterized in great detail (Figures 3.24 and 3.25; Vanneste, Mienert, & Bünz, 2006; Winkelmann, Jokat, Niessen, Stein, & Winkler, 2006; Winkelmann, Geissler, Schneider, Stein, & Schenke, 2008a). The slide involved more than 2,400 km^3 of sedimentary material and affected an area exceeding 10,000 km^2, and the runout length exceeds 275 km at a total drop height of more than 3,750 m. The event left the highest headwalls (up to 1,400 m) so far reported for glacier-fed continental margins and probably worldwide. These numbers put the Hinlopen/Yermak Slide among the largest exposed submarine slides (Winkelmann, Jokat, Niessen, Stein, & Winkler, 2005, 2006; Vanneste et al.,

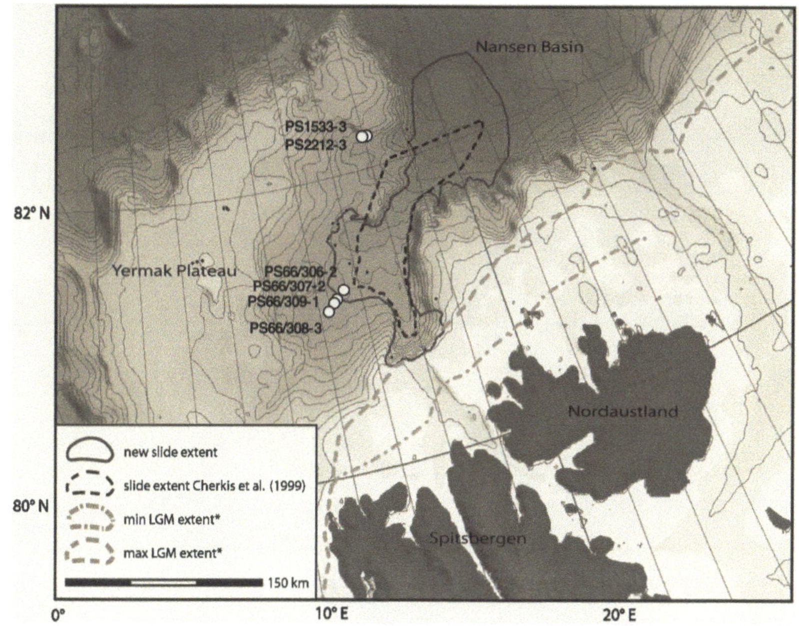

Figure 3.24 Overview map of the Hinlopen/Yermak Mega-Slide region with location of core stations and slide extent. Note that the north-eastern border represents a minimum extent (based on Winkelmann et al., 2006). Minimum and maximum extent of LGM ice sheet (*) according to Landvik et al. (1998). Cores shown on PARASOUND profile of Figure 3.25 are marked as open circles.

2006; Winkelmann & Stein, 2007). Based on the AMS^{14}C dating of slide-related turbidites in key Core PS66/309, the Hinlopen/Yermak mega-slide event occurred ~30 Cal. kyr BP (Figure 3.25), probably resulting from a partial shelf collapse at the termination of a cross-shelf trough (Winkelmann et al., 2006). The minor events (younger debris flows) have not been dated yet.

In general, the slides are mainly erosional in their upper reaches and depositional downslope. They originate in multiple headwalls on the upper continental slope associated with across-slope tensional fractures and longitudinal fractures (Figure 3.26; Kenyon, 1987; Laberg & Vorren, 2000). In the lower depositional area, large isolated blocks, separated by a more smooth sea floor, may occur. In the Hinlopen/Yermak Slide, tens of rafted blocks reaching extensions of more than 3 km and up to 450 m relief are present within the main debris body inside the Sophia Basin (Figure 3.26). The processes of slide failure and down-slope debris transport are very complex, and the whole slide event may contain several subevents including debris flows and turbidity currents, as also shown for the Hinlopen/Yermak Slide (Winkelmann et al., 2006). The triggering process of these large-scale slide events are often uncertain. Processes considered to be involved in triggering major slope failure events include rapid sediment deposition during glaciation, ice-loading effects, gas hydrate decomposition, and earthquake activity (e.g., Vorren et al., 1998; Laberg & Vorren, 2000; Mienert, Posewang, & Lukas, 2001, Mienert et al., 2005; Atakan & Ojeda, 2005; Bryn, Berg, Forsberg, Solheim, & Kvalstad, 2005; Solheim,

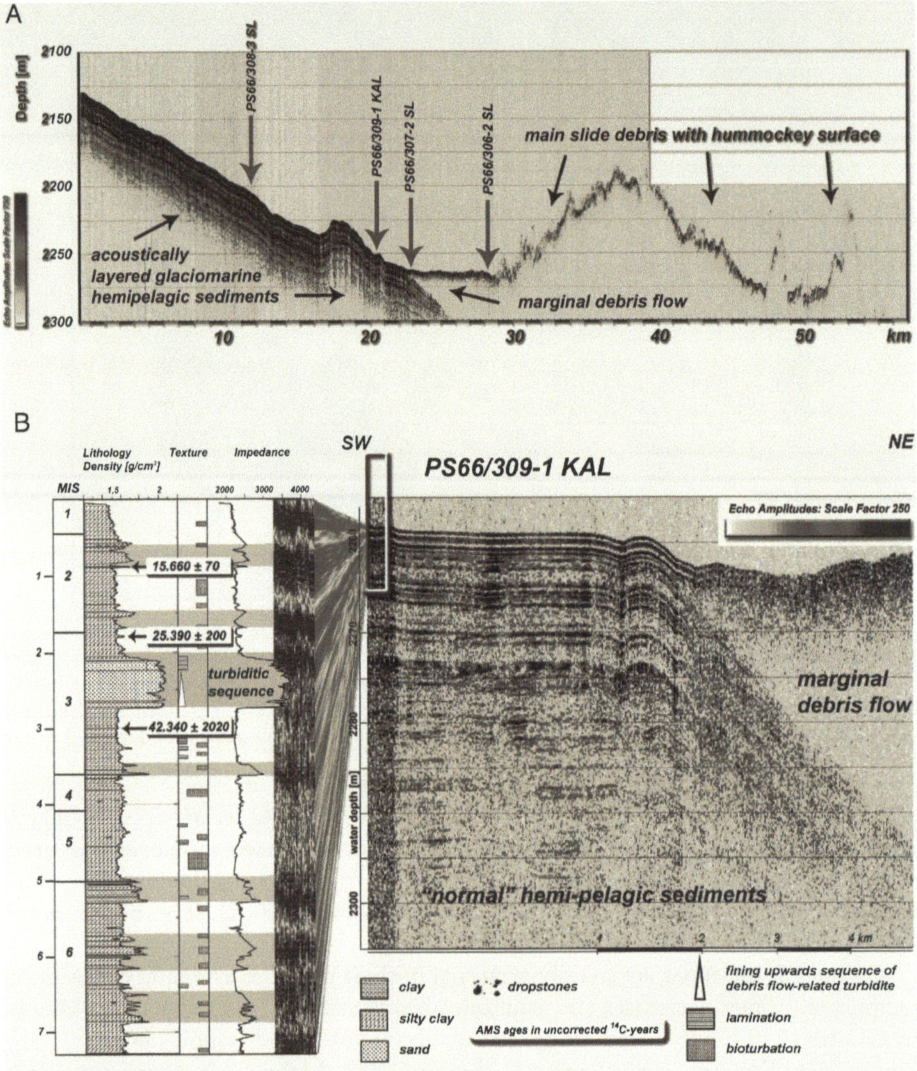

Figure 3.25 (A) Typical PARASOUND profile across the western margin of the Hinlopen/Yermak Slide in the southwestern Sophia Basin. Note the consistent transition from pushed-up slide debris into the marginal debris flow. An enlargement of the western part of the profile is shown in (B). (B) Close-up of PARASOUND profile showing marginal debris flow pinching out into hemipelagic sediments and position, lithology, density, texture, and acoustic impedance of core logging of key core PS66/309-1 KAL (giant gravity corer, 81°11.22N, 12°59.07E, 2270 m water depth, 7.29 m recovery; see Stein, 2005 for details). AMS radiocarbon ages above and below the debris flow-associated turbidite are depicted as ^{14}C years BP, uncorrected for the marine reservoir effect. Marine Isotope Stages (MIS) are based on preliminary correlation of magnetic susceptibility from core logging to cores with existing isotopic stratigraphy. For position of profile and core, see Figure 3.24 (from Winkelmann et al., 2006).

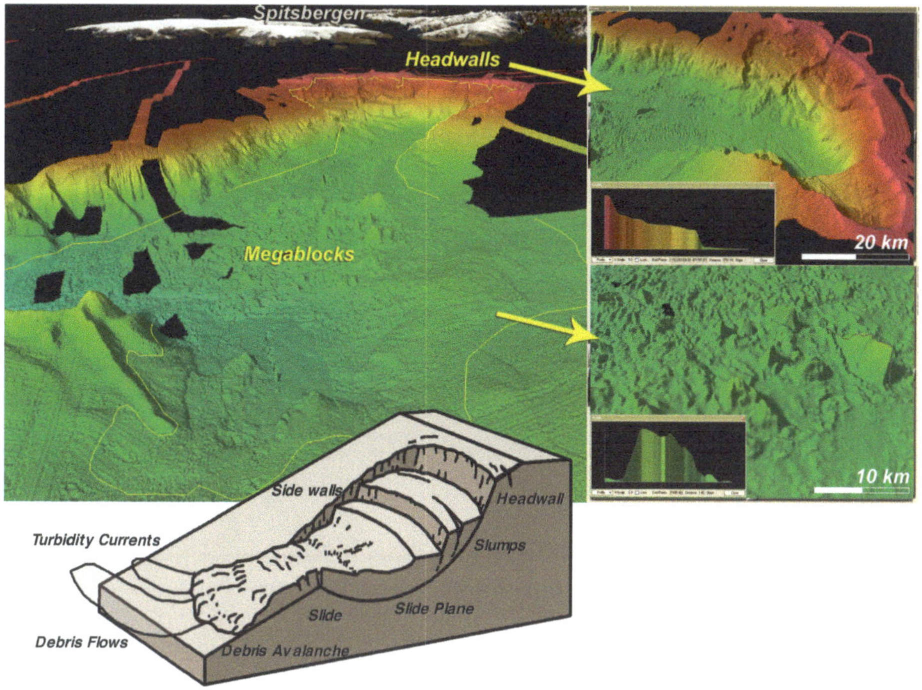

Figure 3.26 Headwalls and blocks of the Hinlopen/Yermak Slide (from Winkelmann, 2007). Inner slide-debris deposition area with tens of megablocks reaching extensions up to 5 km and 450 m relief. Multiple headwalls show combined headwall heights of up to 1,600 m (scales refer to centre of images). The 3D figures were produced, combining high-resolution bathymetry sets of AWI Bremerhaven/Germany and UiT Tromsø/Norway (Vanneste et al., 2006; Winkelmann et al., 2006).

Bryn, Sejrup, Mienert, & Berg, 2005; Winkelmann et al., 2006; Paull et al., 2008). For the Hinlopen/Yermak Megaslide, Winkelmann and Stein (2007) propose that this slide has been the consequence of the rapid onset of late Weichselian glaciation resulting in a drastic sea-level drop near 30 Cal. kyr BP (*ca.* 50 m in <1 kyr; Lambeck, Yokoyama, & Purcell, 2002), asymmetrical ice loading, and a fore bulge development. As the final trigger, a strong earthquake positioned below or close to the SE-Sophia Basin is assumed (see Winkelmann et al., 2006, 2008a and Winkelmann & Stein, 2007 for details).

3.3.3. Submarine Channel-System Formation

3.3.3.1. Norwegian-Greenland Sea

Major submarine channel systems are described for the Norwegian-Greenland Sea (Figures 3.23 and 3.27). Well developed submarine channels occur on the Northeast Greenland margin, where they extend for ~300–400 km down the continental slope and into the abyssal depths of the Greenland Basin (Mienert et al., 1993; Dowdeswell et al., 1996, 2002; Ó Cofaigh et al., 2002, 2004; Taylor, Dowdeswell, & Kenyon, 2003; Wilken & Mienert, 2006). On the Norwegian

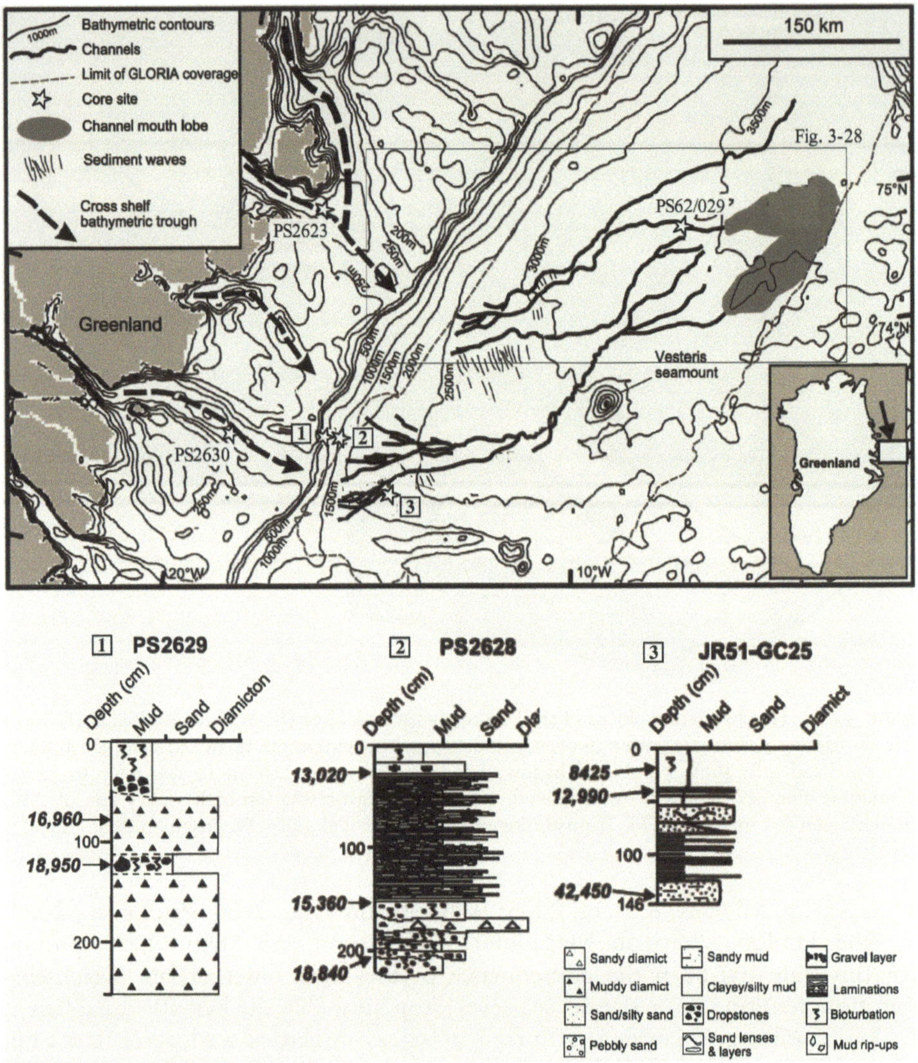

Figure 3.27 Bathymetry of the Greenland Basin and adjoining continental shelf, Northeast Greenland and major submarine geological features (channel systems, sediment waves, and channel-mouth lobes), and lithology of three selected sediment cores (for location of cores see map) (from Ó Cofaigh et al., 2004, supplemented). Location of sediment core PS62/029-4 is also shown. Open rectangle marks close-up shown in Figure 3.28. The major bathymetric troughs on the continental shelf are marked by arrows. Inset shows the location of the study area in Northeast Greenland.

margin, channel systems are confined to the Inbis Channel and Andøya Canyon/ Lofoten Channel (Figure 3.23; Vorren et al., 1998; Laberg, Vorren, Dowdeswell, Kenyon, & Taylor, 2000; Ó Cofaigh, Dowdeswell, & Kenyon, 2006). In general, both the Lofoten and Greenland Basin channel systems are similar in terms of their length, presence of channel-mouth lobes, presence of turbidites and sediment waves within the channel and on the channel mouth lobe, and the presence of braiding where the channels reach the abyssal plain (Ó Cofaigh et al., 2004, 2006). In contrast to the much better studied trough mouth fans, however, there are still relatively few detailed investigations of polar submarine channels and much remains to be learned about the processes of channel formation and architecture (Dowdeswell et al., 2002; Ó Cofaigh et al., 2004, 2006).

The most extensively studied area of submarine channels in the Norwegian-Greenland Sea occurs on the Northeast Greenland margin between ~72°N and 75°N (Figure 3.27). Between 1992 and 2002, several expeditions have been carried out to investigate this major channel system (RV *Livonia* in 1992, Mienert et al., 1993; RV *James Clark Ross* in 1994 and 2000, Ó Cofaigh et al., 2004; RV *Polarstern* in 2000, 2001, and 2002; Krause & Schauer, 2001; Fahrbach, 2002; Lemke, 2003). The distribution and morphology of these channel systems have been defined by marine-geophysical investigations, that is, GLORIA 6.5 kHz side-scan sonar, multibeam swath bathymetry (RV *James Clark Ross*: Kongsberg-Simrad EM120 system; RV *Polarstern*: Hydrosweep D2), and sub-bottom profiling (RV *James Clark Ross*: TOPAS system; RV *Polarstern*: PARASOUND system) (Mienert et al., 1993, 1995; Dowdeswell et al., 2002; Frahm, 2003; Frahm, Hohmann, & Matthiessen, 2003; Matthiessen et al., 2003; Ó Cofaigh et al., 2004). Based on these surveys, there are four main channel systems which extend from water depths between ~1,600 m (southern systems) and 2,700 m (northern systems) to the abyssal depths of the Greenland Basin and which in dimensions are up to ~100-m deep, 4-km wide, and have braided and sinuous reaches (Figures 3.27 and 3.28). Towards the abyssal plain, dimensions of channels decrease (Figure 3.28), and in the deepest part of the basin, the well-defined channels disappear and are replaced by channel-mouth lobes (Figure 3.27) probably composed of more sandy sediments as suggested from sub-bottom profiling records (Ó Cofaigh et al., 2004). Upslope, lenses of acoustically homogeneous sediments were recorded in TOPAS and PARASOUND profiles, which are interpreted as debris-flow deposits (Figure 3.28A; Matthiessen et al., 2003 and unpublished data; Ó Cofaigh et al., 2004).

Sediment cores taken from the upper-middle continental slope in front of the southern channel system contain massive diamictons (Core PS2629, 855 m water depth) and thinly interbedded, massive muddy diamict, massive to bioturbated sandy mud and massive mud (Core PS2628, 1,694 m water depth) (Figure 3.27; Stein et al., 1995; Evans et al., 2002). The uppermost part of the cores are characterized by hemipelagic bioturbated sediments. The diamictons are interpreted as debris flows, sourced either directly from an ice margin positioned at the shelf edge or from iceberg-rafting followed by downslope remobilization (Ó Cofaigh et al., 2002, 2004). Further downslope, laminated sediments characterized by couplets of coarse silt or fine sand fining upward into fine silt/clay are dominant (Figure 3.29). Very often, very sharp contacts are obvious at the coarser-grained base, occasionally also cross-bedding occurs. These laminated sediments are interpreted as fine-grained distal turbidites.

A key question is still the origin of these channels. Here, different processes are under discussion. Mienert et al. (1993) and Dowdeswell et al. (1996, 2002) proposed that the channels relate to the downslope flow of dense-water and

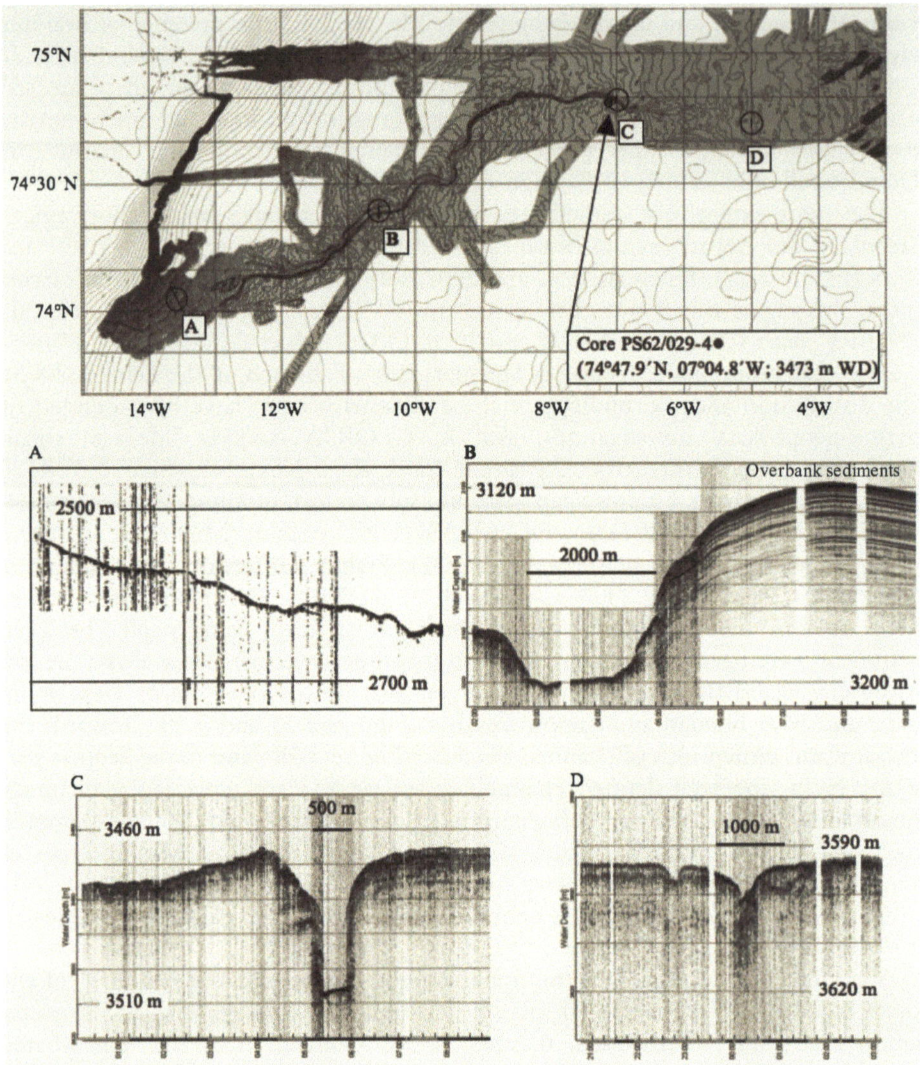

Figure 3.28 Detailed hydrosweep map from the northern part of the channel system (see Figure 3.27 for location) and selected PARASOUND profiles from the study area (for location of profiles see map). The location of sediment core PS62/029-4 is also shown (from Frahm, 2003; Lemke, 2003; Matthiessen et al., 2003).

turbidity currents originating from sea-ice formation on the shelf and upper slope. Ó Cofaigh et al. (2004), on the other hand, favouritize a formation under glacial conditions. Very recently, Wilken and Mienert (2006) proposed that debris flows and channel formation are related to major glaciations with varying ice stream behaviour and extents across the outer and inner shelves.

In order to understand the processes controlling the formation of the channel system, that is, to infer when the channel systems in the Greenland Basin was last active, dating of the transition between downslope mass-wasting, as indicated by the

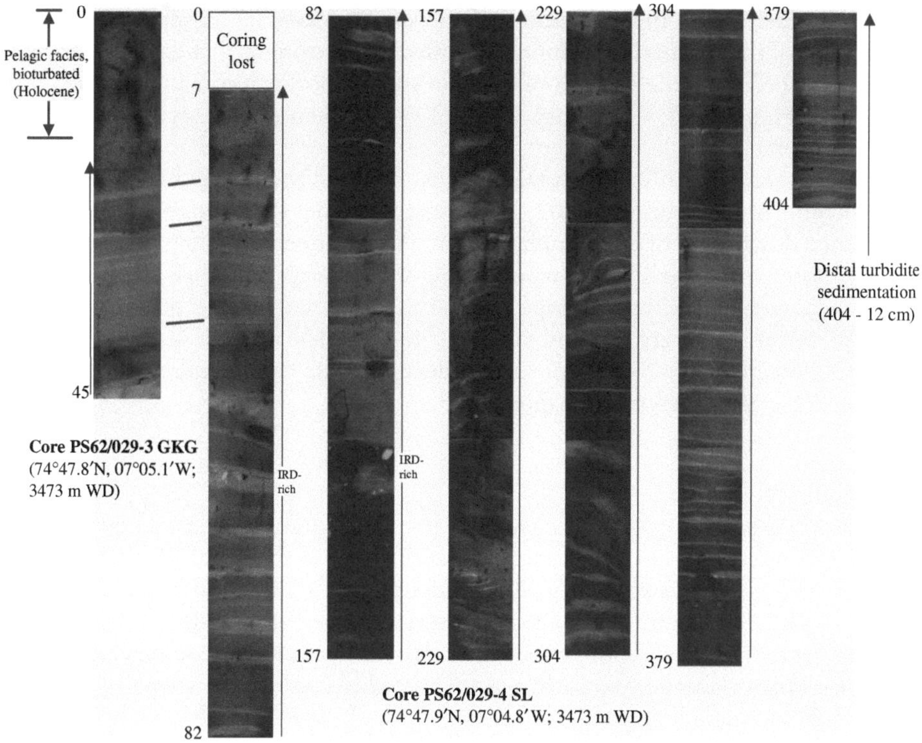

Figure 3.29 X-ray photographs of giant box corer Core PS62/029-3 GKG and the entire sediment sequence of Core PS62/029-4 SL (74°47.9′N, 07°04.8′W; 3,473 m WD; for location see Figure 3.27), dominated by turbidite sedimentation. Abundant fining-upward cycles of silt (light colours) to clay (grey to dark grey colours) facies; partly cross-bedded intervals between 180 and 290 cm indicating higher current regime. The upper 13 cm are composed of bioturbated silty clays, probably representing the last 13 ka (see text). Enrichments of IRD in Sections 1 and 2 of Core PS62/029-4 are indicated. Figure compiled from single X-ray photographs of 1 cm × 10 cm × 25 cm thick sediment slabs; photographs downloaded from PANGAEA (http://www.pangaea.de/PHP/Cores.php?B=PS&C=PS62&S=PS62/029&ID=133390).

presence of turbidites and debris-flow deposits, and the overlying, most recent interval of hemipelagic sedimentation, is of major importance. AMS^{14}C-dated sediment cores (PS2628 and JR51-GC25) indicate that turbidity-current activity in the southernmost channel system had ceased by ~13 ka (Figure 3.27), and sedimentation rates decrease by an order of magnitude after this time (Ó Cofaigh et al., 2004; see also Wilken & Mienert, 2006). At the same time, the ice sheets retreated from the continental shelves around the Norwegian-Greenland Sea (e.g., Funder et al., 1998; Landvik et al., 1998; Mangerud et al., 1998). This cessation of turbidity-current activity implies that the channels were last active under glacial conditions and that the most recent major episode of mass-wasting terminated during the initial stages of deglaciation, which started at ~15.3–15.7 ka at this part of the Greenland continental margin (Nam, Stein, Grobe, & Hubberten, 1995; Stein, Nam, Grobe, & Hubberten, 1996; Evans et al., 2002). Furthermore, a glacial-influence for the formation of the channel systems is supported by the relationship of the submarine channels to bathymetric troughs that extend across the continental shelf from the mouths of many of the major fjords of East and Northeast Greenland (Figure 3.27; Ó Cofaigh et al., 2004; Wilken & Mienert, 2006). These troughs generated by glacial erosion at times when the ice sheet reached the shelf (e.g., Bartek et al., 1997; Ó Cofaigh et al., 2002; Ottesen, Dowdeswell, Rise, Rokoengen, & Henriksen, 2002), terminate at the shelf edge and channels commence downslope of the trough mouths. An ice sheet advanced onto the outer shelf or close to the shelf break would allow the delivery of debris and sediment-laden meltwater to the upper slope, providing a source for the debris flows and turbidity currents that are documented on the continental slope of the Greenland Basin (Ó Cofaigh et al., 2004). For further details on the extent and evolution of the Greenland Ice Sheet refer to Chapter 6.2.3.

3.3.3.2. Arctic Ocean

In the Amundsen Basin, a major channel system was identified, extending from the Lincoln Sea at the Greenland-Canadian margin (84°N) to the North Pole (Figure 3.30; Kristoffersen, Sorokin, Jokat, & Svendsen, 2004b). As cited in Svindland and Vorren (2002), Svendsen (1997) already described (in his unpublished Thesis) such channels on the deep-sea floor of the Amundsen Basin, with a geometry one would expect for channels serving as pathways for turbidity currents flowing east from the Greenland continental margin towards the North Pole, influenced by Coriolis force. The paper by Kristoffersen et al. (2004b) is mainly based on three seismic transects which show that the gentle continental slope north of Greenland adjacent to the Lomonosov Ridge is constructed by levee deposits from the main channel of a submarine fan. In seismic transect AWI-91098 crossing the fan close to the North Pole, the uppermost sequences AB-7 and AB-8 exhibit upward convex internal horizons (Figure 3.30; Jokat et al., 1995). The position of the submarine channels at the highest elevation of the abyssal plain as well as internal units of the sequence, particularly the lowermost one tapering off towards either side, suggests that sequences AB-7 and AB-8 represent the distal lobe of a submarine fan (Kristoffersen et al., 2004b). These authors have proposed to name this fan the "North Pole Submarine Fan", and the main channel the "NP-28 Channel" in recognition of the contribution of the Russian ice stations to the exploration of the deep Arctic basins.

According to Kristoffersen et al. (2004b), main processes developing the North Pole Submarine Fan were increased sediment input to the margin of Canada and North Greenland and topographic focussing of gravity driven flows, triggering turbidity currents via the NP-28 Channel into the deep Amundsen Basin. The latter may have created the observed levee deposits by over-bank flow. These processes were probably most active during glacial maxima or deglaciation, that is, times of maximum sediment input, when ice streams from North Greenland and Ellesmere Island may have reached the Lincoln Sea and brought sediments (close) to the shelf edge (England, 1999; Kristoffersen et al., 2004b). Concerning the timing of development of the submarine fan, Kristoffersen et al. (2004b) infer that

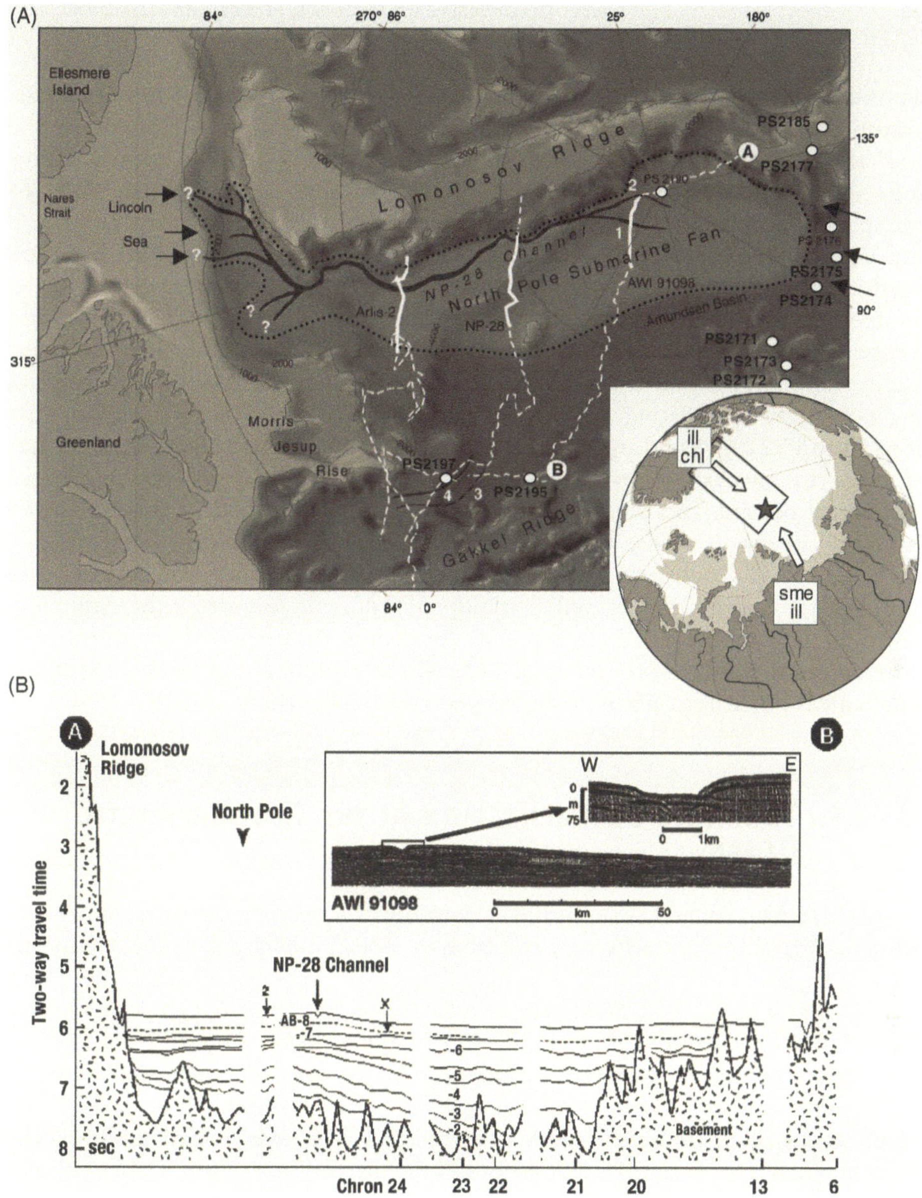

Figure 3.30 (A) Map of the Amundsen Basin with the locations of seismic profiles and observed crossings of submarine channels. Major sediment pathways are indicated by a bold black line and the outline of the North Pole Submarine Fan as grey area surrounded by black dotted line. Tributary paths from the Canadian-Greenland continental slope are inferred from the position of bathymetric downslope depressions. In addition, Polarstern cores characterized by major occurrence of turbidites (based on lithological core description from Fütterer, 1992 and Svindland & Vorren, 2002) are shown. Black arrows indicate assumed sediment input from the Greenland-Canadian and Siberian margin, characterized by illite–chlorite (ill-chl) and smectite–illite (sme-ill) clay–mineral asemblages, respectively (for clay–mineral data of potential source areas see Chapters 4.3.2 and 5.1.2). Bathymetry from International Bathymetric Map of the Arctic Ocean (Jakobsson et al., 2003a). (B) Multichannel seismic reflection line crossing the NP-28 Channel and line drawing of the multichannel seismic section across Amundsen Basin (AWI-91098). Sequence nomenclature from Jokat et al. (1995). Location of transect A-B is shown in (A) (from Kristoffersen et al., 2004b, supplemented).

deposition of most of sequences AB-7 and AB-8 is related to input from increased erosion during the full scale NHG after 2.5 Ma.

Besides seismic profiles, further support for turbidity-current activity derives from sediment cores. Kastenlot Core PS2190-1 taken at the North Pole, for example, is dominantly composed of fine-grained turbidites mostly ranging in thickness between ~5 and 25 cm each (Figure 3.31; see Section 3.4 for more details). Kristoffersen et al. (2004b) interpret this sequence as deposition of hemipelagic clays repeatedly interrupted by over bank flow from turbidity current events which are related to sediment input from the Greenland-Canadian continental margin. Furthermore, these authors suggest that rhythmic alternation of silt and clay dominating the lithology in a second core PS2176, 245 km away from Core PS2190-1 and outside the proposed submarine fan (Figure 3.30), must be distal to any fan lobes extending from either the Greenland or Siberian continental margin, and the most important sediment transport in this part of the basin may occur in a near bottom nepheloid layer (see Hunkins, Thorndike, & Mathieu, 1969). Considering more detailed grain-size and clay-mineral data from cores PS2176 and PS2190 as well as additional sediment cores from the Amundsen Basin, the terrigenous sediment input in the central Amundsen Basin seems to be more complex and Kristoffersen et al.'s (2004b) interpretation in terms of sediment source areas has at least partly to be revised (see next section).

3.4. Turbidite Sedimentation in the Central Arctic Ocean

In the deep Arctic basins, turbidity currents may become the major process for sediment input. Turbidites are very common in most (probably all) of the Arctic deep basins (e.g., Hunkins et al., 1969; Campbell & Clark, 1977; Goldberg, 1983; Darby et al., 1989; Fütterer, 1992; Stein et al., 1994a; Grantz et al., 1996; Grantz, Phillips, & Jones, 1999; Svindland & Vorren, 2002). In the Canada Abyssal Plain and the Amundsen Basin, for example, turbidites may be the dominant facies of the Holocene as well as Pleistocene sedimentary sequences (Campbell & Clark, 1977; Grantz et al., 1996, 1999; Svindland & Vorren, 2002). In the Nansen Basin, specific (kaolinite-rich) clay–mineral assemblages and the preservation of more labile organic-carbon compounds in surface sediments (see Chapters 5.1.2 and 5.2) suggest that turbidite sedimentation may even be active during recent times (Stein et al., 1994a).

Because most studies of sediment cores from the Arctic Ocean are mainly devoted to palaeoenvironmental reconstructions (see Chapter 6) and for this type of studies undisturbed (hemi-) pelagic sediment sections are needed, however, sediment cores containing turbidites are much less investigated. One of the few more detailed studies of turbidite sedimentation in the central Arctic Ocean, namely the Amundsen Basin, has been carried out by Svindland and Vorren (2002). These authors selected two key cores (PS2176-3 and PS2190-1) out of several turbidite-containing sediment cores recovered in the Amundsen Basin during *Polarstern* Expedition ARK-VIII/3 (Figures 3.30 and 3.32; Fütterer, 1992) for their detailed

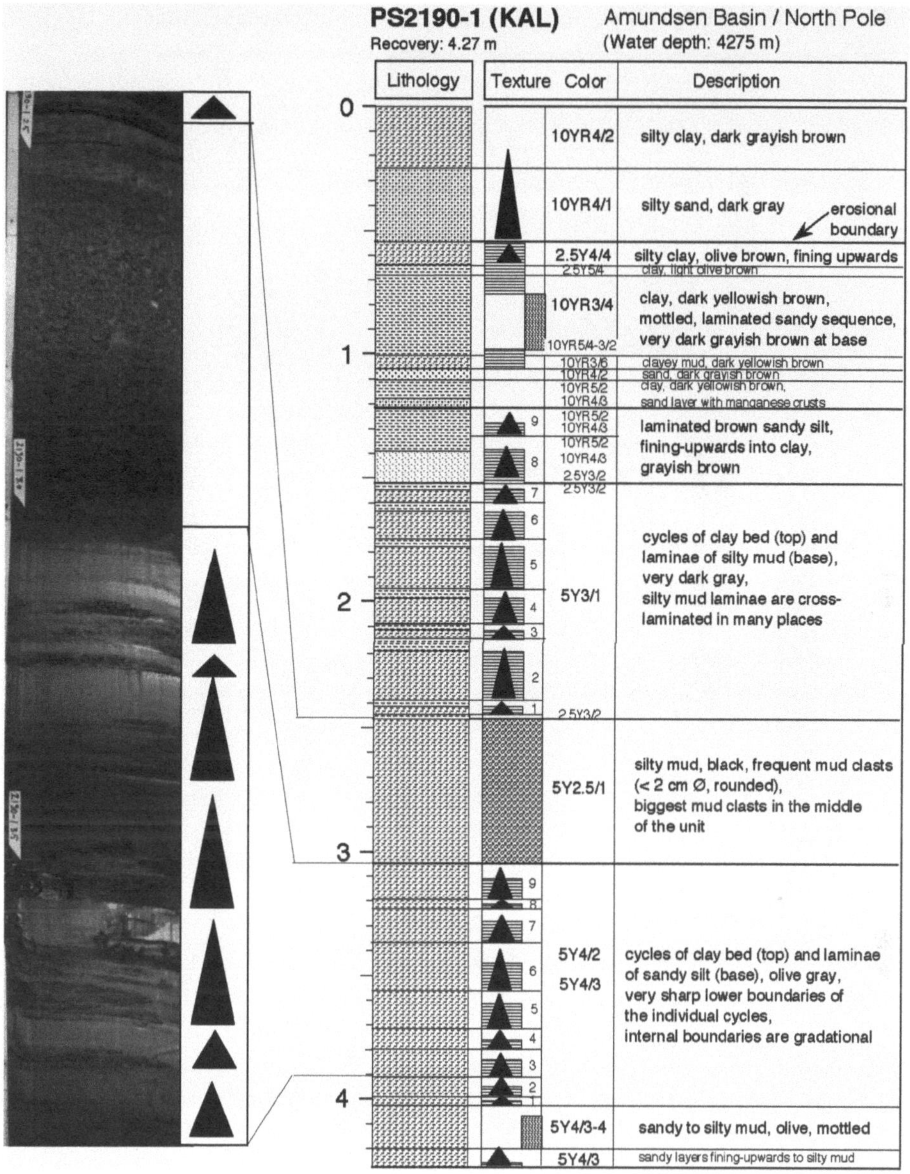

Figure 3.31 Visual core description of North Pole Core PS2190-1 (from Fütterer, 1992, supplemented). Black triangles indicate turbidites. In addition, a core black and white photograph of the depth interval 245–390 cm is shown.

facies analysis of the turbidites, including grain-size measurements (sand–silt–clay as well as SediGraph analysis of the silt fraction), sediment structure, and clay mineralogy. These data are compared with records from a non-turbidite influenced sediment core from the Lomonosov Ridge (PS2185).

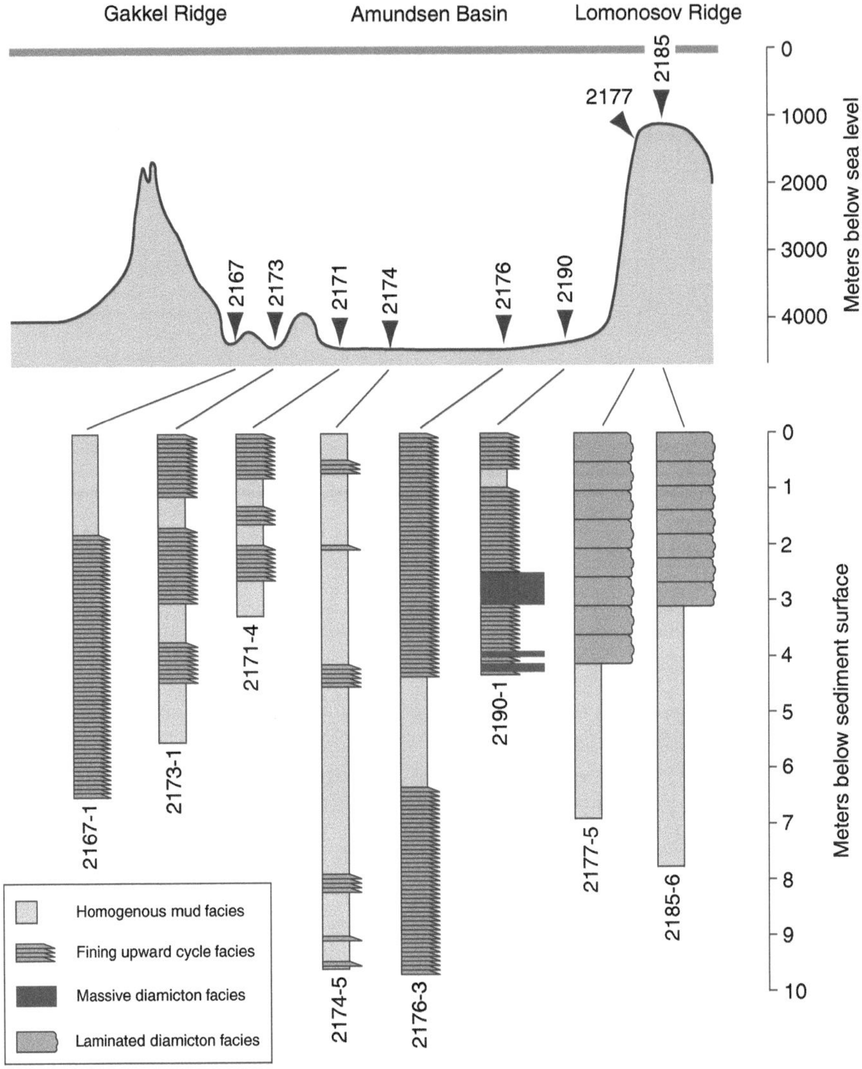

Figure 3.32 Schematic bathymetric profile from the Gakkel Ridge through the Amundsen Basin to the Lomonosov Ridge. Core locations are indicated, and general sedimentary facies of the cores are shown (from Svindland & Vorren, 2002).

On the basis of visual core description and smear-slide analysis carried out onboard *Polarstern* during Expedition ARK-VIII/3, fining-upward sequences were identified to be a dominant lithology in most of the sediment cores from the central Amundsen Basin (Figure 3.32; Fütterer, 1992). The proportions of as well as the grain-size range within fining-upward lithologies, however, vary significantly between the cores. From smear-slide estimates most of these sequences seem to be composed of sand, fine sand, sandy silt, or sandy clay, changing gradually into silt or clay. Later, more detailed grain-size analyses indicate that at least in the two sediment cores PS2176-3 and PS2190-1 studied by Svindland and Vorren (2002), the shipboard smear-slide estimates significantly overestimated the sand content. In both cores, the sand content of the fining-upward facies is predominantly very low, ranging between 0% and 2%, and the content of silt or clay individually of a sample may be as high as 90% or below 10%, alternating between top (~90% clay) and bottom (~90% silt) of the cyclic beds (Figure 3.33; Svindland & Vorren, 2002).

On the basis of sedimentary structures such as cross-bedding and the silt–clay cycles, Svindland and Vorren (2002) conclude that the sediment closely resembles the facies T0–T8 (Stow & Shanmugam, 1980) and E1–E3 (Piper, 1978) of fine-grained turbidites (Figure 3.34, columns E and F) and infer that this fining-upward facies represents distal turbidite deposition. Furthermore, this sediment facies lacks or only contains very little IRD (Figure 3.33), indicating that deposition from floating ice is negligible, or IRD input has been masked by rapid turbidite sedimentation.

As mentioned above, Kristoffersen et al. (2004b) suggested that the source of the turbidites is the Northern Greenland/Canadian continental margin. Clay mineral data indicative for sediment sources (see Chapter 5.1.2), however, tell a different story. In both cores PS2176-3 and PS2190-1 there is a significant change in clay–mineral assemblages at ~420 cm and 100 cm depth, respectively, from a smectite-dominated assemblage in the lower part to an illite–chlorite assemblage in the upper

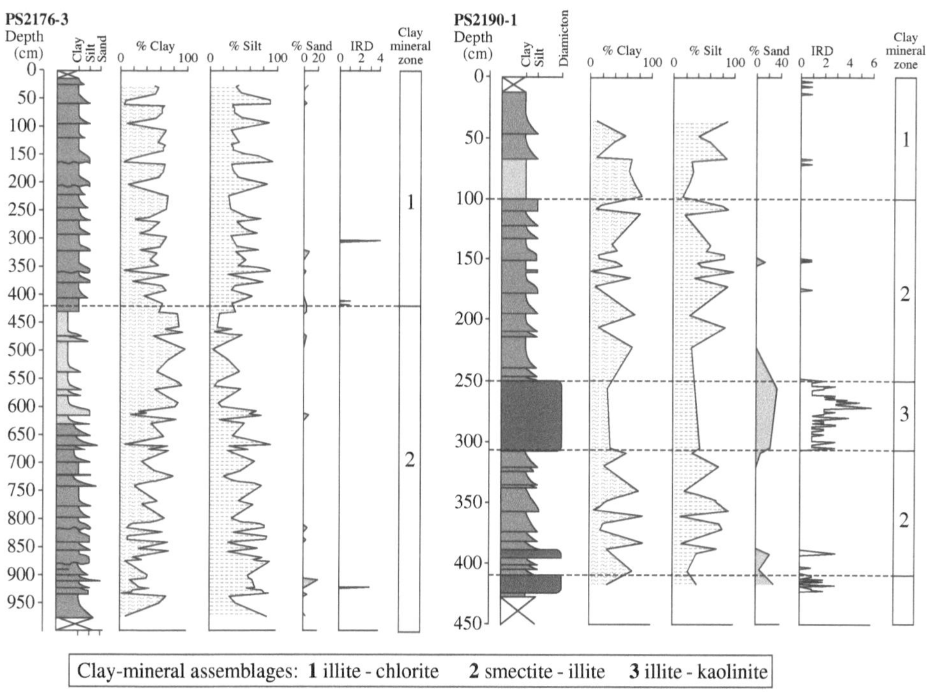

Figure 3.33 Lithology, grain-size distribution, IRD frequency, and clay–mineral assemblages of cores 2176-3 and PS2190-1. Shading pattern in the lithostratigraphical column to the left indicates sedimentary facies (see Figure 3.32 for legend) (modified from Svindland & Vorren, 2002).

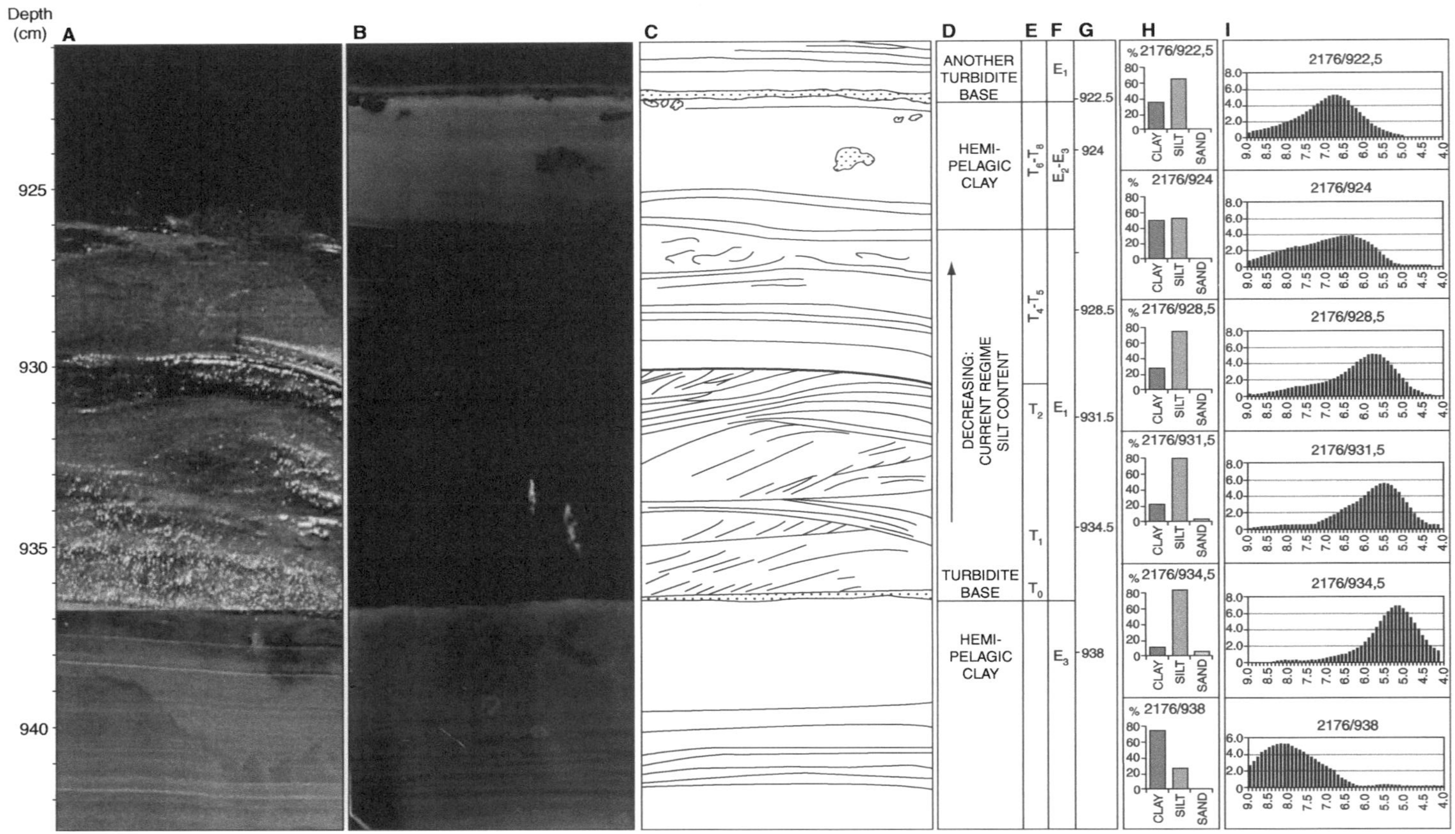

Figure 3.34 Selected sample of fining upward cycle facies, core 2176-3, 921–943 cm: (A) photo, (B) X radiograph, (C) sketch of sedimentary structures, (D) interpretation, (E) classification of turbidite cycle according to Stow and Shanmugam (1980), (F) classification of turbidite cycle according to Piper (1978), (G) sample locations, (H) histograms showing distribution of clay (<2 μm), silt (2–63 μm), and sand (63 μm–2 mm), and (I) histograms showing grain-size distribution (Phi values) of silt (from Svindland & Vorren, 2002).

part of the cores (Figure 3.33; Svindland & Vorren, 2002). This may indicate that during the deposition of the fining-upward cycle facies in the lower part of the cores bottom current sediment transport was dominated by turbidity currents from the Siberian continental margin (i.e., the western Laptev Sea) characterized by elevated smectite values, whereas in the uppermost parts of the cores, sediment input by turbidity currents from northern Greenland characterized by high illite–chlorite values (see Figure 3.30), was predominant. From the clay–mineral association, however, the eastern Laptev Sea could also be a possible source area of the upper turbidites.

Svindland and Vorren (2002) infer that the shift in clay mineral content from smectite-dominated to illite–chlorite facies in both cores, is synchronous, and they correlate this shift to a similar distinct shift in clay mineralogy found at ~50 cm in the pelagic core PS2185-6 on Lomonosov Ridge (Spielhagen et al., 1997, 2004). Based on the revised stratigraphy of Core PS2185-6 (Jakobsson et al., 2000a; Spielhagen et al., 2004; see Chapter 6.1.5), this shift in clay mineralogy occurs at ~50 ka (see Chapter 6.3, Figure 6.57A). If the correlation of the cores and the sychronity of the clay–mineral shifts are correct, this would imply that during the late Weichselian (upper MIS 3 and MIS 2) a sediment input from northern Greenland/Ellesmere Islands was dominant, whereas during the middle Weichselian and earlier, sediment input from Siberia was dominant.

In Core PS2190-1, a massive diamicton facies mainly composed of mud clasts, occurs in 250–305 cm depth (Figures 3.31 and 3.33). Here, Svindland and Vorren (2002) suggest that the massive diamicton facies is a result of downslope transport from the adjacent Lomonosov Ridge as submarine debris flows (Figure 3.32). In the Lomonosov Ridge cores, also intervals of diamicton facies with abundant mud clasts were found (Fütterer, 1992; Kassens & Thiede, 1994). This explanation seems to be less probable because the diamicton facies at Core PS2190-1 is characterized by a kaolinite–illite facies (Figure 3.33) and high organic-carbon contents of >1%, whereas the sediments of Core PS2185-6 have lower kaolinite contents and organic-carbon values are <0.2% (see Chapter 6.4.2 for organic-carbon values). Thus, a major and rapid glacio-marine sedimentation event from sea ice or icebergs (probably originating from the central Barents Sea/Franz Josef Land area; see Chapters 5.1 and 6.3.1) might be the most important sedimentary process generating the mud-clast-rich diamicton. Similar sediment pellets or mud clasts were described by Goldschmidt, Pfirman, Wolllenburg, and Henrich (1992) who concluded that the pellets were generated as cryoconites in sea-ice or icebergs.

Sedimentary sections dominated by turbidites display significantly increased sedimentation rates in comparison to the normal pelagic sedimentation. For the Amundsen Basin, Svindland and Vorren (2002) determined high sedimentation rates of up to 25 cm ky^{-1} in the turbidite sequences. For a sediment core obtained from the Canada Abyssal Plain, even sedimentation rates of ~145 cm ky^{-1} were measured (Grantz et al., 1999). That means, sedimentation rates are 20 to more than 100 times higher than those typical for normal pelagic sediments of the central Arctic (see Chapter 6.1, Table 6.1). Unfortunately, the data base on exact age determinations (needed for calculation of sedimentation rates) and basin-wide distribution maps of turbidite-influenced areas are very limited. Champbell and

Clark (1977), for example, concluded that up to 68% (average of 27%) of the cored sediments in the southeastern portion of the Canada Abyssal Plain adjacent to the narrow Canadian shelf and the Mackenzie cone were deposited by turbidity currents (Darby et al., 1989). Northward, the abundance of turbidites decreases significantly to a maximum of 6.5% of cored sediments in the northwestern part of the Canada Abyssal Plain. Thus, despite its importance for sediment accumulation in the deep basins of the Arctic Ocean, turbidite-influenced areas and sequences are under-represented in available sedimentary and organic carbon budget estimates (see Chapter 6.4.4).

CHAPTER FOUR

Proxies Used for Palaeoenvironmental Reconstructions in the Arctic Ocean

A recent review of proxies commonly used in palaeoceanography for reconstruction of surface- and deep-water chracteristics, depositional environment, etc. is published by Hillaire-Marcel and de Vernal (2007). In this chapter, an overview of geological proxies and methods used for reconstruction of modern and ancient environmental conditions in the Arctic Ocean is presented (Table 4.1). Some more general background information as well as examples of using these proxies in Arctic Ocean environmental studies are given. It should be mentioned that a number of described proxies have their main strength outside the polar ice-covered regions. Nevertheless, they are included here because they yield important information about (palaeo-) environmental conditions in the marginal ice-covered (Subarctic) zones or are used for palaeoenvironmental reconstructions of the old pre-glacial Arctic Ocean. For further more detailed discussion of selected proxies, it will be referred to Part III of this book where results of case studies dealing with reconstructions of modern and ancient Arctic Ocean environment are presented.

4.1. Lithofacies Concept

The most detailed information about the depositional environment (including sea-ice and iceberg sediment input, water-mass characteristics, etc.) and sedimentary processes (e.g., current influence, redepostion, etc.) and their changes in space and time can be obtained using a combined facies analysis of sediment cores (e.g., Clark et al., 1980; Darby et al., 1989; Henrich, Kassens, Vogelsang, & Thiede, 1989; Scott, Mudie, Baki, MacKinnon, & Cole, 1989; Henrich, Wagner, Goldschmidt, & Michels, 1995; Phillips & Grantz, 2001; Dowdeswell & Ó Cofaigh, 2002). Such a sediment facies analysis may include lithology, sedimentary structures (visual core description and X-ray photographs), grain size, coarse-fraction analysis, mineralogy (microscopy and/or XRD), micropalaeontological proxies (assemblages of foraminifers, diatoms, palynomorphs, etc.), organic carbon (OC) content and composition, and carbonate content and dissolution records.

The lithofacies concept has been described by Henrich et al. (1989) in a detailed study of the sedimentary facies of glacial-interglacial cycles in the Norwegian Sea during the last 350 ka. Using this approach, the authors were able to identify distinct changes in IRD input, carbonate productivity and dissolution, and bottom-water circulation (Figure 4.1), and to interpret the data in terms of a conceptional palaeoenvironmental model for glacial, deglacial, and interglacial phases (Figure 4.2). In general, interglacial sediments are of brownish colours, rich in biogenic carbonate (high content of foraminifers), low amounts of OC, and strongly

Table 4.1 Proxies Used for Reconstruction of (Palaeo) Environment and Sedimentary Processes in the (Sub-) Arctic Ocean. A Restricted Number of References of Arctic Studies are Listed for Each Proxy; for More References as well as References for General Background, See Specific Chapters.

Environment/Processes	Proxies in sediments	References (examples)
Sea-surface temperature	Planktonic foraminifers	Pflaumann et al. (2003) and Sarnthein et al. (2003b)
	Dinoflagellates	de Vernal et al. (2001)
	Alkenones	Rosell-Melé (1998, 2001)
	TEX_{86}	Sluijs et al. (2006, 2008)
	HBIs	Belt et al. (2007)
Surface-water salinity	Stable isotopes	Bauch et al. (1995) and Spielhagen and Erlenkeuser (1994)
	Fresh-water diatoms	Bauch and Polyakova (2003) and Polyakova (2003)
	C37:4 alkenones	Rosell-Melé (2001)
	δD of alkenones	Pagani et al. (2006)
Euxinic conditions	Trace elements	Lipinski et al. (2003) and Mutterlose et al. (2003)
	Pyrite framboids	Langrock et al. (2003a) and Boucsein and Stein (2008)
	Biomarker (isorenieratane)	Sluijs et al. (2006)
Primary production	Planktonic foraminifers	Volkmann (2000a, 2000b)
	Benthic foraminifers	Wollenburg and Kuhnt (2000) and Wollenburg et al. (2001)
	Calcareous nannoplankton	Gard and Backman (1990) and Gard (1993)
	Marine diatoms	Koç et al. (2002) and Polyakova (2003)
	Dinoflagellates	Kunz-Pirrung (1999) and Matthiessen et al. (2000)
	Biogenic opal concentration	Schubert (1995) and Nürnberg (1996)
	Barium concentration	Nürnberg (1996)
	$\delta^{13}C_{org}$ and $\delta^{15}N$	Schubert et al. (2001) and Schubert and Calvert (2001)
	Marine biomarker	
	Short-chain *n*-alkanes, sterols	Yunker et al. (1995) and Fahl and Stein (1999)
	GDGTs (crenarchaeol)	Sluijs et al. (2006, 2008)
Sea-ice cover	Sea-ice diatoms	Polyakova (2003)
	Planktonic foraminifer	Sarnthein et al. (2003a)
	Biomarkers (HBIs)	Belt et al. (2007)

Sediment-source indicators and transport by sea ice and/or icebergs	Grain-size distribution	Bischof (2000) and Spielhagen et al. (2004)
	Composition of lithic grains	Spielhagen (1991) and Phillips and Grantz (2001)
	Chemistry of Fe grains	Darby and Bischof (1996) and Darby (2003)
	Clay minerals	Naidu and Mowatt (1983) and Wahsner et al. (1999)
	Heavy minerals	Naugler et al. (1974), Levitan, Tarasov, Bourtman, & Kukina (1999) and Behrends (1999)
	Bulk mineralogy (XRD)	Vogt (1997) and Vogt et al. (2001)
	Major, minor, and trace elements	Rachold (1999) and Viscosi-Shirley, Pisias, & Mammone (2003b)
	Nd, Rb, and Sr isotopes	Eisenhauer et al. (1999) and Tütken, Eisenhauer, Wiegand, & Hansen (2002)
Carbonate dissolution	Dissolution index foraminifers	Henrich et al. (1989) and Pagels (1991)
River discharge/terrigenous input	Clay minerals	Rossak et al. (1999) and Stein et al. (2004a)
	Heavy minerals	Behrends et al. (1999) and Peregovich et al. (1999)
	Magnetic susceptibility	Dittmers et al. (2003) and Stein et al. (2004a)
	Freshwater diatoms	Polyakova (2003) and Polyakova and Stein (2004)
	Palynomorphs	Kunz-Pirrung (1999) and Matthiessen et al. (2000)
	Total organic carbon	Stein and Fahl (2000, 2004a, 2004b)
	δ^{13}Corg	Müller-Lupp et al. (2000) and Naidu et al. (2000)
	C/N ratios	Naidu (1985) and Stein and Fahl (2004a, 2004b)
	Rock-Eval parameter	Stein and Fahl (2000, 2004a, 2004b)
	Maceral composition	Boucsein and Stein (2000) and Boucsein et al. (2000)
	Terrigenous biomarker	
	Long-chain *n*-alkanes	Yunker et al. (1995) and Fahl and Stein (1999)
	Sterols	Fahl and Stein (1997, 2007)
	Lignin	Schubert (1995) and Lobbes et al. (2000)
	GDGTs (BIT Index)	Hopmans et al. (2004) and Brinkhuis et al. (2006)
	Compound specific δ^{13}C	Birgel et al. (2004)
Debris flows, slumps	Lithofacies	Kleiber et al. (2000) and Evans et al. (2002)
	Acoustics, seismics	Vorren et al. (1998) and Kleiber et al. (2001)
Glacigenic sedimentation	Lithofacies	Henrich et al. (1989) and Evans et al. (2002)

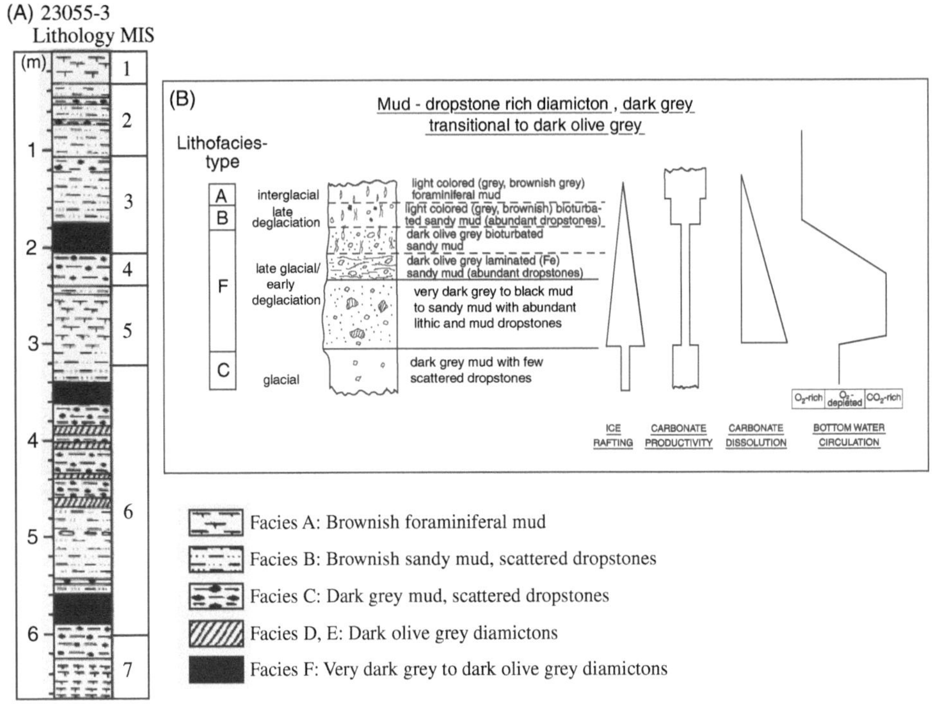

Figure 4.1 (A) Lithofacies distribution in Core 23055-3 from the Vøring Plateau (68°25.4′N, 04°01.3′E; water depth 2,311 m). MIS = Marine Isotope Stages; (B) Compilation of general lithofacies successions deposited during late glacial to interglacial periods, as obtained from a sediment facies study of sediment cores from the Norwegian Sea (from Henrich et al., 1989).

bioturbated. Glacial sediments, on the other hand, are mainly grey to dark grey silty muds with low carbonate content (low amount of foraminifers) and increased OC values. Concerning the IRD input, maximum values were recorded during the late glacial/deglacial transition (Figure 4.1), a phenomenon also observed in the central Arctic Ocean (e.g., Phillips & Grantz, 2001; see Chapter 6.3). Extensive reworking of organic matter rich sediments during glacial advance resulted in deep-sea deposition of terrigenous organic matter. Decreased deep-water renewal and marine-OC oxidation caused carbonate dissolution at the seafloor and, immediately after incipient burial, oxygen-deficient environment within the sediment, causing the dark grey colours (Henrich et al., 1989). Estimates of the degree of carbonate dissolution can be recorded by conventional fragmentation indices of *Neogloboquadrina pachyderma* sin., the only species which occurs in both glacial and interglacial sediments of the High Northern Latitudes (see Chapter 6), and by sensitive Scanning Electron Microscopy (SEM)-based dissolution indices (Henrich, 1986, 1989; Henrich et al., 1989; Pagels, 1991). The SEM approach is based on recognition of four stages of structural breakdown (Dissolution Index D_0 to D_3) during progressive dissolution of two morphotypes of *N. pachyderma* sin., the

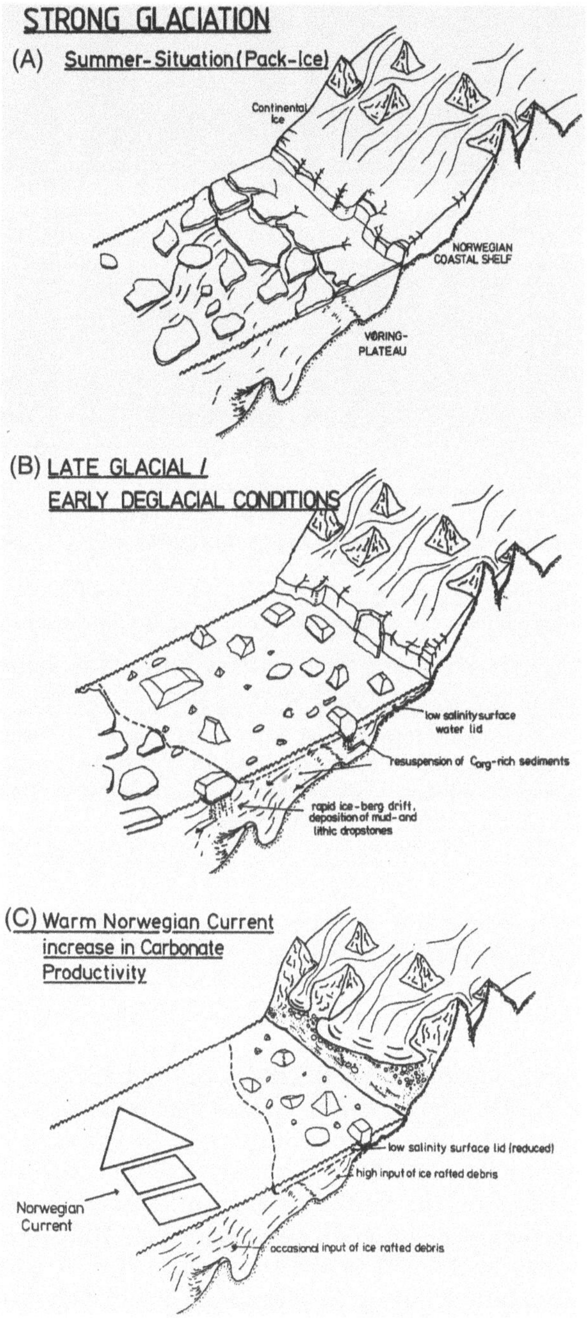

Figure 4.2 Conceptual palaeoenvironmental model. (A) Strong glaciation reflecting advance of continental ice on the Norwegian coastal shelf associated with offshore packice drift; (B) late glacial-early deglacial strong iceberg drift induced by surges of the marine-based parts of the continental ice; surface-water connection with the North Atlantic may be partially blocked off by large iceberg barriers; and (C) intrusion of the Norwegian Current with retreat of iceberg drift progressively closer to the coastal regions and offshore increase in carbonate productivity (from Henrich et al., 1989).

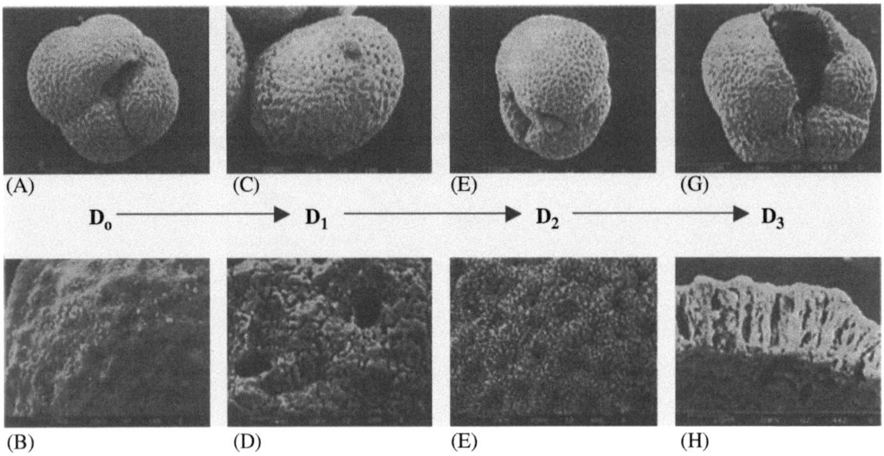

Figure 4.3 SEM dissolution indices on *N. pachyderma* sin., reticulate morphotype (from Henrich et al., 1989). (A and B) Well-preserved reticulate test (D_0); (C and D) incipient dissolution at the surface of the test (D_1); (E and F) strong dissolution at the surface of the test (D_2); (G and H) strong dissolution starting from the interior parts of the test resulting in fragmentation (D_3).

A numerical dissolution index can then be calculated using the following equation:

$$D = (D_0 * n_1 + D_1 * n_2 + D_2 * n_3 + D_3 * n_4)/(n_1 + n_2 + n_3 + n_4)$$

where D = numerical dissolution index of *N. pachyderma* sin., D_0, D_1, D_2, D_3 = steps of dissolution (0 = unaffected to 3 = strong dissolution), and n_1, n_2, n_3, n_4 = number of tests, with dissolution index subscript. This dissolution index can be calculated for the reticulate morphotype (presented here) and the cristalline morphotype of *N. pachyderma* sin., resulting in a numerical composite dissolution index (see Henrich et al., 1989 for details).

reticulate form and the crystalline form of *N. pachyderma* sin. (Figure 4.3; see Henrich et al., 1989 for details).

The distinct alternation of brown faunal-rich and grey nearly abiotic lithological units are also described for the central Arctic Ocean, probably representing interglacial/glacial cycles (Figure 4.4; e.g., Jakobsson et al., 2000a; Phillips & Grantz, 2001; Polyak, Curry, Darby, Bischof, & Cronin, 2004). The brown beds, including the surficial Holocene interval, contain low to moderate amounts of IRD, faunal remnants, and Mn oxides, whereas the grey beds corresponding to glacial periods are almost unfossiliferous and largely fine-grained, but may contain prominent IRD layers near the top and/or bottom of grey units that often extend into the adjacent brown interglacial unit (Figure 4.4; Phillips & Grantz, 2001; Darby, Polyak, & Bauch, 2006). The Mn cycles may reflect variation in the degree of sediment oxidation resulting from changes in ventilation of bottom waters, although the exact mechanisms for these changes are still speculative (Jakobsson et al., 2000a; Polyak et al., 2004). Furthermore, the dark grey sediments (mainly correlating with glacial intervals) may provide a first-hand correlation tool for sediment cores in the Arctic Ocean (e.g., Jakobsson et al., 2000a; Stein et al., 2001; Stein, 2005).

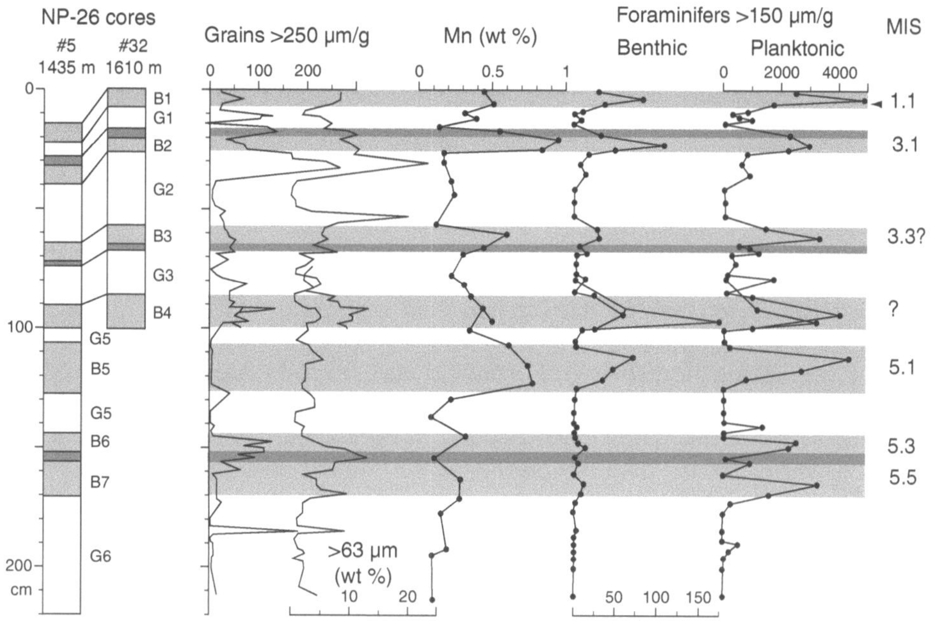

Figure 4.4 Example of the stratigraphy of bottom sediments from cores NP-26 from the Mendeleev Ridge, western Arctic Ocean (from Polyak et al., 2004, supplemented). Indexes to the right of lithologic columns show lithologic units (B, brown; G, grey). Brown (interglacial) beds are shaded; darker shading shows pink-white detrital-carbonate layers used as stratigraphic markers. Proposed MIS stratigraphy of the interglacials are shown at the right-hand side (Darby et al., 2006). For location see Figure 6.54.

In sediment facies analysis, X-ray imagery of sediment sections has become a common tool in accessing the internal structure of soft marine sediments. Standard slices (10 × 25 × 1 cm) of sediment continuously taken downcore from all long gravity and kastenlot cores as well as from short cores, may already be processed onboard. The structures identified on the photographs may reflect marine sedimentary environments including, for example, bioturbated to homogeneous pelagic glacio-marine deposits with varying amounts of IRD, laminated sediments representing distal turbidites, debris flows, etc. (see Chapter 3.3.3, Figure 3.29).

4.2. Grain-Size Distribution

Grain-size distribution of marine sediments may allow to obtain information on transport processes such as ice transport and bottom-current influence.

4.2.1. Grain-Size Distribution and Bottom-Current Speed

A well-established proxy for estimating palaeo-bottom-current velocity is the non-cohesive "sortable-silt" fraction (10–63 µm) introduced by McCave, Manighetti, and Robinson (1995a) (for recent review of strengths and limits of this proxy, see McCave & Hall, 2006). Bottom-current estimates based on the sortable-silt fraction are very limited in the ice-covered High Northern Latitudes and restricted to the Fram Strait area (Hass, 2002; Birgel & Hass, 2004) and the Lomonosov Ridge close to the Greenland continental margin (Nørgaard-Pedersen, Mikkelsen, Lassen, Kristoffersen, & Sheldon, 2007a). In this area (as well as in the glacial North Atlantic; Manighetti & McCave, 1995; McCave, Manighetti, & Beveridge, 1995b), grain-size variations might be influenced by IRD input variations, independent of variations in bottom-current speed. Different approaches for correction of the current-sensitive sortable-silt fraction for IRD input are discussed by Manighetti and McCave (1995) and Hass (2002).

In a sedimentological study of a sediment core from the Yermak Plateau (Core PS2837; 81°13.99′N, 02°22.85′E; 1,042 m water depth), Birgel and Hass (2004) used the sortable silt (10–63 μm) mean diametre (SS) to assess relative current-speed variations during the last deglaciation (Figure 4.5). In this area, the so-called Yermak Slope Current (YSC), a contour current at an average water depth of 1,000–1,500 m, directly affects sedimentation at the location of the core (e.g., Rudels et al., 2000). Current speed at the core location is suggested to be too low for the transport of terrigenous sand-sized materials (Fahrbach et al., 2001, based on data from an array of current-metres located on a transect at 79°N across Fram Strait). Thus, the bulk terrigenous sand fraction and part of the silt fraction are assumed to be ice-rafted. In order to significantly reduce the influence of ice-rafted silt on the silt fraction (i.e., the fraction used for palaeo-current-speed analysis), the predominantly ice-rafted sand fraction was related to the predominantly current-sorted silt fraction (Figure 4.5B; Hass, 2002; Birgel & Hass, 2004). Using the regression equation calculated from the two variables, a steady-state condition was calculated. Deviations from the steady state (ΔSS) in the coarse direction were interpreted as higher current speed, whereas deviations in the fine direction suggest current speed lower than average (Figure 4.5B and C; see Hass, 2002, for a detailed description of this method). The method applied sharpens existing trends but usually does not produce trends when there are none visible in the original data.

The sortable silt mean (SS) and in particular the IRD-corrected value ΔSS in the record of Core PS2837 suggest that colder periods, as visible in the GISP 2 ice-core, generally correspond with periods of lower bottom-current strength (Figure 4.5C; Birgel & Hass, 2004). During Heinrich Event 1 and the Younger Dryas (YD) cooling event, for example, bottom currents significantly decreased. At those time intervals, meltwater probably extended over the North Atlantic and the Nordic Seas and reduced thermohaline overturn (e.g., Broecker et al., 1988; Broecker & Denton, 1989; Bond et al., 1993; Keigwin & Lehman, 1994). Near 14.7 ka, bottom currents dramatically increased for a short period of time before they collapsed to the lowest values in the record. The initial warming at the beginning of the Bølling interstadial most likely controlled this process. It can be assumed that the initial warming was paralleled by intense thermohaline overturn that also accelerated northbound intermediate-depth currents, transporting newly formed NSDW into the Arctic Ocean. The rapid warming at the onset of the Bølling interstadial triggered intense melting processes that may have caused deep-water production to

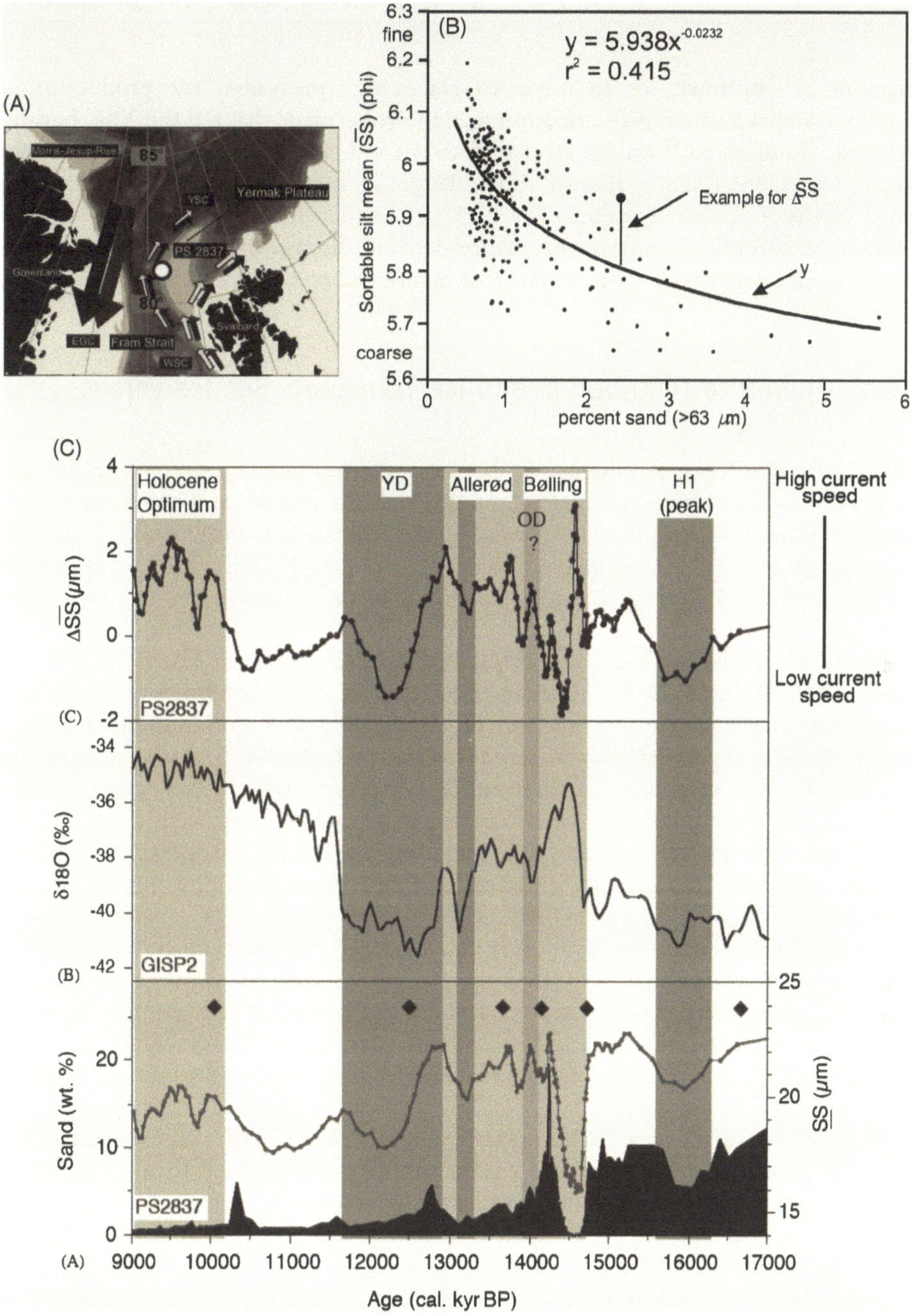

Figure 4.5 (A) Location of Core PS2837 and general circulation pattern in the Fram Strait area. EGC, East Greenland Current; WSC, West Spitsbergen Current; YSC, Yermak Slope Current. (B) Correlation between sortable silt SS and percentages of sand at Core PS2837 (Hass, 2002). (C) (A) Sand fraction (= IRD, black area plot); grey curve: sortable silt mean grain-size SS as measured (unweighed 5 pt. running mean). Black diamonds: ^{14}C age points (calendar years) by Nørgaard-Pedersen et al. (2003). (B) $\delta^{18}O$ of Greenland ice-core GISP 2 (Grootes et al., 1993). (C) IRD-corrected sortable silt mean ΔSS of Core PS2837 (unweighed 5 pt. running mean). Negative values indicate lower, positive values indicate higher current-speed events (Hass, 2002). YD, Younger Dryas; OD, Older Dryas; H1, phase equivalent to Heinrich Event 1 peak phase (from Hass, 2002; Birgel & Hass, 2004).

temporarily shutdown, or to move to places that prevented the production of northbound deep water. As a consequence, the water mass that fed the YSC became reduced, as suggested from the clear shift to the sedimentation of finer grains at the location of Core PS2837 (Figure 4.5C; Birgel & Hass, 2004). Near 10.3 ka, the speed of the YSC accelerated to the high values that characterize early Holocene interglacial conditions. For further more detailed discussion and interpretation of the grain-size records of Core PS2837, it is referred to Hass (2002) and Birgel and Hass (2004).

4.2.2. Grain-Size Distribution and Ice Transport: Sea Ice versus Icebergs

IRD in Arctic Ocean sediments can be derived from icebergs and/or sea ice and occurs throughout glacial and interglacial intervals. Its amount and composition may vary significantly on a spatial as well as temporal scale. A first-order proxy to identify ice transport in deep-sea sediments is the grain-size distribution. Here, many authors use the amount of sand (and gravel) fraction, but the limits on the fractions varies from study to study, that is, >63, >250, >500, >1,000, and >2,000 μm (e.g., Hebbeln, Henrich, & Baumann, 1998; Nørgaard-Pedersen et al., 1998; Nam & Stein, 1999; Andrews, 2000; Andrews & Principato, 2002; Spielhagen et al., 2004).

A very simple and quick method to count IRD >2 mm (related to iceberg transport; see next paragraph) has been developed by Grobe (1987). Instead of sieving the sample and weighing the amount of coarse fraction, X-radiographs which are standard preparations from each core for structural analysis, were used. The X-radiographs were placed on a light table and using a graph paper the particles >2 mm in diametre were continuously counted in 1 cm intervals (Figure 4.6A). Using a radiograph width of 10 and 1 cm interval, the amount of sediment counted represents ~10 cm^3. Comparison with grain-size data obtained by the more time-consuming sieving method shows that both methods give very similar results (Figure 4.6B).

For distinguishing between sea-ice and iceberg transport it is generally accepted that very coarse-grained material >250 μm, that is, coarse-sand-, gravel-, and pebble-sized particles, are mainly restricted to iceberg transport (e.g., Clark & Hanson, 1983; Baumann et al., 1995; Elverhøi et al., 1995a; Dowdeswell et al., 1998; Hebbeln et al., 1998; Nørgaard-Pedersen et al., 1998; Spielhagen et al., 2004). In sediment cores from the central Arctic Ocean, Spielhagen et al. (2004) determined a pronounced positive correlation ($r = 0.6$) between the contents of the >63 μm fraction and the undoubtedly iceberg-rafted >250 μm fraction in the coarse-grained, foraminifer-poor layers, which suggests icebergs as the predominant transport agents also for >63 μm fraction in these sediments (see Chapter 6.3.1 for more details). The presence of large (over-sized) clasts of several centimetres in diametre in ice-distal deep-sea sediments (Figure 4.7B) are an explicit indication of transport by icebergs containing basal glacial, englacial, and supraglacial sediments. In laminated or bedded sediments, these large dropstones may cause depressed laminae directly underneath the clast (Figure 4.7A; Domack & Domack, 1992).

Clark and Hanson (1983) carried out detailed grain-size analyses of Arctic deep-sea sediments and identified — based on grain-size data — different types of

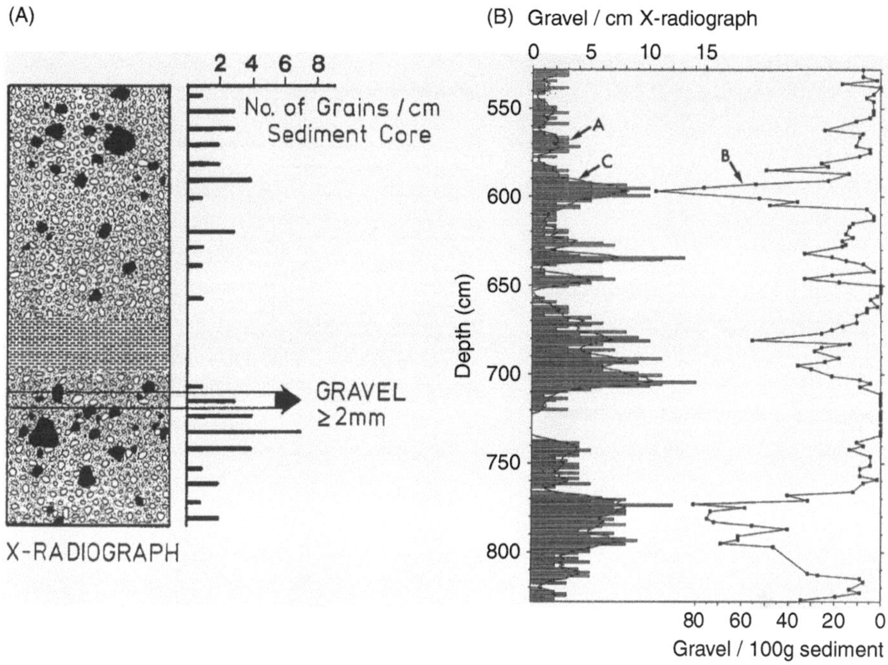

Figure 4.6 (A) Schematic drawing of a X-Ray photograph taken from a 1 cm thick, 10 cm wide, and 25 cm long sediment slab and used to determine the distribution of IRD. On the photograph, the gravel content (>2 mm) was counted in horizons of 1 cm width and plotted as number of grains per centimetre in a diagram. These sediment slabs for X-Ray photography are routinely taken onboard *Polarstern* during sampling of sediment cores as a standard method. (B) Comparison of the number of grains counted in X-ray radiographs (A) with the results of a sieve analysis (B). (C) Shows 7-point running averages of the results of (A) (from Grobe, 1987).

sediments, related to iceberg and sea-ice transport, and ocean-current influence (Figure 4.8). In their study, they defined an Arctic deep-sea sediment called "type III", which contains generally high percentages of fine silt and clay and which was attributed to rafting and release by drifting sea ice. When looking at the grain sizes of sediment entrained in modern sea ice, also a clear dominance of clay- and silt-sized material is obvious (Wollenburg, 1993; Nürnberg et al., 1994; Dethleff, 2005; see Chapter 3.1, Figures 3.3 and 3.4), which may support Clark and Hanson's (1983) interpretation of the "type III" sediments.

As shown by Dethleff (2005), the silt and clay texture distributions of Laptev and Beaufort sea-ice sediments well match the "type III" fine fraction distribution of Arctic deep-sea deposits classified by Clark and Hanson (1983) as ice-transported material, though the "type III" sediment has a relative low at 6–7 phi, and a relative high at 8–9 phi compared to both sea-ice sediment distributions (Figure 4.9). Clark and Hanson (1983) explained the "type III" low at 6–7 phi by particle removal subsequent to post-depositional current reworking of the sediment surface. The pronounced relative high at 8–9 phi in the "type III" sediment may be attributed to the fact that (1) this grain-size spectrum is of limited occurrence in Laptev and Beaufort sea-ice sediments, and also in Laptev shelf source sediments, and/or

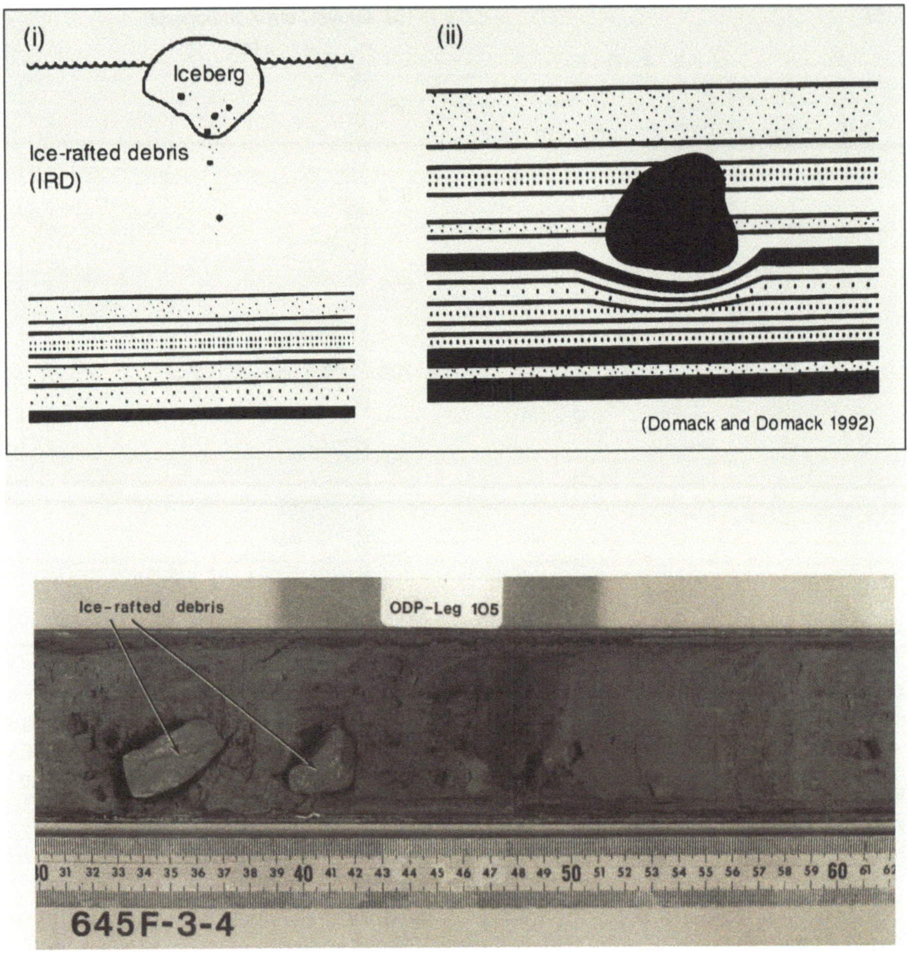

Figure 4.7 Schematic figure showing sedimentation of large-sized dropstones and resulting sediment structure (according to Domack & Domack, 1992). Photograph of large-sized dropstones in a section from ODP Site 645, Baffin Bay (Srivastava et al., 1987).

(2) particles released from melting sea ice in the central Arctic Ocean will be mixed with other fine-grained material of different origins such as icebergs, atmosphere, and river discharge (Dethleff, 2005).

In contrast to sea-ice sediments, the grain-size distribution pattern of the whole clay to sand fraction of iceberg-transported material is dominated by coarse silt and sand (Figure 4.8; "type IV"; Clark & Hanson, 1983).

When using grain-size distributions as proxy for iceberg transport, however, one should have in mind possible limitations of this approach:

- Diamict sediments retrieved in a 7–11 cm diametre core do not provide a complete nor adequate representation of the grain-size spectra of the diamict because very large clasts/particles of several centimetres to decimetres will not be

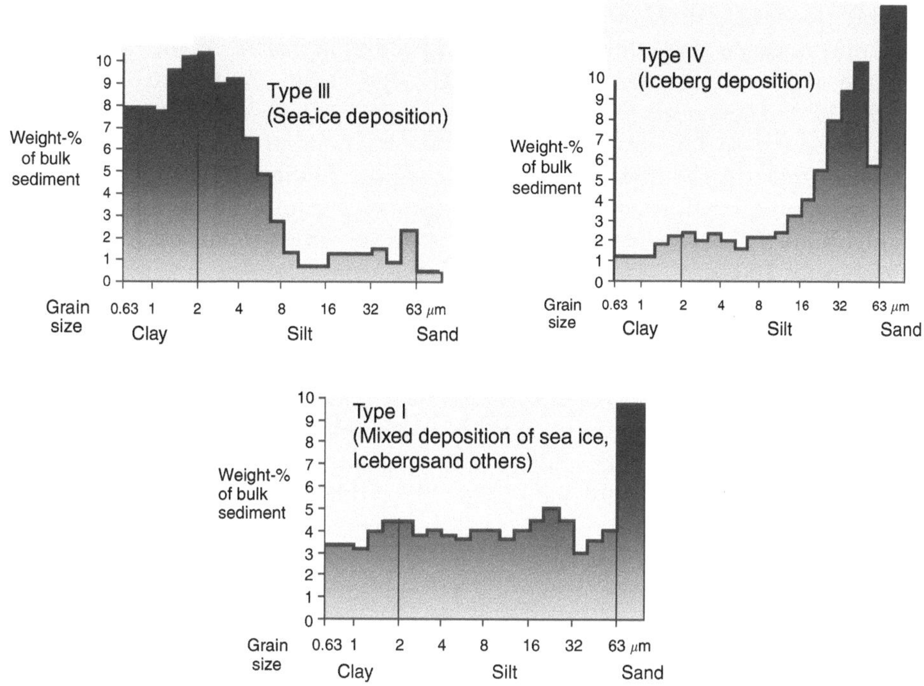

Figure 4.8 Different sediment-size types in central Arctic Ocean sediments based on Coulter counter analysis of the silt and clay fractions, and interpretation in terms of transport processes (Clark & Hanson, 1983; redrawn by R. Spielhagen).

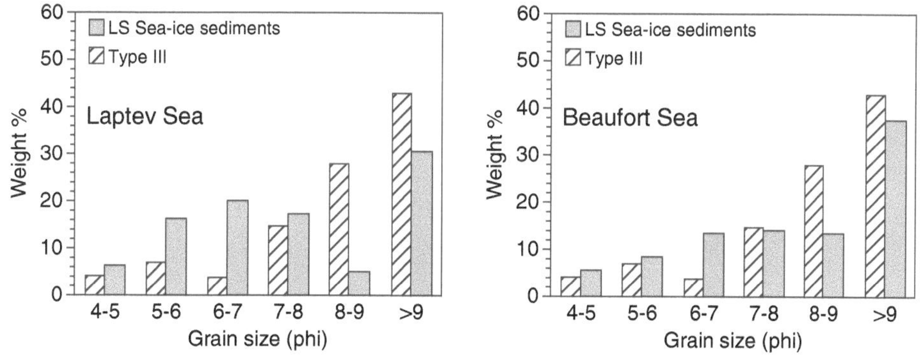

Figure 4.9 Bar diagrams comparing averaged grain-size data >4 phi (<63 mm; in phi steps) of sea-ice sediments from the Laptev Sea (Dethleff, 2005) and the Beaufort Sea (Reimnitz et al., 1998) with Arctic deep-sea sediment "type III" graphically adapted from Clark and Hanson (1983), which was interpreted as ice-transported, silty-clayey deposits (from Dethleff, 2005).

sampled. Most of the grain-size studies are done on the sand fraction, that is, the matrix of the diamicton. But even with a complete grain-size analysis of clasts of all sizes in marine sediment cores, there is a sampling bias because of the coring operation itself (Andrews & Principato, 2002).

- Other processes, such as bottom current can also cause winnowing of fine fraction and accentuate the coarse-grained fraction in marine sediments, resulting in the apparent enhancement of the IRD signal (e.g., Anderson & Andrews, 1999; Ó Cofaigh et al., 2002).
- Down-core variations of different sand-size fractions in the same core do not always parallel each other (see also Chapter 6.2.3, Figure 6.49). That means, different interpretations of IRD events might be determined from the same set of samples purely on the basis of the choice of the sand-size cut-off (Andrews & Principato, 2002).
- At glaciated continental slopes, diamict layers of variable thickness may be deposited by debris flows. Here, considering the sediment facies is important for the interpretation of grain-size data in terms of IRD input. Based on lithology and grain-size data obtained from a sediment core from the lower continental slope of East Greenland near 73°N, for example, debris-flow diamicts and turbidites dominate the core, and the distinct spikes in the coarse fraction >2 mm below 200 cm depth record individual debris-flow pulses rather than discrete iceberg-rafting events (Figure 4.10; Ó Cofaigh et al., 2002; Evans et al., 2002; see Chapter 6.2.3 for more details).
- Single isolated "dropstones" may be related to other than iceberg-transport processes such as rafting by vegetation ("algal rafting") (Emery, 1963; Gilbert, 1984, 1990).

Thus, while IRD texture gives a first-hand indication of the type of sedimentation regime, it can be misleading and needs to be supplemented by other data, such as surface texture analysis of sand grains by SEM (e.g., Jansen, Sjølholm, Bleil, & Erichsen, 1990; Helland & Holmes, 1997; Dunhill, 1998; \St. John, 2008), determination of sediment sources (see later description), and sediment facies. Considering texture in combination with lithology and sediment provenance provides a much sounder interpretation of sea ice versus iceberg rafting.

4.3. Proxies for Sources and Transport Processes of Terrigenous Sediments

The identification of source areas of sediments entrained in sea ice and icebergs based on the IRD composition, gives important information about the pathways of sea-ice and iceberg and, thus, the surface circulation pattern in the Arctic Ocean. Whereas today sea ice is the dominant sediment transport agent, during Pleistocene times icebergs from the continental icesheets surrounding the Arctic Ocean delivered coarse-grained IRD to the Arctic Ocean (see Chapter 6). Because the landmasses surrounding the Arctic Ocean are composed of different geological terraines characterized by a very specific mineralogical and chemical signature (see Chapter 5.1.1), mineralogical, geochemical, and lithological data obtained from sediments can be used to identify source areas of the IRD. If the IRD

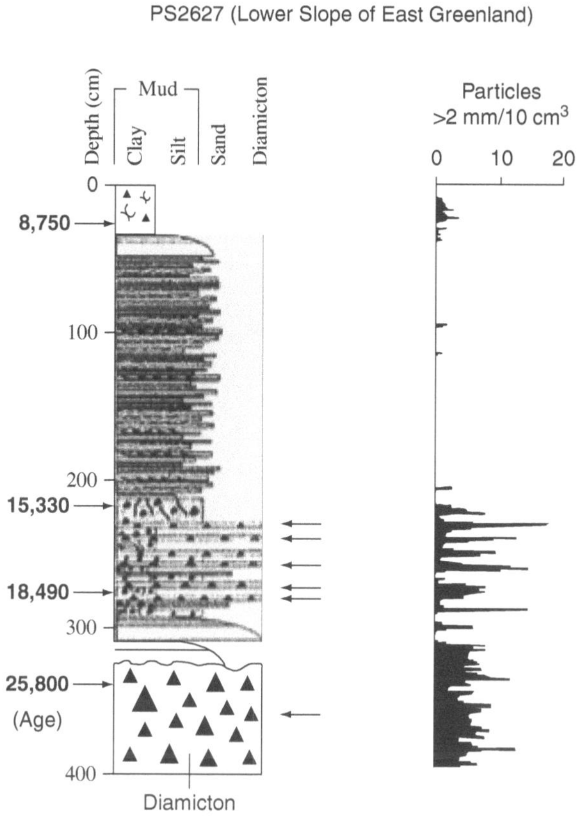

Figure 4.10 Lithology of Core PS2627 and percentages of gravel >2 mm (determined according to Grobe, 1987; see Figure 4.6). Black arrows indicate debris flow pulses (diamictons). For core location, see Chapter 6.2.3, Figure 6.53 (from Ó Cofaigh et al., 2002, supplemented).

composition can be related to specific source areas of these icebergs, information about both past circulation patterns as well as locations where icesheet existed and calved into the Arctic Ocean, may be obtained.

4.3.1. Composition of Coarse Fraction

The most commonly used approach is a study of the composition of the lithic fragments by microscopy (e.g., Spielhagen, 1991; Bischof, Clark, & Vincent, 1996; Phillips & Grantz, 2001). Phillips and Grantz (2001), for example, studied the gravel-sized (iceberg-transported) clasts from a transect across the central Arctic Ocean including the Amerasian Basin influenced by the Beaufort Gyre and the Eurasian Basin influenced by the Transpolar Drift. The samples were taken as vertical slides from the entire stratigraphic section of box cores, representing composite samples of ice-rafted clasts from the Holocene and late Pleistocene, that is, including sediments deposited in both interglacial and glacial environments. Seven clasts types were identified: sandstone and siltstone, quartz and chert, limestone and dolostone, metamorphic, igneous, coal, and unknown. Based on the abundances of these clasts, there is a clear distinction possible between the clast suites from the Amerasian and Eurasian basins (Figure 4.11; Phillips & Grantz, 2001).

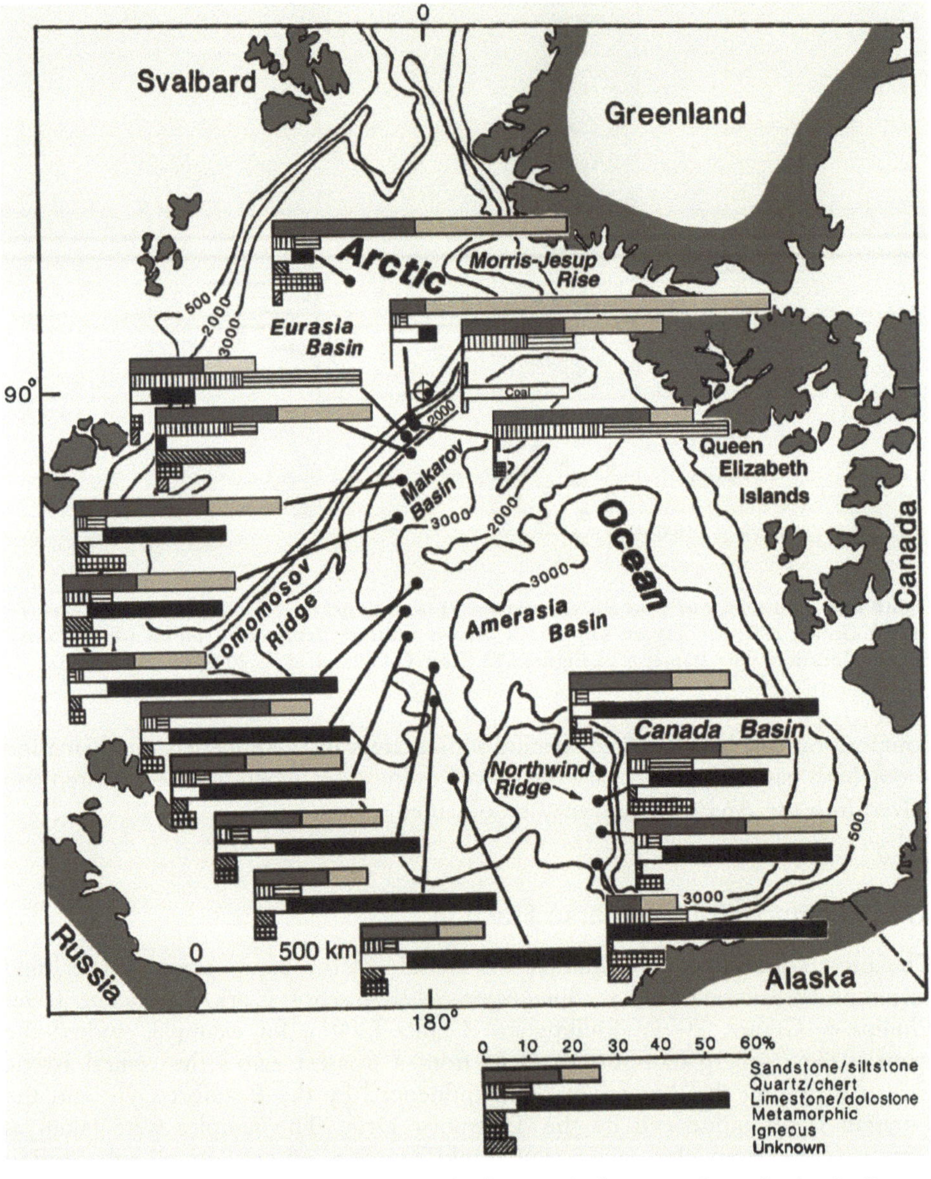

Figure 4.11 Composition of clasts >2.0 mm in diameter in box cores from the Arctic Ocean. Depth contours in metres (from Phillips & Grantz, 2001).

The clasts from the Amerasian Basin are characterized by a dominance of dolostones/limestones that originated in the carbonate-rich Palaeozoic terranes of the Canadian Arctic Islands (see Chapter 5.1.1) transported by icebergs to the core sites in the clockwise Beaufort Gyre. The total concentration of clasts decreases northward by 98% along the trend of the Beaufort Gyre from the southeastern Canada Basin to the Makarov Basin (Figure 4.12; Phillips & Grantz, 2001). Across the Makarov Basin flank of the Lomonosov Ridge, the concentration increased again by a factor of three, and siltstone, sandstone and siliceous clasts become dominant in the Eurasian Basin (Figures 4.11 and 4.12). The source of these clasts is

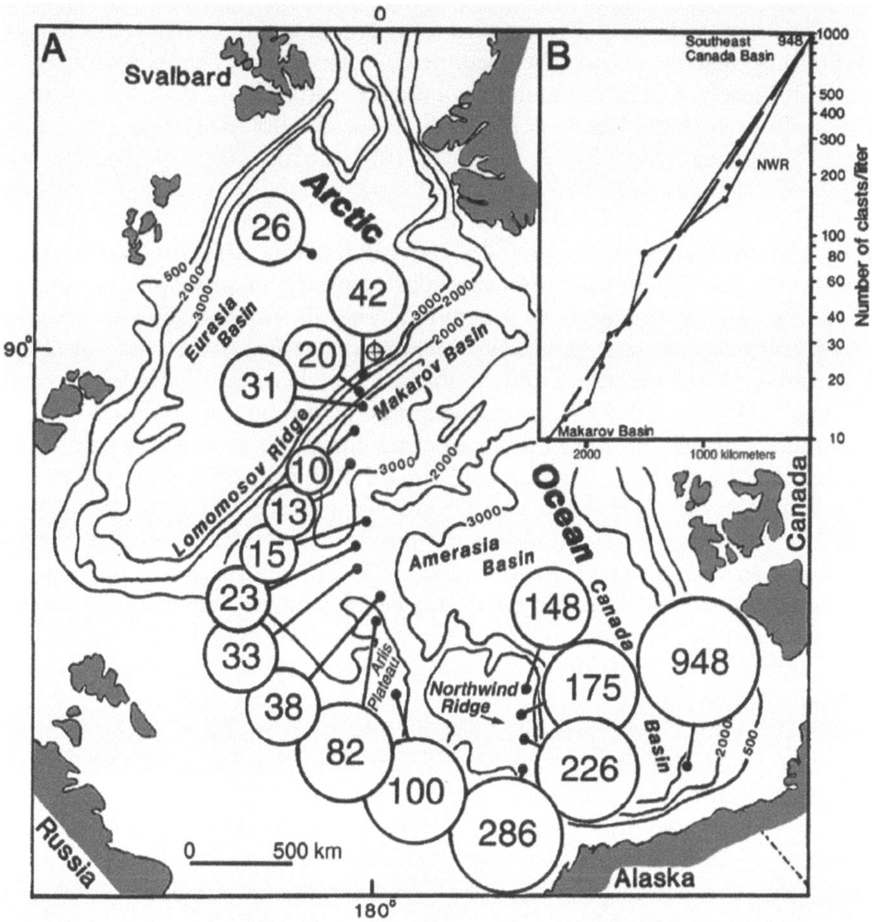

Figure 4.12 (A) Number of clasts $>2.0\,\mathrm{mm\,l^{-1}}$ of sediment in box cores from the Arctic Ocean. (B) Semilog plot of clast/liter versus sample distance from the southeast Canada Basin near the Canadian Arctic Islands to the Makarov Basin in the central Arctic Ocean, showing a logarithmic decrease in clast abundance that resembles a glacial ice decay-gravel sedimentation function (see text). The logarithmic decrease in clast abundance with distance indicates that the source of the sediment-laden icebergs in the Beaufort Gyre was northwestern Canada. Depth contours in metres (from Phillips & Grantz, 2001).

less certain, but bedrock distribution and the distribution of glaciation in northern Eurasia point to the Taymyr Peninsula-Kara Sea region (Phillips & Grantz, 2001). Based on the abruptness of the change in both composition and concentration of clasts on the Makarov Basin flank of the Lomonosov Ridge, Phillips and Grantz (2001) suggest that the boundary between the Beaufort Gyre and the Transpolar Drift has been relatively stable during the last interglacials (see Chapter 6.3 for more detailed discussion).

4.3.1.1. Lithic and Fe oxide grains as source indicator of sea-ice and iceberg sediments

A very intensive study of >300 samples from potential circum-Arctic source areas of IRD, using shallow-marine shelf sediments, inter-island channel sediments, and glacial tills closely representing nearby bedrock formations, was carried out by Darby and co-workers (Darby & Bischof, 1996; Bischof & Darby, 1997; Darby, Bischof, Spielhagen, Marshall, & Herman, 2002; Darby, 2003; Darby & Bischof, 2004). These samples were analysed for petrographic and mineralogic composition of detrital grains >250 μm and for the mineralogy and geochemistry of individual opaque Fe oxide minerals in the 45–250 μm fraction. For the coarse-fraction analysis more >300 grain types were distinguished microscopically, and nine different Fe oxide minerals (fresh ilmenite, altered ilmenite, magnetite, magnetite with other phases, haematite, ferric-ilmenite, titano-haematite, titanomagnetite, and chromite) were identified and confirmed by chemical composition using a microprobe. As a result, the source samples could be grouped into 38 (41) geographically distinct groups of unique petrographic and chemical composition by clustering and discriminant function analysis (DFA) (Figure 4.13; Darby et al., 2002; Darby, 2003). To match grains to a source area, for each grain in a sea-ice or iceberg sample the petrographic and chemical composition was determined and compared to all source area composition groups using DFA. This approach first introduced by Darby and Bischof (1996), provides more specific source information than those commonly used (see Chapter 6.3.1, for Pleistocene IRD records and palaeoceanographic implications). Furthermore, the geochemistry of individual detrital Fe minerals has the advantage that it can be used for identification of the sources of both iceberg and sea-ice debris (Darby et al., 2002; Darby, 2003; Darby & Bischof, 2004). More details of method are described in Darby and Bischof (1996), Darby et al. (2002), and Darby (2003).

Darby (2003) used the geochemical fingerprint of Fe oxide minerals grains entrained in modern Arctic Ocean sea ice to determine more precisely the sources of these sea-ice sediments. For this study, sea-ice sediment samples collected from ice floes in the western Arctic Ocean (i.e., Beaufort Sea, Chukchi Sea, Chukchi Borderland, and central Arctic Ocean; Figure 4.13) during the past 30 years were analyzed. Based on these data, Ellesmere Island and especially the Bank Island shelf and the Laptev Sea (Figure 4.13, source areas 3, 8, and 18, respectively) were identified as major source areas of the sea-ice samples (Figure 4.14; Darby, 2003). In the Canadian Arctic, ice floes drift west with the Beaufort Gyre. Darby's results

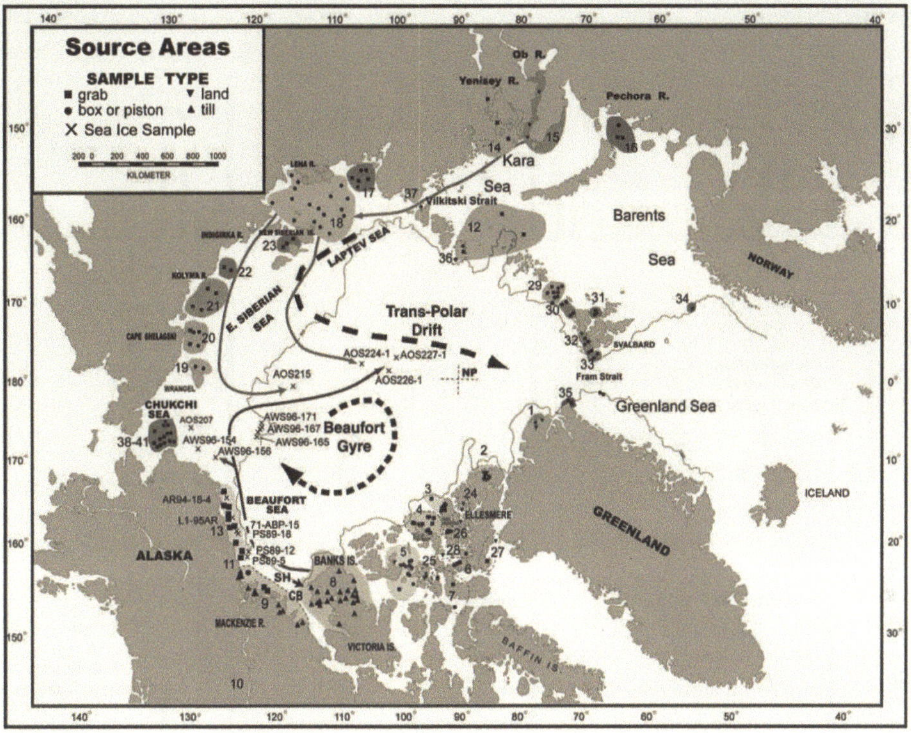

Figure 4.13 Circum-Arctic source areas (1–41) defined by unique mineralogical/geochemical source compositions (Darby et al., 2002; Darby, 2003), sea-ice sample locations, and general sea-ice circulation (bold dashed arrows) in the Arctic Ocean. The proposed drift trajectories (grey arrows) shown are for the general areas where floes were sampled and are based both on the Fe oxide grain sources and buoy back trajectories with known or general surface currents. Samples from Sachs Harbor (SH) are too close together to show. The Cape Bathurst (CB) Polynya generally occurs in the coastal waters between Banks Island and the Mackenzie Delta (from Darby, 2003).

not only substantiate the importance of the Laptev Sea as source of sea ice, already known from clay and heavy-mineral data determined in sea-ice and surface sediments from the Eurasian Arctic (e.g., Nürnberg et al., 1994; Pfirman et al., 1997; Behrends et al., 1999; Dethleff et al., 2000; see Chapter 5.1.2 and 5.1.4), but even show the presence of ice from the Siberian shelves (Laptev Sea and Kara Sea) in the Beaufort Sea off Alaska (Figure 4.14). This indicates the importance of either trans-Arctic transport and deposition of Siberian grains onto shallow North American shelves where they can later be entrained in sea ice, or co-mingling of Siberian and North American ice floes in the Beaufort Gyre or west-central Arctic Ocean. In either case, ice floes originating in shallow Siberian seas are a regular occurrence in the Beaufort Gyre, to be explained by a Transpolar Drift shifted more towards North America (Bischof & Darby, 1997; Darby, 2003; Darby & Bischof, 2004; see Chapter 6.3.1 for more details).

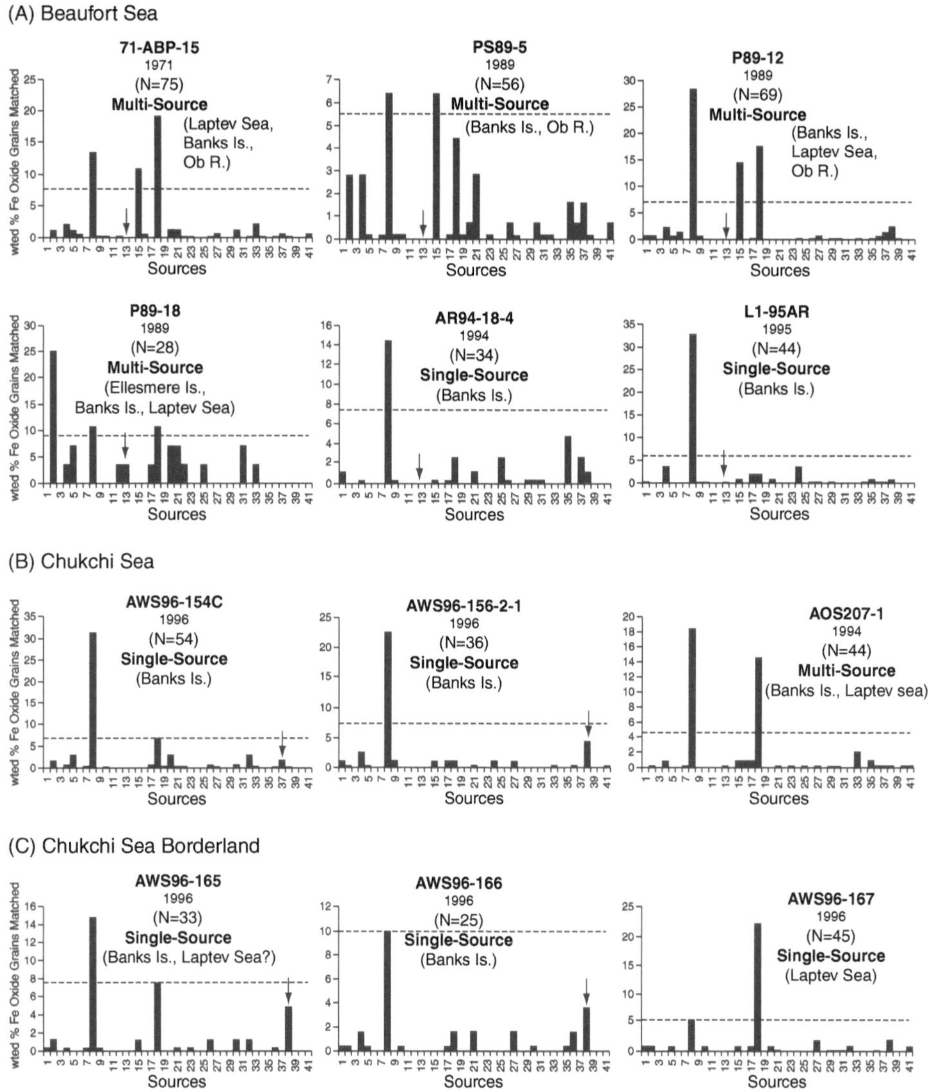

Figure 4.14 The weighted percentage of Fe oxide grains from sea-ice floes matched to circum-Arctic source areas (from Darby, 2003). For sample location see Figure 4.13. (A) Source matches for six floes sampled in the Beaufort Sea. Arrows show the local northern Alaska shelf source area (source area 13). The Banks Island source (source area 8) is significant in every sample. (B) Source matches for three floes sampled in the northern Chukchi Sea. Arrows indicate the local Chukchi Sea source. (C) Source matches for three floes sampled in the Chukchi Borderland area in 1996. Arrows indicate the local Chukchi Sea source. (D) Source matches for four floes sampled in the west-central Arctic in 1994.

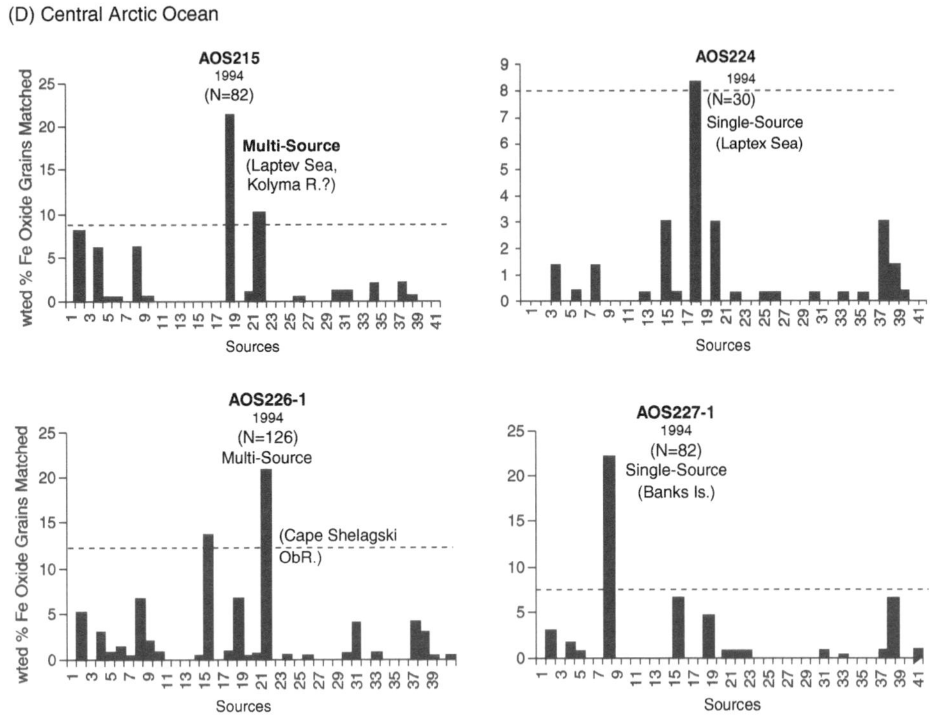

Figure 4.14 (*Continued*)

4.3.2. Bulk, Clay, and Heavy-Mineral Assemblages

Mineralogical data, that is, bulk-, clay-, and heavy-mineral data are also often and successfully used as indicators for sediment sources, transport pathways, and processes in the Arctic Ocean and its marginal seas (e.g., Darby, 1975; Naidu & Mowatt, 1983; Stein, Grobe, & Wahsner, 1994c; Stein et al., 2004a; Levitan et al., 1996; Levitan et al., 1999; Levitan, Bourtman, Demina, Chudetsky, & Schoster, 2005; Vogt, 1997; Behrends et al., 1999; Wahsner et al., 1999; Müller & Stein, 2000; Schoster et al., 2000; Vogt, Knies, Spielhagen, & Stein, 2001; Viscosi-Shirley, Mammone, Pisias, & Dymond, 2003a; Dethleff, 2005; Krylov et al., 2008b). Clay and bulk mineralogy are determined by XRD technique using the clay fraction separated by Atterberg method and the ground bulk sediment sample, respectively. Whereas the major part of clay-mineral studies on Arctic Ocean sediments are based on the $<2\,\mu m$ fraction, a large number of the early studies carried out by Russian scientists are based on the $<1\,\mu m$ fraction. This has to be considered when comparing the different data sets because specific clay minerals such as smectites are enriched in the very fine fraction (e.g., Vogt, 1997 and references therein). Heavy minerals, on the other hand, are determined by counting particles of the heavy subfractions 32–63 μm and/or 63–125 μm (separated from the bulk sediment using sodium metatungstate with a density of $2.83\,g\,cm^{-3}$ or bromoform with a density of $2.89\,g\,cm^{-3}$ under a microscope. Main heavy minerals often determined in Arctic Ocean sediments and used as sediment source indicator, include amphibole (hornblende), pyroxenes (clinopyroxene and orthopyroxene), epidote, garnet, and opaque minerals (illmenite, magnetite, haematite) (e.g., Lapina, 1965; Naugler et al., 1974; Zauderer, 1982; Levitan et al., 1996; Levitan, Arnold, Bourtman, Ivanova, & Marina, 2000; Behrends et al., 1999).

In order to give credit to pioneer studies already performed by Russian and American scientists in the 1960s in the Eurasian Arctic (Lapina, 1965; Holmes, 1967; Lisitzin, 1972; Silverberg, 1972), mineralogical data from Laptev Sea surface sediments are shown here as an example (Figure 4.15). During the last decade, numerous international studies were carried out in the Eurasian Arctic, following this pioneer work. The general picture from the 1960s/1970s, however, remains the same (see later description), although more detailed information on sediment sources and transport became available with the new data from the Laptev/Kara Sea (e.g., Levitan et al., 1996, 1999; Levitan, Lavrushin, & Stein, 2007; Vogt, 1997; Behrends, 1999; Behrends et al., 1999; Peregovich, 1999; Rossak, Kassens, Lange, & Thiede, 1999; Wahsner et al., 1999; Schoster et al., 2000; Stein et al., 2004a; Dethleff, 2005). A more detailed discussion of sediment sources based on different mineralogical data available now for the entire Arctic Ocean and its marginal seas is presented in Chapter 5.1.1.

Based on these pioneer data, the Laptev Sea can be separated into a western and eastern provenance area of terrigenous sediments. The distribution maps of the different heavy minerals in the surface allow the following classification (Lapina, 1965, presented in Stein & Korolev, 1994): The eastern Laptev Sea is characterized by high amounts of epidote and amphibole (Figure 4.15A and 4.15B), transported by the Lena River into the Laptev Sea. The suspension of the Khatanga River draining into the western Laptev Sea, on the other hand, is dominated by pyroxenes (Figure 4.15C). The high amounts of ore minerals (ilmenite and magnetite) in the central part of the Laptev Sea, at the coast of the Taymyr Peninsula, and south of the New Siberian Islands (Figure 4.15D) may indicate erosional processes in very shallow-water environments. Based on the occurrences of the different heavy minerals, three major heavy-mineral provinces can be identified: an epidote–pyroxene assemblage ("Khatanga-assemblage"), an epidote–amphibole assemblage ("Lena-assemblage"), and an ilmenite–magnetite assemblage (Figure 4.15E; Lapina, 1965).

Because the published heavy-mineral data are based on different grain-size classes, that is, the 32–63, 50–100, and 63–125 μm fractions, Behrends (1999) performed a systematic study of heavy-mineral distributions in Laptev Sea surface sediments. As main result, there is a quite good agreement between the different data sets. Thus, for a general classification of heavy-mineral distributions in the Arctic Ocean as source indicator it seems to be reliable to use data sets based on different grain-size classes.

In addition to the heavy-mineral composition, the suspensions of the Lena and Khatanga also show differences in the clay mineralogy (Figure 4.15F) as well as quartz and feldspar contents, already identified in the pioneer studies mentioned earlier (Figure 4.15G and H). The Lena material appears to contain relatively high amounts of feldspar and a clay-mineral assemblage dominated by illite, whereas the

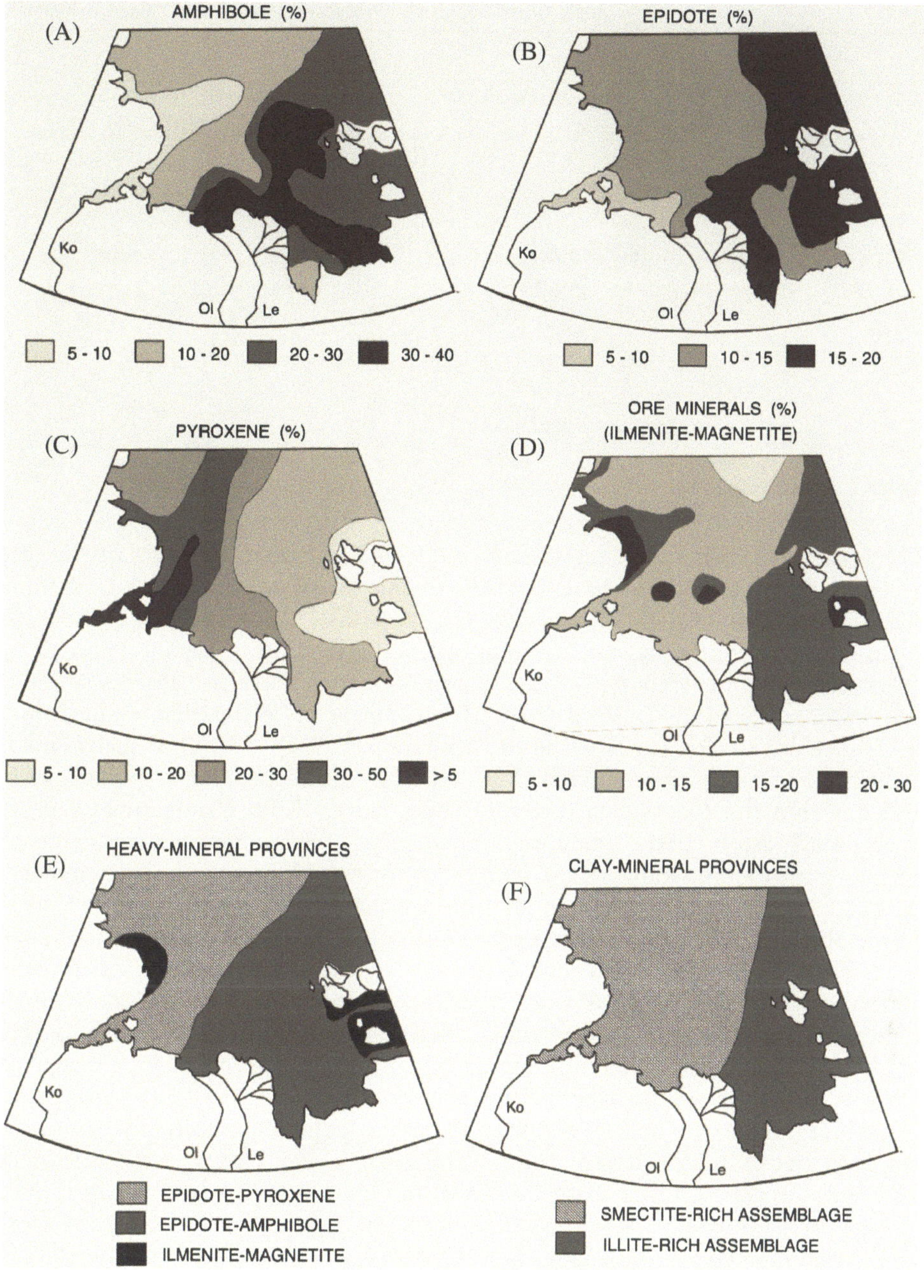

Figure 4.15 Distribution maps of heavy minerals, clay minerals, quartz, and feldspars in surface sediments from the Laptev Sea (from Stein & Korolev, 1994, based on Lapina, 1965; Silverberg, 1972; Wollenburg, 1993). Le, Lena; Ol, Olenek; Ko, Kotuy (Kathanga). (A–E) Heavy minerals; (a) amphibole, (b) epidote, (c) pyroxene, (d) ore minerals, and (e) heavy-mineral provinces (according to Lapina, 1965). In the grain-size fraction 0.05–0.10 mm, heavy minerals (density $>2.9\ g\,cm^{-3}$), and light minerals (density $<2.9\ g\,cm^{-3}$) were separated using bromoform. For heavy-mineral determinations, ~300–400 grains were counted; the results are expressed in percentage values of the total heavy-mineral fraction 50–100 μm. (F) Major clay-mineral assemblages (according to Silverberg, 1972; Wollenburg, 1993). (G) Distribution maps of quartz and (H) feldspars in surface sediments from the Laptev Sea (according to Lapina, 1965). Unfortunately, the raw data and sample locations of these distribution maps were not available. For data source and further details in methods of determination of heavy and light minerals, see Lapina (1965).

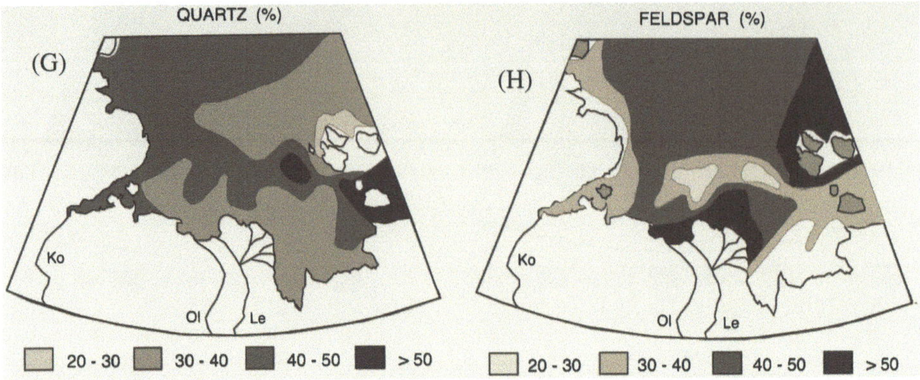

Figure 4.15 (*Continued*)

Khatanga material is more quartz-rich and contains higher amounts of smectite related to the Siberian flood basalts of the Putoran Massif (see Chapter 5.1 for more details).

In order to get more quantitative numbers of XRD mineral assemblages in Arctic Ocean sediments for identifying sources areas, a quantitative phase analysis of XRD data was introduced by C. Vogt (Vogt, 1997; Vogt et al., 2001). The evaluation of XRD data was carried out by using the software packet QUAX (Quantitative phase analysis with X-ray powder diffraction) originally developed for crystalline rocks drilled within the Continental Deep-Drilling Project KTB (Lauterjung, Will, & Hinze, 1985; Stroh, 1988; Lauterjung, 1994). Vogt (1997) optimized this approach for soft Arctic Ocean sediments and could show that in addition to the main minerals (quartz, feldspars, dolomite, calcite, etc.) also reliable data for clay and heavy minerals can be obtained. The easy preparation of powder specimen of samples (which allows to handle a large number of samples) and the objective determination of the bulk mineralogy by QUAX is a major advantage in comparison to the more time-consuming and highly specialized mineralogical investigations such as clay-mineral and heavy-mineral determinations. Details on background, analytical procedure, and limits of use of QUAX are described in Vogt (1997).

As an example, the bulk mineralogy of surface sediment samples from potential source areas of the Canadian Arctic/Northern Greenland as well as surface sediments from the Canada Basin are shown in Figure 4.16. The information to be obtained from this figure is very similar to that of Figure 4.11 based on the more time-consuming coarse-fraction analysis. One of the most obvious characteristics of the Canadian Arctic is the high abundance of carbonate minerals (calcite and dolomite). Carbonate minerals, on the other hand, are more or less absent in the Eurasian Arctic (Vogt, 1997; see also Chapter 5.1.3 for more detailed discussion).

4.3.3. Major, Minor, Trace, and REE Elements

While numerous studies have described modern clay and heavy-mineral distributions in Arctic sediments (see Chapter 5.1), much less is known of their

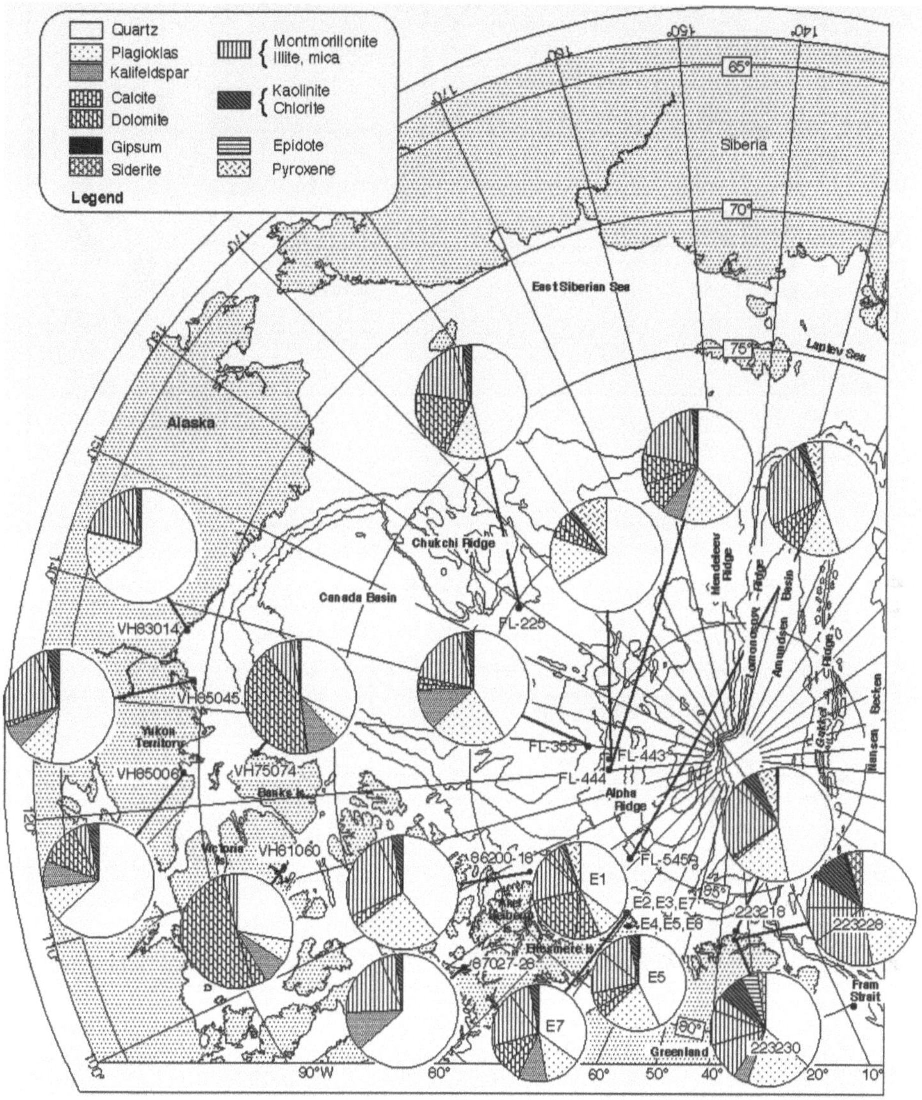

Figure 4.16 Bulk mineralogy data obtained by XRD analysis of sediments from potential Canadian source areas and surface sediments from the Canada Basin (from Vogt, 1997).

multi-element chemistry (e.g., Rachold, 1999; Schoster et al., 2000; Viscosi-Shirley et al., 2003a, 2003b; Schoster, 2005). Like the mineralogy also the chemical composition of the sediments strongly depends on the source-rock composition. Rachold (1999), for example, studied the chemical composition of suspended particulate matter (SPM) in the rivers Khatanga, Lena, and Yana, all draining into the Laptev Sea (Figure 4.17). In total, 31 different major, minor, and REE elements were determined. The most prominent differences were observed in the Khatanga

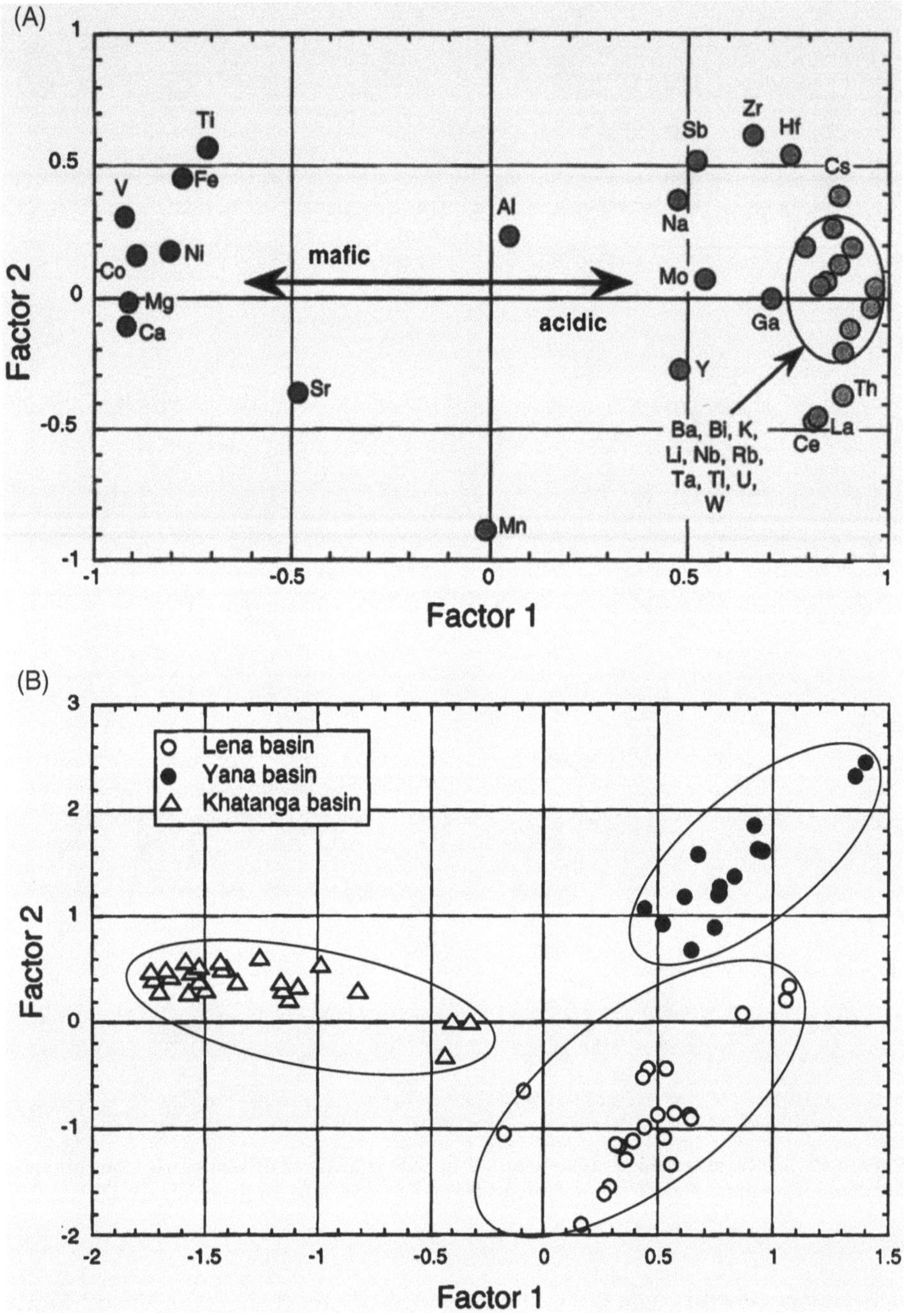

Figure 4.17 Statistical evaluation of chemical composition of suspended matter from the Laptev Sea and Khatanga, Lena, and Yana rivers (from Rachold, 1999, supplemented). (A) Factor loadings of principle component analysis (64 elements, 31 chemical elements). Factor 1 (x-axis, 59.6% of the variance) includes high positive factor loadings for Ba, Bi, Ce, Cs, Hf, K, La, Li, Mo, Na, Nb, Rb, Sb, Ta, Th, U, W, Y, and Zr, and high negative factor loadings for Ca, Co, Fe, Mg, Ni, Ti, and V. Factor 2 (y-axis, 11.8% of the variance) is characterized by high positive factor loadings for Cs, Hf, Sb, Ti, and Zr, and a negative Mn factor loading. (B) Factor scores of principle component analysis (64 elements, 31 chemical elements). Factor 1 (x-axis) clearly separates the Khatanga Basin, whereas factor 2 differentiates between Lena and Yana basins.

SPM, which is strongly influenced by basaltic rocks of the Putoran Massif (see Chapter 5.1.1, Figure 5.3). Thus, the SPM is enriched in elements associated with basalts, for example, Co, Cu, Mg, Ni, and Ti (Rachold, 1999). To consider all measured elements for characterizing the SPM, but to reduce the number of variables, a principle component (factor) analysis has been applied. The general variations in the chemical composition of the SPM can be explained by two factors accounting for 71.4% of the total variance. Factor 1 discriminates elements associated with basalts and consequently separates the Khatanga SPM from the Lena and Yana SPM. Factor 2 describes the Yana Basin and includes elements that are enriched in coarse-grained material and elements originating from granitic intrusions which formed ore-deposits in the Yana catchment (Figure 4.17; Rachold, 1999).

Schoster et al. (2000) also determined higher Ni/Al- and Ti/Al ratios in Kara-Sea surface sediments, related to sediment input from the basalts of the Putoran Massif. These authors used the Ni/Al ratio to trace the sediment input from the Kara-Laptev seas into the central Arctic (Figure 4.18).

4.3.4. Sr, Rb, and Nd Isotopes

Sr, Rb, and Nd isotopic ratios (i.e., $^{87}Sr/^{86}Sr$, $^{87}Rb/^{86}Sr$, and $^{143}Nd/^{144}Nd$) are suitable tracers for provenance studies of continental detritus (e.g., Goldstein, O'Nions, & Hamilton, 1984; Goldstein & Jacobsen, 1988; Grousset et al., 1988; Winter et al., 1997). Thus, they may be used as proxies to obtain information on sediment sources and transport processes of detrital marine sediments (e.g., Grousset et al., 1988; Revel et al., 1996; Hemming et al., 1998). In a series of studies, these tracers have also been used for identification of sediment provenance and transport-pathway reconstruction in the Arctic Ocean (Winter et al., 1997; Rachold et al., 1998; Eisenhauer et al., 1999; Tütken et al., 2002). When interpreting Sr- and Rb-isotope ratios in terms of provenance, however, the grain-size influence on these ratios must be considered, and only well-defined grain-size fractions should be compared (Figure 4.19). The ε_{Nd} values, on the other hand, seem not to be fractionated between mineralogically different silt- and clay-size fractions. Hence, the ε_{Nd} values of all three grain-size fractions (<2, 2–20, 20–63 μm) are almost identical in one sample within the error range (Figure 4.19). The difference in ε_{Nd} values between the two bulk (<63 μm) samples is three times larger than the analytical error (0.3 ε_{Nd} units) and, therefore, indicates a variation in sediment provenance between these two samples (Tütken et al., 2002). Chemical alteration which may also influence the isotope ratios, is negligible in the Arctic environment (Eisenhauer et al., 1999; Tütken et al., 2002).

In Figure 4.20, ε_{Nd} values and $^{87}Sr/^{86}Sr$ ratios of Arctic Ocean sediment samples and potential circum-Arctic source regions are shown (Tütken et al., 2002). From this figure, it is obvious that the Eurasian shelf sediments as natural mixed samples of the drained continental crust in the vast Siberian hinterland have a narrow range of $^{87}Sr/^{86}Sr$ ratios of 0.713–0.719 and ε_{Nd} values between −8.3 and −13.4. These numbers, for example, are quite different from those of other source terranes like Greenland or Svalbard. The flood basalts from the Putoran Massif, an important

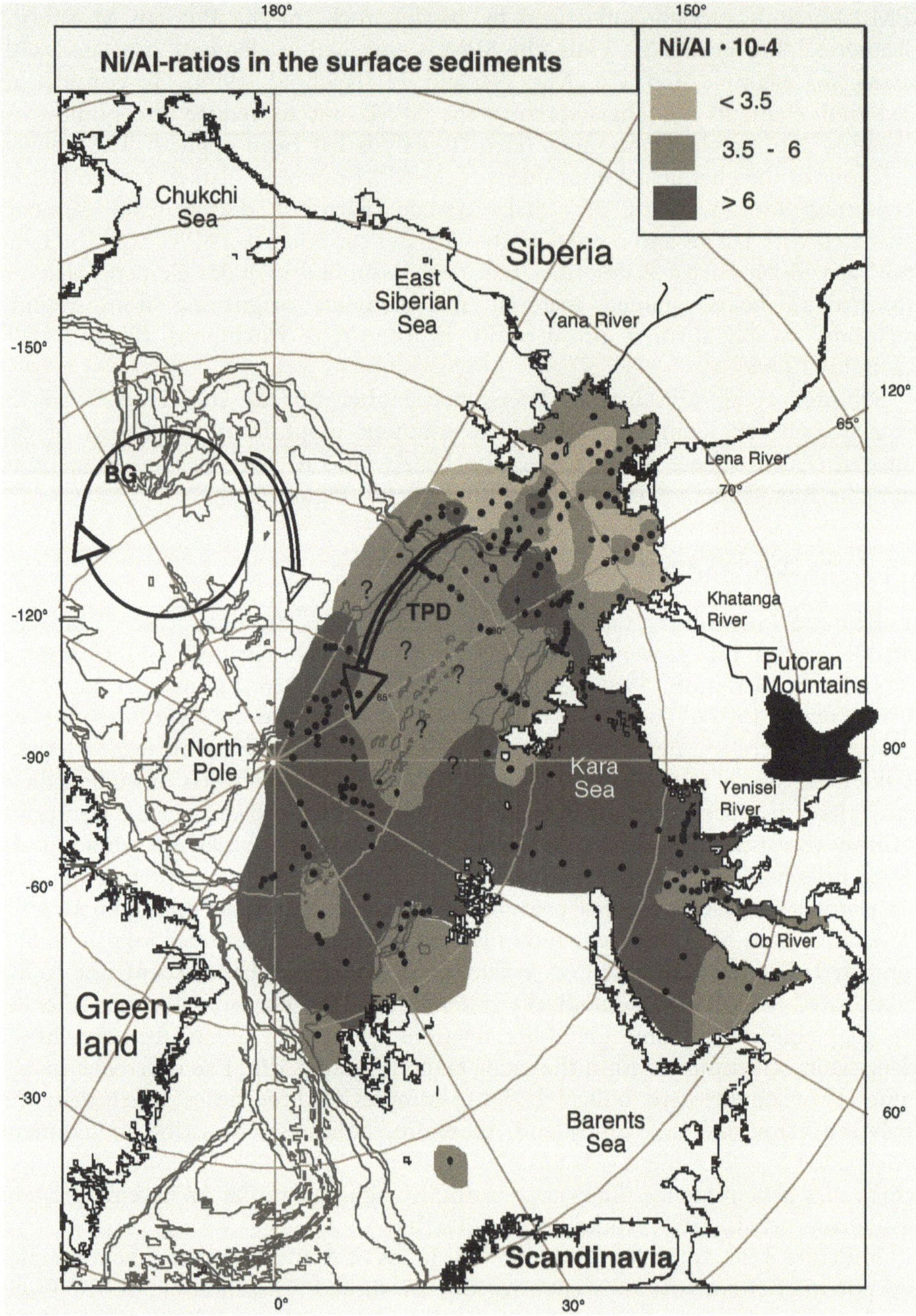

Figure 4.18 Distribution map of Ni/Al ratios in Arctic Ocean surface sediments (from Schoster et al., 2000). The occurrence of the flood basalts of the Putoran Massif in the hinterland is shown.

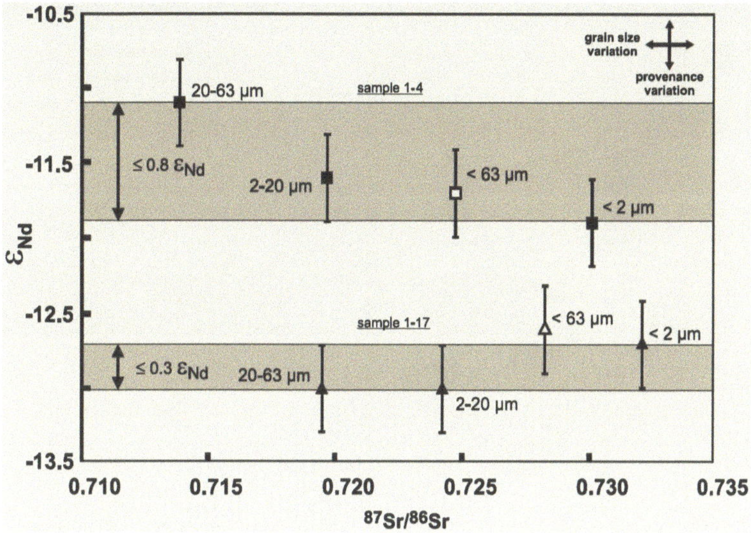

Figure 4.19 ε_{Nd} versus $^{87}Sr/^{86}Sr$ diagram of the clay, fine silt, coarse silt (filled symbols) and $<63\,\mu m$ (open symbols) grain-size fractions of two sediments from Core PS1533 (for core location, see Figure 3.28). Square, sample 1–4 ($\sim$15 ka, MIS 2), triangle, sample 1–17 ($\sim$71 ka, MIS 5). Grey-shaded bars are the within sample variability of the ε_{Nd} values between the grain-size fractions of each sample, indicating the possible grain-size effect (from Tütken et al., 2002).

source for sediments in the Kara Sea as well as western Laptev Sea (see Chapter 5.1.1), have a very distinct Sr and Nd isotope signature (Figure 4.20; $^{87}Sr/^{86}Sr$ of $\sim$0.705 and ε_{Nd} of $\sim$0; Sharma, Basu-Asish, & Nesterenko, 1992). Owing to the input of this basaltic material, there is a strong difference in the isotopic signature of the sediments from the eastern and western Laptev Sea. Eastern Laptev Sea sediments dominantly controlled by the Lena River SPM which has $^{87}Sr/^{86}Sr$ ratios of 0.716–0.717 (Rachold et al., 1998; Rachold, 1999), have a mean $^{87}Sr/^{86}Sr$ ratio of 0.717 and a mean ε_{Nd} value of -12.5 (Eisenhauer et al., 1999). Western Laptev Sea sediments containing more weathering products from the Siberian flood basalts, on the other hand, have lower $^{87}Sr/^{86}Sr$ ratios and higher ε_{Nd} values compared to the eastern Laptev Sea (mean $^{87}Sr/^{86}Sr$ ratio of 0.715 and a mean ε_{Nd} value of -8.7; Eisenhauer et al., 1999) (Figure 4.20).

The separation of Lena and Khatanga river-discharge SPM as well as eastern and western Laptev Sea surface sediments is also clearly reflected in a $^{87}Sr/^{86}Sr$ versus $^{87}Rb/^{86}Sr$ diagram (Figure 4.21; Eisenhauer et al., 1999). The surface sediments from the eastern Laptev Sea fall within an envelope provided by the Lena SPM ("LME" – Lena mixing envelope). The sediments from the western Laptev Sea fall between the LME and the FBE ("FBE" – Flood basalt envelope, defined by the Khatanga SPM), suggesting that these sediments are a mixture of Lena and Khatanga material. Mixing calculations based on $^{143}Nd/^{144}Nd$ ratios show that $\sim$75% of the western Laptev Sea sediments originate from the Lena drainage area whereas $\sim$25% of the sediments originated from the Siberian flood basalt province (Eisenhauer et al., 1999).

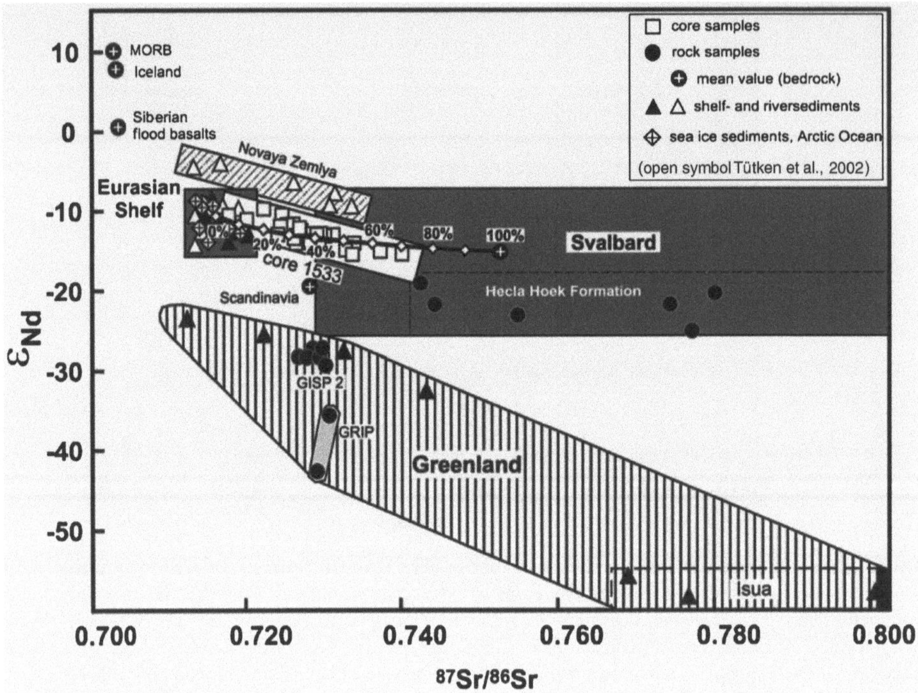

Figure 4.20 ε_{Nd} versus $^{87}Sr/^{86}Sr$ diagram summarizing the isotope compositions of the Core 1533 sediments in relation to whole rock, river, shelf, and sea-ice sediment samples covering the circum-Arctic region (from Tütken et al., 2002; for location of Core 1533 see Figure 3.24). Eurasian shelf sediment data are from Tütken et al. (2002) and Eisenhauer et al. (1999); Svalbard data from Johansson, Gee, Bjørklund, and Witt-Nilsson (1995), Johansson et al. (2000) and Johansson and Gee (1999); Greenland data from McCulloch and Wasserburg (1978), Goldstein and Jacobsen (1988), and Weis, Demaie, Souchez, Gow, and Meese (1997); Scandinavian data from Miller, O'Nions, Hamilton, and Welin (1986); Novaya Zemlya data from Tütken et al. (2002); MORB and Iceland data from Revel, Sinko and Grousset (1996) and references therein; and Siberian flood basalt data from Sharma et al. (1992).

Figure 4.21 (A) Sample locations of the Laptev Sea area. Circles mark surface sediments and rhombs mark ice-rafted sediment (IRS) samples. In addition, sampling locations of suspended particulate matter (SPM) in the Khatanga River (open squares) and Lena River (circled black dots) are shown. For Laptev Sea stations, filled black symbols mark sediments which are isotopically related to Lena derived SPM. Open symbols represent sediment samples which are isotopically a mixture of Lena SPM, Khatanga SPM and Kara Sea sediments. Samples marked with "x" represent sediments with the largest $^{87}Rb/^{86}Sr$ and $^{87}Sr/^{86}Sr$ ratios. Samples marked with "xx" indicate sediments with lowest $^{87}Rb/^{86}Sr$ and $^{87}Sr/^{86}Sr$ ratios. Arrows mark major water outflows from the Lena delta. Note that the most drastic changes of the Rb/Sr systematics occur close to the Lena mouth. (B) $^{87}Sr/^{86}Sr$ ratios as a function of their corresponding $^{87}Rb/^{86}Sr$ ratios. Most data fall within or between two arbitrarily defined envelopes following the slopes provided by the Lena and Khatanga SPM. LME, Lena Mixing Envelope; FBE, Flood Basalt Envelope. These envelopes or "pseudo-isochrons" represent provenance and average age of the corresponding river drainage areas. Samples marked with -x- correspond to the finest grain-size fraction and samples marked with -xx- correspond to the coarsest grain-size fraction (from Eisenhauer et al., 1999).

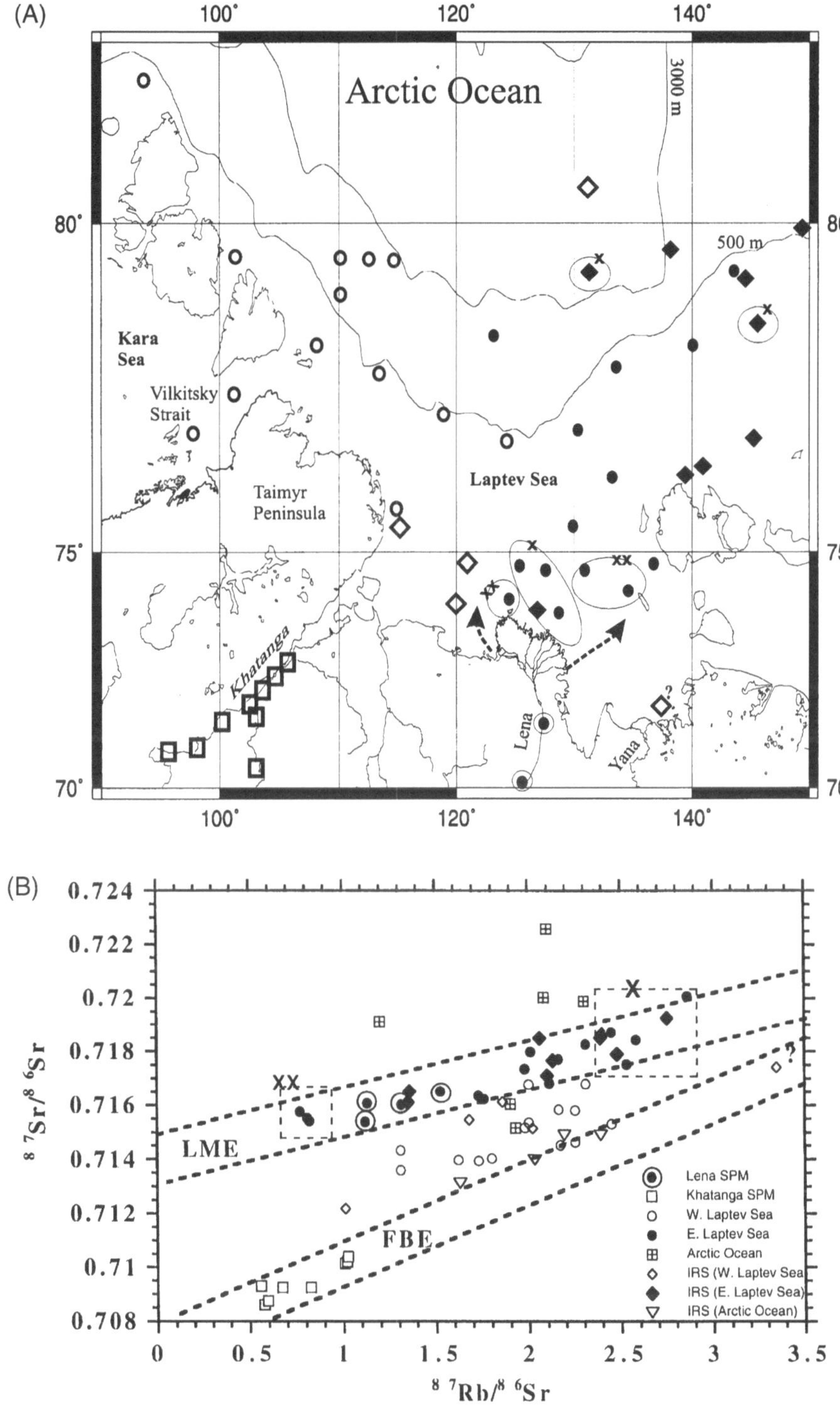
(A)
100°
120°
140°
Arctic Ocean
3000 m
80°
500 m
Kara Sea
Vilkitsky Strait
Taimyr Peninsula
Laptev Sea
75°
Khatanga
Lena
Yana
70°
(B)
0.724
0.722
0.72
0.718
0.716
0.714
0.712
0.71
0.708
$^{87}Sr/^{86}Sr$
LME
FBE
Lena SPM
Khatanga SPM
W. Laptev Sea
E. Laptev Sea
Arctic Ocean
IRS (W. Laptev Sea)
IRS (E. Laptev Sea)
IRS (Arctic Ocean)
0
0.5
1
1.5
2
2.5
3
3.5
$^{87}Rb/^{86}Sr$

The variation in Sr and Nd isotope ratios can be used to reconstruct the climate-related changes in sediment provenance and transport pathways (Tütken et al., 2002; Haley, Frank, Spielhagen, & Fietzke, 2008) in a similar way as other mineralogical proxies like clay and heavy minerals (see Chapter 6.2.2 for discussion).

4.3.5. Magnetic Susceptibility

Changes in magnetic susceptibility (MS) are normally controlled by variations in the content of ferrimagnetic matter (mostly magnetite, titanomagnetite or maghemite), because ferrimagnetic minerals have a significantly higher susceptibility ($k = +10^{-2}$) than most other common minerals (-10^{-6} to $+10^{-6}$) (Thompson & Oldfield, 1986). Volcanic rocks contain significantly higher amounts of ferrimagnetic minerals than other rock types. Consequently, the MS can be used to distinguish between different source areas of terrigenous sediments (e.g., Niessen & Weiel, 1996; Kleiber & Niessen, 2000; Dittmers et al., 2003; Stein et al., 2003a, 2004a). Because sediment cores may exhibit a strong gradient in porosity and because MS of water is nearly zero, in high-porosity intervals the MS signal carried by the minerals would be underestimated. Thus, a porosity correction of the measured MS values might be useful (see Niessen, Jarrard, & Bücker, 1998 for details). This enables estimates of amounts of MS carrier grains, independently of the porosity.

Physical properties (i.e., MS, density, and p-wave velocity) of whole sediment cores can be measured using a GEOTEK "Multi-Sensor-Core-Logger" (MSCL). The system provides data of density, P-wave velocity and MS (e.g., Schultheiss, Mienert, & Shipboard Scientific Party, 1987; Schultheiss & McPhail, 1989; Weaver & Schultheiss, 1990; Weber, Niessen, Kuhn, & Wiedicke, 1997; Rothwell, 2006 and references therein). These data may give high-resolution records of the variability of sediment composition which may be related to climate-controlled changes in sediment input, source, and transport processes. In addition to the use as sediment source proxy, MS (as well as the other physical properties) may provide a first-order tool for correlation (and dating) of sediment cores (e.g., Kleiber et al., 2000; Stein et al., 2001, 2003a).

As an example, MS data from southern Kara Sea surface sediments (Figure 4.22) and a sediment core (BP99-04/7) obtained directly off the Yenisei Estuary (Figure 4.23; for location see Figure 4.22) are presented in the following. The Kara Sea is characterized by its strong influence of river discharge, mainly related to the major rivers Ob and Yenisei (see Chapter 2.5). This huge water and suspended matter discharge by these rivers is clearly reflected in the siliciclastic as well as biogenic fractions of surface sediments (see also Chapters 5.1.2 and 4.5.4). Here, MS values allow to distinguish between Ob and Yenisei sediment input. A major source of the suspended matter of the Yenisei River are the widespread Triassic plateau basalts and tuff deposits of the Putoran Massif (see Chapter 5.1.1, Figure 5.3; Duzhikov & Strunin, 1992). About 70% of the modern annual Yenisei discharge is contributed by the main tributaries Nizhnyaya Tunguska and Podkamennaya Tunguska rivers, draining this area (Meade et al., 2000). The relationship between riverine input from basaltic source areas and MS values is clearly reflected in surface sediments from the Kara Sea (Figure 4.22). The Yenisei-influenced sediments have very high MS values, whereas the sediments supplied by the river Ob draining the

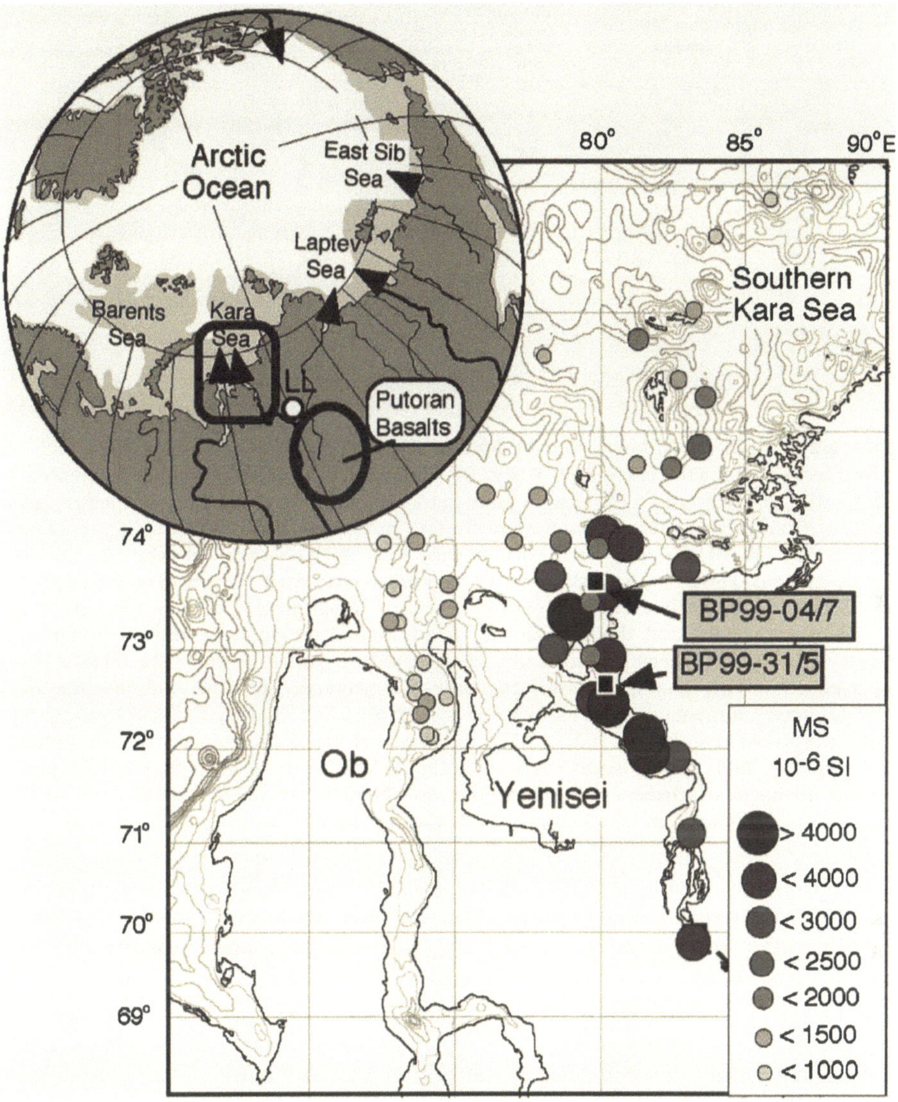

Figure 4.22 Porosity-corrected magnetic susceptibility values (10^{-6} SI) in surface sediments. Location of sediment cores BP99-04/7 and BP99-31/5 are indicated by black squares. The overview map of northern Eurasia and the Arctic Ocean indicate the major Eurasian rivers draining into the Arctic Ocean (black triangles), the study area, the occurrence of Triassic flood basalts of the Putoran Massif (Duzhikov & Strunin, 1992), and the location of the Lama Lake (LL) (from Stein et al., 2003a).

Siberian lowlands are characterized by very low MS values (Dittmers et al., 2003; Stein et al., 2003a, 2004a).

Based on the reliable correlation between measured Yenisei discharge values (period 1950–1995) and MS values determined in near-surface sediments (Core BP99-31/5; see Figure 4.22 for location) from the maximum accumulation rate area

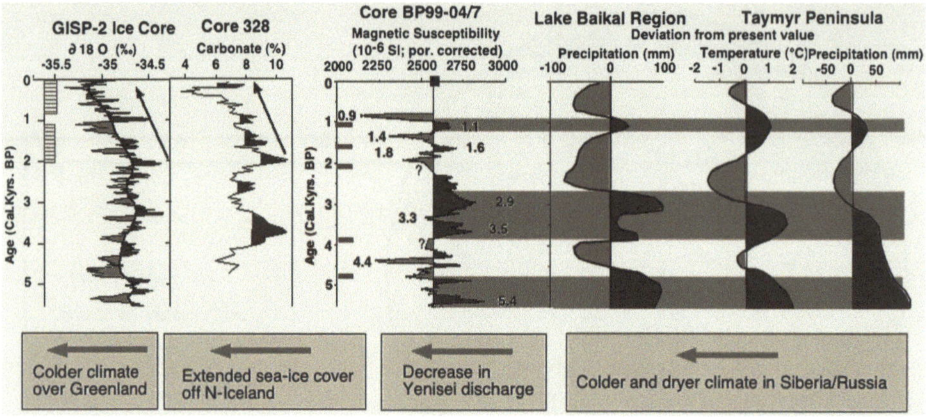

Figure 4.23 High-resolution record of porosity-corrected magnetic susceptibility (10^{-6} SI) of Core BP99-04/7 for the last ~5.5 Cal. kyr BP (based on Stein et al., 2004a, supplemented; for location of core, see Figure 4.22). Black square marks susceptibility value determined in surface sediments from this area. Numbers close to the MS minima and maxima are ages in calendar kiloyears BP. Question marks in the MS record indicate data gaps between the measured 1 m core sections. The BP99-04/7 MS record is compared with the δ^{18}O record of the GISP-2 ice core (Grootes et al., 1993); and the deviation of mean annual precipitation in the Lake Baikal area (Vorobyeva, 1994) and the deviation of mean annual temperature and precipitation on Taymyr Peninsula (Andreev & Klimanov, 2000), based on pollen spectra. The GISP-2 record is a 9-point moving average record. In addition, carbonate maxima and minima in an AMS^{14}C-dated sediment core off Northern Iceland (Core 328; 65°57.4′N, 21°33.17′W; water depth 96 m), interpreted as signal for reduced and increased sea-ice cover, respectively, are indicated (Andrews et al., 2001). Hatched grey vertical bars at the left-hand side of the figure mark periods of major advances of glaciers in western Norway (Nesje et al., 2001). Location of AMS^{14}C datings in the BP04/7 record is shown as black horizontal bars.

of the central Yenisei Estuary (Stein et al., 2003a), the variability of MS values in the sedimentary record of Core BP99-04/7 was used as a proxy for the Holocene variability in Yenisei River discharge and climatic conditions of the hinterland (Stein et al., 2003a, 2004a). The upper half of this MS record (i.e., the last ~5.5 Cal. kyr BP) is presented in Figure 4.23, showing a distinct short-term cyclicity. Maxima in MS are interpreted as periods of increased discharge of suspended matter from the Putoran Massif. Elevated temperature and precipitation values probably resulted in enhanced weathering and erosion and, thus, increased supply of MS-rich basaltic weathering products. In addition to the short-term centennial variability, a lower-frequency MS variability seems to be obvious. On this scale, maxima centred around 5.4 and 2.5–3.5 Cal. kyr BP. These maxima appear to correlate with the more general climatic curves of air temperature and precipitation on Taymyr Peninsula as well as in the Lake Baikal area (Figure 4.23; Vorobyeva, 1994; Andreev & Klimanov, 2000). The major decrease in MS values, that is, a decrease in river discharge, starting near 2.5 Cal. kyr BP and being more pronounced during the last ~2 Cal. kyr BP, may be related to a significant drop in air temperature and mean annual precipitation, as indicated in pollen records from Taymyr Peninsula (Andreev & Klimanov, 2000), Lake Baikal (Vorobyeva, 1994), and Lama Lake in the Putoran Plateau (Hahne & Melles, 1997; Hubberten, Andreev, Kuhnke, Melles, & Siegert, 2001).

The general decrease in river discharge and climate in Siberia near 2.5 Cal. kyr BP coincides with a cooling trend over Greenland indicated in the GISP-2 Ice Core (Grootes, Stuiver, White, Johnsen, & Jouzel, 1993), as well as an advance of glaciers in western Norway (Nesje, Matthews, Dahl, Berrisford, & Andersson, 2001) (Figure 4.23). At the same time, the sea-ice cover off Northern Iceland increased as suggested from decreasing carbonate contents used as proxy for primary production and sea-ice cover (Figure 4.23; Andrews, Helgadottir, Geirsdottir, & Jennings, 2001). This centennial to millennial variability of climate proxies may reflect natural cyclic climate variations which can be seen in context with the interannual and interdecadal environmental changes recorded in the High Northern Latitudes over the last decades, such as, for example, the NAO/AO pattern (Hurrell, 1995; Thompson & Wallace, 1998; Dickson et al., 2000; Peterson et al., 2002). Positive NAO/AO phases bringing warm and wet air to the Russian Arctic and causing increased surface temperatures, precipitation, and weathering, may have been the trigger for increased riverine input from the Putoran Massif throughout the Holocene.

4.4. Trace Elements Used for Palaeoenvironmental Reconstruction

Trace elements (metals) are proxies used in palaeoceanographic studies to reconstruct surface-water productivity and bottom-water characteristics (see references below) as well as in (bio)geochemical studies related to the flux and diagenesis of (labile) organic matter at the water/sediment interface and in near-surface sediments (e.g., Pedersen, Waters, & Macdonald, 1989; Thomson et al., 2001; Arctic Ocean: Gobeil, Macdonald, & Sundby, 1997; Gobeil, Sundby, Macdonald, & Smith, 2001; Aldahan et al., 2000). Both type of studies, however, are still very rare in Arctic Ocean research. A short summary of the application of redox-sensitive elements to OC sediment geochemistry with some information on background of approach and use in Arctic Ocean studies, is described in Stein and Macdonald (2004c).

In palaeoceanographic research, barium is an important proxy as indicator for palaeoproductivity, as shown in numerous studies from the world oceans (e.g., Dehairs, Chesselet, & Jedwab, 1980; Bishop, 1988; Dymond, Suess, & Lyle, 1992; Gingele & Dahmke, 1994; Wehausen & Brumsack, 1998). For the Arctic Ocean, however, data on productivity related biogenic barium are very limited and restricted to surface sediments (Nürnberg, 1996). In general, the very low barium concentrations and the high detrital input in Arctic Ocean sediments make an interpretation of the barium values in terms of productivity difficult, although some correlation to other productivity proxies such as marine biomarkers, OC content, and benthic foraminifers, seems to be obvious in restricted areas (see Nürnberg, 1996 for more details). Thus, in this chapter it will be concentrated on trace elements as proxy for water-mass characteristics, that is, especially the oxygenation of deep (bottom) waters.

General information on bottom-water oxygenation can be obtained from geochemical proxies such as the OC/sulfur and OC/iron/sulfur relationships (see Chapters 4.7.2 and 7.1.2, respectively). In this context, specific redox-sensitive trace elements to be determined by X-ray fluorescence (XRF) and inductively coupled plasma mass spectrometry (ICP-MS) analysis (see e.g., Brumsack, 1980; Schnetger, 1997; Nijenhuis, Brumsack, & de Lange, 1998; Lipinski, Warning, & Brumsack, 2003 for background and analytical approach), are important proxies for reconstruction of depositional environment and related palaeoceanographic conditions in more detail. The geochemical behaviour of these trace elements was a major focus of numerous studies dealing with the formation of modern and ancient OC-rich sediments deposited in upwellling environments as well as anoxic basins (e.g., Brumsack, 1980, 1986, 1989; Hatch & Leventhal, 1992; Calvert & Pedersen, 1992; Piper, 1994; Warning & Brumsack, 2000). In a number of studies, it could be demonstrated that the high enrichments of redox-sensitive and sulfide-forming elements such as Mo, V, U, Ag, Cd, and Zn in various OC-rich sediments (Figure 4.24) have been related to anoxic bottom-water conditions during deposition (e.g., Brumsack, 1980, 1986; Jacobs, Emerson, & Skei, 1985, 1987; Hatch & Leventhal, 1992; Piper, 1994; Nijenhuis et al., 1998; Crusius & Thomson, 2000). In recent sediments of euxinic basins like the Black Sea or the Framvaren Fjord, for example, Mo concentrations are significantly higher than in suboxic sediments like those from the upwelling-influenced Gulf of California or the Arabian Sea (Brumsack, 1986; Jacobs, Emerson, & Huested, 1987; Crusius, Calvert, Pedersen, & Sage, 1996). According to Piper (1994) the accumulation rate by diffusion from an oxic or suboxic water column probably is much lower than by precipitation or adsorption of the reduced species in an anoxic water column. Brumsack (1980) showed through budget calculations that seawater represents the most important source for trace metals (except for Zn) in Cenomanian/Turonian black shales.

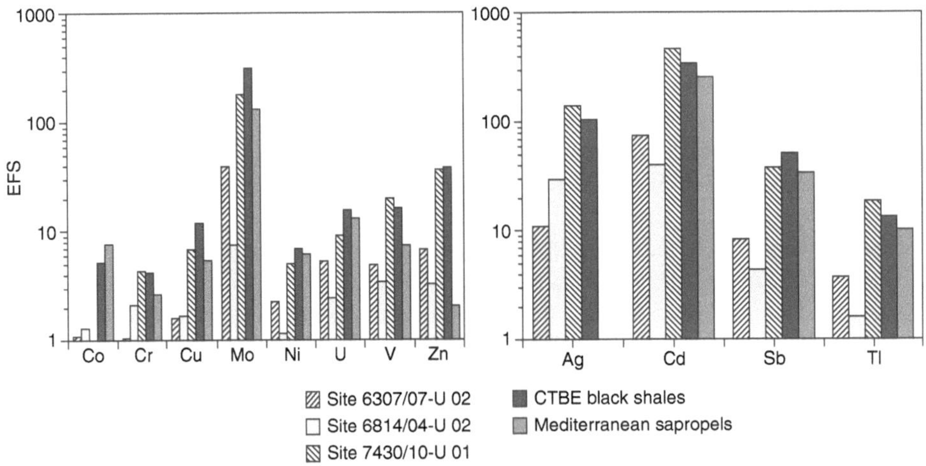

Figure 4.24 Comparison of mean trace-element enrichment factors relative to "average shale" (Wedepohl, 1971, 1991) for upper Jurassic/lower Cretaceous black shales of Cores 6307/07-U-02 and 6814/04-U-02 (Norwegian Sea) and Core 7430/10-U-01 (Barents Sea), Mediterranean sapropels, and Cenomanian/Turonian Boundary Event (CTBE) black shales (from Lipinski et al., 2003 and references therein). For location of cores, see Figure 7.5. Enrichment factor (element/ Al_{sample})/(element/$Al_{average\ shale}$).

As outlined in Lipinski et al. (2003), rhenium (Re) is regarded as one of the most promising palaeo-redox indicator elements as its concentration is extremely low in the continental crust ($0.5\,ng\,g^{-1}$; Crusius et al., 1996) and in oxic sediments ($<0.1\,ng\,g^{-1}$; Koide et al., 1986), but is significantly enriched in OC-rich sediments deposited under euxinic conditions such as the Black Sea ($43\,ng\,g^{-1}$; Table 4.2) (Koide et al., 1986; Colodner et al., 1993; Colodner, Edmond, & Boyle, 1995; Crusius et al., 1996). The Re/Mo ratio was used by Crusius et al. (1996) to distinguish between euxinic and suboxic conditions during sedimentation. According to these authors, low ratios close to the recent seawater value of 0.8×10^{-3} likely indicate euxinic conditions (such as typical for the Black Sea), whereas high ratios presumably indicate suboxic conditions owing to the preferential enrichment of Re over Mo in low-sulfide environments (such as typical for the Japan Sea and the Arabian Sea) (Table 4.2).

Based on these results, Lipinski et al. (2003) used trace elements for the characterization of the depositional environment of upper Jurassic/lower Cretaceous black shales from the Norwegian Sea (Cores 6307/07-U-02, Hitra Basin, and 6814/04-U-02, Lofoten Ridge) and the Barents Sea (Core 7430/10-U-01) (see Chapter 7.1.2, Figure 7.5 for location of cores). At all three locations, the redox-sensitive trace elements display a significant enrichment relative to "average shale" (Wedepohl, 1971, 1991), which are, except for Co, quite similar to the enrichments found in Mediterranean sapropels and, especially, Cenomanian/Turonian black shales (Figure 4.24). Trace metal enrichments are generally most pronounced at the Barents Sea and Hitra Basin sites, whereas trace elements are significantly lower at the Lofoten Ridge site. Thus, the high concentrations of trace elements in these OC-rich sediments from the Barents Sea and Hitra Basin sites most likely are related to water column anoxia, which is responsible for the effective removal of elements from seawater, whereas the shallower Lofoten Ridge location might have been situated in an oxygen minimum layer (Lipinski et al., 2003). In general, this

Table 4.2 Average Re and Mo Contents and Re/Mo Ratios of Norwegian Sea (Hitra Basin, Lofoten Ridge) and Barents Sea, upper Jurassic/lower Cretaceous, Black Shales, Cenomanian/Turonian Boundary Event (CTBE) Black Shales, Mediterranean Sapropels, and Recent Anoxic and Suboxic Sediments from the Black Sea, Japan Sea, and Pakistan Margin (from Lipinski et al. (2003, supplemented)).

Reference	a				b	c		
	Hitra Basin 6307/07	Lofoten Ridge 6814/04	Barents Sea 7430/10	CTBE Black shales	Mediterranean Sapropels	Black Sea	Japan Sea	Pakistan Margin
Re ($ng\,g^{-1}$)	83	20	200	210	330	43	12	31
Mo ($\mu g\,g^{-1}$)	48	8	133	145	76	29	1.3	1.7
Re/Mo ($\times 10^{-3}$)	1.9	4.1	2.6	5.3 (d)	2.9	1.5	9	19

Notes: Reference code (a) Lipinski et al. (2003); (b) Warning and Brumsack (2000); (c) Crusius et al. (1996); and (d) Brumsack (written comm. 2007).

interpretation is supported by the Re/Mo ratios. Samples from the Hitra Basin and Barents Sea have low average Re/Mo ratios of 1.9×10^{-3} and 2.6×10^{-3}, indicating more euxinic depositional conditions. In the black shales from the Lofoten site, on the other hand, the average Re/Mo ratio is $\sim 4.1 \times 10^{-3}$, which is closer to the ratio of determined in recent suboxic sediments from the Japan Sea (Table 4.2; Lipinski et al., 2003).

For a more detailed discussion of the trace element data from the Norwegian/Barents Sea black shales in relationship to other palaeoenvironmental proxies, see Chapter 7.1.2.

4.5. Micropalaeontological Proxies and Their (Palaeo-) Environmental and Stratigraphical Significance

Micropalaeontological proxies most commonly found and investigated in the Arctic Ocean sediments are foraminiferal tests. Other micropalaeontological groups often used for palaeoenvironmental reconstructions are calcareous nannofossils, ostracods, diatoms, radiolarians, and palynomorphs. In addition to their significance as palaeoenvironmental proxies, they are partly also important for correlating and dating sediment cores.

4.5.1. Foraminifers

4.5.1.1. Distribution and variability of planktonic foraminifers

Calcite shells of planktonic foraminifers, major components in pelagic sediments deposited above the carbonate compensation depth (CCD), are important tools for modern and ancient oceanographic reconstructions (e.g., Bé & Tolderlund, 1971; Vincent & Berger, 1981; Kennett & Srinivasan, 1983; Hemleben, Spindler, & Anderson, 1989). There is a general agreement that abundance and composition of planktonic foraminifers are influenced surface-water temperature, surface-water salinity, food supply, and predation. In the Arctic Ocean, diversity and distribution of planktonic foraminifers are largely controlled by the permanent sea-ice cover, the high freshwater discharge, and nutrient and food availability (e.g., Volkmann, 2000a, 2000b, and references therein).

Planktonic foraminifers in the Arctic Ocean are represented largely by one species, *Neogloboquadrina pachyderma* sin. (e.g., Carstens & Wefer, 1992; Kohfeld, Fairbanks, Smith, & Walsh, 1996; Carstens, Hebbeln, & Wefer, 1997; Volkmann, 2000a, 2000b). Towards the central Arctic Ocean as well as in the western Greenland Sea influenced by the cold East Greenland Current, abundances of *N. pachyderma* sin. increased to >99% (Figure 4.25; Pflaumann et al., 2003). In the eastern Fram Strait, some higher abundances of the sub-polar species *Turborotalita quinqueloba* and *N. pachyderma* dex. (up to 8% and 3–4%, respectively, in the 125–250 μm fraction) were found in the surface sediments, attributed to the advection of warm Atlantic water and seasonally ice-free conditions (Figure 4.26A; Volkmann, 2000a, 2000b).

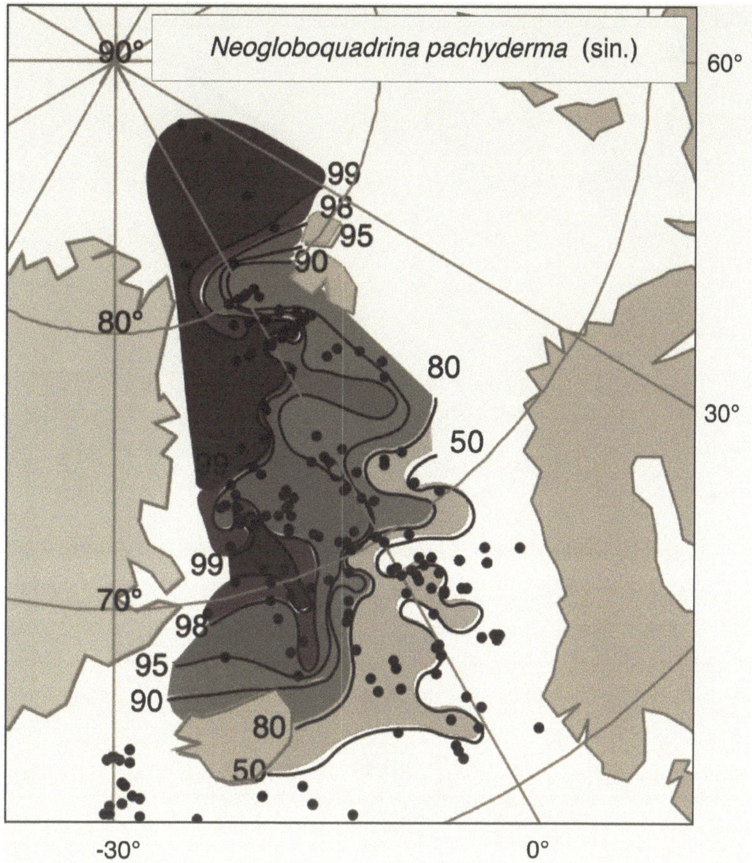

Figure 4.25 Spatial distribution of percentage of *Neogloboquadrina pachyderma* (sin.) in total planktonic foraminifera of modern pelagic sediments in the Nordic Seas (from Pflaumann et al., 2003, supplemented).

In surface sediments, the planktonic foraminiferal assemblages in the Arctic Ocean show significant enrichments of *N. pachyderma* in comparison to multinet tows, predominantly caused by selective dissolution of thin-shelled species such as *T. quinqueloba*. While significant to dominant amount of living *T. quinqueloba* were found in the eastern Arctic Ocean during various years (Carstens & Wefer, 1992; Carstens et al., 1997) and in the eastern Fram Strait and western Barents Sea, this species is rare to absent in the $>125\,\mu m$ fractions of the surface sediments due to carbonate dissolution (Figure 4.26B; Volkmann, 2000a).

Under a perennial ice cover, abundances of planktonic foraminifers are about a factor of 10 lower than in open water and ice-marginal regions (Carstens & Wefer, 1992; Volkmann, 2000a, 2000b). Moreover, the preferred habitat depth of Arctic polar and sub-polar planktonic foraminifers is $\sim$100–300 m in open water and ice-marginal settings (i.e., in the upper Atlantic-Water layer). In the central Arctic Ocean with its permanent sea-ice cover, planktonic foraminifers are forced, due to

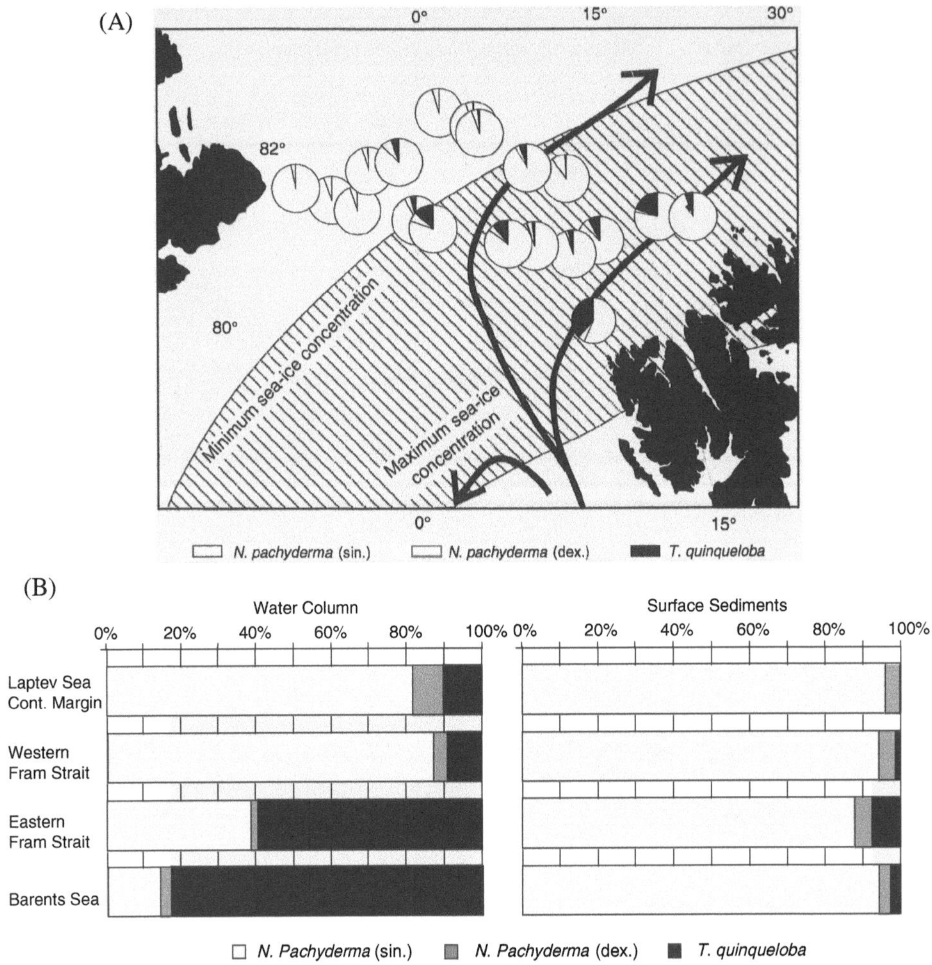

Figure 4.26 (A) Relative abundances of planktonic foraminifers in surface sediment samples from the Fram Strait and (B) Comparison between the relative abundances of planktonic foraminifers in the water column and in surface sediments for different Arctic Ocean areas (from Volkmann, 2000a, 2000b, supplemented). Sea-ice concentrations are according to Romanov (1995). Grey arrows indicate the Atlantic-water recirculation and inflow through the Fram Strait after Rudels et al. (2000).

limited food availability and reduced light penetration, to dwell in the low-salinity and cold polar water that characterize the upper 100 m (Carstens & Wefer, 1992; Carstens et al., 1997; Volkmann, 2000a; Nørgaard-Pedersen et al., 2003). Flux records of planktonic foraminifers may be used as a relative productivity proxy reflecting nutrient supply and light penetration coupled to the degree of ice cover (Hebbeln & Wefer, 1991).

The abundance of foraminifers (predominantly *N. pachyderma* sin.) shows distinct temporal fluctuations, with maxima during the interglacial periods and strongly decreasing values during glacial times, which is probably associated with drastic reduction in organic productivity and increased sea-ice cover (e.g., Darby et al., 1989; Darby, Bischof, & Jones, , 1997; Darby et al., 2006; Poore, Ishman, Phillips, & McNeil, 1994; Phillips & Grantz, 1997; Polyak et al., 2004; Spielhagen et al., 2004). This pattern is especially evident in the Western Arctic, where glacial foraminiferal numbers decrease to nearly zero (Figure 4.4; Polyak et al., 2004; Darby et al., 2006). In the Eurasian Basin, on the other hand, maxima in planktonic foraminifers may correlate with glacial intervals, interpreted as episodes of high productivity ("High Productivity Events, HPEs") presumably related to the advection of North Atlantic waters and adiabatic winds flowing off the Barents Sea Ice Sheet to create polynyas (Hebbeln, Dokken, Andersen, Hald, & Elverhøi, 1994; Dokken & Hald, 1996; Knies, Vogt, & Stein, 1999; Spielhagen et al., 2004 and references therein). It is assumed that the advection of warm Atlantic water to the Norwegian-Greenland Sea (NGS) triggered the growth of the Svalbard/Barents Sea Ice Sheet (SBIS) (Hebbeln et al., 1994). For a more detailed discussion it is referred to Chapter 6.2.2.

4.5.1.2. Planktonic foraminifers and transfer-function technique

Based on microfossil assemblages such as planktonic foraminifers, quantitative estimates of surface-water characteristics are possible. Here, the Climate Long-Range Investigation, Mapping and Prediction project (CLIMAP) was a pioneer approach presenting the first distribution maps of seasonal SST of the LGM North Atlantic and the global ocean at "18 ^{14}C ka" (McIntyre et al., 1976; CLIMAP, 1981). These reconstructions were based on census data of various microfossil assemblages from deep-sea sediment cores that were calibrated to the SST of modern surface water via a set of transfer functions on the basis of principal component analysis (CABFAC) and a multiple regression approach (Imbrie & Kipp, 1971; Kipp, 1976). About 20 years later, a new reconstruction of peak glacial SSTs in the Atlantic using census counts of planktonic foraminifers and the Maximum Similarity Technique Version 28 (SIMMAX-28) modern analog technique, was carried out ("Glacial Atlantic Ocean Mapping-GLAMAP"; Pflaumann et al., 2003; Sarnthein et al., 2003b). This new reconstruction not only used a different transfer technique, but is also based on a more precise age definition and age control of the LGM time slice and on improved calibration of the microfaunal census data to modern SST at 10 m water depth. The original data base of the SIMMAX modern analog technique (Pflaumann, Duprat, Pujol, & Labeyrie, 1996) was enlarged to 947 modern core-top samples, with new samples added especially in the northernmost North Atlantic, that is, the Fram Strait area. Here, the species richness of the planktonic foraminifera fauna is strongly reduced and nearly monospecific, which has necessitated an increase in the number of counted specimens to more than 1,000 per sample to differentiate species portions beyond 99.5% (Figure 4.25; Pflaumann et al., 2003; Sarnthein et al., 2003b). Modern winter and summer SST at 10 m water depth used for calibration were extracted from the World Climatological Atlas (Levitus & Boyer, 1994). For further details on background, methodology, and limitations of the SIMMAX approach, it is referred to Pflaumann et al. (1996, 2003).

(A)

(B)

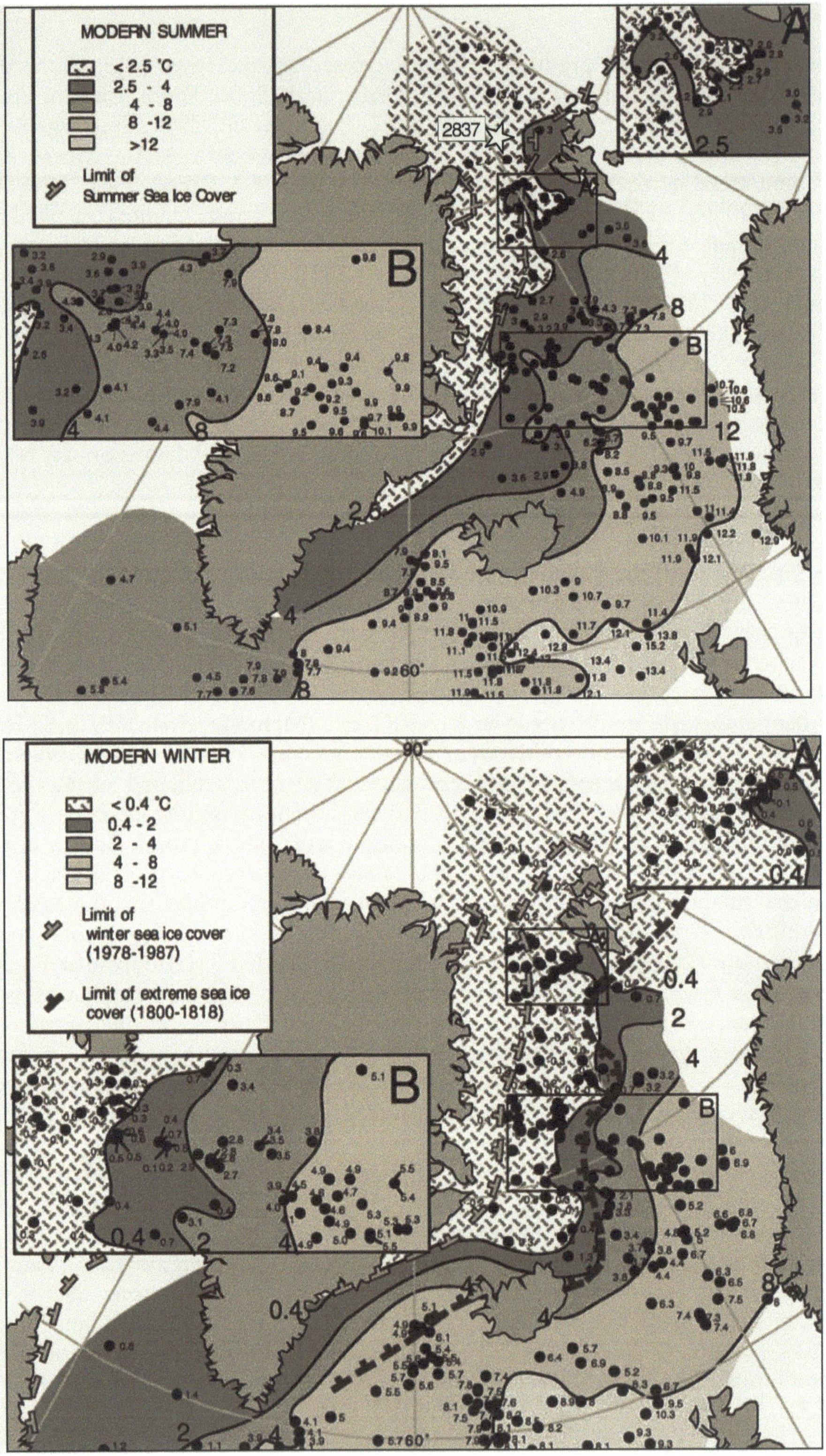

MODERN SUMMER
< 2.5 °C
2.5 - 4
4 - 8
8 -12
>12
Limit of
Summer Sea Ice Cover
2837
A
B
MODERN WINTER
< 0.4 °C
0.4 - 2
2 - 4
4 - 8
8 -12
Limit of
winter sea ice cover
(1978-1987)
Limit of extreme sea ice
cover (1800-1818)
90°
60°

The SIMMAX approach of SST reconstructions also allows estimates of the palaeo-sea-ice cover in the North Atlantic and its seasonal variability for the LGM. To reconstruct the maximum extent of sea ice, Sarnthein, Pflaumann, and Weinelt (2003a) developed a new proxy derived from a (SIMMAX-based) modern threshold SST calibrated to modern satellite imagery of $>50\%$ sea-ice coverage (Gloersen et al., 1992). These SST threshold estimates were defined for the two seasonal ice extremes as a conservative measure to derive the largest possible sea-ice cover per analogy to the sea ice observed during summers and winters of 1978–1987 (Gloersen et al., 1992), as well as to the extreme ice extent of the Little Ice Age (LIA) winters 1800–1818 (Kellogg, 1980). According to this conservative approach, today SST of $>2.5°C$ characterize ice-free conditions during summer while SST of $>0.4°C$ characterize ice-free conditions during winter (based on sea-ice distribution from 1978–1987) (Figure 4.27). When compared to sea-ice conditions of the LIA winters, the threshold winter SST estimates increase to 0.75°C (Figure 4.27B; Sarnthein et al., 2003a). This approach holds true for almost 100% of all sites seaward of the modern sea-ice margin. Many sites seaward of the sea-ice cover, however, also produce SST estimates that are lower than the chosen threshold values. Thus, this empirical approach leads to conservative, fairly robust estimates of the maximum extension of sea ice (or conversely, of the minimum extent of ice-free regions) during past summer and winter times (Sarnthein et al., 2003a). That means, the actual LGM sea-ice cover (with an ice density of $>50\%$) was probably more reduced than the reconstructions shown in Figure 4.28.

In general, the new reconstruction by the GLAMAP group (Pflaumann et al., 2003; Sarnthein et al., 2003a, 2003b) based on a set of more than 60 well-dated sediment cores, shows that the LGM North Atlantic was characterized by an extreme seasonal variability of climate (Figure 4.28). During glacial summers, sea ice retreated up to the Arctic Ocean and the western Fram Strait. This is in contrast to the CLIMAP Project Members (1981) reconstruction that postulated a complete perennial sea-ice cover. Thus, according to the GLAMAP reconstruction most of the Nordic Seas remained ice-free, which may have resulted in small-scale, but more or less continual deep-water formation during the shorter glacial summers (e.g., Voelker, 1999; Sarnthein et al., 2001). Furthermore, ice-free conditions in the Fram Strait area would have also led to seasonally enhanced evaporation, and thus contribute to the continued glacial build-up of large icesheets in Scandinavia and, in particular, on the Barents Shelf (e.g., Hebbeln et al., 1994; Vorren & Laberg, 1996).

Figure 4.27 Modern limits (bold broken lines) of $>50\%$ sea-ice concentration (satellite radiometric observations of Gloersen et al. (1992)) in the northern North Atlantic during (A) summer (July–September) as compared to SIMMAX-based estimates of modern summer SST at 10 m water depth, and (B) winter (January–March) and as reconstructed for the Little Ice Age (Kellogg, 1980) in comparison to SIMMAX-based estimates of modern winter SST at 10 m water depth (from Pflaumann et al., 2003; Sarnthein et al., 2003a, 2003b, supplemented). Large numbers label isotherms (°C). Open star indicates location of Polarstern Core PS2837.

(A)

(B)

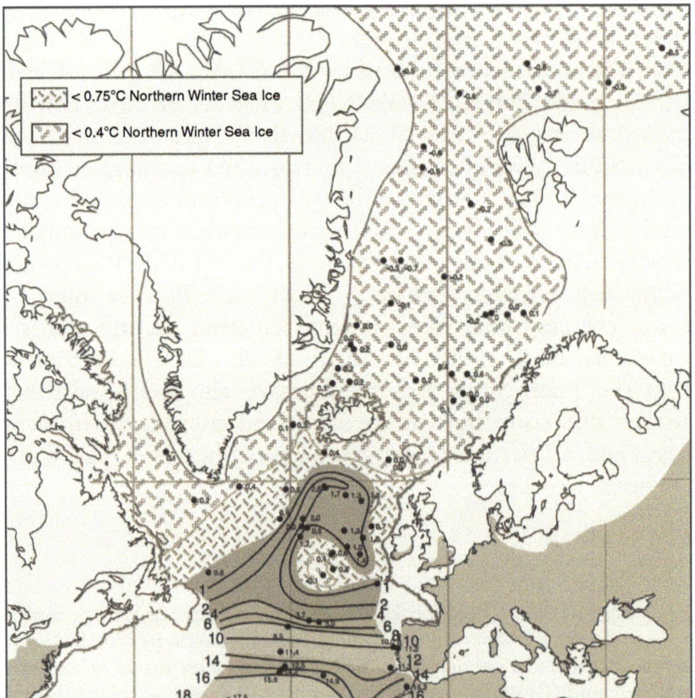

Figure 4.28 GLAMAP 2000 SST reconstruction of the glacial North Atlantic and Fram Strait area. (A) Northern summer, (B) northern winter (from Sarnthein et al., 2003a, 2003b; Pflaumann et al., 2003, supplemented). Arrows indicate major current directions (solid for cold, open/white for warm). The extension of the LGM icesheets is shown. Open star indicates location of Polarstern Core PS2837.

4.5.1.3. Distribution and variability of benthic foraminifers

In contrast to the planktonic foraminifers, the benthic foraminiferal fauna is characterized by a high species richness. Although the controls on the habitats of benthic foraminifers in the Arctic Ocean are not completely understood, distinct patterns in the modern distribution of foraminiferal assemblages are obvious, reflecting a bathymetric zonation, changes in water masses, surface-water productivity, and/or sea-ice conditions (Vilks, 1969, 1989; Lagoe, 1977; Hald, Danielsen, & Lorentzen, 1989; Scott & Vilks, 1991; Hald & Steinsund, 1992; Ishman & Foley, 1996; Polyak & Mikhailov, 1996; Wollenburg & Mackensen, 1998; Wollenburg & Kuhnt, 2000; Polyak et al., 2002b; Wollenburg, Kuhnt, & Mackensen, 2001; Wollenburg, Knies, Mackensen, 2004).

Based on a detailed study of faunal composition, standing stock and diversity of living benthic foraminifers from the central Arctic Ocean, Wollenburg and Mackensen (1998) came to the conclusion that in general the distribution of foraminiferal associations in the Arctic Ocean is not controlled by different water masses as previously suggested (e.g., Scott, Mudie, Baki, MacKinnon, & Cole, 1989; Scott & Vilks, 1991; Bergsten, 1994). Instead they proposed that mainly biological factors control the distribution of certain species and associations. That means, in the Arctic Ocean as one of the least productive areas in the world, standing stock, diversity, and the distribution of faunal associations are predominantly controlled by the availability of food and competition. The abundance of living benthic foraminifers (standing stock) and estimated marine-OC flux are positively correlated. Highest standing stock values correspond to seasonally ice-free more-productive areas (such as the area around Svalbard influenced by warm Atlantic water) and the adjacent continental slope and basin reflecting down-slope organic matter transport (Figure 4.29). The number of living foraminifers under the permanent ice-cover is one order of magnitude lower than in seasonally ice-free areas of the same water depth (Wollenburg & Mackensen, 1998). Thus, other factors such as bottom-current activities, water mass, and sediment type seem to be of minor importance, acting only as modifying agents in the seasonally ice-free areas.

According to Wollenburg and Mackensen (1998), most of the living associations are represented by an almost similar dead association. Owing to the extensive carbonate dissolution in arctic and subarctic shelf and slope areas (e.g., Steinsund & Hald, 1994) predominantly calcareous living associations are often dominated by agglutinated species in the dead associations, whereas the opposite is found in deep-sea environments where loosely agglutinated species dominate the living associations. Because of the low fossilization potential of these species, the corresponding dead associations are often characterized by calcareous taxa (Wollenburg, 1995; Wollenburg & Mackensen, 1998).

In the modern Arctic Ocean the abundance of "Atlantic species" (*Epistominella pusilla, Pullenia bulloides, Discorbinella bertheloti*) is highest close to Fram Strait, and decreases eastwards (Figure 4.30; Wollenburg et al., 2001). These species are relatively abundant in sediments influenced by the WSC, but are not present where this water mass has lost much of its heat and salt because of mixing, and sinks below a 200-m-thick low-salinity surface layer on its way to the Siberian

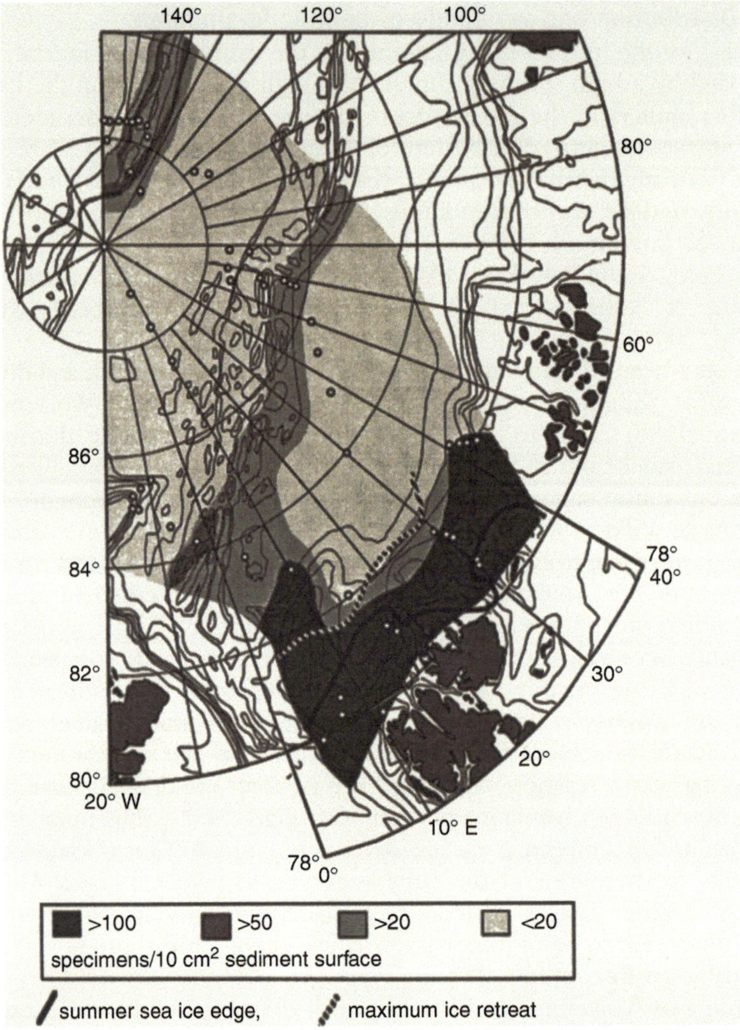

Figure 4.29 Standing stock $>125\,\mu m$ (number of Rose Bengal stained specimens from the sediment surface down to 15 cm sub-bottom depth) (from Wollenburg & Mackensen, 1998).

shelves. *Cassidulina teretis* with its circum-Arctic distribution in water depths between 200 and 1,400 m and coincidence with the distribution of the warmest water masses, is also used as an indicator for the presence of Atlantic water (e.g., Scott & Vilks, 1991; Bergsten, 1994; Polyak & Solheim, 1994; Hald et al., 1999; Lubinski, Polyak, & Forman, 2001). However, it is not clear which parametres restrict the habitat of "Atlantic species" to areas under the influence of the WSC (i.e., temperature, salinity, availability of food, etc.; see earlier) (Wollenburg et al., 2001, 2004).

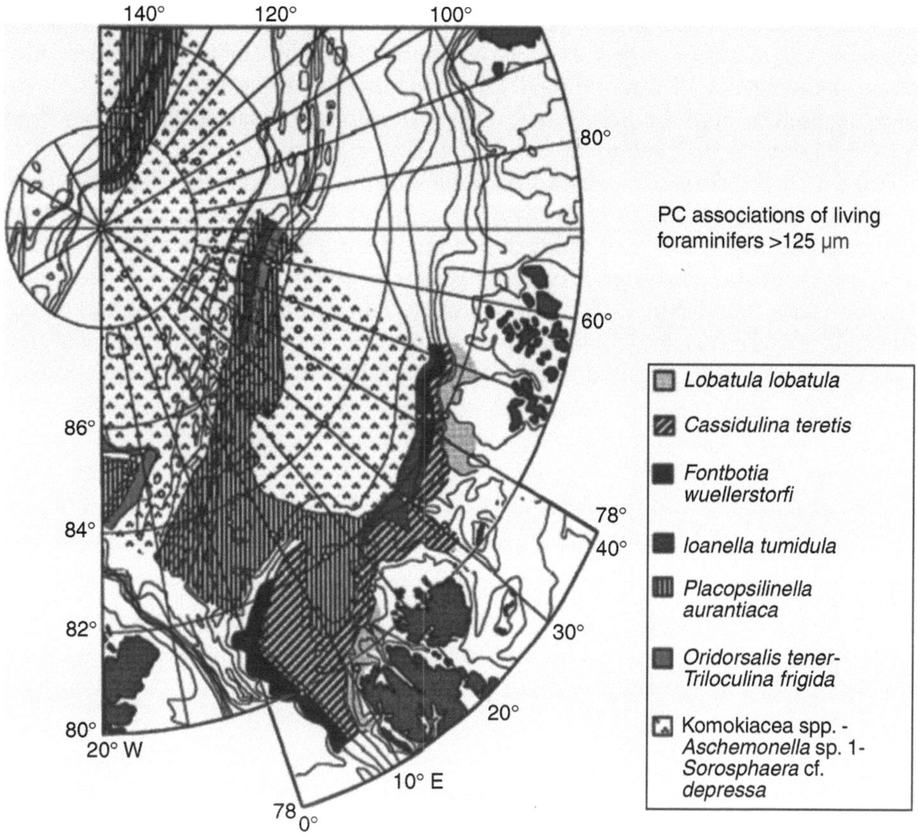

Figure 4.30 Distribution of living benthic foraminiferal associations, highest PC (Principal component) values of the >125 µm fraction (from Wollenburg & Mackensen, 1998).

4.5.1.4. Benthic foraminifers and transfer-function technique: Estimates of productivity

Loubere (1994) developed a transfer function to estimate palaeoproductivity from relative abundance data of benthic foraminiferal species (e.g., Loubere & Fariduddin, 1999; Loubere, 2000), which then was applied to the Arctic Ocean by Wollenburg and Kuhnt (2000) and Wollenburg et al. (2001), although adapted to the high seasonality of productivity in High Northern Latitudes. Owing to the high seasonality of productivity and variable water depths of samples these analyses differ in the statistical method used, correspondence instead of factor analysis, and by using OC fluxes instead of primary production values for comparison, from those of Loubere and colleagues (Wollenburg & Kuhnt, 2000; Wollenburg et al., 2001). Correspondence analysis reveals only one meaningful factor (CAF1), which explains 41% of the total variance of data. CAF1 correlates very well with the OC flux calculated from published Arctic primary production data using the equation of Betzer et al. (1984) (Figure 4.31A; Wollenburg & Kuhnt, 2000). This correlation between modern dead foraminiferal CAF1 values and the estimated OC flux can then be used as transfer function for the calculation of palaeo-OC fluxes from the CAF1 record in a sediment core. In a next step, palaeoproductivities can be calculated from estimated palaeo-OC fluxes using the equation of Betzer et al. (1984). When interpreting the absolute values of calculated productivity, one should have in mind that these may differ depending on the equations used for OC flux estimates from surface-water productivity (Suess, 1980; Betzer et al., 1984; Pace, Knauer, Karl, & Martin, 1987; or Schlüter, Sauter, Schäfer, & Ritzrau, 2000), although the trends in palaeoproductivity values are very similar for all the different equations. For a more detailed discussion of background and limits of the approach

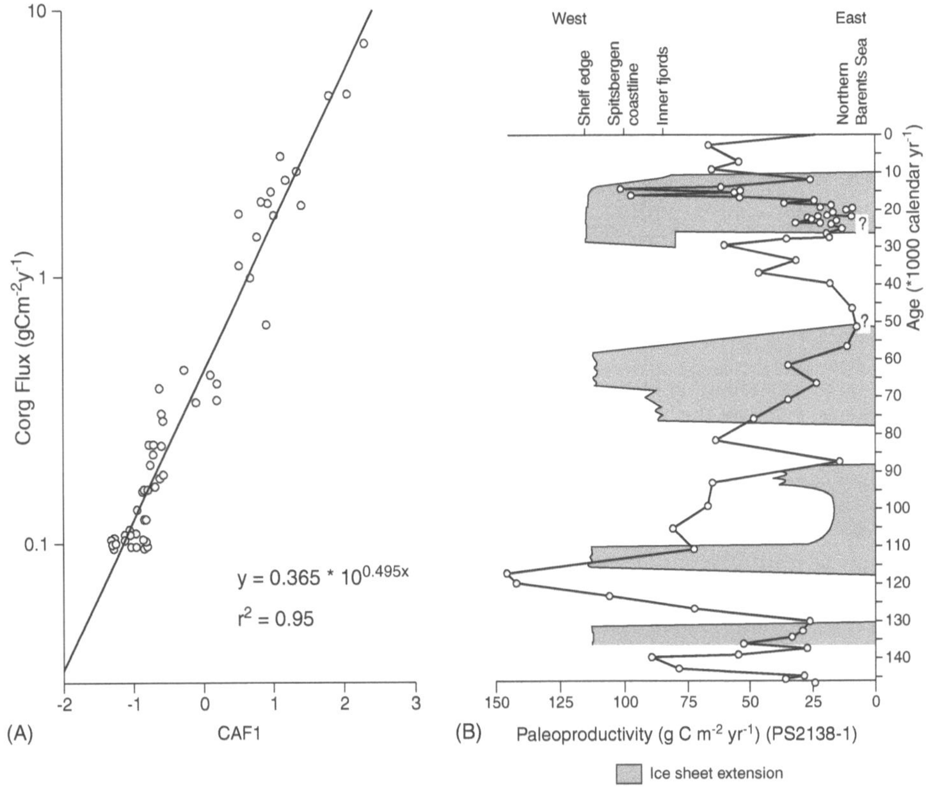

Figure 4.31 (A) CAF1 sample values of modern dead foraminifera obtained from correspondence analysis, in relation to carbon flux (from Wollenburg & Kuhnt, 2000), and (B) a 150 ka palaeoproductivity record from Core PS2138-1 (for location see Figure 6.54), based on benthic foraminifers and compared to ice-sheet extensions (Wollenburg et al., 2001). Ice-sheet extension according to Mangerud et al. (1998). For background and approach of estimating palaeoproductivity using benthis foraminifers see Wollenburg and Kuhnt (2000) and Wollenburg et al. (2001, 2004).

of palaeoproductivity estimates based on benthic foraminifers, it is referred to Wollenburg et al. (2001, 2004).

Using this approach, Wollenburg et al. (2001, 2004) determined the palaeoproductivity and its change during the last ~150 ka in sediment cores from the northern Barents Sea continental margin (Figure 4.31B, Core PS2138). From the record of this Core PS2138 it is obvious that calculated palaeoproductivity values generally correlate with the growth and decay of the Svalbard Ice Sheet (Elverhøi et al., 1995a; Knies et al., 1999). Highest palaeoproductivity occurred in interglacial periods and at the termination of interstadials to stadials within MIS substages 6.3, 5.5, 5.3, 5.1, 3.2, and Termination Ia. These productivity maxima are related to distinct ice retreat and partly coincide with the increased inflow of temperate saline Atlantic water into the Arctic Ocean (Wollenburg et al., 2001). During glacial periods, on the other hand, productivity was significantly reduced (Figure 4.31B), indicating increasing sea-ice cover.

4.5.1.5. Foraminifers and stratigraphy

Planktonic and benthic foraminifers are not only palaeoenvironmental proxies, but are also of major importance for correlating and dating sediment cores. In late Quaternary sections from the high latitudes (e.g., Nordic Seas, Fram Strait, Barents Sea, etc.), for example, the benthic foraminifer *Pullenia bulloides* is present in few stratigraphic intervals, but distinct abundance maxima occur only in upper MIS 1 and above MIS 5.5 (Haake & Pflaumann, 1989; Struck, 1997; Wollenburg et al., 2001).

Despite all the problems encountered in Arctic Ocean biostratigraphy, compositional and abundance variations often show coherent patterns over wide distances. As outlined in Backman, Jakobsson, Løvlie, Polyak, and Febo (2004), for example, Ishman, Polyak, and Poore (1996) proposed that systematic variations among benthic foraminiferal assemblages in intermediate to deep (>1,000 m) environments may be more useful for core correlations than the lithostratigraphic zonation of Clark et al. (1980) (see Chapter 6.3), simply because the variability among distinctive faunal assemblages yields more easily recognized signals in comparison to the lithostratigraphic variability in sediment texture (i.e., grain size).

Based on benthic foraminifer data, cores from the Northwind Ridge (NWR 5) and the Lomonosov Ridge (96/12-1pc), located some 1,600 km apart and both taken from comparable water depths (1,089 and 1,003 m) to eliminate a bathymetric bias in comparing benthic foraminifers, may be correlated (Figure 4.32; Backman et al., 2004). As outlined by Backman et al. (2004), the proposed correlation is based on maxima or the presence of a few rare foraminiferal species at certain stratigraphic levels: The zone of abundance of *Bulimina aculeata* is confined to foraminiferal maxima F2-F3 in Core 96/12-1pc (Jakobsson et al., 2001) and M3-L1 in Core NWR 5 (Poore et al., 1994). This peak zone has been observed also at the corresponding stratigraphic position in other cores from the Northwind and Mendeleev Ridges (Polyak, 1986; Ishman et al., 1996). Similarly, *Oridorsalis tener* occurs in significant amounts only in F3 and L1 and upcore, whereas *Pullenia* spp. is present only in F7 and K2 (Figure 4.32). Furthermore, the boundary between predominantly calcareous and almost exclusively arenaceous faunas that occurs

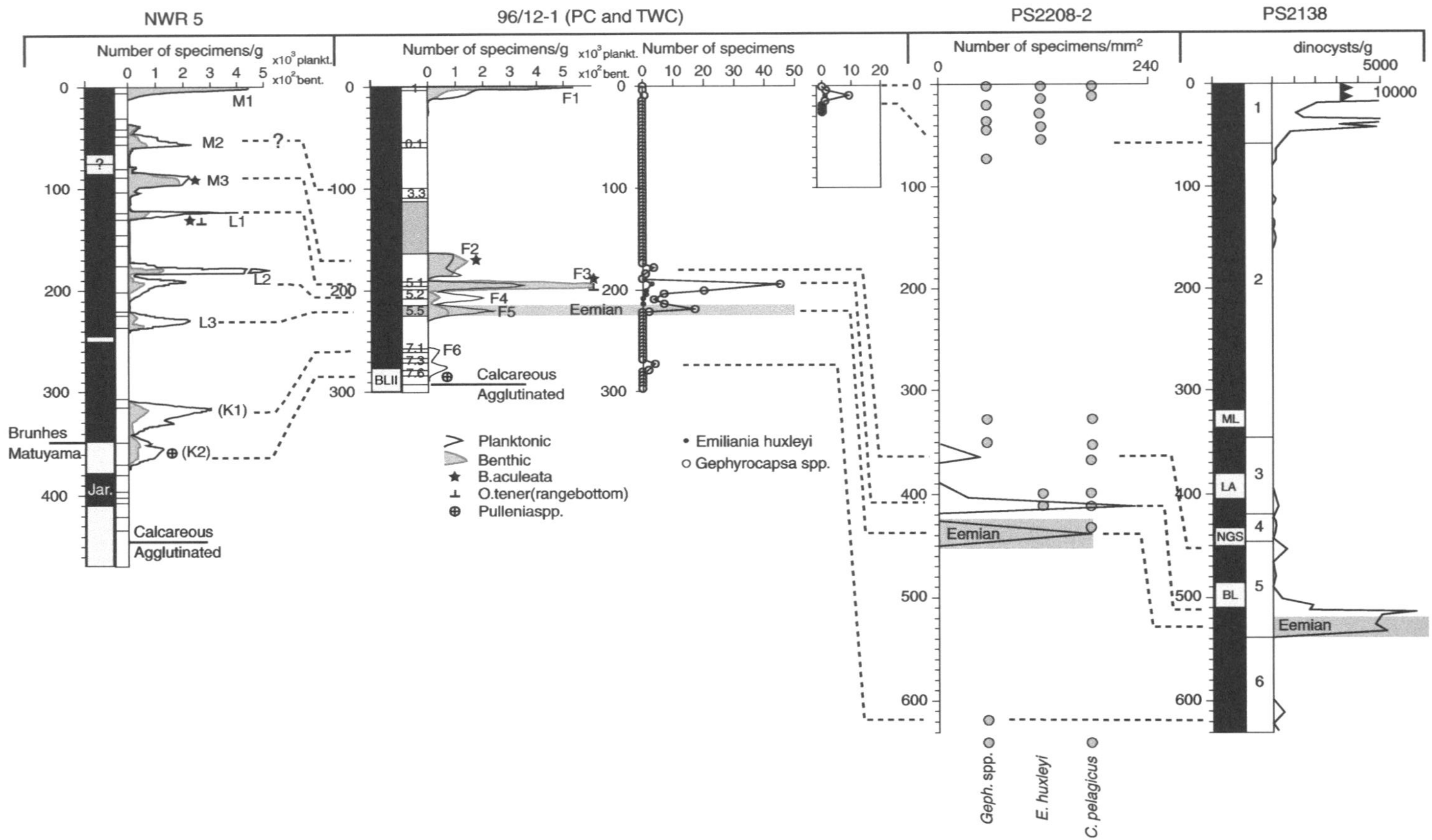

Figure 4.32 Correlation of foraminiferal stratigraphies in cores 96/12-1 (Jakobsson et al., 2001; Backman et al., 2004) and NWR 5 (= PI88-P5) (Poore et al., 1993, 1994), and nannofossil data from Core PS2208-2 (Gard, 1993) and dinocyst data from Core PS2138-1 (Matthiessen et al., 2001) (from Backman et al., 2004). Interglacial brown-mud units are shaded; criss-cross pattern in NWR 5 column shows detrital-carbonate layers. Curves show numbers of planktonic ($*10^3$) and benthic ($*10^2$, shaded) $>150\,\mu m$ foraminifers per gram sediment. Letters in foraminiferal-peak indices in NWR 5 correspond to Clark's et al. (1980) lithologic units (see also Poore et al., 1994). Indices within unit K (in parentheses) are added. Selected biostratigraphic markers are shown (see text). The position of the Brunhes/Matuyama boundary in NWR 5 reflects the interpretation of Poore et al. (1993, 1994). For location of cores, see Figure 6.54.

at 287 cm in Core 96/12-1pc and at 445 cm in Core NWR 5, seems to be another useful stratigraphic marker (Figure 4.32; Backman et al., 2004; see also Cronin, Smith, Eynaud, O'Regan, & King (2008) for the ACEX record).

When comparing the correlation based on foraminifer data with the palaeomagnetic record, it is obvious that a prominent polarity change occurs between peaks F6 and F7 in Core 96/12-1pc and between the corresponding peaks (K1-K2) in Core NWR 5 (Figure 4.32; Backman et al., 2004). The agreement between correlations based on independent bio- and magnetostratigraphic records supports on one hand the proposed correlation. On the other hand, this may last doubts on the absolute age model based on magnetostratigraphy, because the change in polarity has been interpreted as Brunhes–Matuyama boundary in Core NWR 5 (Poore, Phillips, & Rieck, 1993) and as prominent excursion within the Brunhes Chron in Core 96/12-1pc (Figure 4.32; Jakobsson et al., 2000a; Backman et al., 2004; see Chapters 6.1 and 6.3 for more detailed discussion).

As suggested by Jakobsson et al. (2001), it seems to be possible to link the fluctuations in foraminifers on Lomonosov Ridge to nannofossil records in the Nansen Basin (Gard, 1993) and dinocycst records from the Barents Sea continental margin (Matthiessen, Knies, Nowaczyk, & Stein, 2001) (Figure 4.34). By using such biostratigraphic signals caused by glacial/interglacial climatic variability, it is possible to correlate cores from the Eurasian margin via the Nansen Basin to the Lomonosov Ridge, and further to the Amerasian margin, a distance exceeding 2,600 km (Backman et al., 2004).

4.5.2. Calcareous Nannofossils

Calcareous nannofossils is an useful proxy for palaeoenvironmental reconstruction (e.g., Gard, 1988, 1993; Gard & Backman, 1990; Baumann, 1990; Baumann, Andruleit, & Samtleben, 2000; Jakobsson et al., 2000a). The presence of nannofossils that are in places abundant, indicates that open-water conditions must have existed in the Arctic Ocean, because nannofossils are derived from coccolithophorids which are photosynthetic algae that cannot live under permanent sea-ice cover (e.g., Gard, 1993). The presence of nannofossils, however, does not imply that sea ice was absent in the central Arctic Ocean but that large leads of open water were present between scatter ice floes. The absence of nannofossils may indicate a more closed sea-ice cover and/or carbonate dissolution.

Furthermore, nannofossils are important as stratigraphic marker. Based on studies of sediment cores from the Norwegian-Greenland Sea, Gard (1988) and Gard and Backman (1990) established a calcareous nannofossil biozonation for the High Northern Latitudes, consisting of 11 zones for the last ~500 ka. Absolute ages of the zones are based on correlation with oxygen isotope records and ages of worldwide nannofossil datum events. The data show a strong variability in nannofossil abundance and composition, related to glacial-interglacial climate cycles (Figure 4.33A). Extending this study to sediment cores from the central Arctic Ocean, Gard (1993) showed that most of the sediment sections were barren

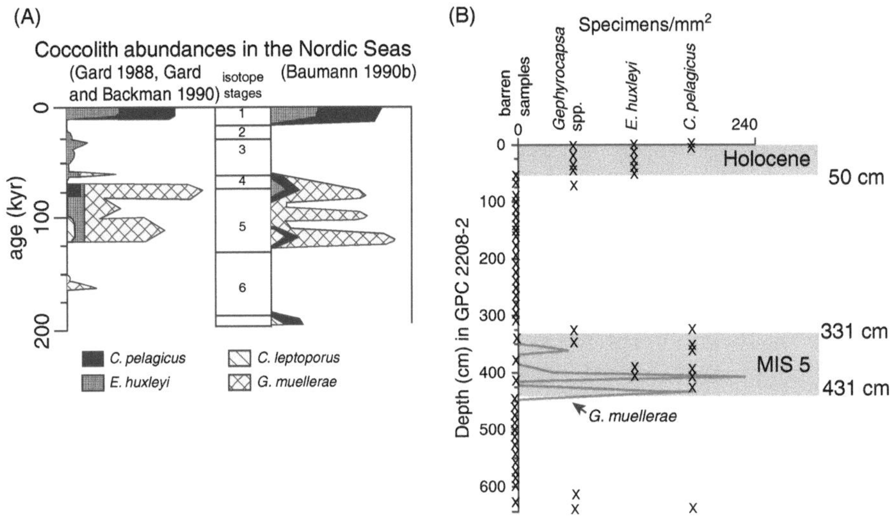

Figure 4.33 (A) Comparison of two standard composite coccolith records from the Nordic Seas, (1) according to Gard (1988) and Gard and Backman (1990), and (2) according to Baumann (1990b) (from Hebbeln et al., 1998). (B) Abundance of *Gephyrocapsa* spp. (dominantly *G. muellerae*), *E. huxleyi*, and *C. pelagicus* in giant piston core PS2208-2 (from Gard, 1993). X indicates presence and absence of data. For location of core, see Figure 6.54.

of nannofossils. If they are present, the abundances are lower than in the Norwegian-Greenland Sea. High abundance of nannofossils is restricted to the present interglacial (Holocene) and the last interglacial (Marine Isotope Stage 5). The Holocene is identified by the dominance of *Coccolithus pelagicus* and *Emiliania huxleyi*, whereas MIS 5 is dominated by *Gephyrocapsa muellerae* with some *E. huxleyi* and other *Gephyrocapsa* spp., as shown in an example from a sediment core from the Nansen Basin (Core PS2208; Figure 4.33B; Gard, 1993). The high-amplitude variations in abundance of *G. muellerae* during MIS 5 may be correlated with warm interstadials (5.1, 5.3, and 5.5) characterized by the maxima and cold stadials (5.2 and 5.4) characterized by the minima.

Diverse assemblages of reworked calcareous nannofossils ranging from the early Jurassic to the Neogene, were also found in the Quaternary sediments in the entire Eurasian Arctic basins where nannofossils are preserved (Gard & Crux, 1994). These are mostly transported from the Eurasian continental shelves and their hinterlands, probably by ice-rafting and/or turbidity currents.

4.5.3. Ostracods

Ostracods, bivalved Crustacea, occur throughout the Arctic Ocean and are common in Quaternary Arctic sediments (Clark, Chern, Hogler, Mennicke, & Atkins, 1990; Mostafawi, 1990; Pak, Clark, & Blasco, 1992; Cronin, Holtz, & Whatley, 1994; Cronin et al., 1995; Taldenkova, Bauch, Stepanova, Dem'yankov, & Ovsepyan, 2005). Furthermore, they have a good potential as palaeoceanographic

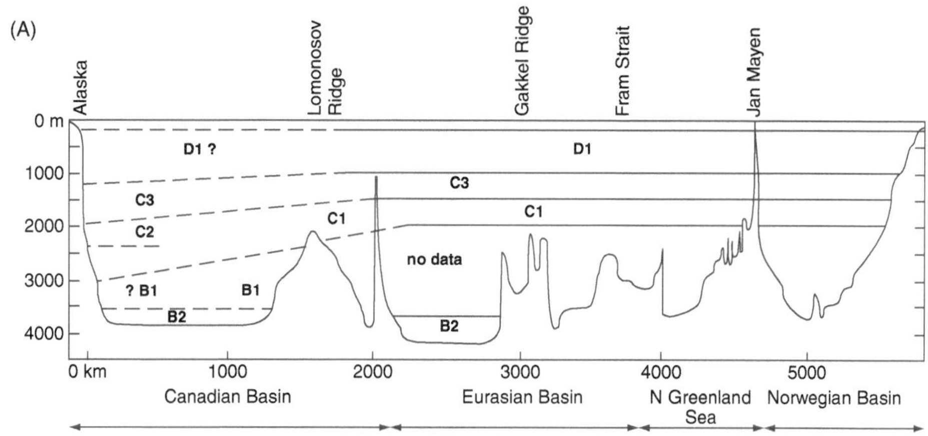

(B)

Biofacies	Area	Water mass	Water depth (m)	Characteristic Ostracodes
A1	Yermak Plateau S. Nansen Basin	Deep Nansen Basin	2500-4100	*Krithe*
B1	Canada Basin Makarov Basin Lomonosov Ridge	Deep Canada Basin	2000-4000	*Cytheropteron bronwynae* *C. hamatum, C. alatum,* *C. carolinae*
B2	Amundsen Basin Nansen Basin	Deep Eurasian Basin	3800-4500	*Krithe, C. bronwynae* *C. carolinae, C. alatum*
C1	Lomonosov Ridge Gakkel Ridge Greenland Sea Chukchi Rise	Upper Deep Eurasian Basin Deep Canada Basin Greenland Sea Intermediate	1000-2000 (Eurasian) 2800-3200 (Canada)	*Polycope, C. carolinae* *C. alatum* *C. hamatum* *Acetabulastoma*
C2	Canada Basin	Upper Deep Canada Basin	1500-2300	*Polycope, C. medistriatum* *C. bronwynae*
C3	Morris Jesup Rise Lomonosov Ridge Greenland Sea	Lower Intermediate, Eurasian, Greenland Sea Intermediate	1000-1400 (Eurasian) 1300-2100 (Canada)	*Polycope, C. bronwynae* *C. carolinae* *Acetabulastoma* *Pseudocythere*
C4	Chukchi Rise Alpha Ridge	Upper Deep Canada Basin Upper Deep Canada Basin	2300-2800	*Polycope, C. medistriatum* *C. bronwynae* *Acetabulastoma* *Parakrithella*
D1	Greenland Sea, Norwegian Sea Kara Sea	Arctic Intermediate (Atlantic Layer)	220-1000	*Eucytherura, Krithe glacialis* *C. inflatum, C. testudo* *Sarsicytheridea punctillata* *C. arcuatum, Rabilimis,* *Cluthia cluthae, S complanata* *Muellerina abyssicola*

Figure 4.34 (A) Cross section of Arctic Ocean and Greenland/Norwegian Seas showing depth distribution of major ostracode biofacies, and (B) Arctic Ocean ostracode biofacies and water masses (from Cronin et al., 1994).

tracers because many species are ecologically sensitive and limited to particular oceanic conditions (Cronin et al., 1994, 1995; Cronin, 1996). The bathymetric and geographic distributions of the most common Arctic and Nordic seas ostracode species indicate there are a variety of factors that control the occurrence and relative

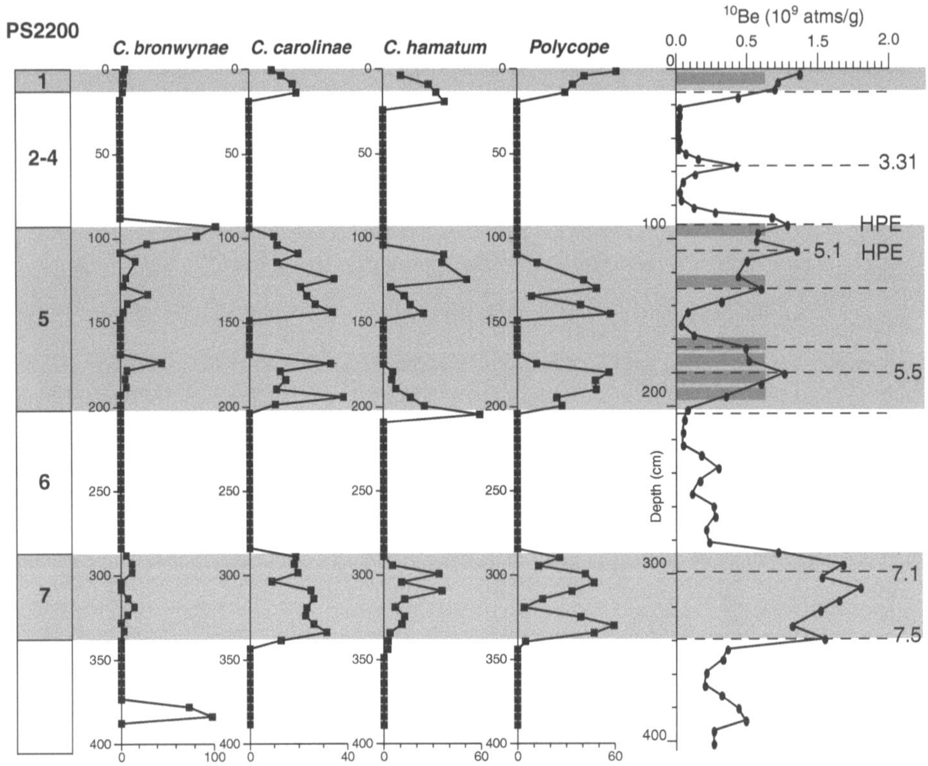

Figure 4.35 Plot of downcore percentages of four ostracode taxa of Kastenlot Core PS2200-5 from Morris Jesup Rise and proposed isotope stratigraphy (MIS 1–7) (from Cronin et al., 1994) and ^{10}Be stratigraphy with isotope substages from the same core (from Spielhagen et al., 2004). HPE indicates "High Productivity Events" based on occurrence of foraminifers, grey bars indicate occurrence of coccoliths. For core location, see Figure 6.54.

abundance of species. These factors include physical barriers like the Lomonosov Ridge, the presence of sea ice and surface-water productivity, salinity, temperature and dissolved oxygen, and availability of food and nutrient resources, and other as yet unknown biotic factors such as competition and predation. In general, however, the modern stratified Arctic Ocean appears to have produced a clear water-depth-related zonation of ostracod species from the shelf to the deep basins that reflects a first-order control of species depth distributions by physico-chemical and/or resource-related characteristics of Arctic water masses (Figure 4.34; Cronin et al., 1994; Cronin, 1996).

As shown in a 300 ka record from Morris Jesup Rise (Core PS2200; Figure 4.35), specific ostracod assemblages reached peak abundances especially during interglacial stages, related to increases in dissolved oxygen and productivity during glacial-interglacial transitions (Cronin et al., 1994). Thus, especially in combination with other proxies such as coccoliths, foraminifers, and ^{10}Be, ostracodes might also be a useful stratigraphic marker (see Chapter 6.1).

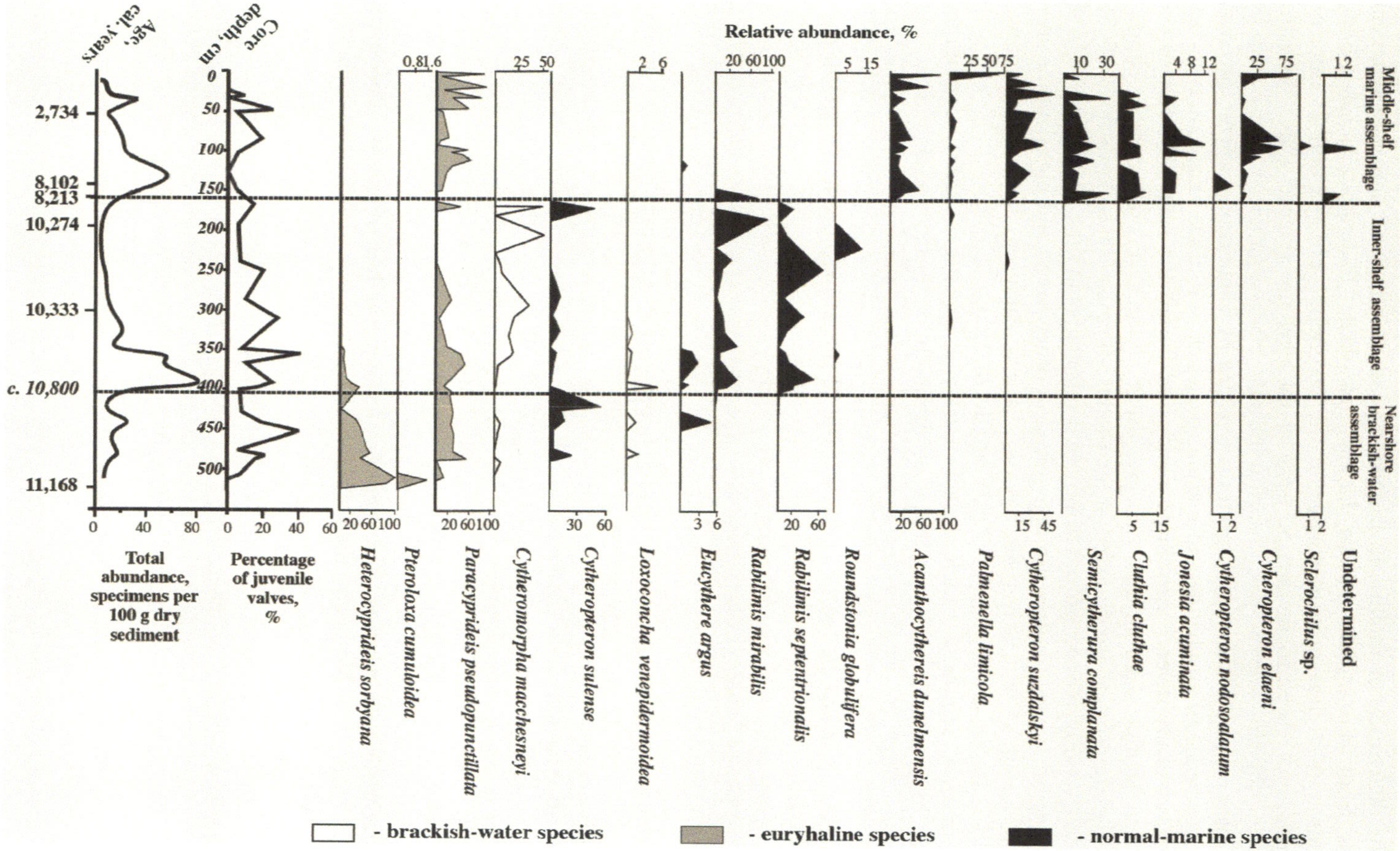

Figure 4.36 Downcore variations in the relative abundance of ostracods in Laptev Sea Core PS-51/138-12 (from Taldenkova et al., 2005). For core location, see Figure 6.54.

In their study of molluscan, ostracodal, and foraminiferal faunas in sediment cores from the middle eastern Laptev Sea shelf, Taldenkova et al. (2005) demonstrate a clear upward change in the taxonomic composition of ostracod assemblages, related to the last post-glacial sea-level rise (Figure 4.36). With the flooding of the Laptev Sea, a taxonomically poor assemblage dominated by euryhaline and brackish-water species is replaced by a transitional assemblage with co-occurrence of euryhaline and marine species at ~10.8 ka. Finally, near 8.2 ka, a taxonomically diverse assemblage dominated by marine species similar to the modern benthic communities, occupies the study area. These trends in benthic faunal composition, also observed in the molluscan fauna (Taldenkova et al., 2005), from survivor or pioneer groups to a more diverse fauna in the course of post-glacial flooding are typical features also recognized in other Arctic and high-latitude areas (e.g., Cronin, 1977, 1981, 1989; Syvitski, Farrow, Atkinson, Moore, & Andrews, 1989; Schoning & Wastegard, 1999; Gordillo & Aitken, 2001).

Simstich, Stanovoy, Bauch, Erlenkeuser, and Spielhagen (2004) determined stable oxygen and carbon isotopes of ostracods in surface sediments as well as in an AMS^{14}C-dated sediment core representing the last ~8 ka, from the southern Kara Sea (Core BP00-07; for location see Figure 6.26). These isotope data could be related to the last post-glacial sea-level rise and Holocene changes in river discharge.

4.5.4. Diatoms

In contrast to sediment records from the Southern Ocean where abundant diatoms have allowed the development of direct proxies to reconstruct past variations in the distribution pattern of sea ice and SST (e.g., Crosta, Pichon, & Burckle, 1998; Armand & Leventer, 2003; Gersonde & Zielinski, 2000; Gersonde et al., 2003; Gersonde, Crosta, Abelmann, & Armand, 2005), the use of diatoms in the Arctic Ocean is much more limited. In subrecent (Quaternary) sediments of the High Northern Latitudes, diatoms are often absent or only occur in very minor amounts, and palaeoceanographic reconstructions are mainly restricted to subarctic areas. Diatom assemblages, for example, have been successfully used for reconstructions of surface-water conditions in the Greenland, Iceland, and Norwegian seas during the last late glacial to Holocene time interval (Koç & Jansen, 1992; Koç, Jansen, & Haflidason, 1993; Koç, Jansen, Hald, & Labeyrie, 1996; Koç, Kristensen, Hasle, Forsberg, & Solheim, 2002; Birks & Koç, 2002; Jiang, Eiriksson, Schulz, Knudsen, & Seidenkrantz, 2005). In a sediment core directly north of Svalbard representing the last ~14 ka, Koç et al. (2002) determined diatom species related to sea ice or the marginal ice zone (MIZ). In this core, diatoms are absent before 10.8 ^{14}C ka, interpreted by the authors as indication that the surface waters in the area were mostly sea-ice covered. For the Sea of Okhotsk, Bering Sea, and subarctic North Pacific, down-core profiles of selected diatom taxa allowed reconstructions of changes in palaeoclimatic and palaeoceanographic conditions, for example, seasonal productivity, sea ice, and surface-water circulation of the glacial and deglaciation periods during the late Quaternary (e.g., Sancetta & Robinson,

1983; Sancetta & Silvestri, 1986; Shiga & Koizumi, 2000; Katsuki, Takahashi, & Okada, 2003; Koizumi, Shiga, Irino, & Ikehara, 2003; Katsuki & Takahashi, 2005).

In pre-Neogene (pre-glacial) sediments from the central Arctic Ocean, however, diatoms may become a very significant microfossil group. Middle Eocene sediments from Lomonosov Ridge, for example, consist of diatom oozes which yield important information on palaeoproductivity, fresh-water discharge, etc. (Backman et al., 2006; see Chapter 7.2.2). Almost pure laminated siliceous oozes with excellently preserved diatoms and silicoflagellates of late Cretaceous age were recovered in two short gravity cores from the Alpha Ridge (Dell'Agnese & Clark, 1994; Firth & Clark, 1998; see Chapter 7.1.3).

In the Eurasian Arctic, diatoms were used as valuable proxy for the reconstruction of late Quaternary environmental conditions in the Arctic shelf areas characterized by strong river discharge and seasonal variability in sea-ice cover and primary production. Freshwater diatoms are an important component of particulate organic matter that is transported onto shelves by rivers, marine diatoms are a major portion of the phytoplankton production in the Kara Sea, and sea-ice diatoms are indicative for the extent of sea-ice cover (e.g., Cremer, 1999; Bauch & Polyakova, 2000, 2003; Polyakova, 2003). Like other micropalaeontological (e.g., Naidina & Bauch, 1999; Matthiessen, Kunz-Pirrung, & Mudie, 2000; Kraus, Matthiessen, & Stein, 2003) and organic-geochemical tracers (e.g., Fahl & Stein, 1997, 1999; Stein & Fahl, 2004a, 2004b) diatom assemblages can be used to identify and characterize different OC sources (i.e., terrestrial/freshwater versus marine) in shelf sediments. Moreover, established linkages between hydrographical parametres (e.g., surface-water salinity and sea-ice conditions) and the composition of surface sediment diatom assemblages (Bauch & Polyakova, 2000, 2003; Polyakova, 2003) can be used to reconstruct palaeoenvironmental conditions (e.g., Bauch & Polyakova, 2003; Polyakova & Stein, 2004).

As an example, diatom data from the southern Kara Sea and the relationship to sea-surface salinity are shown in Figure 4.37 (Polyakova, 2003; Polyakova & Stein, 2004). Direct comparison of summer surface-water salinity in the Ob and Yenisei estuaries and the inner Kara Sea shelf and relative abundances of freshwater diatoms in the surface sediment indicate a direct correlation between these two parametres. The entire investigated area covers an average summer salinity range 0.5–28.5. The highest relative abundances of freshwater diatoms ($>90\%$) are characteristic for the outer Ob and Yenisei estuaries with salinities < 10. Seaward, the overall decrease of relative abundances of freshwater diatoms is obviously related to the decreasing influence of the riverine water. Abundances of $\sim 60\%$ and 40% seem to correspond to the 11 and 12 isohalines. The steep decrease of relative abundances of freshwater diatoms down to 10% correlates with a salinity range 13–15, thus indicating this as the northern border of the main area influenced by riverine waters (Figures 4.37 and 4.38). The same trend is also shown in the abundances of chlorophycean algae genus *Pediastrum* spp. (Figure 4.38; see later description). Northward of this boundary, the occurrence of freshwater diatoms in surface sediments is obviously connected to other processes, for example melting of riverine ices and sediment entrainment into sea ice (e.g., Kempema, Reimnitz, & Barnes, 1989; Abelmann, 1992; Lisitzin, 2002).

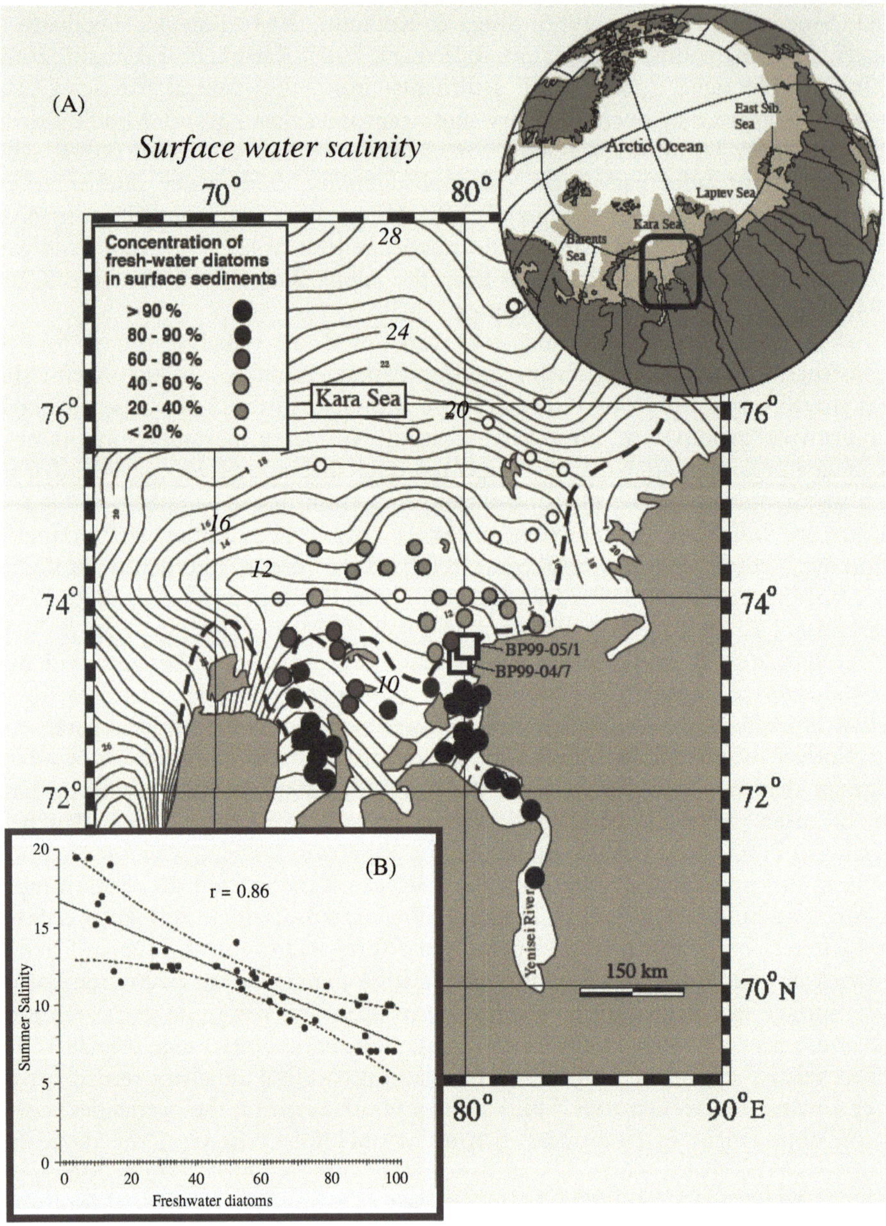

Figure 4.37 (A) Distribution patterns of relative abundance of freshwater diatoms (%) in the surface sediments of the Kara Sea and the outer Ob and Yenisei estuaries (from Polyakova, 2003; Polyakova & Stein, 2004). Salinity contour lines (numbers in italics) represent the mean inter-annual summer surface-water salinity distribution in the Kara Sea based on 50-years measurements (after Dmitrenko, Gribanov, Volkov, & Kassens, 1999). (B) Correlation between relative abundances of freshwater diatoms and the summer surface water salinity pattern (from Polyakova, 2003; Polyakova & Stein, 2004). A linear regression between average summer surface-water salinity and relative proportions of freshwater diatoms reveals a correlation coefficient of $r = 0.86$ for a salinity range of $\sim$ 5–17. Stippled lines indicate 2 sigma range.

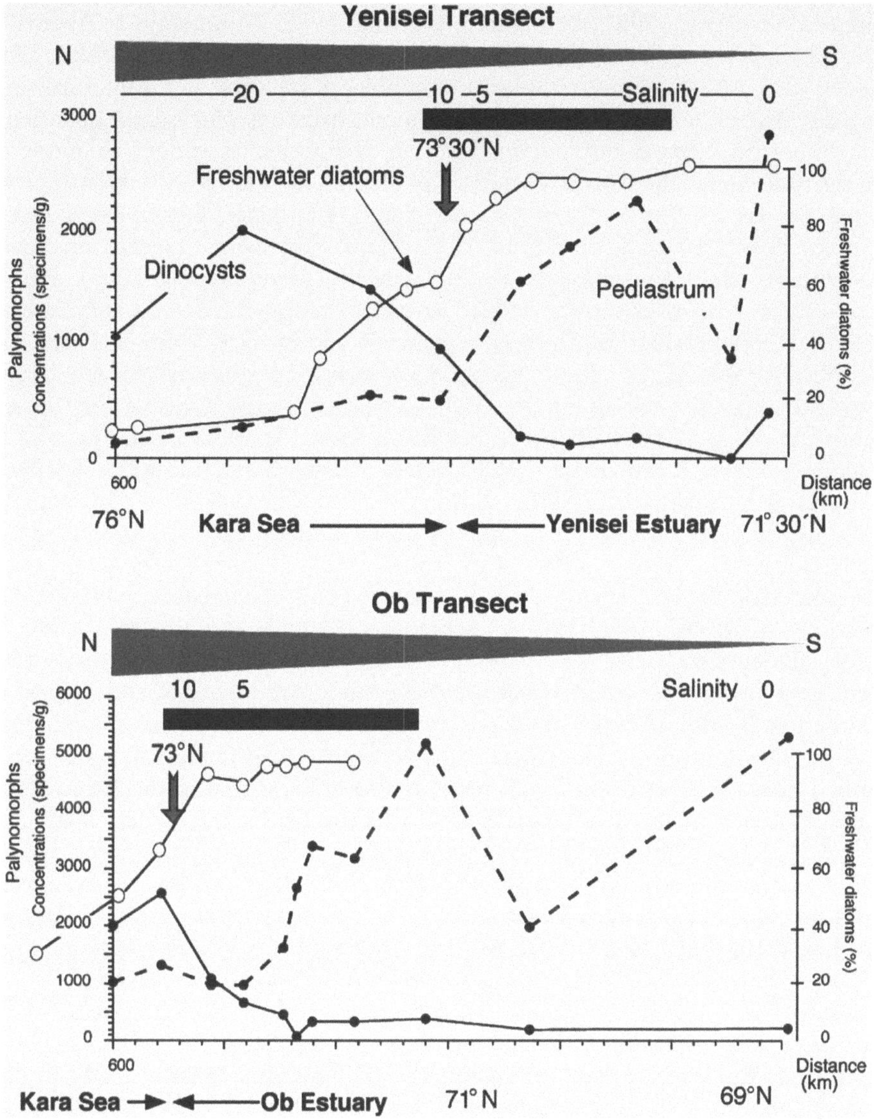

Figure 4.38 Distribution of freshwater diatoms, dinoflagellate cysts and *Pediastrum* spp. along the Ob and Yenisei transects (from Stein et al., 2004a). Salinity data are from Burenkov and Vasilkov (1995). Dark grey bars indicate location of marginal filter.

4.5.5. Radiolaria

Radiolaria, marine planktonic protozoans, are widely distributed in the world oceans, and comprise surface- and deep-living species. Based on studies of plankton tows, sediment traps, and surface sediments, radiolarian assemblages are adapted to specific environmental conditions such as water mass and its vertical hydrographic structure, nutrients, temperature, salinity, etc. (e.g., McMillen & Casey, 1978;

Kling & Boltovskoy, 1995; Abelmann & Gowing, 1997; Nimmergut & Abelmann, 2002; Abelmann & Nimmergut, 2005; Itaki, Ito, Narita, Ahagon, & Sakai, 2003; Okazaki et al., 2003). Thus, radiolarian assemblages are very useful micropalaeontological proxies for palaeoceanographic reconstructions, including quantitative estimates of SST based on transfer-function technique, as shown in numerous investigations from the world oceans (e.g., Hays, Imbrie, & Shackleton, 1976; Lozano & Hays, 1976; Matul, 1995; Abelmann, Brathauer, Gersonde, Sieger, & Zielinski, 1999; Dolven & Bjørklund, 2001; Matul, Abelmann, Tiedemann, Kaiser, & Nürnberg, 2002; Gersonde et al., 2005; Matul & Abelmann, 2005). For the Arctic Ocean proper, however, detailed studies of radiolaria are still very rare and limited to investigations of the water column and surface sediments (see later description). Well-dated fossil records to be used for palaeoceanographic reconstructions, are missing for the Arctic Ocean and only available for subarctic regions such as the Sea of Okhotsk, the Bering Sea, the Norwegian Sea, and the Labrador Sea (e.g., Matul et al., 2002; Hays & Morley, 2003; Tanaka & Takahashi, 2005; Matul & Abelmann, 2005).

In the modern Arctic Ocean and its adjacent marginal seas, several tens of radiolarian species were reported from plankton and surface sediment samples (Bernstein, 1932; Hülsemann, 1963; Tibbs, 1967; Kruglikova, 1989, 1999; Bjørklund, Cortese, Swanberg, & Schrader, 1998; Bjørklund & Kruglikova, 2003; Itaki et al., 2003). The most comprehensive compilation of radiolarian assemblages in surface sediments from the central Arctic basins and ridges was published by Bjørklund and Kruglikova (2003). For the Chukchi and Beaufort seas, the depth distributions of radiolarian fauna were examined quantitatively in depth-stratified plankton tows from 4 or 5 intervals above 500 m and in surface sediments from various depths between 163 and 2907 m (Figure 4.39; Itaki et al., 2003). Itaki et al. (2003) could show a clear relationship between radiolarian fauna and the vertical water structure of the Canadian Basin composed mainly of three water masses, that is, Arctic Surface Water (ASW), Arctic Intermediate Water (AIW), and Canadian Basin Deep Water (CBDW) (see Chapter 2.2).

The radiolarian assemblage in the ASW <150 m is characterized by the dominance of *Amphimelissa setosa* and common to abundant *Actinomma boreale/leptoderma* group, *Pseudodictyophimus gracilipes* and *Spongotrochus glacialis*, in both plankton tows and surface sediments (Figure 4.39). This assemblage is similar to those in the plankton and surface sediments of the Barents, Iceland, and Greenland seas (Kruglikova, 1989; Swanberg & Eide, 1992; Bjørklund & Kruglikova, 2003). As outlined in Itaki et al. (2003), all locations where *A. setosa* dominated the radiolarian assemblage (i.e., the Chukchi, Beaufort, Barents, Iceland, and Greenland seas), were taken near the summer ice edge. Furthermore, these authors found that *A. setosa* reached a maximum in the cold halocline layer, the lower part of the ASW, near the freezing point. On the other hand, in the central Arctic Ocean near the North Pole, characterized by thick sea-ice cover during the year, *A. leptoderma* dominated the assemblage, but *A. setosa* was absent (Kruglikova, 1989; Bjørklund & Kruglikova, 2003; Itaki et al., 2003). Based on this distribution pattern and the fact that in the Greenland Sea *A. setosa* and *P. gracilipes* are dominant in the ice edge assemblage related to chlorophyll-a and phaeopigments (Swanberg & Eide, 1992), Itaki et al.

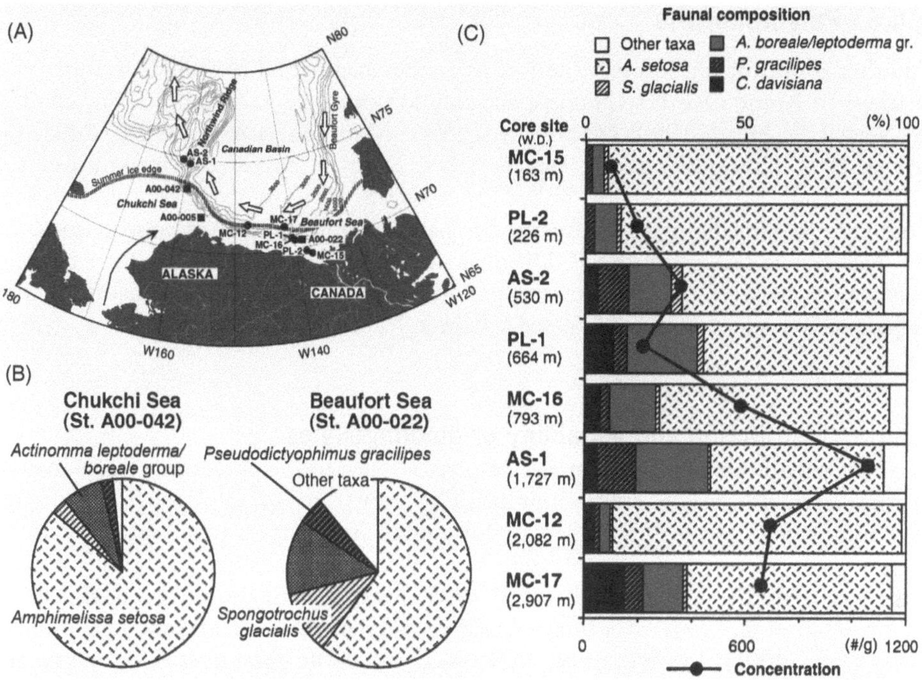

Figure 4.39 (A) Map of the Chukchi and Beaufort seas showing the locations of plankton tows (solid squares) and surface sediment stations (solid circles). Open and solid arrows indicate the circulation of the Beaufort Gyre and the inflow of Pacific water, respectively. Dashed line indicates the mean summer ice edge. (B) Compositions of living radiolarian assemblages through the upper 500 m of the water column at stations A00-042 (Chukchi Sea) and A00-022 (Beaufort Sea). (C) Depth-arranged total radiolarian concentrations (line graph) and their faunal composition (bar graph) in surface sediments from the Chukchi Sea (AS-1 and AS-2) and the Beaufort Sea (MC-12, MC-15, MC-16, MC-17, PL-1, and PL-2) (from Itaki et al., 2003).

(2003) suggest that the dominance of *A. setosa* resulted from both the high primary production and low temperature at the ice edge area.

In the samples studied by Itaki et al. (2003), *Ceratocyrtis historicosa* commonly occurred in water depths of 300–500 m, closely corresponding to the relatively warm upper AIW with water temperatures 0.5–1.0°C, originating from the North Atlantic Ocean and Norwegian Sea. The radiolarian species *Cycladophora davisiana* was absent in the plankton tows (< 500 m) and restricted to surface sediments below 500 m of water depth (Figure 4.39), suggesting that this species mainly lives in the lower AIW or CBDW. High abundance of *C. davisiana* has been only reported from surface sediment and plankton studies of the Sea of Okhotsk (Morley & Hays, 1983; Nimmergut & Abelmann, 2002), with highest standing stock values and percentages in the 200–1,000 m deep Intermediate Water, a very cold and highly oxygenated water mass originated from the shelf areas. These characteristics may be also applied to the lower AIW and CBDW, which are probably composed of cold and highly oxygenated shelf waters. Thus, this may support that the occurrence of a deep, cold, and well-oxygenated water mass is at least one important factor controlling the depth distribution of *C. davisiana* (Itaki et al., 2003; *cf.* Nimmergut & Abelmann, 2002).

4.5.6. Palynomorphs

Another group of microfossils used as palaeoenvironmental as well as stratigraphic markers in Arctic Ocean sediments, are palynomorphs (e.g., Mudie, 1985; Mudie, de Vernal, & Head, 1990; Matthiessen, 1995; Matthiessen et al., 2000, 2001; de Vernal et al., 2001; Naidina & Bauch, 1999; Kraus et al., 2003; Brinkhuis et al., 2006). Most of these organic-walled microfossils have an excellent preservation potential, making them so important for palaeoenvironmental reconstructions. In the Palaeogene section of the IODP drill site on Lomonosov Ridge, for example, palynomorphs were the main microfossil group available for precise dating of the sequence (Backman et al., 2006; see Chapter 7.2.2).

4.5.6.1. Distribution and variability of dinoflagellates

Aquatic palynomorphs are remains of phyto- and zooplankton that may live in marine and freshwater environments. The most important group of marine palynomorphs, also used for palaeoenvironmental and palaeoclimate reconstructions in the Arctic Ocean and circum-Arctic regions, are dinoflagellates and their resting cysts (e.g., Mudie, 1992; Matthiessen, 1995; Head, 1996; de Vernal et al., 2001; Matthiessen et al., 2001; Matthiessen and de Vernal, 2001; Brinkhuis et al., 2006; Sluijs et al., 2006). In contrast to siliceous or carbonate microfossils, dinocysts are generally well preserved in sediments affected by dissolution because they are composed of highly resistant refractory organic matter. Furthermore, dinoflagellates may thrive despite extremely cold conditions, and they occupy a wide range of marine environments with respect to SST, SSS, and sea-ice cover (SIC) (e.g., de Vernal et al., 2001). Depending on the sea-surface water characteristics, the abundances of specific dinocysts taxa vary significantly. As an example, distribution maps of two taxa are shown in Figure 4.40. *Islandinium minutum* is a polar to sub-polar taxon that is common to abundant in environments with low summer SST $<5^{\circ}C$ and an extensive SIC (Dale, 1996; de Vernal et al., 2001; Matthiessen & Knies, 2001). Its occurrence is abundant in the northern Baffin Bay, Kara Sea, and Laptev Sea areas as well as in the western Greenland Sea influenced by the cold East Greenland Current. The second taxon is the more cosmopolitan taxon *Operculodinium centrocarpum* which is dominant in the North Atlantic Ocean and the Barents Sea and may characterize the pathway of relatively warm Atlantic waters in this region (e.g., Matthiessen, 1995; Rochon, de Vernal, Turon, Matthiessen, & Head, 1999).

In studies of sediment cores from the northern North Atlantic and Eurasian continental margin, increased abundance of *O. centrocarpum* is used as proxy for warm Atlantic-water inflow (e.g., Matthiessen, 1995; Matthiessen et al., 2001; Matthiessen & Knies, 2001). Maximum abundances of *O. centrocarpum* in a sediment core NE of Svalbard, for example, revealed clear evidence of interglacial conditions with SST similar to present or even warmer and increased Atlantic-water inflow during the early MIS 5 (Figure 4.41; Matthiessen & Knies, 2001). This interpretation is supported by simultaneous peak abundances of "Atlantic water" benthic foraminifers indicating a strong advection of water masses from the Norwegian Sea (Wollenburg et al., 2001).

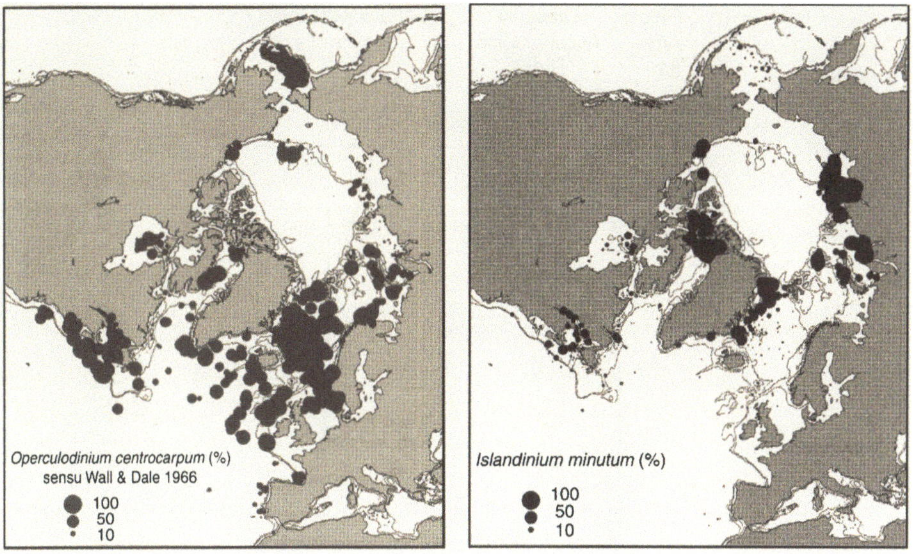

Figure 4.40 Distribution maps of (A) *Operculodinium centrocarpum* sensu Wall & Dale 1966 and (B) *Islandinium minutum* (from de Vernal et al., 2001).

Using large data matrixes of dinocyst assemblages in modern Arctic and Subarctic sediments (based on 677 sediment surface samples and 30 dinocysts taxa) and their statistical correlation with SST, SSS, and SIC, even quantitative reconstructions of past sea-surface conditions are possible by means of modern analogue techniques (Mudie, 1992; de Vernal, Turon, & Guiot, 1994; de Vernal et al., 2001; Rochon et al., 1999).

As an example for reconstruction of SST, SSS, and SIC, records from a sediment core taken on the eastern Barents Sea shelf, an area close to the Polar Front but still influenced by Atlantic water (Figure 4.42; Core PL-96-112). Close to the core location, modern August SST and SSS are 7.8 $\pm$1.9°C and 34.8 $\pm$0.2 (de Vernal et al., 2001). The dinocysts assemblages of Core PL-96-112 representing the last ~8.5 Cal. ka, are generally dominated by *O. centrocarpum* and the cyst of *Pentapharsodinium dalei* and only show a moderate variability in the percentages of the main taxa (Figure 4.42; de Vernal et al., 2001; Voronina, Polyak, de Vernal, & Peyron, 2001). The most prominent up-core change is the increase of *P. dalei* cysts at the expense of *O. centrocarpum*. Voronina et al. (2001) interpret this change as a generally closer position of the Polar Front in the younger part of the record. The reconstruction of sea-surface conditions based on modern analogue technique indicates relatively warm SST and limited sea-ice extent, related to a significant inflow of Atlantic water into the southeastern Barents Sea, during most time of the last ~8.5 Cal. kyr BP (Figure 4.42). The SST variability is mostly within the range of modern SST (7.8 $\pm$1.9°C). Especially warm and stable conditions probably occur between 8 and 5 Cal. ka. Distinct cooling pulses were recorded at ~8.1, 5.0, 3.5, and 2.5 Cal. kyr BP. The 8.1 Cal. kyr BP cooling may correlate with the widespread 8.2 Cal. kyr BP cooling event detected in many high-latitude areas

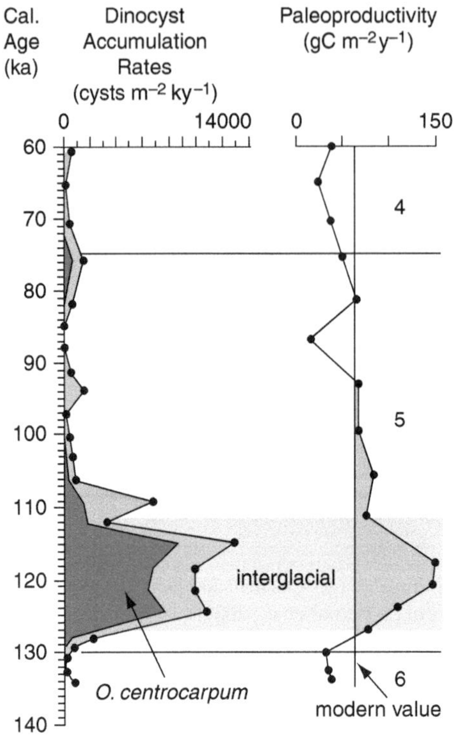

Figure 4.41 Dinocyst accumulation rates and palaeoproductivity estimated from benthic foraminiferal assemblages (Wollenburg et al., 2001) in MIS 6 to MIS 4 of Core PS2138-1, Barents Sea continental margin (from Matthiessen & Knies, 2001). For core location, see Figure 6.54.

(e.g., Alley et al., 1997). The SST values at Core PL-96-112 are significantly higher than those determined in a sediment core from the western Barents Sea (Core M23258; Figure 4.42 and 6.76 for location), that is, an area with a stronger influence of Atlantic water, where SST was estimated based on planktonic foraminifers and the SIMMAX technique (see Chapter 4.5.1 for background). For the same time interval, most of the SIMMAX summer SST values vary between 2 and 5°C (Sarnthein et al., 2003c; see Chapter 6.4.3, Figure 6.93). These differences are difficult to explain.

High abundances of dinocysts, especially in combination with OC records, may also be used as proxy for increased primary production (Aksu, de Vernal, & Mudie, 1989; Stein & Stax, 1991; Hillaire-Marcel et al., 1994; Mudie & Rochon, 2001; Hamel, de Vernal, Gosselin, & Hillaire-Marcel, 2002).

4.5.6.2. Distribution and variability of chlorophycean algae

Among the freshwater palynomorphs, the chlorophycean algae *Pediastrum* and *Botryococcus* are most widespread in limnic and marine sediments throughout the Arctic (Matthiessen et al., 2000). These algae have been recorded in recent Arctic

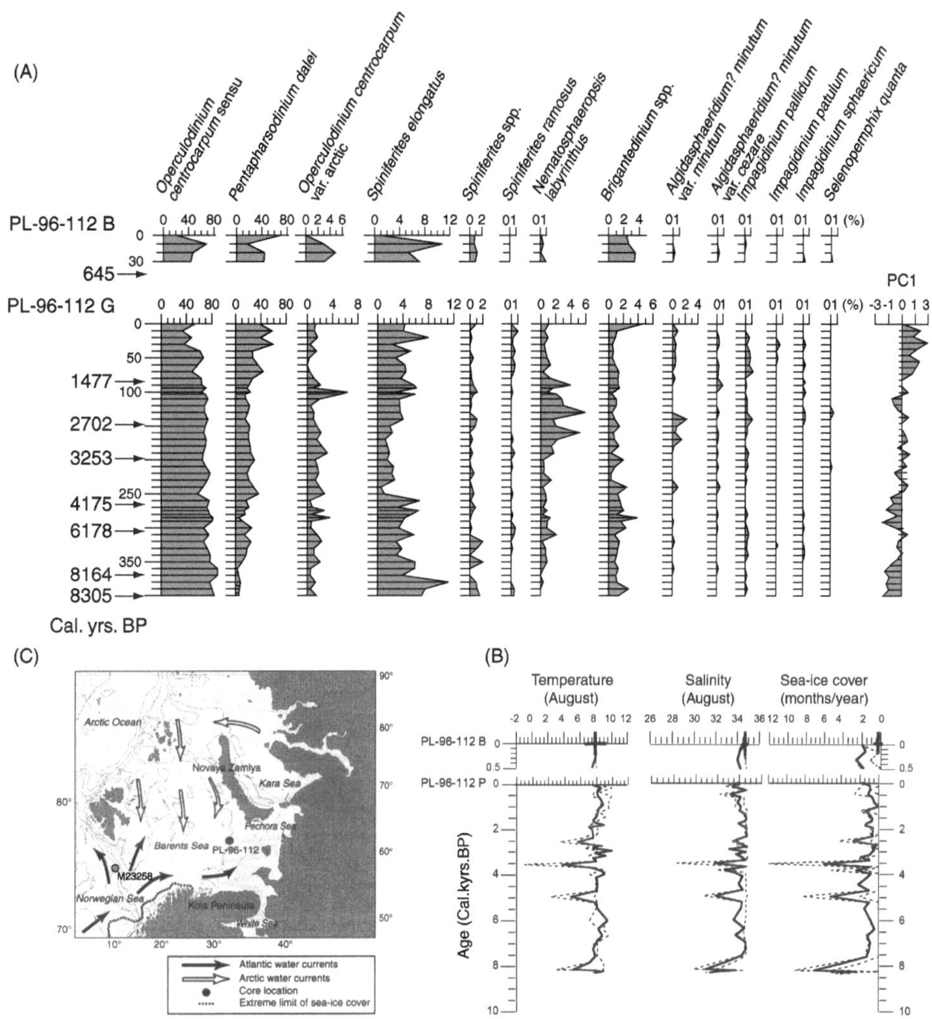

Figure 4.42 (A) Dinocyst taxa percentages (plotted versus depth) and (B) reconstruction of sea-surface conditions from dinocyst distribution (plotted versus calendar kiloyears) in the gravity (G) and box (B) Cores PL-96-112; (C) map of the Barents Sea showing core location and major current systems (from de Vernal et al., 2001, supplemented). Solid lines in (B) represent the most probable reconstructed values (weighted averages of the five best analogues) within a confidence interval delimited by minimum and maximum values (dotted lines) from the five best analogues. In addition, location of Core M23258 is shown.

Ocean sediments (Mudie, 1992; Kunz-Pirrung, 1999; Matthiessen, 1999; Matthiessen et al., 2000), circum-Arctic Holocene and late Weichselian lake deposits and shelf sediments (e.g., Fredskild, 1973; Hill, Mudie, Moran, & Blasco, 1985; Kunz-Pirrung, 1998; Hahne & Melles, 1999; Naidina & Bauch, 1999), and Pliocene and Pleistocene marine sediments from the High Northern Latitudes (Mudie, 1985, 1989; de Vernal & Mudie, 1989). Thus, these freshwater palynomorphs have a high potential as proxy to identify freshwater (riverine) input into the Arctic Ocean in marine sedimentary sections and its change through time. As an example, distribution maps of *Pediastrum* spp. and *Botryococcus cf. braunii* in surface sediments from the Beaufort Sea strongly influenced by Mackenzie River discharge, and the Kara Sea strongly influenced by discharge of the rivers Ob and Yenisei, are shown in Figure 4.43 (Matthiessen et al., 2000). The distribution of these algae is clearly related to the discharge of freshwater and suspended matter by the large rivers onto the shelf. In transects from the estuaries towards the open Kara Sea, the high absolute abundances of the chlorophycean algae genus *Pediastrum* spp. are characteristic for the Ob and Yenisei estuaries with salinities < 10 (Figure 4.38). Seaward, the overall decrease of abundances of *Pediastrum* spp. is related to the decreasing influence of the riverine water. The same picture is also reflected in the abundances of freshwater diatoms (Figure 4.38). The steep decrease in freshwater species occur at ~73°N in the Ob Estuary and at ~73°30′N in the Yenisei Estuary, paralleled by an increase in marine dinoflagellates (Figure 4.38; Matthiessen et al., 2000; Stein et al., 2004a). This border approximately corresponds to the outer boundary of the "marginal filter" in the Kara Sea (see Chapter 2.5.2, Figure 2.20).

4.5.6.3. Distribution and variability of terrestrial palynomorphs

Terrestrial palynomorphs include pollen and spores. Numerous studies on Arctic and Subarctic shelf sediments as well as lake sediments illustrate the potential of pollen to reconstruct palaeoclimate change in the adjacent coastal region and to link the terrestrial and marine palaeoclimate records (e.g., Kulikov & Khitrova, 1982; Mudie, 1982; Mudie & Short, 1985; Hill et al., 1985; Mudie & McCarthy, 1994; Naidina & Bauch, 1999, 2001; Levac, de Vernal, & Blake, 2001; Kraus et al., 2003; Premke-Kraus, 2008). Also for palaeoenvironmental reconstruction of the early pre-glacial Arctic Ocean, terrestrial palynomorphs are of major significance (Brinkhuis et al., 2006; Sluijs et al., 2006; see Chapter 7.2.2 for discussion).

The modern vegetation in the Siberian coastal area and the adjacent hinterland is characterized by a latitudinal succession from south to north (Figure 4.44): Boreal forests and forest tundra are replaced by Subarctic and Arctic tundra and finally by polar desert covering most islands (Aleksandrova, 1980, 1988; Atlas Arktiki, 1985; Walter & Breckle, 1994). Pollen assemblages in sediments reflect these vegetation zones. Thus, studies of pollen assemblages and their spatial and temporal fluctuations in sediment cores from shelf or lake areas may give information on changes in vegetation and palaeoclimate. Since the type of vegetation depends upon the duration and intensity of positive summer temperatures and the amount of precipitation, palynological records — if calibrated for the modern environment — can be used to give even quantitative numbers on air temperature and precipitation (e.g., Hahne & Melles, 1997; Andreev & Klimanov, 2000; Hubberten et al., 2001; see also Figure 4.23). As an example, a palynological record from Lama Lake, Taymyr Peninsula, representing the late Weichselian–Holocene time interval, is shown in Figure 4.44 (Hahne & Melles, 1997). Based on these records, the late Weichselian was characterized by a cold and dry climate leading to strongly reduced vegetation. During the Pleistocene/Holocene transition, a distinct step-wise

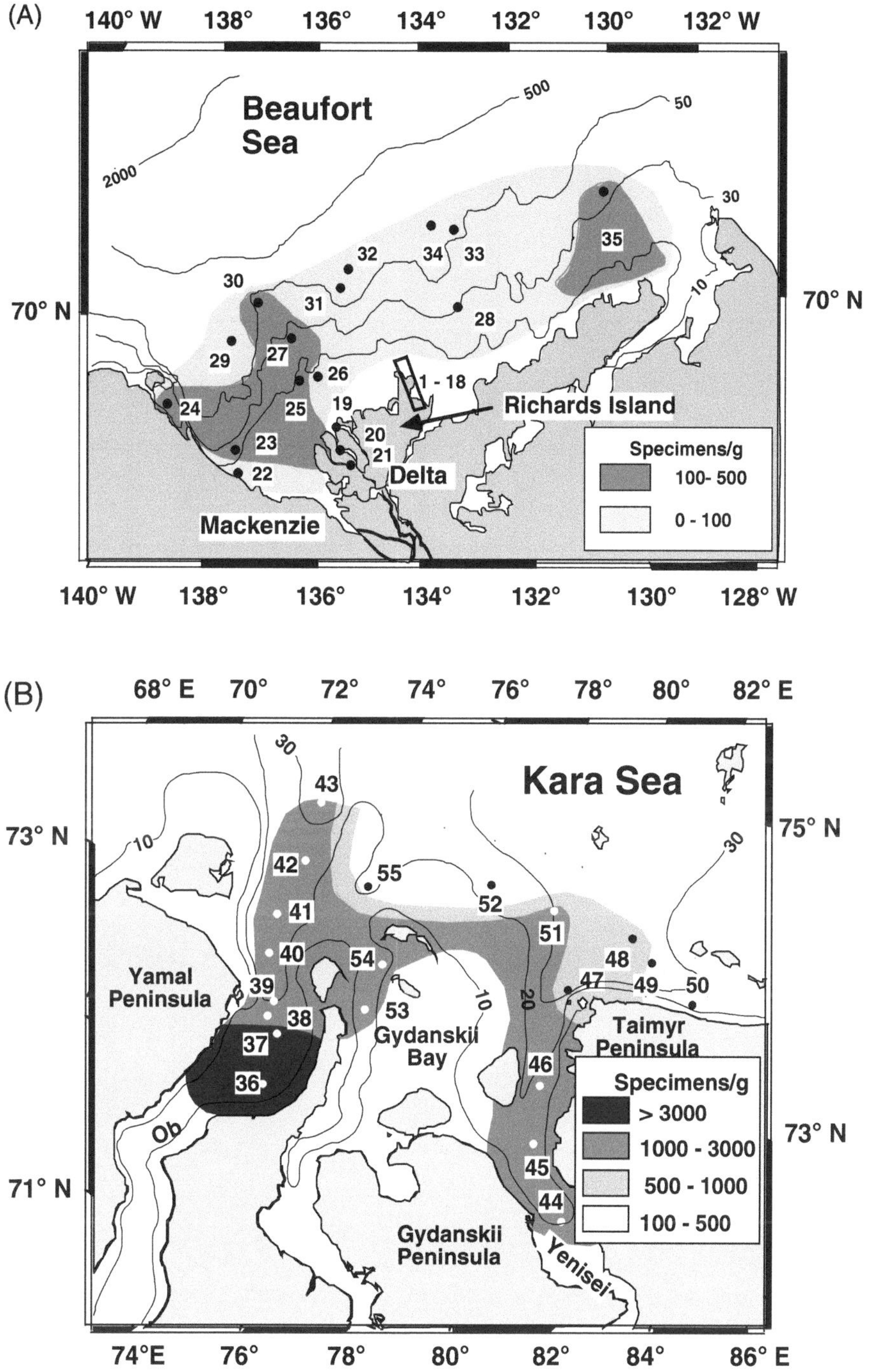

Figure 4.43 Sediment concentrations of *Pediastrum* spp. and *Botryococcus cf. braunii* in the (A) Beaufort Sea and (B) Kara Sea (from Matthiessen et al., 2000). Depth contours in metres are indicated.

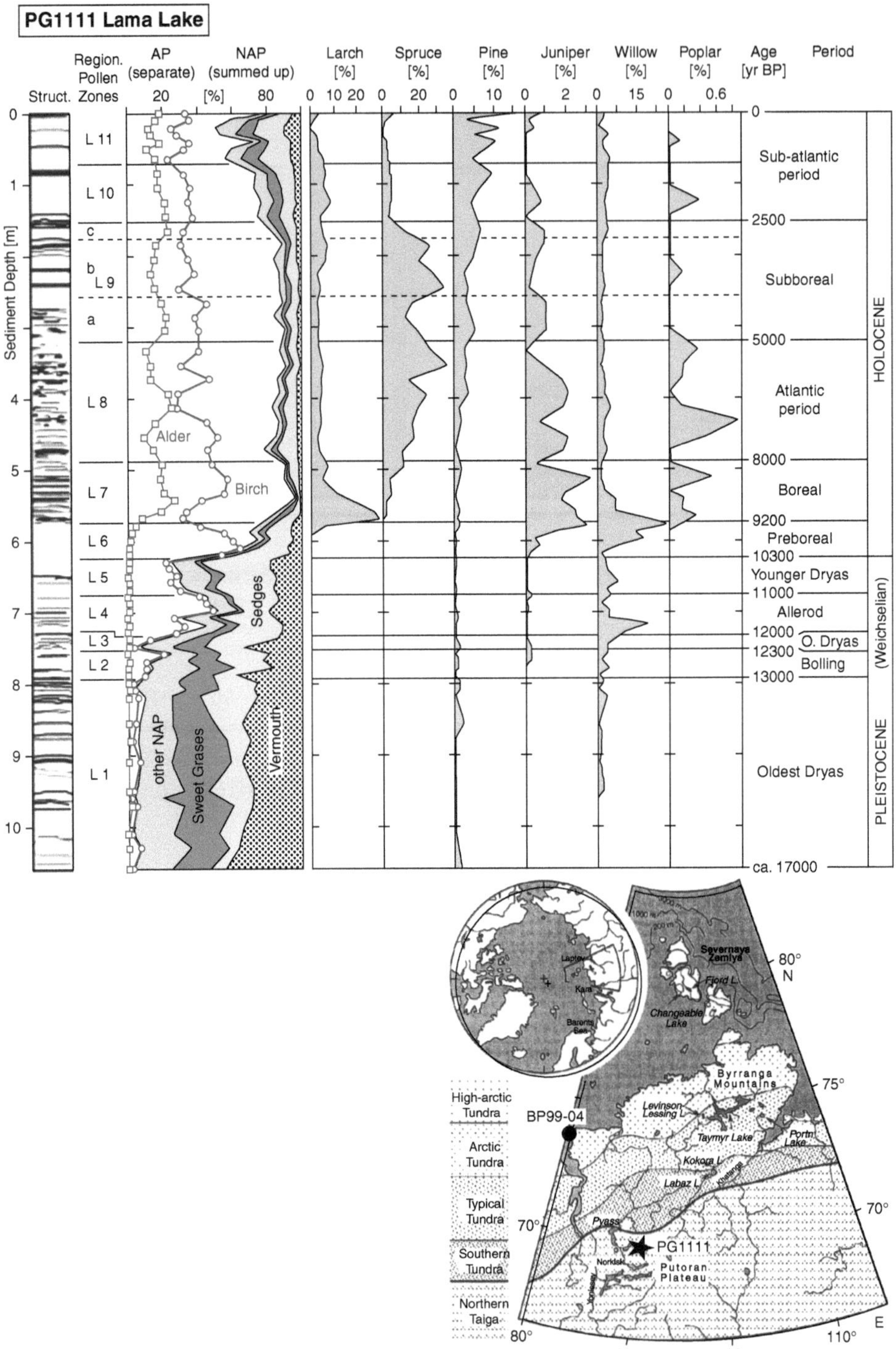

Figure 4.44 Map showing current vegetation zones, and pollen profile of sediment core PG1111 from the central part of the Lama Lake (Hahne & Melles, 1999, supplemented). The chronology is based on correlations of the regional Pollen Assemblage Zones (PAZ) with radiocarbon-dated pollen records on land, supported by ^{210}Pb measurements in near-surface sediments.

warming is obvious, culminating in the Early Holocene (Boreal) Climatic Optimum (see e.g., Duplessy, Ivanova, Murdmaa, Paterne, & Labeyrie, 2001; Andrews et al., 2003). Warmer intervals occur in the Bølling, Allerød, and Preboreal, interrupted by colder intervals during the Older and YD. During the last ~3,000 years, a significant climatic deterioration took place, which correlates with a decrease in Siberian River discharge (Stein et al., 2003a, 2004a; see Figure 4.23).

4.6. Stable Isotopes of Foraminifers

The oxygen isotope ($\delta^{18}O$) record determined in foraminifer shells depend on (1) the isotopic composition of the water mass (controlled by the continental ice-sheet volume and freshwater input) and (2) the water temperature. In general, $\delta^{18}O$ records of benthic foraminifers from marine sediment cores reflect the variability of continental ice volume through time ("ice effect") because temperature changes in great water depths are supposed to be minimal in the late Quaternary glacial–interglacial cycles (e.g., Shackleton, 1967; Imbrie et al., 1984). $\delta^{18}O$ records of planktonic foraminifers, on the other hand, are controlled by both changes in ice effect and temperature. In low- and mid-latitudes, benthic and planktonic $\delta^{18}O$ records show a typical "global" pattern of light interglacial and heavy glacial values (e.g., Imbrie et al., 1984; Martinson et al., 1987; Lisiecki & Raymo, 2005). Thus, these records can be used as palaeoenvironmental proxy as well as stratigraphic tool. In contrast to the global open-ocean $\delta^{18}O$ stratigraphy, isotope records from the central Arctic Ocean may be strongly superimposed by local/regional input of isotopically lighter freshwater (i.e., meltwater or river water).

The isotopic composition of river water is highly depleted in $\delta^{18}O$ relative to marine waters as well as to sea ice. Thus, the Arctic marginal seas characterized by strong river discharge show a distinct gradient in $\delta^{18}O$ of surface waters from the river mouths towards the open ocean (Figure 4.45A; Bauch, Erlenkeuser, & Andersen, 2005). There is an excellent correlation between salinity and oxygen isotopic composition of the surface waters (Figure 4.45B; Bauch et al., 1995, 2005; Bauch, Carstens, & Wefer, 1997). Therefore, the $\delta^{18}O$ composition of the water (and foraminifers living in this water mass) is a good measure for the amount of river-runoff within the water column.

Oxygen isotopic composition of the planktonic foraminiferal species *N. pachyderma* (sin.) appears to be in equilibrium with the isotopic composition of the surface water (e.g., Charles & Fairbanks, 1990). In the modern central Arctic Ocean, this species lives mainly in the uppermost water column (50–100 m; Carstens & Wefer, 1992), where temperatures are near the freezing point all across the basin (see Chapter 2.2). Thus, the oxygen isotope composition of its carbonate shell is almost entirely controlled by the isotopic composition of the ambient water mass. Owing to the correlation between salinity and oxygen isotopic composition of the surface waters (Bauch et al., 1995), variations of the $\delta^{18}O$ values of *N. pachyderma* (sin.) mostly reflect the regional salinity variations caused by the

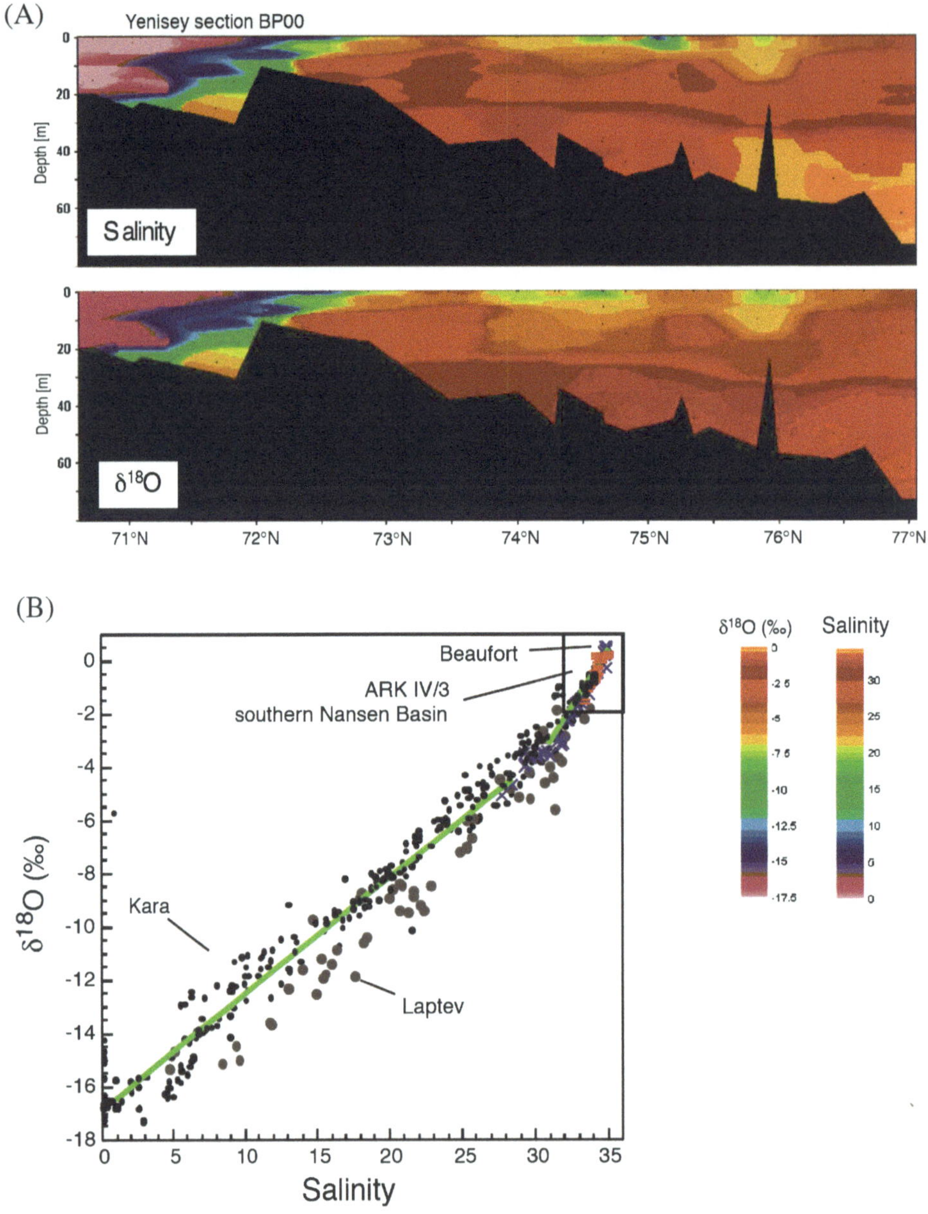

Figure 4.45 (A) Salinity and $\delta^{18}O$ data on a section from the Yenisei Estuary northwards to 77°N (*Akademik Boris Petrov* Expedition, 2000; Stein & Stepanets, 2001). This figure was plotted using Ocean Data View (Schlitzer, 2001). (B) $\delta^{18}O$ versus salinity for stations in the Laptev Sea (grey dots), the Beaufort Sea (crosses; Macdonald, Paton, Carmack, & Omstedt, 1995a, 1995b), the Kara Sea (BP99, BP00, BP01; black dots), and data from ARK IV/3 (Bauch et al., 1995) located in the southern Nansen Basin (squares) (from Bauch et al., 2005).

inflow of low saline, ^{18}O-depleted river water (Spielhagen & Erlenkeuser, 1994; Bauch et al., 1997; Volkmann, 2000a).

Arctic Ocean surface sediments reveal a zonation of $\delta^{18}O$ values in *N. pachyderma* (sin.), which traces the salinity distribution in the uppermost water layer (0–100 m) (Figure 4.46 compare with Chapter 2.2, Figure 2.7), that is, boundaries of defined $\delta^{18}O$ zones are almost parallel to the isohalines in the Eurasian part of the Arctic Ocean (Spielhagen & Erlenkeuser, 1994). In the Amerasian Basin, oceanographic and $\delta^{18}O$ data are too scarce to allow a similar classification. As outlined in Spielhagen and Erlenkeuser (1994), the seaward decrease in $\delta^{18}O$ values (Figure 4.46) opposes an increase of sea-surface salinity at the Laptev Sea continental margin. In this area, which is ice-free in its southern part during the summer season, the thickness of the pronounced low-salinity layer is $\sim$50 m (Fütterer, 1994). The distribution of foraminiferal $\delta^{18}O$ values from surface sediments of this area suggests a habitat change of *N. pachyderma* (sin.) from deeper, saline waters in the southern, ice-free part of this area to the low salinity layer in the ice-covered northern zone. The proposed habitat change of *N. pachyderma* (sin.) in this area of variable ice cover resembles observations reported by Carstens and Wefer (1992) in the southern Nansen Basin.

Spielhagen and Erlenkeuser's (1994) interpretation of the isotope data obtained from surface sediments is also supported by isotope data on living specimen of *N. pachyderma* (sin) from the same areas (Volkmann, 2000a; Volkmann & Mensch, 2001). Under a permanent SIC as typical in the western Fram Strait and the outer Laptev Sea, *N. pachyderma* (sin.) lives in the Arctic halocline and reveals $\delta^{18}O$ values <2‰. In the eastern Fram Strait and northeast of Servernaya Semlya, $\delta^{18}O$ values are between 2.5–3.7‰ due to the deeper average depth of habitat mostly in the Atlantic waters. For a more detailed discussion of isotope data in relationship to salinity, see Spielhagen and Erlenkeuser (1994) and Volkmann and Mensch (2001).

The modern distribution of planktonic foraminiferal $\delta^{18}O$ in the Arctic Ocean, except for the area near the Fram Strait that is affected by warm Atlantic water, suggests that down-core $\delta^{18}O$ variations will primarily reflect the history of freshwater budget rather than temperature changes. Thus, these planktonic $\delta^{18}O$ records of Arctic and Subarctic sediment cores often deviate from the "global" pattern because additional ^{18}O-depleted meltwater has entered the surface ocean during glaciations and especially in the glacial terminations. That is, in contrast to "global" records, the isotope curves may be characterized by heavier values in interglacial/interstadial intervals and light peaks associated with glacial beds (Polyak et al., 2004; Spielhagen et al., 2004). Several of these meltwater events were found in Weichselian records from the Subarctic Norwegian–Greenland Sea (e.g., Jones & Keigwin, 1988; Köhler & Spielhagen, 1990; Sarnthein, Pflaumann, Ross, Tiedemann, & Winn, 1992; Stein et al., 1996; Hald, Dokken, & Mikalsen, 2001) as well as in the central Arctic Ocean (e.g., Stein et al., 1994a; Nørgaard-Pedersen et al., 1998, 2003; Poore, Osterman, Curry, & Phillips, 1999; Polyak et al., 2004; Spielhagen, et al., 2004; Spielhagen, Erlenkeuser, & Siegert, 2005), correlated to the history of icesheets on Scandinavia, the Barents Sea, and/or Greenland (for further discussion, see Chapter 6).

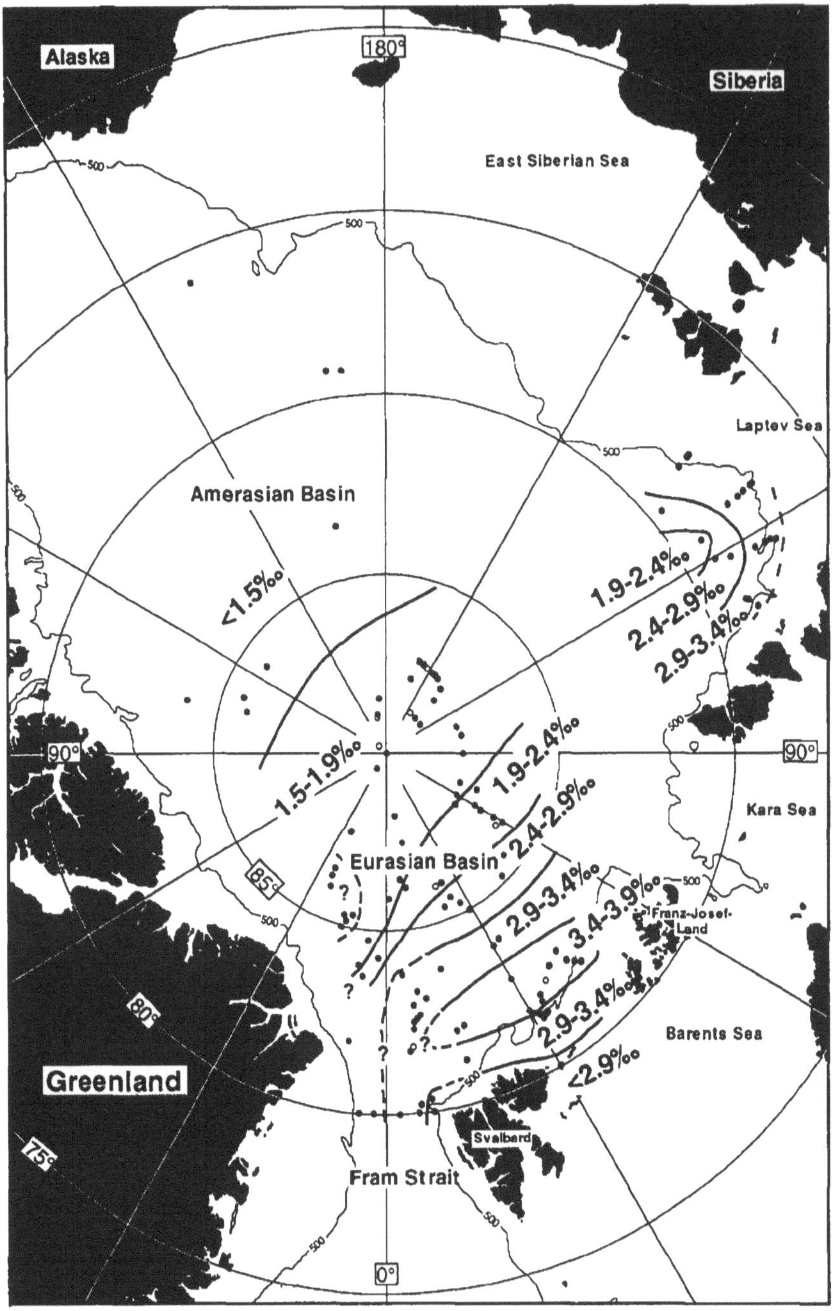

Figure 4.46 Defined zones of $\delta^{18}O$ values from *N. pachyderma* (sin.) in sediment surface samples (from Spielhagen & Erlenkeuser, 1994). Dots mark values matching the $\delta^{18}O$ zone range defined. Open circles mark values out of this range.

Meltwater spikes observed in the Arctic planktonic $\delta^{18}O$ records often coincide with a low $\delta^{13}C$ signal (e.g., Stein et al., 1994a; Nørgaard-Pedersen et al., 1998, 2003; Poore et al., 1999; Spielhagen et al., 2004). This feature is ascribed to reduce ventilation of surface waters caused by strong stratification from the surficial low-salinity freshwater lid (see Chapter 6.3.5).

4.7. Organic-Geochemical Proxies for Organic-Carbon Source and Palaeoenvironment

In this chapter, an overview about the background of different organic-geochemical proxies (more commonly) used for palaeoenvironmental reconstructions from Arctic Ocean sediments is presented. These proxies give information about the origin of the OC preserved in the sediments, needed to interpret the OC data in relation to transport pathways, depositional environment and (palaeo-) climatic/environmental changes as well as for the calculation of flux, remineralization and burial rates of marine and terrigenous OC. For more detailed information on organic-geochemical proxies and their use in (palaeo-) environmental reconstructions, the reader is referred to Romankevich (1984), Tissot and Welte (1984), Stein (1991a), Peters and Moldowan (1993), Tyson (1995), Meyers (1997), Wagner and Dupont (1999), Rullkötter (2000), and Stein and Macdonald (2004a) and further references therein.

Information on OC sources in sediments, that is, estimates of the marine and terrigenous proportion of the organic matter, and on the (palaeo-) environment can be obtained by means of (1) organic-geochemical bulk parameters (C/N ratios, $\delta^{13}C_{org}$ values, Rock-Eval parameters; C/S ratios), (2) maceral composition, and (3) specific biomarker distributions (Figure 4.47). Biomarkers may also allow reconstructions of past SST, SSS, and SIC as well as oxygenation of water masses.

4.7.1. Organic-Geochemical Bulk Parameters and Organic-Carbon Sources

4.7.1.1. Carbon/nitrogen ratios

Carbon/nitrogen (C/N) ratios measured by elemental analysis are often used to distinguish between marine (algal) and terrigenous (higher plant) organic matter in marine sediments. C/N ratios of marine organic matter (mainly phytoplankton and zooplankton) are around 6, whereas terrigenous organic matter (mainly from higher plants) has C/N ratios of >15 (e.g., Bordowskiy, 1965a, 1965b; Scheffer & Schachtschabel, 1984; Hedges et al., 1986). However, many studies on Quaternary marine sediments have shown that interpretation of C/N ratios is not always straightforward and should be done very carefully. It has to be considered, for example, that the C/N ratios given above represent "C_{org}/N_{org}" ratios and that, for simple technical reasons, the C/N ratios of sediments are commonly calculated from the measured total OC and total (i.e., organic and inorganic) nitrogen contents.

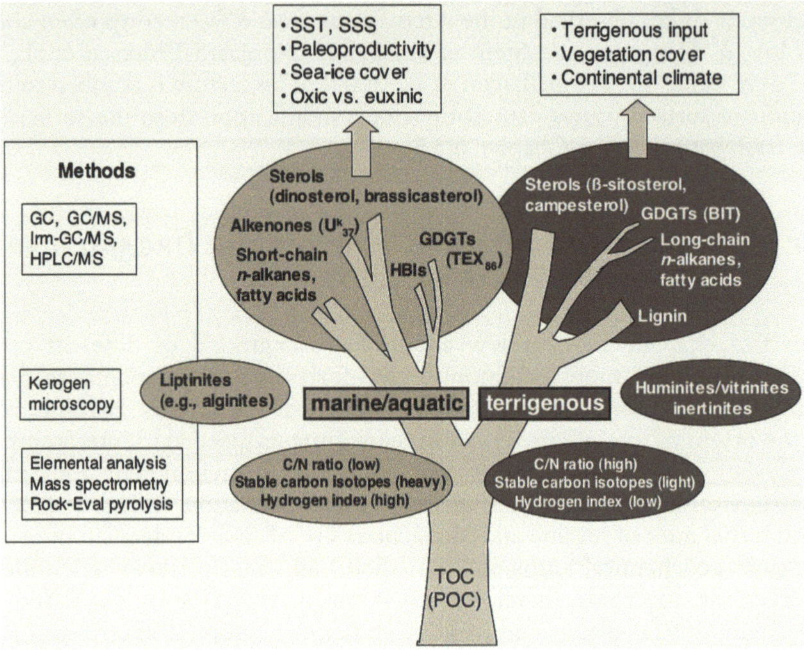

Figure 4.47 Schematic figure indicating organic-geochemical methods and different organic-geochemical bulk parameters, macerals, and biomarkers used for identification of organic-carbon composition and for reconstruction of palaeoenvironment (based on K. Fahl, unpublished course script, 1995; modified and supplemented).

In general, inorganic nitrogen concentrations are small in comparison to those of organic nitrogen, justifying this approach. If the OC content is low (<0.5%), however, the inorganic nitrogen may become a major proportion of the total nitrogen. Furthermore, inorganic nitrogen is bound as ammonium in clay minerals (N_{inorg}), especially in the clay-mineral illite (Stevenson & Cheng, 1972; Müller, 1977). A first-order estimate of the amount of the inorganic nitrogen might be obtained from a total OC versus total nitrogen plot (Figure 4.48).

For surface sediments from the central Arctic containing a high proportion of inorganic nitrogen (N_{inorg}) bound as ammonium in clay minerals (Stein et al., 1994c), for example, calculated C/N ratios are quite low (5–8). This would point to a dominance of marine OC. After correction for N_{inorg}, the "C_{org}/N_{org}" ratios lie between 8 and 15 (Figure 4.48A), indicating a more mixed marine/terrigenous source for the organic matter. This agrees much better with hydrogen index and maceral data, indicating a high proportion of terrigenous OC being present in the surface sediments (Stein et al., 2004b, 2006a; B. Boucsein, unpublished data, 2006). The first-order correction for N_{inorg} based on the OC/N_{tot} plot is well supported by measured N_{org} values (Figure 4.48A; Schubert & Calvert, 2001).

In the Eurasian shelf seas strongly influenced by high river discharge, the OC is predominantly of terrigenous origin (e.g., Stein & Fahl, 2004a, 2004b). This is also reflected in the high C/N ratios. In the southern Kara Sea and estuaries, for

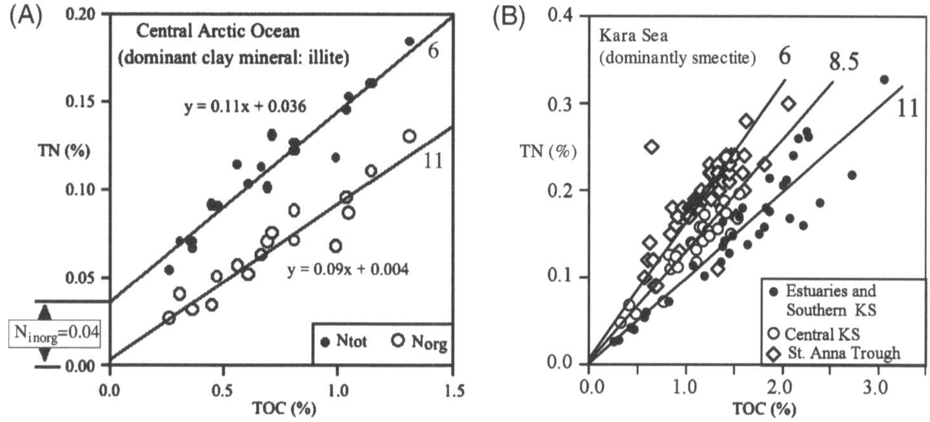

Figure 4.48 Total organic carbon (TOC) versus total nitrogen (N_{tot}) correlation in surface sediments from (A) the central Arctic Ocean (data from Schubert & Calvert, 2001) and (B) the southern Kara Sea (from Stein & Macdonald, 2004b). The estimated amount of inorganic nitrogen (N_{inorg}) is indicated. For the central Arctic Ocean sediments, the correlation between TOC and organic nitrogen (N_{org}) is also shown.

example, high C/N ratios between 9 and 14 were determined (Figure 4.49). As shown in a OC versus N_{tot} diagram, Kara Sea surface sediments probably do not contain significant amounts of inorganic nitrogen (due to the dominance of the clay mineral smectite; Stein et al., 2004a) as implied by the negligible intercept ($<0.01\%$ N_{inorg}) at OC = 0 (Figure 4.48B). Towards the open Kara Sea, with increasing distance from the estuaries, a decrease in amount of terrigenous matter is indicated by decreasing C/N ratios (Figure 4.49). In the St. Anna Trough area of the northern Kara Sea, the relative amount of marine organic matter seems to be significantly higher than in the southern Kara Sea, as suggested from the C/N ratios <7. Exceptions are sites close to Franz Josef Land where local inputs of terrigenous organic matter are indicated by the higher C/N ratios (Figure 4.49). The same picture is obvious from the hydrogen index values (see later description). Probable source areas of the terrigenous organic matter are the OC-rich Mesozoic rocks which out-crop on the islands (e.g., Elverhøi et al., 1995a).

In high-productivity settings, high C/N ratios atypical for algal-source OC were determined in Mediterranean sapropels, late Neogene OC-rich sediments from upwelling areas as well as Cenomanian–Turonian black shales, all characterized by the dominance of marine OC (Meyers, 1997; Twichell, Meyers, & Diester-Haass, 2002). In these sediments, the high C/N ratios are explained by the fact (1) that algae are able to synthesize lipid-rich OC during times of abundant nutrient supply, that is, high primary productivity, and/or (2) that during sinking, partial degradation of algal OC may selectively diminish N-rich proteinaceous components, and raise the C/N ratio (Meyers, 1997). A similar situation, that is, increased preservation of algae-type OC and increased primary production, coinciding with high C/N ratios, was also discussed for the Palaeogene central Arctic Ocean (Stein et al., 2006a; see Chapter 7.2.2 for more details).

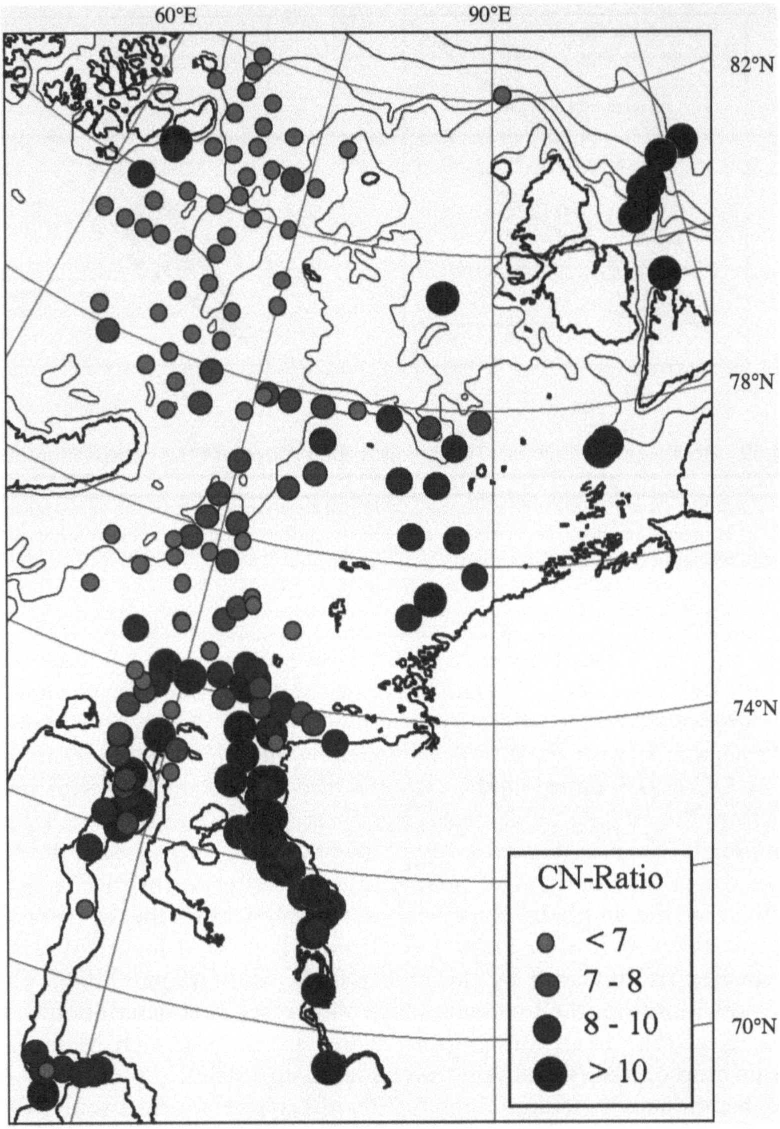

Figure 4.49 Distribution of C/N ratios in Kara Sea surface sediments (from Stein & Fahl, 2004a).

Recent studies from the marginal, modern Arctic Ocean demonstrated the potential of using relative amounts of inorganic nitrogen (N_{inorg}) and organic nitrogen (N_{org}) to track inputs of terrigenous and aquatic/marine organic matter to marine sediments (Winkelmann & Knies, 2005; Knies, Brookes, & Schubert, 2007a). This approach is based on the observation that N_{inorg} values (defined as $N_{inorg}/N_{tot}\%$) make up to 70% of the total nitrogen fraction along Arctic shorelines. In near-surface sediments, the negative correlation of % N_{inorg} with $\delta^{13}C_{org}$

values ($R^2 = 0.65$) implicate that fixed ammonium is land-derived (Knies et al., 2007a).

4.7.1.2. Stable carbon isotopes of organic matter

A common approach for estimating the relative proportions of marine (MarOC) and terrigenous OC (TerrOC) in marine sediments is to assume two endmembers of different isotopic composition (e.g., Hedges, Clark, & Cowie, 1988; Jasper & Gagosian, 1990; Prahl, Ertel, Goñi, Sparrow, & Eversmeyer, 1994; Goñi et al., 2000; Schubert & Calvert, 2001). Assuming a linear mixing of terrigenous and marine organic matter, TerrOC or MarOC can be calculated using the following equation:

$$\text{TerrOC}\ (\%) = \frac{\delta^{13}\text{C}_{\text{sample}} - \delta^{13}\text{C}_{\text{mar}}}{\delta^{13}\text{C}_{\text{terr}} - \delta^{13}\text{C}_{\text{mar}}} \times 100$$

or

$$\text{MarOC}\ (\%) = \frac{\delta^{13}\text{C}_{\text{sample}} - \delta^{13}\text{C}_{\text{terr}}}{\delta^{13}\text{C}_{\text{terr}} - \delta^{13}\text{C}_{\text{mar}}} \times 100$$

where $\delta^{13}C_{terr}$ and $\delta^{13}C_{mar}$ are the isotope values of the terrigenous and marine OC endmembers. In general, light isotopic $\delta^{13}C_{org}$ values of −26 to −28‰ are typical of terrigenous organic matter (i.e., land plants using the C_3 pathway of photosynthesis), and heavy $\delta^{13}C_{org}$ values of −20 to −22‰ are given as characteristic for marine organic matter (marine algae) in low- to mid-latitudes (Meyers, 1994, 1997, and further references therein).

Using this approach for Arctic studies, the terrigenous endmember can be constrained with quite confidence. Land plants using the C_4 pathway for carbon fixation and, thus, characterized by heavy $\delta^{13}C_{org}$ values of about −14‰ (O'Leary, 1988; Meyers, 1994), are of very minor importance in high-latitude areas (Teeri & Stowe, 1976). $\delta^{13}C_{org}$ values determined in particulate organic matter from the Lena River dominating the riverine input into the Laptev Sea, for example, vary between −25.7 and −28.8‰ with an average of −27.1‰ (Rachold & Hubberten, 1999), for Ob and Yenisei draining into the Kara Sea, $\delta^{13}C_{org}$ values of −28 to −28.7‰ and −26.5‰, respectively, were measured (Fernandes & Sicre, 2000; Krishnamurthy, Machavaram, Baskaran, Brooks, & Champs, 2001; Fahl et al., 2003). For the Mackenzie/Beaufort area, −26.5‰ to −27‰ was determined for the terrigenous endmember (Goñi et al., 2000; Naidu et al., 2000). The modern riverine input onto the shelf and its decrease towards the open sea is very well reflected in the distribution of $\delta^{13}C_{org}$ values of surface sediments in the Arctic marginal seas as shown in an example for the Beaufort Sea (Figure 4.50; Macdonald et al., 2004a, 2004b).

The marine endmember is much less well constrained in the Arctic Ocean environment. In high latitudes, $\delta^{13}C_{org}$ values of marine phytoplankton may approach terrestrial values, ranging between −16.7 and −30.4‰ (e.g., Rau, Sweeney, & Kaplan, 1982; Rau, Takahashi, & Des Marais, 1989; Rau, Sullivan, & Gordon, 1991; Goericke & Fry, 1994). This high variability can be related to several

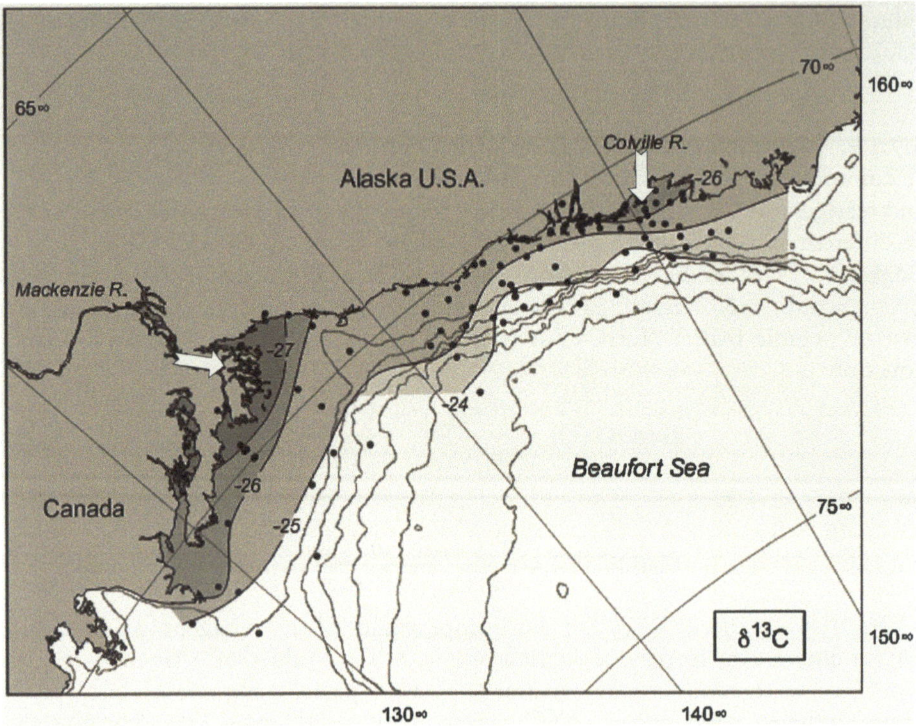

Figure 4.50 A contour plot showing the distribution of $\delta^{13}C$ values for surface sediments collected on the Beaufort shelf and slope (from Macdonald et al., 2004b, based on data from Naidu et al., 2000, supplemented). Open arrows indicate riverine input.

factors such as higher concentration of dissolved CO_2 at low surface-water temperatures, cell growth rate, cell size, and cell membrane CO_2 permeability (e.g., Rau, Takahashi, & Des Marais, 1989; Rau, Riebesell, & Wolf-Gladrow, 1997; Goñi, Ruttenberg, & Eglinton, 1997). Furthermore, $\delta^{13}C_{org}$ values in sea-ice algae may reach heavy values of −15 to −8‰ (Gibson, Trull, Nichols, Summons, & McMinn, 1999). This high variability in the marine endmember often makes it difficult to determine OC sources in marine sediments from sedimentary $\delta^{13}C_{org}$ values.

One possibility to estimate the marine $\delta^{13}C_{org}$ endmember may be obtained from a correlation between $\delta^{13}C_{org}$ values and the sum of long-chain (C_{27}, C_{29} and C_{31}) *n*-alkanes (mg g^{-1} OC), a specific biomarker indicative of vascular plant debris (see later description for details) (Figure 4.51B; Jasper & Gagosian, 1990; Fernandes & Sicre, 2000; Knies, Hald, Ebbesen, Mann, & Vogt, 2003). The almost exclusive terrestrial origin for these biomarker compounds, that is, the absence of the odd high molecular weight *n*-alkanes in the marine organic matter, is a crucial assumption in this approach (Fernandes & Sicre, 2000). Then, the marine $\delta^{13}C_{org}$ endmember can be tentatively predicted from the x-intercept (for $y = 0$) of the linear regression of long-chain *n*-alkanes (mg g^{-1} OC) versus $\delta^{13}C_{org}$ values in the marine sediments. For the Kara Sea surface sediments, Fernandes and Sicre (2000)

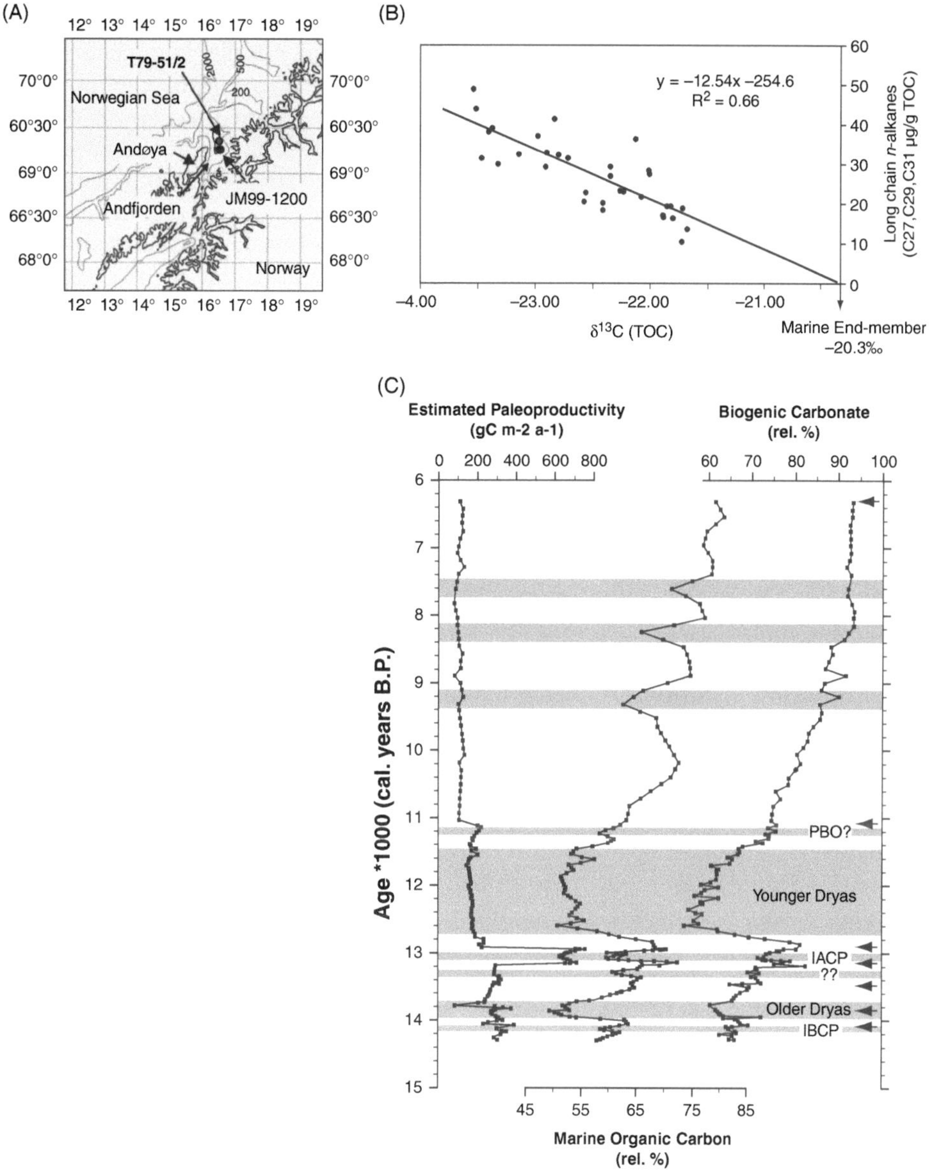

Figure 4.51 (A) Bathymetric map of the Andfjorden area (off Northern Norway) including locations of cores JM99-1200 and T79-51/2. (B) Cross plot of the sum of long-chain (C_{27}, C_{29}, C_{31}) *n*-alkanes (µg/g TOC) and the $\delta^{13}C$ isotopic signature of the total organic matter ($\delta^{13}C_{org}$) in Core JM99-1200. By approaching the regression to zero, a $\delta^{13}C_{org}$ marine endmember of −20.3% was estimated. (C) Palaeoproductivity estimates based on Equation (4.2) in gC m^{-2} y^{-1} (see Table 4.3), relative proportions of marine organic carbon, and relative biogenic (terrigenous-free) carbonate percentages (Core JM99-1200) shown against calendar years. Black arrows mark available AMS^{14}C datings. Multiple cooling events during the last deglaciation and the early/middle Holocene are marked. IBCP, Intra Bølling Cool Period; IACP, Intra Allerød Cool Period; PBO, Preboreal Oscillation (from Knies et al., 2003).

determined a very heavy marine $\delta^{13}C_{org}$ endmember of −16‰. Knies et al. (2003) used the same approach in their study of a deglacial–middle Holocene record of biogenic sedimentation and palaeoproductivity changes from the northern Norwegian continental shelf (Figure 4.51A) and determined a marine $\delta^{13}C_{org}$ endmember of −20.3‰ (Figure 4.51B). Using the $\delta^{13}C_{org}$ endmember mixing model with a marine endmember of −20.3‰ and a terrigenous endmember of −27‰, Knies et al. (2003) obtained a high-resolution record of the variability of marine OC (Figure 5.51C), interpreted by the authors as palaeoproductivity changes in surface-water masses and biogenic sedimentation on the shelf off northern Norway. Multiple cooling events during the last deglaciation appear to correlate with minima in biogenic sedimentation and surface-water productivity, with the most drastic drop occurring during the Younger Dyas. The marine OC record has a distinctly higher time resolution than a palaeoproductivity record calculated from the same core (Figure 4.51C), because for the latter mean linear sedimentation rates for specific intervals have to be used (for calculation procedure see Chapter 4.7.2). Thus, the productivity record reflects only general trends (see Knies et al., 2003 for detailed discussion and further references).

In summary, one should be aware of these difficulties when using $\delta^{13}C_{org}$ values as proxy for OC sources, and interpret the data with caution, or better still, apply the isotope approach together with other proxies for OC source (Goñi et al., 2000; Schubert & Calvert, 2001).

To evaluate the proportion of marine OC and the influence of primary production on the OC accumulation in the central Arctic Ocean, Schubert, Stein, and Calvert (2001) used carbon stable isotopes in combination with nitrogen stable isotopes. There are several general possibilities to explain the variability in carbon and nitrogen isotope distributions (Miyaka & Wada, 1967; Wada et al., 1987; Altabet & Francois, 1994; Francois & Altabet, 1992; Farrell, Pedersen, Calvert, & Nielsen, 1995): (1) Isotopic measurements of nitrogen and carbon represent a variable contribution of terrigenous and marine organic material to the sediment sample, (2) nitrogen isotopes increase due to higher nutrient utilization of the phytoplankton in surface waters and the related heavier carbon isotopic composition of a sample represents the higher contribution of marine organic matter, that is higher productivity of the surface waters, and (3) nitrogen isotopes in the sediment sample increase due to higher nutrient utilization of the phytoplankton in surface waters and the OC isotopic composition is related to changes in the pCO_2 of the surface waters. For details of this approach it is referred to Schubert et al. (2001), Schubert and Calvert (2001), and Knies, Mann, Popp, Stein, and Brumsack (2008).

4.7.1.3. Rock-Eval parameters

The so-called "Rock-Eval" parameters are useful indicators for the characterization of the composition of the OC fraction (i.e., to estimate the amount of terrigenous and marine proportions) in OC-rich (OC >0.5%), immature sediments. The Rock-Eval pyrolysis method is originated from the Petroleum Geology (e.g., Espitalié et al., 1977; Tissot & Welte, 1984; Peters, 1986), but also used very successfully in palaeoceanography research during the last about two decades, in the world oceans including the Arctic Ocean and in time intervals from the Jurassic to the Quaternary (e.g., Arthur, Dean, & Stow, 1984; Dean, Arthur, & Claypool, 1986; Stein, Rullkötter, & Welte, 1986; Stein, 1991a, 2007; Wagner, 1999, 2002; Stein et al., 2001, 2006a).

The pyrolysis is conducted on bulk sediment samples to determine (1) the amount of hydrocarbons already present in the sample (S1 peak in milligram hydrocarbons per gram sediment), (2) the amount of hydrocarbons generated by pyrolytic degradation of the kerogen during heating of up to 550°C (S2 peak in milligram hydrocarbon per gram sediment), (3) the amount of carbon dioxide generated during heating of up to 390°C (S3 peak in milligram carbon dioxide per gram sediment), and (4) the temperature of maximum pyrolysis yield (T_{max} value in °C) (Tissot & Welte, 1984; Peters, 1986). As indicator for the composition of the organic matter (kerogen type), hydrogen (HI), and oxygen index (OI) values were calculated and shown in a "van-Krevelen-type" diagram (Figure 4.52). The HI value corresponds to the quantity of pyrolyzable hydrocarbons (S2) per gram TOC (mgHC gC^{-1}), the OI value corresponds to the quantity of carbon dioxide (S3) per gram TOC (mgCO_2 gC^{-1}). In immature sediments, HI values of <100 mgHC gOC^{-1} are typical of terrigenous organic matter (kerogen type III), whereas HI values of 300–800 mgHC gOC^{-1} are typical of aquatic (algae-type) organic matter (kerogen types I and II) (Figure 4.52; Tissot & Welte, 1984). Based on the measured parameters S1, S2, and S3, additional parameters can be calculated. These are the "Genetic Potential" (S1+S2) and the "Production Index" (S1/(S1+S2)), proxies which give further information about the OC source, source-rock potential, presence of hydrocarbons, and maturity (for further details, see Tissot & Welte, 1984; Peters, 1986; see also Chapter 7.2.2). The T_{max} value from Rock-Eval pyrolysis can be used as maturity indicator. T_{max} values <435°C is an indication for fresh immature organic matter, whereas T_{max} values >435°C point to the presence of more mature and/or refractive organic matter.

When using HI values it has to be considered that this proxy only allows a very general classification of the origin of the organic matter and does not allow to give more precise estimates of the proportions of terrigenous and marine organic matter. Nevertheless, HI values may also allow a first-order quantitative estimate of the abundance of marine or terrigenous OC, as shown by the correlation with kerogen microscopy data (Figure 4.53). Based on the comparison of HI values and maceral composition in immature sediments obtained within the Deep Sea Drilling Project (DSDP), for example, Stein et al. (1986) showed that in samples with HI values around 100 mgHC−gOC^{-1} the relative proportions of terrigenous OC vary between 60% and 100%. Data from arctic and Subarctic areas fit quite well into this correlation plot (Figure 4.53; Stein, Littke, Stax, & Welte, 1989, 2006).

Bulk sediment samples with low OC contents (<0.5%) may yield spuriously low HI values due to adsorption of hydrocarbons onto clay mineral surfaces

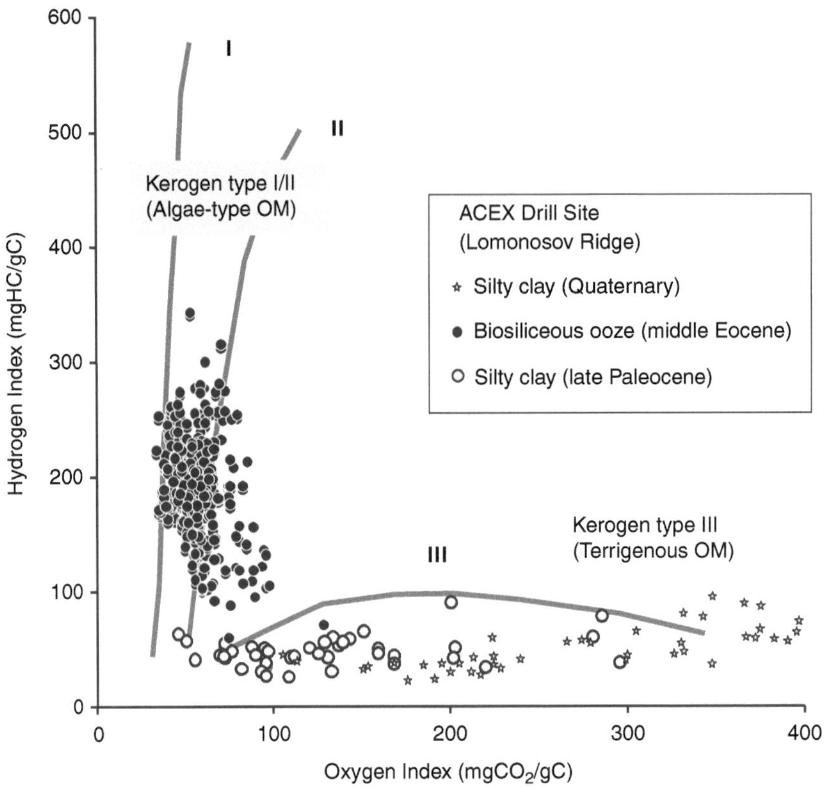

Figure 4.52 Hydrogen index versus oxygen index ("van-Krevelen-type") diagrams for Quaternary, Eocene, and Palaeocene sediments from the ACEX drill site on Lomonosov Ridge, IODP Expedition 302 (Stein, 2007)

("mineral matrix effect"; Espitalié, Makadi, & Trichet, 1984), overestimating the terrigenous proportion of OC in marine sediments. In a plot of the Rock-Eval S2 peak versus TOC, very often a positive TOC value for zero S2 yield is obtained in immature sediments (Figure 4.54). This background value reflects dead carbon (DC) that does not contribute to the S2 pyrolysate yield. For a better classification of the OC type, the HI value can be re-calculated as DC-free $HI' = S2/(TOC-DC) * 100$ (e.g., Langford & Blanc-Valleron, 1990; Cornford, Gardner, & Burgess, 1998; Calvert, 2004; Stein et al., 2006a).

In the Arctic Ocean, Rock-Eval parameters were used to characterize OC sources in surface sediments (Stein et al., 1994c; Birgel & Stein, 2004; Stein & Fahl, 2004a, 2004b; see Chapter 5.2.7) and late Quaternary sediments (Stein & Fahl, 2000; Stein et al., 2001; see Chapter 6.4) from the Eurasian continental margin and adjacent deep-sea basins as well as in a Palaeogene section from Lomonosov Ridge (Figure 4.52; Stein et al., 2006a; Stein, 2007; see Chapter 7.2.2) and upper Jurassic/lower Cretaceous records from the Barents Sea (Langrock et al., 2003a, 2003b;

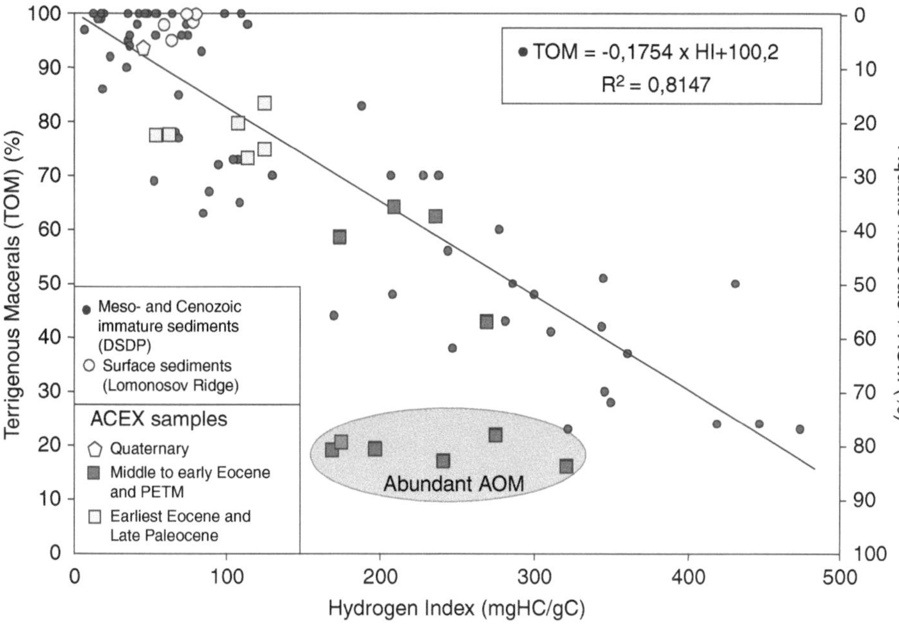

Figure 4.53 Relationship between hydrogen index and maceral composition obtained from kerogen microscopy (TOM, terrigenous macerals as sum of the macerals vitrinite/huminite, inertinite, detritus, that is, vitrinite/huminite and inertinite with grains < 10 μm, and terrigenous liptinites in percentage of total macerals). Regression is based on data determined in immature Mesozoic and Cenozoic sediments (DSDP samples) (Stein et al., 1986). Furthermore, new maceral data from Lomonosov Ridge surface sediments (open circles) and ACEX sediments (squares) also mainly containing immature organic matter (Stein et al., 2006a), were included. Some data points from the middle to early Eocene and PETM event (grey-shaded area) fall out of the correlation due to the very high content of amorphous organic matter (AOM). Following Mann et al. (2008) suggesting that significant amounts of the AOM in the ACEX samples is probably of terrigenous origin as based on the negative correlation between aquatic/algal macerals and % N_{inorg} (see Knies et al., 2007a for background); however, this would increase the terrigenous proportion significantly and shift the data points close to regression line.

Langrock & Stein, 2004; see Chapter 7.1.2). These data clearly indicate the general dominance of terrigenous OC in Quaternary sediments from the central Arctic Ocean and the Eurasian shelves, related to the low primary production and strong river discharge.

In the Eocene OC-rich sediments from Lomonosov Ridge, on the other hand, aquatic (marine and/or freshwater-derived) OC may become dominant (Figure 4.52A), probably caused by increased preservation rate under euxinic conditions (Stein et al., 2006a). Based on the correlation with maceral data, ~40–70% of the OC is of aquatic origin (Figure 4.53). Using the S1, S2, and S3 peaks and the T_{max} values in the Palaeogene record from the Lomonosov Ridge, also information about the hydrocarbon potential, maturity, and possible presence of hydrocarbons became available (Stein, 2007; see Chapter 7.2.2).

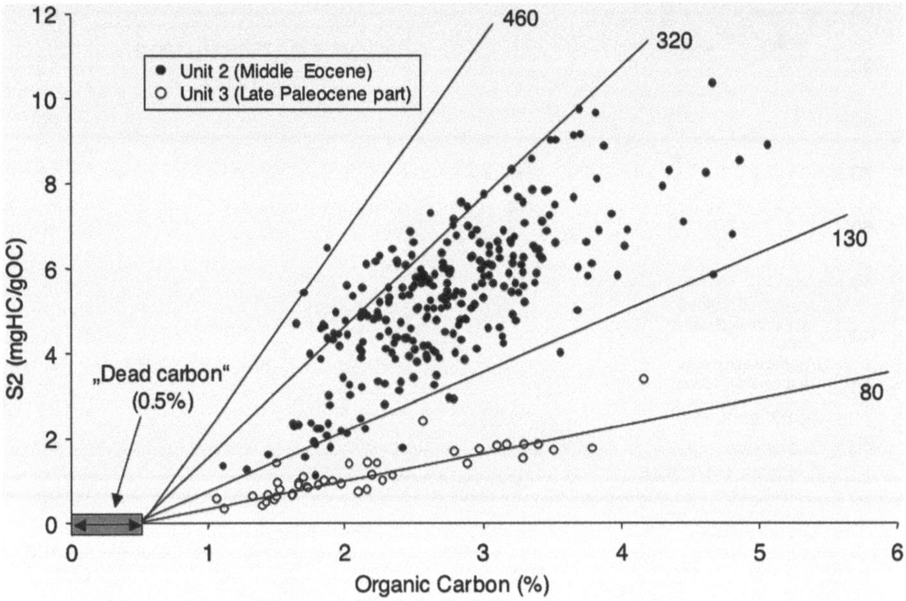

Figure 4.54 Relationship between Rock-Eval S2 peak and total organic carbon (TOC) for middle Eocene and upper Palaeocene sediments of the ACEX sequence (modified from Stein et al., 2006a). The positive TOC value for zero S2 yield reflects dead carbon (DC = 0.5%) that does not contribute to the S2 pyrolysate yield. For a better classification of the organic carbon type, the HI value can be re-calculated as dead carbon-free HI′ = S2/(TOC-DC) $*$ 100 (e.g., Langford & Blanc-Valleron, 1990; Cornford et al., 1998; Calvert, 2004). Given numbers show average hydrogen index values, considering the dead organic carbon content.

4.7.2. Organic-Geochemical Bulk Parameters and Palaeoenvironment

4.7.2.1. Total organic carbon and sedimentation rate

Based on (marine) OC and sedimentation-rate data derived from recent to subrecent sediments deposited in different pelagic to hemi-pelagic, non-turbiditic environments, it may be possible to distinguish between open-marine low- and high-productivity environments and to recognize OC deposition under anoxic deep-water conditions in areas of low sedimentations rate (Figure 4.55; see Stein, 1990, 1991a for background and details).

Under oxic deep-water conditions, a positive correlation exists between (marine) OC content and sedimentation rate (Field A-A′ in Figure 4.55; Heath, Moore, & Dauphin, 1977; Müller & Suess, 1979; Stein, 1990, 1991a; Stein & Littke, 1990). In central open-ocean environments, very low OC contents and low sedimentation rates are typical (Field A), whereas in high-productivity areas (i.e., upwelling areas) very high OC contents and high-sedimentation rates occur (Field A′). In open-ocean oxic environments the positive correlation can be explained by the fact that high-sedimentation rates favour the preservation of OC by reducing its residence time in zones of bioturbation and oxic decomposition. In upwelling areas, increased productivity and flux of organic matter is the dominant factor controlling the OC accumulation.

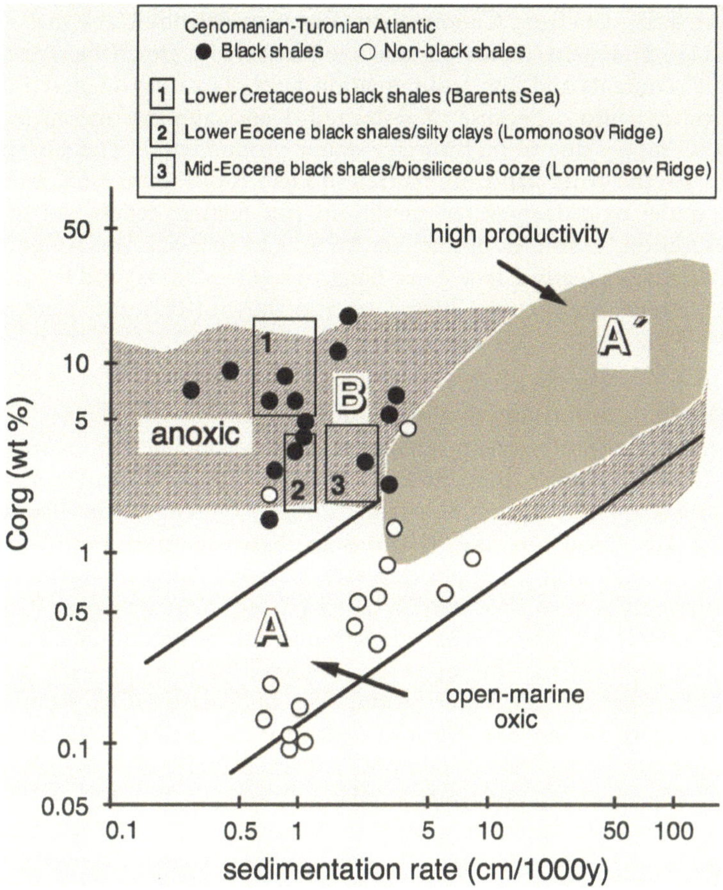

Figure 4.55 Correlation between sedimentation rates and (marine) organic carbon content as based on surface and near-surface sediments from the world oceans (Stein, 1986, 1990) which allows to separate three different environments: (A) open-marine oxic, (A′) high productivity, and (B) anoxic. In addition, data from Cenomanian/Turonian sediments from the Atlantic (Stein, 1986, 1990) and Eocene black shales from Lomonosov Ridge (see Stein, 2007 and further references therein) are shown.

Under anoxic deep- and bottom-water conditions, very high OC contents may occur together with high as well as low sedimentation rates. There is no positive correlation between OC content and sedimentation rate (Field B in Figure 4.55). For example, in the modern Black Sea, maximum OC contents coinciding with minimum surface-water productivity, occur in areas with minimum as well as maximum sedimentation rates (Stein, 1991a, based on Ross, Degens, & MacIlvaine, 1970; Shimkus & Trimonis, 1974; Izdar, Konuk, Ittekkot, Kempe, & Degens, 1983). An OC enrichment under such ("Black-Sea-type") conditions is possible because of a high preservation rate of (marine) organic matter in an anoxic water mass (Demaison & Moore, 1980), although the surface-water productivity and the sedimentation rate may be low.

In Figure 4.55, data from Cenomanian/Turonian Atlantic black shales and non-black shales are presented, indicating that the black shales are characterized by high (marine) OC contents and low sedimentation rates. This may suggest a deposition under anoxic conditions because of restricted deep-water circulation (rather than increased OC supply due to high surface-water productivity). The OC-poor non-black shales, on the other hand, are plotted in Field A of Figure 4.55, interpreted as deposition under oxic deep-water conditions (for further details, see Stein et al., 1986). In addition, examples of ancient Arctic Ocean black-shale-type sediments from the Barents Sea (early Cretaceous; Langrock et al., 2003a) and the Lomonosov Ridge (Palaeogene; Stein et al., 2006a) are also shown (for some more discussion, see Chapter 7.1.2 and 7.2.2, respectively).

4.7.2.2. Organic carbon and palaeoproductivity

A first-order estimate of palaeoproductivity may be obtained from (marine) OC records (Müller & Suess, 1979; Stein, 1986; Sarnthein, Winn, & Zahn, 1987; Sarnthein et al., 1992; Knies & Mann, 2002). The equations for estimating productivity are based on the relationship between measured surface-water productivity, estimates of decomposition rates of organic matter during its transfer from the surface water to the seafloor based on sediment trap data (Suess, 1980; Betzer et al., 1984), and the OC accumulation in surficial sediments. Furthermore, it has to be mentioned that the equations obtained by this approach describe the relationship between surface-water productivity and marine OC accumulation in oxic environments. Below, two equations used for estimating Arctic and Subarctic palaeoproductivity under oxic conditions, are listed (Table 4.3; Equation (4.1) of Stein, 1986; Equation (4.2) of Knies & Mann, 2002).

Based on the correlation between accumulation rates of (marine) OC and primary productivity in recent anoxic environments such as the Black Sea, for example, Bralower and Thierstein (1984) estimated that ~2% of the primarily produced OC is preserved and accumulated in the sediments under anoxic

Table 4.3 Equations for Used for Estimating Arctic and Subarctic Palaeoproductivity under Oxic and Anoxic (Euxinic) Conditions from Organic-carbon Data.

Oxic environments

$$PP = 5.31 * MOC * DBD^{0.71} * LSR^{0.07} * z^{0.45} \quad (4.1) \quad \text{(Stein, 1986)}$$

$$PP = \left(\frac{MOC \times 0.378 \times DBD \times LSR \times z^{0.63}}{(1 - (1/0.037LSR^{1.5} + 1))}\right)^{0.71} \quad (4.2) \quad \text{(Knies \& Mann, 2002)}$$

Anoxic environments

$$PP = 5 * MOC * LSR * DBD \quad (4.3) \quad \text{(Bralower \& Thierstein, 1984)}$$

Notes: PP, paleoproductivity ($gC\,m^{-2}yr^{-1}$); MOC, marine organic carbon content (%); DBD, dry-bulk density ($g\,cm^{-3}$); LSR, linear sedimentation rate ($cm\,ky^{-1}$); and z, (palaeo) water depth of seafloor in metres.

conditions. Using this relationship, these authors derived the formula (4.3) for estimating palaeoproductivities in anoxic environments (Table 4.3).

For Subarctic environments, Equation (4.1) has been used to reconstruct long-term changes in productivity in Baffin Bay during Neogene/Quaternary times as well as short-term glacial/interglacial productivity cycles in the Labrador Sea (Stein, 1991a, 1991b; Stein & Stax, 1991). Knies et al. (2003) published a high-resolution record of palaeoproductivity based on Equation (4.2), for the last deglacial-Holocene time interval in the eastern Norwegian Sea (see Figure 4.51C). The Bralower and Thierstein (1984) approach has been used to estimate palaeoproductivity in the Palaeogene section of the IODP Expedition (ACEX) drill site on Lomonosov Ridge, characterized by euxinic conditions (Stein, 2007; Knies et al., 2008; see Chapter 7.2.2 for discussion).

For further details including deduction of the equations, comparisons of the different approaches, and limitations of use it is referred to Müller and Suess (1979), Bralower and Thierstein (1984), Stein (1986, 1991a), Sarnthein et al. (1987, 1992), and Knies and Mann (2002).

4.7.2.3. Organic carbon versus sulfur and anoxia

The organic carbon/sulfur (C/S) ratio may be used to get first-order information about the oxygenation of bottom water (Leventhal, 1983). For Quaternary normal marine fine-grained detrital sediments a positive correlation between pyritic sulfur and OC with a mean C/S ratio of ~2.8 exists (Goldhaber & Kaplan, 1974; Berner & Raiswell, 1983; Berner, 1984, 1989). In euxinic environments such as the modern Black Sea, however, H_2S already occurs in the water column, and framboidal pyrite can be initially formed, resulting in an excess of sulfur in the C/S diagram (Figure 4.56). C/S ratios of the sediments are very low (typically <1), and a positive S intercept is obvious (Leventhal, 1983). Similar information on the oxygenation of bottom water can be obtained from the TOC-Fe-S relationship (Brumsack, 1988; Dean & Arthur, 1989; Hofmann, Ricken, Schwark, & Leythaeuser, 2000; Langrock et al., 2003b; see Chapter 7.1.2 for some more details). The C/S approach — in combination with other proxies for anoxia such as fine lamination and occurrence of small-sized pyrite framboids (Wilkin et al., 1996, 1997) — has been used by Stein et al. (2006a) for palaeoenvironmental reconstructions of the Palaeogene Arctic Ocean based on ACEX sediments. In the example of Figure 4.56, the changes from oxic to euxinic to oxic conditions across the PETM Event are obvious, also supported by other data (see Chapter 7.2.2 for further discussion). Very recently, the same approach (however, extended by sulfur stable isotope values) was also applied on ACEX sediments by Ogawa, Takahashi, & Yamanaka (2008).

4.7.3. Maceral Composition and Organic Carbon Sources

Macerals, organic particles determined by microscopy under incident and fluorescent light, allow a precise and quantitative distinction of marine and terrigenous organic matter in sediment samples. In general, macerals are distinguished into the three main groups vitrinite/huminite, inertinite and liptinite, and several subgroups,

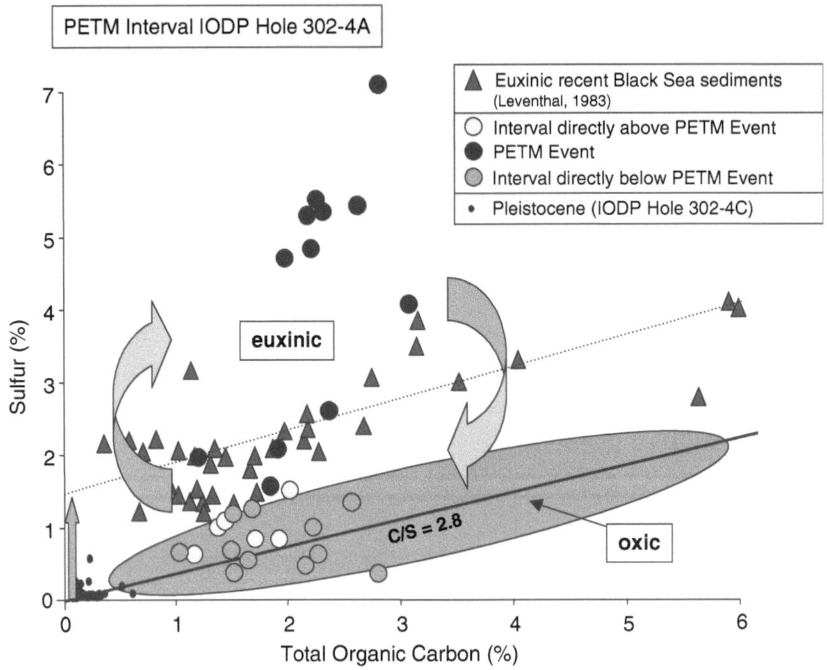

Figure 4.56 Plot of the total organic carbon versus (pyritic) sulfur (C/S diagram), indicating a positive correlation for oxic environments (according to Goldhaber & Kaplan, 1974; Berner & Raiswell, 1983; Berner, 1984). Sediments from the euxinic environment of the modern Black Sea (black triangles; Leventhal, 1983), on the other hand, are characterized by an excess of sulfur in the C/S diagram. In addition, ACEX data across the Palaeocene/Eocene Thermal Maximum (PETM) indicating an evolution from oxic (pre-PETM) to euxinic (PETM) to oxic (post-PETM) conditions, and from Pleistocene sediments (oxic conditions), are shown (data from Stein et al., 2006a). For further explanation see text.

according to the nomenclature described by Taylor et al. (1998). The classification is mainly based on organic petrography studies of coals and sedimentary rocks. Macerals are broadly disperse and generally not abundant in marine sediments (see Littke, 1993 and Wagner, 1999, for introduction of maceral approach to immature marine sediments). Identification and quantification of small-sized organic particles in marine sediments is therefore restricted by various factors, such as size, abundance, and nature of macerals. The recognition of individual macerals is restrained by the maximum resolution of light microscopy (1 μm particle diametre). Based on a comprehensive set of microscopic data Littke (1993) conclude that in most oxic marine sediments more than 80% of the total OC occurs in form of discrete, optically resolvable particles. On the other hand, this percentage may be considerably smaller if marine OC dominates, that is <10% in the upwelling high-productivity areas off Oman and Peru (Lückge, Lallier-Vergès, & Littke, 1996).

Despite some concerns against quantitative optical approaches (i.e., organic petrography and palynology), there are a significant number of studies covering the whole range of old, mature rocks to modern, immature sediments, which show that quantitative optical data convincingly correlate to organic-geochemical proxies (e.g., Stein et al., 1986, 1989, 2006a; Jones, 1987a; Stein & Littke, 1990; Wagner & Henrich, 1994; Wagner & Hölemann, 1995; Boucsein, Fahl, & Stein, 2000; Boucsein, Knies, & Stein, 2002; for reviews, see Littke, 1993; Tyson, 1995; Wagner, 1999, 2002).

For the studies of marine sediments from high latitudes, the classic maceral concept was modified. With reference to the environment and different biological sources, terrigenous and marine macerals were distinguished as shown in Table 4.4 (Boucsein & Stein, 2000). Vitrinite/huminite, inertinite, detritus (vitrinite/huminite and inertinite with grains <10 μm) and specific liptinites (sporinite, cutinite, and fresh-water alginite) are counted as terrigenous particles, whereas marine particles are represented mainly by lamalginite, dinoflagellate cysts, and marine liptodetrinite. Particles <10 μm showing fluorescence are counted as "terrigenous-marine liptodetrinite" because a precise source indication is not possible due to the small particle size.

Based on the abundances of terrigenous macerals (i.e., vitrinite/huminite, inertinite, terrigenous liptinites, detritus, freshwater alginites) and marine macerals (lamalginite, dinoflagellate cysts, marine liptodetrinite), the amount of marine versus terrigenous OC in the Laptev and Kara Sea surface sediments were estimated quantitatively (Figure 4.57; Boucsein, Fahl, Siebold, & Stein, 1999; Boucsein & Stein, 2000). In the southern part of these seas, the sediments contain OC that is >90% of terrigenous origin. Furthermore, close to the river mouths minor but

Table 4.4 Modified Classification of Macerals (Boucsein & Stein, 2000).

Maceral group	Maceral	Origin
Terrigenous macerals	Huminite Vitrinite Inertinite Detritus	 Land plants → fragments <10 μm
	Coal fragments	Coal
	Liptinite: Sporinite Cutinite Suberinite	 Spores/Pollen Cuitcles Bark/Cork
	Freshwater — Alginite	Algae for example, chlorococcalean algae: *Botryococcus or Pediastrum*
Marine macerals	Lamalginite Dinoflagellate cysts Marine liptodetrinite	Marine algae → fragments <10 μm
Liptodetrinite, terrigenous-marine	Liptinite <10 μm	Fragments of terrigenous and marine lipitinites

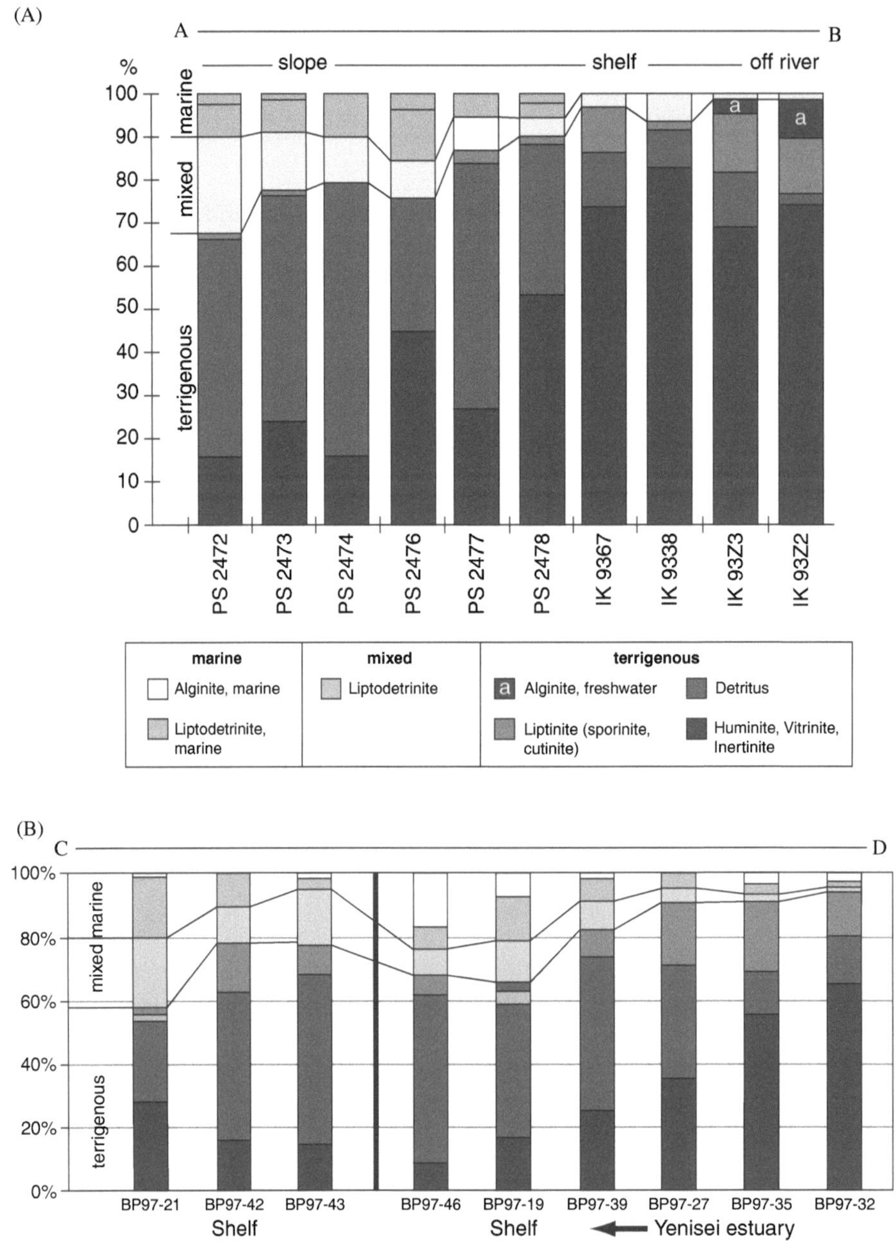

Figure 4.57 Maceral composition in surface sediments from transects (A) across the Laptev Sea continental margin (Boucsein & Stein, 2000) and (B) in the southern Kara Sea/Yenisei Estuary area (Boucsein et al., 1999). For the location of transects see Figures 5.18 and 5.17, respectively.

significant amounts of fresh-water algae were found (3 to almost 10%), indicating aquatic, land-based primary production (Boucsein & Stein, 2000). Towards the north, with increasing distance from the land, the amount of terrigenous macerals decreases, whereas marine macerals occur in significant amounts. However, even on the open shelf, terrigenous OC remains dominant, reaching values of ~70 to >85%. At the upper Laptev Sea continental slope, marine OC may contribute as much as 20–30% of the total OC. A south–north trend is also obvious in the grain size of the terrigenous organic particles. Towards the north, the size of the huminite/vitrinite and inertinite particles decreases while the amount of detritus <10 μm increases, which is caused by a stronger fragmentation and sorting due to long-distance transport (Figure 4.57; Boucsein & Stein, 2000).

4.7.4. Biomarker and Organic Carbon Sources

Whereas the organic-geochemical bulk parameters described above are important proxies for identifying the origin of organic matter in a more general sense (e.g., Stein, 1991a; Meyers, 1994, 1997), specific biomarkers — determined by gas chromatography (GC), gas chromatography/mass spectrometry (GC/MS), high-performance liquid chromatography/mass spectrometry (HPLC/MS), and isotope ratio monitoring gaschromatography/mass spectrometry (Irm-GC/MS) techniques provide more precise information about the different marine and terrigenous proportions of the OC fraction in marine sediments, (e.g., Volkman, Eglinton, Corner, & Forsberg, 1980; Volkman, Barrett, Dunstan, & Jeffrey, 1993; Volkman et al., 1998; Brassell, Eglinton, Marlowe, Pflaumann, & Sarntheim, 1986; Prahl & Muehlhausen, 1989; Sikes, Farrington, & Keigwin, 1991; Yunker, Macdonald, Veltkamp, & Cretney, 1995; Yunker, Belicka, Harvey, & Macdonald, 2005; Hedges, Keil, & Benner, 1997; Fahl & Stein, 1999; Goñi et al., 2000; Villanueva et al., 2001; Hopmans et al., 2004; Pancost & Boot, 2004). Besides marine and terrigenous OC sources, the main topic of this chapter, bacteria may become a significant OC source in marine sediments. Contributions of bacterial sources, for example, can be recognized by specific hopanoid-type compounds (Rohmer, Bouvier-Nave, & Ourisson, 1984) or the abundance of iso- and anteiso-fatty acids (Leo & Parker, 1966; Perry, Volkman, Johns, & Bavor, 1979; Kaneda, 1991). For the Arctic Ocean with its low primary production and high terrigenous (fluvial) input, especially organic-geochemical proxies to be used for the characterization and quantification of terrigenous OC input and burial are of special interest. More detailed discussions on the use of biomarkers as OC source indicators in Arctic Ocean sediments are published in Yunker, Macdonald, Cretney, Fowler, and McLaughlin (1993), Yunker et al. (1995), Yunker et al. (2005), Fahl and Stein (1997, 1999), Goñi et al. (2000), Belicka, Macdonald, and Harvey (2002); Belicka, Macdonald, Yunker, and Harvey (2004), and Birgel, Stein, and Hefter (2004).

4.7.4.1. Biomarker for terrigenous OC

Long-chain odd-carbon-numbered *n*-alkanes are the most widely found terrestrial compounds in marine sediments and are derived from the surface waxes of terrestrial vascular plant material (e.g., Eglinton & Hamilton, 1963; Prahl & Muehlhausen, 1989). In Arctic Ocean studies, long-chain *n*-alkanes were used to characterize the terrigenous OC input by the major rivers into the shelf seas: Beaufort Sea (Yunker et al., 1991, 1995, 2005; Macdonald et al., 2004a, 2004b), East Siberian Sea (Petrova, Batova, Zinchenko, Kursheva, & Narkevskiy, 2004), Laptev Sea (Fahl & Stein, 1997, 1999; Fahl et al., 2001; Stein & Fahl, 2004b), and Kara Sea (Fernandes & Sicre, 2000; Fahl et al., 2003; Stein & Fahl, 2004a; Fahl & Stein, 2007). As example, *n*-alkane data from Kara Sea surface sediments are presented in Figure 4.58. The predominantly terrigenous origin of OC in the southern Kara Sea and, especially in the estuaries, is clearly reflected in the high concentrations of long-chain *n*-alkanes (C_{27}+C_{29}+C_{31}) of ~350–410 μg gOC^{-1}. Towards the open Kara Sea, with increasing distance from the estuaries, a decrease in amount of terrigenous OC is indicated by decreasing long-chain *n*-alkanes. In the St. Anna Trough area in the northern Kara Sea, minimum concentrations of <150 μg gOC^{-1} are reached, suggesting that the relative amount of marine OC seems to be significantly higher than in the southern Kara Sea. This is also supported by organic bulk parameters such as C/N ratios (Figure 4.49; Stein & Fahl, 2004a).

Based on the absolute concentration of these long-chain *n*-alkanes, even quantitative estimates of the proportion of terrigenous organic matter in coastal environment might be possible (Pacific Ocean: Prahl & Carpenter, 1984; Arctic Ocean: Fernandes & Sicre, 2000; Stein & Fahl, 2004a, 2004b; Fahl & Stein, 2007). This approach to quantify the proportion of terrigenous organic matter originated from the observation that at the mouth of the Columbia River to the Washington coast, the concentrations of sum of long-chain *n*-alkanes C_{25}–C_{31} in μg gOC^{-1} were fairly constant despite of the large variability in OC contents (ratio = 277 ±87 μg gOC^{-1}; Prahl & Carpenter, 1984). Based on this observation, the authors developed a simple binary model assuming that (1) long-chain *n*-alkanes and terrestrial OC ("ALKOC"; Figure 4.59) are exported at the mouth of the Columbia River to the Washington coast in a fixed ratio and (2) a change in the ALKOC ratio determined in coastal surface sediments is caused by dilution of terrestrial OC with marine OC that does not contain long-chain *n*-alkanes (Prahl & Carpenter 1984; Prahl & Muehlhausen, 1989; Prahl et al., 1994). Accordingly, the percentage of terrigenous OC (TerrOC) at each station can be calculated from the concentration of long-chain *n*-alkanes in μg gOC^{-1} normalized to the ALKOC ratio:

$$\text{TerrOC (\%)} = \frac{(\text{sum of long-chain } n\text{-alkanes } (\mu\text{g gSed.}^{-1})/\text{OC (g g Sed.}^{-1}))_{\text{sample}}}{\text{ALKOC}} \times 100$$

Fernandes and Sicre (2000) successfully applied this approach for Kara Sea surface sediments strongly influenced by supply of terrigenous OC by the Ob and Yenisei rivers. Following Fernandes and Sicre (2000), TerrOC values were quantified in surface sediments from the Kara and Laptev seas (Stein and Fahl, 2004a, 2004b) and in Holocene records from the Kara Sea (Fahl & Stein, 2007). In the latter, for example, accumulation rates of terrigenous OC were calculated based on long-chain *n*-alkanes, and used for reconstruction of climate-related Holocene change in Yenisei River discharge (see Chapter 6.4.1).

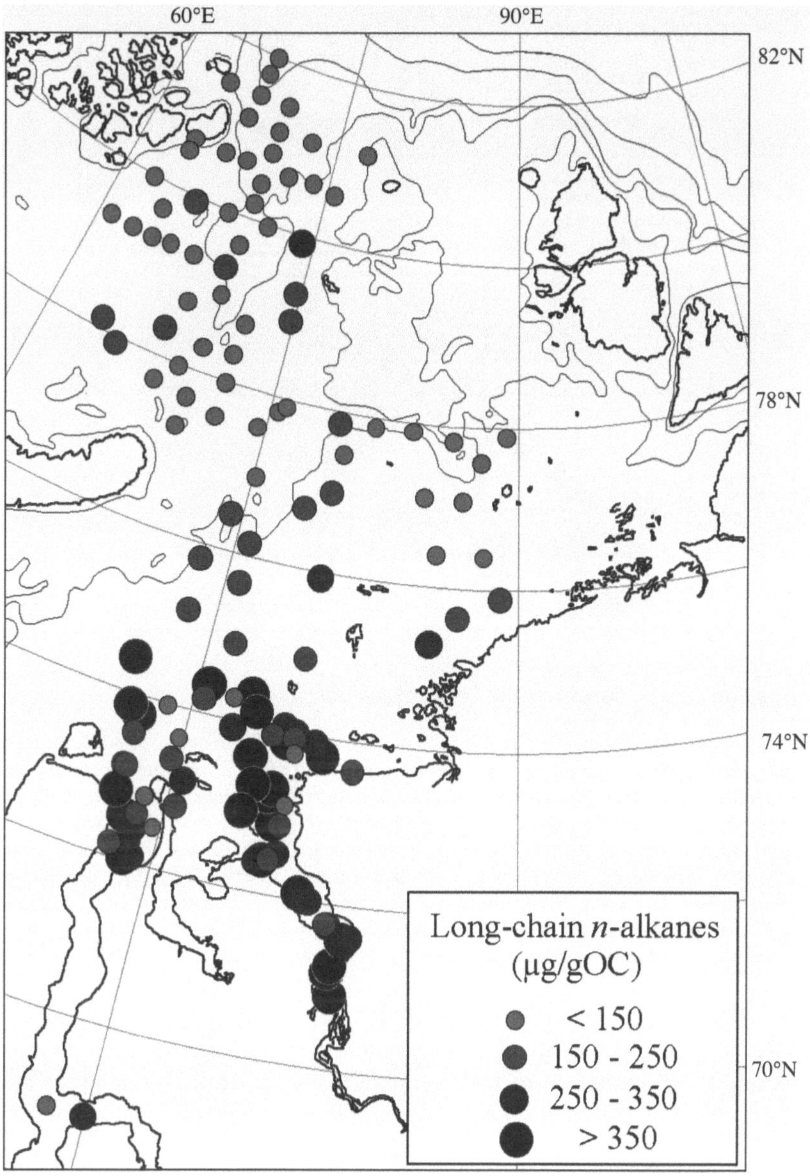

Figure 4.58 Distribution of long-chain *n*-alkanes (C_{27}+C_{29}+C_{31} in $\mu g\,gOC^{-1}$) in Kara Sea surface sediments (from Stein & Fahl, 2004b).

Concerning the "*n*-alkane approach" to estimate the proportions of terrigenous and marine OC, it is important to note that the ALKOC ratio seems to vary from region to region and has to be empirically evaluated for each coastal system (Figure 4.59; Stein et al., 2004b). Using the sum of C_{27}–C_{31} *n*-alkanes, ALKOC ratios of 369 and 750 were calculated for the Ob/Yenisei River and for the Lena River, respectively (Stein & Fahl, 2004a, 2004b), whereas for the Mackenzie River no correlation between *n*-alkanes and OC is obvious (Figure 4.59; Macdonald et al.,

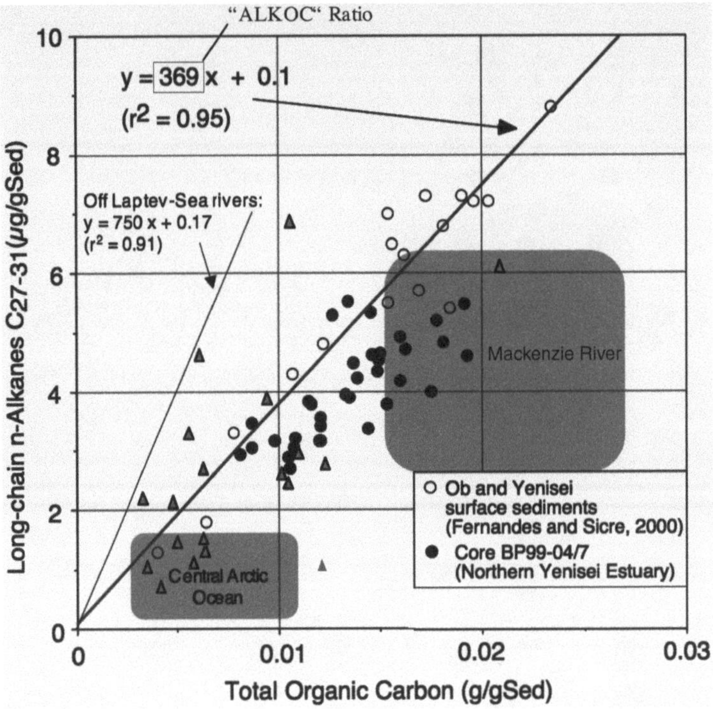

Figure 4.59 Correlation between absolute concentrations of long-chain *n*-alkanes (sum of C_{27}+C_{29}+C_{31} in µg gSed.$^{-1}$) and organic carbon content (g gSed.$^{-1}$) in surface sediments from the Ob and Yenisei estuaries, Kara Sea (open circles; data from Fernandes & Sicre, 2000). In addition, data points from Kara Sea Core BP99-04/7 (solid circles; Stein & Fahl, 2004a) and the correlation line of data points from the Laptev Sea (Stein & Fahl, 2004b) are shown. The grey areas mark the fields of data points from the Mackenzie River and central Arctic Ocean areas, respectively. Triangles are data from Amundsen Basin Core PS2174-5 (Schubert, 1995), suggesting a OC source from the Laptev and/or Kara Sea (see Chapter 6.4.2). Figure from Stein et al., 2004b, supplemented; for further explanation see text.

2004a, 2004b). This implies that the ALKOC ratio cannot simply be used for estimating the relative proportion of terrigenous organic matter in the central Arctic Ocean where the importance of terrigenous OC input from the different marginal seas (characterized by different ALKOC ratios) is generally not known and may vary from area to area.

Other biomarker proxies for the input of higher land plants are long-chain even-carbon-numbered fatty acids and *n*-alcohols (e.g., Meyers, 1997) as well as specific C_{29} sterols (e.g., campesterol and ß-sitosterol) (e.g., Huang & Meinschein, 1976). Based on the relative abundance of C_{29} sterols compared to C_{27} and C_{28} sterols, an estimate of the terrestrial input might be possible (Huang & Meinschein, 1976). According to Volkman (1986), however, C_{29} sterols have also shown to be ubiquitously occurring in marine algae, that is, they may not be purely restricted to terrestrial input. Here, compound-specific carbon isotope measurements may help to distinguish, attribute, and deconvolute distinct (marine or terrigenous) sources of the organic compounds (e.g., Schefuß, Ratmeyer, Stuut, Jansen, & Sinninghe Damsté, 2003; Birgel et al., 2004).

One of the best proxies for vascular plant OC input is lignin (e.g., Hedges & Mann, 1979; Goñi & Hedges, 1992; Goñi et al., 2000). From the specific lignin phenols, an even more detailed identification of different terrigenous sources is possible (e.g., Kattner et al., 1999). Lignin data determined in central Arctic Ocean sediments, however, are very rare (Schubert, 1995; Lobbes et al., 2000).

Very recently, Hopmans et al. (2004) proposed a new tracer for terrestrial OC in sediments based on the analysis of specific tetraether lipids (glycerol dialkyl glycerol tetraethers, GDGTs) using HPLC/MS. These GDGTs (Figure 4.60) were identified in terrestrial as well as marine sediments, however, branched GDGTs appear clearly to be predominant in terrestrial soils and peats (structures VII–IX), whereas structurally related isoprenoid GDGTs are predominant in marine planktonic

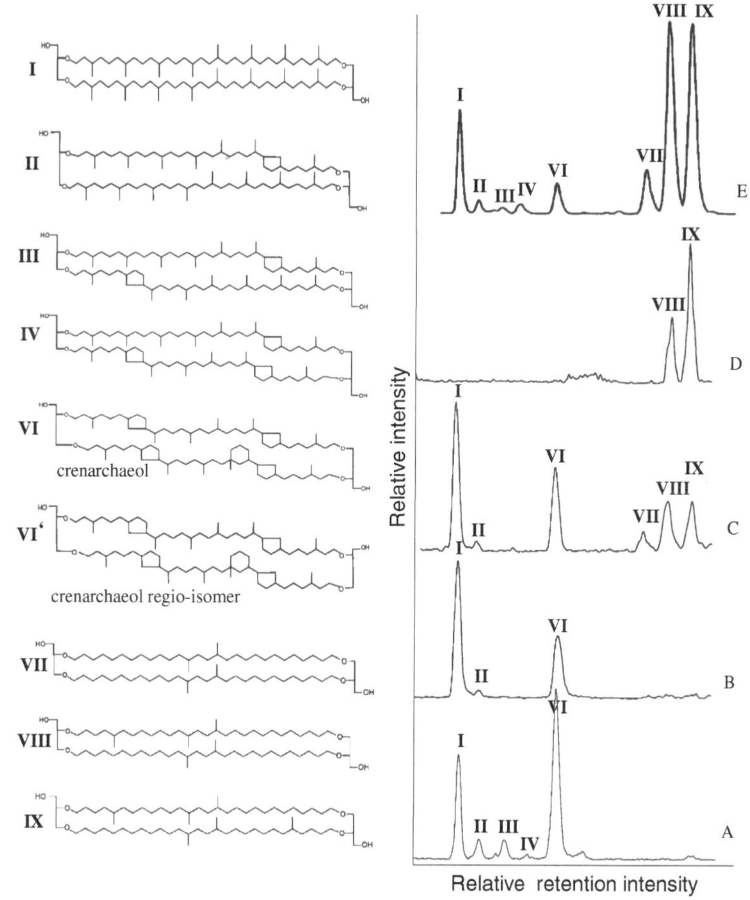

Figure 4.60 Chemical structures of glycerol dialkyl glycerol tetraether (GDGT) membrane lipids (left-hand side, I to IX) (Weijers et al., 2006), and HPLC/MS base peak chromatograms of tetraether lipids in (A) core-top sample from the Congo River basin, (B) water column sample of the Wadden Sea; (C) surface sediment from the Mok Bay, Texel, the Netherlands; (D) soil from a deciduous tree forest, Texel; and (E) France-15, semi-natural (grazed) grassland. Roman numbers refer to structures shown in left-hand side of the figure. (A–D) from Hopmans et al. (2004), (E) from Weijers et al. (2006).

archaea (structure VI; "crenarchaeol") (Sinninghe Damsté, Hopmans, Pancost, Schouten, & Geenevasen, 2000; Sinninghe Damsté et al., 2002a; Sinninghe Damsté, Schouten, Hopmans, van Duin, & Geenevasen, 2002b; Weijers, Schouten, Spaargaren, & Sinninghe Damsté, 2006 and further references therein). Compound I is a less specific GDGT but also predominantly derived from planktonic archaea (Schouten, Hopmans, Pancost, & Sinninghe Damsté, 2000). The isoprenoid GDGTs are also used for SST reconstructions (see Chapter 4.7.5). Based on these findings, Hopmans et al. (2004) defined the "*Branched and Isoprenoid Tetraether (BIT)*" index, which is the ratio between the abundance of terrestrially derived tetraether lipids (Compounds VII, VIII, and IX) versus crenarchaeol (Compound VI):

$$\mathrm{BIT} = \frac{[\mathrm{VII} + \mathrm{VIII} + \mathrm{IX}]}{[\mathrm{VII} + \mathrm{VIII} + \mathrm{IX}] + [\mathrm{VI}]}$$

According to this definition, the BIT index can reach values ranging from 0, representing no branched GDGTs, to 1, representing no crenarchaeol. In a globally distributed set of surface sediments from very different environments, it could be shown that the BIT index correlates very well with the terrestrial (fluvial) input (Hopmans et al., 2004). For example, BIT values clearly follows the plume of the Congo River outflow into the Angola Basin, with BIT values ranging from 0 in open-marine surface sediments to 0.91 for the sediments taken directly at the mouth of the Congo River. In samples from the highly productive upwelling zone of the Peru continental margin, far away from riverine input, BIT values are (close to) 0.

As described by Hopmans et al. (2004) there are several advantages to the use of tetraether lipids compared to other molecular and bulk proxies.

(1) The terrestrial compounds have a similarly functionalized chemical structure as their isoprenoid counterparts and are therefore likely to be degraded at similar rates during sediment diagenesis.
(2) The branched GDGTs are derived from seemingly ubiquitous organisms living in soils and peats and are thus not selective for particular vegetation types or climates. Similarly, crenarchaeol is ubiquitously present in the marine water column and marine and lake sediments (Schouten et al., 2000; Schouten, Hopmans, Schefuß, & Sinninghe Damste', 2002; Powers et al., 2004).
(3) The branched GDGTs as well as crenarchaeol were found in substantial amounts in sediments up to at least the Cretaceous (Kuypers, Pancost, Nijenhuis, & Sinninghe Damsté, 2002). Thus, reconstruction of terrestrial input in ancient immature marine sediments based on the BIT index may be possible at least up to the Cretaceous.

It should be mentioned here, however, that Weijers et al. (2006) found crenarchaeol being unambiguously also present in terrestrial soils. Nevertheless, the BIT values of all the soils they studied are still >0.9, that is, substantially higher than BIT values from open-marine settings. That means, the abundance of crenarchaeol (VI) in soils relative to the branched GDGTs (VII–IX) is very low compared to the marine environment. Thus, Weijers et al. (2006) concluded that, despite the presence of crenarchaeol in soils, the BIT index as proxy for terrigenous OC input in the marine environment is still applicable, although with a somewhat narrower range than previously suggested.

For the Arctic Ocean strongly influenced by river discharge and, thus, high input of terrigenous OC (see Chapter 5.2, Figure 5.23), BIT index data are unfortunately missing yet (but might be a promising approach for future research) — with one exception. In their studies of palaeoenvironmental change across the Palaeocene–Eocene Thermal Maximum (PETM) and the so-called Eocene "Azolla Freshwater Event" on Lomonosov Ridge, Sluijs et al. (2006, 2008) and Brinkhuis et al. (2006), respectively, used the BIT index to get information about terrigenous OC input (see Chapter 7.2.2, for a more detailed discussion of the Palaeogene palaeoenvironment in the central Arctic Ocean).

4.7.4.2. Biomarker for marine OC

Besides alkenones (see later) short-chain C_{17} and C_{19} *n*-alkanes are often used as indicator for marine phytoplankton (e.g., Blumer, Guillard, & Chase, 1971; Kollatukudy, 1976; Fahl & Stein, 1997). These compounds, however, may also be synthesized by freshwater algae (e.g., Eglinton & Hamilton, 1963; Venkatesan, Ruth, Steinberg, & Kaplan, 1987). Dinosterol is well known as a biosynthetic product of dinoflagellates (Boon et al., 1979; De Leeuw, Rijpstra, Schenck, & Volkman, 1983; Volkman et al., 1993, 1998), and brassicasterol reflects the presence of diatoms that occur in both marine and freshwater environments (e.g., Robinson, Eglinton, Brassell, & Cranwell, 1984; Thiel, 1993; Yunker et al., 1995). Thus, especially in environments strongly influenced by fluvial discharge, such as the Beaufort, Laptev, and Kara seas shelves, these biomarkers may derive from both aquatic (freshwater) algae as well as from marine phyto- and zooplankton. Consequently, a pronounced algal signature is not necessarily marine in origin, and all data have to be interpreted cautiously (see Fahl & Stein, 1999 for discussion).

In surface sediments from Fram Strait and across the Yermak Plateau, molecular and isotopic organic-geochemical methods were used to determine the composition, distribution, and origin of extractable aliphatic lipids (*n*-alkanes, *n*-alkanols, fatty acids) (Birgel et al., 2004; Kierdorf, 2006). In the eastern Fram Strait area influenced by the WSC and characterized by partly significantly elevated OC values (Figure 4.61A), Birgel et al. (2004) were able to assign most compounds to three lipid pools, representing (i) primary production (marine phytoplankton, sea-ice algae), (ii) secondary inputs (by feeding of zooplankton, benthic organisms, and bacteria on the former), and (iii) terrestrially derived contributions (Figure 4.61B). The first two compound groups dominated, but varied significantly in relation to the environment, and were highest at the MIZ and along the permanently ice-covered western flank of the Yermak Plateau. In contrast, compounds attributable to a terrestrial source were of only minor importance in terms of absolute concentrations and less variable, but showed increasing relative proportions from an average of 8–14% at and southwards of the MIZ up to 27–33% on the Yermak Plateau and towards the central Arctic Ocean, probably caused by a decrease

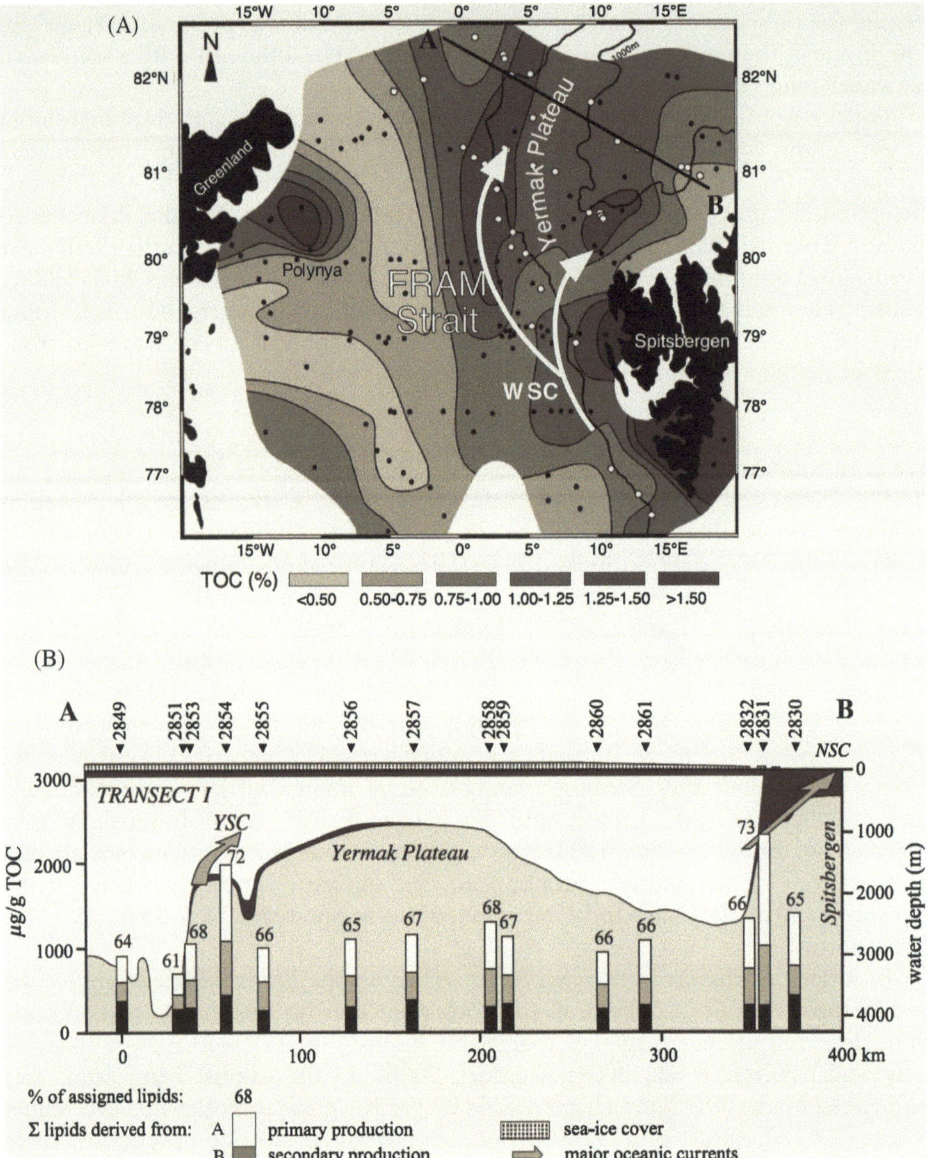

Figure 4.61 (A) Total organic carbon (TOC) content of surface sediments in the Fram Strait (from Birgel et al., 2004, including data from Hebbeln & Berner, 1993; Stein et al., 1994c; Notholt, 1998; Kierdorf, 2006). WSC, West Spitsbergen Current. Line A–B indicates location of transect (B). (B) Total lipids of different sources in surface sediments across the Yermak Plateau (from Birgel et al., 2004). For location of Transects see (A). The thickness of ice cover and water column parts enriched in suspension (adapted from Rutgers van der Loeff, Meyer, Rudels, & Rachor, 2002) is shown schematically and not scaled in vertical extension. The numbers given to the columns represent percentages of lipid pools A, B, and C of the total lipid amounts.

in primary and secondary production (Birgel et al., 2004). When interpreting these data, however, one should have in mind that the lipid fractions only represent a very minor fraction of the total OC. Using organic HI values as a rough proxy for the origin of total OC, terrigenous OC seems to be dominant also in this area (see Chapter 5.2.7, Figure 5.22; Stein et al., 1994c, 2004b; Birgel & Stein, 2004).

To quantify the importance of the three main organic matter inputs involving marine, continental, and ancient reworked organic matter in a sedimentary record from the western Barents Sea (Core M23258; for location see Figure 4.42), Martrat, Grimalt, Villanueva, van Krefeld, and Sarnthein (2003) developed a mass balance model based on the down-core quantification of the C_{37} alkenones as proxy for the marine OC proportion, the odd long-chain *n*-alkanes as proxy for the allochthonous/terrigenous OC proportion, and the unresolved complex mixture of hydrocarbons (UCM) as proxy for the proportion of old, reworked OC. This approach has also been successfully used in studies of sediments from the Atlantic (Müller, Cepek, Ruhland, & Schneider, 1997; Villanueva, Grimalt, Cortijo, Vidal, & Labeyrie, 1997; Sicre et al., 2000; Villanueva et al., 2001) and Indian Oceans (Schubert et al., 1998). According to this model,

$$\mathrm{TOC} = \mathrm{OC}_{\mathrm{reworked}} + \mathrm{OC}_{\mathrm{allochthonous}} + \mathrm{OC}_{\mathrm{marine}} \qquad (4.4)$$

Furthermore, these authors assume that the concentration of each biomarker is linearly related to its associated contribution and that the coefficients K_r, K_a, and K_m remain constant throughout the core.

$$\mathrm{OC}_{\mathrm{reworked}} = K_r * [\mathrm{UCM}] \qquad (4.5)$$

$$\mathrm{OC}_{\mathrm{allochthonous}} = K_a * [n\text{-alkanes}] \qquad (4.6)$$

$$\mathrm{OC}_{\mathrm{marine}} = K_m * [\mathrm{alkenones}] \qquad (4.7)$$

Using Equations (4.5) to (4.7), Equation (4.4) can be expressed as:

$$\mathrm{TOC} = K_r * [\mathrm{UCM}] + K_a * [n\text{-alkanes}] + K_m * [\mathrm{alkenones}] \qquad (4.8)$$

Based on the biomarker concentrations in Core M23258, Martrat et al. (2003) came to the following solution:

$$\mathrm{TOC} = 130 * [\mathrm{UCM}] + 4,628 * [n\text{-alkanes}] + 3,654 * [\mathrm{alkenones}] \qquad (4.9)$$

Using Equation (4.9), the main features of the TOC record of Core M23258 (TOC measured) can be reconstructed quite well (Figure 4.62A and B; for more details on deduction of the equations, error calculations, and limitations of use, see Martrat et al., 2003). As main result of the study by Martrat et al. (2003), the predominance of reworked organic matter is characteristic of Termination Ia (up to 70%), continental organic matter was dominant during the Bølling-Allerød (B-A) and YD periods (~85%), and a strong increase of marine organic matter occurred in the Holocene (between 50% and 75%) (Figure 4.62C).

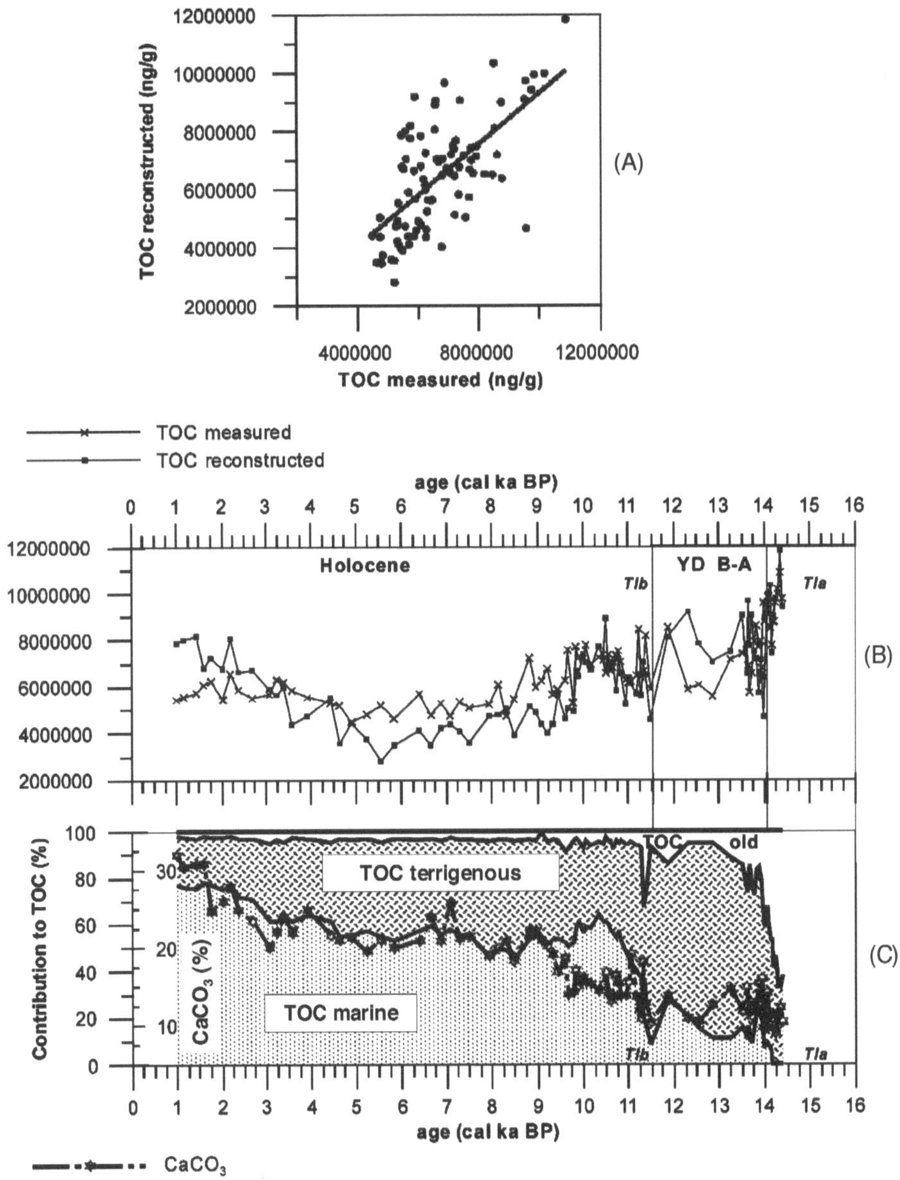

Figure 4.62 (A, B) Comparison of measured versus reconstructed TOC, (C) Percentage of $CaCO_3$ (right axis) and estimation of the marine, allochthonous/terrigenous, and old/reworked contributions to TOC (left axis) at Core M23258 (for location of Core, see Figure 6.76), as defined in the model described in the text (from Martrat et al., 2003). YD, Younger dyras; B-A, Bølling-Allerød; TIa and Tib, Termination Ia and Ib.

4.7.5. Biomarker as Proxies for Temperature, Salinity, Sea-Ice Cover, and Water-Mass Oxygenation

4.7.5.1. Biomarker and sea-surface temperatures

The reconstruction of ocean (and lake) temperatures is of overall importance for understanding past climatic changes. Besides transfer-function techniques based on microfossil assemblages (see Chapter 4.5) and Mg/Ca ratios (e.g., Hastings, Russell, & Emerson, 1998; Lear et al., 2000; Kisakürek, Eisenhauer, Böhm, Garbe-Schönberg, & Erez, 2008), very promising organic-geochemical (palaeo-) temperature tools are the so-called U^K_{37}, TEX_{86}, and HBI proxies. Whereas the first approach is very well established and widely used in numerous palaeotemperature reconstructions, the letter two have been developed more recently (see later description).

The alkenone approach (U^K_{37}). A widespread technique for estimating past SST is based on long chain, unsaturated ketones (alkenones) which are found in marine sediments throughout the world ocean (e.g., Brassell et al., 1986; Rosell-Melé, Carter, & Eglinton, 1994; Rosell-Melé et al., 2001; Müller, Kirst, Ruhland, von Storch, & Rosell-Melé, 1998; Bendle & Rosell-Melé, 2004; Moros et al., 2004). This method evolved from the observation that certain microalgae of the class *Prymnesiophyceae*, notably the marine coccolithophorids *Emiliania huxleyi* and *Gephyrocapsa oceanica* (e.g., Volkman et al., 1980a; Marlowe, Brassell, Eglinton, & Green, 1984; Conte, Eglinton, & Madureira, 1992), and presumably other living and extinct members of the family *Gephyrocapsae* (Marlowe, Brassell, Eglinton, & Green, 1990; Müller et al., 1998), have or had the capability to synthesize alkenones whose extent of unsaturation changes with growth temperature (Marlowe et al., 1984; Brassell et al., 1986; Prahl & Wakeham, 1987). Alkenones were found in samples as old as Cretaceous (Farrimond, Eglinton, & Brassell, 1986; Yamamoto, Ficken, Baas, Bosch, & de Leuw, 1996; Brassell, Dumitrescu, & ODP Leg 198 Shipboard Scientific Party, 2004) and Eocene age (Marlowe et al., 1984; Dzvonik, 1996). In order to permit a direct comparison between water (growth) temperature and alkenone unsaturation, Brassell et al. (1986) introduced the ketone unsaturation index U^K_{37} which is calculated from the relative abundances of C_{37} methyl alkenones containing 2–4 double bonds:

$$U^K_{37} = [C_{37:2} - C_{37:4}]/[C_{37:2} + C_{37:3} + C_{37:4}]$$

As the tetra-unsatured alkenone ($C_{37:4}$) which becomes more significant with decreasing temperature (e.g., Rosell-Melé et al., 1994) and/or reduced salinity (e.g. Rosell-Melé, Jansen, & Weinelt, 2002; Sicre, Bard, Ezat, & Rostek, 2002), is not always present in particulate matter and sediments, a simplified version of the index ($U^K_{37'}$) is usually employed (Prahl & Wakeham, 1987):

$$U^K_{37'} = [C_{37:2}]/[C_{37:2} + C_{37:3}]$$

Globally, the Müller et al. (1998) core-top calibration

$$U^K_{37'} = 0.033\ \mathrm{SST} + 0.044 \quad \text{or} \quad \mathrm{SST} = [U^K_{37'} - 0.044]/0.033$$

which is based on the correlation with annual average temperatures between 60°S and 60°N, is considered the standard in the field (Figure 4.63). This calibration is

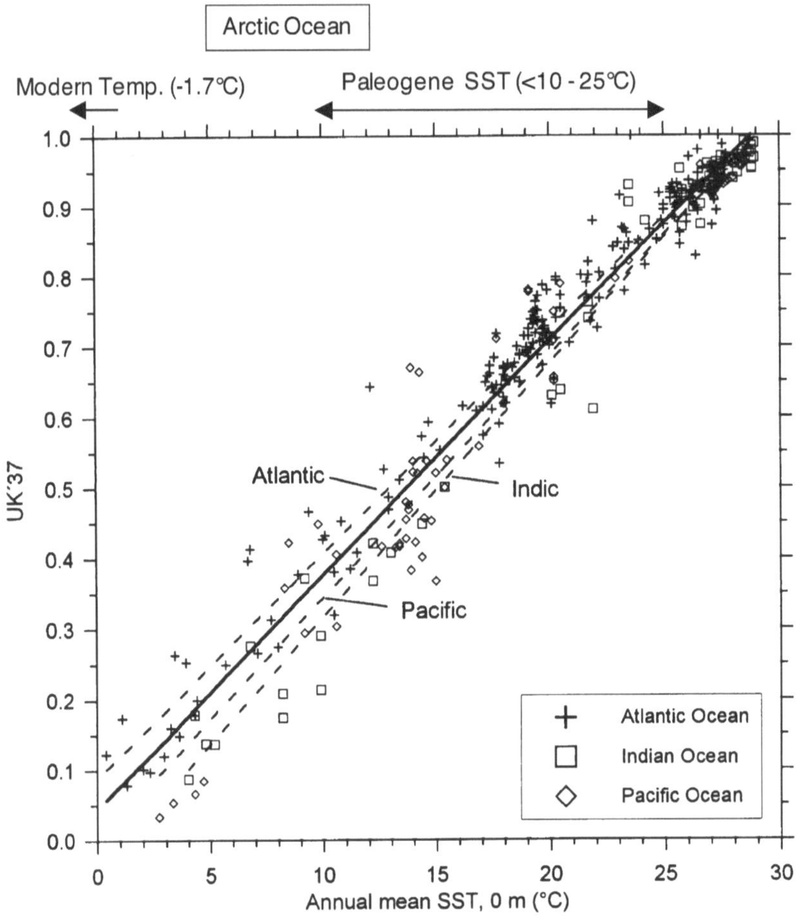

Figure 4.63 Relationships between $U^{K}_{37'}$ and annual mean SST (0 m) for surface sediments from the global ocean between 60°S and 60°N (from Müller et al., 1998, supplemented). The temperature range estimated for the Palaeogene Arctic Ocean (~10–25°C) is shown (see Chapter 7.2.2).

notably equivalent to that originally obtained on both culture and *in situ* data by Prahl and Wakeham (1987). If seasonal calibration is needed, other correlation equations might be more appropriate (Sikes & Volkman, 1993; Sikes, Howard, Neil, & Volkman, 2002).

The optimum temperature range for using the alkenone approach is between 10°C and 25°C, that is, well in the central part of the calibration curve (Figure 4.63). In the central Arctic Ocean, however, modern (interglacial) SST is < 0°C (see Chapter 2.2), and even during glacial times SST cannot become colder. Thus, SSTs of the (Neogene–Quaternary) Arctic Ocean are distinctly outside the calibration curve, and high-latitude SST reconstructions based on alkenones are restricted to Subarctic areas such as the Norwegian Sea where higher SSTs of 5–10°C occur due to the influence of the warm Norwegian Current (e.g., Rosell-Mélé, 2001). For the pre-glacial Palaeogene Arctic Ocean characterized by much warmer SST, however, the U^{K}_{37} index approach has been used on ACEX material (Weller & Stein, 2008; see Chapter 7.2.3).

The TEX_{86} approach. Besides alkenones, the distribution of a number of other lipids present in organisms, especially in their membranes, are known to vary with growth temperatures. In several studies, the membrane lipids of marine *Crenarchaeota*, ubiquitous microorganisms that make up 20–30% of the picoplankton in present-day oceans, have been identified (e.g., DeRosa & Gambacorta, 1988; Schouten et al., 2000; Sinninghe Damsté et al., 2000, 2002a, 2002b). These lipids consist of GDGTs with 0–4 cyclopentane rings and, in one case, a cyclohexane ring (for molecular structures, see Figure 4.60). As shown in a study of marine surface sediments from the world oceans, these organisms adjust the composition of their tetraether membranes lipids in response to SST (Schouten et al., 2002, 2003; Schouten, Hopmans, & Sinninghe Damsté, 2004; Weijers et al., 2006). This response is quantified as the so-called TEX_{86} (tetraether index of 86 carbon atoms), calculated by dividing the quantity of each isomers (II–VI′; see Figure 4.60) according to:

$$TEX_{86} = \frac{[III + IV + VI']}{[II + III + IV + VI']}$$

Based on the original calibration data set based on 43 samples, the TEX_{86} shows a linear relationship for a temperature range from 0°C to 28°C (Figure 4.64A; Schouten et al., 2002; Wuchter, Schouten, Coolen, & Sinninghe Damsté, 2004):

$$TEX_{86} = 0.015\ T + 0.29\ (r^2 = 0.92)\ \text{or}\ T = (TEX_{86} - 0.29)/0.015$$

Thus, it can be used to determine SST from the tetraether membrane composition preserved in sediments and sedimentary rocks. Laboratory experiments support that temperature is a major controlling factor for the GDGT distribution of marine *Crenarchaeota* and that, therefore, TEX_{86} values are mainly determined by temperature. The comparison of the correlation line from the incubation series with the core-top correlation shows, however, that there is a difference in the intersection to the *y*-axes, which is mainly determined by the relative amount of the crenarchaeol regio-isomer (Wuchter et al., 2004 and further references therein).

Very recently, Kim, Schouten, Hopmans, Donner, and Sinninghe Damsté (2008) established a more extensive, global calibration for TEX_{86} based on 287 surface sediment samples. They found a linear relationship with temperature between 5°C and 30°C, but not below 5°C. Using 223 data points, they proposed a new calibration model for the temperature range between 5°C and 30°C (for discussion of regional differences see Kim et al., 2008):

$$TEX_{86} = 0.0178\ T + 0.192(r^2 = 0.935)\ \text{or}\ T = (TEX_{86} - 0.192)/0.0178$$

For the central Arctic Ocean, the TEX_{86} approach (Schouten et al., 2002) has been successfully used for the reconstruction of the pre-glacial Cretaceous and

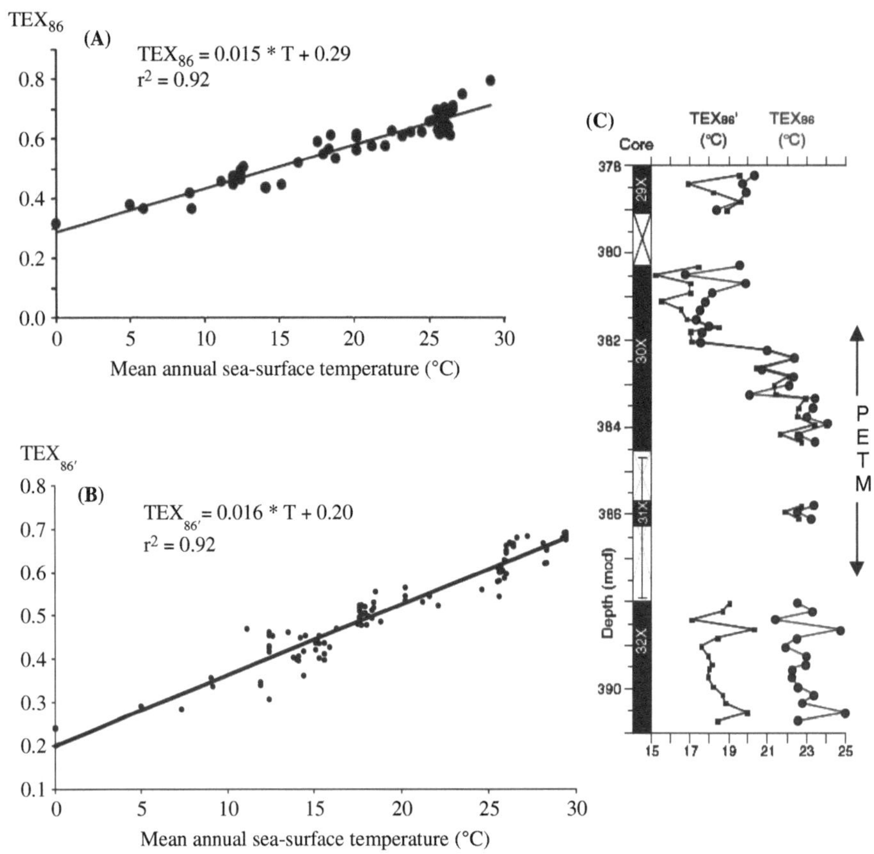

Figure 4.64 (A) Updated correlation line of TEX_{86} in core top samples with annual SST according to Schouten et al. (2002) (from Wuchter et al., 2004); (B) calibration of $TEX_{86'}$ to mean annual sea-surface temperature (from Sluijs et al., 2006); and (C) comparison between $TEX_{86'}$ and TEX_{86} SST across the Palaeocene-Eocene Thermal Maximum (PETM) of IODP Hole 302-4A from Lomonosov Ridge (from Sluijs et al., 2006). See text for further explanations.

Palaeogene time intervals (Jenkyns, Forster, Schouten, & Sinninghe Damsté, 2004; Sluijs et al., 2006; see Chapter 7.2.2). Sluijs et al. (2006), however, used an alternative TEX_{86} proxy, the $TEX_{86'}$. This different approach is based on the finding that the concentration of GDGT-3 (isomer IV in Figure 4.60), almost always the least abundant GDGT lipid in marine sediments, was unusually high in intervals with a high BIT index of 0.5–0.7 (see Chapter 7.2.2). Since terrestrial OC also contains isoprenoid GDGT lipids with cyclopentane rings (Weijers et al., 2006), it is likely that the high terrestrial input in this interval may have disturbed the marine signal. The high abundance of GDGT-3 resulted in unusually high TEX_{86} values and, thus, probably to too high SST values (Figure 4.64C). To circumvent this problem, the $TEX_{86'}$ was introduced which has the same definition as TEX_{86}

except that isomer GDGT-3 was removed from the denominator:

$$\mathrm{TEX}_{86'} = \frac{[\mathrm{III} + \mathrm{VI'}]}{[\mathrm{II} + \mathrm{III} + \mathrm{VI'}]}$$

The $\mathrm{TEX}_{86'}$ was determined for 104 marine surface sediments and found to correlate very well with annual mean SST (Figure 4.64B):

$$\mathrm{TEX}_{86'} = 0.016 \times \mathrm{T} + 0.20\ (r^2 = 0.93) \quad \text{or} \quad \mathrm{T} = (\mathrm{TEX}_{86'} - 0.20)/0.016$$

This equation was used to convert $\mathrm{TEX}_{86'}$ into temperature values (Figure 4.64C).

The fact that cyclopentane-containing isoprenoid GDGTs (isomers II–IV; Figure 4.60) were — although in low concentrations — also found in soils, gives some limitation to the use of TEX_{86} as SST proxy. Especially in coastal or lake environments with a high fluvial input, TEX_{86} SSTs may be biased by this process. At these places, therefore, quantification of the relative terrigenous OC input with the BIT index is required to determine a possible bias in the TEX_{86} proxy (Weijers et al., 2006). This is an important aspect for SST reconstructions in the Arctic Ocean strongly influenced by fluvial (OC) input. Furthermore, it has to be considered, that the GDGT-producing *Crenarchaeota* may live deeper down ($\sim$100 m) and are not dependent on light for growth (e.g., Könnecke et al., 2005; Wuchter, Schouten, Wakeham, & Sinninghe Damsté, 2005; Huguet et al., 2007). Thus, TEX_{86} may represent a subsurface temperature signal in contrast to $\mathrm{U}^{\mathrm{K}}_{37}$ representing surface-water temperature (see also discussion in Chapter 7.2.2). Based on a detailed study of archaeal lipid distributions from globally distributed samples of freshwater, marine, and hypersaline suspended particulate matter, Turich et al. (2007) stated that the lipids used in the TEX_{86} proxy may not record temperature alone, but may also reflect changes in nutrient concentrations, oceanographic conditions, and archaeal ecology.

For further detailed discussion of strengths and limits of the TEX_{86} approach, the reader is referred to Schouten et al. (2002, 2003, 2004), Wuchter et al. (2004, 2005), Herfort, Schouten, Boon, and Sinnighe Damsté (2006), Weijers et al. (2006), Huguet et al. (2007), Turich et al. (2007), and Kim et al. (2008).

The HBI temperature proxy. Highly branched isoprenoid (HBI) alkenes in phytoplankton with numbers of unsaturations that co-vary with temperature, have been discovered (Wraige et al., 1997; Rowland et al., 2001). C_{25} and C_{30} highly branched isoprenoid (HBI) alkenes, known as haslenes and rhizenes, respectively, are biosynthesized by a limited number of diatom genera, *Haslea* spp., *Rhizosolenia* spp., *Pleurosigma* spp., and some *Navicula* spp., and also preserved in the marine sediments (Rowland & Robson, 1990; Volkman, Barrett, & Dunstan, 1994; Belt, Allard, Masse', Robert, & Rowland, 2000, 2001; Massé, Belt, Rowland, & Rohmer, 2004; Sinninghe Damsté et al., 2004). In culturing experiments of *Haslea ostrearia*, Rowland et al. (2001) demonstrated a strong dependence of the extent of unsaturation to the growth temperature. Whereas tetra- and tri-unsaturated haslenes are produced as the major isomers at 25°C and 15°C, respectively, at 5°C diunsaturated haslenes (hasladienes) are biosynthesized as the major isomers. Although this proxy has not been applied in palaeotemperature reconstructions yet, it may be developed to become a promising new palaeotemperature proxy, especially for high latitudes with SST $<<10°C$ where the use of the other SST proxies (i.e., U^{k}_{37} and TEX_{86}) is very limited (Sachs, Pahnke, Smittenberg, & Zhang, 2008).

4.7.5.2. Biomarker and (palaeo-) salinity

Along with temperature, salinity is the most important oceanographic parameter. The reconstruction of past salinity may give, for example, information about water mass stratification and circulation as well as freshwater discharge (see Chapter 7.2). Recent developments in organic geochemistry have produced two promising palaeosalinity proxies: the relative abundance of tetra-unsaturated C_{37} alkenones ($C_{37:4}$), and the hydrogen isotopic composition (D/H) of organic compounds produced by phytoplankton and plants (e.g., Sachs et al., 2008 and references therein).

$C_{37:4}$ Alkenones and salinity. In contrast to the di- and tri-unsaturated alkenones, known to be useful molecules to reconstruct past SST (see earlier), $C_{37:4}$ is rare or absent outside of polar and sub-polar waters and lake sediments. For the Northern North Atlantic, the increased relative abundance of $C_{37:4}$ is coincident with the introduction of increasing scatter in the $U^{K}_{37'}$ to SST-relationship. Moreover, at concentrations of % $C_{37:4}$ >5% a closer correlation of %$C_{37:4}$ with SSS than with SST has been suggested (Rosell-Melé et al., 1994; Rosell-Melé, Weinelt, Sarnthein, Koc, & Jansen, 1998; Rosell-Melé et al., 2001, 2002; Sicre et al., 2002). A $C_{37:4}$-salinity dependence was also found in the low-salinity waters of the Sea of Okhotsk (Seki, 2005) and the Bering Strait area (Harada, Shin, Murata, Uchida, & Nakatani, 2003), whereas no such correlation, for example, occurs in the Southern Ocean region (Sikes & Sicre, 2002). Furthermore, the established % $C_{37:4}$-salinity relations vary significantly between oceanic regions (Table 4.5). Thus, the applicability of % $C_{37:4}$ as a salinity proxy may be restricted to certain cold, low-salinity marine environments (Sachs et al., 2008).

Table 4.5 %$C_{37:4}$-Salinity Relationships in Different Ocean Basins. $C_{37:4}$ Expressed as Relative Abundance of $C_{37:4}$ to the Total Abundance of $C_{37:2}$, $C_{37:3}$, and $C_{37:4}$.

Ocean Basin	Linear regression	R2	References
North Atlantic and Nordic seas	%$C_{37:4}$ = 146.9 (±10.6)−4.1S (±0.3)	0.69	Rosell-Melé et al. (2002)
North Atlantic	%$C_{37:4}$ = 48.1S+1,691	0.78	Sicre et al. (2002)
Bering Strait	%$C_{37:4}$ = 11.7S+397.6	0.76	Harada et al. (2003)

Source: From Sachs et al. (2008).

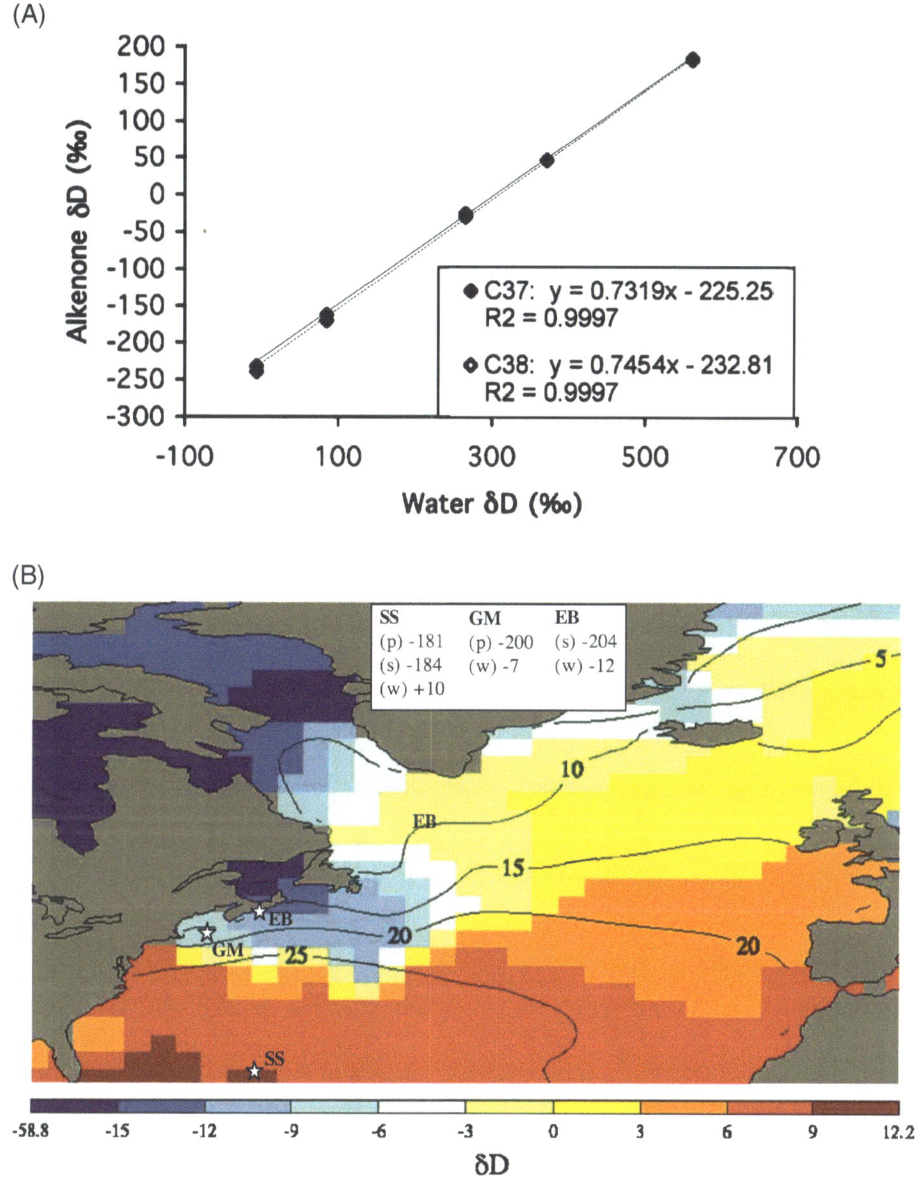

Figure 4.65 (A) Hydrogen isotopic composition of C_{37} (solid symbol and line) and C_{38} (open symbol, dashed line) alkenones from *Emiliania huxleyi* cultures grown at five deuterium enrichments. A linear regression of each data set yields R2 values >0.99, providing strong support for the use of alkenone δD values as a proxy for water δD values. The per mil enrichment factor derived from the intercept of the regression for the C_{37} and C_{38} alkenones, is −225‰ and −233‰, respectively. (B) Summer (July–September) SST (contour lines) and surface water D (contour colours) values in the North Atlantic Ocean; isotope data from Schmidt, Bigg, and Rohling (1999). In the white box, δD values of water (w), particulate alkenones (p), and near-surface sedimentary alkenones (s) from the Sargasso Sea (SS), Gulf of Maine (GM), and Emerald Basin (EB) are listed (from Englebrecht & Sachs, 2005).

Hydrogen isotope ratios of organic compounds and (palaeo-) salinity. All aquatic photosynthetic organisms obtain hydrogen for lipid biosynthesis from the surrounding water. Therefore, hydrogen isotope ratios (δD) of organic compounds synthesized by such organisms directly reflect the δD of the environmental water where they grow in, albeit offset by biochemical fractionation processes. These biochemical fractionation factors ($\varepsilon_{\text{lipid/water}}$) have been determined for a variety of organisms, lipid classes, and distinct biomarkers, such as *n*-alkanes, fatty acids, *n*-alcohols, sterols, and alkenones (Sessions, Burgoyne, Schimmelmann, & Hayes, 1999; Sauer, Eglinton, Hayes, Schimmelmann, & Sessions, 2001; Paul, 2002; Sachse, Radke, & Gleixner, 2004; Englebrecht & Sachs, 2005). δD values in algal sterols and C_{37} alkenones from marine sediments, for example, reliably reflect the δD composition of the water in which they were produced, but are depleted relative to that of the water by about -201 ± 10‰ (Sauer et al., 2001) and -193 ± 3‰ (Figure 4.65; Englebrecht & Sachs, 2005), respectively. These δD values from alkenones in the field were $\sim$32‰ more enriched than predicted from the aggregate *E. huxleyi* culture experiments (Figure 4.65; -225‰). Englebrecht and Sachs (2005) explain this difference by the fact that the field results are more likely to be representative of wild populations of coccolithophorids than those obtained from batch cultures. Furthermore, sedimentary applications show that δD of organic compounds does not only record modern climate variability (Sachse et al., 2004), but is also a suitable proxy for the reconstruction of δD values from past environmental waters and related changes in palaeoceanographic circulation patterns (Andersen et al., 2001; Sauer et al., 2001; Huang, Bryan, Yi, & Thompson, 2002; Paul, 2002; Englebrecht & Sachs, 2005; Sachs et al., 2008).

Salinity perturbations in the Arctic Ocean, induced for example within periods of increased freshwater discharge, massive ice melting or meltwater inflows from glaciers (e.g., during Heinrich Events or glacial–interglacial transitions), can be expected to have introduced a large variability in the source-water δD available to ancient aquatic organisms, and this should be recorded in the δD of specific biomarkers determined in the sedimentary record. Pagani et al. (2006) have successfully used the δD approach to record changes in hydrology, including surface-water salinity and precipitation, in the Palaeogene Arctic Ocean. Their results implicate different water sources for long- and short-chain *n*-alkanes, with δD values of C_{17} *n*-alkane recording the δD of Arctic surface waters, and δD values of C_{27} and C_{29} *n*-alkanes reflecting the hydrogen isotopic composition of precipitation in the Arctic region (see Chapter 7.2.2).

4.7.5.3. Biomarker and (palaeo-) sea-ice cover

The Arctic sea-ice cover and its change through time and space is for several reasons an important key parameter in the (palaeo-) climate system, for example, due to its influence on the earth's albedo, water mass stratification, and primary production (see Chapters 1.1 and 2.3). Despite this importance, detailed information about the extent and variability of palaeo-sea ice is still very sparse. Foraminifers and diatom assemblages and related transfer-function techniques, for example, are common

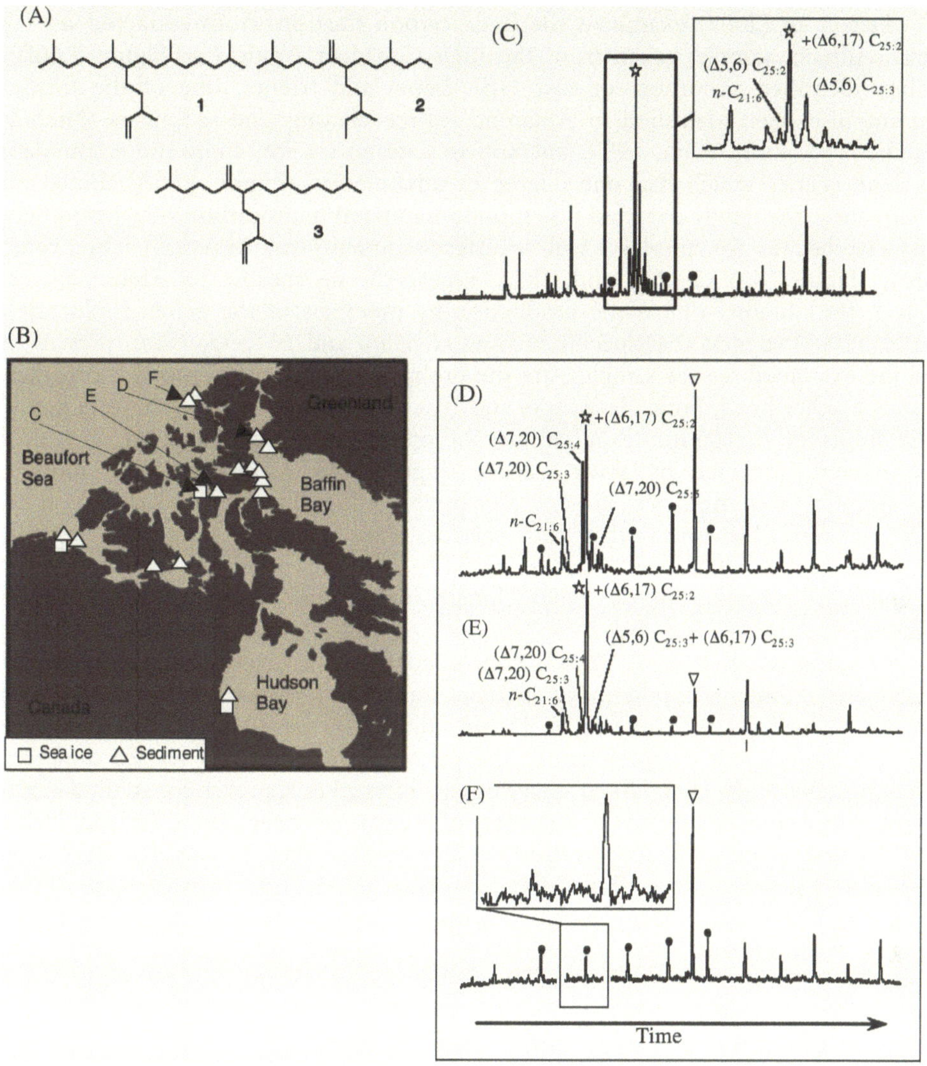

Figure 4.66 (A) Chemical structures of (1) IP_{25}, (2) C_{25} HBI monoene, and (3) C_{25} HBI diene. The double bond in IP_{25} is located at C_{23-24}; (B) map of sampling locations for sea ice and sediment samples of HBI study; location sites of samples (C–F) are indicated; (C) partial total ion current chromatograms of hydrocarbon fractions from extracted Arctic sea ice (McDougall Sound); (D–F) partial total ion current chromatograms of hydrocarbon extracts from Arctic sediments. The peaks labelled as star and open triangle are due to IP_{25} and the internal standard, respectively. Filled circles denote n-alkanes, C_{20}–C_{24}. Note, the absence of IP_{25} in (F) (A–F modified from Belt et al., 2007).

approaches, but limited in use in the Arctic Ocean (see Chapter 4.5.1 and 4.5.4). Here, a novel biomarker approach which is based on the determination of sea-ice diatom specific isoprenoids (C_{25} HBIs), seems to be a major step forward in filling this important gap of knowledge.

Belt et al. (2007) examined the hydrocarbon fractions from extracted sea ice and sediment samples taken from the high Canadian Arctic (see Figure 4.66B). They identified a number of C_{25} HBI dienes and trienes, one of the former having also been identified in Antarctic sea-ice diatoms and sediments (Nichols et al., 1988; Johns et al., 1999) but not, to date, in sea ice, and a more abundant haslene isomer which had one degree of unsaturation (Figure 4.66A). Based on their study, the newly detected C_{25} monounsaturated hydrocarbon (IP_{25}) is present in all studied sea-ice samples as well as surface sediments from seasonally ice-covered areas (Figures 4.66C–66E) and, thus, appears to be specific to Arctic sea ice. Given the ubiquity of haslene production by members of the genus *Haslea*, Belt et al. (2007) suggest that species such as *H. vitrea* and *H. crucigeroides*, present in all the examined sea-ice samples, are the probable sources of this novel biomarker. On the other hand, both *Haslea* spp. and IP_{25} were absent or, at least, were below the limits of detection, from samples of open-water phytoplankton and areas permanently covered by thick sea ice (Figure 4.66F). In a selected number of samples taken from a sediment core representing the last ~9 ka, Belt et al. (2007) could also show that this palaeo-sea-ice proxy is applicable at least throughout the Holocene. Whether the IP_{25} are also preserved in longer sediment cores representing the older climate history, has to be proven in future studies.

In summary, Belt et al. (2007) concluded that the newly characterized HBI monoene represents a potentially specific, stable, and sensitive proxy measure of Arctic sea ice and its presence in Arctic sediment cores may well prove to be a reliable indicator of the presence of the palaeo-ice edge and of sea-ice duration. Very recently Massé et al. (2008) used this new biomarker approach for an ultra-high-resolution reconstruction of the variability of the sea-ice cover around Iceland during the last millennium and its correlation with historical data. Future IP_{25} studies on well-dated sediment cores from high-sedimentation-rate key areas (e.g., the Fram Strait/Yermak Plateau area) will give new insight in the short-term climate variability and its driving forces.

4.7.5.4. Biomarker and oxygenation of water masses

As outlined in Chapter 4.7.2, organic-geochemical bulk parameters may give a first-order information on water-mass anoxia. Specific biomarker, that is, isorenieratane and related isorenieratene derivates, which are exclusively synthesized by green sulfur bacteria (*Chlorobiaceae*), may even allow to reconstruct the extension of euxinic conditions within the water column (Sinninghe Damsté, Wakeham, Kohnen, Hayes, & de Leeuw, 1993; Koopmans et al., 1996; Sinninghe Damsté & Köster, 1998). The *Chlorobiaceae* are photoautotrophic organisms that are strictly anaerobic and require both light and H_2S to thrive, and their presence attests to an anoxic water layer reaching into the photic zone, and euxinic conditions (Koopmans et al., 1996). Isorenieratane and related isorenieratene derivates were found in numerous (sub-) recent and ancient sedimentary sections such as, for example, in Holocene sediments from the Black Sea (Sinninghe Damsté et al., 1993) and in Cretaceous black shales from the Atlantic Ocean (Sinninghe Damsté &

Köster, 1998; Nijenhuis, Bosch, Sinninghe Damsté, Brumsack, & de Lange, 1999), and used there successfully as indication for euxinic condition expanding into the photic zone. In the Arctic Ocean, this biomarker approach allowed to identify euxinic conditions reaching the photic zone during short events at or close to the Palaeocene/Eocene boundary, that is, during the global PETM Event (Sluijs et al., 2006) and the Elmo Event (Weller & Stein, 2008) (see Chapter 7.2.2 for more details).

Part 3: The Marine-Geological Record

CHAPTER FIVE

Modern Environment and Its Record in Surface Sediments

5.1. Terrigenous (Non-Biogenic) Components in Arctic Ocean Surface Sediments: Implications for Provenance and Modern Transport Processes

Factors controlling terrigenous particle input, transportation, and accumulation on the shelf, the adjacent continental slope and the deep sea of the Arctic Ocean are river discharge, coastal erosion, sea ice and icebergs, ocean currents (i.e., surface and bottom currents, transport within the nepheloid layer), gravitational flows (i.e., turbidity currents, debris flows, slumps, slides), and, although of minor importance, aeolian supply (Figure 5.1). Transport processes strongly depend on the grain size of particles. Fine-grained sediments such as clay minerals, for example, are very mobile and can be transported within the water column over long distances into the deep ocean basins. On the other hand, transport by sea ice and icebergs becomes more important for heavy minerals. Thus, a combination of different mineralogical proxies is useful to distinguish between different transport processes (see Chapters 3 and 4 for more details).

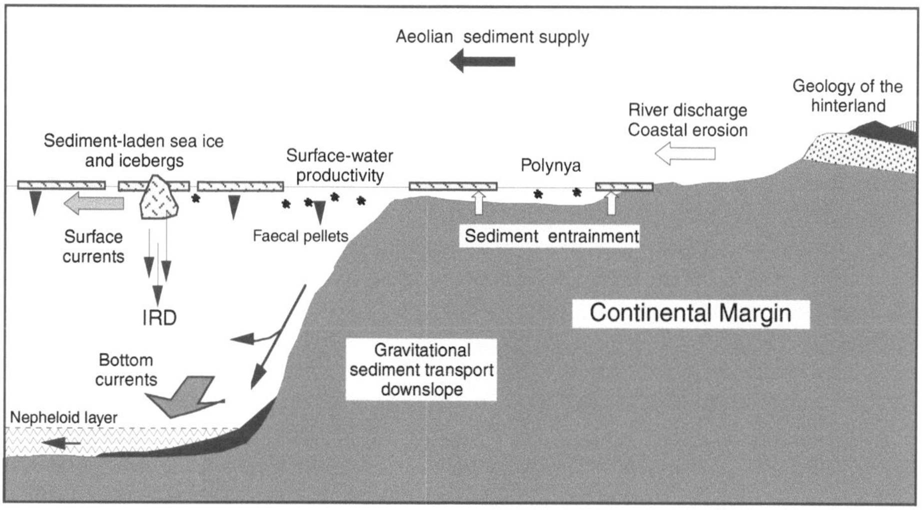

Figure 5.1 Processes controlling terrigenous sediment supply in the Arctic Ocean (modified from Stein & Korolev, 1994).

The main purpose of this chapter is to review data on clay minerals, bulk (light) minerals, and heavy minerals of surface sediments from the marginal seas and the adjacent central Arctic Ocean. Based on the distribution patterns of these minerals, potential source areas and transport processes will be discussed. In order to interpret the composition of terrigenous sediments in terms source indicator, a knowledge of the geology of the surrounding landmasses and its characteristics is needed.

5.1.1. The Geology of the Surrounding Arctic Ocean Landmasses and Its Characteristics

The Arctic Ocean is surrounded by very different geologic terrains (see Chapter 1.3, Figure 1.10; Zonenshain & Natapov, 1989) which are characterized by a very different rock lithologies and composition. Thus, spatial trends in the lithology, mineralogy, and element chemistry of terrigenous surface sediments of the Arctic shelves may reflect variability in the composition of the sediment's source rocks. Here, a few examples from geographically distinct key areas for terrigenous sediment input (especially by sea ice and/or icebergs), characterized by source rocks of a very specific, unique petrographic and chemical composition, are presented.

In the Canadian Arctic, Palaeozoic (largely dolomitized) limestones are widespread (Figure 5.2; Okulitch, 1991; Phillips & Grantz, 2001). Carbonate rocks (partly Mg-rich calcite) are also typical for northern Greenland (Trettin, 1991 and references therein). These lithologies are characteristic for the near-coastal glacigenic sediments (glacial tills) of these areas as well. Thus, dolomite and Mg-rich calcite and their rock fragments are key indicators for identifying sediment input from the Canadian Arctic (e.g., Clark et al., 1980; Darby et al., 1989; Bischof et al., 1996; Vogt, 1997; Phillips & Grantz, 2001). On the other hand, a widespread occurrence of basaltic rocks are restricted to the Eurasian Arctic. Here, the flood basalts of the Putoran Massif (Figure 5.3; Choubert & Faure-Muret, 1976; Duzhikov & Strunin, 1992; Vyssotski, Vyssotski, & Nezhdanov, 2006) are the most important and characteristic source rock. Basalt and its weathering products are characterized by very specific minerals and elemental composition such as clinopyroxene, plagioclase, and smectite as well as, for example, high Ni/Al ratios in the elemental composition, which can be used as tracers for sediment input from the Kara Sea and western Laptev Sea (e.g., Levitan et al., 1996; Wahsner et al., 1999; Schoster et al., 2000; see Chapter 4.3).

Similar to the examples from the Canadian Arctic, North Greenland and the Kara/Laptev Sea, also other areas surrounding the Arctic Ocean show some characteristics which are useful for identifying source areas in the marine records (see later description).

5.1.2. Clay Minerals: Distribution and Significance

In general, records of the clay mineral composition of marine surface sediments may give important information about source areas and transport mechanisms of terrigenous material as well as the climate of the source areas (see Chamley, 1989 for a synthesis). The clay-minerals illite and chlorite are common constituents of plutonic and metamorphic rocks. Mobilized during physical weathering, these

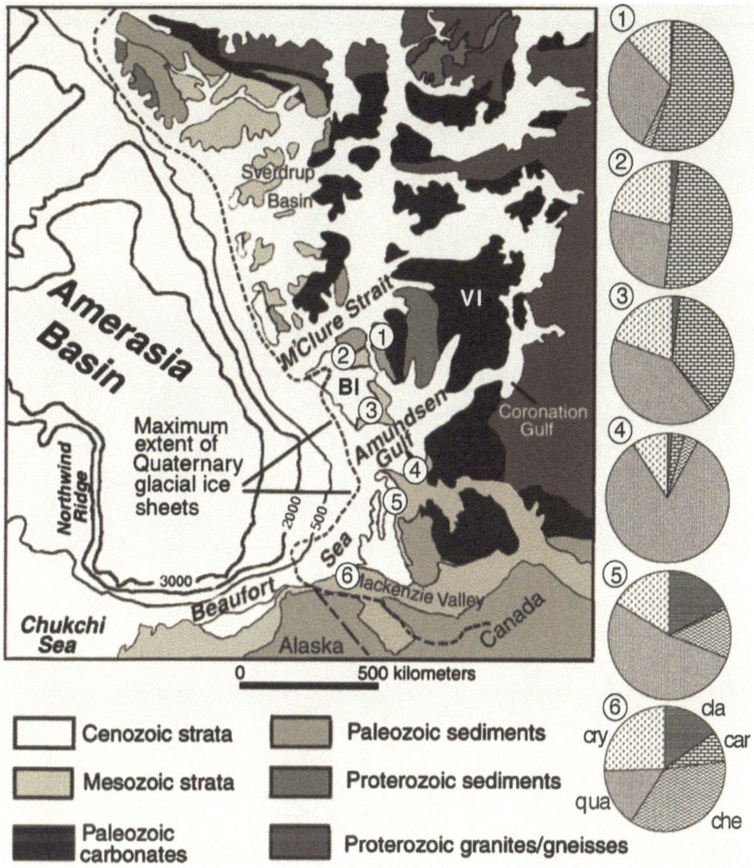

Figure 5.2 Generalized geologic map of the Canadian Arctic Islands, northwestern Canada, and Alaska showing that outcrops of Palaeozoic dolostones and limestones in this area supplied abundant carbonate clasts to the Arctic Ocean (from Phillips & Grantz, 2001, supplemented). The maximum Pleistocene glacial ice edge from Prest (1984) and bedrock geology from Okulitch (1991). In addition, the mineralogical composition of till samples (grain percentage of $>250\,\mu m$ fraction) from Canadian Arctic (locations 1–6) are shown (data from Bischof et al., 1996). Abbreviations used in this figure: cry, crystalline; cla, clastics; car, carbonates; che, cherts; qua, quartz.

minerals are typically found in high-latitude marine sediments. Kaolinite, on the other hand, forms under hot humid conditions and is abundant in the tropics. Kaolinite, however, is also found in polar regions in sedimentary deposits that were formed either under past warmer and wetter climatic conditions than currently exist or at low latitudes and later displaced northward through plate motion. Derived from the weathering of volcanic rocks, smectite is a good indicator of volcanic sediment sources. In marine sediment cores these data may be used to reconstruct the palaeoceanographic and palaeoclimatic changes through time, as shown in numerous studies from different parts of the world ocean (e.g., Janecek & Rea 1983; Stein, 1985; Stein & Robert, 1985; Ehrmann, Melles, Kuhn, & Grobe, 1992; Ehrmann & Mackensen, 1992; Robert & Kennett, 1992, 1994).

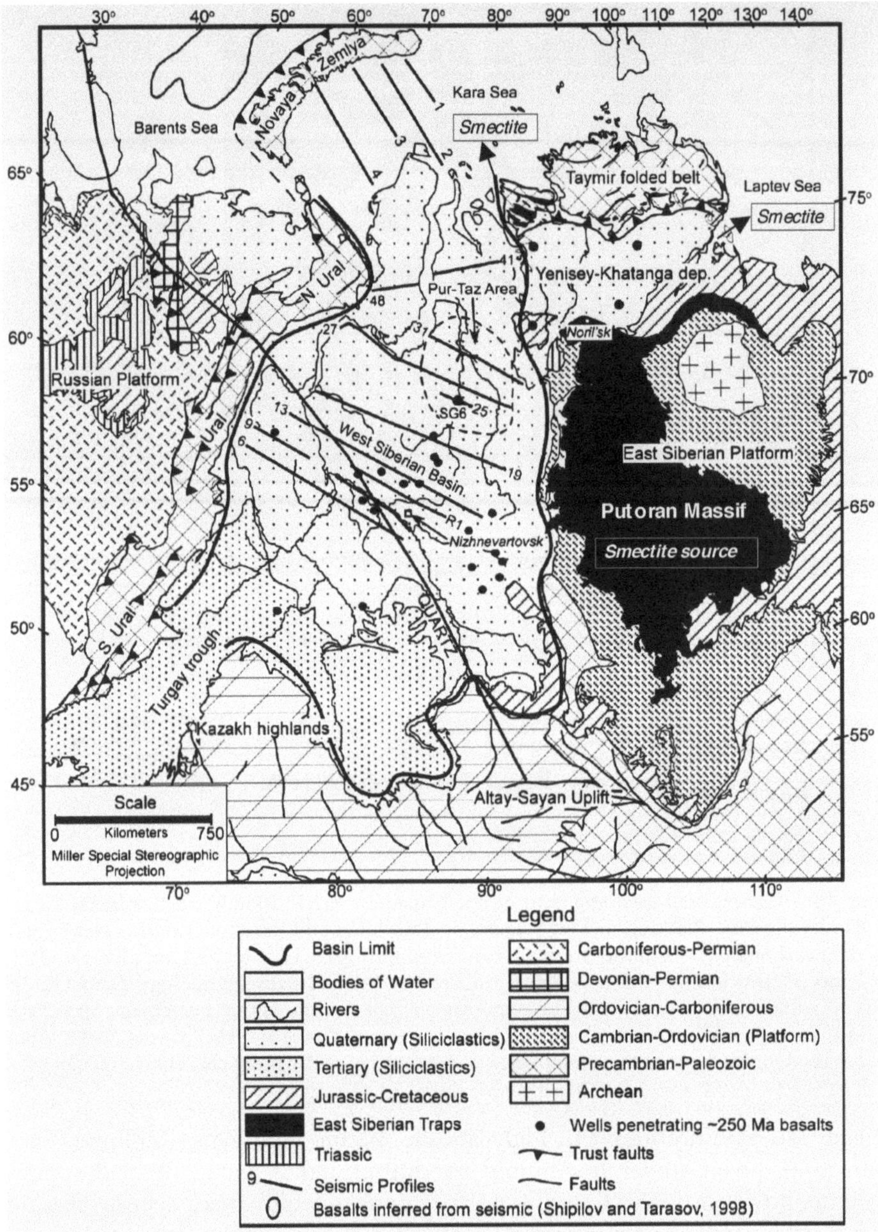

Figure 5.3 Simplified geologic map of the West Siberian Basin and surrounding areas, and source area and transport of smectite (from Vyssotski et al., 2006, supplemented, based on Choubert & Faure-Muret, 1976). Locations of seismic profiles and key wells studied by Vyssotski et al. (2006) are shown.

In polar and subpolar regions, where a cold climate was dominant at least during Neogene and Quaternary times, physical weathering processes dominate and chemical and diagenetic alterations are negligible, the clay-mineral association in marine sediments is a valuable indicator to identify origin and transport pathways of terrigenous sediments.

5.1.2.1. Distribution of clay minerals

In general, the clay mineralogy of recent Arctic Ocean sediments reflects the source mineralogies of the landmasses and shelf areas surrounding the Central Arctic Ocean basins (Naidu, Mowatt, Hawkins, & Hood, 1975; Naidu, Creager, & Mowatt, 1982; Naidu & Mowatt, 1983; Mowatt & Naidu, 1987; Darby et al., 1989; Stein et al., 1994c; Wahsner et al., 1999; Viscosi-Shirley et al., 2003a). Illite is the dominant clay mineral (mostly >50%), followed by chlorite (15–30%). Kaolinite and smectite concentrations are highly variable on the shelves, whereas their distribution is relatively uniform with values <20% in the Central Basin (see Chapter 5.1.5, Table 5.1). Differences in clay-mineral associations between various source areas are obvious and these can be used as source rock indicators.

5.1.2.2. Illite

Illite, a typical clay mineral for cold regions, is most abundant and is delivered by all shelf areas surrounding the Arctic Basin (Figure 5.4A). The maximum illite concentrations north of Svalbard and on the Morris Jesup Rise indicate a transport of illite-rich material from northern Greenland and Svalbard to this area. Transport of terrigenous sediments from the southwestern Canadian Arctic, especially from northern Greenland onto the Morris Jesup Rise, is also supported by the occurrence of sand-sized grains of detrital carbonate (Bischof et al., 1996; Vogt, 1997). Elverhøi, Pfirman, Solheim, and Larsen (1989) have shown that illite is a major component on the northern Barents Sea shelf. Further illite sources for the Barents Sea are Quaternary moraines on Kola Peninsula which contain up to 60% illite, sediments from the White Sea (Kalinenko et al. 1974; Saukel, 2006), and the Pechora River. In the eastern part, Novaya Zemlya is a source area for illite-rich sediments (Figure 5.4A; Nürnberg, Levitan, Pavlidis, & Shelekhova, 1995). The relatively low illite concentrations in the Kara Sea result from dilution with the high smectite component supplied by the rivers Ob and Yenisey (see later description). In the Laptev Sea, high concentration of illite are typical for the eastern Laptev Sea, mainly delivered by the rivers Lena and Yana (Rossak et al., 1999). In the western Laptev Sea, on the other hand, illite concentration is significantly lower due to smectite input by the river Khatanga and inflow from the Kara Sea (see later description). In the East Siberian Sea, coastal abrasion and riverine input by the Indigirka and Kolyma rivers are the main illite sources. Modern soils and soil-forming rocks of the Yana-Indigirka lowlands contain up to 70% illite (Kalinenko, Shelekhova, & Wahsner, 1996). Other illite sources are the Mesozoic rocks on the banks of the big rivers, metamorphic rocks, sandstones, and shales in the Chukchi

Table 5.1 Average Values of Clay Minerals (CM) and Selected Heavy Minerals (HM).

Area	Illite	Chlorite	Smectite	Kaolinite	Amphibole	Clinopyroxene	References
Laptev Sea							
Khatanga SPM (Western LS)	9	8	83	0	9	27	CM:1, HM: 2
Lena SPM (Eastern LS)	54	22	16	8	22	8	CM:1, HM: 2
Yana SPM (Eastern LS)	67	29	0	4	5	2	CM:1, HM: 2
SS (Western Shelf)	40	22	26	14	21	29	CM:4, HM: 2
SS (Eastern Shelf)	49	21	19	11	30	13	CM:4, HM: 2
SIS (Western LS)	29	15	31	25	22	31	CM:1, HM: 3
SIS (Eastern LS)	49	13	12	26	31 (off Lena)	10 (off Lena)	CM:1, HM: 3
Kara Sea							
Ob SPM	28	15	43	14	10	31	CM:2, HM: 2
Yenisei SPM	23	18	44	15	10	48	CM:2, HM: 2
SS (Southwestern Kara Sea)	38	15	38	10	20	10	CM:4, HM: 5
SS (Southeastern Kara Sea)	19	12	60	9	8	36	CM:4, HM: 6
Area around Franz Josef Land	21	14	23	29	7	35	CM:4, HM: 7
St. Anna Trough	36	22	25	17	15	25	CM:4, HM: 7
Barents Sea	56	20	12	12	10	8	CM:4, HM: 10
East Siberian Sea	69	20	4	8	25	10	CM:4, HM: 2
Chukchi Sea	50	23	20	7	18	15	CM:8, HM: 2
Beaufort Sea	59	19	11	11	12	12	CM:9, HM: 2
Central Arctic	53	23	9	15	22	14	CM:1, HM: 2

Notes: References 1 (Dethleff et al., 2000), 2 (Schoster et al., 2000), 3 (Behrends et al., 1999), 4 (Wahsner et al., 1999), 5 (Levitan et al., 1996), 6 (Levitan et al., 2005), 7 (Levitan et al., 1999), 8 (Viscosi-Shirley et al., 2003a), 9 (Naidu & Mowatt, 1983), and 10 (based on Figure 5.13). LS, Laptev Sea; SS, surface sediments; SIS, sea-ice sediments; SPM, suspended particulate matter.

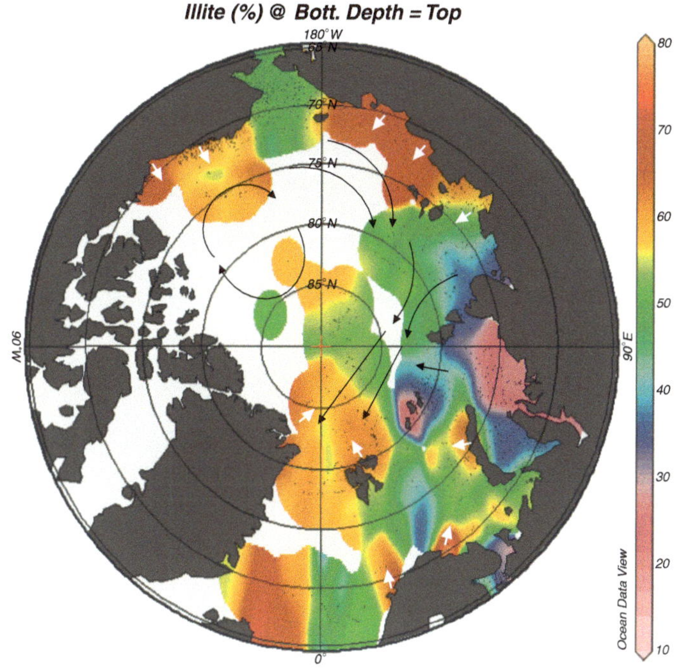

(A)

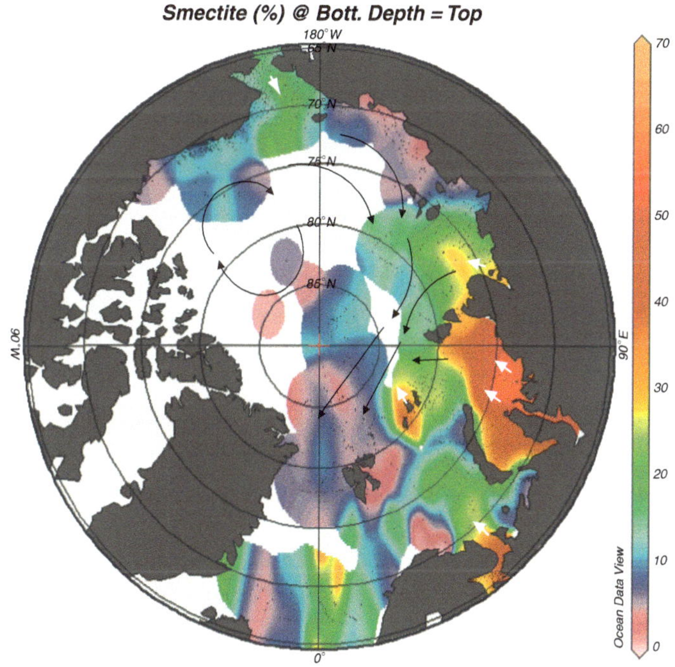

(B)

Figure 5.4 Clay-mineral distribution in Arctic Ocean surface sediments (A) Illite, (B) smectite, (C) chlorite, and (D) kaolinite. Surface-water circulation patterns (i.e., mainly the Beaufort Gyre and the Transpolar Drift) are shown by black arrows, main input areas by white arrows. Maps have been compiled using the Ocean Data View software (Schlitzer, 2001) and data from Naidu and Mowatt (1983), Darby et al. (1989), Berner (1991), Stein et al. (1994c), Wahsner and Shelekova (1994), Nürnberg et al. (1995), Kalinenko et al. (1996), Levitan et al. (1996), Wahsner et al. (1996, 1999), Rossak et al. (1999), Viscosi-Shirley et al. (2003a), Stein et al. (2004a), and Saukel (2006).

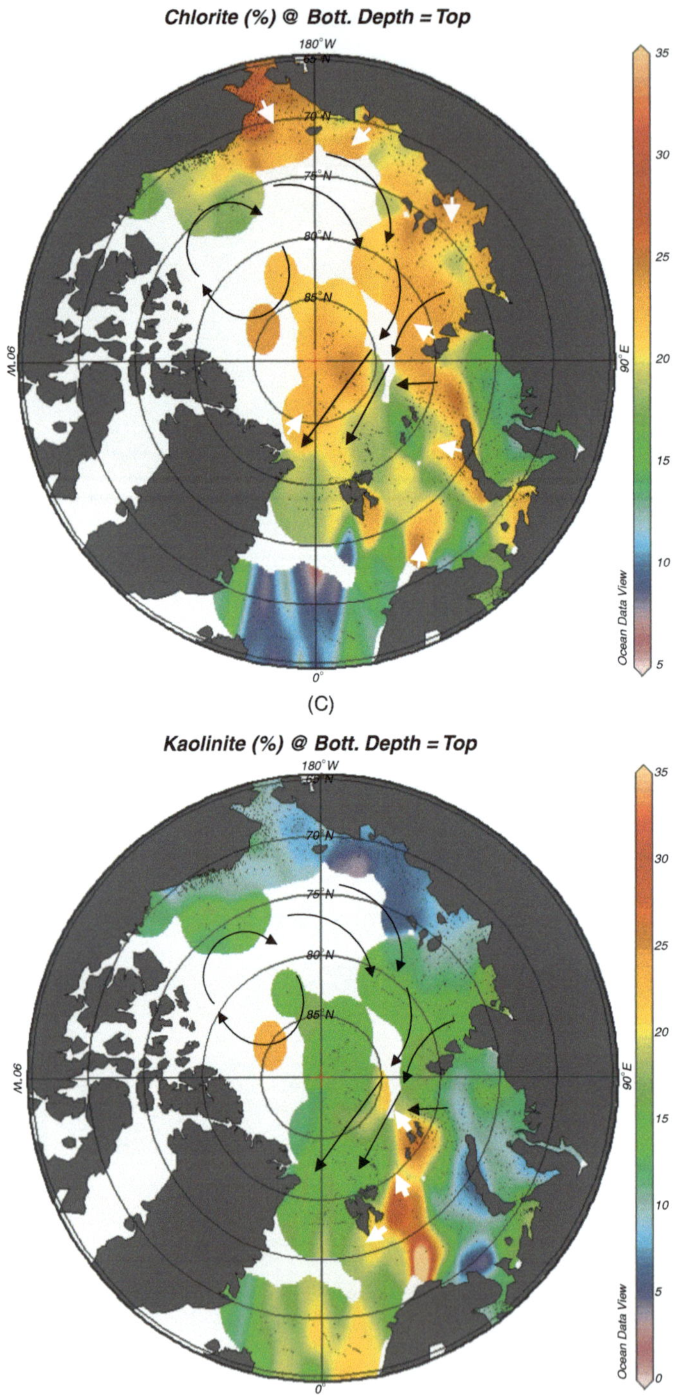

Figure 5.4 (*Continued*)

region, and granitoid massives from the Chukchi Peninsula (Kalinenko et al., 1996). In the Amerasian part of the Arctic Ocean higher illite values (up to 70%) were found along the Arctic coast of Alaska and on the Chukchi Rise (Figure 5.4A; Naidu et al., 1975; Clark et al., 1980).

5.1.2.3. Smectite

The importance of the clay-mineral smectite in the Arctic environment as source indicator was already described by Silverberg (1972), Wollenburg (1993), Nürnberg et al. (1994), Stein et al. (1994c), and Wahsner et al. (1999). Maximum concentration of smectite was recorded in the southern Kara Sea and the Ob and Yenisei estuaries. Here values of >60% are reached (Figure 5.4B; Wahsner et al., 1999; Stein et al., 2004a). The smectite maxima result from erosion and weathering of extensive flood basalts of the Putoran Massif of the Siberian Hinterland (Figure 5.3; Duzhikov & Strunin, 1992; Vyssotski et al., 2006). The Yenisei River with its tributaries drains this area and transports smectite into the Kara Sea. This process is supported by high concentrations of clinopyroxene, also a typical weathering product of basaltic material (Silverberg, 1972; Levitan et al., 1996; Vogt, 1997; Behrends et al., 1999; see Chapter 5.1.4, Figure 5.13B) and high magnetic susceptibility values (Chapter 4.3.5, Figure 4.22) in southern Kara Sea surface sediments, both reaching maximum values directly off the Yenisei River.

Relatively high values of smectite were also measured in western Laptev Sea (Figure 5.4B). These elevated smectite values can be related to two processes. First, a source for smectite enrichment in the western part of the Laptev Sea is sediment input by the Khatanga River which also drains the flood basalts of the Putoran Massif (Müller, 1999; Rossak et al., 1999). Similar to the situation in the Kara Sea (see earlier), clinopyroxene (Chapter 5.1.4, Figure 5.13B; Lapina, 1965; Stein & Korolev, 1994) and magnetic susceptibility (Niessen & Weiel, 1996) also show high values. Second, a distinct suspension transport exists from the eastern Kara Sea (characterized by high smectite values) towards and through the Vilkitsky Strait into the western Laptev Sea (Figure 5.4B; Wahsner et al., 1999).

Elverhøi et al. (1989) related the smectite occurrence in the northern Barents Sea to an extra-basinal source. Sea-ice sediments (SIS) from eastern shelf areas, enriched in smectite, may be transported within the Transpolar Drift entering the Barents Sea between Svalbard and Franz Josef Land. High smectite concentrations occur in surface sediments in the Franz Josef Land area and on the western slope of the St. Anna Trough. Weathering of basaltic rocks on the islands leads to smectite enrichment in soils and sea-floor sediments around the Franz Josef Land islands (Figure 5.4B; Nürnberg et al., 1995; Wahsner, Tarasov, & Ivanov, 1996).

The East Siberian Sea is impoverished in smectite (Figure 5.4B). In the Chukchi Sea, significant amounts of smectites of almost 30% were determined (Naidu et al., 1982; Naidu & Mowatt, 1983; Wahsner et al., 1999; Viscosi-Shirley et al., 2003a). Based on smectite concentrations in surface sediments of rivers flowing into the Chukchi, East Siberian and Bering seas, as well as regional trends in geology, ocean currents, and shelf sediment smectite distributions, Naidu et al. (1975, 1982) and Naidu and Mowatt (1983) conclude that the high smectite contents in Chukchi Sea surface sediments are derived from Siberian and Alaskan volcanic rocks, discharged to the Bering Sea and transported northward through the Bering Strait into the Chukchi Sea. Also the Colville Delta contains 10–30% smectite (Naidu & Mowatt, 1983).

5.1.2.4. Chlorite

Chlorite distribution in Arctic Ocean surface sediments is relatively uniform (Figure 5.4C), and no shelf areas with especially high values were identified in the Eurasian part of the Arctic. Mean values of chlorite vary between 14% and 25% (Table 5.1). The Novaya Zemlya Island is identified as a local chlorite source by slightly higher chlorite concentrations along their western coasts (Nürnberg et al., 1995) and south of the St. Anna Trough. In the Laptev Sea a slightly higher chlorite component is delivered by the rivers Lena and Yana (Rossak et al., 1999). In the Chukchi Sea, local enrichments of chlorite (up to 34%) exist in the delta areas of some rivers (Figure 5.4C; Naidu et al., 1982; Naidu & Mowatt, 1983). In contrast to these relatively low chlorite concentrations in the Eurasian Arctic, higher values (up to 47%) were identified along the Canadian Continental Rise and along the American coast (up to 37%; Clark et al., 1980).

5.1.2.5. Kaolinite

Potential source areas for kaolinite in Arctic Ocean sediments are very limited (Figure 5.4D). Main source areas for kaolinite occur in the central Barents Sea and on Franz Josef Land (Figure 5.4D; Birkenmajer, 1989; Elverhøi et al., 1989). Local enrichments in the Barents Sea are related to outcrops of Triassic and Jurassic sedimentary rocks south of Svalbard and on some shallow banks. The high kaolinite concentrations in surface sediments of the Franz Josef Land area are caused by erosion of Mesozoic sediments on the islands. In all other shelf seas, kaolinite occurs only in minor amounts and plays only a minor role in the clay-mineral characterization of surface sediments. In the Laptev Sea, the kaolinite originates from sediments of some Jurassic and Cretaceous clastic series on the Siberian Platform, forming the drainage area of Anabar and Olenek rivers (Rossak et al., 1999). Horizons of pure kaolinitic clay occur beneath the Upper Jurassic coal deposits of east Siberia (Kalinenko et al., 1996). In the Amerasian Arctic, some Mesozoic and Cenozoic strata along the north coast of Alaska and Canada show higher kaolinite concentrations up to 25% (Darby, 1975; Naidu & Mowatt, 1983; Dalrymple & Maass, 1987). Minimum kaolinite values of $<8\%$ are predominant in the East Siberian Sea (Figure 5.4D).

5.1.2.6. Clay-mineral assemblages and reconstruction of transport pathways

As described above, clay-mineral data can be used to distinguish among the shelf areas by the amounts of smectite, illite, and kaolinite in surface sediments (Figure 5.5). The

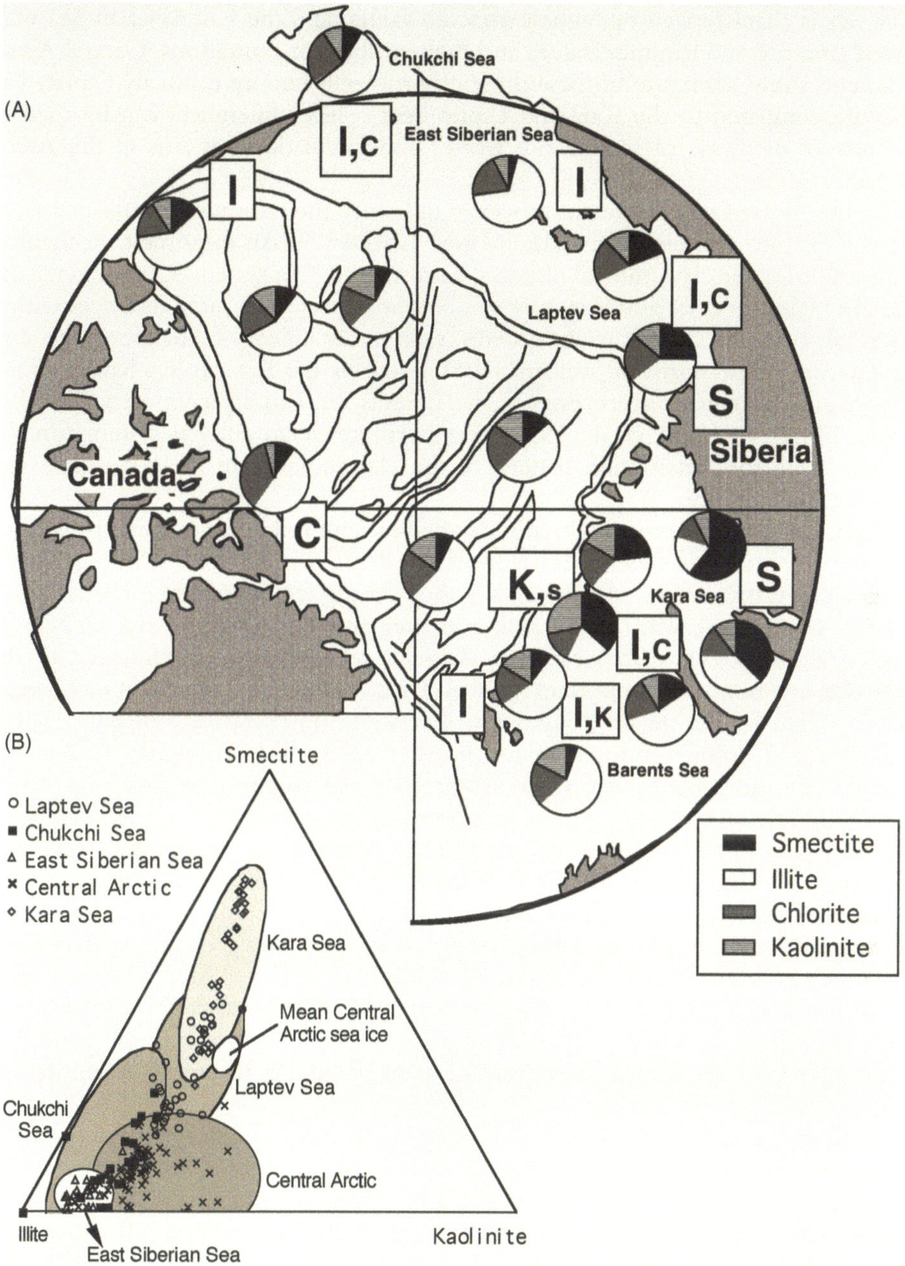

Figure 5.5 Summary of average clay-mineral distribution in Arctic Ocean surface sediments (from Wahsner et al., 1999). (A) Map and (B) triangle diagram of average clay-mineral distribution for the Arctic marginal seas and central Arctic Ocean.

Kara Sea is characterized by highest smectite values, and the East Siberian Sea with lowest smectite and kaolinite values and highest illite concentrations. Central Arctic sediments show a mixture of these three clay minerals but are distinctly depleted in smectite compared to the Kara and Laptev seas. These differences can be used to reconstruct pathways of terrigenous matter and circulation patterns in the Arctic Ocean.

In the Arctic Ocean different processes influence the transport of sediment from the shelves into the deep Arctic Basins (see Chapter 3). An important mechanism responsible for the clay-mineral dispersal pattern is current system. Clay minerals are extremely fine-grained and therefore can be transported over large distances within the water column. Major current systems, such as the inflow of Atlantic waters into the Central Arctic north of Svalbard (WSC) and via the Barents Sea have a strong impact on sedimentation processes in the Barents Sea and adjacent areas (Elverhøi et al., 1989; Nürnberg et al., 1995). Furthermore, local current systems on the shelves shape the distribution pattern of the clay minerals in these marginal seas (Levitan et al., 1996).

Additional processes in high-latitude shelves responsible for sediment distribution are the formation of sea ice and associated dense brines. Cold, saline, and well-oxygenated water masses, formed on the shelf during ice formation (Schauer et al., 1997a; Schauer et al., 1997b) may sink over the continental margin into the Nansen Basin (Aagaard et al., 1985) carrying clay-rich suspension from the shelf to the deep-sea environment. Important transport tracks for the suspension-rich water masses might be the major troughs northwest and northeast of Svalbard and the St. Anna and Voronin troughs in the northern Kara Sea. Turbidity currents are another common mechanism to transport clay-rich suspensions from the Arctic shelves into the deep basins. This is supported by the occurrence of a large number of clayey-silty distal turbidites recorded in the Nansen and Amundsen basins (see Chapter 3.4). Higher concentrations of kaolinite in turbidites in the Nansen Basin reflect the supply from the Franz Josef Land slope (Stein et al., 1994c).

Sediment transport with drifting sea ice is of importance in the Arctic Ocean (see Chapter 3.1), and clay minerals might be useful proxies to identify source areas of the sea ice (e.g., Letzig, 1995; Pfirman et al., 1997; Dethleff, 2005). For example, both SIS and shelf sediments from the Laptev Sea show very similar averaged clay mineral percentages which, however, substantially differ from the Beaufort Sea (Figure 5.6; Dethleff, 2005).

It is suggested that the Transpolar Drift ice is a significant agent contributing large amounts of sediments to the deep-sea environment (Wollenburg, 1993; Nürnberg et al., 1994; Pfirman et al., 1997; Dethleff et al., 2000). Sediments sampled from sea ice of the Siberian Branch of the Transpolar Drift show high smectite contents of 15–60% (Figure 5.7; Nürnberg et al., 1994; Dethleff et al., 2000), similar to surface sediments from the Laptev Sea shelf (up to 45% smectite) and from the Kara Sea (up to 70% smectite). This suggests that first the Laptev Sea and second the Kara Sea are potential source areas of the siliciclastic material included in the sea ice of the southern branch of the Transpolar Drift. The Polar Branch of the Transpolar Drift, however, transports SIS with significantly low smectite contents (Figure 5.7). This is supported by reconstructions of origin and

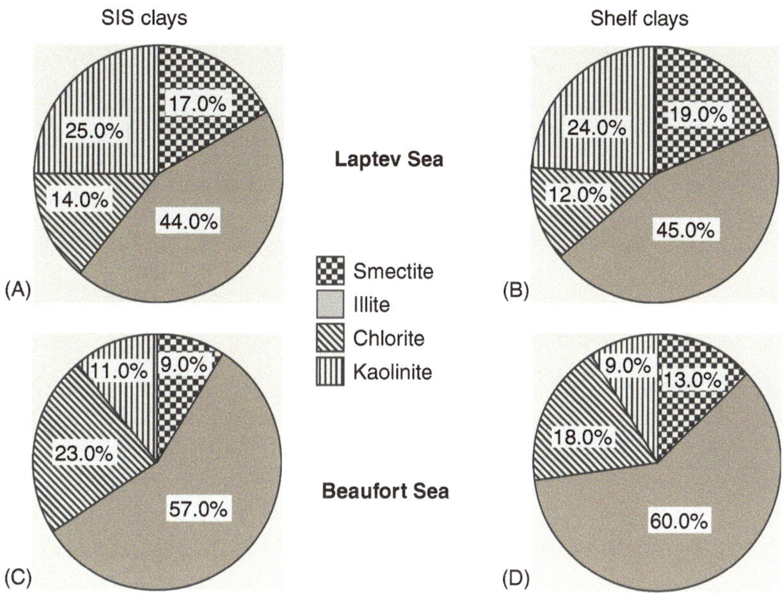

Figure 5.6 Average clay-mineral assemblages in sea-ice sediments (SIS) and shelf sediments from the Laptev Sea (A and B; Dethleff, 2005) and the Beaufort Sea (C and D; Reimnitz et al., 1998) (from Dethleff, 2005).

trajectory of the drifting sea ice identifying the New Siberian Islands in the Laptev Sea and the Central Kara Plateau as two regions from which sediment-laden sea ice is exported into the Transpolar Drift (Figure 5.8; Pfirman et al., 1997).

In contrast to the sea ice, the underlying surface sediments of the Eurasian Basin show only a very low content of smectite (Figure 5.9; Stein et al., 1994c; Wahsner et al., 1999). The data suggest, that today sediment particles in sea ice from the southern branch of the Transpolar Drift, released by melting, are probably not the dominant source for the underlying bottom sediments of the basin. Other processes, such as turbidity and ocean currents may be also important and overprint the sediment input by sea ice (Stein & Korolev, 1994). The sizes of the clay minerals, even flocculated, are so small that they remain within the water column for long periods and even weak currents can transport the settling clays over long distances away from the point of release, producing a mixed clay-mineral signal in the bottom sediments. On Lomonosov Ridge where sediments are elevated from turbidites and likely more of pelagic origin, on the other hand, the smectite values of surface sediments come more close to the sea-ice sediment values (Figure 5.9).

Further to the south, that is, in the Fram Strait area where melting processes are intensified due to the influence of the warm WSC, illite is the predominant clay mineral and smectite is of minor importance in sea ice as well as surface sediments (Figures 5.4A and B). This may be related to sediment input via the Polar Branch of the Transpolar Drift with sediments originated from the eastern Laptev Sea, the east Siberian Sea, and even the Canadian Arctic (Figures. 5.7 and 5.8; Pfirman et al.,

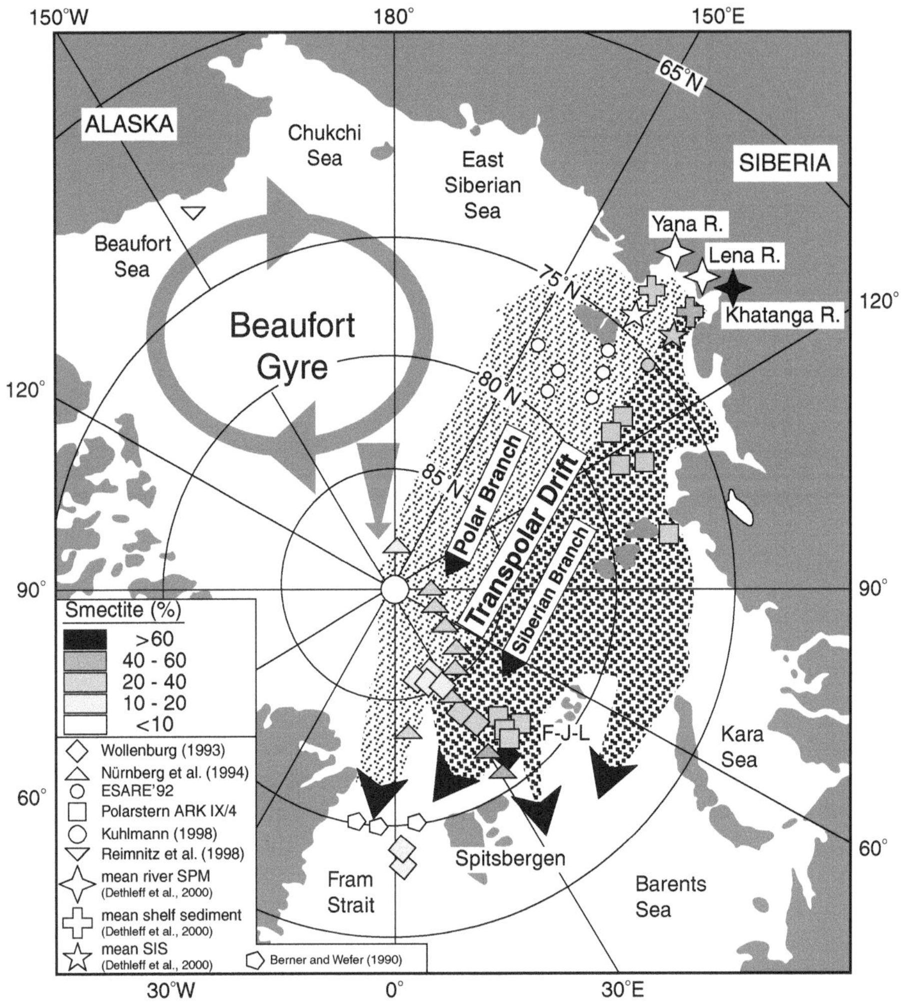

Figure 5.7 Smectite percentages in sea-ice sediments (SIS) from the Laptev Sea, the central Arctic Ocean, and Fram Strait, indicating potential transport pathways of sediment-laden sea ice. In addition, average smectite percentages of sea-ice sediments from the Beaufort Sea (9% smectite) as well as Laptev Sea surface sediments and river-suspended particulate matter (SPM) are shown. Main surface-water circulation systems, that is, the Beaufort Gyre and the Transpolar Drift with its Siberian Branch and its Polar Branch are indicated (from Dethleff et al., 2000, supplemented).

1997). This is also supported in sediment trap data from the Fram Strait area, showing a clear dominance of illite with >70% (Figure 5.10A–C; Berner & Wefer, 1990; Berner, 1991). Kaolinite increases from ~10% in the western trap FS3 (Figure 5.10B) to over 20% in the eastern trap SP1 (Figure 5.10C). The higher kaolinite values in the eastern trap probably indicate transport of kaolinite-rich suspension derived from the Barents Sea shelf within the WSC (see Figure 5.4E; Elverhøi et al., 1989; Berner & Wefer, 1990). At the MIZ primary production is

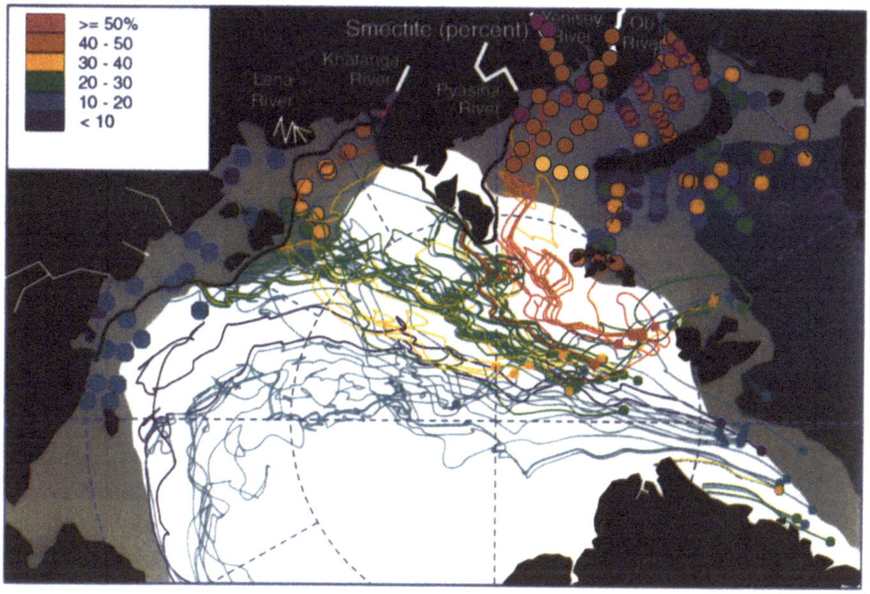

Figure 5.8 Reconstructed backward trajectories for sea-ice samples (small coloured dots), indicated by lines. Colour of dots and lines are related to smectite concentrations. Concentrations of smectite of surface sediments on the shelves are incitated by large coloured symbols. The heavy black line on the shelf indicates the location of the fast ice border, where the trajectories were terminated. The smectite concentration of the sea-ice samples is shown to generally, but not always, match the characteristics of the shelf source area. Clay mineralogy data from Wollenburg (1993), Nürnberg et al. (1994, 1995), Wahsner and Shelekova (1994), and Stein et al. (2004a). The perennial ice pack in the Arctic Basin is shown as the white area on the map, the marginal ice zone as light grey, and open-water conditions as dark grey. Figure taken from Pfirman et al. (1997), supplemented by new clay-mineral data from the Kara Sea (Stein et al., 2004a).

enhanced by algae blooms (e.g., Berner & Wefer, 1990). Increased grazing of copepods lead to strongly enhanced sedimentation of fine-grained particles via fecal-pellet production (Alldredge & Silver, 1988). With this process, fine-grained sediment, just released from the sea ice, is transported to the seafloor close to the actual position of the MIZ, as reflected in the increased kaolinite contents as well as in maximum fluxes of lithogenic, biogenic opal, and OC fluxes observed in sediment traps in the eastern Fram Strait close to the MIZ (Berner & Wefer, 1990; Hebbeln & Wefer, 1991; Hebbeln & Berner, 1993).

5.1.3. Bulk Mineralogy: Distribution and Significance

Bulk minerals most commonly used for indication of source rocks of terrigenous sediments in the Arctic Ocean include quartz, feldspars (i.e., plagioclase and kalifeldspar), and detrital carbonates (e.g., dolomite) (Figure 5.11; see also Figure 4.16), the most abundant minerals (besides clay minerals) in Arctic Ocean sediments. These data can be obtained by microscopic analysis (e.g., Spielhagen,

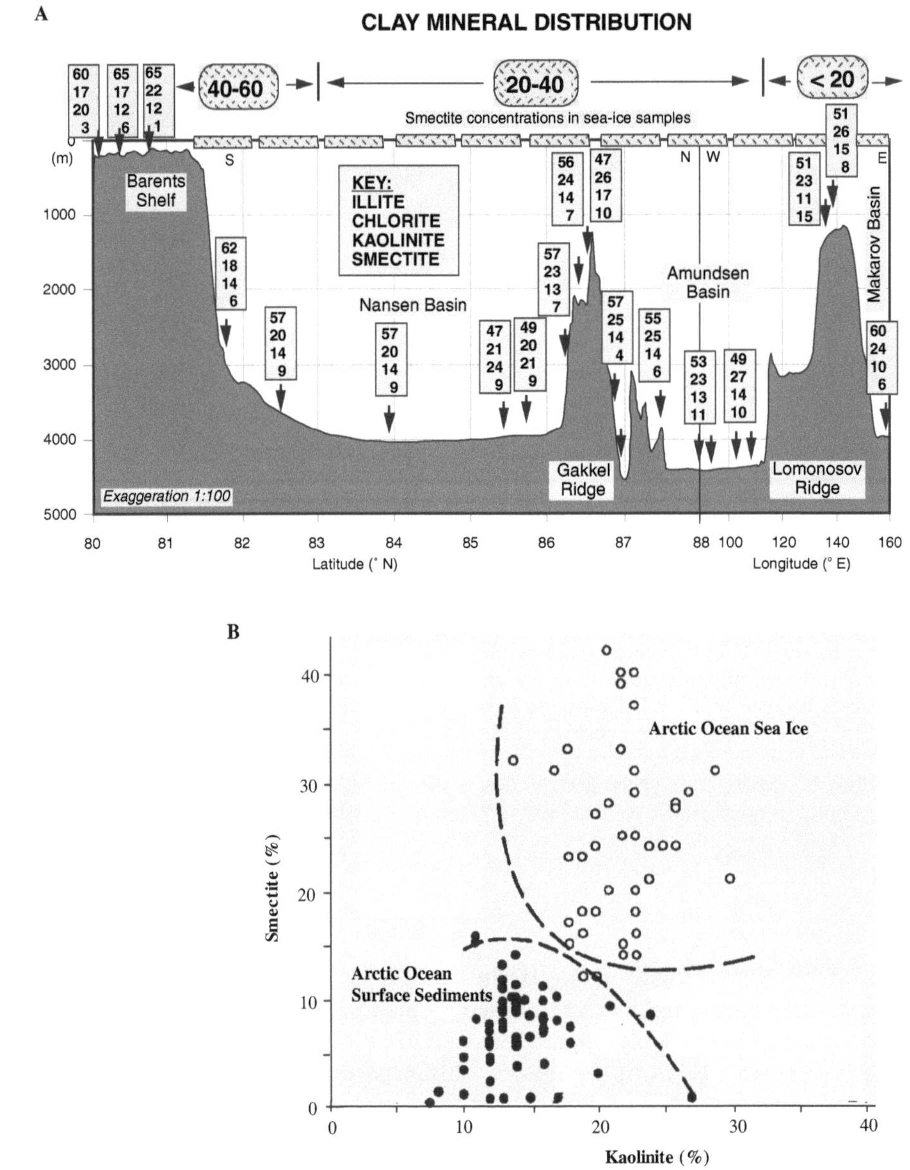

Figure 5.9 (A) Clay-mineral percentages in surface sediments from a transect from the Barents Sea to the Lomonosov Ridge and in sea-ice sediment samples from the same area (Wahsner et al., 1999). For location of transect see Figure 5.4. (B) Kaolinite versus smectite diagram of surface sediments from the Eurasian Basin (Stein et al., 1994c) and sea-ice sediments (Nürnberg et al., 1994).

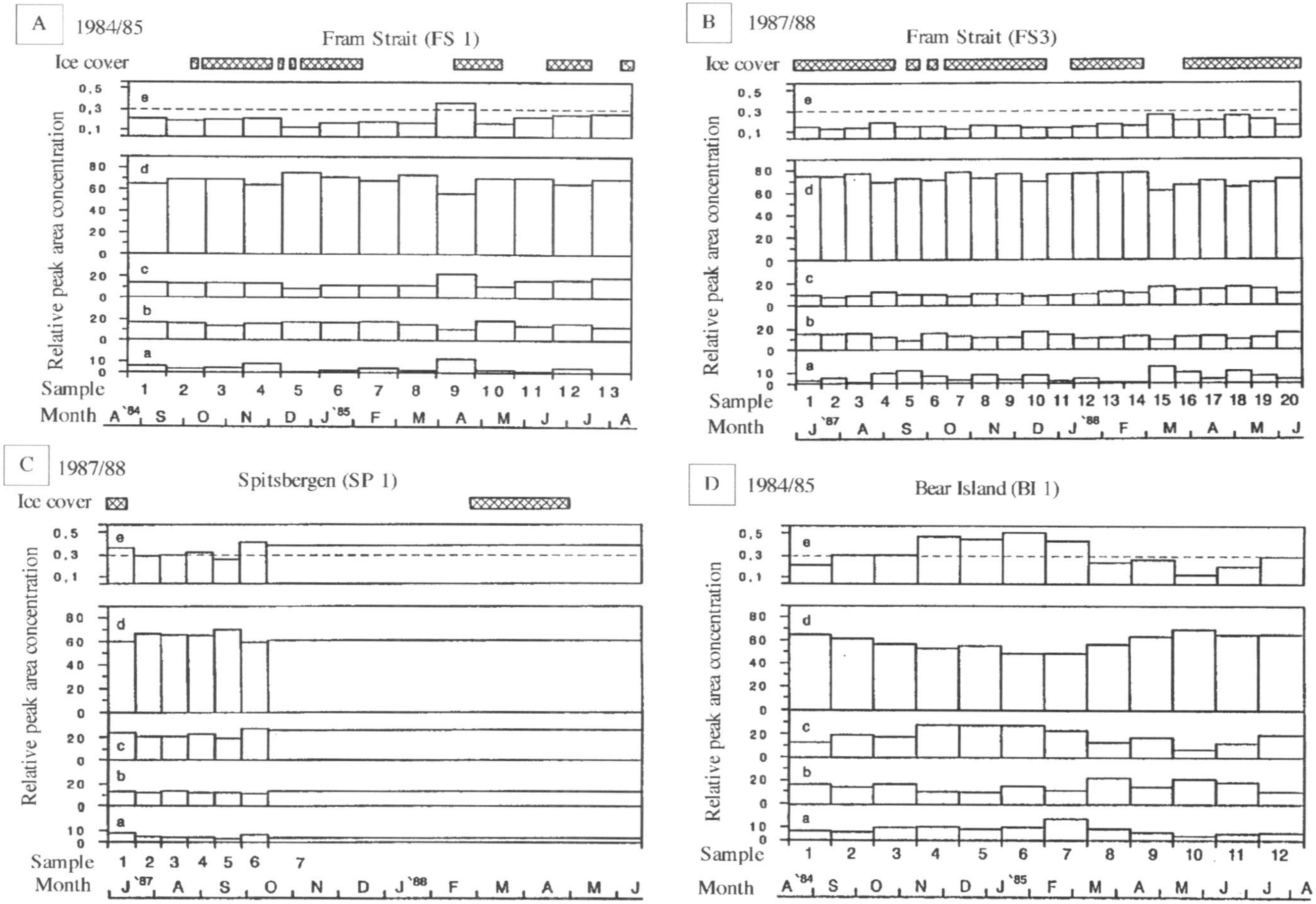

Figure 5.10 Clay-mineral composition of clay fraction (< 2 μm) in samples from sediment traps, representing a 12 months cycle each (from Berner & Wefer, 1990; Berner, 1991, supplemented). (A) Trap Fram Strait 1 (FS1), (B) trap Fram Strait 2 (FS2), (C) trap Spitsbergen 1 (SP1), and (D) Bear Island 1 (BI1). a, smectite (17Å); b, kaolinite (7Å); c, chlorite (4.7Å); d, illite (10Å); e, kaolinite/illite ratio. For location and data of sediment traps, see Chapter 2.4, Figure 2.16 and Table 2.4, respectively.

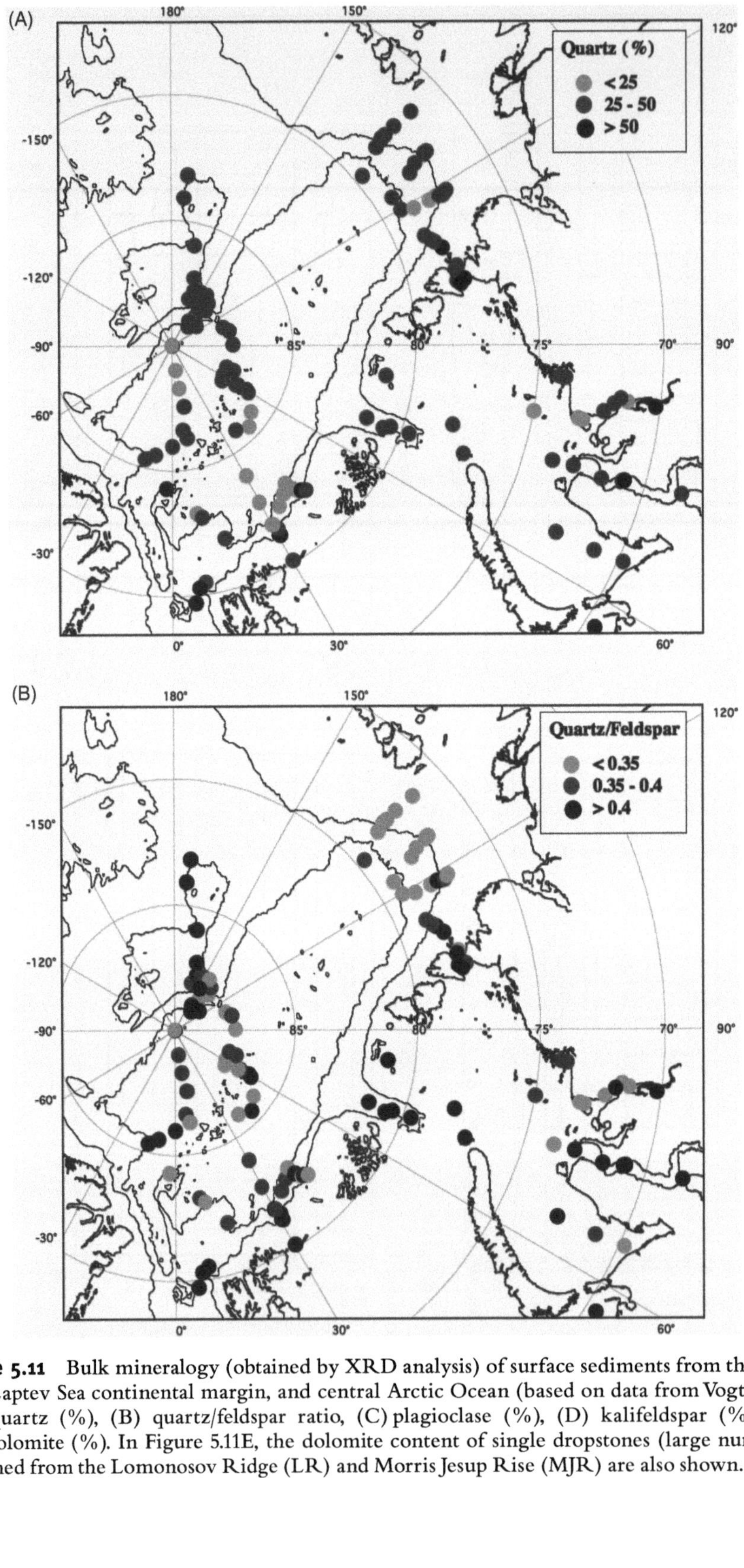

Figure 5.11 Bulk mineralogy (obtained by XRD analysis) of surface sediments from the Kara Sea, Laptev Sea continental margin, and central Arctic Ocean (based on data from Vogt, 1997). (A) quartz (%), (B) quartz/feldspar ratio, (C) plagioclase (%), (D) kalifeldspar (%), and (E) dolomite (%). In Figure 5.11E, the dolomite content of single dropstones (large numbers) obtained from the Lomonosov Ridge (LR) and Morris Jesup Rise (MJR) are also shown.

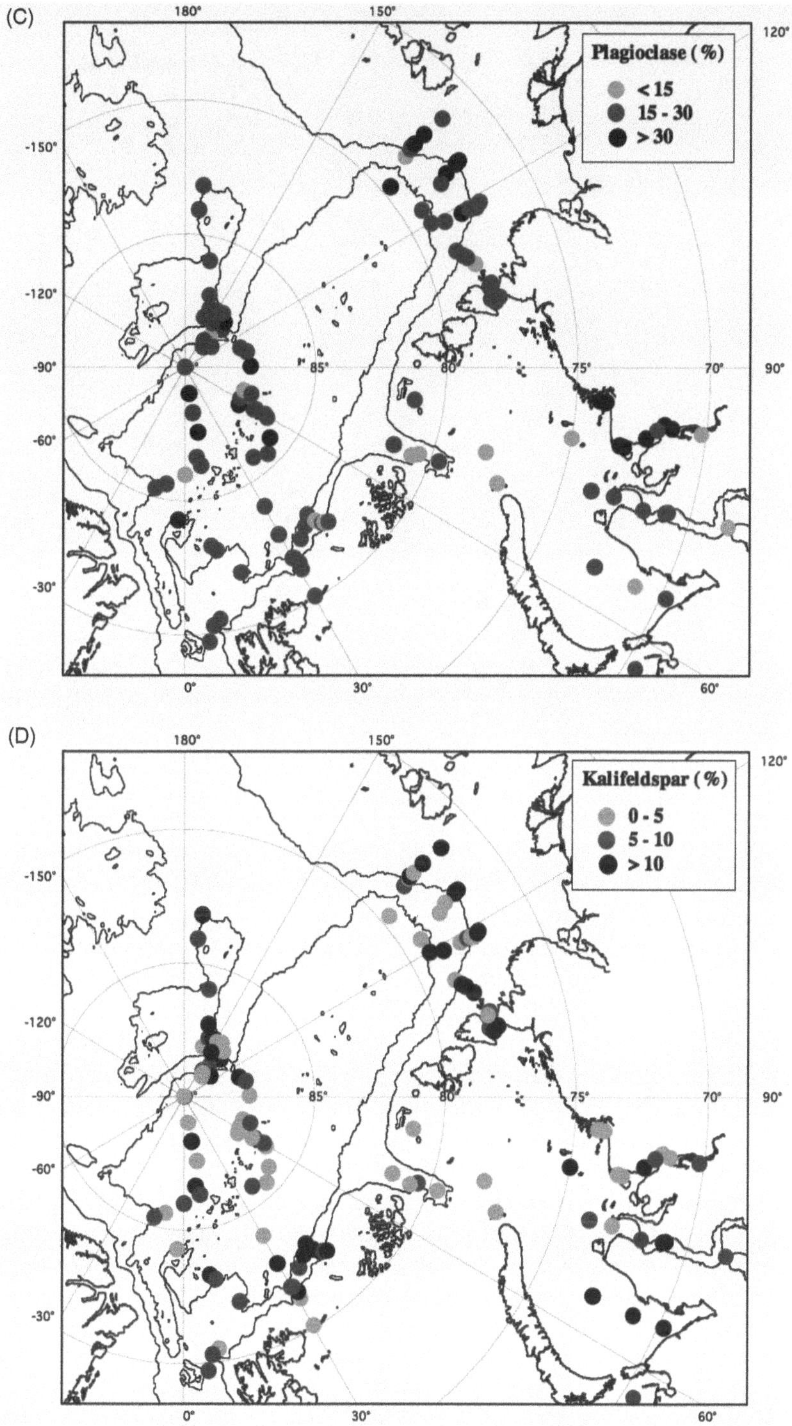

Figure 5.11 (*Continued*)

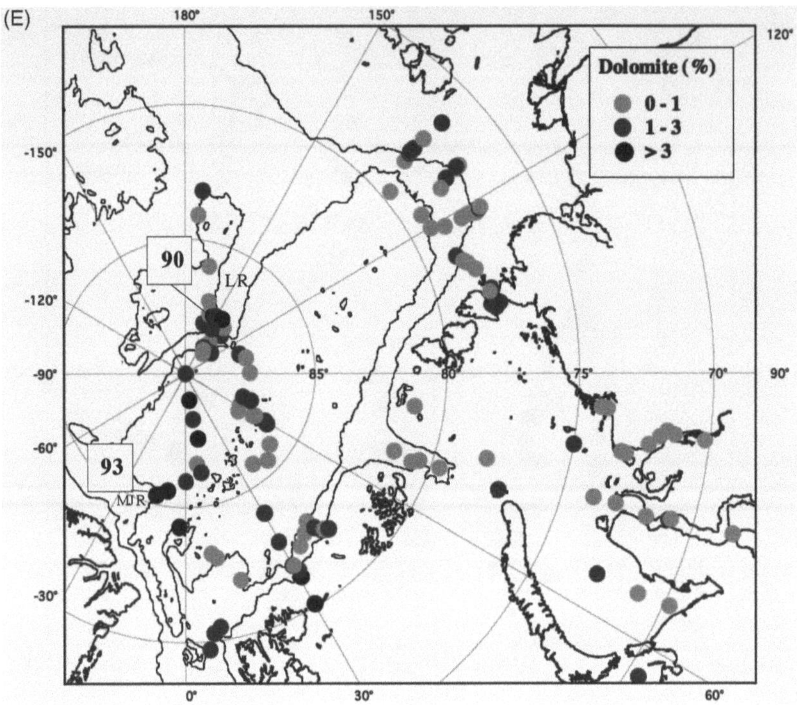

Figure 5.11 (*Continued*)

1991; Kubisch, 1992; Bischof et al., 1996; Levitan et al., 1996; Nørgaard-Pedersen et al., 1998, 2003; Bischof, 2000; Phillips & Grantz, 2001; see also Chapter 4.2) or X-ray diffraction analysis (e.g., Vogt, 1997; Vogt et al., 2001; Nørgaard-Pedersen, Mikkelsen, Lassen, & Kristoffersen, 2007b). Bulk mineralogy X-ray diffraction data of Arctic Ocean sediments are still limited. Here, the most comprehensive data base including X-ray diffraction analyses of bulk samples, silt fraction as well as clay fraction, has been produced by Vogt (1997; see Chapter 4.2).

Although the mineral quartz is ubiquitous in the Arctic reaching very high values >50% in the Canadian Arctic (Figure 4.16) as well as the Eurasian Arctic (Figure 5.11), a combination with other minerals such as feldspars may yield important information about source areas and transport processes. As shown by Vogt (1997), for example, surface sediments from potential circum-Arctic source areas and basins can be characterized by very specific kalifeldspar/plagioclase (Kfs/Plg) and quartz/feldspar (Qz/Fsp) ratios. Based on these ratios, a major proportion of the surface sediments of the Eurasian Basin seem to be derived from the Laptev Sea and the Barents Sea continental slope (Figure 5.11B and 5.12; Kfs/Plag 0.2–0.7, Qz/Fsp 0.3–0.5), which is also supported by clay- and heavy-mineral data (see Chapter 5.1.2 and 5.1.4). In contrast, most sediments from the Canadian Archipelago and the Kara Sea have much higher Qz/Fsp ratios (mostly 0.4 to >2; Vogt, 1997). It has to be mentioned, however, that a more comprehensive data base from other circum-Arctic regions as well as the central Arctic (e.g., similar to

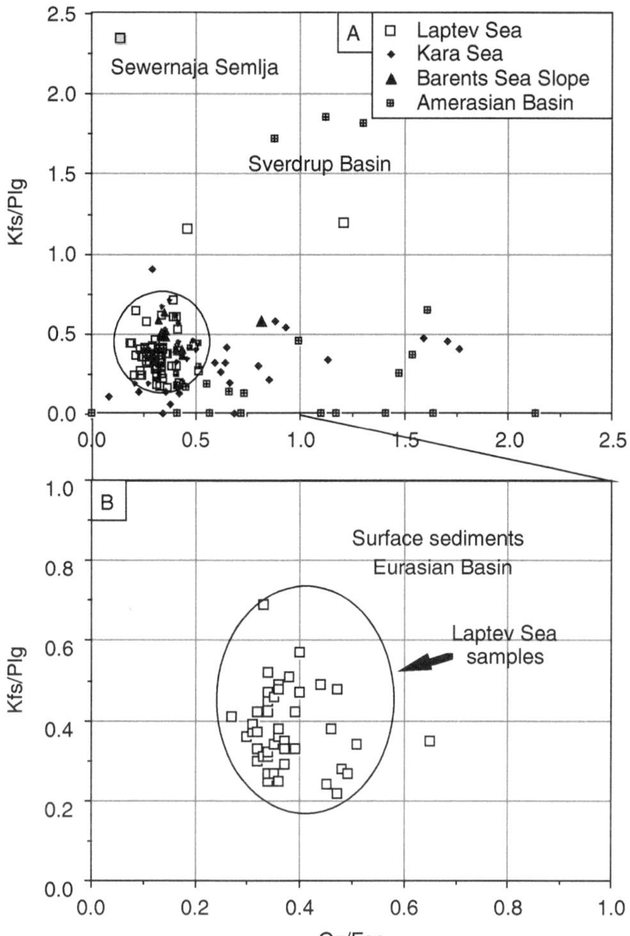

Figure 5.12 (A) Kalifeldspar/plagioclase (Kfs/Plg) versus quartz/feldspar (Qz/Fsp) diagram in surface sediments from the Laptev Sea, Kara Sea, Barents Sea slope, and Amerasian Basin. The majority of the Latev Sea samples fall into the circle. (B) Kalifeldspar/plagioclase (Kfs/plg) versus quartz/feldspar (Qz/Fsp) diagram in surface sediments from the Eurasian Basin. The circle indicates the ratios of the Laptev Sea samples shown in Figure (5.12A). Most of the samples from the Eurasian Basin fall into this field, suggesting a Laptev Sea sediment source (from Vogt, 1997).

that available for Fe-minerals; Darby, 2003; see Chapter 4.3) is still needed to prove the use of this proxy as reliable source indicator.

As outlined in Chapter 4.3, detrital carbonate, especially dolomite, is one of the best proxies for the identification of sediments derived from the Canadian Arctic. This is reflected in the X-ray bulk mineralogy data from circum-North American/ Canadian coastal areas and the Amerasian Basin where values of up to 50% and 20%, respectively, were determined (Figure 4.16; Vogt, 1997). In the Eurasian Arctic, on the other hand, detrital carbonate/dolomite is more or less absent (Vogt, 1997).

Dolomite concentrations in surface sediments from the Eurasian Basin are of very minor significance (0–3%; Figure 5.11E), suggesting that a major input of sediments from the Canadian Arctic can be excluded. Only on the Lomonosov Ridge and in a restricted area from the western Amundsen Basin towards the Morris Jesup Rise higher dolomite values of up to 17% are reached (Figure 5.11E). For single grains (dropstones) from the Lomonosov Ridge and Morris Jesup Rise, X-ray analysis showed that these grains may contain >90% dolomite, indicating sediment input by ice rafting from the Canadian Arctic via the Beaufort Gyre (Figure 5.11E; Vogt, 1997). The dolomite distribution obtained by X-ray diffraction method is also supported by microscopy data obtained from heavy-mineral analysis. Dolomite contents are almost negligible in the Eurasian Arctic, whereas in the Beaufort and Chukchi seas values of 16% are determined (Behrends, 1999; Behrends et al., 1999).

5.1.4. Heavy Minerals: Distribution and Significance

The data base for heavy-mineral distributions in Arctic Ocean surface sediments shows large spatial differences. Russian scientists recognized the high potential of heavy minerals as source indicator for terrigenous sediments very early. Thus, a large data base is available for the Eurasian marginal seas (e.g., Belov & Lapina, 1961; Lapina, 1965; Naugler, 1967; Silverberg, 1972; Naugler et al., 1974; Gurevich, 1995; Lisitzin, 1996; Levitan, Nürnberg, Pavlidis, & Shelekova, 1994; Levitan et al., 1996, 1999, 2005, 2007 and further references therein). Within a major Russian–German research project in the Laptev Sea (Kassens et al., 1999), further heavy-mineral data from the Laptev Sea was produced by Behrends et al. (1996, 1999); Behrends (1999); Hoops (1999); Peregovich (1999), and Peregovich, Hoops, and Rachold, (1999). For the Beaufort and Chukchi seas (Luepke & Escowitz, 1989; Lisitzin, 1996) as well as the central Arctic Ocean (Darby et al., 1989; Behrends, 1999; Behrends et al., 1999), on the other hand, the data base is still very limited.

In the following, distribution maps of the main heavy minerals often used as source indicator, that is, amphibole (hornblende), clinopyroxene, epidote, and granat, are presented (Figure 5.13).

5.1.4.1. Amphibole

From the distribution map of amphiboles in Arctic Ocean surface sediments, it is obvious that the East Siberian Sea and, especially, the eastern Laptev Sea are the main source areas (e.g., Naugler, et al., 1974; Behrends et al., 1999). There, high amphibole values of >30 to 50% are reached (Figure 5.13A). Increased amphibole contents were also determined in the southern Barents Sea with maxima north of Kola Peninsula. In the Kara Sea and, especially, in the northern Barents Sea amphibole contents are significantly lower with typical values of 8–15% and 5–10%, respectively (Figure 5.13A). In the central Arctic Ocean, the amphibole distribution shows values between 10% and 20%.

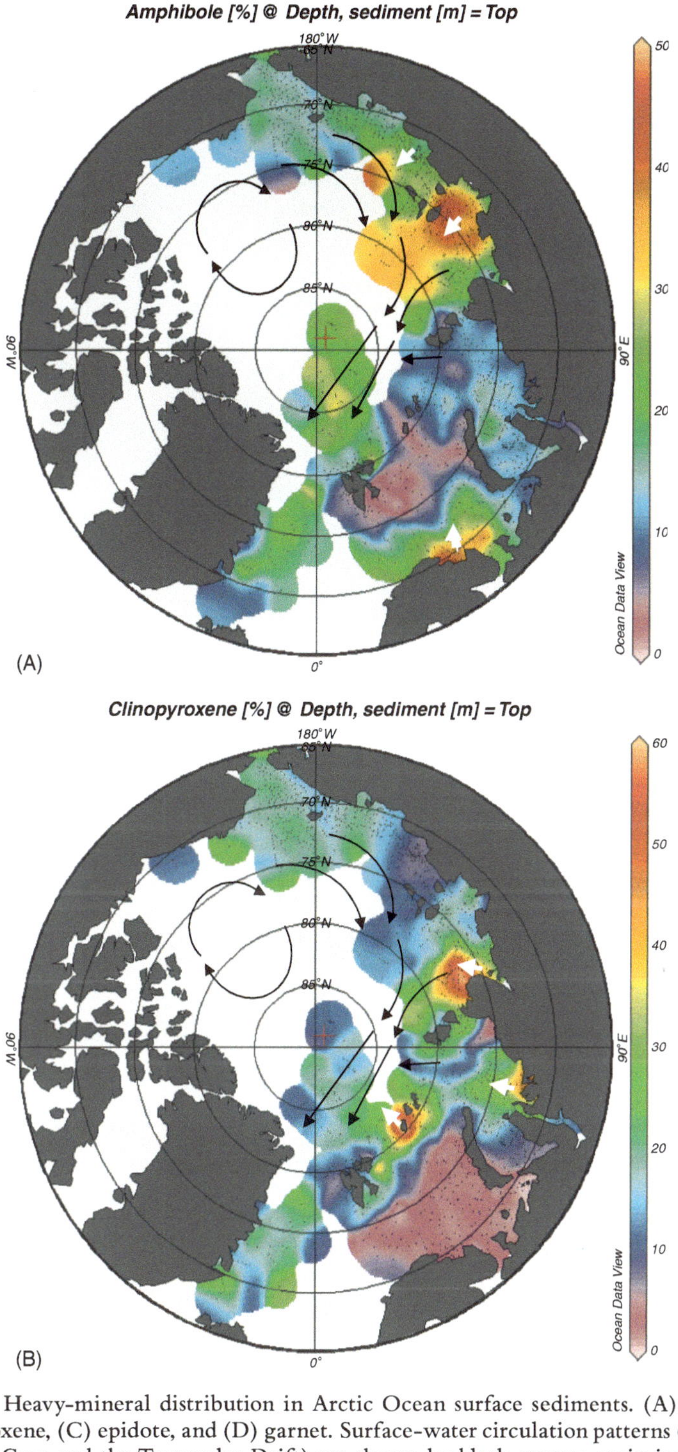

Figure 5.13 Heavy-mineral distribution in Arctic Ocean surface sediments. (A) amphibole, (B) clinopyroxene, (C) epidote, and (D) garnet. Surface-water circulation patterns (i.e., mainly the Beaufort Gyre and the Transpolar Drift) are shown by black arrows, main input areas by white arrows. Maps have been compiled using the Ocean Data View software (Schlitzer, 2001) and data from Naugler (1967), Naugler et al. (1974), Behrends (1999), Behrends et al. (1999), Peregovich (1999), Levitan et al. (1996, 1999, 2005), and Andreeva (www.pangaea.de/search?q = Andreeva), Klenova (www. pangaea.de/search?q = Klenova), Kulikov (www. pangaea.de/search?q = Kulikov), and Pavlidis (www.pangaea.de/search?q = Pavlidis).

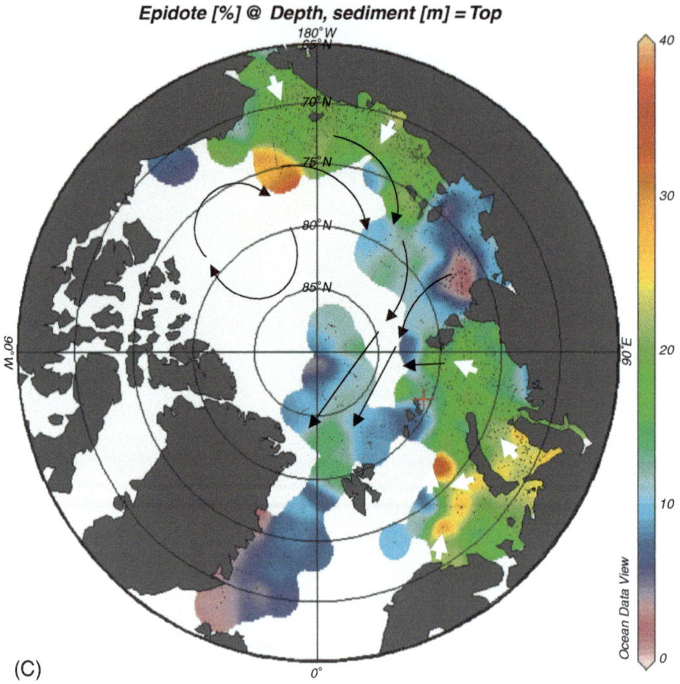

(C)

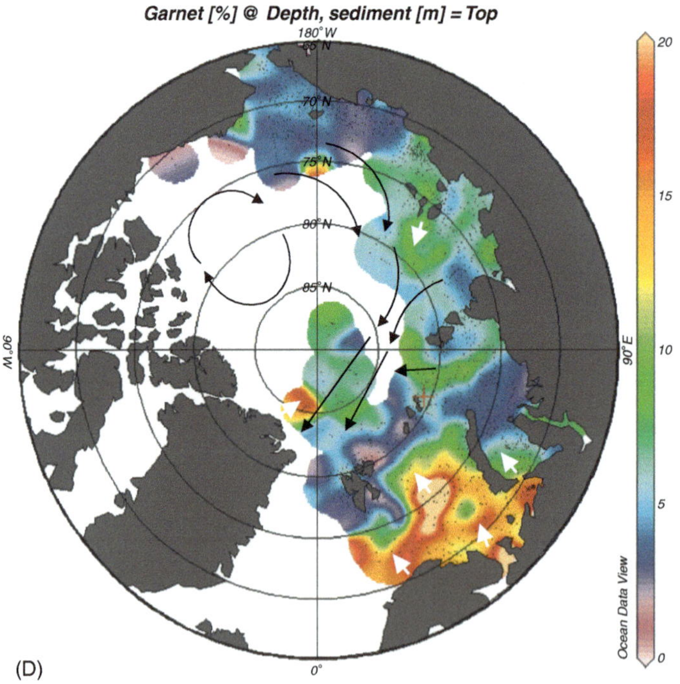

(D)

Figure 5.13 (*Continued*)

5.1.4.2. Clinopyroxene and orthopyroxene

The pyroxenes are dominated by clinopyroxene varying between 5% and >50% (Figure 5.13B), whereas orthopyroxene are generally very low in Arctic Ocean surface sediments (<5%; Behrends et al., 1999). Main source areas for clinopyroxene mainly related to the riverine sediment input of weathering products of the Putoran Massif (Figure 5.3) are the western Laptev Sea, the southeastern Kara Sea, and the area around Franz Josef Land. The Barents Sea, on the other hand, is characterized by minimum clinopyroxene values (Figure 5.13B).

5.1.4.3. Epidote

Epidote occurs in high abundance in the East Siberian Sea, Chukchi Sea, (southwestern) Kara Sea, and eastern Barents Sea, where values of 15–30% are determined (Figure 5.13C). Low values of <10% are typical for Laptev Sea surface sediments. In the central Arctic Ocean as well as along the Eurasian continental margin and Fram Strait area also low epidote values of 15–15% are dominant.

5.1.4.4. Garnet

Garnet seems to be a heavy mineral of secondary importance in most of the Arctic Ocean sediments with contents <10%, with two exceptions (Figure 5.13D). In the central and southern Barents Sea, garnet contents display values of 15–20%. Another area with increased garnet content is the area north of northern Greenland.

Because heavy minerals — in contrast to clay minerals — cannot be transported over long distances via ocean currents, transport by ice rafting seems to be a major process (see Chapter 3.1). Concerning the distribution of heavy-mineral assemblages in central Arctic surface sediments, sediment input via sea ice from the Laptev Sea (and East Siberian Sea) and transport within the Transpolar Drift is most obvious. Underneath the Siberian Branch of the Transpolar Drift, high clinopyroxene values can be related to the western Laptev Sea, whereas amphibole concentrations are higher underneath the Polar Branch, related to sediment input from the eastern Laptev Sea and/or East Siberian Sea (Figure 5.13A and 5.13B; Behrends et al., 1999). That means, the heavy-mineral assemblages generally support the interpretation of the clay-mineral data (see Figure 5.7).

5.1.5. Summary: Mineralogy of Surface Sediments and Sediment Source Areas

As outlined in the preceding subchapters and summarized in Figure 5.14, the different Arctic Ocean marginal seas are characterized by specific mineral assemblages which can be used to identify source areas and transport pathways. The average values of clay- and heavy-mineral assemblages in surface sediments are listed in Table 5.1. Based on these data, the following general statements concerning sediment source areas can be made.

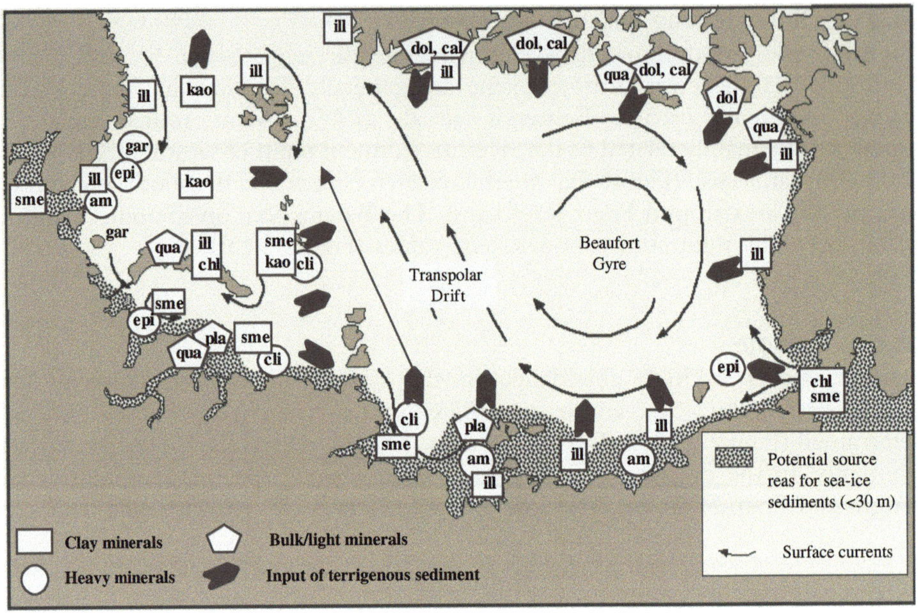

Figure 5.14 A map summarizing key minerals of Arctic Ocean shelf sediments and adjacent landmasses (based on Stein et al., 1994c; Vogt, 1997; Wahsner et al., 1999).

- Sediment input from the Canadian Arctic (and northern Greenland) is characterized by high detrital carbonate values (dolomite and calcite); illite is the dominant clay mineral. Dolomite is more or less absent in sediments from the Eurasian Arctic.
- Sediments from the East Siberian Sea have very high illite concentrations and high contents of amphibole.
- The eastern Laptev Sea sediments are characterized by high illite and amphibole values, whereas the western Laptev Sea sediments are characterized by high smectite and clinopyroxene values.
- Kara Sea surface sediments are similar to the western Laptev Sea sediments with elevated smectite and clinopyroxene values. In the Kara Sea, the absolute maximum of smectite concentrations in Arctic Ocean sediments are determined, reaching values $>60\%$.
- Around Svalbard and in the eastern Barents Sea influenced by input from Nowaja Semlja, illite, chlorite, and quartz are most important minerals, whereas in the central Barents Sea and on/around Franz Josef Land maximum kaolinite values occur.

Important transport agents for terrigenous material today seem to be sea ice and, within the basins, turbidity currents (see Chapters 3.1 and 3.4). The mineralogical assemblages described above and, especially, the Fe-oxid proxies (see Chapter 4.2; Darby, 2003) can be used to identify changes in terrigenous sediment input from different source areas in sediment cores, which then can be related to changes in extent of glacial ice sheets and oceanic circulation patterns (see Chapter 6.2 and 6.3).

5.2. Organic-Carbon Content: Terrigenous Supply versus Primary Production

The amount and composition of particulate OC preserved in the sedimentary records are controlled by different factors such as terrigenous input, primary production, transformation processes in the water column and at the seafloor as well as bulk sedimentation rates. Furthermore, sediment grain size will also be an important factor controlling OC concentrations, based upon the covariance between sediment mud and OC contents generally observed (Blackburn, 1987; Keil, Tsamakis, Fuh, Giddings, & Hedges, 1994; Mayer, 1994; Stein & Macdonald, 2004a and further references therein). All these factors are highly variable in the different areas and environments of the Arctic Ocean. Although the ice-covered Arctic Ocean is of low productivity, the OC content in the surface sediments is generally high (0.4–2%; Figure 5.15) compared to other world open ocean areas

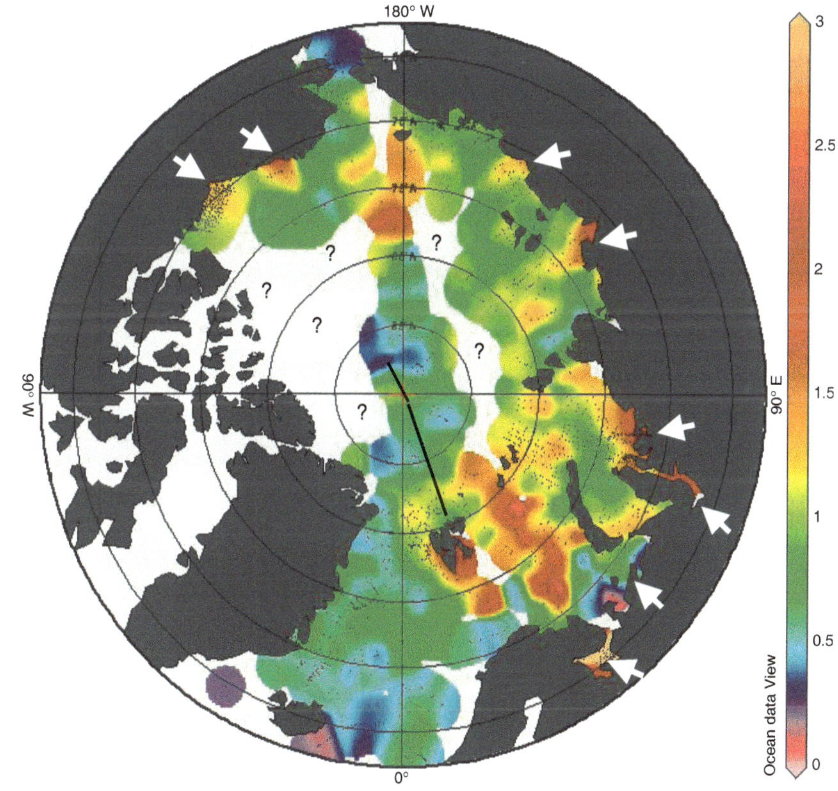

Figure 5.15 Distribution map of organic carbon content in Arctic Ocean surface sediments. Map has been compiled using the Ocean Data View software (Schlitzer, 2001) and data from Boström et al. (1984), Hebbeln and Berner (1993), Stein et al., 1994c, Naidu et al. (2000, 2004), Stein and Fahl (2000), Stein et al., (2003a), Birgel and Stein (2004), Macdonald et al. (2004b), Petrova et al. (2004), Vetrov and Romankevich (2004), and Kierdorf (2006), and Pavlidis (www.pangaea.de/search?q=Pavlidis). Black line indicate location of transect shown in Figure 5.22.

where OC contents of <0.5% are most typical (e.g., Suess, 1980; Romankevich, 1984). Looking at the overview distribution map of Figure 5.15, however, it becomes obvious that the data base for the various marginal seas and the central Arctic Ocean is very different. Much more data, for example, are available from the Beaufort, Laptev, and Kara seas than for the East Siberian Sea and the central Arctic Ocean. For the shelf area of the Canadian Archipelago as well as large parts of the central Arctic, no data at all are available.

In this chapter, a summary of OC distribution and its controlling processes in surface sediments from the central Arctic Ocean and its marginal seas is presented. For a more discussion of the Arctic Ocean OC cycle, see Stein and Macdonald (2004a) and Macdonald et al. (2008) and references therein.

5.2.1. Barents Sea

OC content in surface sediments of the Barents Sea ranges from <0.1% to >3% with an average of 1.1% (Figure 5.16; Vetrov & Romankevich, 2004 and references therein). The OC distribution in the Barents Sea sediments depends strongly on sediment grain size and bottom topography. Mean OC contents in the clayey sediments are 1.2–1.4%, whereas in the more sandy sediments values of 0.2–0.4% are typical. Highest OC contents (~3%) were determined in clay-rich sediments of the troughs and depressions as well as near the foot of submarine rises (>2%). In the coastal zone, strong tidal currents lead to the deposition of mainly coarse-grained sediments, whereas fine-grained sediments are transported to deep water. As a result, in the southern open part of the Barents Sea adjacent to the continent, sediments are represented mostly by sands containing OC content<0.5 and 0.5–1.0%.

$\delta^{13}C_{org}$ values, C/N ratios, and biomarker data all indicate a mixed allochtonous-autochtonous (terrigenous/marine) origin of organic matter in surface sediments of the Barents Sea with a predominance of marine organic matter in most areas (Vetrov & Romankevich, 2004). In areas more affected by river run-off such as, for example, the Pechora Sea with its input by the Pechora River, terrigenous OC may become more significant, but even in this southeastern part of the Barents Sea the marine proportion remains dominant. In average, ~70% of the OC in surface sediments of the modern Barents Sea has a marine origin (Vetrov & Romankevich, 2004). The predominance of marine OC is explained by the influence of warm Atlantic-water inflow causing reduced sea-ice cover and increased primary production (see Chapter 2).

5.2.2. Kara Sea

In the southern Kara Sea, the highest OC values of 1.5–2.5% occur in the Ob and Yenisei estuaries and submarine channels towards the north (Figure 5.17; Stein & Fahl, 2004a). The shallower areas outside the channels are characterized by significantly lower OC values (<1%). This distribution pattern is related to the riverine suspended matter input. The estuaries, functioning as marginal filter (see Chapter 2.5, Figure 2.20), capture and accumulate most of the fine-grained

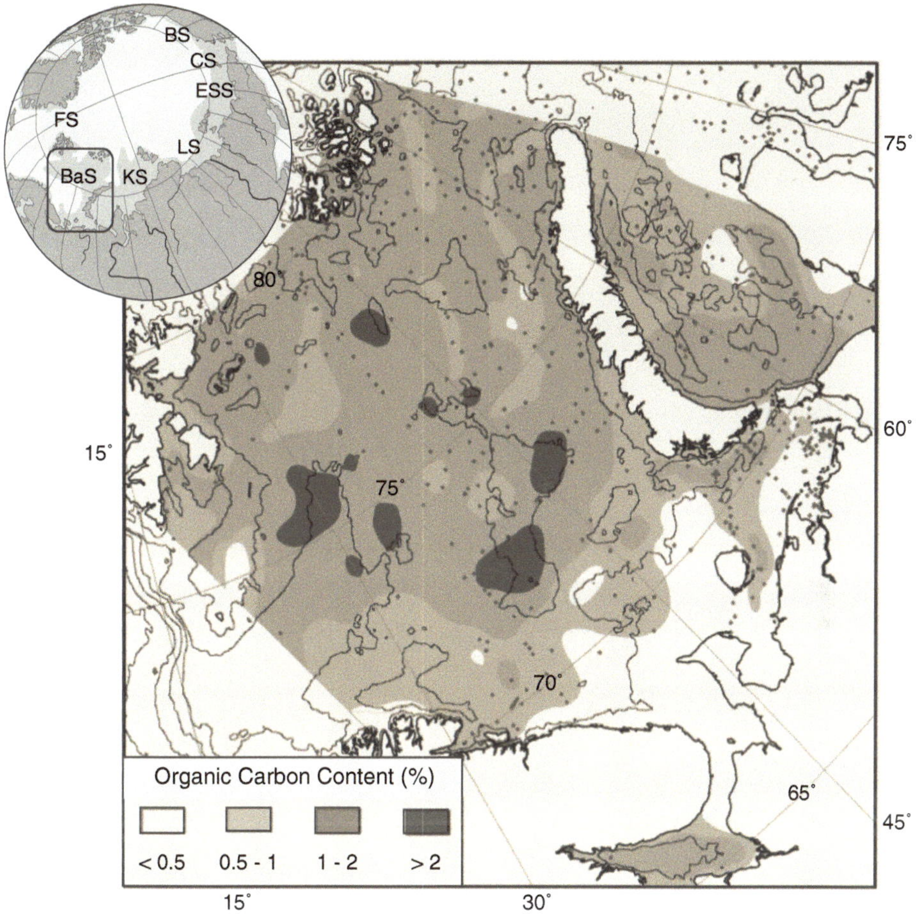

Figure 5.16 Distribution of organic carbon in surface sediments in the Barents Sea (Vetrov & Romankevich, 2004, supplemented).

suspended sediment including the organic matter (Lisitzin, 1995). This is reflected in absolute OC maxima >2% measured in the central estuaries. In the northern Kara Sea, the OC distribution is very variable (Figure 5.17). Elevated OC values of 1–2% occur in the St. Anna Trough and the Voronin Trough, separated by low OC values in the Central Kara Plateau area. Maximum OC values in the St. Anna Trough are concentrated in its eastern central part. High OC values of >1.5% also occur within a narrow rim around Franz Josef Land and in the deeper part (water depth >300 m) of the area between Franz Josef Land and Novaya Zemlya. The former maximum is probably related to input of terrigenous matter from Franz Josef Land (see later description), the latter related to OC input from the Barents Sea via Atlantic-water inflow (Figure 5.17).

The predominantly terrigenous origin of OC in the southern Kara Sea and, especially, in the estuaries is reflected in high C/N ratios >10 (Figure 4.49), low

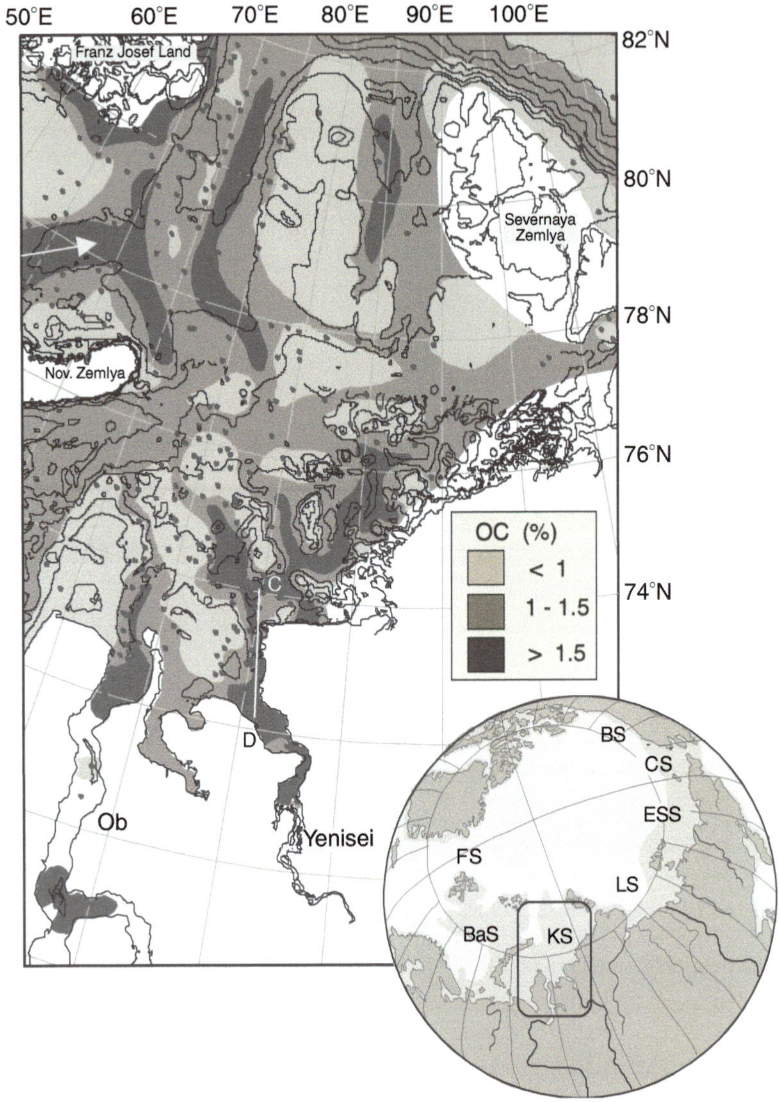

Figure 5.17 Distribution of organic carbon in surface sediments in the Kara Sea (Stein & Fahl, 2004a, supplemented). C-D indicates transect of maceral data shown in Figure 4.57B.

hydrogen index values < 100 mgHC gOC^{-1} and high concentrations of long-chain n-alkanes (C_{27}+C_{29}+C_{31}) of ~350–410 µg gOC^{-1} (Figure 4.58) (Stein & Fahl, 2004a) as well as the light $\delta^{13}C_{org}$ values of about −27‰ (Yenisei) and −28 to −28.7‰ (Ob) (Fernandes & Sicre, 2000; Krishnamurthy et al., 2001; Fahl et al., 2003). Based on n-alkane and kerogen microscopy data, ~80–100% of the OC is of terrigenous origin (Figure 4.57B; Boucsein et al., 1999; Stein & Fahl, 2004a). Towards the open Kara Sea, with increasing distance from the estuaries, a decrease in amount of terrigenous OC is indicated by decreasing C/N ratios, increasing $\delta^{13}C_{org}$ values, and decreasing long-chain *n*-alkanes. Based on the C/N ratios as well as the low hydrogen index values, a dominance of terrigenous OC is implied for the central Kara Sea up to 78°N and in the eastern Kara Sea. A transport of riverine material towards the north and northeast probably occurs via eastward-flowing (near-bottom) currents through the submarine channels. This mechanism is also reflected in the OC distribution pattern (Figure 5.17).

In the St. Anna Trough area of the northern Kara Sea, the relative amount of marine OC seems to be significantly higher than in the southern Kara Sea, as suggested from the HI values $> 100\,mgHC\,gOC^{-1}$, C/N ratios <7 (Figs. 4.48b and 4.49), and long-chain *n*-alkane concentrations of $<150\,\mu g\,gOC^{-1}$ (Figure 4.58) (Stein & Fahl, 2004a). Exceptions are sites close to Franz Josef Land where local inputs of terrigenous OC are indicated by higher C/N ratios and low HI values. Probable source areas of the terrigenous OC are the OC-rich Mesozoic rocks which outcrop on the islands (Elverhøi et al., 1995a).

In average, ~90–95% (5–10%), 85% (15%), 80% (20%), and 75% (25%) of the OC in surface sediments from the estuaries, southern, central, and northern Kara Sea, respectively, are probably of terrigenous (marine) origin (for data base of estimates, see Stein & Fahl, 2004a). The dominance of terrigenous OC is related to the high input by rivers and coastal erosion (see Chapter 2.5 and 2.7).

5.2.3. Laptev Sea

In the Laptev Sea, OC values vary between ~0.5% and 2.3% (Figure 5.18; Stein & Fahl, 2000, 2004b). Maximum OC values of up to 2% and more occur in the vicinity of the eastern Lena Delta, off the Olenjok river mouth and southwest of the New Siberian Islands, and the central part of the lower Laptev Sea continental slope. In the eastern and middle Laptev Sea, these OC maxima follow submarine channels which can be traced back to the river mouths. These submarine channels are the main modern (late Holocene) depocenter in the inner Laptev Sea, where fine-grained riverine material accumulates (e.g., Kuptsov & Lisitzin, 1996). The OC maximum along the lower continental slope (Figure 5.18) is probably related to the inflow of Atlantic-water masses laterally transporting (OC-enriched) suspended matter (see, e.g., Knies et al., 2000; Stein et al., 2001) and/or input from the Laptev Sea shelf.

The predominantly terrigenous origin of the organic matter of the Laptev Sea is shown in the organic-geochemical bulk parameters, maceral composition as well as biomarker data. Areas of high OC concentration commonly correspond to low HI values ($<100\,mgHC\,gOC^{-1}$), light stable carbon isotope composition ($\delta^{13}C_{org} = -26.5‰$; Müller-Lupp et al., 2000; Fahl et al., 2001), and high C/N ratios of 10–25 (using a correction for inorganic nitrogen; see Stein & Fahl, 2004b) pointing to a terrigenous origin. The lightest $\delta^{13}C_{org}$ values recorded near the mouth of the Lena River are in good agreement with the composition of the particulate organic matter of the Lena itself (−27.1‰; Rachold & Hubberten, 1999). The dominantly terrigenous source is also verified by the distribution of odd long-chain *n*-alkanes C_{27}–C_{31} (Fahl & Stein, 1997). The highest concentrations of

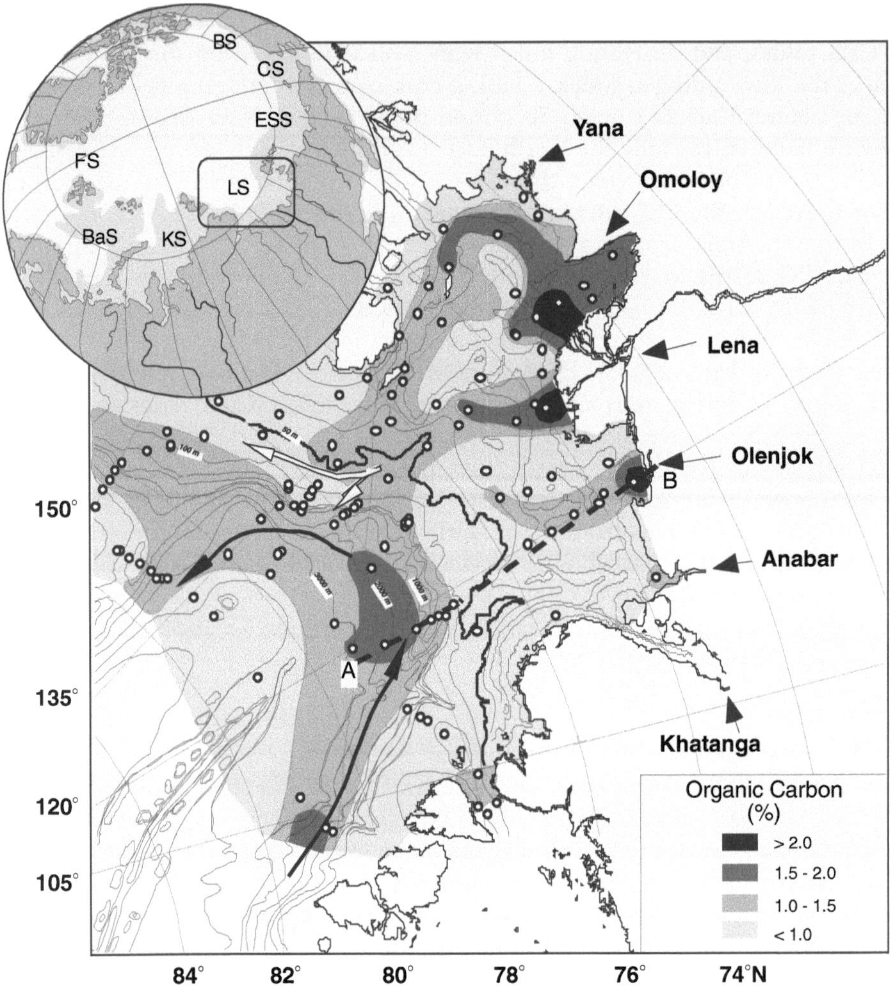

Figure 5.18 Distribution of organic carbon in surface sediments in the Laptev Sea (Stein & Fahl, 2004b, supplemented). A-B indicates transect of maceral data shown in Figure 4.57A.

these biomarkers, which are generally accepted as an indicator for a terrigenous higher plant source (see Chapter 4.7.4), were found in the vicinity of the eastern Lena Delta as well as off the rivers Anabar, Olenek, and Yana. There, the *n*-alkane concentrations reach values of $\sim$600 to $>$1,000 $\mu g\,gOC^{-1}$. Towards the continental slope, the concentration of long-chain *n*-alkanes decrease to values of $<$250 $\mu g\,gOC^{-1}$ (Stein & Fahl, 2004b), indicating a decreasing terrigenous influence.

A very similar picture can be obtained from maceral data. In the inner Laptev Sea $\sim$95% of the OC is of terrigenous origin, increasing to even 97–99% directly off the river mouths (see Chapter 4.7.3, Figure 4.57A; Boucsein & Stein, 2000;

Stein & Fahl, 2004b). Towards the north, with increasing distance from the land, the amount of terrigenous macerals decreases. At the outer Laptev Sea shelf and the upper continental slope, however, still 85% and 70–80% of the OC, respectively, is terrigenous. As in the Kara Sea, the dominance of terrigenous OC is related to the high input by rivers and coastal erosion (see Chapters 2.5 and 2.7).

5.2.4. East Siberian Sea

In general, OC content varies between 0.5% and 1.5% OC (Figure 5.19; Petrova et al., 2004). The higher values (1.5–2%) are associated with sediments near the Indigirka and Kolyma river mouths, whereas the lowest values (<0.5%) are associated with sandy sediments adjacent to islands of the western shelf, in local wash and re-deposition zones, and in sediments of continental shelf.

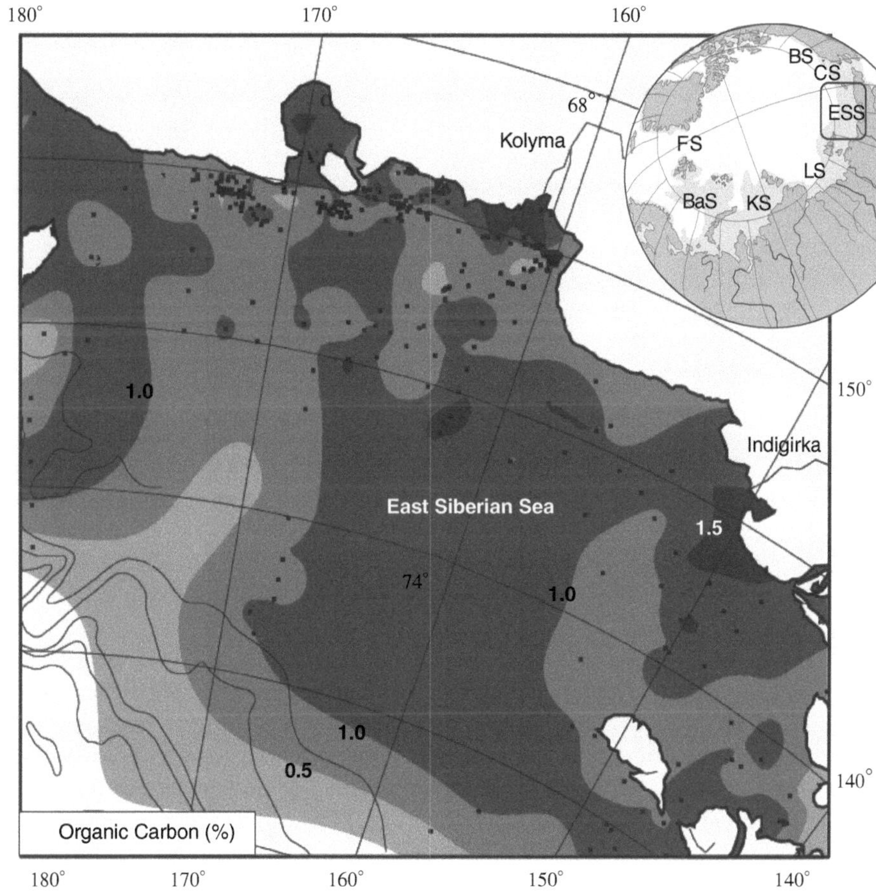

Figure 5.19 Distribution of organic carbon in surface sediments in the East Siberian Sea (Petrova et al., 2004, supplemented).

Unfortunately, organic-geochemical bulk parameters to be used as OC source indicator are not available for the East Siberian Sea. A limited number of biomarker data, however, allow a general identification of the sources (Petrova et al., 2004). The composition of *n*-alkanes in estuarine/deltaic sediments of the East Siberian Sea (e.g., of the Indigirka River) shows a strong predominance of high-molecular odd-number compounds (*n*-C_{22-37} up to almost 90% of the total *n*-alkanes) indicating dominantly terrigenous organic material. Towards the open shelf the long-chain *n*-alkanes decrease on cost of the short-chain *n*-alkanes, related to an increased influence of marine production of organic material. Along the shelf edge and slope, the proportion of short-chain *n*-alkanes becomes the dominant proportion of the *n*-alkanes (~65% of the total *n*-alkanes), suggesting increased biological productivity.

Based on the data from Petrova et al. (2004), Stein and Macdonald (2004b) roughly estimated that ~70% (30%) of the OC in surface sediments of the East Siberian Sea is of terrigenous (marine) origin. Terrigenous input by river discharge and coastal erosion (see Chapter 2.5 and 2.7) and primary production related to Pacific-water inflow, respectively, are the main processes controlling OC accumulation in the East Siberian Sea.

5.2.5. Continental Margin of the North Bering-Chukchi Sea

The concentrations of OC in surface sediments vary widely at the continental margin of the North Bering-Chukchi Sea, ranging from 0.1% to 2.8% (Naidu, Cooper, Grebmeier, Whitledge, & Hameedi, 2004). In general, the OC values are between 0.5% and 1.5%, that is, similar to those of most world shelves (Figure 5.20; Premuzic, Benkovitz, Gaffney, & Walsh, 1982; Romankevich, 1984). The exceptions are the northwest and southeast Gulf of Anydyr and the northwest Chukchi Sea where locally higher (1.5–2.8%) OC values were determined (Grebmeier, 1993).

The OC sources in surface sediments of the North Bering-Chukchi Sea are discussed in detail by Blackburn (1987) and Naidu et al. (1993, 2000, 2004), using the distribution patterns of $\delta^{13}C_{org}$ and C/N as proxies. In general, in the Bering Sea $\delta^{13}C_{org}$ progressively increases from −24‰ in the coastal region in the east, especially off the Yukon River and Norton Sound, to −20‰ in the west within the Gulf of Anadyr. Further inshore along the Gulf margin and the Anadyr Bay and Anadyr Estuary, the sediments are distinguished by the lowest $\delta^{13}C_{org}$ values (mean −26.6‰) (Naidu et al. 1993, 2000). Likewise, in the Chukchi Sea a progressive increase in $\delta^{13}C_{org}$ across the shelf is observed from the east to about the central shelf, and then farther west the values decrease to −22‰ or less near the East Siberian coast. Within the entire study area a progressive cross-shelf decrease from the east to west in the C/N ratios is clearly observed, with relatively higher values concentrated off major fluvial systems (Naidu et al. 1993, 2004). The general lateral increase in $\delta^{13}C_{org}$ and corresponding decrease in C/N ratios are explained by a net increase in marine- versus terrigenous-dominated particulate OC in the sediments.

In average, ~50% of the OC in surface sediments of the North Bering Chukchi Sea is estimated to be of terrigenous origin (Naidu et al., 2004; Stein & Macdonald,

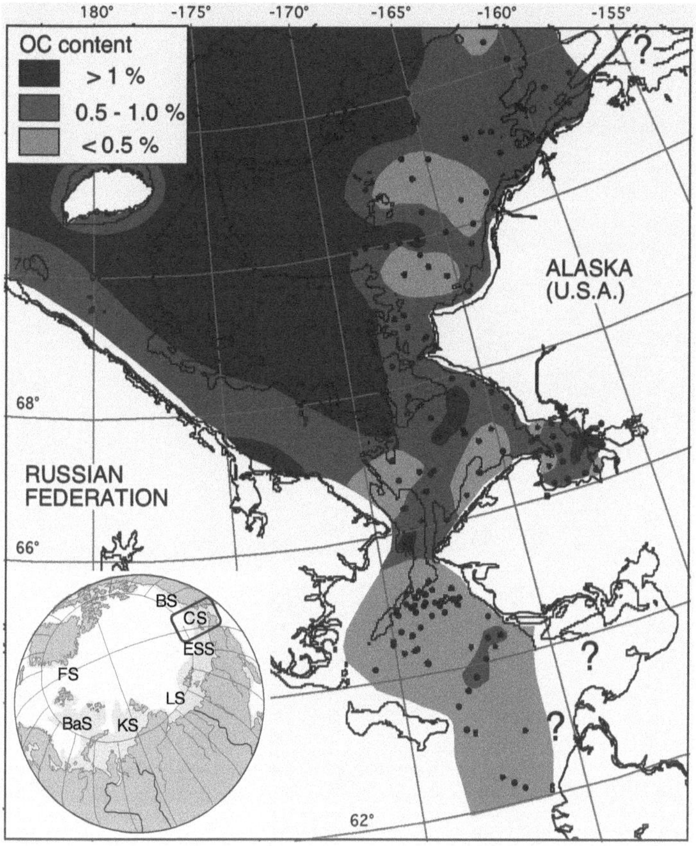

Figure 5.20 Distribution of organic carbon in surface sediments in the North Bering-Chukchi Sea (Naidu et al., 2004, supplemented).

2004b). The high proportion of marine OC is probably related to increased primary production due to the inflow of nutrient-rich Pacific water.

5.2.6. Beaufort Sea

The distribution of OC in the Beaufort Sea is strongly influenced by inputs from river discharge, especially for the Mackenzie Shelf which receives a disproportionately large portion of its fine-grained inorganic sediments from the Mackenzie River. OC contents within this region display a clear spatial pattern with the highest OC values observed closer to the Mackenzie River mouth (1.25–1.5%) and lower OC values in sediments from the slope (1–1.25%) (Figure 5.21). In general, high OC percentages are associated with fine sediments (Macdonald et al., 2004b). For the Alaskan Shelf, the OC data base is much more limited than for the Mackenzie Shelf. OC contents of Alaskan slope sediments usually exceed 1% with the highest value being 1.7% OC. Although there is much scatter in the shelf

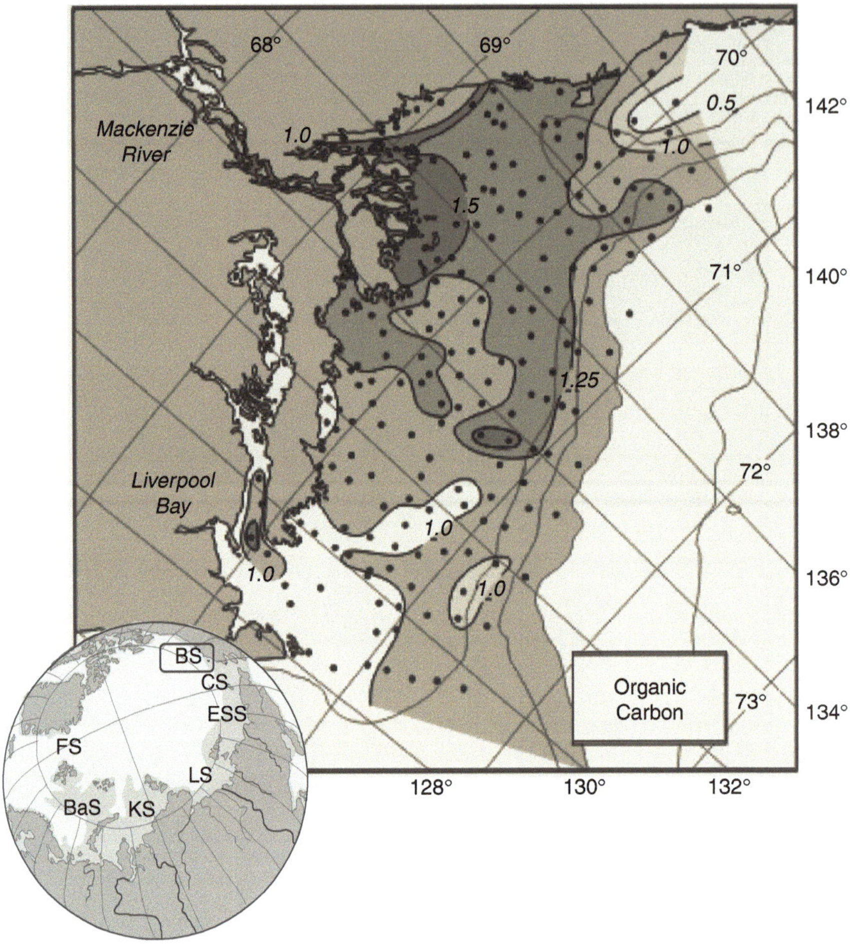

Figure 5.21 Distribution of organic carbon in surface sediments in the Beaufort Sea (Macdonald et al., 2004b, supplemented).

samples, they generally contain lower OC concentrations, most of them ranging between 0.25% and 0.9% OC (Macdonald et al., 2004b).

Based on $\delta^{13}C$ values and C/N ratios in surface sediments, information about the OC sources in the Beaufort Sea is available (Naidu, 1985; Naidu et al., 2000; (Macdonald et al., 2004b and references therein). C/N ratios vary predominantly between 6 and 14 suggesting mixtures of terrigenous and marine OC with no obvious relationship between OC and C/N values. For the Alaskan margin, the distribution of C/N ratio shows a broad spatial trend, with values decreasing from 14 in the inshore to 9 on the shelf and 8 in the slope region (Naidu, 1985). The first-order interpretation of the spatial distribution of $\delta^{13}C$ suggests strong terrestrial inputs of OC near shore with progressively greater contribution from marine primary production towards the outer shelf and interior ocean (see Figure 4.50). Furthermore, there is an east to west broad trend in sediments accumulated along the outer Beaufort shelf, manifested as a greater ^{13}C enrichment (i.e., less negative values) toward the west (Macdonald et al., 2004b). In summary, the C/N ratio and $\delta^{13}C$ data indicate that the Beaufort Sea shelf sediments predominantly are admixtures of terrigenous OC (C/N ~10 to 20, $\delta^{13}C$ ~ −27‰) and marine OC (C/N ~6–7, $\delta^{13}C$ ~ −24‰ to −23.4‰). Based on these data, Macdonaold et al. (2004b) estimated that almost all of the sediments contain over 30% terrigenous OC and the majority over 50%. These estimates are supported by biomarker data which also point to a system strongly influenced by terrigenous OC inputs (river input or coastal erosion) augmented by marine OC (Venkatesan, Kaplan, & Ruth, 1983; Yunker et al., 1991, 1993, 1995, 2005; Steinhauer & Boehm, 1992; Yunker & Macdonald, 1995; Goñi et al., 2000; (Macdonald et al., 2004b). Based on OC budget calculations, even >90% of the OC buried in near-surface delta and shelf sediments off the Mackenzie River is of terrigenous origin (Macdonald et al., 2004b), clearly reflecting the importance of fluvial OC supply.

Using ^{14}C isotopic mass balance, Goñi et al. (2005) demonstrated that the majority (~70%) of the particulate organic matter exported by the Mackenzie River and deposited in surface sediments of the Beaufort Shelf, is probably ancient OC. According to these authors, the source of this ancient terrigenous particulate OC is likely to be a combination of fossil carbon (i.e., bitumen and/or kerogen) eroded from sedimentary rocks in the drainage basin, and highly degraded soil carbon with long residence times bound in permafrost. These results are consistent with the findings from the Siberian Arctic, where old ages for sedimentary OC have been measured in the eastern region of the Siberian continental shelf off the Khatanga, Lena, and Indigirka rivers (Guo et al., 2004).

5.2.7. Central Arctic Ocean

According to the limited OC data base for central Arctic Ocean surface sediments (e.g., Belicka et al., 2002; Stein et al., 2004b), only a very general picture of the OC distribution with major gaps is available. Based on the existing data, the continental slopes are characterized by OC contents of ~1%. In the major basins such as the Canada Basin, the Amundsen Basin, and the Makarov Basin, OC values of 0.7–1% are typical whereas on the central Gakkel Ridge as well as the Lomonosov Ridge, the OC contents seem to be lower (0.3–0.6%) (Figure 5.22). Towards the Eurasian continental margin, OC contents increase to 0.8–1.3% in surface sediments over the Lomonosov Ridge, with higher values more typical for the Amundsen Basin side (Figure 5.18). Although there is no clear correlation between OC content and grain size, OC contents >0.5% occur mainly in fine-grained sediments with a silt plus clay content of >90% (Stein et al., 2004b). These fine-grained sediments are concentrated in the basins, whereas on the oceanic ridges and plateaus higher fractions of sand tend to be observed.

A likely cause for the high OC content of Arctic surface sediments is the relatively high contribution of terrigenous organic matter derived from the surrounding marginal seas (Stein et al., 2004b and further references therein). This

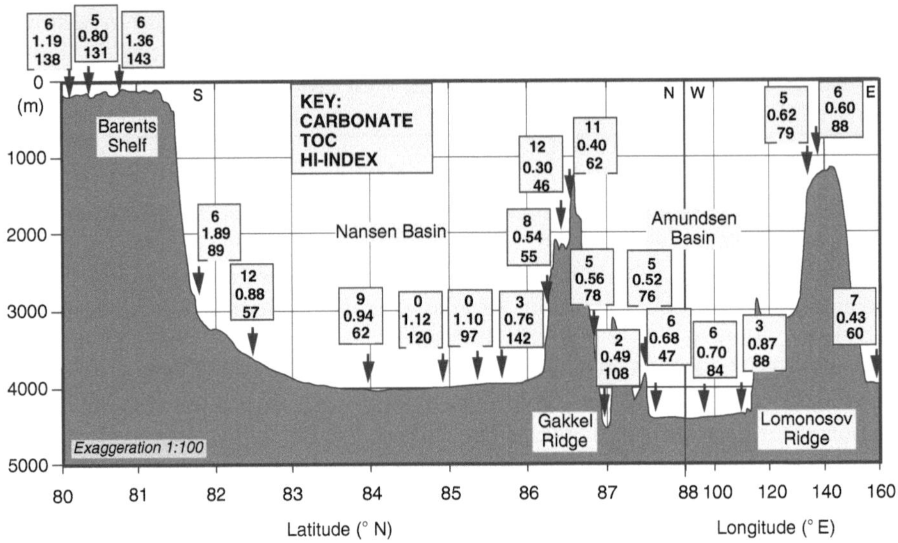

Figure 5.22 Transect across the Eurasian Arctic Ocean (for location of transect, see Figure 5.15) showing organic carbon, carbonate, and hydrogen index values in surface sediments (data from Stein et al., 1994c).

is supported by very low hydrogen index values <100 mgHC gC^{-1} (Figure 5.22). Only close to the Svalbard and Barents Sea continental margin, where elevated HI values of 150 mgHC gOC^{-1} were determined (Figure 5.22), pointing to some increased preservation of marine OC. These increased marine OC values are probably caused by increased (ice edge) surface-water productivity due to reduced sea-ice cover and increased nutrient supply, both triggered by the inflow of warm Atlantic water via the WSC. The somewhat higher HI values determined in surface sediments from the central Nansen Basin are related to turbidity currents causing rapid burial and increased preservation of marine OC (Stein et al., 1994c, 2004b).

C/N ratios (corrected for the amount of inorganic nitrogen; see Chapter 4.7.1 and Stein et al., 2004b) vary between 8 and 15 (Figure 4.48A) indicating a mixed marine/terrigenous OC source. Across the Lomonosov Ridge, close to the Laptev Sea continental margin, C/N ratios are generally higher on the Amundsen Basin side and lower on the Makarov Basin side (Stein et al., 2004b). This may reflect a higher terrigenous influence in the Amundsen Basin than in the Makarov Basin. In the Amerasian Basin, C/N ratios of ~8–15 also suggest a mixed marine/terrigenous OC source for this part of the Arctic (Cranston, 1997).

Although only based on a very limited number of samples from the Lomonosov Ridge, maceral data support the HI (see Chapter 4.7.1) and C/N data, clearly indicating the predominance of terrigenous OC in central Arctic Ocean surface sediments. The proportion of the terrigenous macerals huminite/vitrinite and inertinite reach values of >90% of the total macerals (Figure 4.53; Stein et al., 2006a; Boucsein & Stein, 2008). Biomarker data, that is, the clear dominance of long-chain *n*-alkanes over short-chain *n*-alkanes, agree with a mainly terrigenous OC source as well (Schubert & Stein, 1997; Belicka et al., 2002).

In contrast to the HI values, C/N ratios, maceral composition, and biomarker data, very heavy $\delta^{13}C_{org}$ values ranging between −21.1‰ and −22.1‰ across the Eurasian Basin (Schubert & Calvert, 2001), may suggest higher amount of marine OC being preserved in the surface sediments. Based on the $\delta^{13}C_{org}$ values, Schubert and Calvert (2001) estimated the maximum terrigenous contribution to be only ~30% of the organic matter in central Arctic Ocean sediments. It has to be considered, however, that the carbon isotope composition of phytoplankton in the Arctic is highly variable. For example, high amount of ice algae representing up to 60% of the primary production in the permanently ice-covered central Arctic Ocean (Sakshaug, 2004) and characterized by very heavy $\delta^{13}C_{org}$ values of

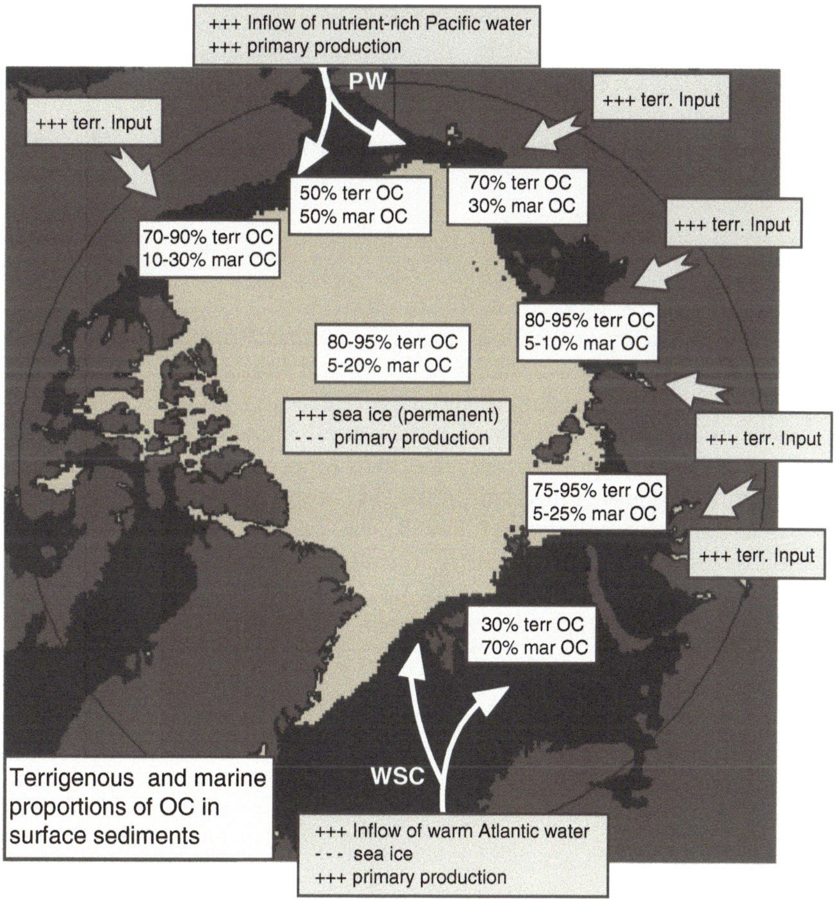

Figure 5.23 Average percentage values of marine and terrigenous organic-carbon proportions in surface sediments from the Arctic marginal seas and central Arctic Ocean (for data base, see Stein & Macdonald, 2004a and references therein).

about −15 to −8‰ (Gibson et al., 1999), may also explain the heavy $\delta^{13}C_{org}$ values measured in the surface sediments. This fact may limit the use of $\delta^{13}C_{org}$ values as a marine/terrigenous source indicator (Schubert & Calvert, 2001; Stein & Macdonald, 2004c).

Another argument against high proportion of marine OC being preserved in near-surface sediments is derived from the comparison of estimated marine OC accumulation rates based on primary production values and OC decomposition rate during its transfer from the surface water to the seafloor (for calculation procedure, see Stein & Macdonald, 2004c), and OC accumulation rates based on sediment core data. For the latter, OC accumulation rates of ~2.5–4.5 mgC $cm^{-2}\,kyr^{-1}$ have been calculated for the central Arctic Ocean (see Chapter 6.4; Stein et al., 2004b). Using a realistic central Arctic Ocean primary production value of 10–20 gC $m^{-2}\,yr^{-1}$ (see Chapter 2.4) and mean sedimentation rate values of 0.5–1 cm kyr^{-1}, on the other hand, the expected maximum accumulation of marine OC would be between ~0.1 and 0.5 mgC $cm^{-2}\,kyr^{-1}$ for these areas. This difference can best be explained by the high proportion of terrigenous organic matter in the Arctic Ocean sediments. The calculations imply that about only 5–20% of the organic matter is of marine origin, that is, 80–95% of the organic matter in central Arctic Ocean sediments is of terrigenous origin (Stein et al., 2004b).

5.2.8. Summary

In summary, it can be stated that the modern Arctic Ocean OC accumulation is strongly controlled by terrigenous input, in contrast to the other world oceans (see Stein & Macdonald, 2004a for more details and references). Owing to the high fluvial and coastal erosion sediment supply, terrigenous OC is predominant in the Kara, Laptev, and Beaufort seas as well as the central Arctic Ocean (Figure 5.23). For the latter, the very low productivity due to the permanent sea-ice cover further strengthens the terrigenous character of the OC in surface sediments. In the Barents Sea as well as the East Siberian and Chukchi seas, on the other hand, a much higher proportion of the OC is of marine origin, caused by increased primary production due to the increased Atlantic-water inflow and reduced sea-ice cover, and increased inflow of nutrient-rich Pacific waters, respectively (Figure 5.23).

CHAPTER SIX

Quaternary Variability of Palaeoenvironment and Its Sedimentary Record

6.1. The Stratigraphic Framework of Arctic Ocean Sediment Cores: Background, Problems, and Perspectives

For the interpretation of geological records in terms of palaeoenvironmental changes versus time, a reliable stratigraphical (chronological) framework is necessary. This framework may be based on (a combination of) AMS^{14}C and U/Th datings (for U/Th approach see Fietzke, Liebetreu, Eisenhauer, & Dullo, 2006), oxygen isotope stratigraphy, palaeomagneto-stratigraphy, biostratigraphy, lithostratigraphy, and logging records. These dating techniques are in standard use in studies from most world ocean areas. For Arctic Ocean sediments, however, several problems are obvious resulting in difficulties in establishing accurate age-depth relationships in the existing sediment cores (see Backman et al., 2004 for summary and further references). Poor preservation of calcareous and biosiliceous microfossil faunas and floras in Arctic cores often precludes application of conventional biostratigraphic and isotopic dating techniques (see later discussions). Therefore, chronological models have largely relied on lithostratigraphy (e.g., Clark et al., 1980), magneto-stratigraphy (e.g., Steuerwald, Clark, & Andrew, 1968; Clark, 1970; Clark et al., 1980; Aksu, 1985; Poore et al., 1993; Frederichs, 1995; Schneider, Backman, Curry, & Possnert, 1996; Phillips & Grantz, 1997; Nowaczyk et al., 2001; Jakobsson et al., 2000a), ^{10}Be records (Eisenhauer et al., 1994; Aldahan, Ning, Possnert, Backman, & Bostrom, 1997; Spielhagen et al., 1997, 2004), amino acids (Sejrup, Gifford, Brigham-Grette, Løvlie, & Hopkins, 1984; Macko & Aksu, 1986), and AMS^{14}C datings (e.g., Markussen, Zahn, & Thiede, 1985; Zahn, Markussen, & Thiede, 1985; Poore et al., 1994; Stein et al., 1994a; Darby et al., 1997; Spielhagen et al., 1997; Nørgaard-Pedersen et al., 2003), and the correlation between these records. However, also these approaches have to be applied with caution when using as single approach because records might be interpreted in different ways (e.g., the palaeomagnetic and ^{10}B records; see later discussions). Others such as amino acids yielded different opposing results. Using amino acids, evidence for high sedimentation rates was found in cores from the Alpha Ridge by Sejrup et al. (1984), whereas Macko and Aksu (1986) proposed an older age and, thus, low sedimentation rates for sediment cores from the same area.

6.1.1. AMS^{14}C Datings

While age models and sedimentation rates in the Arctic Ocean over the entire Pleistocene are still under discussion (Backman et al., 2004; see later discussions), the last glacial and Holocene are quite well documented by radiocarbon chronology (e.g., Stein et al., 1994a, 1994b, 2003a; Darby et al., 1997, 2006; Nørgaard-Pedersen et al., 1998, 2003; Poore et al., 1999; Polyak et al., 2004; Levitan & Stein, 2007). In general, sedimentation rates are an order of magnitude higher in the Fram Strait and continental margin areas than in the central Arctic Ocean (Figure 6.1A and B; Table 6.1). Average Holocene sedimentation rates in the central Arctic of $\sim$0.5–1 cm kyr^{-1}, increasing towards the continental margins to 5 to >10 cm kyr^{-1} (Figure 6.2). In areas influenced by turbidites, sedimentation rates may significantly increase (e.g., Grantz et al., 1999; Svindland & Vorren, 2002; see Chapter 3.4). For

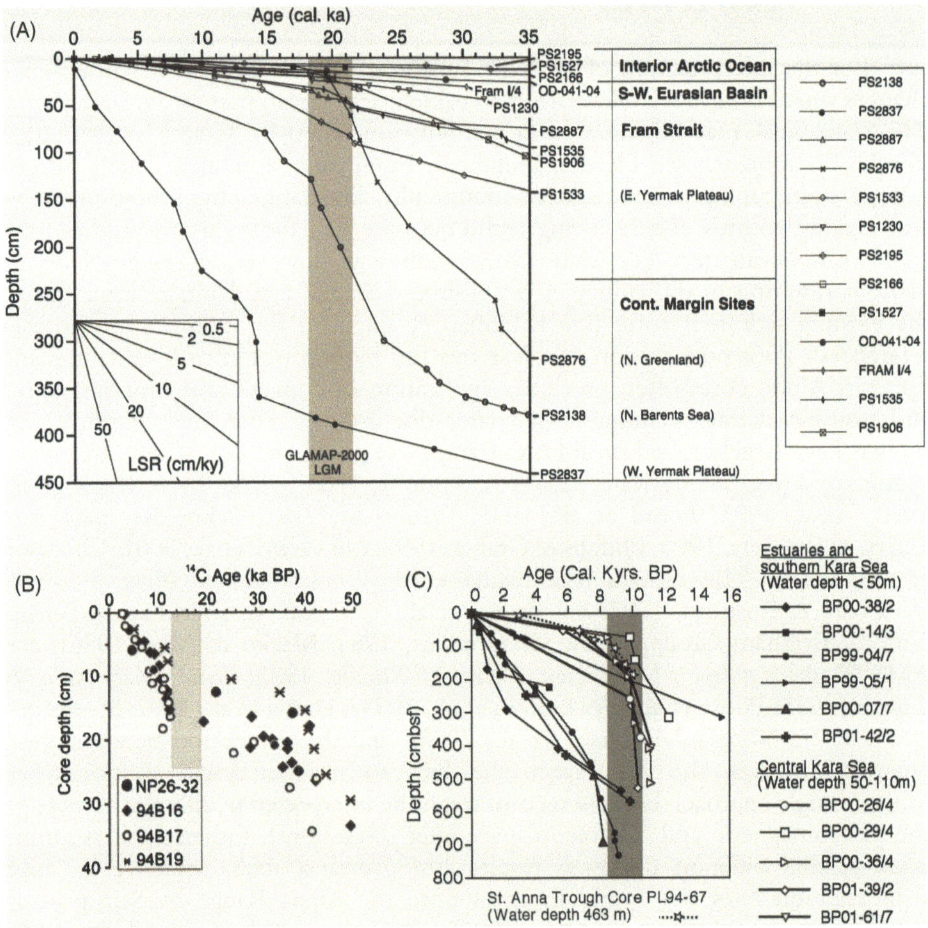

Figure 6.1 Age depth diagrams of Arctic Ocean sediment cores, based on AMS^{14}C datings. (A) Arctic Ocean and adjacent continental margin (Nørgaard-Pedersen et al., 2003) (B) Mendeleev Ridge (Polyak et al., 2004) (C) Kara Sea (Stein et al., 2003a).

Table 6.1 Compilation of Sedimentation Rates of MIS-1 (Holocene) and MIS-2 time intervals from AMS^{14}C-Dated Sediment Cores. References of the Data are Listed in the Last Column. WD, Water Depth.

Station	Latitude	Longitude	WD (m)	MIS 1 (Holocene)	MIS 2	Reference
Fram Strait (north of 80°N)/Yermak Plateau						
PS51/ 309	81°11.2′N	12°59.1′E	2,269	4	8	Winkelmann and Stein (2007)
PS2212	82°04.2′N	15°51.2′E	2,550	1.8		Nørgaard-Pedersen (1997)
PS2419	80°27′ N	13°40′W	412	6	nd	Notholt (1998)
PS2424	80°02′ N	05°44′W	445	2.9	41.2	Notholt (1998)
FRAM 1/4	84°30′ N	08°59′N	3,820	1	1.9	Zahn et al. (1985)
PS 2423a	80°02′N	05°26′W	829	0.8	7.2	Notholt (1998)
PS 2422a	80°01′N	04°13′W	1,938	0.3	nd	Notholt (1998)
PS 2837	81°13′N	02°22′E	1,042	22.6	4.1	Nørgaard-Pedersen et al. (2003)
PS 55/ 100	80°29′N	02°56′E	1,538	20	6.2	Birgel and Stein (2004)
Morris Jesup Rise						
PS2200	85°19.6′N	14°00.0′W	1,074	0.5	0.3	Nørgaard-Pedersen (1997)
Barents Sea slope						
PS2138	81°32.1′N	30°35.6′E	995	5.4	21.2	Knies et al. (2000)
PS2446	82°23.8′N	40°54.5′E	2,022	10	>30	Knies et al. (2000)
P191-AR-JPC5	81°07.1	43°25.9	463	22.7	nd	Lubinski et al. (2001)
Laptev Sea slope						
PS51/ 154-11	77°15.6′N	120°36.6′E	270	18	nd	Bauch et al. (2001)
PS2778-2	77°58.7′N	113°03.9′E	341	38	nd	Stein and Fahl (2000)
PS2458-4	78°10.0′N	133°23.9′E	983	34	nd	Spielhagen et al. (2005)
PS2742-5	80°47.3′N	103°48.9′E	1,890	22	nd	Stein and Fahl (2000)
PS2741	81°06.0′N	105°22.0′E	2,400	11.9	2	Knies et al. (2000)
East Siberian Sea slope						
PS2763	80°16.9′N	150°26.1′E	1,591	12.6	10.2	Stein et al. (2001)
Chukchi Rise						
92BC17	76°05.2′N	164°50.2′W	402	2	nd	Darby et al. (1997), Grantz et al. (1999)
Northwind Ridge						
88BC6	74°35.9′N	158°02.2′W	954	3.2	nd	Grantz et al. (1999)
88BC22	75	161°60′W		3.4	nd	Darby et al. (1997)
P1-92-AR-P2	73°57′N	161°32′W	369	9	nd	Polyak et al. (2007)

Table 6.1 *(Continued)*

Station	Latitude	Longitude	WD (m)	MIS 1 (Holocene)	MIS 2	Reference
Mendeleev Ridge						
94BC16	80°20.3′N	178°42.7′W	1,533	1.1	0.5	Darby et al. (1997), Grantz et al. (1999)
94BC17	81°15.91′N	178°58.1′E	2,217	1.5	0.3	Darby et al. (1997), Grantz et al. (1999)
94BC19	82°26.8′N	178°45.5′E	2,400	0.8	<0.3	Poore et al. (1999)
NP26-32	79°19.4′ N	178°04′W	1,610	1		Polyak et al. (2004)
Lomonosov Ridge						
GreenICE Core 10	84°48.8′N	74°17′W	1,040	0.3		Nørgaard-Pedersen et al. (2007a, 2007b)
PS2177	88°02.2′N	134°55.1′E	1,388	0.9	0.3	Nørgaard-Pedersen (1997)
PS2185	87°31.9′N	144°08.4′E	1,073	0.7	0.3	Nørgaard-Pedersen (1997)
94BC28	88°52.3′N	140°22.8′E	2,015	1.3	0.1	Darby et al. (1997)
PS2757	81°09.6′N	140°12.1′E	1,230	5.2	5.8	Stein et al. (2001)
Gakkel Ridge						
PS1527						Köhler (1992)
PS2163	86°14.5′N	59°12.9′E	3,040	0.7	<0.7	Stein, Schubert, Vogt, and Fütterer (1994b)
PS2166	86°51.6′N	59°45.9′E	3,636	0.9	0.4	Nørgaard-Pedersen, (1997)
PS2206	84°16.7′N	02°30.3′W	2,993	0.7	0.5	Stein et al. (1994b)
Nansen Basin						
OD-041-04	84°01.8′N	11°14.3′E	3,344	0.7	0.4	Nørgaard-Pedersen et al. (1998, 2000)
Amundsen Basin						
PS2170	87°35.8′N	60°53.7′E	4,226	0.9	0.5	Stein, et al. (1994b)
PS2192	88°15.7′N	9°52.7′E	4,375	4	nd	Nørgaard-Pedersen (1997)
PS2195	86°13.7′N	9°35.6′E	3,873	0.6	0.3	Nørgaard-Pedersen (1997)
Makarov Basin						
PS2761	81°11.5′N	150°29.8′E	2,640	2	3.4	Stein et al. (2001)
PS2180	87°37.6′N	156°40.5′E	4,005	0.8	<0.4	Nørgaard-Pedersen (1997)
Wrangel Abyssal Plain						
94BC20	83°10.6′N	174°5.9′E	3,145	1	0.2	Darby et al. (1997)
Canada Abyssal Plain						
88PC10	74°43.5′N	156°08.5′W	3,899	145	nd	Grantz et al. (1999)
93BC13	75°48.3′N	155°07.6′W	3,831	1	nd	Grantz et al. (1999)

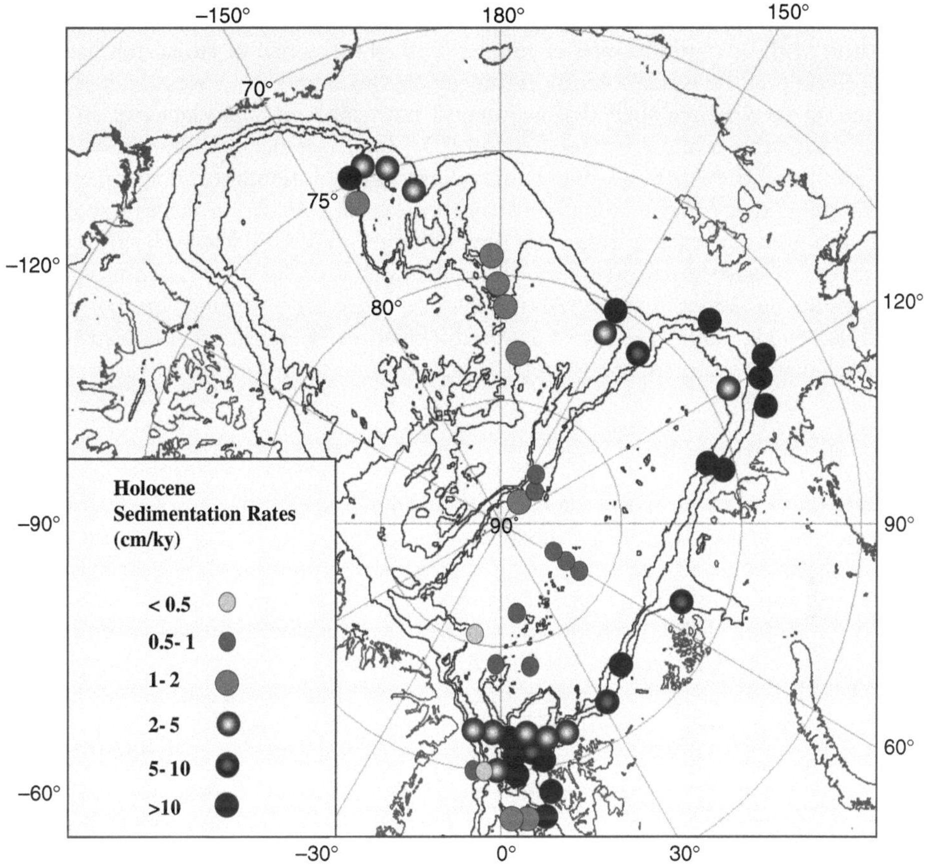

Figure 6.2 Distribution of average Holocene sedimentation rates in the central Arctic Ocean. For data base and references see Table 6.1. For distribution maps of average sedimentation rates for MIS 5, MIS 4, MIS 3, MIS 2 and MIS 1 it is referred to Levitan and Stein (2007).

a sediment core obtained from the Canada Abyssal Plain, for example, even an average Holocene sedimentation rate of $\sim$145 cm kyr^{-1} was measured (Grantz et al., 1999). On the continental shelves, especially in areas proximal to the sources of major terrigenous (fluvial) input such as the Kara and Laptev seas, very high average Holocene sedimentation rates of 50–>100 cm kyr^{-1} were determined. Here, maximum sedimentation rates of several m kyr^{-1} were reached during the last deglaciation at $\sim$11 to 8 ka, that is, at times when the broad shelf areas were flooded after the last glacial (Figure 6.1C; Bauch, Kassens, Naidina, Kunz-Pirrung, & Thiede, 2001a, Bauch et al., 2001b; Stein & Fahl, 2000; Stein et al., 2001; 2003a, 2004a).

During the last glacial, sedimentation rates display a much larger variability, reaching tens of centimetre per thousand years during episodes of massive iceberg discharge, such as along the Barents Sea continental margin (Figure 6.1A; Knies et al., 2000), but decreasing to very low values or even forming a hiatus of up to several thousand years during the glacial maximum in the central Arctic Ocean basins (e.g., Stein et al., 1994a,

1994b; Nørgaard-Pedersen et al., 1998, 2003; Poore et al., 1999; Polyak et al., 2004). This drop in the deposition rates is coupled with the absence of faunal remnants and may indicate that the water column in this region was capped by a very thick and solid pack ice or even an ice shelf that suppressed both biological productivity and lithic deposition (Darby et al., 2006; see Section 6.3.3 for more detailed discussion).

When using the AMS^{14}C-based chronology and sedimentation rates, one should have in mind that for the calculation of absolute ages in Arctic Ocean sequences routinely a reservoir correction of −440 yr (Mangerud & Gulliksen, 1975) was used. Due to the relative isolation of the deep Arctic Basins, however, actual reservoir time could be larger in the Arctic Ocean, especially during glacial periods (Backman et al., 2004; see also Sarnthein, Grootes, Kennett, & Nadeau, 2007). Thus, calculated sedimentation rates for glacial intervals might be too low.

6.1.2. Biostratigraphy

Accurate biostratigraphy based on the calibration of biostratigraphic events (i.e., first and last appearance datums) to magneto-stratigraphy and/or orbitally tuned time scales, an approach well established for low and mid latitudes (e.g., Berggren, Kent, & Van Couvering, 1985; Berggren, Kent, Swisher III, & Aubry, 1995; Shackleton, Berger, & Peltier, 1990; Shackleton, Crowhurst, Hagelberg, Pisias, & Schneider, 1995; Shackleton, Crowhurst, Weedon, & Laskar, 1999; Hilgen, 1991; Backman & Raffi, 1997), is more or less not developed for central Arctic Ocean records because (i) taxonomically well defined and firmly age-calibrated biostratigraphic marker events only occur rarely in Arctic Ocean cores, (ii) biosiliceous groups are not preserved in upper Cenozoic cores, and (iii) calcareous faunas and floras are taxonomically impoverished and occur discontinuously in the cores (Backman et al., 2004). One of the few marker species providing accurate age information and also found in central Arctic Ocean sediments is the nannofossil species *Emiliana huxleyi*. This species, with a first appearance datum at 0.26 Ma (Thierstein, Geitzenauer, Molfino, & Shackleton, 1977), has only been found in the Holocene and the penultimate interglacial in Arctic cores (Baumann, 1990; Gard, 1993; Jakobsson et al., 2001; see Chapter 4.5.2, Figure 4.33). These interglacial intervals may also be identified by foraminifers (e.g., Jakobsson et al., 2001; Wollenburg et al., 2001) and dinoflagellates (e.g., Matthiessen et al., 2001) (see Figures 4.32 and 4.41). Despite all the problems encountered in Arctic Ocean biostratigraphy, compositional and abundance variations of species (e.g., specific benthic foraminifers, ratio benthic versus planktonic foraminifers, boundary between predominantly calcareous and almost exclusively arenaceous foraminiferal faunas) often show coherent patterns over wide distances (see Figure 4.32) and may be used as relative stratigraphic markers (e.g., Herman, 1974; O'Neill, 1981; Polyak, 1986; Scott et al., 1989; Ishman et al., 1996; Backman et al., 2004 and further references therein; see also Chapter 4.5).

6.1.3. Oxygen Stable Isotope and ^{10}Be Stratigraphy

The stable isotope method which still gives reliable information on stratigraphy in the Fram Strait and northeastern Svalbard continental margin/Yermak Plateau area (e.g., Köhler & Spielhagen, 1990; Vogelsang, 1990; Köhler, 1992; Knies & Stein, 1998; Knies et al., 1999, 2000; Knies & Vogt, 2003) as well as in Arctic Ocean AMS^{14}C-dated near-surface sediment records (e.g., Zahn et al., 1985; Stein et al., 1994a, 1994b; Poore et al., 1999; Nørgaard-Pedersen et al., 1998, 2003), is constrained by the discontinuous and limited distribution of foraminiferal carbonate in the pre-MIS 6 central Arctic Ocean sediment cores (e.g., Scott et al., 1989). Furthermore, reduced surface water salinities resulting from large river discharge and meltwater events, strongly influence the oxygen isotope records generated from planktonic foraminifers (e.g., Aksu, Mudie, Macko, & de Vernal, 1988; Stein et al., 1994a, 1994b; Nørgaard-Pedersen et al., 1998, 2003; Spielhagen et al., 2004, 2005; see Section 6.3.5 for more details). Thus, the correlation between oxygen isotope curves obtained from central Arctic Ocean cores and low- to mid-latitude oxygen isotope curves representing global climate records, generally do not yield reliable age information. Here, however, ^{10}Be records may help to identify glacial/ interglacials and, thus, may allow to obtain a stratigraphic framework, as shown by the correlation between oxygen isotope and ^{10}Be records in the Norwegian Sea and on Yermak Plateau (Figure 6.3; Eisenhauer et al., 1994; Knies, 1999).

In general, ^{10}Be concentrations are low in glacial upper Quaternary sediments, whereas in interglacial sediments ^{10}Be concentrations are high. The background of this pattern is that the flux of the cosmogenically produced ^{10}Be from the uppermost water column to the sediment is enhanced by sedimentation of clay minerals and bioproduction (see Sharma, Mahannah, Moore, Ku, & Southon, 1987; Kusakabe et al., 1987), both processes that are intensified during interglacials in the sub-Arctic and Arctic Ocean (Eisenhauer et al., 1994; Gard, 1993). The deposition of fine-grained particles act by scavenging as a carrier to the seafloor. Accordingly, coarse-grained, IRD-rich sequences reveal intervals with very low ^{10}Be concentrations. With this approach, a stratigraphy of Arctic Ocean sediment cores for which no oxygen isotope records are available, could be obtained (Figure 6.3; Eisenhauer et al., 1994). When using ^{10}Be records as stratigraphical tool, however, one should have in mind that without additional independent age control the interpretation ^{10}Be is not unequivocal (Spielhagen et al., 1997, 2004; see Section 6.1.5).

6.1.4. Magneto-Stratigraphy: Reversals and Excursions of Inclination and Palaeointensity

Palaeomagnetic age models are based on interpretation of changes in measured intervals of uniform geomagnetic polarity directions that are caused by time-dependent changes in Earth's magnetic field. Palaeomagnetic stratigraphy, however, strongly relies on the interpretation of zones with negative inclination that may represent either complete reversals (i.e., chrons or subchrons, see Cande & Kent, 1992, 1995) or partial excursions of the geomagnetic field (e.g., Champion, Lanphere, & Kuntz, 1988; Nowaczyk, 1991; Nowaczyk, Frederichs, Eisenhauer, & Gard, 1994; Langereis, Dekkers, de Lange, Paterne, & van Santvoort, 1997; Lund, Stoner, Channell, & Acton, 2006 and further references therein) (Figure 6.4, Table 6.2). Reversals occur over time scales of a few hundred thousand years, whereas excursions commonly last $<<10$ kyr. Thus, different interpretations of the

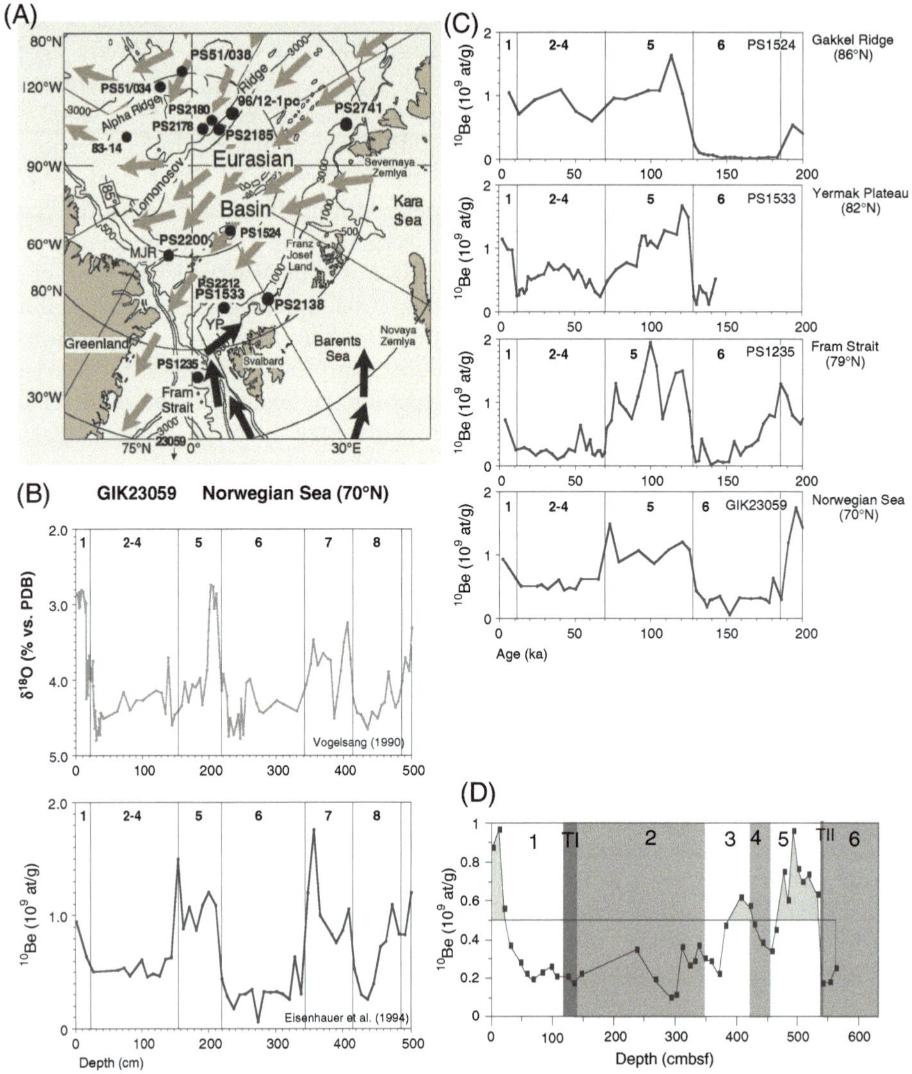

Figure 6.3 (A) Location of sediment cores discussed in this subsection. (B) Oxygen stable isotope and ^{10}Be records of a sediment core from the Norwegian Sea at 70°N (Eisenhauer et al., 1994). (C) ^{10}Be records of sediment cores along a transect between 70°N and 86°N. Marine isotope stages (MIS) are indicated (Eisenhauer et al., 1994). (D) ^{10}Be record of sediment Core PS2138-1 from the northern Barents Sea continental margin; MIS 1-6 and terminations TI and TII are indicated (Knies, 1999). From this core, a detailed oxygen stable istope record is also available, supporting the ^{10}Be stratigraphy (Knies et al., 1999 2000, 2001; see Section 6.3.1, Figure 6.30).

magnetic boundaries may result in significantly different age models, as shown, for example, by Frederichs (1995) for sediments cores from the Morris Jesup Rise and the Lomonosov Ridge (see Section 6.1.5). The most intensive work on geomagnetic excursions in the Arctic Ocean and NGS area has been carried out by

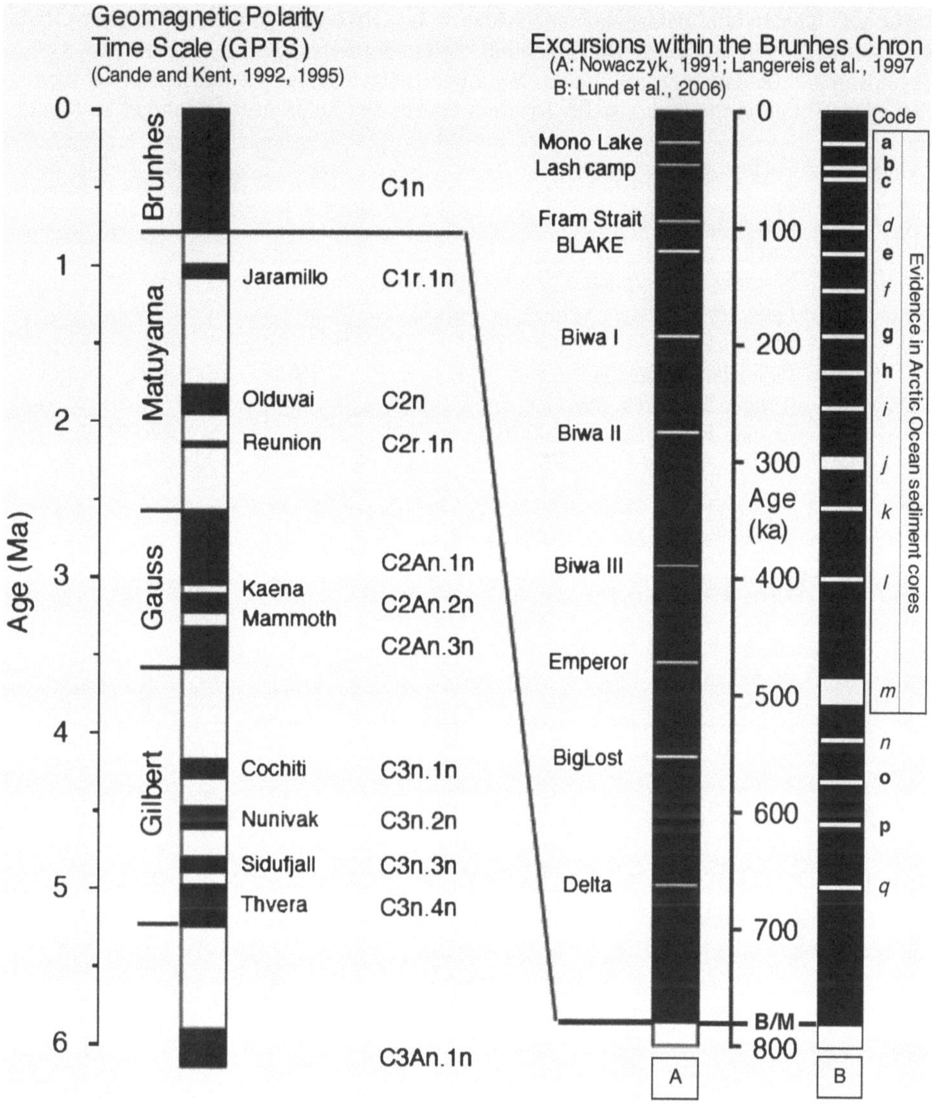

Figure 6.4 Geomagnetic polarity time scale (Cande & Kent, 1992, 1995) and established geomagnetic excursions within the Brunhes Chron (Backman et al., 2004, supplemented). A, Excursions according to Nowaczyk (1991), Langereis et al. (1997); B, Excursions according to Lund et al. (2006). For codes a–q (in bold well-documented excursions) see Table 6.2.

Nowaczyk and co-workers (Nowaczyk, 1991; Nowaczyk & Baumann, 1992; Nowaczyk et al., 1994, 2000, 2001; Nowaczyk & Frederichs, 1999; Nowaczyk & Knies, 2000). Geomagnetic excursions were also discussed for Arctic Ocean sediment cores by Løvlie, Markussen, Sejrup, and Thiede (1986), Schneider et al. (1996), and Jakobsson et al. (2000a).

Table 6.2 Summary of palaeomagnetic Evidence for Existence/Age of Brunhes Chron Excursions (Lund et al., 2006, supplemented). Age in brackets (240 ka) according to Thompson and Goldstein (2006). Italics, More Poorly Documented Excursions; Underlined, Excursions Found in the (sub-) Arctic. For Data Source See References in Lund et al. (2006).

Code	Excursion	Stage	Estimated age (ka)
a	Mono Lake/3α	3	33 ± 1
b	Laschamp/3β	3	41 ± 1
c	Norwegian-Greenland Sea/4α	4	61 ± 2
d	*Fram Strait I/5α*	*5.2/5.3*	*~ 100*
e	Blake/5β	5.5	123 ± 3
f	*Baffin Bay/Fram Strait II/6α*	*6.2/6.3*	*~ 160*
g	Iceland Basin/7α (Jamaica? Biwa II?)	6.6/7.1	~ 190
h	Pringle Falls/Simmer Lake II	7.5	~ 220 (240)
i	*CR0/8α (Biwa III ?)*	*8*	*~ 260*
j	*9α*	*8/9*	*~ 290–310*
k	*CR1/9β*	*9*	*~ 330*
l	*11α*	*11*	*~ 400*
m	*CR2/13α*	*13*	*480–510*
n	*14α*	*14*	*~ 535*
o	Big Lost/CR3 (Emperor ?)	15	~ 575
p	La Palma/15β	15	~ 605
q	*17α*	*17*	*~ 665*

When using geomagnetic excursion for establishing a chronology for Arctic Ocean sediment cores one should have in mind, that (i) the number of excursions found in sediment cores within the Brunhes time interval increased up to 17 during the last decade (Table 6.2) (Langereis et al., 1997; Lund et al., 1998, 2006), (ii) not all excursions were found in all cores, (iii) some of the excursions have a poor age control, and (iv) the absolute ages of several excursions were significantly revised during the last years. The Biwa I, II, and III events, for example, were dated to 180 and 295 ka, respectively (Kawai, Yaskawa, Nakajima, Torii, & Horie, 1972; no age given for Biwa III), 160, 310, and 380 ka, respectively (Kawai, 1984), and 110, 200, and 250 ka, respectively (Machida, Arai, & Yokoyama, 1991). Due to these questionable absolute ages, Langereis et al. (1997) and Lund et al. (2006) do not include the Biwa events in their list of reliably determined and well-dated reversal excursions. Langereis et al. (1997), however, mentioned that the redated excursions Biwa II (~200 ka) and Biwa III (~250 ka) have similar ages as the Jamaica (Iceland Basin?) and CR0 events, respectively (Table 6.2). On the other hand, these Biwa events (in combination with other age fix points; see later discussions) are used as important age markers for dating of Arctic Ocean sediment cores by Nowaczyk (1991), Frederichs (1995), and Jakobsson et al. (2000a). According to Lund et al. (2006), the most reliable and most accurately dated excursions within the Brunhes Chron are the Mono Lake, Laschamp, NGS, Blake, Iceland Basin, Pringle Falls, Big Lost, and La Palma events (see Table 6.2). For Arctic Ocean sediments, examples from two well-dated sediment cores recovered from the continental margin north (east) of Svalbard (Cores PS2138-1 and PS2212-3; for location see Figure 6.3A) are presented in Figure 6.5, containing the Mono Lake, Laschamp, NGS, and Blake events (Nowaczyk & Knies, 2000; Nowaczyk et al., 2001).

Due to the uncertainties of interpretation of magnetic inclination changes described earlier, therefore, it is crucial to acquire independent age control to guide the interpretation of the palaeomagnetic polarity pattern in order to establish an accurate chronology (e.g., Jakobsson et al., 2000a, 2001; Backman et al., 2004).

Another very promising palaeomagnetic stratigraphy tool is the determination of the relative palaeointensity variations of the geomagnetic field (Tauxe, 1993). Geomagnetic palaeointensity records from marine sediment cores have been shown to contain a global signal suitable for fine-scale correlation (e.g., Meynadier, Valet, Weeks, Shackleton, & Hagee, 1992; Guyodo & Valet, 1996; Channell, Hodell, & Lehman, 1997, Channel, Stoner, Hodell, & Charles, 2000; Laj, Kissel, Mazaud, Channell, & Beer, 2000; Stoner, Channell, & Hillaire-Marcel, 1998; Stoner, Channell, Hillaire-Marcel, & C. Kissel, 2000), at least for the last glacial cycle. High-resolution palaeointensity records available beyond 200 ka also indicate that fine-scale features are correlative (e.g., Hayashida, Verosub, Heider, & Leonhardt, 1999; Channell & Kleiven, 2000; Channell et al., 2000; Channell, Mauzaud, Sullivan, Turner, & Raymo, 2002; Channell, Curtis, & Flower, 2004; see Lund et al., 2006 for summary). For example, the palaeointensity record for the MIS 9–11 interval (300–400 ka) from the Iceland Basin (ODP Sites 983 and 984) can be correlated to the sub-Antarctic South Atlantic (ODP Site 1089) at suborbital (millennial) scale (Stoner, Channell, Hodell, & Charles, 2003). As variations in geomagnetic palaeointensity control atmospheric production of ^{10}Be and ^{36}Cl isotopes, and the flux of these isotopes is readily measurable in ice cores, palaeointensity records in marine cores provide an independent link between marine sediment and ice core records (e.g., Mazaud, Laj, & Bender, 1994). Frank et al. (1997) have shown that 10^4–10^5 yr variability in ^{10}Be production rate, as determined from globally distributed deep-sea cores over the last 200 kyr, can be matched to sediment palaeointensity data. This observation and the similarity of globally distributed palaeointensity records indicate that much of the variability in palaeointensity records is globally correlative.

Nowaczyk et al. (2001) also used this approach for Arctic Ocean sediments. Within this study three long sediment cores from the Makarov Basin (PS2178-3, PS2178-5, and PS2180-2; for location of cores see Figure 6.3A) have been selected for detailed palaeomagnetic and rock magnetic analyses. Using these magnetic proxies, an excellent correlation between the three cores was possible and a stack of ChRM inclinations and relative palaeointensity versus composite depth has been created (Figure 6.6; see Nowaczyk et al., 2001 for details). The sediments are dominated by normal polarity including short excursions, suggesting that most of the record is of Brunhes age. The relative palaeointensity variations display a well-documented succession of pronounced minima and maxima that can be correlated to published reference curves (Figures 6.7 and 6.8; Nowaczyk et al., 2001 and further references therein). However, there are still two different interpretations of

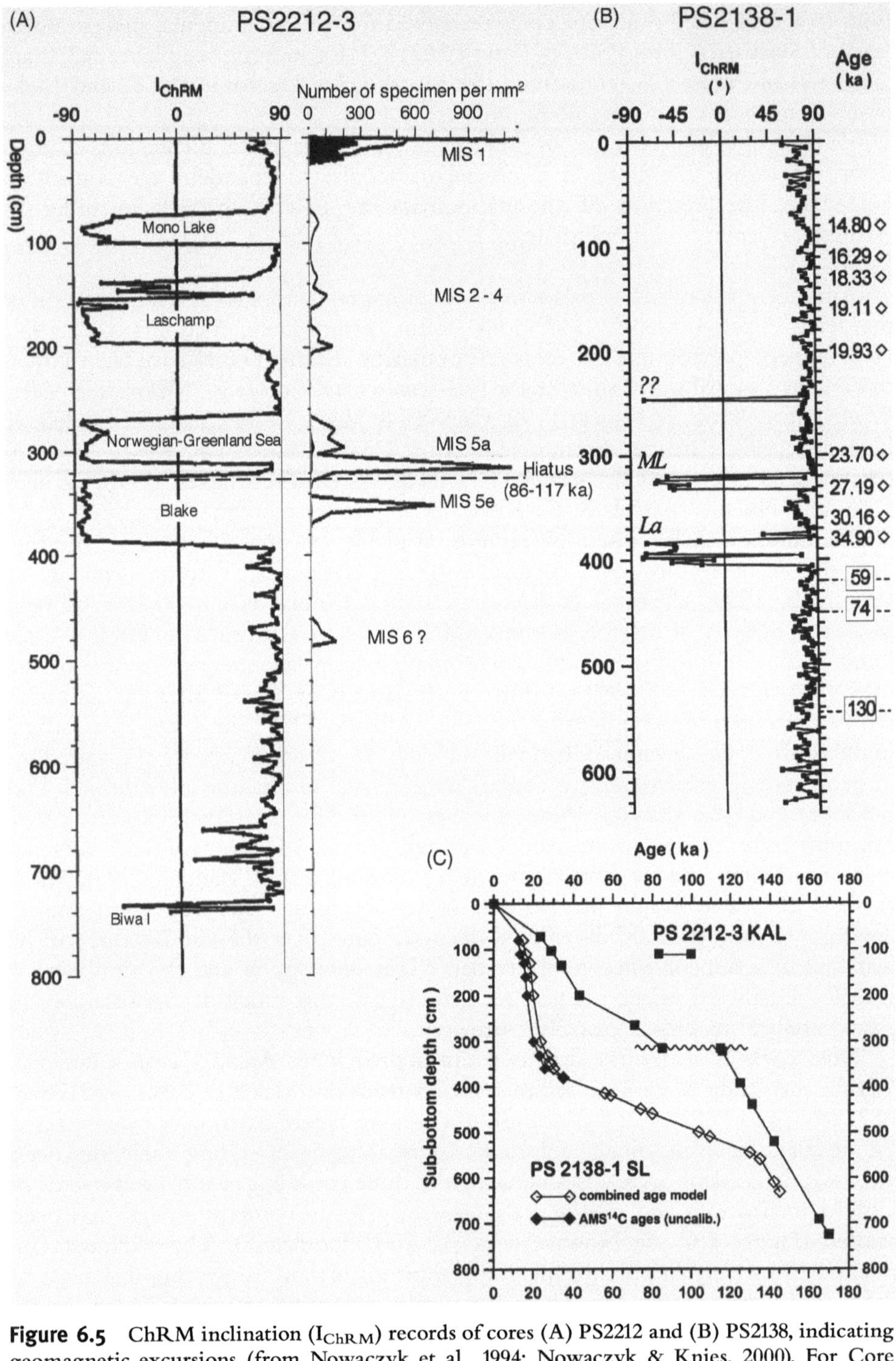

Figure 6.5 ChRM inclination (I_{ChRM}) records of cores (A) PS2212 and (B) PS2138, indicating geomagnetic excursions (from Nowaczyk et al., 1994; Nowaczyk & Knies, 2000). For Core PS2212, also abundances of coccoliths *E. huxleyi* and *Gephyrocapsa* spp. (solid line), and *C. pelagicus* (black), and approximate levels of MIS 1 through 6 are shown. (C) Age-depth diagram of cores PS2138 and PS2212. For core location see Figure 6.3A.

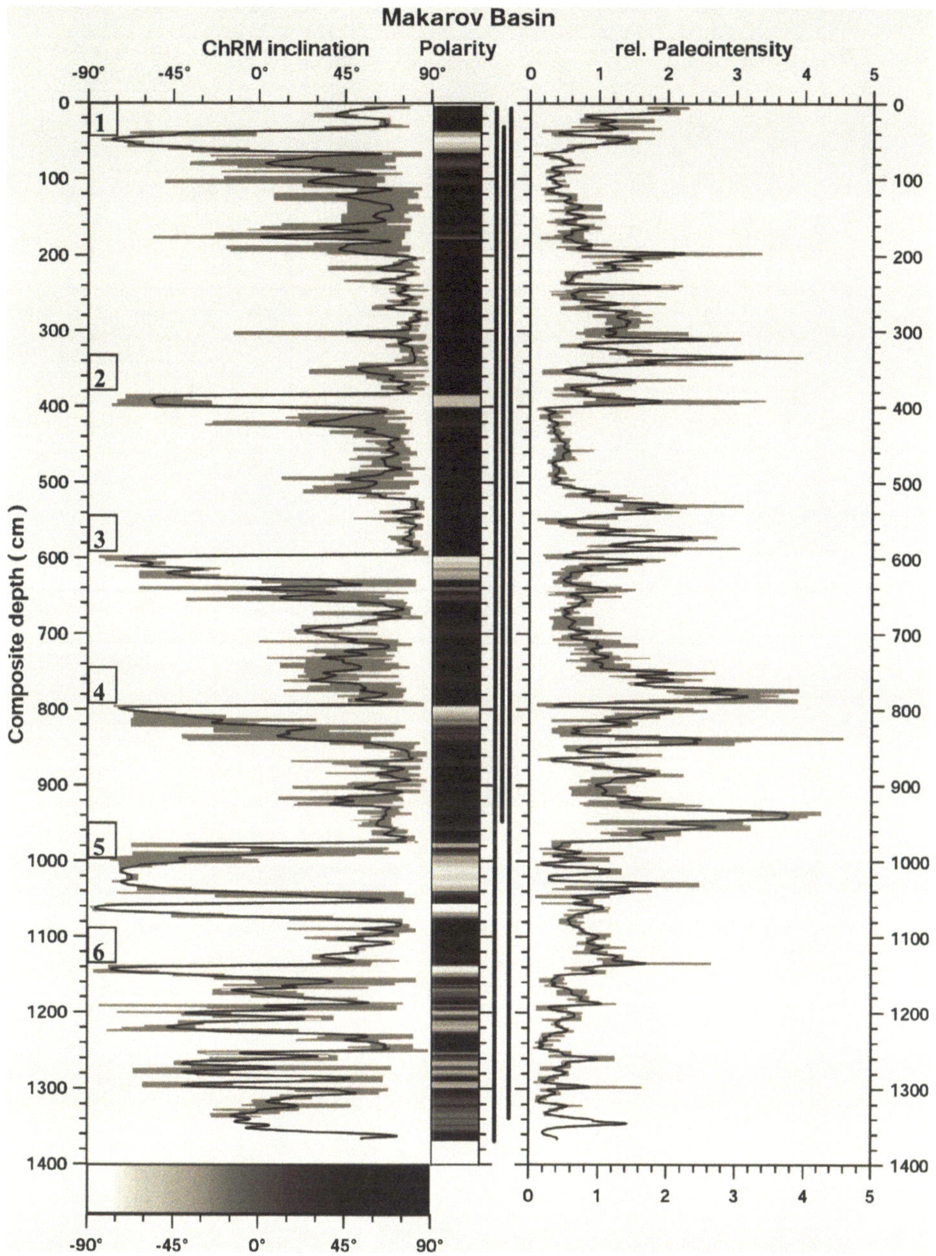

Figure 6.6 Stacks of the ChRM inclinations and relative palaeointensity estimates as function of composite depth, obtained from sediment cores from the Makarov Basin (from Nowaczyk et al., 2001). The shadded areas underlying both curves indicate maximum deviations of the individual records from the stacked curves. Vertical lines in the middle indicate the depth intervals covered by cores PS2178-3 (left line), PS2178-5 (middle line), and PS2180-2 (right line). According to the grey scale bar in the bottom left, the indication pattern was converted into grey values, yielding the "polatity" pattern in the middle. Numbers in boxes mark intervals of reversed magnetizations as discussed in the text. For core location see Figure 6.3A.

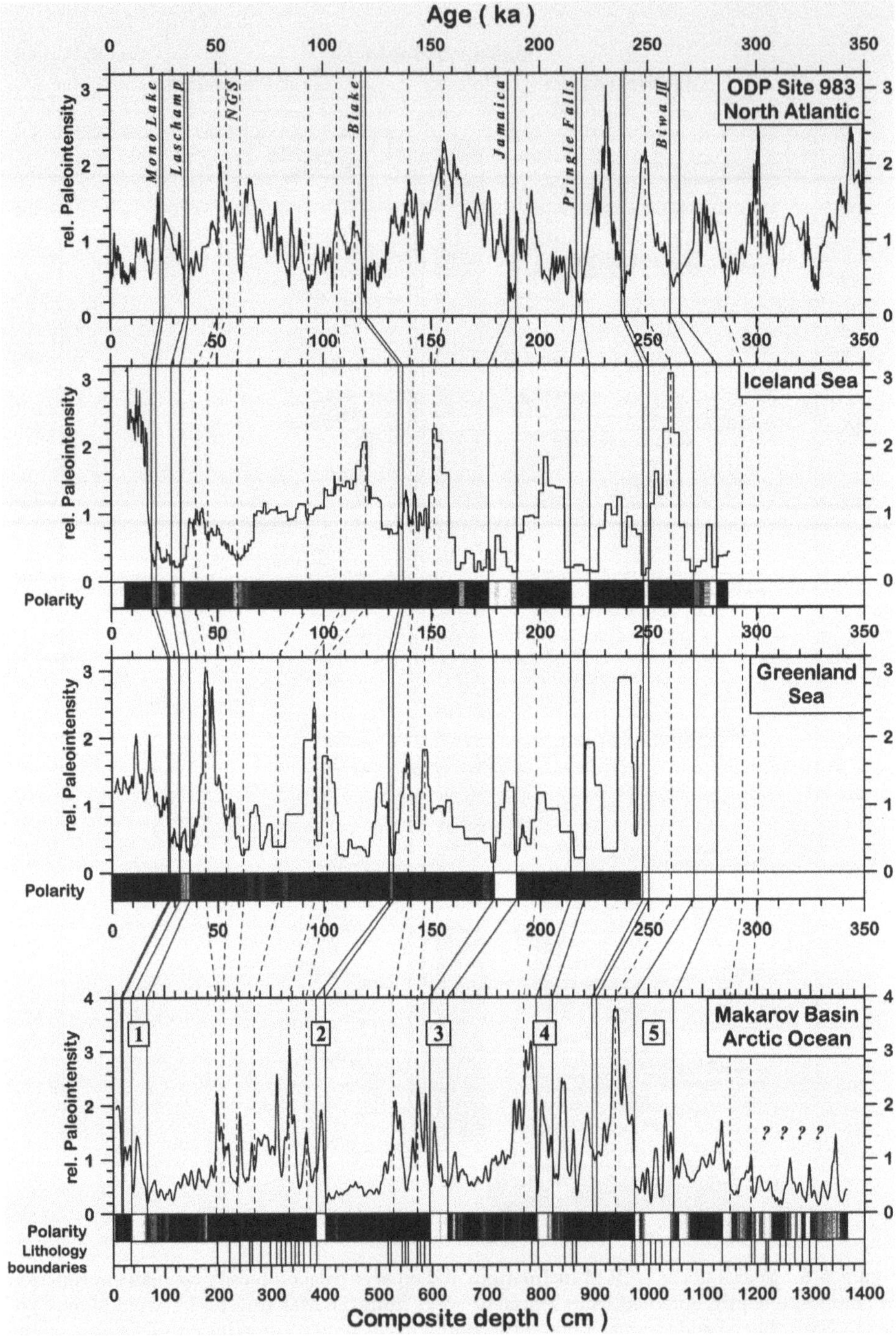

Figure 6.7 Age model 1: correlation of the palaeointensity record from the Makarov Basin, plotted versus composite depth, to other records from the Greenland Sea (after Nowaczyk, 1997), the Iceland Sea (Nowaczyk & Frederichs, 1999), and the North Atlantic, ODP Site 983 (Channell et al., 1997), all plotted versus age (from Nowaczyk et al., 2001). Names in the North Atlantic graph indicate geomagnetic excursions; NGS, Norwegian-Greenland-Sea excursion. Numbers in boxes in the Makarov Basin graph mark intervals of reversed magnetizations discussed in the text.

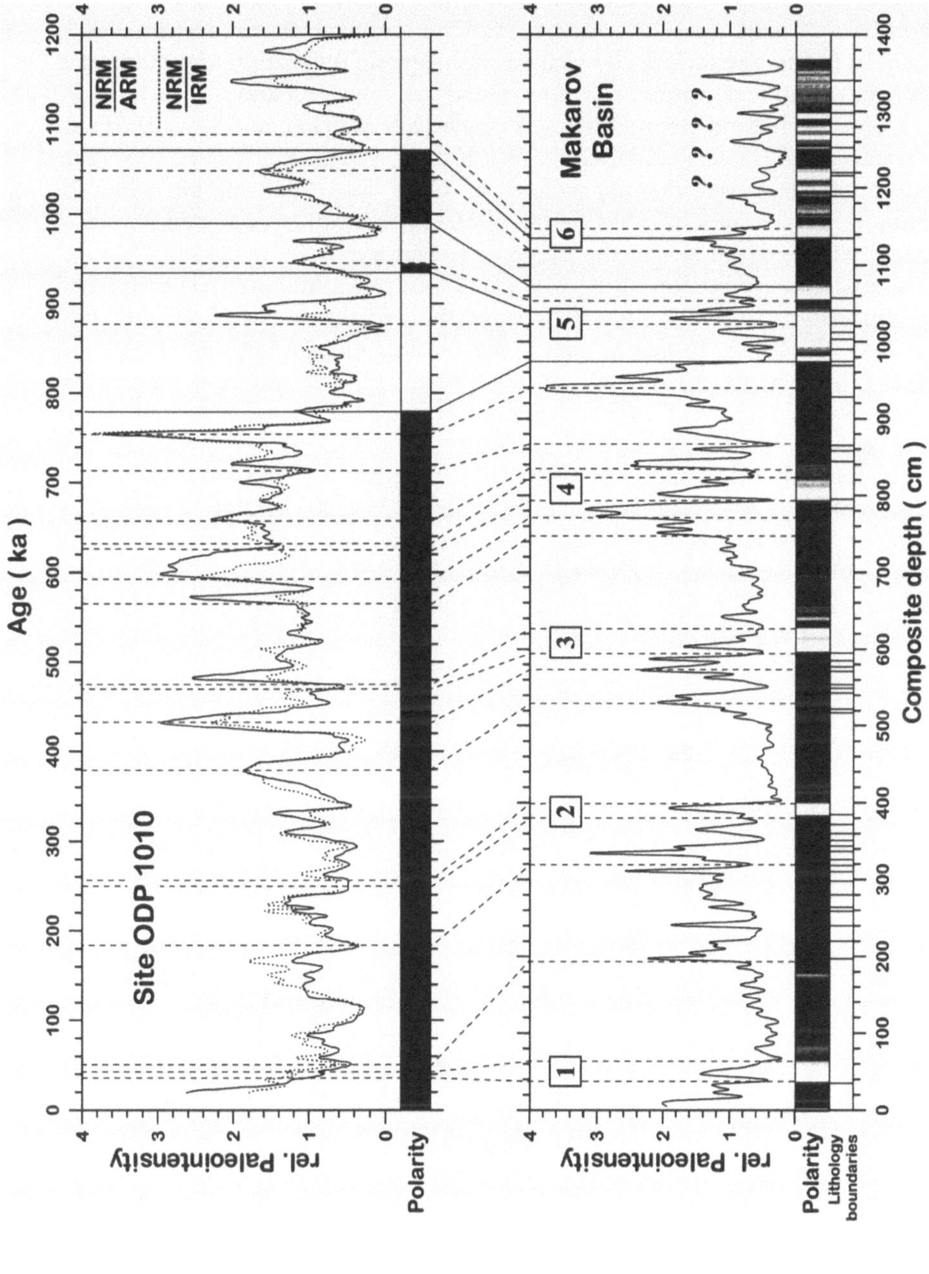

Figure 6.8 Age model 2: correlation of the palaeointensity record from the Makarov Basin with a record from the northeast Pacific, ODP Site 1010 (Hayashida et al., 1999) (from Nowaczyk et al., 2001). Numbers in boxes in the Makarov Basin graph mark intervals of reversed magnetizations discussed in the text.

the palaeomagnetic records possible, resulting in different age models and, thus, different sedimentation rates (see later discussions).

Age model 1 is based on the correlation of the Makarov Basin record to comparable data sets from the North Atlantic ODP Site 983 (Channell et al., 1997) and from the Nordic Seas (Nowaczyk, 1997; Nowaczyk & Frederichs, 1999) (Figure 6.7). In this age model, the changes in magnetic inclination are interpreted as excursions, and excursions 1–5 are related to the Laschamp (40 ka), Blake (118 ka), Jamaica/Iceland Basin (190 ka), Pringle Falls (220 ka), and Biwa III/Fram Strait events (see Table 6.2). Below ~1,170 cm composite depth, no clear correlation of the palaeointensity record to ODP Site 983 can be achieved. According to this age model 1, an age of ~350–400 ka can be estimated, resulting in a mean sedimentation rate in the range of 3–4 cm kyr^{-1}.

An alternate interpretation of the Makarov Basin palaeointensity record is based on the correlation to ODP Site 1010 from the Northeast Pacific (Figure 6.8; Hayashida et al., 1999). In age model 2, the polarity change between 1,000 and 970 cm is related to the Brunhes/Matuyama Boundary (780 ka), and the normal intervals between 1,135 and 1,070 cm and 1,050 and 1,040 cm are related to the Jaramillo and Kamikatsura events, respectively. According to age model 2, four excursions were found in the Brunhes Chron. Excursions 1 and 2 are correlated with the Laschamp (40 ka) and Biwa III/Fram Strait (255–265 ka), respectively. The identification of excursion 3 (470 ka) is not clear, whereas excursion 4 probably correlates with the La Palma event (615 ka) (Table 6.2; Nowaczyk et al., 2001 and references therein). Using age model 2, the bottom of the Makarov Basin record has an age of at least 1.2 Ma (Figure 6.9).

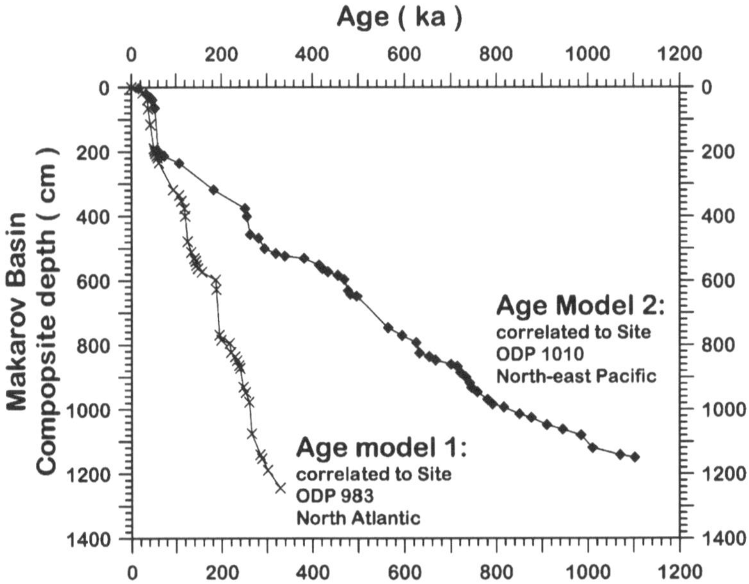

Figure 6.9 Age-depth diagram based on age models 1 and 2; see Figures 6.7 and 6.8 (from Nowaczyk et al., 2001).

From an age-depth diagram it is obvious that both age models display highly variable sedimentation rates which are almost identical in the upper 2 m composite depth (because here AMS^{14}C ages and the interpretation of the palaeomagnetic results are the same), however differ significantly below 2 m (Figure 6.9; Nowaczyk et al., 2001). According to age model 1 and age model 2, average sedimentation rates are ~5 cm kyr^{-1} and only ~1 cm kyr^{-1}, respectively. Based on the new interpretation of the ^{10}Be record of kastenlot Core PS2178-5 (Spielhagen et al., 2004; see later discussions), age model 1 seems to be the more realistic one.

6.1.5. Changing Stratigraphies of Central Arctic Ocean Sediment Cores During the Last Decades

The difficulties in Arctic Ocean stratigraphy may explain why over decades, there has been a long-lasting and still ongoing controversial discussion whether the central Arctic Ocean sediments were deposited under a scenario characterized by very low sedimentation rates or a scenario characterized by significantly higher sedimentation rates. The low sedimentation rate scenario suggesting Plio–Pleistocene rates that vary between ~0.04 and 0.4 cm kyr^{-1} is mainly derived from cores obtained from ridges in the Amerasia Basin and implies that the majority of cores presently available extend well into Plio–Pleistocene (e.g., Steuerwald, et al., 1968; Clark, 1970; Clark et al., 1980; Aksu & Mudie, 1985; Scott et al., 1989). The high sedimentation rate scenario, on the other hand, proposes rates that vary from about one to a few cm/ky and is mainly based on cores from ridges and basins in both the Amerasian and Eurasian parts of the central Arctic Ocean (e.g., Gard, 1993; Stein et al., 1994a, 1994b; Jakobsson et al., 2000a, 2001; Nowaczyk et al., 2001). This discussion has been summarized in a synthesis paper by Backman et al. (2004).

6.1.5.1. Amerasian Basin

A widely applied standard lithostratigraphy was developed by Clark et al. (1980), which has provided a common stratigraphic framework for many investigators to correlate sediments across large areas of the Arctic Ocean (Figures 6.10 and 6.11). This framework is based on a detailed sedimentological study of several hundreds of short sediments cores collected from Ice Island T-3 (Fletcher's Ice Island or Ice Station Bravo) in the Amerasia Basin between 1952 and 1974 (Weber & Roots, 1990; for drift path of T-3 see Figure 6.10A). In their study, Clark et al. (1980) established 13 lithostratigraphic units A to M, which include silty and arenaceous lutites, and carbonate-rich, pinkish-white layers, with variable characteristic contents of quartz-feldspar, detrital grains (carbonate maxima; Darby, 1975), foraminifers, and Fe-Mn particles (Figure 6.10C). The content of sand-sized material (enriched in units C, F, H, J, L, and parts of M) and the pink-white layers were considered to be the key sedimentary characteristic used for correlation of these lithostratigraphic units.

In cores recovered from a drifting ice camp during the Canadian Expedition to Study the Alpha Ridge (CESAR) in 1983 (Jackson et al., 1985; see Figure 1.6), Clark's lithostratigraphic units A to M were also identified, and the lithostratigraphic succession was even expanded by three new lithostratigraphic units A1, A2, and A3

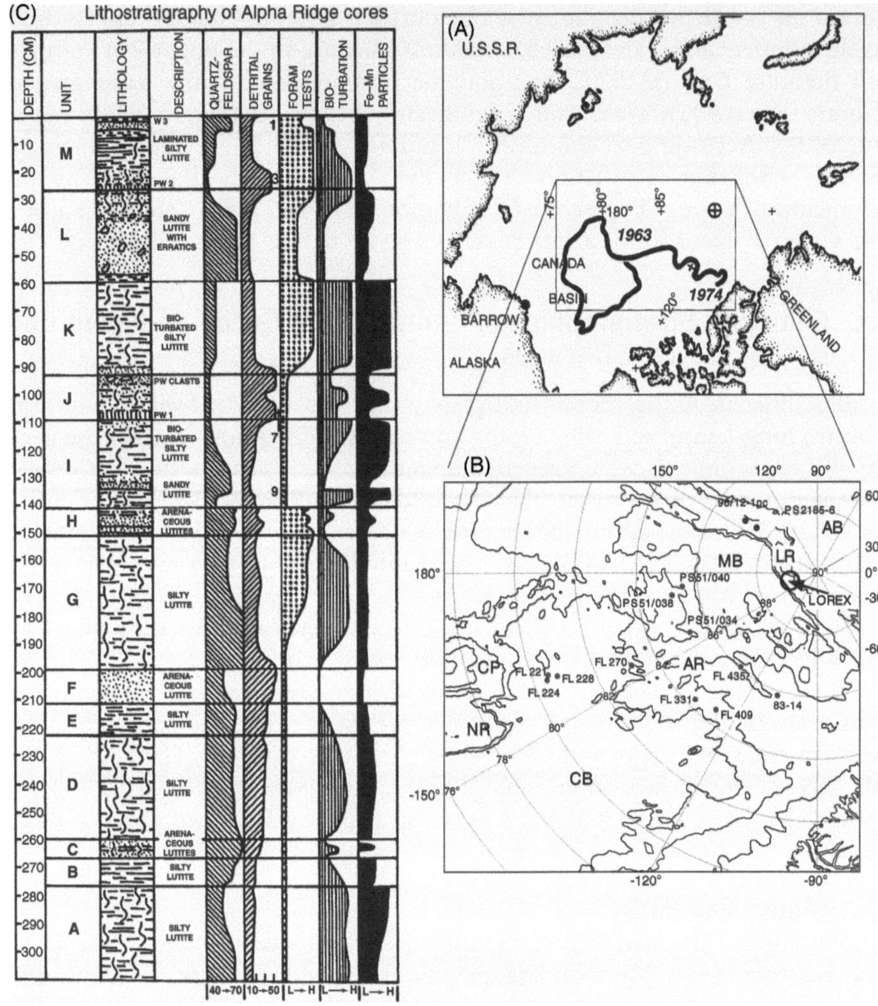

Figure 6.10 (A) Track of ice island T-3 drift. All cores studied by Clark et al. (1980) were taken along the track of this drift (from Clark et al., 1980, supplemented) and (B) location map of cores discussed in the text. (C) Lithostratigraphic units A–M in the central Arctic Ocean. Sedimentary parameters shown are percentages of quartz-feldspar, total detrital grains, foraminifer tests, level of bioturbation, and Fe-Mn particles (modified from Clark et al., 1980). pw, pink white layer; w, white layer. Abbreviations used in the figure: AB, Amundsen Basin; LR, Lomonosov Ridge; MB, Makarov Basin; AR, Alpha Ridge; CB, Canada Basin; CP, Chukchi Plateau; NR, Northwind Ridge.

recovered in CESAR Core 83-14 (Mudie and Blasco, 1985; for location of core, see Figure 6.10B). A similar lithologic sequence was identified in the LOREX cores recovered from Lomonosov Ridge (Morris, Clark, & Blasco, 1985; see Figure 6.10B, for location of LOREX area). The lithostratigraphic units recognized in the LOREX cores correlate with the lithostratigraphic units K, L, and M of the Amerasian Basin.

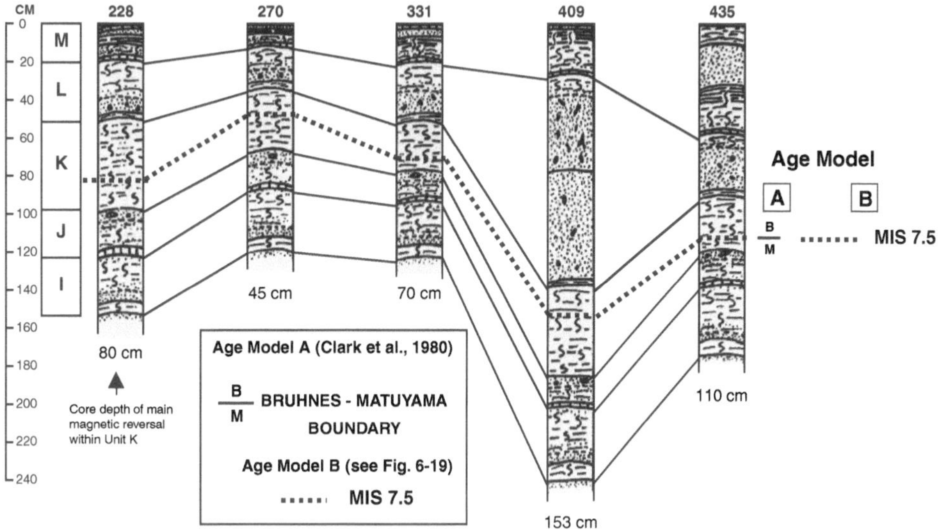

Average sedimentation rates (Unit M, L, and upper K)

	FL 228	FL 270	FL 331	FL 409	FL 435
Model A	0.11 cm/ky	0.06 cm/ky	0.09 cm/ky	0.19 cm/ky	0.14 cm/ky
Model B	0.35 cm/ky	0.2 cm/ky	0.3 cm/ky	0.65 cm/ky	0.46 cm/ky

Figure 6.11 Lithostratigraphic correlation of units I to M between five selected cores (Clark et al., 1980; for location see Figure 6.10B). The Brunhes-Matuyama Boundary in unit K according to the interpretation of Clark et al. (1980) is shown (age model A). In addition, an alternate age model (B) proposed here is shown (see Figure 6.19). Average sedimentation rates based on models A and B are also indicated.

To obtain a stratigraphic framework, the lithostratigraphic units were correlated to palaeomagnetic records determined in these cores. This first chronology of sediment cores from the Amerasia Basin was based on the assumption that zones with negative inclination represented genuine polarity reversals (e.g., Steuerwald et al., 1968; Clark, 1970; Clark et al., 1980). At that time, geomagnetic excursions have not been considered as an alternative to polarity reversals when interpreting palaeomagnetic data in sediment cores (see later discussions). Thus, the first encountered down-core zone with negative inclination was interpreted to be the Brunhes/Matuyama boundary. As result, units M through E, D through B, and A were interpreted to belong to the Brunhes and Matuyama chrons, Gauss Chron, and Gilbert Chron, respectively (Figure 6.12A; Clark et al., 1980). The main reversal within unit K was correlated with the Brunhes/Gauss boundary, resulting in very low sedimentation rates of $\sim < 0.05–0.2\,\mathrm{cm\,kyr^{-1}}$ (Figure 6.11).

In the CESAR cores, a similar palaeo-magnetic age model (at least for the upper half, i.e., units E–M) was established (Figure 6.12B), supporting interpretations in favour of low sedimentation rates (Aksu, 1985; Mudie & Blasco, 1985;

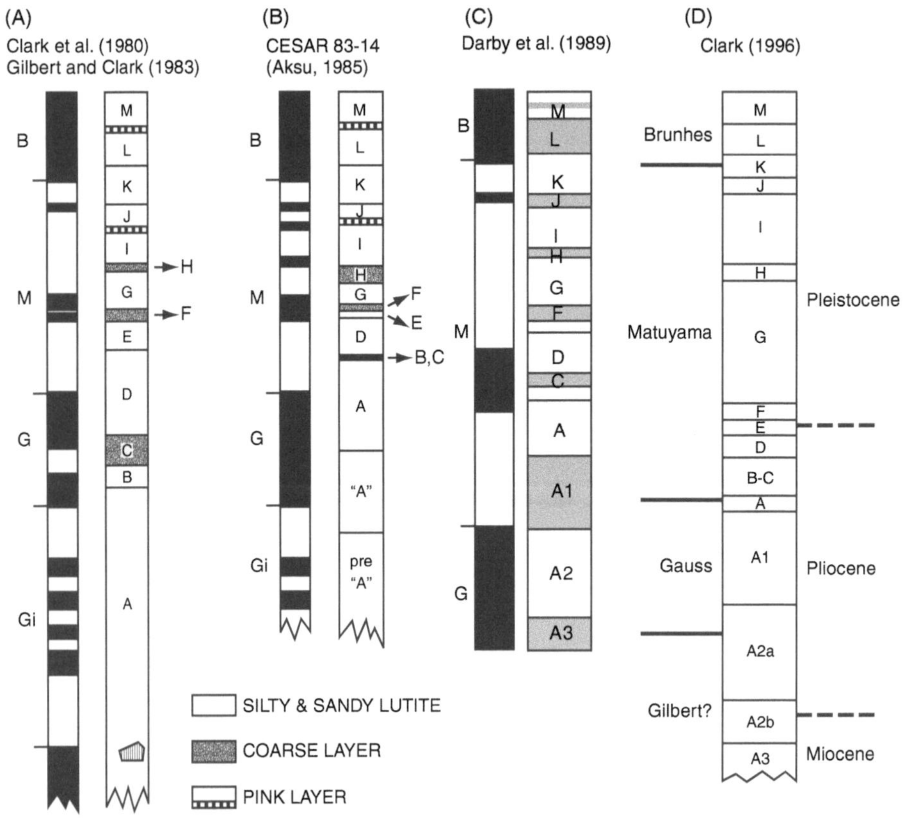

Figure 6.12 Alternate models for correlation of lithostratigraphic units (see Figure 6.10C) with magnetostratigraphy. (A) Correlation according to Clark et al. (1980), Gilbert and Clark (1983), and Minicucci and Clark (1983); (B) correlation based on CESAR cores (Aksu, 1985); (C) revised age model according to Darby et al. (1989); and (D) revised age model according to Clark (1996).

Aksu et al., 1988; Scott et al., 1989). There is, however, a significant discrepancy in the correlation of the older part of the record. Whereas Clark et al. (1980) proposed that the Matuyama–Gauss boundary occurs in unit D (Figure 6.12A), the polarity change is within unit A in the interpretation by Aksu (1985) (Figure 6.12B). According to Aksu (1985), an evaluation of this discrepancy was complicated because only a summary polarity magneto-stratigraphy was presented in Clark et al. (1980) and solid palaeomagnetic data of the cores were not available.

As outlined in Backman et al. (2004), the palaeomagnetic age model of Clark et al. (1980) was challenged by Jones (1987b), who cautioned that the quality of the palaeomagnetic data set varies widely from samples analysed with full stepwise demagnetization and six-spin measurements to no demagnetization and only one-spin measurements. By using only those cores that had been demagnetized up to 50 Oe and that could be correlated with Clark's lithostratigraphic units, and by adding some new records, Jones (1987b) proposed a revised magneto-stratigraphy of the Alpha Ridge cores. Darby et al. (1989) also critically discussed the palaeomagnetic age model of Clark et al. (1980) and published a re-interpretation of the magnetostratigraphy, based on the papers by Aksu (1985), Clark (1970), Clark et al. (1980), and Jones (1987b) (Figure 6.12C). Clark (1996) used another interpretation of the palaeomagnetic record which partly considered the Jones (1987b) results (Figure 6.12D).

These different interpretations of the palaeomagnetic records from the Amerasian Basin display the difficulties in Arctic Ocean stratigraphy. The situation becomes even more complicated when considering that the magnetic reversals may also be interpreted in terms of geomagnetic excursions (see later discussions), as already mentioned by Darby et al. (1989) in their synthesis paper.

6.1.5.2. Lomonosov Ridge

During the *Polarstern* Expedition ARK-VIII/3 in 1991, several undisturbed long sediment cores were recovered in the central Arctic Ocean (Fütterer, 1992). The sediment cores from Lomonosov Ridge are characterized by (cyclic) changes in colour and sedimentary facies (Fütterer, 1992). Here, the 8.2 m long kastenlot Core PS2185-6, in combination with the GKG Core PS2185-3 representing the undisturbed uppermost 35 cm of the section (together used as "Core PS2185"), was selected as key core for several palaeoceanographic and sedimentological studies (e.g., Stein, Schubert, Grobe, & Fütterer, 1994d; Spielhagen et al., 1997, 2004; Behrends, 1999; Svindland & Vorren, 2002). The prerequisite for all of these studies was (and is) a stratigraphic framework. This, however, has been changed significantly between 1992 and today, as outlined in the following.

The first preliminary age model of Core PS2185 was presented at the International Conference on Arctic Margins (ICAM) 1992 and published in the proceedings volume (Stein et al., 1994d). This (ICAM-92) age model was based on carbonate, TOC, and preliminary oxygen isotope records (Figure 6.13); AMS^{14}C datings, ^{10}Be records, and a magnetostratigraphy were not available at that time (see later discussions). In this age model, the boundary between glacial MIS 6 and interglacial MIS 5 was suggested to be at ~245 cm core depth, based on a carbonate maximum coinciding with the occurrence of coccoliths (Gard, unpublished data, 1992, later published in Gard, 1993) and the underlying occurrence of coal fragments. It was assumed that the coccoliths found at 186 and 233 cm core depth, may represent MIS 5, although not proven at that time (see Gard, 1993). The interpretation of coal fragments as MIS 6 indicator was based on Bischof, Koch, Kubisch, Spielhagen, and Thiede (1990) who found significant amount of coal fragments originated from Siberia, in MIS 6 sediments in the Norwegian-Greenland Sea. Using the preliminary ICAM-92 age model, it was suggested that (1) the upper ~5 m of the core may represent MIS 1–MIS 7, (2) the colour cycles present throughout the core are probably glacial/interglacial cycles, with the olive grey to greyish brown sediments are representing glacial intervals and the dark brown to brown sediments are representing interglacial intervals, and (3) the whole record of Core PS2185 represents the last ~450 ka (Stein et al., 1994d). Based on this age model, a mean sedimentation rate for the interval MIS 1–MIS 5 of 1.9 cm kyr^{-1} was calculated.

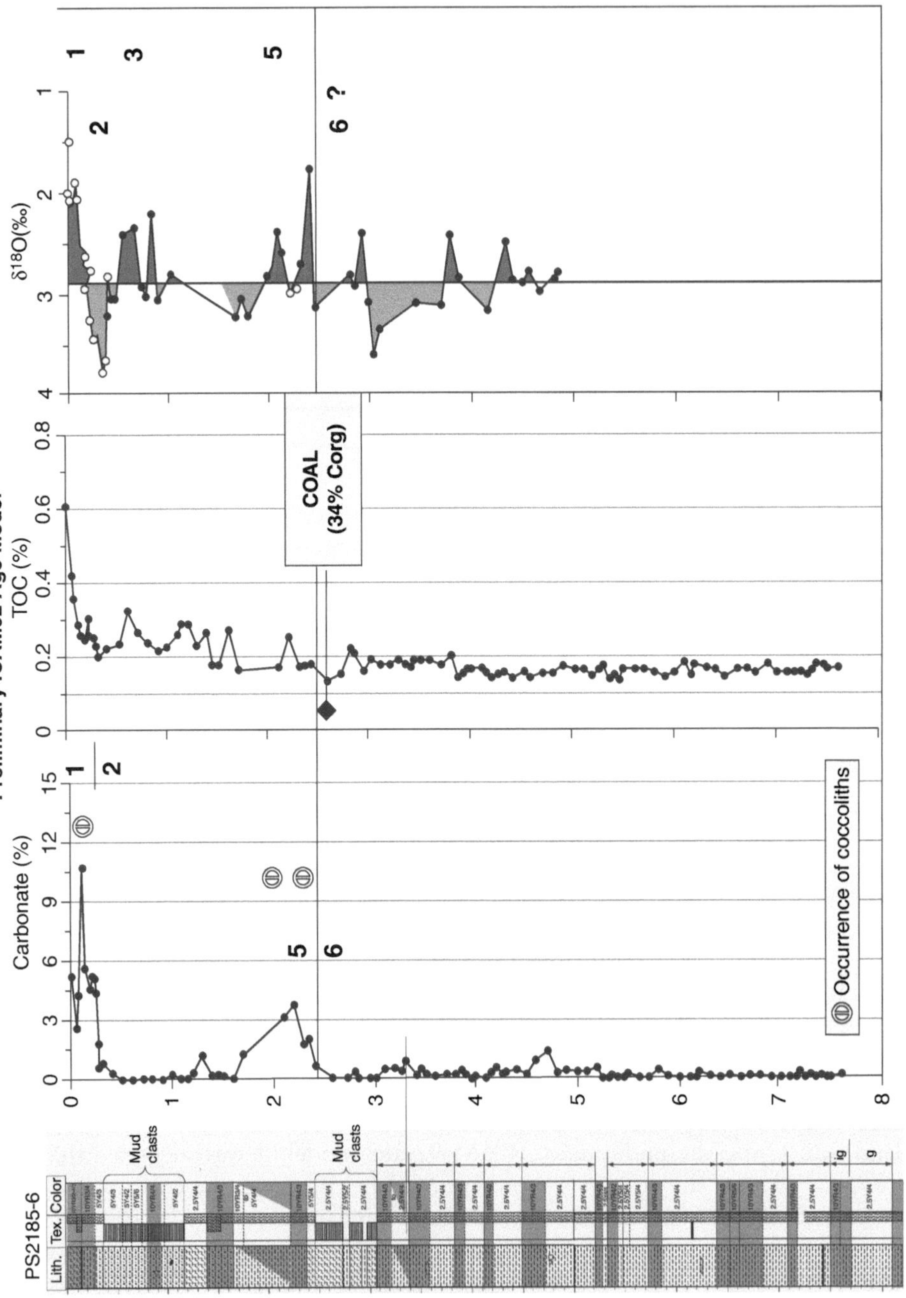

Figure 6.13 ICAM-92 age model of Core PS2185 (Stein et al., 1994d). Lithostratigraphy (according to Fütterer, 1992), carbonate content, TOC content (with occurennce of coal), and preliminary oxygen stable isotope record determined on planktonic foraminifcra *N. pachyderma* sin. Isotope data from Stein and Spielhagen (unpublished data, 1992, later included in Spielhagen et al., 2004), occurrence of coccoliths from Gard (unpublished data, 1992, later published in Gard, 1993). Arrows in the lithology column indicate glacial (g)/interglacial (i) cycles, with brownish intervals (Munsell coulor code 10YR 4/3) representing interglacials/interstadials.

In 1995, Frederichs published magneto-stratigraphies of four sediment cores recovered during the *Polarstern* Expedition 1991 (Fütterer, 1992). The palaeomagnetic record of Core PS2185 (see Figure 6.3A for location) displays distinct variations between intervals with positive (normal) and negative (reversed) inclinations. Because an accurate chronology of this core was not available at that time, Frederichs (1995) presented in his study two palaeomagnetic (PM) age models of this core (Figure 6.14), without giving any preference for one of these models. Model PM-95-1 interpreted the intervals of negative inclinations as full reversals, that is, chrons and subchrons. That means, the upper 342 cm were correlated with the Brunhes Chron, and the lower boundaries of the Matuyama and Gauss Chrons were suggested to be at 450 and 631 cm, respectively (Figure 6.14A). Based on Model PM-95-1, Core PS2185 represents the last ~5 Ma (Figure 6.14A). Model PM-95-2 interpreted the intervals of negative inclinations as geomagnetic excursions, following Nowaczyk (1991) (Figure 6.14B). Using Model PM-95-2, the lowermost part of the core has an age of ~0.5 Ma. Depending on the age model, inferred sedimentation rates differ by at least one order of magnitude, that is, <0.1–0.5 cm kyr^{-1} (Model PM-95-1) and 1–3 cm kyr^{-1} (Model PM-95-2) (Figure 6.14). Frederichs' Model PM-95-2 agrees quite well with the preliminary ICAM-92 age model.

About two years later Spielhagen et al. (1997) presented a multidisciplinary approach for an improved stratigraphy of Core PS2185, using AMS^{14}C datings, ^{10}Be to detect glacial–interglacial changes in sediment Core PS2185 (Figure 6.15). In this age model, the large-scale chronology of Core PS2185 was based on magneto-stratigraphy using Model PM-95-1 of Frederichs (1995) (Figure 6.14A). Then, the ^{10}Be record of Core PS2185 showing a strong variability which can be correlated to climate cycles in sub-Arctic and Arctic Ocean cores (Eisenhauer et al., 1994; see earlier) was used to identify interglacial intervals. Thus, Spielhagen et al. (1997) interpreted peak ^{10}Be concentrations in Core PS2185 between 0 and 350 cm to represent interglacial MIS 1, 3–5, 7, 9, 11, 13, 15, 17, and 19 (Figure 6.15). This (Spielhagen-97) age model resulted in a low average sedimentation rates for the Brunshes Chron of ~0.5 cm kyr^{-1}. Further support for this age model came from AMS^{14}C datings of the uppermost 30 cm giving infinite ages (>38 ka) below 20 cm (Spielhagen et al., 1997). Based on this age model and the interpretation of the terrigenous coarse fraction, Spielhagen et al. (1997) proposed major glaciation in northern Siberia and increased IRD input during glacial MIS 16, 12, 10, and 6 (Figure 6.15).

Seven years later, Spielhagen et al. (2004) published a revised age model of Core PS2185, following the new chronological concept of Jakobsson et al. (2000a). These authors presented a chronology of an Arctic Ocean core (96/12-1pc), recovered from Lomonosov Ridge close to the location of Core PS2185 (see Figure 6.3A). This age model is based on palaeomagnetism, nannofossil biostratigraphy, and cyclic variability in colour and manganese content. In their approach, Jakobsson et al. (2000a) proposed that the Mn and colour variability follows low-latitude oxygen isotope oscillations. Based on this correlation with the low-latitude isotope record, they could transfer the depth scale of the colour cycles into an age scale (see Figure 6.16 for the upper ~3 m of the record; Jakobsson et al., 2000a, 2001). This Mn- and colour-derived time scale is

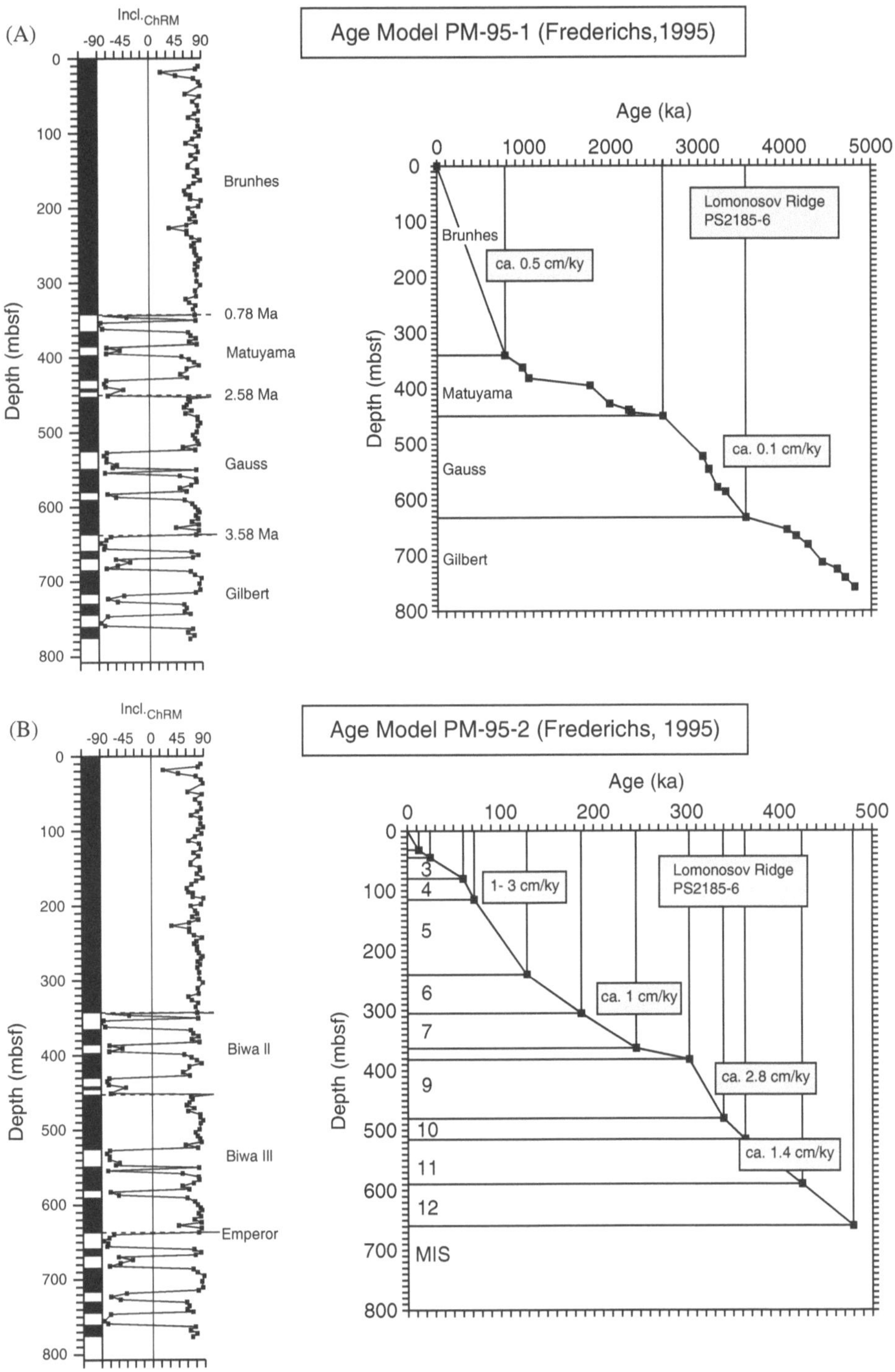

Figure 6.14 Age-depth diagrams of Core PS2185-6 (from Frederichs, 1995). (A) Age model PM-95-1, interpreting polarity changes as reversals; (B) age model PM-95-2, interpreting polarity changes as excursions. Average sedimentation rates for both age models are indicated.

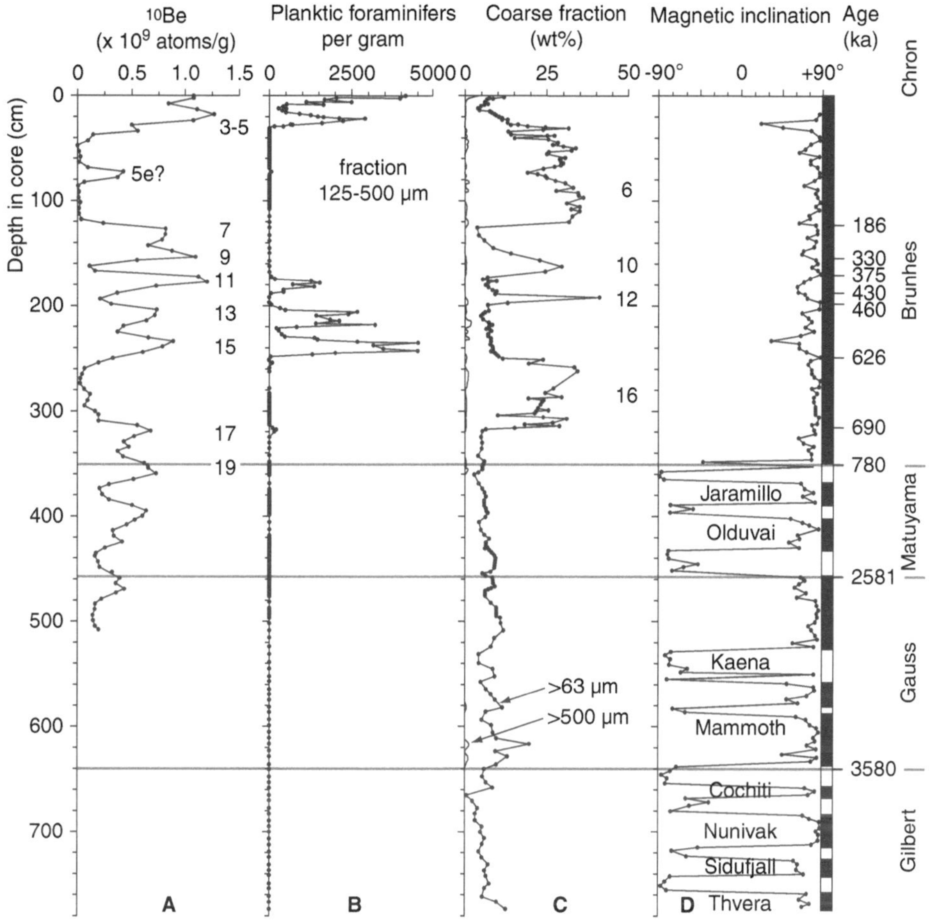

Figure 6.15 Core PS2185 age model of Spielhagen et al. (1997). (A) ^{10}Be concentrations; (B) planktic foraminifer abundances; (C) grain sizes; and (D) magnetic inclination. Numbers in (A) and (C) indicate proposed oxygen isotope stages (modified from Spielhagen et al., 1997).

independently supported by the palaeomagnetic record if the polarity changes determined in the core are related to geomagnetic excursions according to Langereis et al. (1997); see also age model PM-95-2 (Figure 6.14B; Frederichs, 1995). With this approach, Jakobsson et al. (2000a) developed an age model for the upper ~6 m of Core 96/12-1pc, probably representing a record down to MIS 21. The resulting sedimentation rate are 2.8 cm kyr^{-1} in the upper ~2 m (MIS 1–4) and between 0.2 and 1.6 cm kyr^{-1} (with an average of 0.5 cm kyr^{-1}) in the lower part of the dated record (Jakobsson et al., 2000a; Backman et al., 2004).

Optically Stimulated Luminescence (OSL) datings of ice-rafted quartz grains in a sediment core close to Core 96/12-1pc confirmed the new age model of Jakobsson et al. (2000a) at least through MIS 6 (Jakobsson, Backman, Murray, & Løvlie,

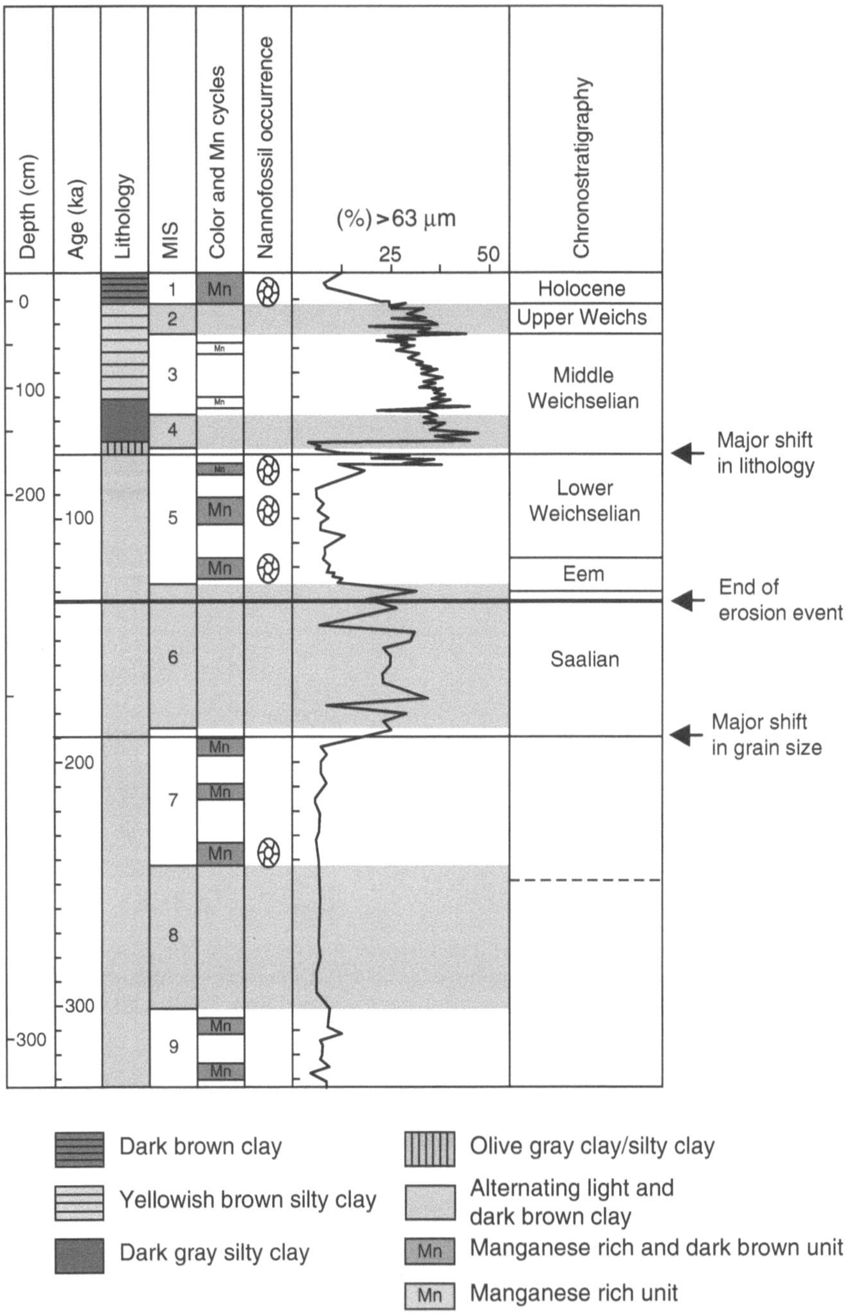

Figure 6.16 Summary of age model and lithology of Core 96/12-1pc from the Lomonosov Ridge (from Jakobsson et al., 2000a and references therein).

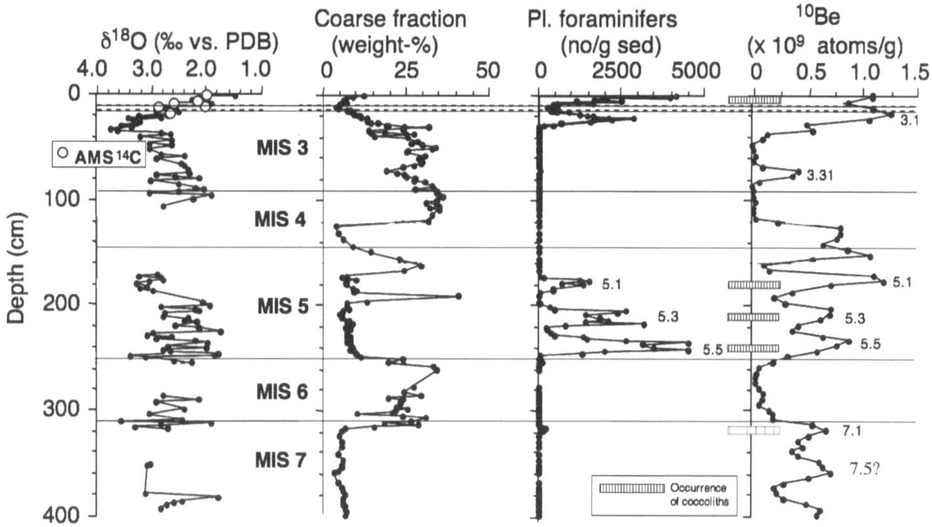

Figure 6.17 Revised age model of Core PS2185 (Spielhagen et al., 2004), based on oxygen isotopes, coarse fraction content, planktonic foraminifer abundances, and ^{10}Be concentrations. Occurrence of coccoliths, MIS stages and substages, and depths of AMS^{14}C datings are indicated.

2003b). Using the coarse fraction records (see Figures 6.15 and 6.16) and MS records, Core 96/12-1pc can be correlated to the record of Core PS2185 very well (Jakobsson et al., 2001), and Jakobsson et al. (2000a) already suggested that the reversed polarity zones represent geomagnetic excursions in both cores.

The revised age model of Spielhagen et al. (2004) is based on the correlation to the Jakobsson et al. (2000a) age model for Core 96/12-1pc, new nannofossil data, AMS^{14}C data, and a re-interpretation of the ^{10}Be and palaeomagnetic records (Figure 6.17). With this new stratigraphy approach for the eastern and central Arctic Ocean, Spielhagen et al. (2004) presented a core-to-core correlation from the central Fram Strait to the Yermak Plateau and the Morris Jesup Rise and across the Lomonosov Ridge to the Alpha Ridge in the Amerasian Basin (Figure 6.18). A key proxy for re-interpreting the ^{10}Be and palaeomagnetic records published in Spielhagen et al. (1997) was the occurrence of specific nannofossils. Here, *G. muellerae* together with *E. huxleyi* unambiguously can be assigned to OIS 5 (e.g., Gard, 1988, 1993; Baumann, 1990; Gard & Backman, 1990). In Core PS2185, nannofossils, mainly *G. muellerae* and *E. huxleyi*, are present at depths of 181–183, 205–217, and 229–245 cm. In addition, the single occurrence of *Coccolithus pelagicus* at 181 cm may indicate MIS 5.1. Spielhagen et al. (2004) used the coccolith-bearing intervals to identify the stratigraphic tie points of isotope substages 5.1, 5.3, and 5.5, supported by maxima in planktonic foraminifers and ^{10}Be (Figure 6.17). At ~310 cm, traces of coccoliths also occur (Gard, 1993), coinciding with a ^{10}Be peak and interpreted as MIS 7.1. This revised new chronology verifying the preliminary ICAM-92 age model (Stein et al., 1994d), resulted in a totally different

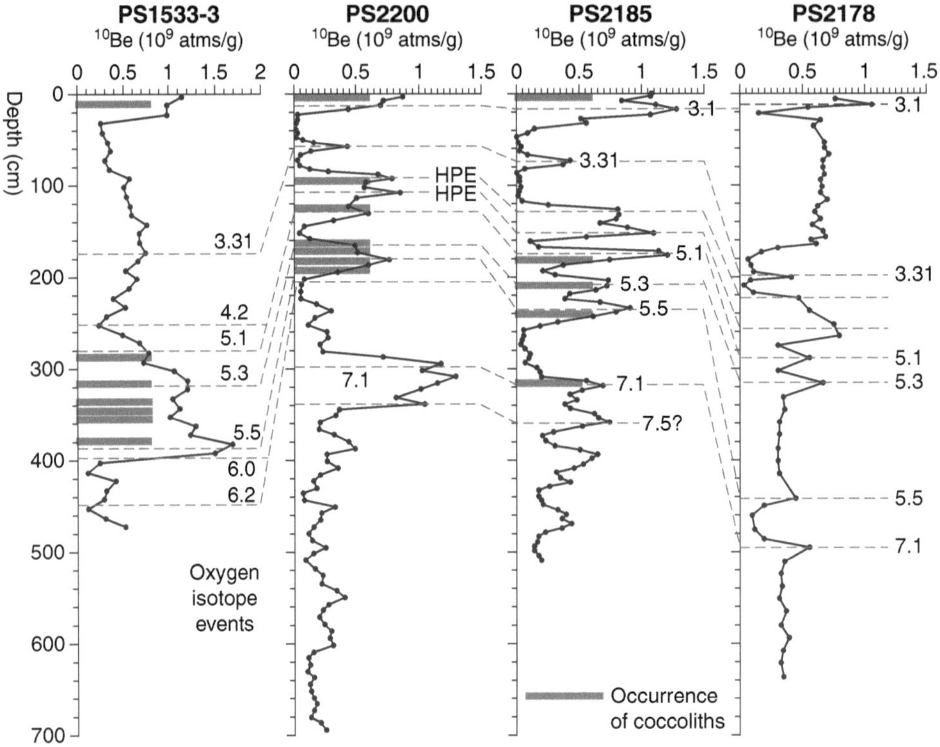

Figure 6.18 ^{10}Be records of sediment cores along a transect from the Yermak Plateau to the Morris Jesup Rise and across the Lomonosov Ridge to the Makarov Basin (from Spielhagen et al., 2004, supplemented). Correlation to MISs (*cf.* Martinson et al., 1987), high production events (HPE) identified in the planktonic foraminifer abundance records of Core PS2200, and occurrence of coccoliths are indicated.

interpretation of the palaeoclimatic record of Core PS2185. The major glaciations in northern Siberia and periods of increased IRD input had to be shifted from glacial MIS 16, 12, 10, and 6 (based on age model Spielhagen et al., 1997) to MIS 6, 5.2, 4, and 3 (Figure 6.17; see Section 6.3.1 for more detailed discussion).

The new age model of Spielhagen et al. (2004) based on a multiproxy approach, is well defined for the upper part of the cores representing MIS 1–7.1, that means for the last ~200 ka (see also O'Regan et al., 2008a for Pleistocene chronology of the ACEX record). For the lower part of section older than 200 ka, these authors also present a correlation of the palaeomagnetic inclination records of cores PS2200 and PS2185 to the corresponding record of Core 96/12-1pc (Jakobsson et al., 2000a). Here, however, the chronology is still open for discussion and strongly depends on the interpretation of the geomagnetic reversals.

In Figure 6.19, an alternate age model for the entire record of Core PS2185 (Age Model PM-2008) is given. The well-established upper part (MIS 1–7.1) is based on Spielhagen et al. (2004). The major geomagnetic excursion coinciding with a ^{10}Be peak and being older than MIS 7.1, however, is interpreted as Pringle Falls Event (MIS 7.5) (see Table 6.2). For the lower part (i.e., below MIS 7.1), a correlation with

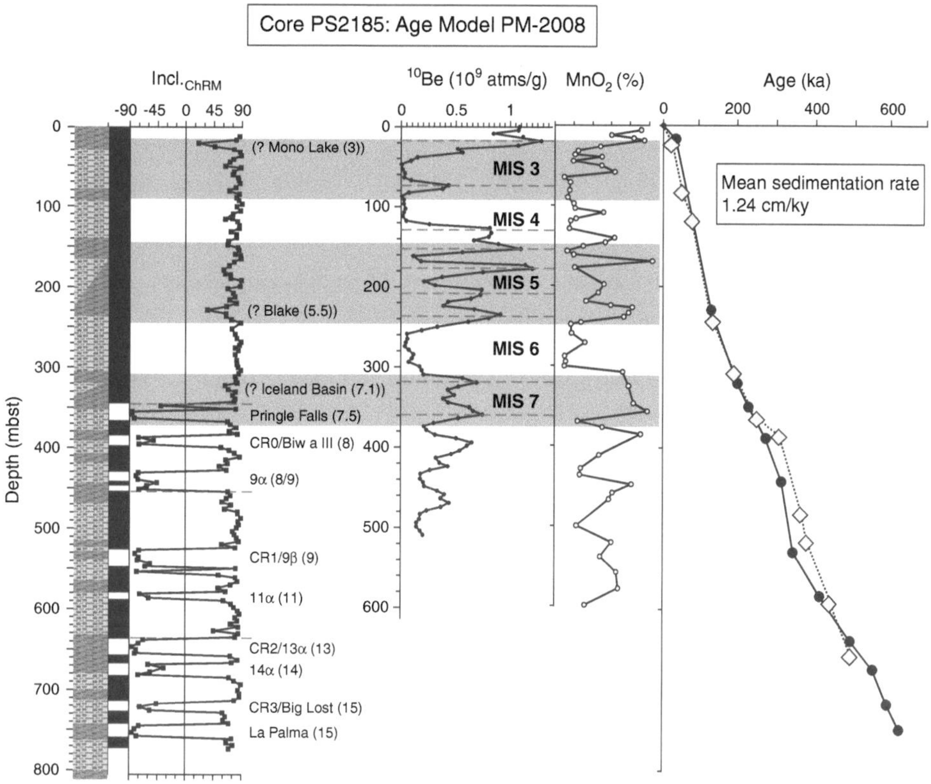

Figure 6.19 Proposed alternate age model PM-2008 of Core PS2185, based on (i) the age model of Spielhagen et al. (2004) for the upper part (i.e., MIS 1–7) and (ii) a re-interpretation of the palaeomagnetic record of Frederichs (1995) and correlation with geomagnetic excursions according to Lund et al. (2006) (see Table 6.2). Lithostratigraphy from Fütterer (1992), ^{10}Be record from Spielhagen et al. (2004), and Mn$_2$O record from Schoster (2005). An age-depth plot based on this age model (solid circles) is shown at the right-hand side, resulting in mean sedimentation rates of 1.24 cm kyr^{-1}. In addition, an age-depth plot based on age model PM-95-2 (open rhombs; see Figure 6.14; Frederichs, 1995) is presented.

the geomagnetic reversals published by Lund et al. (2006) has been used (see Table 6.2). Using this age model, also most of the brownish intervals characterized by high Mn values would fall into interglacials (Figure 6.19), which is in agreement with the interpretation of Jakobsson et al. (2000a). As seen from the age-depth plots, this age model is close to the age model PM-95-2 of Frederichs (1995) (Figure 6.19). Furthermore, all age models including the first preliminary ICAM-92 age model (Stein et al., 1994d) but except the Spielhagen et al. (1997) age model, are similar for the upper ~200 ka, supporting the reliability of these age models for this time interval.

6.1.6. Summary

In summary, it can be stated that (i) sediment cores from the Amerasian Basin and the Eurasian Basin can be correlated very well based on micropalaeontological proxies (Figure 4.32; Backman et al., 2004), palaeomagnetic proxies (Spielhagen et al., 2004), and ^{10}Be records (Figure 6.18; Spielhagen et al., 2004), and (ii) a consistent age model for the time interval down to MIS 7 can be obtained (Backman et al., 2004). Backman et al. (2004) interpreted the first down-core zone of negative inclination in the Lomonosov Ridge Core 96/12-1pc as the Biwa II excursion at 295 ka (age given by Kawai et al., 1972, however, revised to 200 ka by Machida et al., 1991; see discussion in Langereis et al., 1997). According to the age model of Spielhagen et al. (2004), this excursion has to be older than MIS 7.1. Thus, following the most recent compilation of geomagnetic excursions of the Brunhes Chron (Table 6.2; Lund et al., 2006), here it is proposed to interpret the first down-core zone of negative inclination as the Pringle Falls Event (MIS 7.5, ~240 Ma according to Thompson & Goldstein, 2006; Figures 6.19 and 6.20). As the Lomonosov Ridge cores can be clearly correlated to the Northwind and Alpha Ridges cores (Figure 4.32; Backman et al., 2004), this requires a re-interpretation of the "old" magneto-stratigraphy of the cores from the Amerasian Basin, in which the correlatable geomagnetic event within lithostratigraphic unit K has been interpreted as the Brunhes/Matuyama boundary at 780 ka (e.g., Clark et al., 1980; Figure 6.11). Using this revised age model, the sedimentation rates increase by a factor of 3 to 4 (Figures 6.11 and 6.20). For the lower part of the record, i.e., below MIS 7.1, however, additional independent age proxies are still needed to improve the chronology.

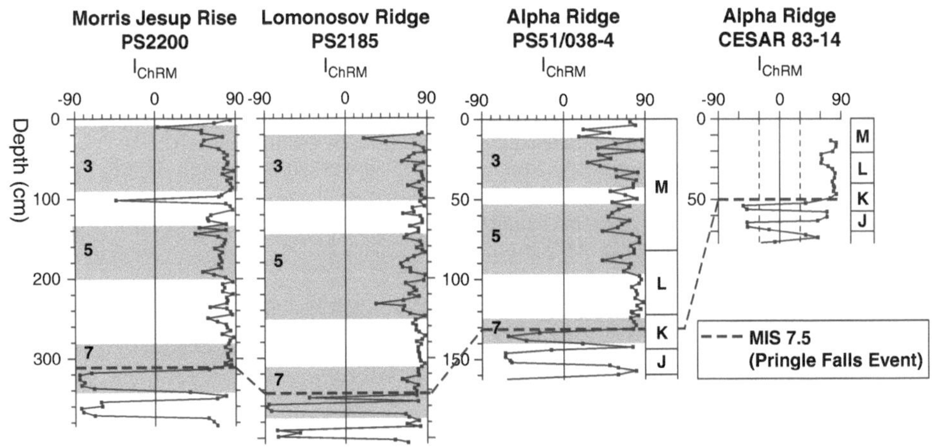

Average sedimentation rates MIS 1-7			
305 cm	342 cm	130 cm	50 cm
1.27 cm/ky	**1.42 cm/ky**	**0.54 cm/ky**	**0.21 cm/ky**

Figure 6.20 Correlation and re-interpretion of the upper part of ChRM inclination (I_{ChRM}) records of cores PS2200, PS2185, and PS51/038-4 (Spielhagen et al., 2004), and CESAR 83-14 (Aksu, 1985), representing the time interval MIS 7–1, and correlation with lithostratigraphic units J to M of Clark et al. (1980). In addition, means sedimentation rates for the time interval MIS 7–1 are shown.

6.2. Variability of Quaternary Ice Sheets and Palaeoceanographic Characteristics: Terrestrial, Model, and Eurasian Continental Margin Records

6.2.1. Extent, Thickness, Volume, and Variability of Northern Hemisphere Ice Sheets: Results from Terrestrial and Modelling Data

The outline of large parts of the major Northern Hemisphere ice sheets for the LGM, for example, the North American, Scandinavian, and British Islands ice sheets, are quite well known (Figure 6.21; e.g., Dyke & Prest, 1987; Dyke et al.,

Figure 6.21 Location map and revised LGM ice extent (from Zweck & Huybrechts, 2005, based on work from Svendsen et al., 1999, 2004; Dyke & Prest, 1987; Dyke et al., 2002; Mangerud et al., 2002). The three main ice sheet systems are plotted in different shades of grey (light, North American ice sheet; medium, Greenland and Iceland ice sheets; dark, Eurasian ice sheet). The dashed line shows the location of the 500 m bathymetric depth of the ocean.

2002; Svendsen et al., 1999, 2004 and references therein). There is also a consensus that the Arctic archipelagos of Svalbard, Franz Josef Land, and Novaya Zemlya were glaciated during the last glacial, and that the western and northern limit of the Eurasian Ice Sheet was located at the continental shelf break, where the ice margin was prevented from further growth into deeper water by processes at the ice–ocean interface, that is, rapid iceberg calving and submarine melting (e.g., Elverhøi et al., 1995a; Andersen, Dokken, Elverhøi, Solheim, & Fossen, 1996; Svendsen et al., 1999, 2004; Knies et al., 2000). However, there is a still ongoing controversial debate about the timing, extent, and thickness of these ice sheets further to the east across the Kara Sea and south towards the coast of Russia.

Since the 1970s, Grosswald and colleagues have pursued the view of a huge LGM continental ice sheet, covering the entire Arctic Eurasia from Scandinavia in the west to Beringia in the far east (Figure 6.22; e.g., Hughes, Denton, & Grosswald, 1977; Grosswald, 1980, 1993, 1998). This ice sheet configuration and reduced versions of it have been used widely as boundary conditions for climate models of the LGM (e.g., Manabe & Broccoli, 1985; Kutzbach & Guetter, 1986; Peltier, 1994). As outlined in Spielhagen et al. (2004), several Russian scientists disagreed with Grosswald and proposed a very limited LGM ice extent in northern Russia and Siberia (e.g., Isayeva, 1984; Arkhipov, Isayeva, Bespaly, & Glushkova, 1986a, 1986b; Biryukov et al., 1988; Dunayev & Pavlidis, 1988), but they achieved only little attention outside of Russia. In 2002, Grosswald still postulated a LGM reconstruction with an ice limit located far to the south on the Russian mainland, arguing that glaciological modelling supports his interpretation (Grosswald & Hughes, 2002). Numerous Russian, Western European, and American field geologists working in the Russian Arctic, on the other hand, maintained that the existing field observations and geochronometric data disprove Grosswald's hypothesis and proposed a much more restricted LGM ice sheet in Eurasia (Dunayev & Pavlidis, 1988; Velichko, Kononov, & Faustova, 1997a; Velichko, Andreev, & Klimanov, 1997b; Velitchko, Dolukhanov, Rutter, & Catto, 1997c; Pavlidis, Dunayev, & Shecherbakov, 1997; Astakhov et al., 1999; Forman, Ingolfsson, Gataullin, Manley, & Lokrantz, 1999; Larsen et al., 1999a; Mangerud, Svendsen, & Astakhov, 1999, 2002; Svendsen et al., 1999; Polyak et al., 2000a). As mentioned in Mangerud et al. (2002), however, Grosswald's idea of a huge circum-Arctic LGM ice sheet (Figure 6.22) is still considered as a viable alternative among researchers from other disciplines of Quaternary science, or by those who have not studied this region themselves.

Figure 6.22 LGM Circum-Arctic ice sheet reconstruction postulated and published by Grosswald and Hughes (2002), showing the Arctic Ice Sheet before and after a LGM basal thawing event (see Hughes, 1996). (A) Ice sheets reconstructed over a largely frozen bed; (B) reconstructed over a largely thawed bed that conserves the volume of ice sheets reconstructed over a largely frozen bed shown in (A). Ice elevations are contoured in 0.2 km intervals. Hatchured lines are ice-shelf calving fronts. Dotted areas are oceans. Parallel lines are lakes dammed by ice sheets or supplied by overflow from ice-dammed lakes. Arrows are directions taken by megafloods (Grosswald, 1999). Bathymetric isopleths are shown for present-day sea level and for 200 and 2,000 m below present-day sea level.

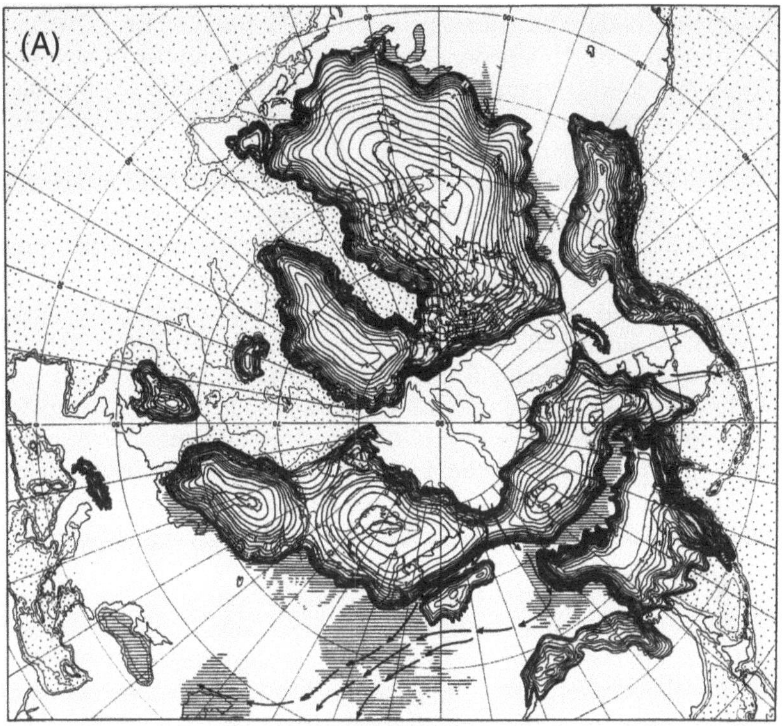
(A)

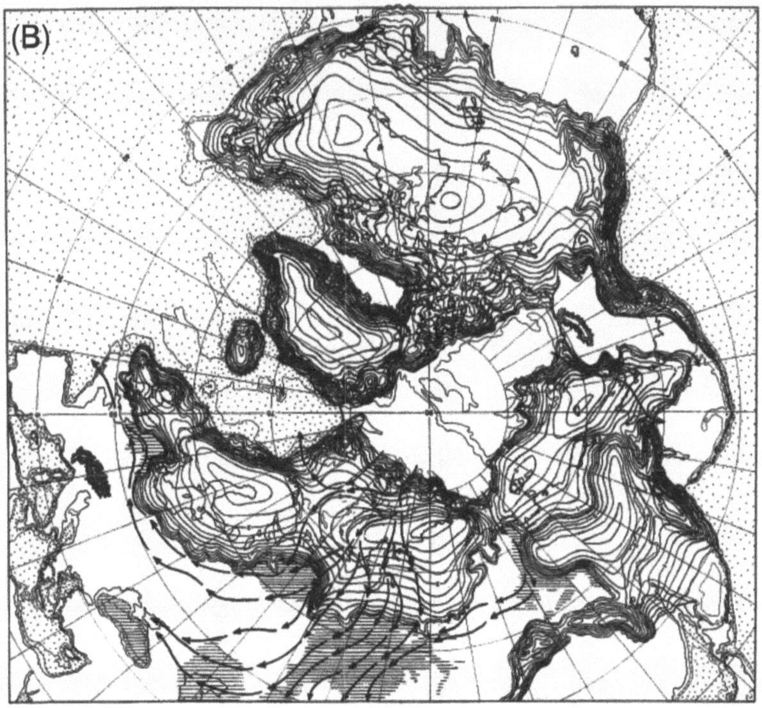
(B)

Field activities undertaken during the QUEEN programme (Thiede, Bauch, Hjort, & Mangerud, 2001; Thiede, 2004) resulted in a number of new geological data sets from terrestrial (and marine) sequences on and around the Eurasian Arctic mainland and archipelagos, providing new evidence of the dimensions and chronology of the Quaternary (and especially LGM) Eurasian ice sheets (e.g., (Svendesen et al., 1999, 2004; Mangerud et al., 2002; Siegert & Dowdeswell, 2004). Mangerud et al. (2002) have described numerous well-dated stratigraphic sequences and geomorphological features from Arctic Russia, which are all dated to 20 ka or older and which are located hundreds of kilometre inside the LGM ice limit proposed by Grosswald (1993, 1998) (Figure 6.23). According to Mangerud et al. (2002), most of the described deposits are composed of easily deformable, loose silt or fine sand, and some contain fragile artefacts and bones, or frozen mammoth carcasses. However, none of the sites shows any sign of glacial overriding during the LGM. At some sites, even deposition of non-glacial sediments and/or formation of ice wedges took place during the LGM. Furthermore, QUEEN field studies revealed permafrost profiles and sediment sequences in modern High Arctic lakes, which indicate that the lowlands of the Taymyr Peninsula, south of Severnaya Zemlya, remained ice-free during the Late Weichselian (Hahne & Melles, 1999; Siegert, Dowdeswell, & Melles, 1999). These are all strong arguments against the huge circum-Arctic LGM ice sheet proposed by Grosswald and in favour for Svendsen et al. (2004)'s reconstruction which is also supported by the marine

Figure 6.23 Map of Northern Russia showing the proposed limits for the Barents-Kara Ice Sheet during the LGM according to Grosswald (1993, 1998) and Svendsen et al. (1999, 2004). Sites for which obtained dates and the stratigraphy or morphology demonstrate that they have not been overrun by a LGM glacier are marked (from Mangerud et al., 2002).

sedimentary records (see Section 6.2.2.). Based on marine geological and geophysical data, the LGM ice sheet limit has been identified on the sea floor off the mainland (Svendsen et al., 1999, 2004; Polyak, Gataullin, Epshtein, Okuneva, & Stelle, 2000a; Polyak, Gataullin, Gainanov, Gladysh, & Goremykin, 2002a; Polyak et al., 2002b; Gataullin, Mangerud, & Svendsen, 2002; Stein et al., 2002; see later discussions).

As a result of earlier as well as QUEEN studies, Svendsen et al. (2004) have reconstructed the maximum limits of the Eurasian ice sheets during four major Quaternary glaciations: (1) the Late Saalian (>140 ka), (2) the Early Weichselian (100–80 ka), (3) the Middle Weichselian (60–50 ka), and (4) the Late Weichselian (LGM) (25–15 ka) (Figure 6.24A–6.24D). The Late Saalian (MIS 6) Ice Sheet (Figure 6.24) was one of the most extensive Quaternary glaciations in this part of the world. At this time, a large ice shelf possibly fringed the Barents-Kara Ice Sheet and may have extended far into the central Arctic Ocean (Polyak et al., 2001; see Section 6.3.2). According to the reconstruction by Svendsen et al. (2004), the northern Svalbard Barents Ice Sheet should also have covered the Yermak Plateau during MIS 6 (Figure 6.24A). Marine data do not support this reconstruction, as the upper ~20 m of the sedimentary record of ODP Site 910 obtained from the crest of the southern Yermak Plateau (80°15.9′ N, 6°35.4′ E; water depth 556 m) and representing the last ~650–800 ka (Flower, 1997; Knies et al., 2007b), do not show any over-consolidation related to grounded ice sheet (Thiede & Myhre, 1996). Below this depth and coincident with an erosional discordance (hiatus), however, shear strength sharply increased, and over-consolidation was attributed to enhanced compaction of sediments beneath an ice sheet extending north from Svalbard prior to MIS 16 (Thiede & Myhre, 1996; Flower, 1997; see Section 6.3.2 for more details).

For the three major Weichselian ice sheet advances following the last interglacial, the results of Svendsen et al. (2004) reveal that the ice sheet development in the more eastern part of the Eurasian Arctic followed a different pattern than over Fennoscandia and along the western margin of the Barents Sea shelf. The Barents-Kara Ice Sheets were getting progressively smaller during the successive glaciations, whereas the dimensions of the Scandinavian Ice Sheet increased through time. Such an asymmetry in the evolution of the Eurasian ice sheets has also already been described by Velichko et al. (1997a).

During the Early and Middle Weichselian glaciations, major ice sheets were centred over the Barents and Kara seas and expanded onto mainland Russia (Figure 6.24B and C). These ice sheets must have blocked north-flowing rivers, such as the Yenisei, Ob, Pechora, and Mezen, and large ice-dammed lakes with reversed outlets, for example, towards the Caspian Sea, were formed south of these ice sheets (Figure 6.25, Table 6.3; Mangerud et al., 2001, 2004). Terrestrial lake deposits in many places prove the formation of large lakes in northernmost Russia west of 90°E (e.g., Astakhov et al., 1999; Houmark-Nielsen et al., 2001; Mangerud et al., 2001, 2004). These lakes were considerably larger than any lake on Earth today, storing water of $>30 \times 10^3$ km^3 (Table 6.3), and probably emptied during deglaciation within few months as suggested from modelling results (see Mangerud et al., 2004 for details). Such a catastrophic outburst with giant freshwater supply into the Barents and Kara seas as well as into the Arctic Ocean proper and via the

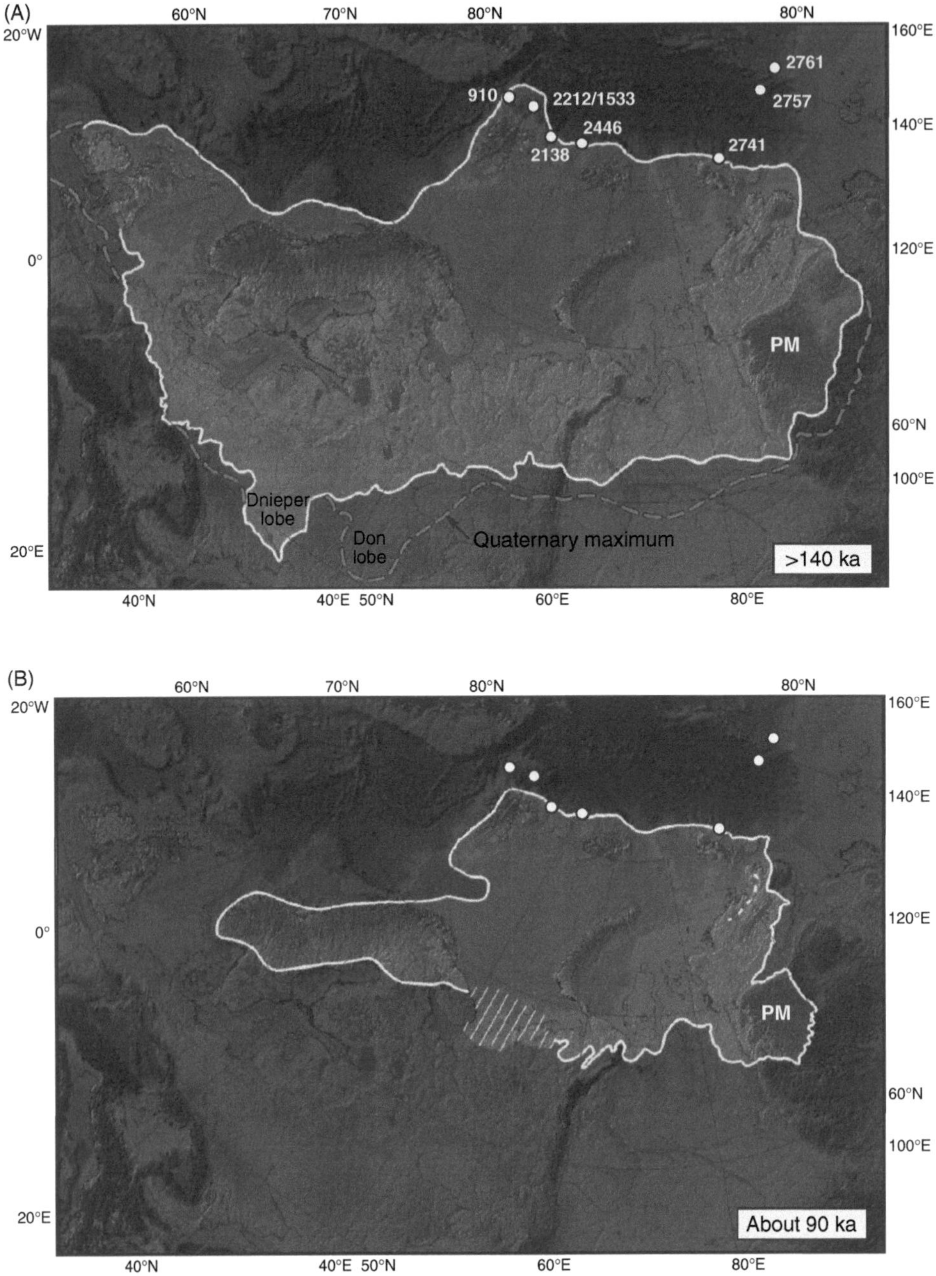
(A)
60°N
70°N
80°N
80°N
20°W
160°E
140°E
120°E
0°
60°N
100°E
20°E
40°N
40°E 50°N
60°E
80°E
910
2212/1533
2446
2138
2741
2761
2757
PM
Dnieper lobe
Don lobe
Quaternary maximum
>140 ka
(B)
60°N
70°N
80°N
80°N
20°W
160°E
140°E
120°E
0°
60°N
100°E
20°E
40°N
40°E 50°N
60°E
80°E
PM
About 90 ka

TPD even into the northern North Atlantic, must have had considerable impact on sea-ice formation in the Arctic Ocean and, probably, deep-water formation in the North Atlantic (Mangerud et al., 2001, 2004; Spielhagen et al., 2004; Knies et al., 2007b). These freshwater outbursts were probably of the same dimension then those described for the Lake Agassiz in North America during the last deglaciation, when the Laurentide Ice Sheet (LIS) retreated and large volumes of water stored in this proglacial lake were episodically released into the oceans (Broecker et al., 1989; Barber et al., 1999; Teller, Leverington, & Mann, 2002; see Section 6.3.5 for discussion).

During the LGM, the southern and eastern flanks of the Barents Kara Ice Sheet terminated on the south-eastern Barents Sea and western Kara Sea shelf (Figure 6.24D), far inside its Early Weichselian maximum extent (Figure 6.24B). During the *Akademik Boris Petrov* Expedition 2001 carried out in the southern Kara Sea, Stein et al. (2002) could map parts of the still unknown southeastern boundary of the LGM ice sheet in great detail (Figure 6.26A). The seafloor in the western part of the study area is strongly furrowed which is interpreted as a relatively fresh (young) subglacial morphology (Figure 6.26C). In numerous profiles this facies has a distinct boundary to the east, which is related to the eastern boundary of the LGM ice sheet. Deeper penetrating chirp profiles exhibit that young morains overlying a pre-Holocene unconformity are associated with the strongly furrowed facies (Figure 6.26B). The influence of an glacial ice sheet is also supported by sediment cores obtained from this area, which are composed of stiff, overconsolidated (?) clays and diamictons (Stein & Stepanets, 2002). East/south of the mapped boundary of the LGM ice sheet, there is no evidence for a LGM ice sheet (i.e., no structures like endmoraines and furrows were recorded in the profiles). Furthermore, the profiling results show a major draining channel system in front of the glacial ice sheet (Figure 6.26C). This channel could be mapped from ~73°N (40 m of water depth) to 76°30′N (120 m of water depth) and appears to be related to the LGM drainage of the Ob river deviated to the east by the LGM Ice Sheet boundary (Figure 6.26A).

Figure 6.24 Reconstructions of the ice-sheet extent in Eurasia during the Late Saalian to Late Weichselian times (from Svendsen et al., 2004; see further references therein for details and data base). (A) Maximum ice-sheet extent during the Late Saalian (*ca.* 160–140 ka). The approximate maximum extent of the Quaternary glaciations (drift limit) is indicated by a dotted line. (B) Ice sheet extent during the Early Weichselian glacial maximum (90–80 ka). The ice front position between the Pechora Lowland and the Kola Peninsula has not yet been defined and the uncertainties are indicated by the hatched field. The dashed line on northern Taimyr marks the retreat stage at the North Taymyr marginal ice zone (NTZ) ~80 ka. (C) Ice sheet extent during the Middle Weichselian glacial maximum (60–50 ka). The ice margin across the southern Kara Sea shelf is tentative and the uncertainty is indicated by the hatched field. The ice limit around the Putoran Massif (PM) shows the Norilsk Stage (Kind, 1974; Isayeva, 1984) that was previously thought to be of Late Weichselian age. (D) Ice sheets at the Late Weichselian glacial maximum (LGM). The presence of restricted valley glaciers on the PM is indicated by hatched lines. In the easternmost Kara Sea the extent of the ice sheet is still under discussion (see text). Note that the glacier distribution over the Bristish Isles, Iceland, Greenland, Alps and/or other mountain areas is not shown on this reconstruction. Furthermore, only the ice limits are shown and not the corresponding ice-dammed lakes (see Figure 6.26; Mangerud et al., 2004). Location of sediment cores discussed in the text is indicated as white circles.

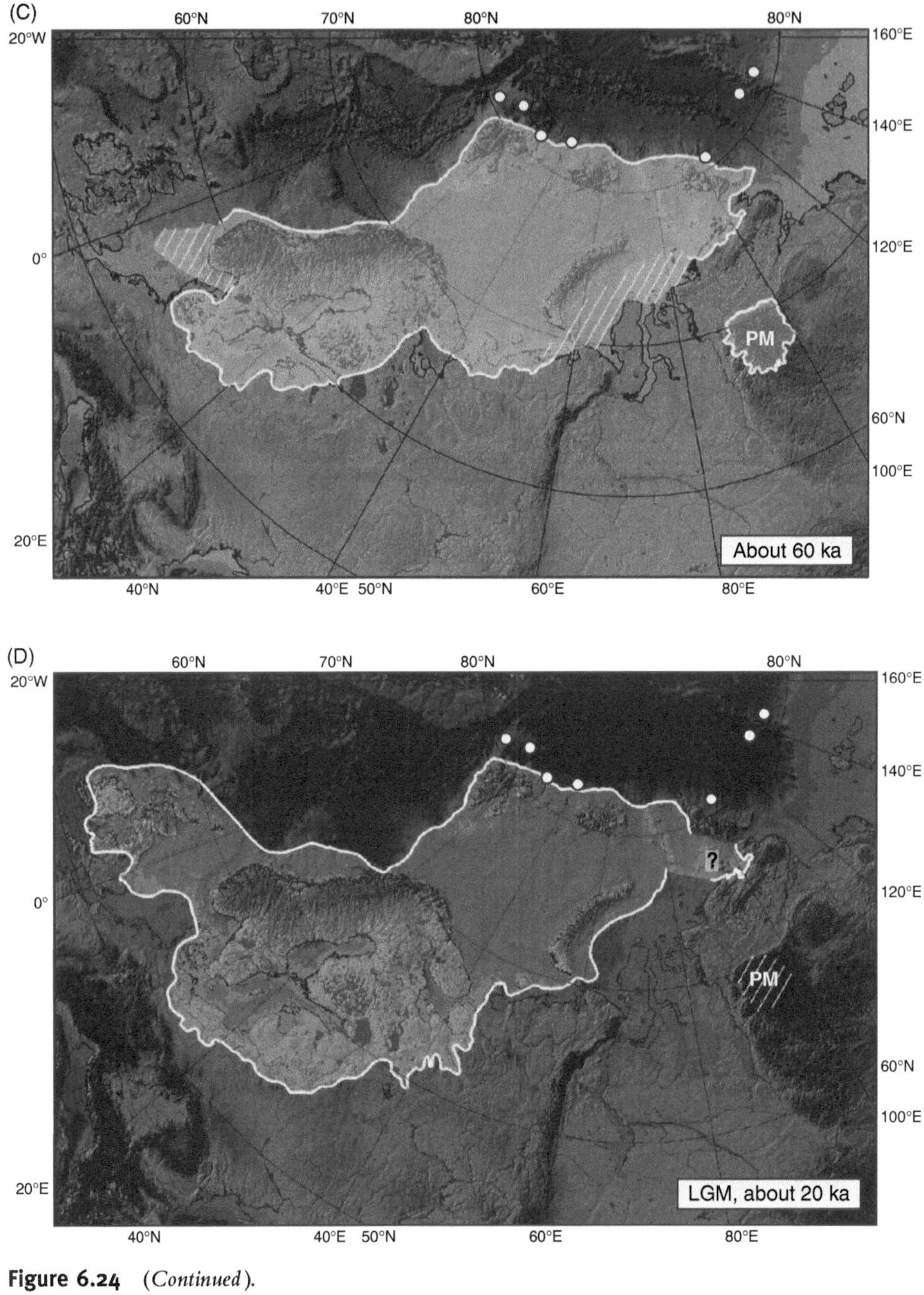

Figure 6.24 *(Continued)*.

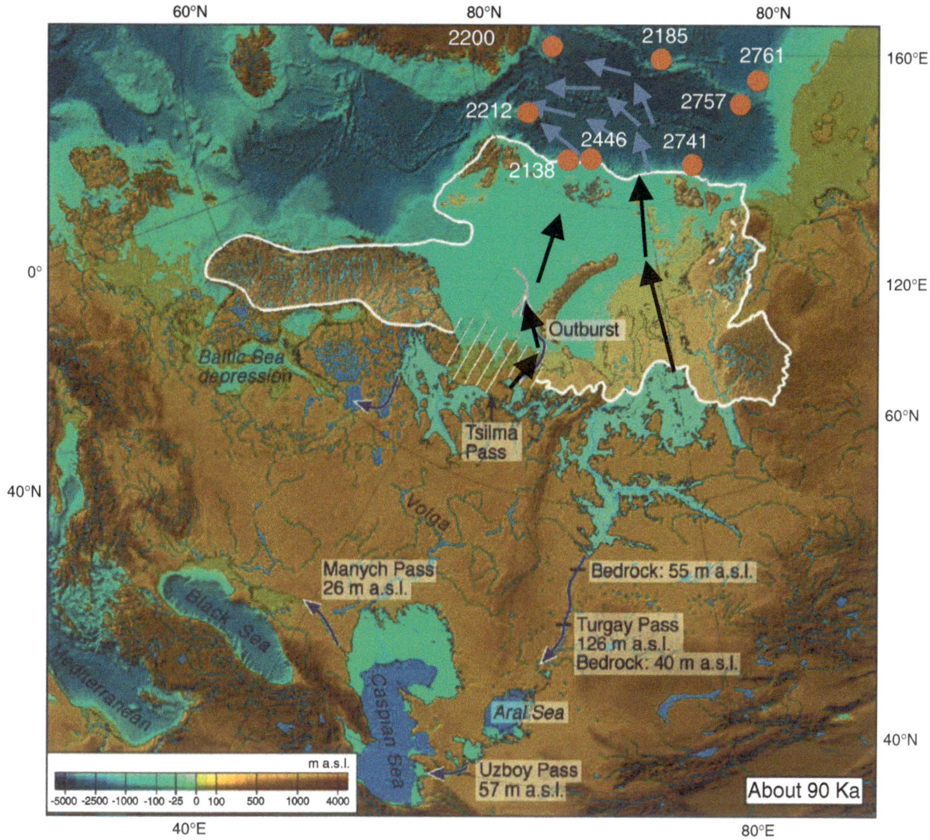

Figure 6.25 Reconstruction of ice-dammed lakes and rerouting of rivers during the early Weichselian, ~90–80 ka (from Mangerud et al., 2004, supplemented). Ice margins are taken from Svendsen et al. (2004). In the hatched area the ice margin position is unknown, probably because it was overrun by the 60 ka ice advance. Stippled line on Taimyr shows a retreat phase damming a lake. Dark arrows show outlets. The arrow in the Barents Sea shows the longest modelled outburst route for Lake Komi, and the corresponding western ice margin. The shorter and more probable routes have the same starting point. Sea level is lowered 50 m (Chappell et al., 1996) without considering any isostatic depression. Light blue arrows in the Nansen Basin indicate pathway of meltwater, red circles cores discussed in the text.

On the northern Kara Sea shelf, the LGM ice sheet may have advanced eastwards and temporarily inundated parts of northwesternmost Taimyr (Figure 6.24D). This ice advance has been explained as a result of surging from the higher parts of the ice sheet at its Barents-Kara Sea interfluve north of Novaya Zemlya (Alexanderson et al., 2002; Svendsen et al., 2004). If correct, the ice masses may have blocked the northward flow of water from the Yenisei and Ob rivers for only a very brief interval, because eroded channels on the Kara Sea shelf contained rivers that flowed towards the Arctic Ocean during most of the LGM

Table 6.3 Properties of Early (90–80 ka) and Middle Weichselian (60–50 ka) Ice-Dammed Lakes in the Eurasian Arctic (from Mangerud et al., 2004 and References Therein). Lake Level is Given Relative to Present Day Sea Level. For the White Sea Basin a Lake at 60–50 ka of the Same Magnitude as at 90–80 ka is Postulated.

Ice dammed lakes	Lake level (m a.s.l.)	Area (1,000 km^2)	Volume (1,000 km^3)	Average depth (m)
Early Weichselian				
Lake Komi in the Pechora Lowland	110–90	76	2.4	32
White Sea Basin	100	218	15	69
West Siberian Plain	60	610	15	24
Taimyr	140	45	2	45
Total			34.2	
Middle Weichselian				
West Siberian Plain	45	881	32	36
Taimyr	80	17	0.4	25

period (Figure 6.26; Stein et al., 2002; Dittmers et al., 2003; Dittmers, Niessen, & Stein, 2008).

6.2.1.1. Modelling results

Whereas the extent of ice sheets can be obtained from geological field work, the thickness and volume of these ice sheets as well as their evolution over time are much harder to reconstruct and generally need to be inferred from indirect evidence and modelling (e.g., Zweck & Huybrechts, 2005 and references therein). Information on time-dependent changes in global ice volume can be obtained from changes in global sea level. These can be inferred, for example, from drilling into tropical coral reefs (Fairbanks, 1989; Bard et al., 1996) or from glacio-isostatic modelling of observations of sea-level change in regions far away from the ice sheets (Yokoyama, Lambeck, de Dekhar, Johnston, & Fifield, 2000; Peltier, 2002; Milne et al., 2002). These studies indicate that the total ice sheet volume change since the LGM has been between 115 and 135 m (Milne et al., 2002), a value well supported by geochemical records of oxygen isotopes from deep-sea sediment cores (e.g., Chappell & Shackleton, 1986; Shackleton, 1987; Waelbroeck et al., 2002). Using this approach, however, the source contributions from individual ice sheets remain undetermined.

A three-dimensional thermo-mechanically coupled ice sheet model of the Northern Hemisphere allowed to reconstruct the Quaternary ice sheets during the last glacial cycle (Huybrechts & T'siobbel, 1997; Huybrechts, 2002; Zweck & Huybrechts, 2003, 2005). As outlined in Zweck and Huybrechts (2005), (1) the model includes basal sliding, internally calculated surface mass balance, glacial isostasy, and a treatment for marine calving, (2) the time-dependent forcing consists of temperature and precipitation anomalies from the United Kingdom

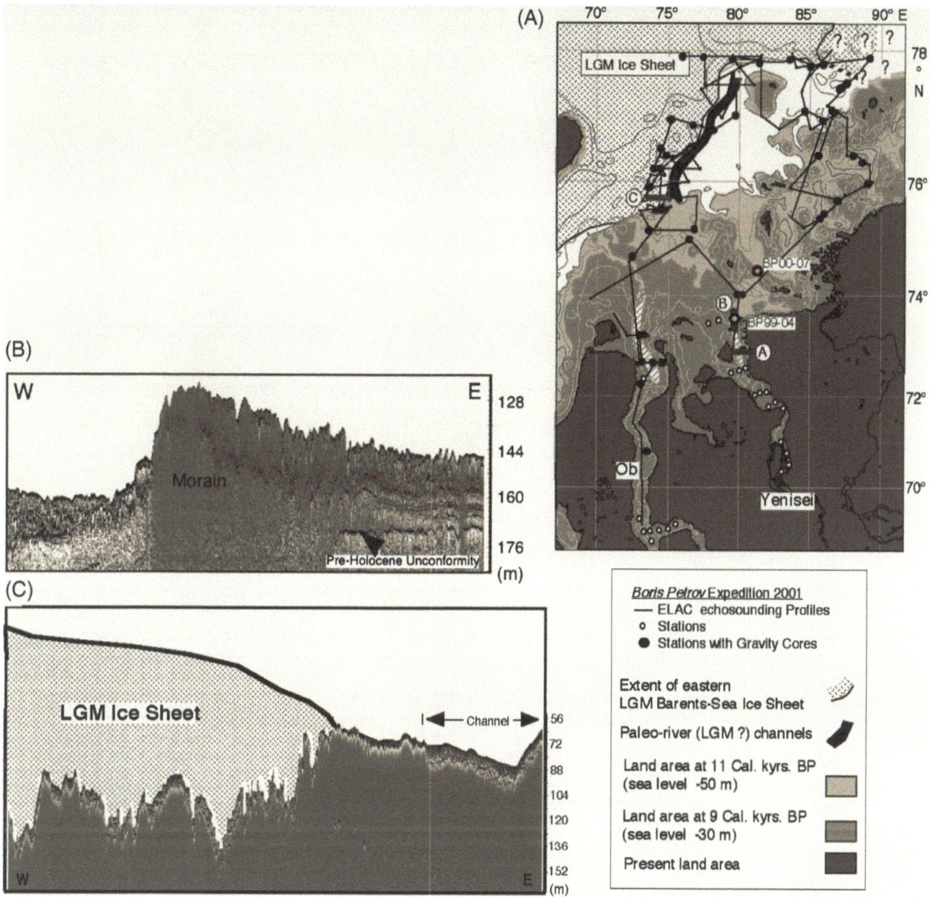

Figure 6.26 Map of the study area of the *Akademik Boris Petrov* Expedition 2001 in the inner Kara Sea and Ob and Yenisei estuaries, presenting (A) location of sampling sites and sediment echograph profiles, and (B and C) two selected acoustic profiles (from Stein et al., 2002). Based on the profiling results, the extent of the southeastern margin of the LGM ice sheet and a LGM palaeo-river channel are shown. Early Holocene (~9 Cal. kyr BP and 11 Cal. kyr BP) land area assuming a 30 and 50 m, respectively, lowered sea level in comparison to today (according to Fairbanks, 1989). The map has been constructed by simply using the 30 and 50 m isolines of the modern bathymetry (and assuming no or negligible isostatic rebound). The location of the ELAC sediment echosounding and GeoChirp profiles from the western study area across the proposed eastern margin of the LGM Barents-Kara Ice Sheet and shown in (B) and (C), respectively, are located at "C" in Map (A). In addition, the location of two ELAC sediment echograph profiles (A and B) presented in Figure 2.21 as well as the location of two AMS^{14}C-dated sediment cores BP99-04/7 and BP00-07/7 for which OC records are presented in Figure 6.87 are shown.

Meteorological Office General Circulation Model scaled to the Greenland Ice Core Project (GRIP) ice core $\delta^{18}O$ record, and (3) model parameters were chosen to best match geomorphological inferences on LGM extent and global eustatic sea-level change. In Figure 6.27, model results of time-dependent predictions over the last

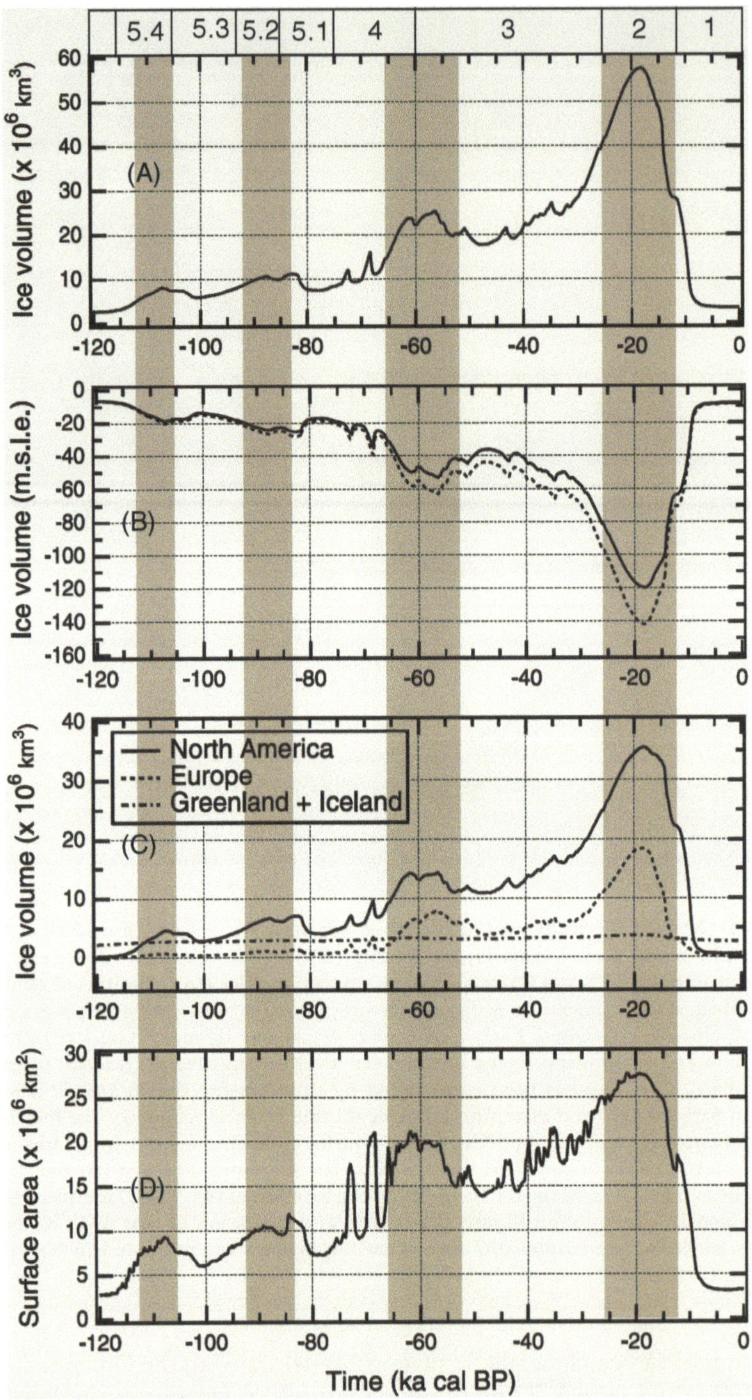
5.4
5.3
5.2
5.1
4
3
2
1
Ice volume (x 10^6 km^3)
(A)
60
50
40
30
20
10
0
-120
-100
-80
-60
-40
-20
0
Ice volume (m.s.l.e.)
(B)
0
-20
-40
-60
-80
-100
-120
-140
-160
Ice volume (x 10^6 km^3)
(C)
North America
Europe
Greenland + Iceland
40
30
20
10
0
Surface area (x 10^6 km^2)
(D)
30
25
20
15
10
5
0
Time (ka cal BP)

glacial cycle using initial conditions of a glaciological steady state ice configuration forced by present-day climate, are shown. The maximum ice volume of $57 \times 10^6\,km^3$ was found at LGM, of which $35 \times 10^6\,km^3$ is stored in North America, $18 \times 10^6\,km^3$ in Eurasia, and another $4 \times 10^6\,km^3$ in Greenland/Iceland (Zweck & Huybrechts, 2005). The total ice volume corresponds to a eustatic sea-level lowering of 110 m after correction for hydro-isostatic displacement. These 110 m can be separated into 82 m stored in the North American ice sheet, and 25 m stored in the Eurasian ice sheet (Figure 6.27). For comparison, today the only two large ice masses that survived the last glacial–interglacial transition, are the Greenland and Antarctic ice sheets with a total volume of $\sim 32 \times 10^6\,km^3$ (Clark & Mix, 2002). As outlined by Huybrechts (2002), their current volume together contains enough ice to raise global sea level by almost 70 m, of which ~60 m would derive from the Antarctic Ice Sheet (Huybrechts, Steinhage, Wilhelms, & Bamber, 2000) and 7 m from the Greenland Ice Sheet (Letréguilly, Huybrechts, & Reeh, 1991).

In order to reconstruct the Weichselian Eurasian Ice Sheet compatible with geological-based limits (Svendsen et al., 1999, 2004), Siegert et al. (1999) used a numerical ice-sheet model with an informal inverse-type procedure in which the ice-sheet model's climate inputs (mean annual precipitation and air temperature) were adjusted until a "fit" between the model's ice-sheet results and the geological evidence was established (see also Dowdeswell & Siegert, 1999; Siegert & Marsiat, 2001; Siegert, Dowdeswell, Hald, & Svendsen, 2001; Siegert & Dowdeswell, 2002). Furthermore, ice-sheet decay was modelled by comparing the bed uplift predicted by the model to the uplift record of Svalbard and Franz Josef Land (e.g., Forman et al., 2004). The result of this modelling procedure is an ice-sheet reconstruction consistent with the geological record, plus a climate and mass balance necessary to build the ice sheet (Siegert & Dowdeswell, 2004). The modelled build-up and decay of the Eurasian Ice Sheet during the late Weichselian are shown in Figure 6.28. At its LGM maximum size, the ice sheet occupied the entire Barents Sea, Scandinavia, and the North Sea, reaching an ice thickness of >2.4 km over Scandinavia and ~1 km across the central Barents Sea, whereas over the Kara Sea only a small ice mass with a maximum thickness of ~300 m developed (Siegert & Dowdeswell, 2004). Fast-flowing ice streams within bathymetric troughs drained

Figure 6.27 Three-dimensional thermomechanically coupled ice sheet modelling of the Northern Hemisphere for reconstruction of the Quaternary Northern Hemisphere ice sheets during the last glacial cycle (from Zweck & Huybrechts, 2005, supplemented). Figure shows the time-dependent change in basic model outputs for a reference model run. For this reference model, model parameters on the basis of good agreement between model output and observations were chosen. Most weight was given to consistency with geomorphological inferences of LGM ice extent, ice-volume estimates from observed sea-level changes, and present-day ice-sheet coverage. (A) Total ice-sheet volume over the last glacial cycle. (B) Total ice-sheet volume converted to change in eustatic sea level from the present day (dashed line) and for geomorphology-corrected sea-level change (for method see Zweck & Huybrechts, 2005) (solid line). (C) Total volumes of ice sheets by region as listed in the legend. (D) Total ice-sheet surface area. Marine isotope (sub-) stages and periods of glacier advance (vertical grey bars) are indicated.

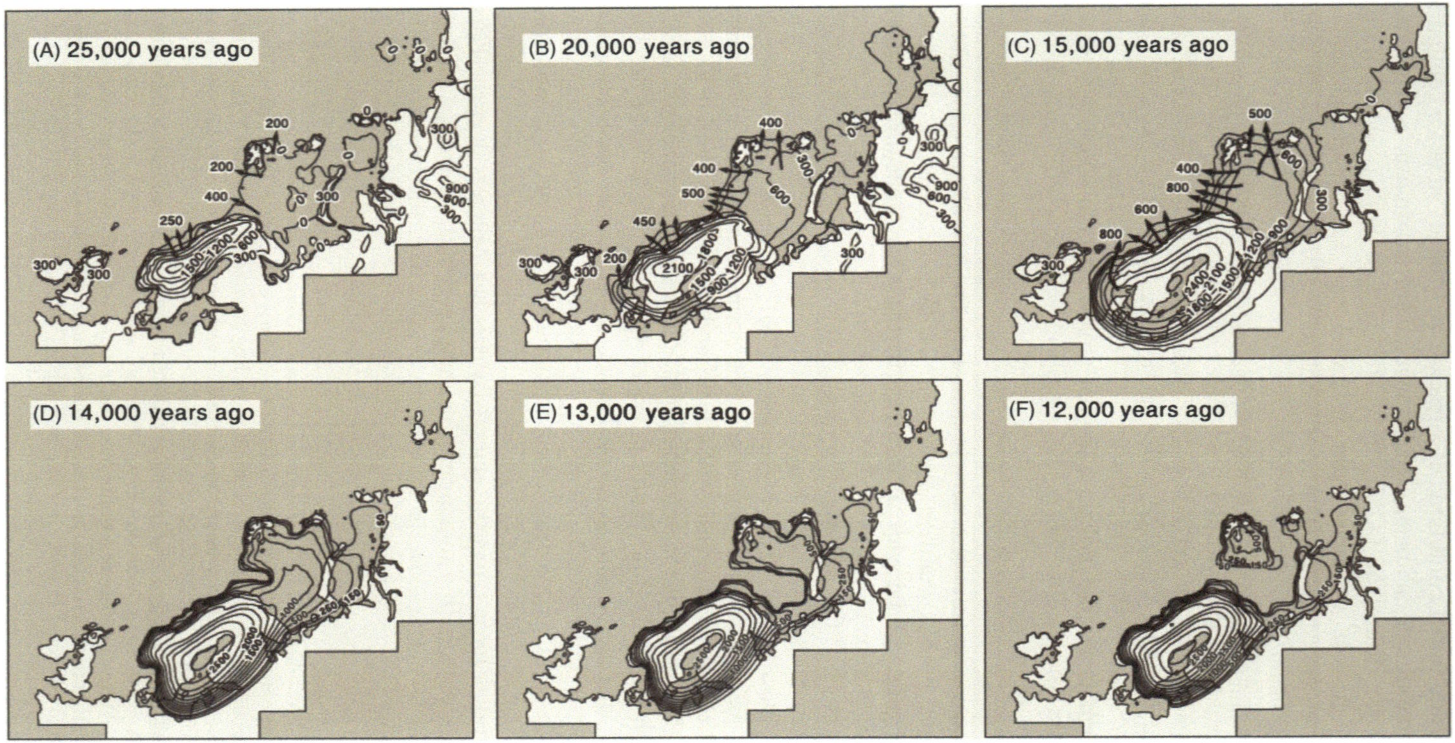

Figure 6.28 Modelled ice-sheet thickness (m) at several time slices during the last glaciation (from Siegert & Dowdeswell, 2004): (A) 25 ka; (B) 20 ka; (C) 15 ka; (D) 14 ka; (E) 13 ka; (F) 12 ka. Arrows and numbers provided in (A–C) correspond to fast-flowing parts of the ice sheet with modelled ice velocities ($m\,yr^{-1}$) at the ice-sheet margin shown.

ice from the central regions of the ice sheet to the marine margin, where iceberg calving and sediment deposition occurred (Figure 6.28).

According to the model results, warm maritime conditions across western Scandinavia and the western Barents Sea resulted in rates of ice accumulation in excess of 300 mm yr^{-1}, caused by the local availability of (a) a moisture source within the seasonally ice-free Norwegian-Greenland Sea and (b) westerly prevailing winds that controlled the storm tracks across this region of the North Atlantic. At the eastern margin of the ice sheet over the Kara Sea, on the other hand, the rate of ice accumulation was <100 mm yr^{-1}, making it similar to the polar desert conditions in central East Antarctica at present (Siegert & Dowdeswell, 2004).

6.2.2. History of Late Saalian to Late Weichselian Glaciations: Sedimentary Records from the Eurasian Continental Margin

A main problem of terrestrial records when used for reconstruction of the glacial history might be the correct dating of the glacial deposits. Furthermore, deposits of older glaciations may have been overprinted or even eroded by younger glacier advances. Thus, these terrestrial records are often incomplete. Here, carefully selected, continuous and dated marine sedimentary records obtained from the continental slopes adjacent to glaciated shelves as well as from the deep-sea basins and ridges have a major advantage, because continental glaciations often leave traces in these marine sediments. As outlined in Chapters 4.2 and 4.3, a number of proxies are available for identifying the input of IRD (i.e., the content of coarse-grained terrigenous detritus) as well as the source areas and pathways of the IRD (i.e., mineralogical and geochemical proxies) (e.g., Clark et al., 1980; Clark, Vincent, Jones, & Morris, 1980, 1984; Darby et al., 1989; Spielhagen, 1991; Elverhøi et al., 1995a; Bischof & Darby, 1997; Mangerud et al., 1998; Nørgaard-Pedersen et al., 1998, 2007a, 2007b; Knies et al., 2000; Knies, Kleiber, Matthiessen, Müller, & Nowaczyk, 2001; Phillips & Grantz, 2001; Spielhagen et al., 1997, 2004; Levitan et al., 2007). The most likely transport agents of the coarse-grained IRD are icebergs breaking off from calving ice sheets at the coastline. Thus, the IRD in deep-sea sediments may provide a direct link to the timing and regional extent of continental glaciations. Furthermore, melt-water discharge related to the decay of glacial ice sheets can be identified by stable isotopes of planktonic foraminifers (e.g., Stein et al., 1994a; Nørgaard-Pedersen et al., 1998; Poore et al., 1999; Spielhagen et al., 2004). Using these proxies in well-dated complete sedimentary sections allow a more precise reconstruction of the history of glacier advance and retreat than terrestrial records do. Marine records, for example, may help to resolve the two major discrepancies concerning the extent and volume of the LGM ice sheet along the Eurasian continental margin discussed earlier. Whereas numerous studies related to the glacial history of the Eurasian and Greenland continental margin were carried out (see later discussions), similar studies in the Arctic proximal to the northern LIS are more or less missing due to the lack of high-quality sediment cores from the permanently ice-covered region off the Canadian Archipelago and the Beaufort Sea continental margin area. Thus, records dealing with the history of the LIS are restricted either to sub-Arctic regions (e.g., the North Atlantic) or the distal central Arctic Ocean (see Section 6.3).

The glaciation curves of the SIS and the SBIS during the Weichselian are quite well known from sections in Scandinavia and on Svalbard as well as from offshore records obtained in the Fram Strait area west of Svalbard and the Norwegian Sea (Figure 6.29; e.g., Hebbeln, 1992; Mangerud & Svendsen, 1992; Baumann et al., 1995; Mangerud, Jansen, & Landvik, 1996, 1998). Based on these studies, the waxing and waning of glaciers correlate well with the IRD input in the deep ocean along the western Norwegian and Svalbard margins. Maximum IRD input occurred particularly during deglaciation phases of the most extended ice-sheets when a broad ice front reached the continental shelf edge, disintegrated, and subsequently melted (Mangerud et al., 1998). During the Weichselian, the glacier front of the SIS passed the coastline of western Scandinavia at least four times, and the waxing and waning of ice seems to follow the 23 kyr precession cycle of the Earth's orbit (Ruddiman & McIntyre, 1981; Mangerud et al., 1996). In comparison

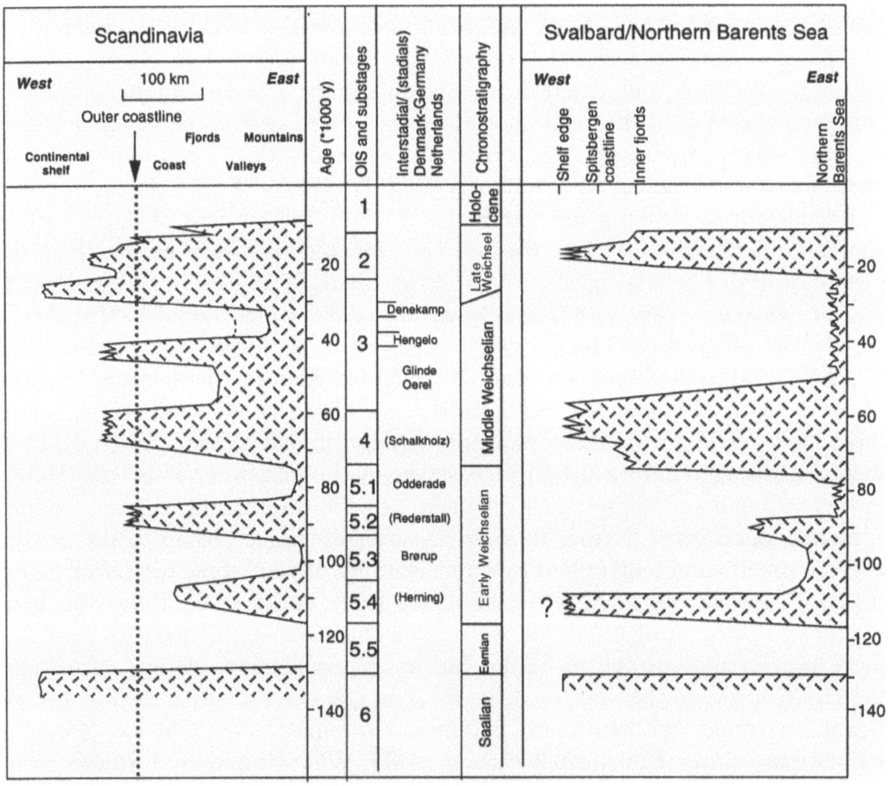

Figure 6.29 The glaciation fluctuations of the Scandinavian Ice Sheet (SIS) and the Svalbard-Barents Sea Ice Sheet (SBIS) during the last glacial/interglacial cycle (from Knies et al., 2001; *cf.* Mangerud et al., 1996, 1998). Based on marine data, the extent of the early Weichselian ice sheet (MIS 5.4) on Svalbard was probably smaller than the Saalian and Middle/Late Weichselian ice sheets, indicated by the question mark (*cf.* Knies et al., 2000, 2001; Winkelmann, 2007; Winkelmann, Stein, Schäfer, & Mackensen, 2008b).

to the SIS, the glaciation curve of the SBIS shows less extensive glacier advances. Significant ice advances reaching the shelf edge west of Svalbard, probably occurred during MIS 6, substage 5.4, MIS 4 and MIS 2, as supported by IRD records along the Svalbard continental margin and adjacent deep-sea area (Hebbeln, 1992; Mangerud et al., 1998). The substage 5.4 ice advance, however, is not documented as a terrigenous input signal in sediment cores north of Svalbard, suggesting that the early Weichselian glaciation on Svalbard might have been less extended than the Saalian and Middle and Late Weichselian glaciations (Winkelmann, 2007; Winkelmann et al., 2008b; see later discussions). Lloyd, Kroon, Boulton, Laban, and Fallick (1996) suggest an additional major glacier advance during substage 5.2, which is, however, only partly consistent with the terrestrial records (Mangerud et al., 1998). According to Landvik, Bolstad, Lycke, Mangerud, and Sejrup (1992), the glaciers probably survived in the northern parts of the Barents Sea during the entire MIS 5. The major glacier advances of SBIS apparently follow the 41-kyr obliquity cycle of the Earth's rotation axis, which at this latitude is more important for the summer insolation than the 23-kyr precession cycle (Mangerud & Svendsen, 1992; Mangerud et al., 1996, 1998).

Towards the east, along the northern Barents Kara Sea continental margin, detailed correlation of IRD records and onshore sections during the Weichselian or even the Saalian are hampered by relatively sparse terrestrial data sets. In these areas, marine records may provide most important information on the history and extension of glaciations (e.g., Polyak et al., 1997, 2000a, 2000b; Knies et al., 1999, 2000, 2001; Lubinski et al., 2001; Kleiber et al., 2000; Knies & Vogt, 2003), which can be used to test or approve ice-sheet reconstructions based on field and modelling studies (e.g., Svendsen et al., 1999, 2004; Grosswald & Hughes, 2002; Mangerud et al., 2002; see discussion earlier).

In this context, selected sediment cores obtained from the Eurasian continental margin (Figure 6.24A) were studied in detail to reconstruct the environmental changes in terms of waxing and waning of the Barents Kara Sea ice sheets and related Atlantic water inflow and sea-ice distribution throughout the last 150 ka (Vogt, 1997; Knies, 1999; Kleiber, 1999; Müller, 1999; Kleiber et al., 2000, 2001; Knies et al., 2000, 2001; Knies & Vogt, 2003). Here, especially Core PS2138-1 located at the Barents Sea continental margin between the Svalbard and Franz Josef Land archipelagos, and Core PS2741-1 located off Severnaya Semlja (Figure 6.24A) are key cores because they allow to distinguish between signal related to the Barents Sea Ice Sheet and the more eastern Kara Sea Ice Sheet. Core PS2138-1, one of the best dated Arctic sediment cores, is dated by using AMS^{14}C, stable isotopes of planktonic foraminifer *N. pachyderma*, palaeomagnetic data (see Figure 6.5), ^{10}Be stratigraphy (see Figure 6.3), and the occurrence of benthic foraminifer *Pullenia bulloides* (Figure 6.30; Knies et al., 2000, 2001; Nowaczyk & Knies, 2000). The age model of Core PS2741-1 was established by AMS^{14}C datings and palaeomagnetic data, and dinoflagellate data (Figure 6.30; Knies et al., 2000, 2001; Matthiessen et al., 2001). As proxies for the reconstruction of the Barents-Kara Sea ice-sheet build-up and decay, lithology, IRD input, clay mineralogy, and bulk accumulation rates were determined in these cores. In addition data from cores PS1533, PS2446-4, and PS2212-3 located in the Yermak Plateau/northern Barents Sea area

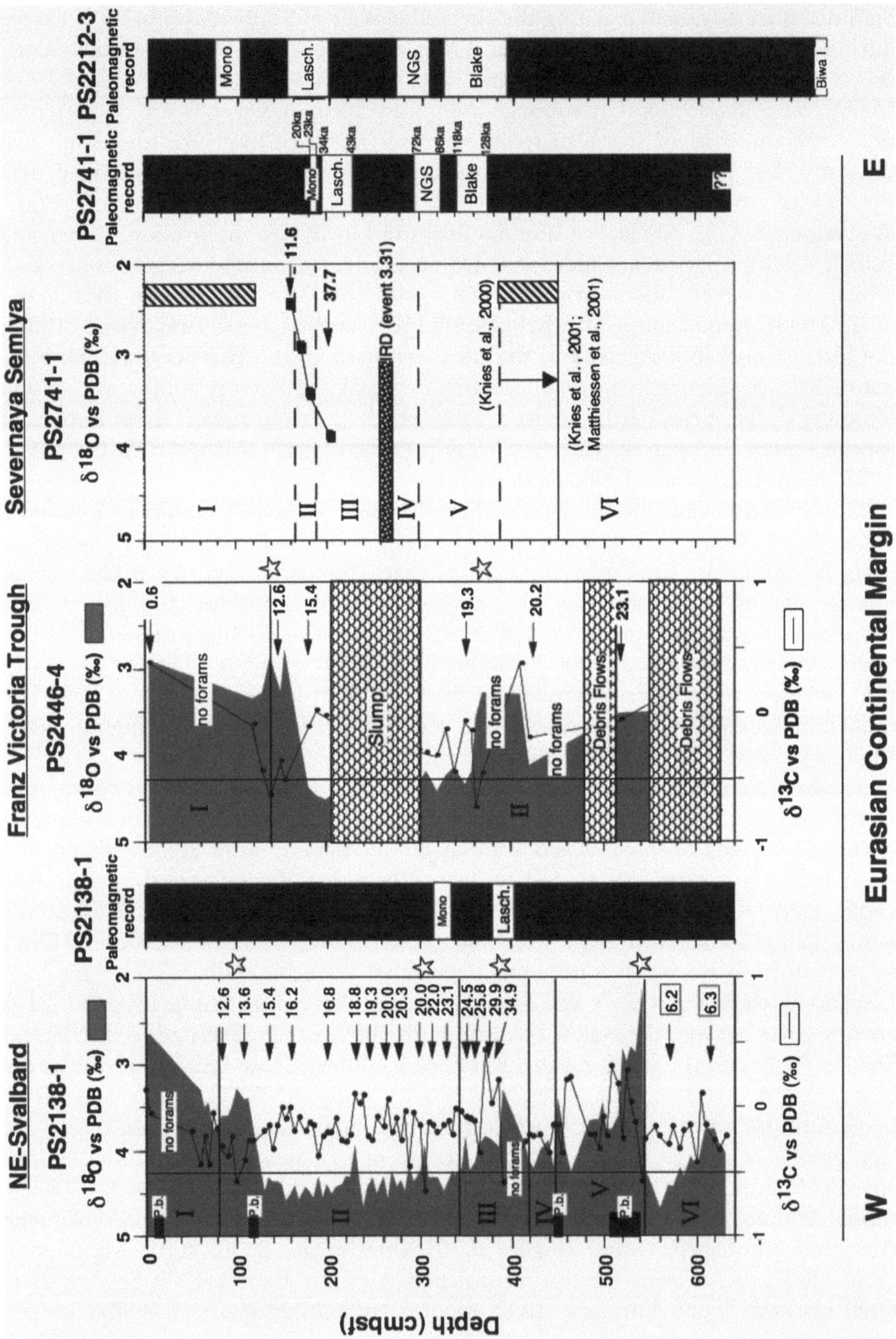
NE-Svalbard
PS2138-1
δ18O vs PDB (‰)
PS2138-1
Paleomagnetic record
Mono
Lasch.
Franz Victoria Trough
PS2446-4
δ18O vs PDB (‰)
no forams
Slump
Debris Flows
Severnaya Semlya
PS2741-1
δ18O vs PDB (‰)
IRD (event 3.31)
(Knies et al., 2000)
(Knies et al., 2001;
Matthiessen et al., 2001)
PS2741-1
Paleomagnetic record
NGS
Blake
PS2212-3
Paleomagnetic record
Biwa I
δ13C vs PDB (‰)
Depth (cmbsf)
Eurasian Continental Margin
W
E

(for location see Figure 6.24A; for stratigraphy of these cores see Figures 6.3, 6.5, and 6.30), and from cores PS2757-8 and PS2761-10 (for location see Figure 6.24A; for stratigraphy of these cores, see Müller, 1999) are presented.

6.2.2.1. Late Saalian to late Weichselian glaciations in the northern Barents Sea

The IRD pattern NE of Svalbard (PS2138-1) recording the history of the northern SBIS, displays a strong variability. The peaks in IRD, however, seem not to be confined to ice volume maxima at MIS 6, 5.4 and 5.2, 4, and 2, but they correlate quite well with deglacial phases on western Svalbard at the MIS 6/5, 4/3, and 2/1 transitions (Figure 6.31; Knies et al., 2001). The Saalian (MIS 6) glaciation probably reached the northern Barents Sea shelf edge as indicated by enhanced bulk accumulation rates of $\sim 10\,\mathrm{g\,cm^{-1}\,kyr^{-1}}$ (Figure 6.31) and high kaolinite contents (Figure 6.32) in the PS2138-1 record. At that time, Mesozoic bedrocks of the northern Barents Sea characterized by high kaolinite contents (Birkenmajer, 1989; see Chapter 5.1.2, Figure 5.4D) and high contents of mature OC (see Section 6.4) were eroded by glacial activity and delivered to the slope by suspension plumes when the maximum ice-sheet extension was reached (Knies et al., 2000, 2001). A first prominent IRD pulse related to the MIS event 6.3 may indicate a destabilization of the ice margin induced by a short-term warming period (Matthiessen et al., 2001). The onset of disintegration of the MIS 6 ice-sheet is recognized at $\sim$134 ka by a single IRD peak, followed by a second IRD spike near 128 ka interpreted as the final retreat from the outer shelf (Figure 6.31).

At Core PS2212-3 located on the Yermak Plateau north of Svalbard, three pulses of coarse-grained sediments were found in the MIS 6 time interval (Figure 6.33; Vogt, 1997), which have very similar characteristics as the contemporaneous sediments at Core PS2138-1. They are characterized by high amounts of smectite and kaolinite (Figure 6.33) as well as elevated terrigenous OC values, interpreted as ice-rafted material originated from the eastern Eurasian shelf area between Franz Josef Land and the western Laptev Sea and transported by the TPD (Vogt, 1997). As mentioned by Knies et al. (2001), a similar scenario of waxing and waning ice-sheets during late MIS 6 is described from the western Scandinavian margin

Figure 6.30 Oxygen and carbon isotope stratigraphy, magneto-stratigraphy, and AMS^{14}C datings of cores PS2138-1, PS2446-4, and PS2741-1 (from Knies, 1999; Knies et al., 2001; Matthiessen et al., 2001; for core location see Figure 6.24A). MIS stages are displayed for each core. Abundances of *P. bulloides* (P.B.) is marked by solid bars in Core PS2138-1. MIS events 6.2 and 6.3 are indicated (*cf.* Martinson et al., 1987). Slumps/debris flows in Core PS2446-4 are labelled. For Core PS2741-1, the occurrence of dinoflagellates (*Operculodinium centrocarpum* and *Impagidinium pallidum*; Matthiessen et al., 2001) are shown as hatched bar. Palaeomagnetic data for cores PS2741-1 and PS2212-3 (*cf.* Nowaczyk et al., 1994) are shown on the right-hand side. Ages for several magnetic events are adopted from Nowaczyk et al. (1994). Distinct IRD-layer in early MIS 3 is correlated to isotope event 3.3.1 according to Martinson et al. (1987) and Nørgaard-Pedersen et al. (1998). Based on dinoflagellates, the MIS 5/6 stage boundary, in contrast to Knies et al. (2000), has been shifted to the base of the concentration maximum of dinocysts at *ca.* 455-cm core depth (Matthiessen et al., 2001). Open stars indicate meltwater events.

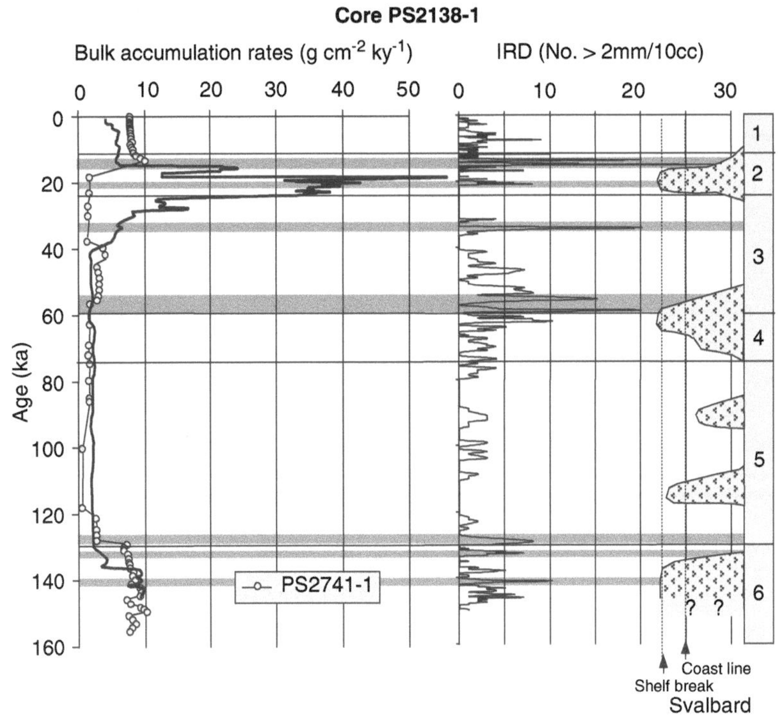

Figure 6.31 Bulk accumulation rates of cores PS2138-1 and PS2741-1, and a IRD record of PS2138-1 versus age (from Knies et al., 2000, 2001). Shaded areas mark distinct IRD peak values. MIS are indicated. At the right-hand side, the extension of the Svalbard Ice Sheet is shown (based on Mangerud et al., 1998).

(Wagner, 1993; Wagner & Henrich, 1994; Baumann et al., 1995; Fronval & Jansen, 1997), suggesting a link in the behaviour of both the SIS and SBIS ice sheets.

The decay of the Saalian ice sheet also resulted in a prominent meltwater pulse near ~131 ka, indicated by a distinct light planktonic $\delta^{18}O$ spike associated with very low $\delta^{13}C$ values of −0.6‰ (Figures 6.30 and 6.34; Knies et al., 2001; Knies & Vogt, 2003; Spielhagen et al., 2004). Data from the Nordic Seas reveal that the Saalian deglaciation penetrated far into MIS 5.5 and moulded the oceanic regime by massive meltwater supply (Fronval & Jansen, 1997; Bauch et al., 1999). Knies and Vogt (2003) infer from the PS2138 data that the Arctic Ocean delivered a major proportion of meltwater during MIS 6/5 transition, at least to the exit of the TPD. This conclusion is further supported by the detection of a major meltwater pulse on the Yermak Plateau (PS1533) during Saalian deglaciation (Eisenhauer et al., 1994).

During MIS 5, moderate input of IRD recorded along the northern Barents Sea margin, and low bulk accumulation rates suggest smaller glacial activity in comparison to MIS 6. After the Eemian (MIS 5.5), five smaller IRD (>2 mm) peaks were recorded between ~112 and 70 ka (Figure 6.31). These IRD peaks are contemporaneous with more prominent peaks in the >250 μm IRD fraction, as shown by data from a more recent high-resolution study carried out for the same

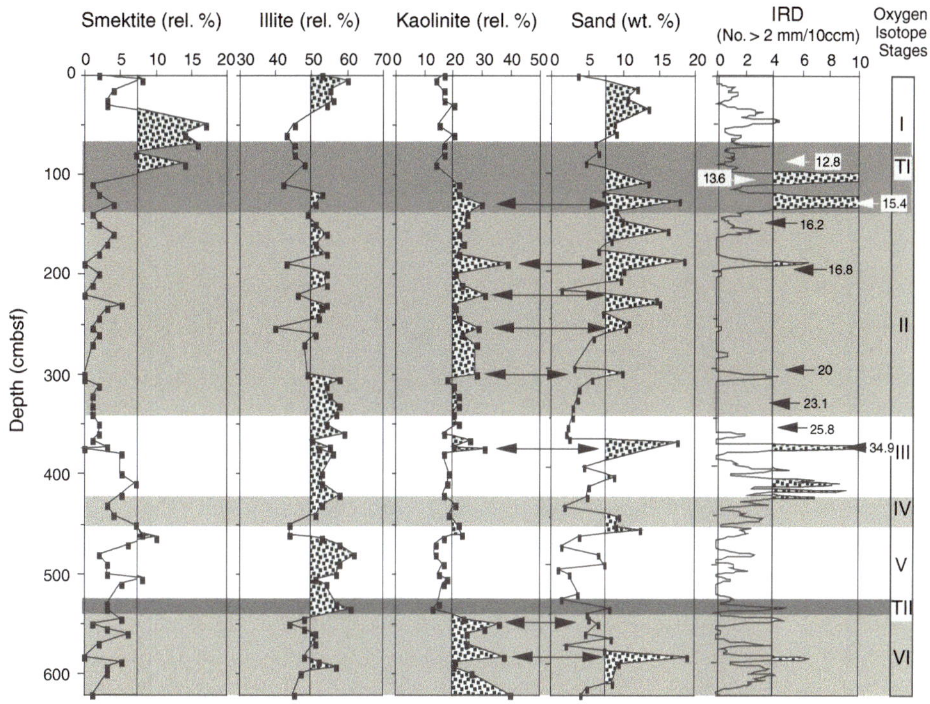

Figure 6.32 Relative contents of smectite, illite, and kaolinite, percentage of sand fraction, and numbers of IRD grains >2 mm per cc versus depth of Core PS2138-1 (from Knies, 1999; Knies et al., 2000). Marine oxygen isotope stages are marked as roman numbers; TI and TII indicate Terminations I and II; numbers at arrows in the IRD record indicate reservoir-corrected AMS[14]C ages.

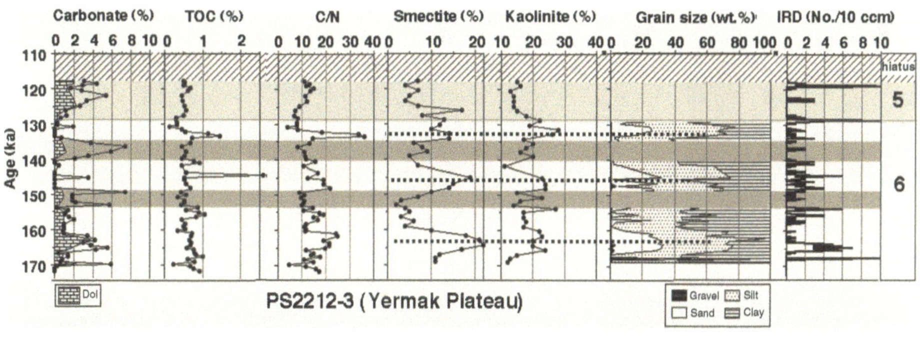

Figure 6.33 Carbonate content (dol, dolomite proportion is indicated), TOC content, C/N ratio, smectite and kaolinite contents, grain-size distribution (sand-silt-clay), and IRD (numbers per 10 cc) determined for the time interval 170–110 ka (MIS 6 to upper MIS 5) of Core PS2212-3 (from Vogt, 1997). For core location see Figure 6.24A.

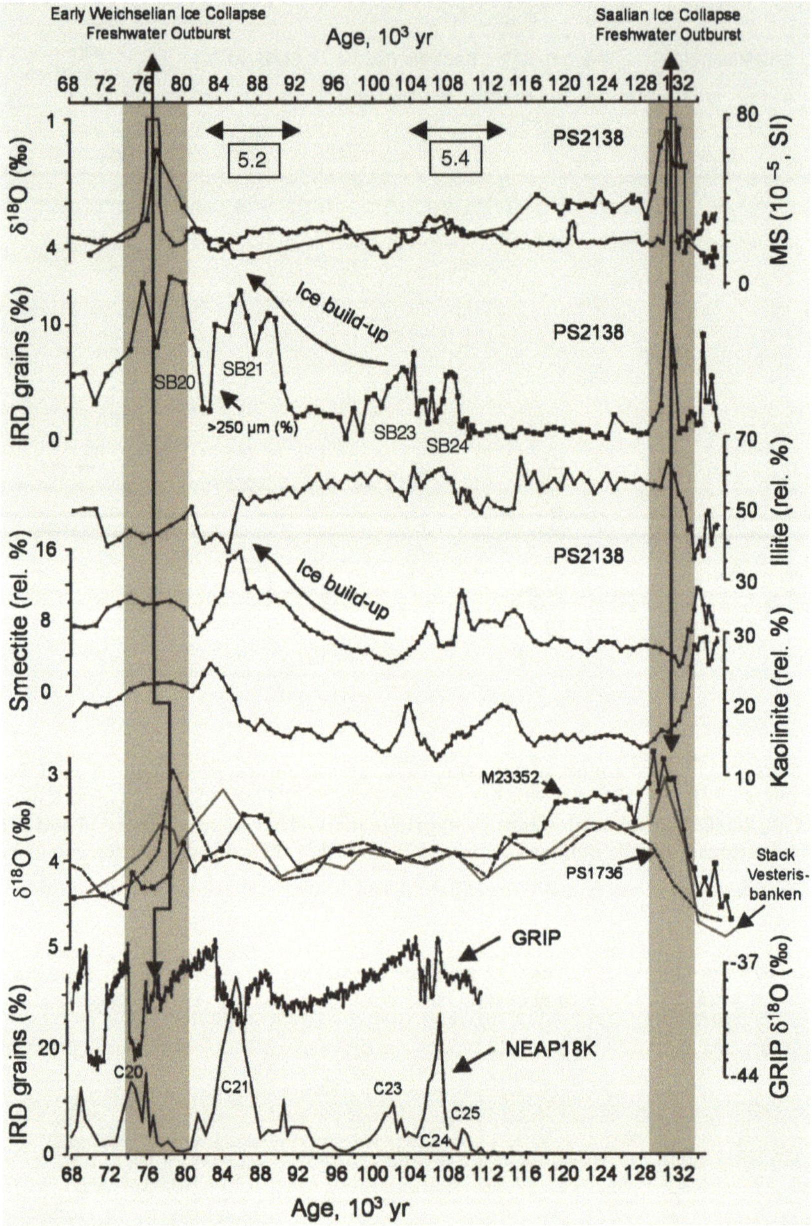

Figure 6.34 Comparison of marine $\delta^{18}O$ records from the western Nordic seas (Jünger, 1994 — PS1736, hatched line; Antonow, Goldschmidt, & Erlenkeuser, 1997 — stack Vesterisbanken, grey thick line; Bauch, Erlenkeuser, Jung, & Thiede, 2000 — M23352, squares), Greenland GRIP ice core (Johnsen et al., 1997), and a North Atlantic IRD record (NEAP18K; Chapman & Shackleton, 1999) versus age in the lower panel with a compilation of $\delta^{18}O$ record, MS (MS volume, 10^{-5} SI; Nowaczyk & Knies, 2000), IRD grains, and clay mineralogy in Core PS2138 versus age in the upper panel (from Knies & Vogt, 2003). IRD events in Core PS2138 corresponding to North Atlatic cooling events C25–C20 (McManus et al., 1994) are abbreviated as SB25–SB20. Arrows depict the prominent meltwater events during Saalian and Early Weichselian deglaciation and indicate the similarities between Arctic Ocean freshwater outburst and sudden atmospheric and oceanic cooling event C20 (McManus et al., 1994).

time interval (Figure 6.34; Knies & Vogt, 2003). According to Knies and Vogt (2003), they coincide with ice-sheet growth in the Barents and Kara seas during the MIS 5.4 and 5.2 (C21), and decay during the MIS 5.4/5.3 and 5.2/5.1 transitions.

The gradual increase in smectite during the large-scale ice-rafting events SB21 and SB20 may suggest the Kara Sea as major contributor to the overall IRD signal (Figure 6.34). This agrees with the proposed Early Weichselian ice-sheet extension in the Kara Sea ~90 ka (Figure 6.24B; Mangerud et al., 2001; Svendsen et al., 2004), when glacially eroded smectite-rich sediments may have been transported by the Siberian branch of the TPD to the northern Barents Sea, and the Kara Sea ice sheet blocked the north-flowing Siberian rivers resulting in the formation of a large ice-dammed lake (Figure 6.25; Mangerud et al., 2001, 2004). The deglaciation is assumed to have occurred during substage MIS 5.1, suggesting a subsequent massive outflow of freshwater into the Arctic Ocean (Siegert et al. 2001).

Contemporaneously with the prominent freshwater pulse within MIS 5.1, distinct peaks in the MS and smectite records (Figure 6.34) support the hypothesis of major fluvial/meltwater discharge from weathering residues of Permo–Triassic Putoran Plateau flood basalts of the northern Siberian hinterland to the Kara Sea/western Laptev Sea (Kleiber & Niessen, 2000; see Chapter 5.1.1, Figure 5.3). Based on studies of the origin and trajectory of Arctic sea ice, Pfirman et al. (1997) pointed to the Kara Sea as a major contributor to enhanced export rates of sediment-laden ice to the southern limb of the TPD (see Chapter 5.1.2, Figure 5.8).

Meltwater events during MIS 5.1 were also recorded on the Yermak Plateau (PS1533; Eisenhauer et al., 1994; see Section 6.3.1), in the Fram Strait (PS1535; Köhler & Spielhagen, 1990), and in the Greenland Basin (PS1736; Jünger, 1994). This may suggest that meltwater originating from the outburst of ice-dammed lakes in northern Siberia and flushing through the Fram Strait potentially created a low-salinity anomaly upon melting in the Fram Strait and possibly further south towards the Greenland Sea (Knies & Vogt, 2003). Such discharges may have contributed to a substantial reduction of NADW formation (Rahmstorf & Ganopolski, 1999). Knies & Vogt (2003) hypothesize that the 77 ka meltwater event probably was a decisive trigger for the rapid cooling during North Atlantic cool event C20 (McManus et al., 1994). Better chronologies for Arctic Ocean sediment cores, however, are needed to test this hypothesis.

The development of a major Middle Weichselian ice sheet (Figure 6.24C) is not clearly reflected in the record of Core PS2138-1. The — in comparison to MIS 6 and MIS 2 — significantly lower bulk accumulation rates values during MIS 4 (Figure 6.31) clearly indicate a lower terrigenous supply and, thus, a more restricted ice-sheet extension and probably a closely packed sea-ice cover (Knies et al., 2000). Slightly increased kaolinite contents near the base of MIS 4 (Figure 6.32), however, may suggest a more extended ice-sheet, probably to the outer coastline, than during MIS 5. Distinct IRD pulses between 57 and 52 ka in PS2138 may reflect a major deglacial event, related to the huge iceberg discharge from a retreating ice-margin from the outer shelf areas (Figure 6.32; Mangerud et al., 1998; Knies et al., 2000, 2001). This event corresponding to MIS event 3.31 and representing the degradation of the Middle Weichselian ice sheet, can be traced far into the central Arctic Ocean as IRD and melt-water signals (Darby et al., 1997; Norgaard-Pedersen et al., 1998;

Spielhagen et al., 2004; see Section 6.3.1). Between the deglaciation event 3.31 and the onset of the last glaciation, three major IRD pulses are recorded northeast of Svalbard (Figure 6.31). At least one (at 34.9 ka) correlates with increased content of kaolinite (Figure 6.32), suggesting an advanced SBIS to the shallow shelf in mid MIS 3 (Knies et al., 2000). These IRD events seem to be contemporaneous with the Heinrich events 5, 4, and 3 in the North Atlantic (Bond et al., 1992).

Due to high sedimentation rates, oscillations of the northern SBIS during the last ~30 ka are preserved on a highly resolved time scale in the records from Core PS2138-1 (Figure 6.35). The final build-up of the SBIS during the early stage of the late Weichselian after 30 ka was enforced by moisture supply due to intruding Atlantic water in combination with lower summer insolation (Hebbeln et al., 1994; Dokken & Hald, 1996; Knies et al., 1999, 2000). The ice sheet probably reached the shelf edge at ~23 ka. At that time, fast flowing ice streams in the Franz Victoria Trough delivered large quantities of sediment to the continental slope, as indicated by a series of debris flows identified in PARASOUND profiles as well as sediment cores (Figure 6.30 and Chapter 3.3, Figures 3.20 and 3.21; Kleiber et al., 2000; see also Laberg & Vorren, 1995). A comparable scenario is not recorded in Core PS2138-1, probably because sedimentation is not directly influenced by a fast-flowing ice stream, but by a more stable ice margin (Knies et al., 2000, 2001; see Chapter 3.3). At Core 2138-1, the maximum extension of the ice sheet is reflected in increased sediment input from the northern Barents Sea, indicated by a distinct increase in bulk accumulation rates up to 50 g cm^{-2} kyr^{-1} (Figure 6.31) and a major increase in kaolinite (Figure 6.32).

Almost contemporaneously, that is, between 27 and 20 ka, a sequence with very specific sedimentological, mineralogical, and organic-geochemical characteristics was identified in sediment cores west of Spitsbergen as well as on Yermak Plateau (Figure 6.36). These sediments contain high amounts of mature terrestrial organic matter, very low carbonate (mainly dolomite), no kalifeldspar, and a very distinct clay mineralogy with extremely low smectite, but high kaolinite percentages accompanied by the occurrence of ordered layered expandable minerals (OLEM) (Andersen et al., 1996; Vogt, 1997; Vogt et al., 2001). Based on these characteristics, the sediments originated from a restricted source area, that is, Spitsbergenbanken S to SE of Spitsbergen where a sequence of lower Cretaceous shales ("hot shale" member) are out-cropping (Figure 6.36). As this event is developed in all studied cores similarly, Vogt et al. (2001) regarded it as a synchronous regional event representing the advanced SBIS and a time of northward transport of fine fraction along the western Svalbard continental slope. The reduction of the horizons thickness from S to N (i.e., tens of centimetre W of Spitsbergen to a few at the Yermak Plateau) may support that currents transported the fine fraction eroded by the advanced SBIS at the Spitsbergenbanken and injected into the intermediate waters of a Palaeo-WSC through dense, suspension-rich bottom-water currents from the Storfjorden Trough (*cf.* Hebbeln et al., 1994; Andersen et al., 1996) towards the north, finally reaching the northern Yermak Plateau (Figure 6.36). Because the OLEM event at Core PS2212-3 coincides with increased IRD values (Figure 6.36), an additional direct transport from the northern margin of the SBIS by ice-rafting towards the core location is probable.

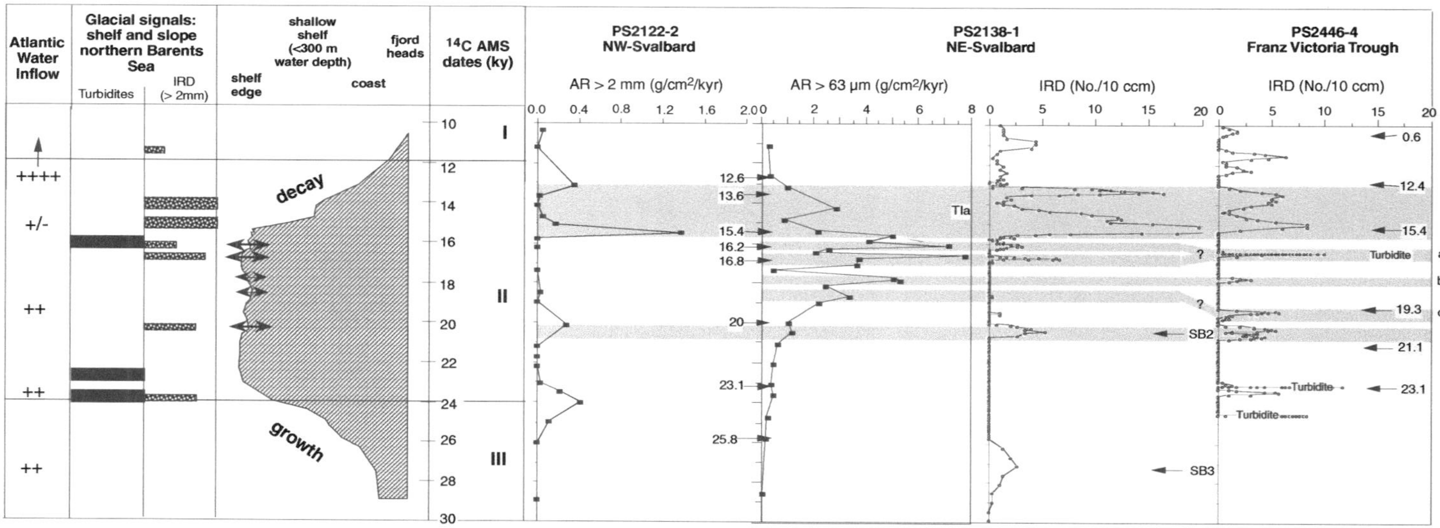

Figure 6.35 The left panel shows extension of the ice sheet on the northern Barents Sea margin during the last glacial/interglacial cycle versus age with respect to Atlantic water inflow (Knies et al., 1999), debris flow occurrences (Kleiber et al., 2000), and IRD peak values; the right panel shows the data corresponding to the model (from Knies, 1999; Knies et al., 2000, 2001). The reconstruction is based on IRD records of cores PS2138-1 and PS2446-4, and accumulation rates of coarse fraction at cores PS2122-2 and PS2138-1. For core location see Figure 6.36. Available AMS^{14}C datings are indicated. SB2 and SB3 (SB, Svalbard/Barents Sea) mark pronounced IRD deposition events contemporary to the North Atlantic Heinrich Events H2 and H3 (*cf.* Bond et al., 1993; Grousset, Pujol, Labeyrie, Auffret, & Boelaert, 2000).

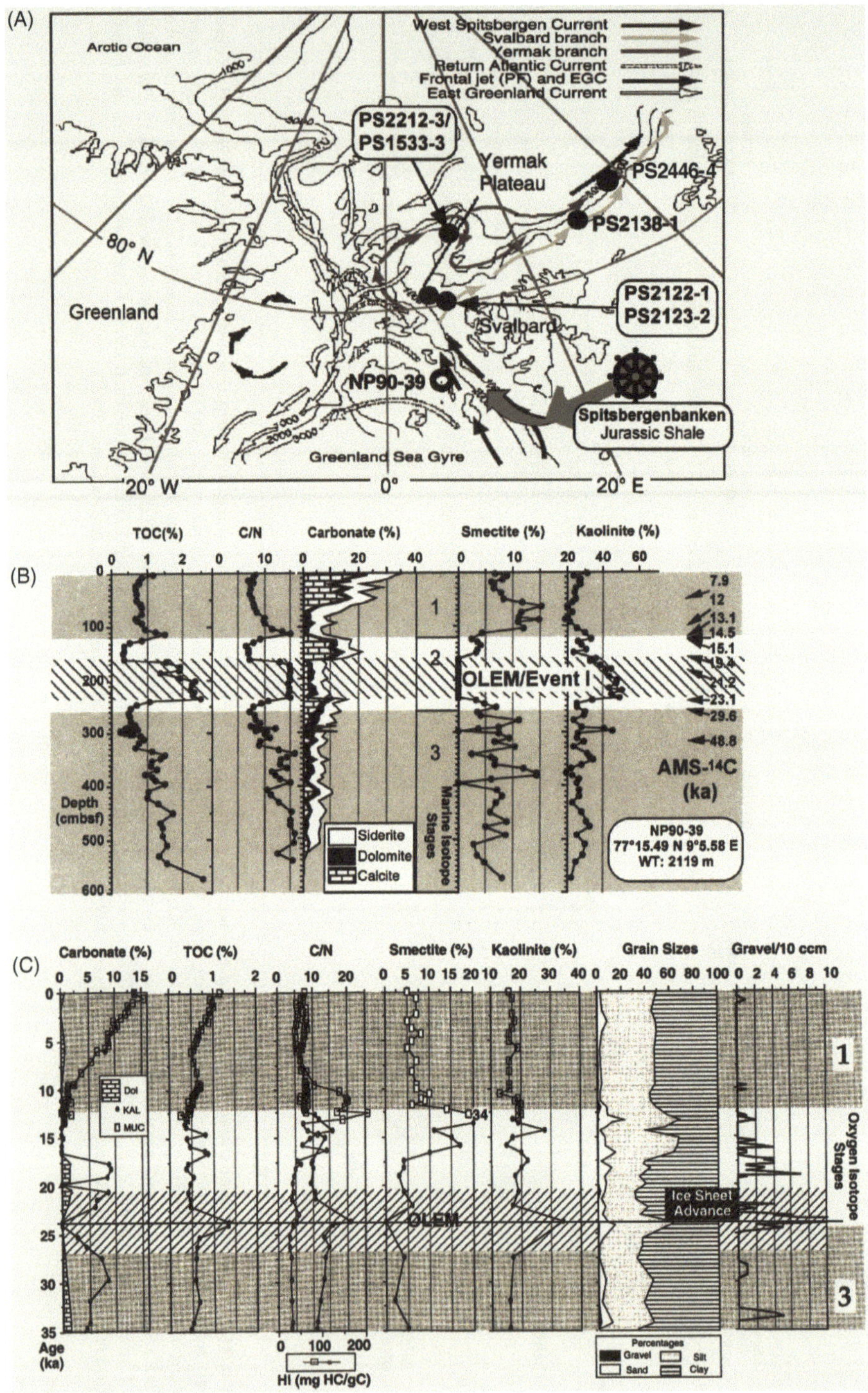

(A)
Arctic Ocean
West Spitsbergen Current
Svalbard branch
Yermak branch
Return Atlantic Current
Frontal jet (PF) and EGC
East Greenland Current
PS2212-3/
PS1533-3
Yermak
Plateau
PS2446-4
PS2138-1
80° N
Greenland
PS2122-1
PS2123-2
Svalbard
NP90-39
Spitsbergenbanken
Jurassic Shale
Greenland Sea Gyre
20° W
0°
20° E
(B)
TOC(%)
C/N
Carbonate (%)
Smectite (%)
Kaolinite (%)
OLEM/Event I
Depth (cmbsf)
Siderite
Dolomite
Calcite
Marine Isotope Stages
AMS-14C (ka)
NP90-39
77°15.49 N 9°5.58 E
WT: 2119 m
(C)
Carbonate (%)
TOC (%)
C/N
Smectite (%)
Kaolinite (%)
Grain Sizes
Gravel/10 ccm
Dol
KAL
MUC
OLEM
Ice Sheet Advance
Oxygen Isotope Stages
Age (ka)
HI (mg HC/gC)
Percentages
Gravel
Sand
Silt
Clay

A first distinct ice break-up occurred at ~20 ka, indicated by a major IRD pulse (Figure 6.35) and a contemporaneous meltwater event (Figure 6.30). Further IRD pulses followed, reflecting the waxing and waning of the SBIS on the outer shelf between 19 and 16.2 ka (Figure 6.35). These IRD events seems to be in good correlation with IRD pulses recorded in the northern North Atlantic (Bond & Lotti, 1995; Fronval, Jansen, Bloemendal, & Johnsen, 1995; Stoner, Channell, & Hillaire-Marcel, 1996) and off East Greenland (Nam et al., 1995; Stein et al., 1996; Nam, 1997; Nam & Stein, 1999; see Section 6.2.3), and may reflect synchronous fluctuations of the Northern Hemisphere ice sheets on a millennial time scale (e.g., Broecker, 1994; Bond & Lotti, 1995). As cited in Knies et al. (2000), however, a sudden collapse of the most extended SBIS could also have occurred in response to an increase in sea level caused by a LIS surge during each Heinrich events, as proposed by McCabe and Clark (1998).

With increasing summer insolation and global sea-level rise a rapid disintegration of the SBIS along the northern Barents Sea continental margin started at 15.4 ka (Ruddiman & McIntyre, 1981; Fairbanks, 1989), as suggested from distinct IRD pulses at that time (Figure 6.35). Contemporaneously, influx of meltwater and poorly ventilated low-saline surface water masses are indicated by prominent $\delta^{13}C$ and $\delta^{18}O$ minima (Figure 6.30; Knies et al., 2000, 2001). This meltwater event can be traced to the central Arctic Ocean and the Fram Strait (e.g., Jones & Keigwin, 1988; Stein et al., 1994a, 1994b; Nørgaard-Pedersen et al., 1998; see Section 6.3.5). Armadas of icebergs and extensive meltwater lids may have prohibited a further decay of the ice sheets because of a significant cooling of the ocean triggered by positive ice–albedo feedback mechanisms and sea-ice formation (e.g., Ruddiman & McIntyre, 1981). A second major IRD pulse at ~13.6 ka are interpreted as increased iceberg calving due to further ice-sheet decay by still rising summer insolation and sea level (Knies et al., 2000, 2001).

The decay of the major LGM SBIS is also well recorded in sediment cores from the Franz Victoria and St. Anna troughs, where stiff diamictons associated with grounded glacier ice during the LGM are overlain by a sequence of laminated clays, silty clays, and sandy layers being deposited in ice-proximal to ice-distal glacial-marine environments (Figure 6.37; Lubinski et al., 1996, 2001; Polyak et al., 1997; Kleiber et al., 2001). An age of >13 ka is inferred for the boundary between the basal diamicton and the overlying sediments in the Franz Victoria Trough and

Figure 6.36 (A) Surface currents in the European sector of the Arctic Ocean and locations of sediment cores PS2122-1, PS2123-2, PS2212-3, PS1533-3, PS2138-1, and NP90-39 (from Vogt et al., 2001 and further references therein). Position of Spitsbergenbanken with Jurassic shale source rock and proposed transport path of fine fraction material after Andersen et al. (1996) (B) Sediment characteristics (TOC and carbonate contents, C/N ratios, and smectite and kaolinite contents) versus depth of Core NP90-39 and radiocarbon dates (Andersen et al., 1996). The occurrence of ordered layered expandable minerals (OLEM) are marked as hatched area. Marine Isotope Stages 1–3 are indicated by bold numbers (from Vogt et al., 2001). (C) Sediment characteristics (TOC and carbonate contents, C/N ratios, smectite and kaolinite contents, grain-size distribution (sand–silt–clay), and IRD > 2 mm in No./10 cc) versus age (0–35 ka) of Core PS2212-3. The occurrence of OLEM are marked. The hatched area indicate the time interval of proposed ice-sheet advance (from Vogt, 1997).

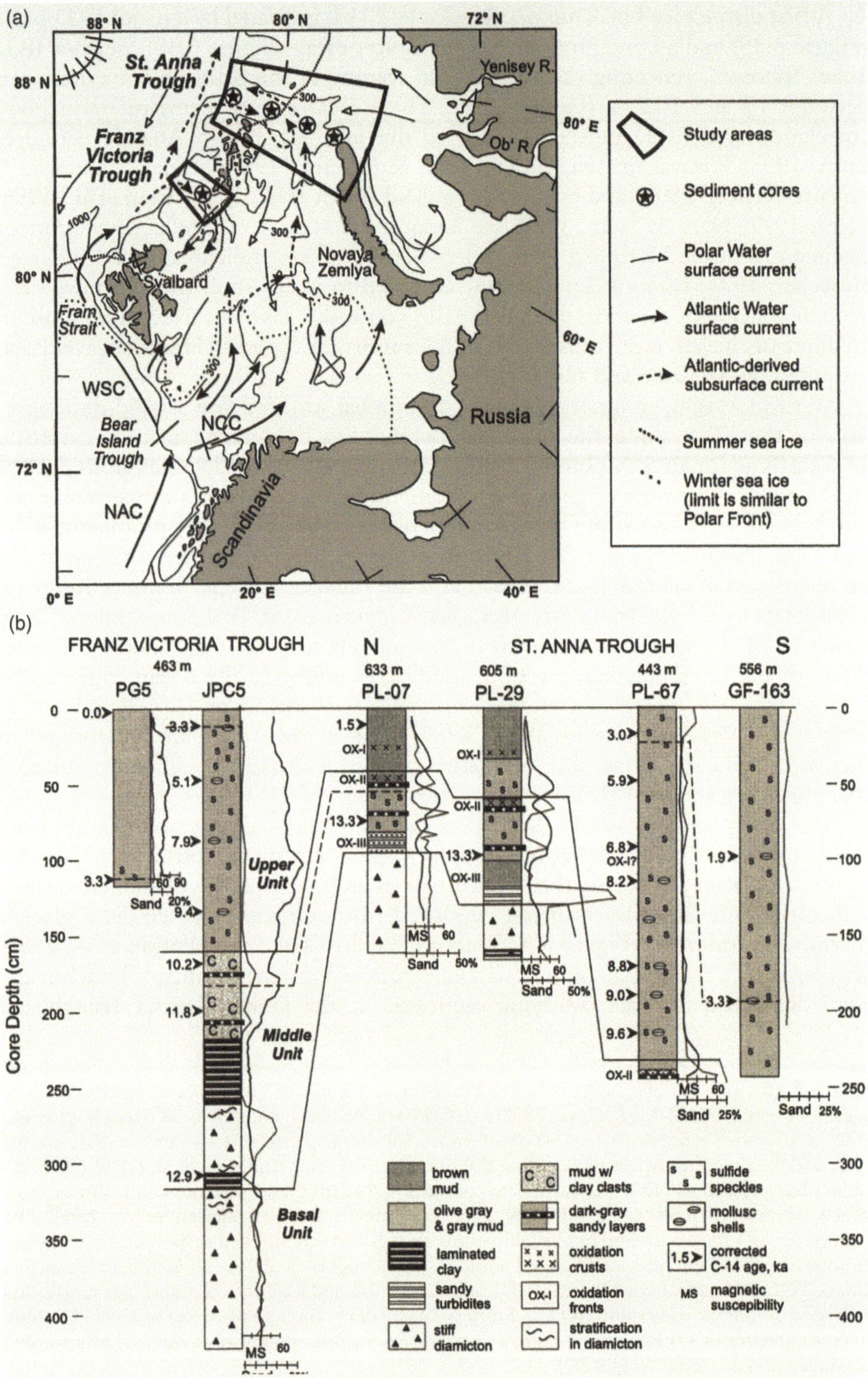
(a)
88° N
80° N
72° N
88° N
80° N
72° N
0° E
20° E
40° E
60° E
80° E
St. Anna Trough
Franz Victoria Trough
Yenisey R.
Ob' R.
Novaya Zemlya
Svalbard
Fram Strait
WSC
Bear Island Trough
NCC
NAC
Scandinavia
Russia
300
1000
Study areas
Sediment cores
Polar Water surface current
Atlantic Water surface current
Atlantic-derived subsurface current
Summer sea ice
Winter sea ice (limit is similar to Polar Front)
(b)
FRANZ VICTORIA TROUGH
N
ST. ANNA TROUGH
S
463 m
633 m
605 m
443 m
556 m
PG5
JPC5
PL-07
PL-29
PL-67
GF-163
Core Depth (cm)
0
50
100
150
200
250
300
350
400
Upper Unit
Middle Unit
Basal Unit
OX-I
OX-II
OX-III
OX-I?
MS
Sand
brown mud
olive gray & gray mud
laminated clay
sandy turbidites
stiff diamicton
mud w/ clay clasts
dark-gray sandy layers
oxidation crusts
oxidation fronts
stratification in diamicton
sulfide speckles
mollusc shells
corrected C-14 age, ka
magnetic suscepibility

northern St. Anna Trough, representing the deglaciation phase and final retreat of the LGM ice sheet from the troughs. The true age, however, might be considerably older than 13 ka in the northern St. Anna Trough (15 ka?), because 30–50 cm of sediment separates the oldest AMS^{14}C-dated samples (13.3 ka) and the diamicton (Figure 6.37). Based on the AMS^{14}C-datings, the age for the final ice-sheet retreat in the Franz Victoria Trough is probably younger than in the St. Anna Trough (Lubinski et al., 2001).

Moderate IRD input found in the Holocene record of Core PS2138-1 may be explained by surging of tidewater glaciers from Nordaustlandet or Franz Josef Land and/or by short-term glacial readvances to the outer coastline (Pfirman & Solheim, 1989; Knies et al., 2000). A distinct IRD peak near 11 ka (Figure 6.31) may indicate an extended ice-sheet along the northern Barents Sea continental margin during the Younger Dryas (YD), an interpretation which remains speculative due to the low stratigraphic resolution of the Holocene record (Knies et al., 2000, 2001).

The growth and decay of the SBIS resulting in changes of the sediment sources, is also reflected in the variations of Sr and Nd isotopic composition in Core PS1533 obtained from the Yermak Plateau (Figure 6.38). As stated by Tütken et al. (2002), these variations can be interpreted as mixtures of Eurasian shelf sediment ($^{87}Sr/^{86}Sr$ ratios of 0.713 to 0.719 and ε_{Nd} values of -8.3 to -13.4) and Svalbard bedrock material ($^{87}Sr/^{86}Sr$ ratios of 0.742 to 1.369 and ε_{Nd} values of -18.6 to -24.6) (see Chapter 4.3.4, Figure 4.20). During glacier advance from Svalbard and intensified glacial bedrock erosion, ε_{Nd} values decrease gradually to a minimum value of -14.9 due to increased input of Proterozoic/Palaeozoic crystalline Svalbard bedrock material. During maximum extensions of the SBIS up to the shelf edge and ice-sheet breakdown of the northern SBIS margin (and/or the Kara Sea Ice Sheet), on the other hand, increasing amounts of Eurasian shelf sediments were supplied by iceberg rafting and shelf-sediment reworking, especially during glacial Terminations I and II (Figure 6.38; Tütken et al., 2002). At that time, ε_{Nd} values in glacial sediments reach maximum values that are comparable to the average value of modern Eurasian shelf and sea ice sediments ($\varepsilon_{Nd} = -10.3$). During interglacials when sediment transport by sea ice probably became more important (e.g., Clark, 1990; Berner & Wefer, 1990; Nørgaard-Pedersen et al., 1998; Darby et al., 2006), Sr and Nd isotopic compositions in Core PS1533 suggest a mixed sediment with input from the Eurasian shelves as well as Svalbard. Based on Sr and Nd mixing calculations Tütken et al. (2002) estimated that during MIS 1, 3, and 5 ~75–55% of

Figure 6.37 Oceanic circulation patterns in the Barents and Kara seas and lithostratigraphy of sediment cores (from Lubinski et al. 2001). (A) Map of the Barents and Kara seas showing simplified bathymetry (300 and 1,000 m isobaths), sediment core sites, study areas, and general circulation pattern (after Loeng, 1991; Loeng et al., 1993; Pfirman, Bauch, & Gammelsrød, 1994; Schauer et al., 1997b); mean sea-ice limits (>12% cover) are from Barry (1989). Three major Atlantic water current systems are labelled: NAC, Norwegian Atlantic Current; NCC, North Cape Current; WSC, West Spitsbergen Current. The position of the winter sea ice limit is similar to that of the Polar Front, which separates cold Polar Water on the surface from warm Atlantic water. (B) Basic lithostratigraphy and correlation of sediment cores from Franz Victoria and St. Anna troughs (for locations of cores see (A)). Based on sediment colour, structure, grain-size, bulk density, and MS, three main lithological units are distinguished. Solid curves, volume MS; dotted curves, sand content.

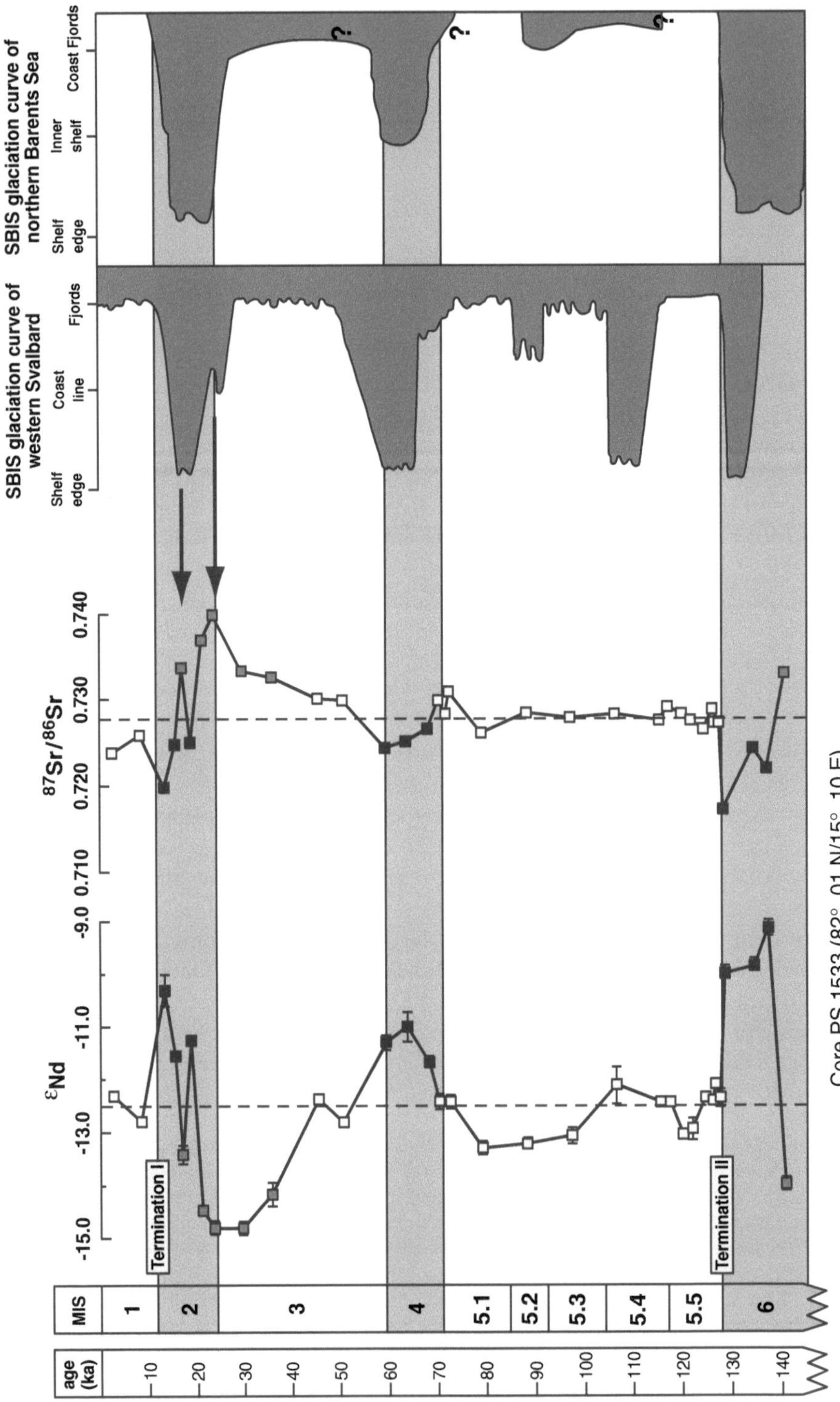
SBIS glaciation curve of northern Barents Sea
Shelf edge
Inner shelf
Coast
Fjords
SBIS glaciation curve of western Svalbard
Shelf edge
Coast line
Fjords
$^{87}Sr/^{86}Sr$
0.710
0.720
0.730
0.740
εNd
-15.0
-13.0
-11.0
-9.0
Termination I
Termination II
MIS
1
2
3
4
5.1
5.2
5.3
5.4
5.5
6
age (ka)
10
20
30
40
50
60
70
80
90
100
110
120
130
140
Core PS 1533 (82° .01 N/15° .10 E)

the sediment is delivered by sea ice via the TPD, whereas the 25–45% originates from local Svalbard sources.

6.2.2.2. Late Saalian to late Weichselian glaciations in the northern Kara to East Siberian Sea area

In comparison to the records from the Svalbard-Barents continental margin, IRD contents in records from the eastern Eurasian continental margin influenced by the fluctuations of ice sheets located in the northern Kara Sea, Severnaya Semlya and East Siberia, are much lower during the last ~180 ka (Figures 6.39 and 6.40; Müller, 1999; Knies et al., 2000, 2001). This may reflect either the presence of relatively stable ice sheets with small fluctuations of the ice margin or stable glacio-marine conditions when ice sheets had a small extent or were even absent (Knies et al., 2000, 2001). At Core PS2741-1, finely laminated sediments and high contents of smectite and kaolinite (Figure 6.39; Müller, 1999; Knies et al., 2000) as well as increased bulk accumulation rates (Figure 6.31) characterize the MIS 6 interval. These sediments may indicate glacially derived sediment input from the Kara Sea, that is, the predominance of smectite and kaolinite suggest bottom transport of suspension-loaded plumes delivered by submarine meltwater discharge from the advanced Saalian Kara Sea Ice Sheet, probably from Franz Josef Land, the St. Anna Trough, and/or the Voronin Trough areas (Vogt, 1997; Müller, 1999; Knies et al., 2000). That means, sedimentation on the lower slope of Severnaya Semlya has probably not been directly affected by the Severnaya Semlya Ice Sheet advance onto the shelf. A distinct IRD spike and peak values of coarse fraction input were recorded near the MIS 6/5 transition, interpreted by Knies et al. (2001) as deglaciation event monitoring the Saalian deglaciation phase of the Severnaya Semlya Ice Sheet, when large amounts of icebergs were released. This is also supported by a distinct increase in illite content (Figure 6.39).

Further to the east, in cores PS2757-8 and PS2761-10, a pronounced IRD spike and increased amount of coarse fraction were also found at the MIS 6/5 transition (Figure 6.40; Müller, 1999). This interval is characterized by high illite contents of

Figure 6.38 Down-core variation in $^{87}Sr/^{86}Sr$ ratio and ε_{Nd} values of the <63 μm sediment fraction in Core PS1533 over the last 140 ka (from Tütken et al., 2002). For location of cores see Figure 6.36. Oxygen isotope stages are indicated. Grey shaded horizontal bands correspond to the glacial isotope stages. Black square, samples from glacial maxima and terminations with maximum SBIS extension onto the shelf and highest shelf sediment input; grey square, samples from SBIS build-up with highest input of Svalbard material; open square, samples from interglacial periods. Vertical dotted lines represent average isotopic values of the core sediments ($\varepsilon_{Nd} = -12.5$; $^{87}Sr/^{86}Sr = 0.728$). SBIS glaciation curves of the northern Barents Sea shelf after Knies et al. (2000) and for western Svalbard after Mangerud et al. (1998). Down-core variation of Sr and Nd isotopic data of Core PS1533 sediments correlates well with the SBIS glaciation curve of the northern Barents Sea, while glacier extensions of the western Svalbard glaciers in MIS 5.4 and 5.2 found by Mangerud et al. (1998) are not reflected in PS1533 sediments. This may support less extended ice sheets on Svalbard during MIS 5.4 and 5.2 (see discussion earlier and Winkelmann et al., 2008b). The two build-up phases of the SBIS in MIS 2 between 27 and 23 ka and 19 and 16 ka, however, seem to be recorded in PS1533 sediments (black arrows). For more details and background see Tütken et al. (2002).

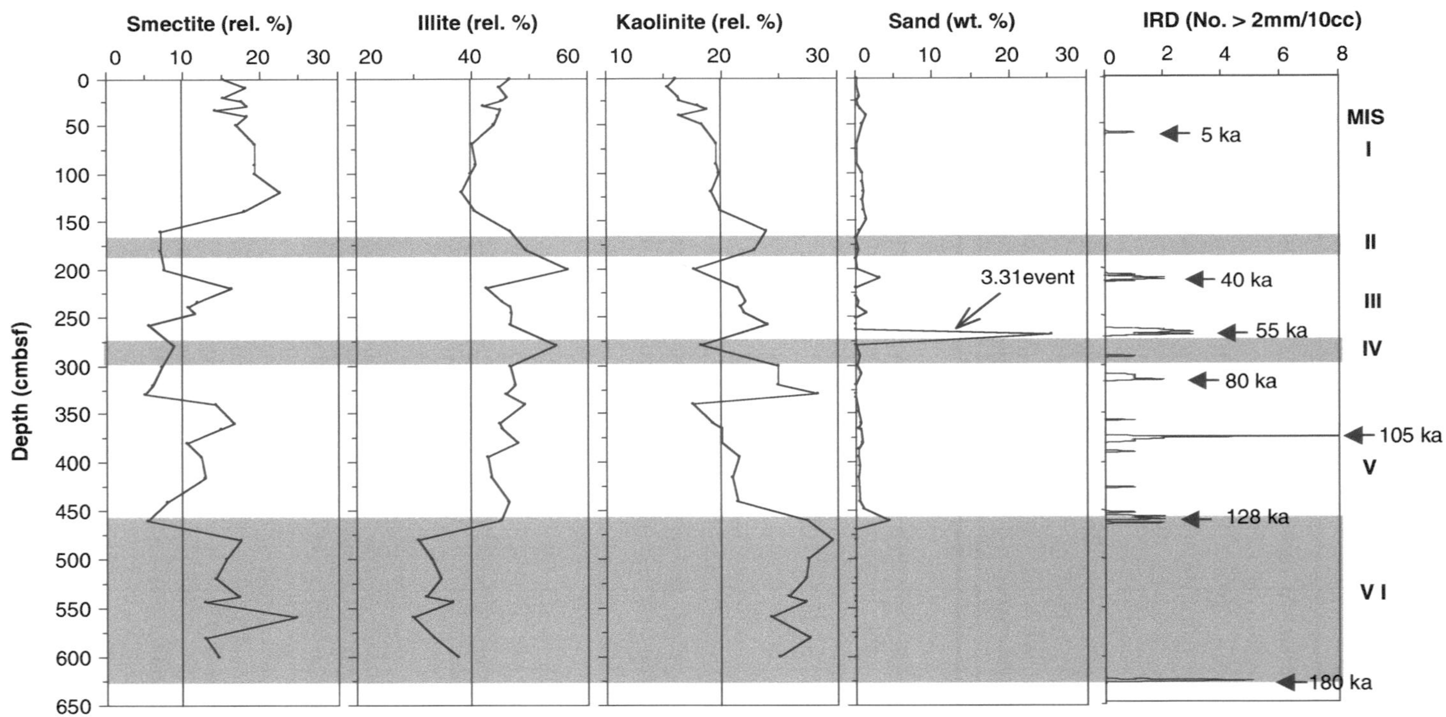

Figure 6.39 Relative contents of smectite, illite, and kaolinite, percentage of sand fraction, and numbers of IRD grains >2 mm per cc versus depth of Core PS2741-1. Marine oxygen isotope stages are marked as roman numbers; TI and TII indicate Terminations I and II. Estimated ages of major IRD peaks are shown (from Müller, 1999; Knies et al., 2000).

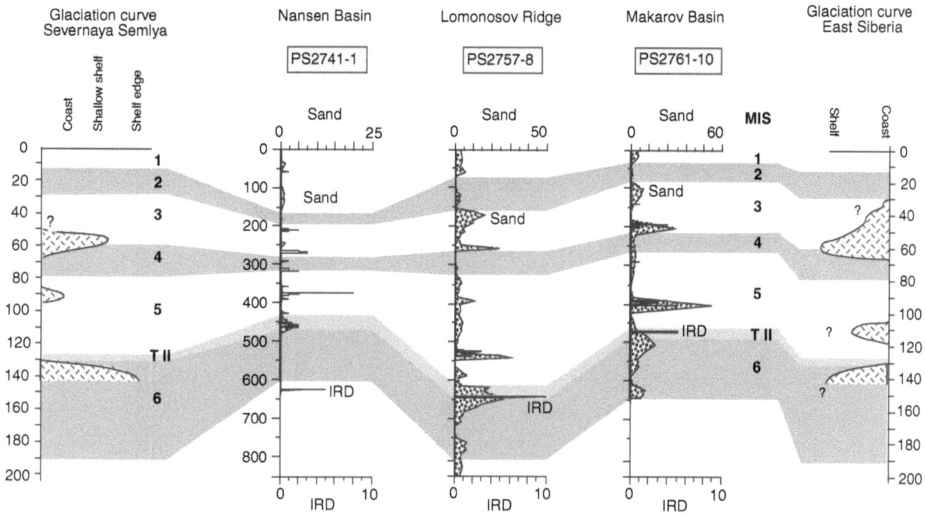

Figure 6.40 Distribution of IRD and sand fractions in cores PS2741-1, PS2757-8, and PS2761-10, and postulated glaciations on Servernaya Semlya (based on Knies et al., 2000) and in East Siberia (from Müller, 1999).

up to 60% and increased amounts of feldspars, epidote, amphibole, and garnet, pointing to the western East Siberian Sea/New Siberian Islands as probable source area (Behrends, 1999; Müller, 1999). These characteristics suggest an ice sheet in the area around the New Siberian Island, collapsed and disintegrated during Termination II (Figure 6.41; Arkhipov, Ehlers, Johnson, & Wright, 1986b, 1995; Müller, 1999).

During MIS 5, two distinct IRD peaks dated — based on linear interpolation — at the MIS transitions MIS 5.4/5.3 near 105 ka and MIS 5.2/5.1 near 80 ka might reflect the deglaciation patterns from two readvances of the Kara Sea/Severnaya Semlya Ice Sheet to the outer shelf during early Weichselian time (Figure 6.39; Knies et al., 2000, 2001). More than one ice advance or local glaciations during the Early Weichselian in the Arkhangelsk region were also described by Houmark-Nielsen et al. (2001), which may support Knies et al.'s (2001) interpretation of the PS2741-1 record from the northern Severnaya Semlya margin. The pronounced IRD peak in cores PS2757-8 and PS2761-10 at ~540 and 400 cmbsf, respectively, may represent the decay of an early Weichselian ice sheet developed during MIS 5.4 (Figure 6.40). Elevated smectite values suggest the glaciated Putoran Massif being a possible source of this IRD material (Müller, 1989).

The Middle Weichselian ice-sheet advance onto the outer shelf off Severnaya Semlya during MIS 4 is indicated by a distinct IRD peak overlain by laminated sediments during MIS 4/3 transition of core PS2741-1 (Figure 6.39). This IRD event correlating with the widespread 3.31 event (see earlier) may reflect the disintegration of the Severnaya Semlya Ice Sheet during the Middle Weichselian MIS 4. Contemporaneously, a similar IRD peak was recorded in cores PS2757-8 and PS2761-10 (Figure 6.40). Furthermore, in both cores the sediments from this

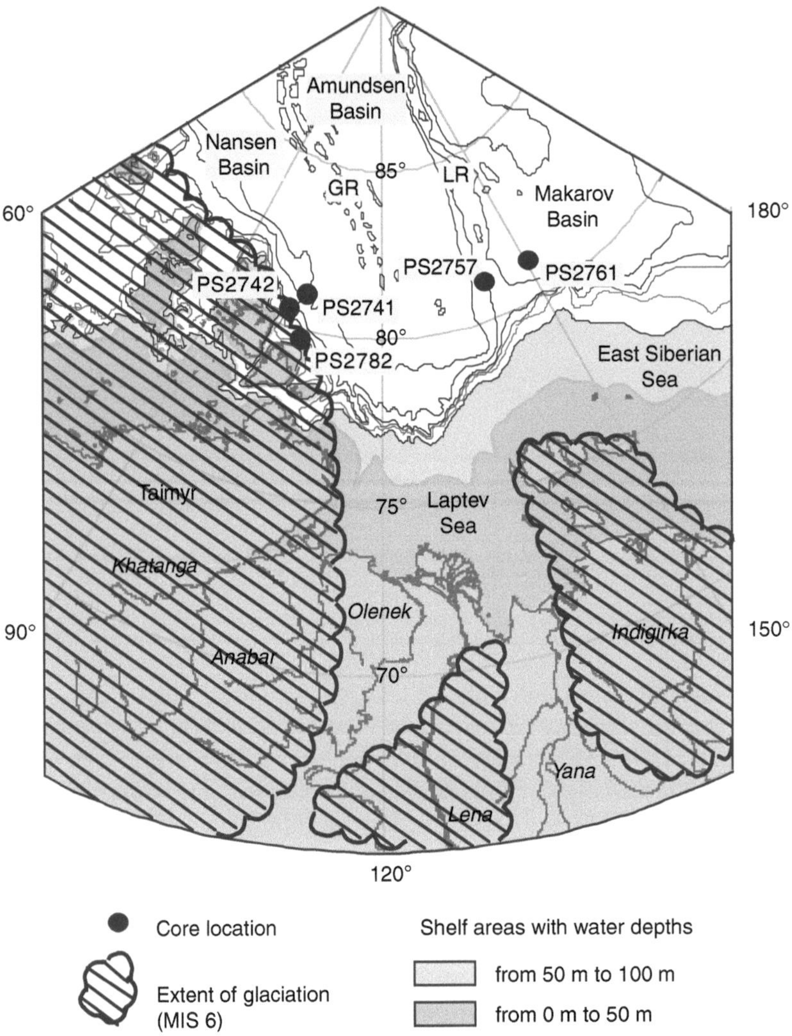

Figure 6.41 Postulated distribution of ice sheets in East Siberia during MIS 6 (from Müller, 1999, according to Arkhipov et al., 1986b, 1995).

time interval are characterized by low smectite values and high quartz and illite values, suggesting a sediment input from the East Siberian Sea and the New Siberian Islands area (Müller, 1999). This may support an extended Middle Weichselian ice sheet in East Siberia, probably similar to the MIS 6 (Saalian) glaciation (Figure 6.41; Arkhipov et al., 1986b, 1995; Müller, 1999).

Further support for an extended Middle Weichselian ice sheet in the area east of Franz Josef Land derives from a detailed PARASOUND profiling survey at the continental slope off Severnaya Semlya and Taymyr Peninsula (Kleiber, 1999; Kleiber et al., 2001; see Figure 3.22). These authors suggest that the ice-proximal facies recorded in these PARASOUND profiles is associated with the Middle Weichselian (MIS 4) maximum extent of an ice sheet on the Kara Sea, Severnaya Zemlya and Taymyr Peninsula (Figure 6.24C; Svendsen et al., 1999, 2004). This interpretation is consistent with evidence from terrestrial and permafrost sequences (Siegert et al., 1999), sediment cores, and high-resolution seismic profiles of lake sediments on the Taymyr Peninsula (Niessen, Ebel, Kopsch, & Federov, 1999), as well as a diamicton in Core PS2782-1 east of Severnaya Zemlya Archipelago with a radiocarbon dating of >44 ka (Figure 6.42), indicating that the last glaciation of the area occurred during MIS 4 (Knies, 1999; Kleiber et al., 2001; Knies et al., 2001). The diamicton is also associated with end moraine structures down to a water depth of 390 m (Figure 6.42). A Saalian age (MIS 6) for the formation of the ice-proximal facies (unit IV, see Chapter 3.3.1, Figure 3.22) is excluded because, after the deposition of unit IV, there is no evidence for typical MIS 5 interglacial deposits indicative for warmer climate and higher sea level than at present in the entire sedimentary succession of the western Laptev Sea (Kleiber et al., 2001).

At ~40 ka, a last significant IRD pulse was recorded in Core PS2741-1 which coincides with elevated smectite and kaolinite contents (Figure 6.39), suggesting an advanced Kara Sea Ice Sheet with sediment input from the Franz Josef Land area and further suspension transport towards the core location.

Except for a small IRD peak at ~5 ka no IRD was found in the record of Core PS2741-1 during the last ~30 ka (Figure 6.39), that is, during the LGM and the following deglaciation. Also the very low IRD values in the shallow Core PS2782-1 (Figure 6.42) do not suggest an ice advance to the outer coastline. The absence of any IRD as well as the very low bulk accumulation rates at Core PS2741-1 (Figure 6.31) during this time interval suggest a perennial sea-ice cover and the absence of a larger northern Kara Sea Ice Sheet (Knies et al., 2000, 2001). Thus, these data rather support the LGM reconstruction by Svendsen et al. (2004), but reject the huge Pan-Arctic Glaciation proposed by Grosswald and Hughes (2002).

At Core PS2742-5 from the upper slope off Severnaya Semlya (see Figure 6.41 for location) a very dark greyish brown to dark brown diamicton with large-sized dropstones of up to 5 cm in diameter was found in 343–360 cmbsf (Weiel, 1997; Müller, 1999; Stein & Fahl, 2000; Stein et al., 2001). This ~20 cm thick diamicton could be tentatively dated to the late Termination Ia, based on a single age control point at 321 cmbsf giving an age of 12.5 ka. It may document the retreat of smaller ice caps on Severnaya Semlya, presumably from the coast line to the inner fjords (Knies et al., 2000; Stein et al., 2001). As suggested by Knies et al. (2000, 2001), however, an alternate explanation for the origin of the thin diamicton in Core PS2742-5 could be iceberg discharge from the retreating marine-based Barents/Kara Sea Ice Sheet in the St. Anna Trough and subsequent transport along the Eurasian Continental Margin with the increasing Atlantic water inflow from the west after 13 ka (Figure 6.37; *cf.* Lubinski et al., 1996, 2001; Polyak et al., 1997; Hald et al., 1999).

The final stage of mountain deglaciation of the Putoran Massif during the earliest Holocene (Termination I) and the subsequent release of huge amounts of

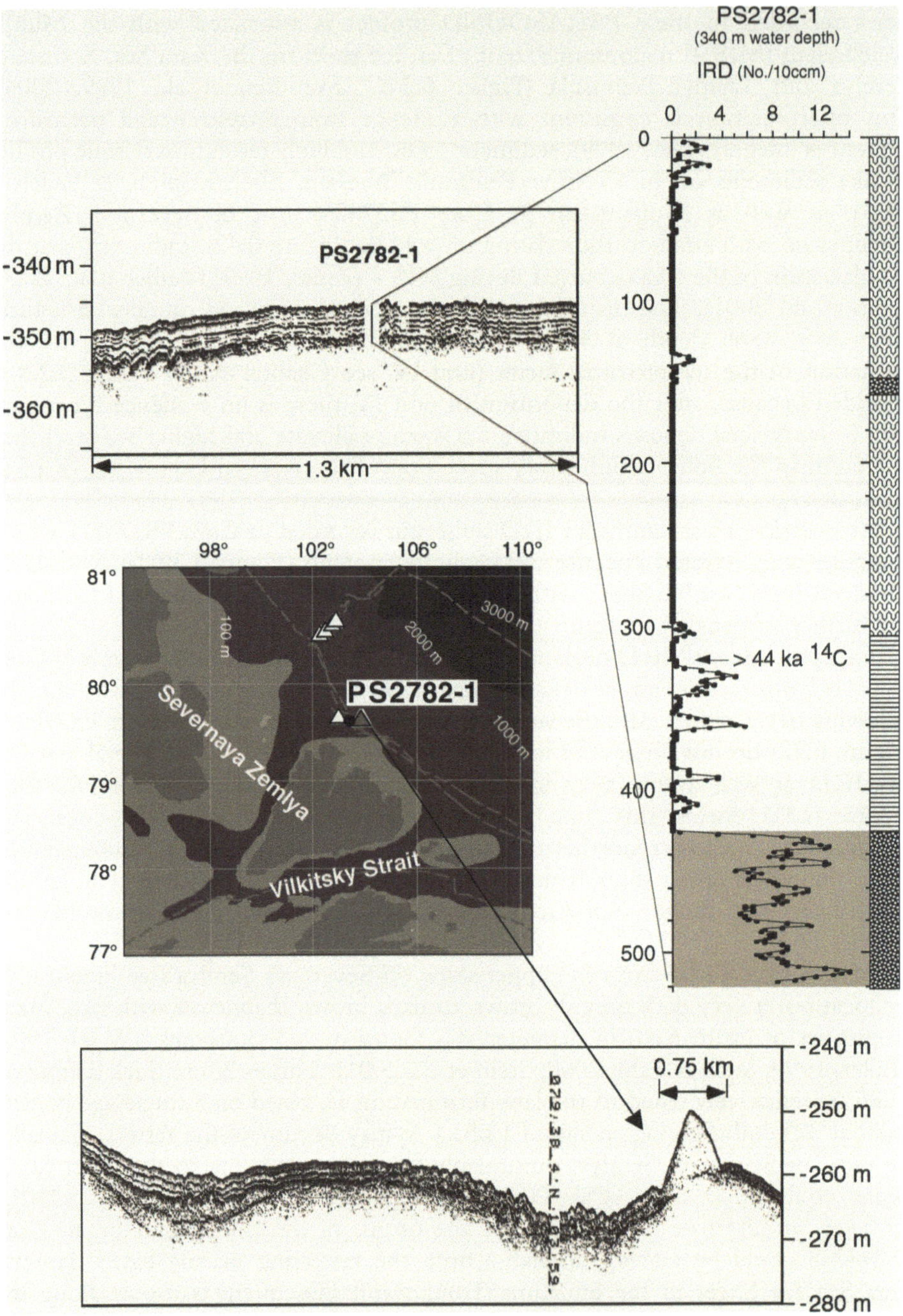

Figure 6.42 Lithology and numbers of IRD grains >2 mm per 10 cc in Core PS2782-1 obtained off Severnaya Zemlya (340 m of water depth) (Knies et al., 2001). An AMS^{14}C dating of >44 ka indicates that the diamicton is of pre-late Weichselian age. In addition, PARASOUND profiles across the core location and across an endmoraine structure found very close to the core location are presented (Weiel, 1997; Niessen, unpublished data, 1997). In the map, core location (solid circle) as well as end moraines found by PARASOUND profiling (triangles) are shown.

fine-grained glacigenic debris enriched in ferrimagnetic minerals into the Yenisei River (draining into the Kara Sea) and the Khatanga River (draining into the Laptev Sea) are reflected in maximum MS values observed in several sediment cores from the eastern Kara Sea and western Laptev Sea (e.g., Kleiber & Niessen, 2000; Stein & Fahl, 2000; Stein et al., 2001, 2003a).

6.2.2.3. Processes controlling glaciations along the Eurasian continental margin

As described earlier, a clear asymmetry in the evolution of the Eurasian ice sheets has been identified by terrestrial and marine data (e.g., Velichko et al., 1997a, 1997b; Svendsen et al., 1999, 2004; Knies et al., 2000, 2001). How could these glaciations be developed along the Eurasian continental margin, and how can the differences in development between western and eastern Eurasia be explained?

In several studies, the connection between the build-up and decay of the Barents/Kara Sea ice sheets, the outflow of meltwater to the central Arctic Ocean, and the inflow of Atlantic water during the last glacial/interglacial cycle has been documented (e.g., Hebbeln et al., 1994, 1998; Dokken & Hald, 1996; Hebbeln & Wefer, 1997; Nørgaard-Pedersen et al., 1998; Knies et al., 1999, 2000). Hebbeln et al. (1994) reported two short periods (27–22.5 ka and 19.5–14.5 ka) of relatively warm Atlantic water advection during the Late Weichselian in the eastern Norwegian-Greenland Sea, which — as a regional moisture source — had a major influence on the SBIS build-up and also caused an increased production of coccoliths and subpolar planktonic and benthic foraminifera (Hebbeln & Wefer, 1997). Recurring ice-free conditions due to Atlantic water advection to the Fram Strait and Norwegian-Greenland Sea are also reported during MIS 6 for at least three time periods (145, 165, and 180 ka) (e.g., Lloyd et al., 1996; Hebbeln & Wefer, 1997). The importance of Atlantic water inflow along the northern Barents Sea continental margin has been discussed by, for example, Polyak et al. (1997), Hald et al. (1999), Knies et al. (1999, 2000, 2001), Lubinski et al. (2001), Matthiessen et al. (2001), and Wollenburg et al. (2001).

The strongest inflow of Atlantic waters is primarily related to interglacial periods of high sea level and absence or low extent of ice sheets on the Barents Sea/Spitsbergen shelves, when Atlantic water could intrude both around Spitsbergen and across the Barents/Kara Sea. This is, for example, clearly demonstrated by the retreat of the Barents Sea Ice sheet and related Atlantic water inflow during the last deglaciation (Polyak et al., 1997). Pronounced concentration maxima of cosmopolitan and temperate–subpolar dinoflagellate cysts along the Eurasian continental margin during the Eemian (MIS 5.5) and the Holocene support maximum inflow of Atlantic waters and seasonally increased production of cysts during peak interglacials (Figure 6.43; Matthiessen et al., 2001; Matthiessen & Knies, 2001). In addition, smaller concentration maxima of dinocysts were determined in MIS 6.3, MIS 5.1, and at the stage MIS 4/3 boundary, indicating in comparison to peak interglacials reduced but still significant Atlantic water inflow at the northern Barents Sea continental margin (Figure 6.43). Furthermore, the influence of

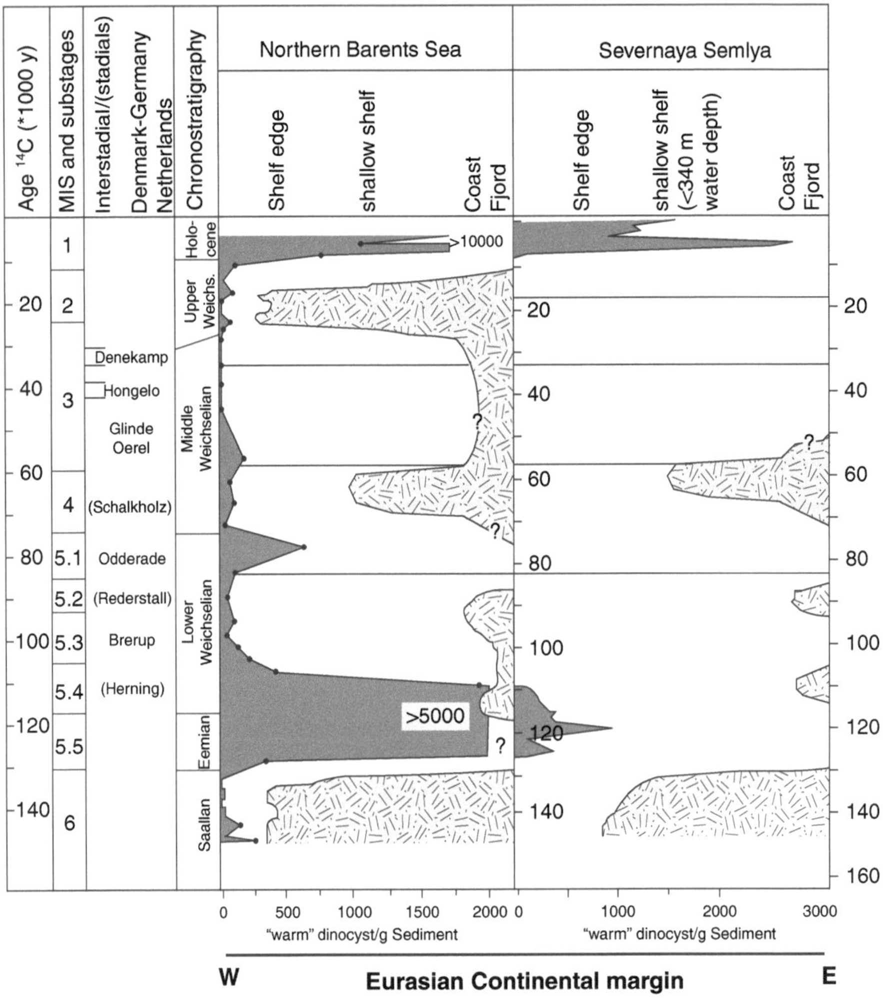

Figure 6.43 Reconstructed ice-sheet advances along the northern Eurasian margin over the last two glacial/interglacial cycles. In addition, abundances of the warm adapted dinoflagellate cysts in Core PS2138-1 (northern Barents Sea) and PS2741-1 (northern Kara Sea) indicating influence of warm Atlantic-derived water masses are displayed (see Matthiessen et al., 2001 for details) (from Knies et al., 2001 and further references therein).

Atlantic (sub-) surface waters generally decreased from the western (high-precipitation) to the eastern (low-precipitation) Eurasian continental margin, in particular during the glacials (Matthiessen et al., 2001; Knies et al., 2001). Atlantic water flowing towards the east in intermediate water depths as contour current, however, can be traced along the Eurasian continental far towards the east, as shown in the smectite records from the Laptev Sea continental slope. High smectite contents originated from the Kara Sea are indicative for an eastwards transport of suspension within the Atlantic layer (Figure 6.44; Müller, 1999). In this far-eastern

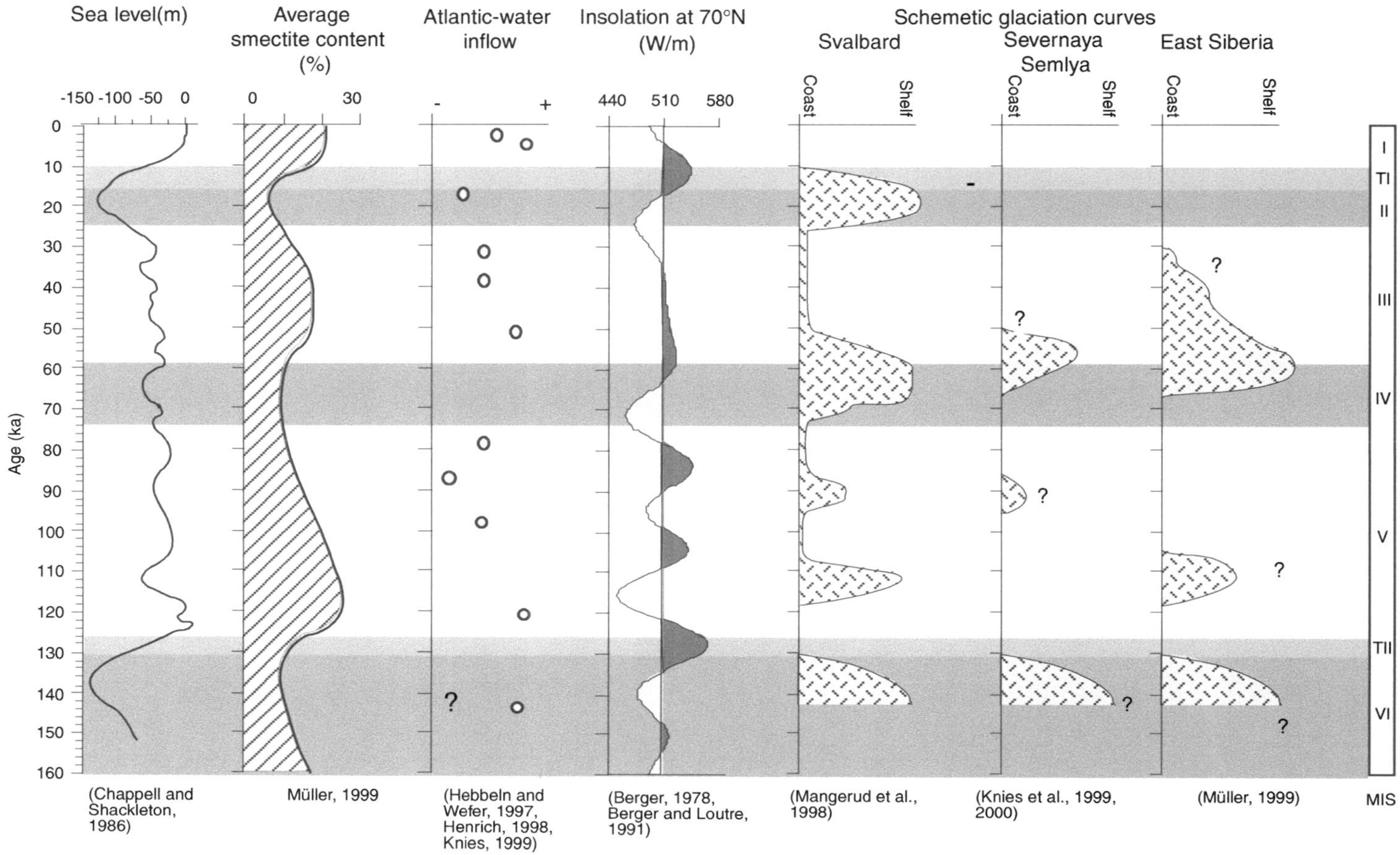

Figure 6.44 Sea level, average smectite record of sediment cores from the Laptev Sea continental slope and adjacent deep sea, Atlantic-water inflow, insolation at 70° N, and extension of glaciation on Svalbard, Servernaya Zemlya, and in East Siberia (from Müller, 1999, based on references given in the figure).

area, however, the water mass is too deep to cause open-water conditions that may act as moisture source.

Whereas biogenic carbonate records are used to trace increased Atlantic water inflow in the Fram Strait area (e.g., Hebbeln et al., 1998; Bauch et al., 1999), the use of such carbonate records are limited along the northern Barents Sea continental margin due to strong carbonate dissolution during interglacials. Dissolution of biogenic calcite in sediments of the eastern and northern Barents Sea has increased during deglacial and interglacial periods due to a combination of Atlantic water influx, the annual formation of sea ice, dense bottom water formation, and surface-water productivity blooms (Steinsund & Hald, 1994). During seasonal sea-ice formation in the Barents Sea brines, that is, highly saline, CO_2- and oxygen-enriched water masses, are ejected by sea ice and descend in troughs and depressions on the eastern and northern Barents Sea margin. In addition, oxidation of marine organic matter highly accumulated near the ice edge (Hebbeln & Wefer, 1991) is enhanced resulting in a further increase of CO_2 and, thus, reinforce dissolution of biogenic calcite (Knies et al., 1999). Wollenburg et al. (2001) explicitly stated that carbonate dissolution affected benthic foraminiferal faunas during the interglacial periods of enhanced productivity (see Chapter 4.5.1, Figure 4.31) and Atlantic water advection in cores PS2138-1 and PS2212-3. This supports the view of Knies et al. (1999) that the almost complete dissolution of biogenic carbonate is linked to the strongest advection of Atlantic waters in MIS 1 and 5.

Although the absence of biogenic carbonate cannot be used for statements regarding the intensity of Atlantic water inflow, the presence of biogenic calcite with values up to 15% in sediments along the northern Barents Sea margin during late MIS 6, mid and late MIS 3, and MIS 2 probably reflect at least some significant Atlantic water influx (Knies et al., 1999). The abundance of the benthic foraminifera *Cassidulina teretis*, a proxy for Atlantic water influx into the Arctic Ocean (see Chapter 4.5.1, Figure 4.30; Wollenburg & Mackensen, 1998; Lubinski et al., 2001), during late MIS 6, MIS 3, and MIS 2 (Wollenburg et al., 2001) also supports that Atlantic water reached the northern Barents Sea margin. The high abundances of (planktonic and benthic) foraminifers point to seasonally ice-free conditions during these glacial periods because the production of zooplankton is largely a function of irradiance and nutrient concentrations (i.e., primary production) and, thus, of open water conditions (Smith, 1995; Wollenburg et al., 2001; see also Chapter 4.5.1, Figure 4.31). Because Atlantic water inflow was significantly stronger during the interglacials MIS 5 and MIS 1 (see earlier), however, Knies et al. (1999) suggest that highest accumulation rates of biogenic calcite of up to $6\,g\,cm^{-2}\,kyr^{-1}$ during MIS 2 along the northern Barents Sea margin, coinciding with increased lithogenic fluxes, are probably not related just to Atlantic water inflow. Following Hebbeln and Wefer (1991) and Kohfeld et al. (1996) who argued that favourable conditions for biological productivity and high lithogenic flux may especially occur in polynyas caused by the upwelling of relatively warm water triggering increased productivity and the release of IRD by melting, Knies et al. (1999) proposed a widespread coastal polynya also for the northern Barents Sea continental margin at least to the northern Franz Victoria Trough (Figure 6.45). Such a coastal polynya, triggered by katabatic winds from the

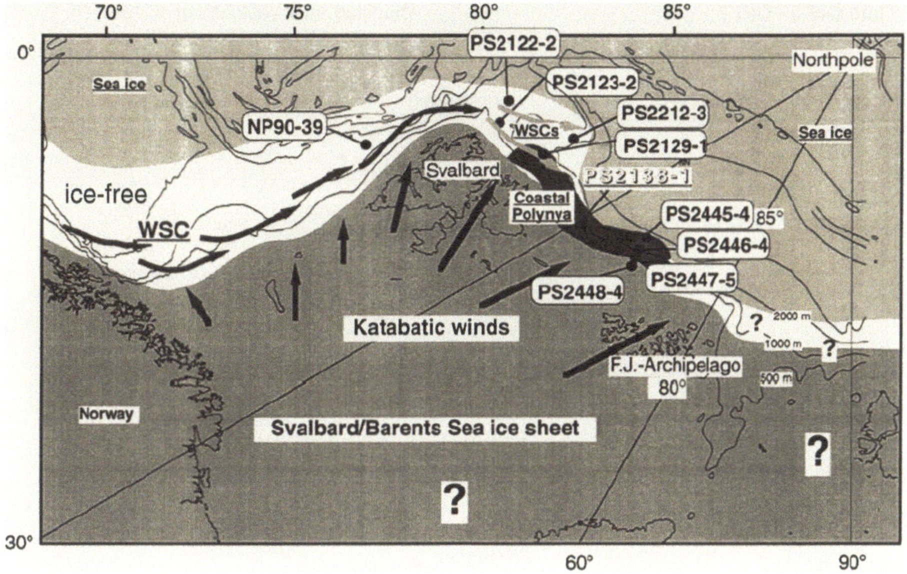

Figure 6.45 Palaeoenvironmental model of the SBIS during LGM (at 19 ka) (from Knies et al., 1999). Extension of ice sheets and Atlantic water advection are based on Vorren et al. (1988) and Hebbeln et al. (1994). All investigated cores (except PS2447 and PS2448) show enhanced abundances of planktonic and benthic foraminifers in the coarse fraction (>63 μm) during late MIS 3 and MIS 2 and can be correlated to PS2138-1. Based on these data, a coastal polynya along the northern Barents Sea continental margin triggered by katabatic winds has been proposed (WSC, West Spitsbergen Current; WSCs, submerging West Spitsbergen Current).

protruding SBIS and the inflow of subsurface Atlantic water masses, may have supported the seasonally ice-free conditions and provided the moisture, which, in combination with moisture bearing storms and low summer insolation, had a major influence on the final ice build-up during the Saalian and Weichselian glaciations along the western Eurasian margin (Figure 6.46).

In low-precipitation areas of eastern Eurasia, on the other hand, the local source of moisture for the atmosphere was probably restricted due to the reduced influence of Atlantic water and a more or less closely packed Sea-Ice Cover over the whole time span (Knies et al., 2000, 2001). Ice advance onto the shelf occurred during MIS 6 and MIS 4 and follows distinct interglacial periods. This confirms the model of the asymmetry of the cryosphere in Eurasia during the last glaciation with a maximally extended ice-sheet along the western margin and an ice-sheet of limited size along the eastern margin (Figure 6.46; Velichko et al., 1997a; Knies et al., 2000, 2001; Svendsen et al., 2004). That means, during the initial cooling following MIS 5, and probably MIS 7, the combined effect of sustained inflow of Atlantic water into the Arctic Ocean and eastward penetration of moisture-bearing cyclones supported major ice build-up during the Saalian and Middle Weichselian glaciations. For MIS 6 and MIS 4, a cyclonic atmospheric circulation over Eastern Siberia has been proposed which allowed transport of moisture towards the east (Figure 6.47; Arkhipov et al., 1986b; Naidina, 1995; Velichko et al., 1997a).

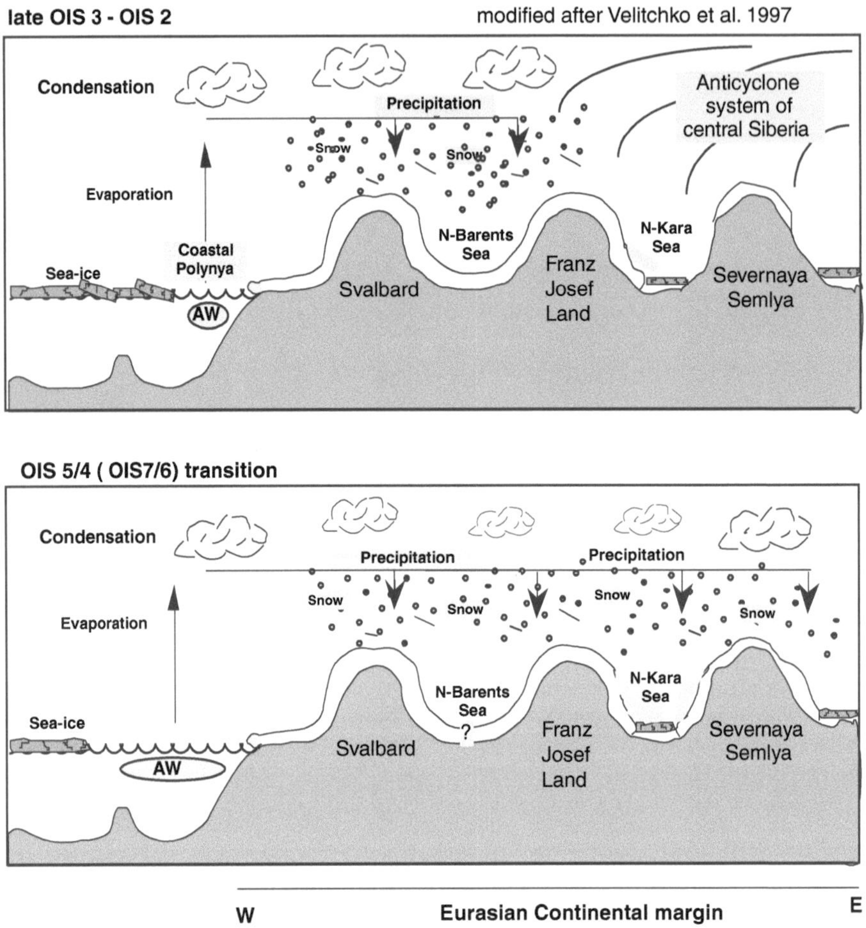

Figure 6.46 Proposed model that explains the environmental conditions along the Eurasian continental margin during MIS 5/4 (7/6?) and MIS 3/2 transitions, respectively (from Knies et al., 2001, modified after Velitchko et al., 1997a, 1997b). No data are available for MIS 7/6 so far. Based on evidences for an extended ice sheet during the Saalian, it is hypothesized that comparable environmental conditions than during MIS 5/4 prevailed. Significant ice-free conditions in the Nordic Seas during the end of MIS 5 (MIS 7?), indicated by enhanced abundances of warm adapted dinoflagellate cysts in PS2138-1 (see Figure 6.43), could have provided enough moisture to build up ice sheets in the Barents and Kara Sea during MIS 4 (MIS 6). In contrast, restricted and time limited areas of open water conditions along the western and northern Eurasian continental margin, indicated by generally low dinoflagellate cysts abundances in PS2138-1 and PS2741-1 (Figure 6.43) and growth of severe anticyclones over Siberia (see Figure 6.47) could have caused build-up of only local ice caps in the Kara Sea during the late Weichselian.

According to these authors, on the other hand, an anticyclonic regime was probably predominant during MIS 2, which blocked the transport of moisture towards the east, preventing the build-up of a major ice sheet in the southern Kara Sea/East Siberian area (Figure 6.47).

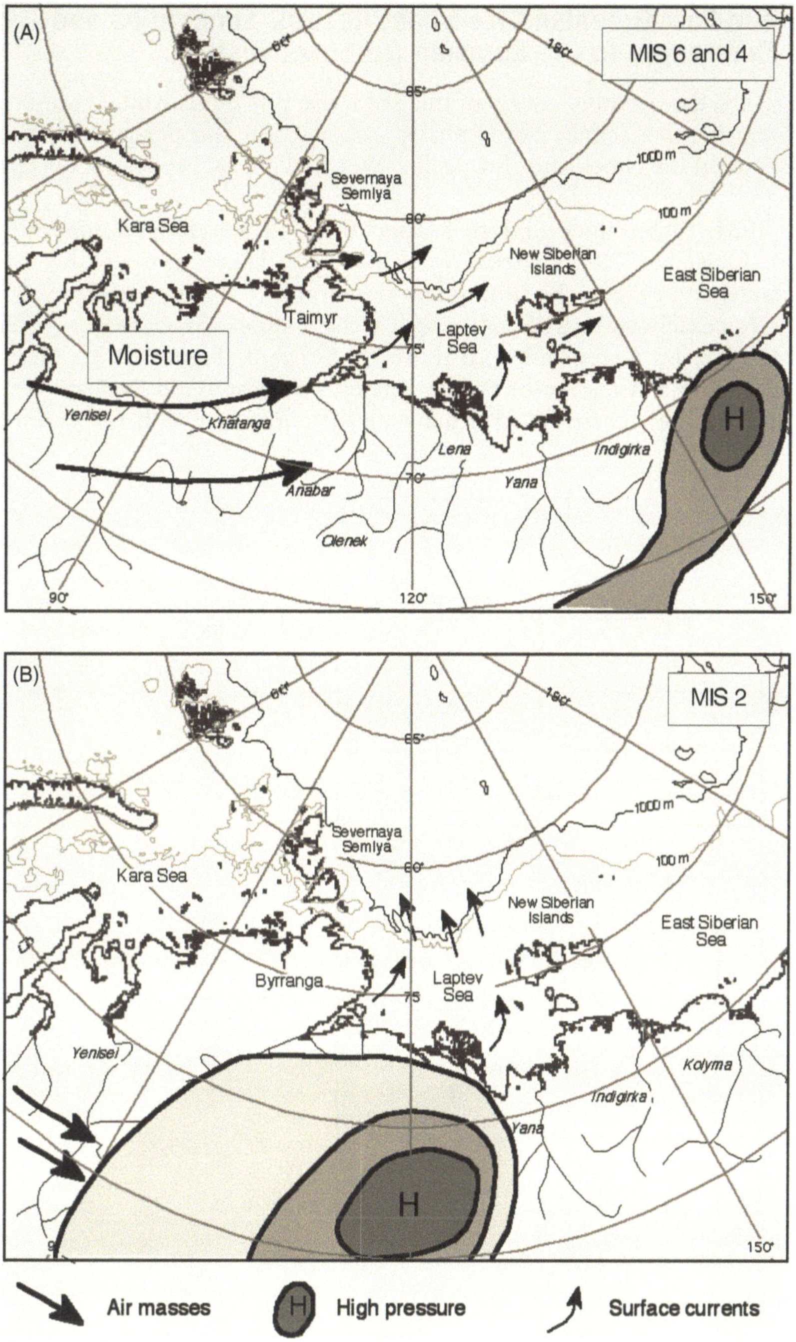

Figure 6.47 Proposed distribution of high-pressure field and related atmospheric circulation over Siberia during (A) MIS 6 and 4, and (B) MIS 2 (from Müller, 1999, according to Naidina, 1995; Velichko et al., 1997a).

6.2.3. The East Greenland Ice-Sheet Record: Similarities and Differences to the Eurasian Ice-Sheet History

In contrast to the dynamic nature of the SBIS, the East Greenland ice sheet margin appears much more stable. Three phases of glaciation and deglaciation have also been identified on Greenland during the Weichselian, but — based on terrestrial data — they involve much less dramatic variations in ice sheet configuration (e.g., Funder, 1984, 1989; Funder, Hjort, & Landvik, 1994, Funder et al., 1998; Funder, Jennings, Q Kelly, 2004; Elverhøi et al., 1998a). There seems to be a general correspondence between the timing and duration of the first post-Eemian glacier advance (Aucellaelv stade/MIS 5.4) as well as the ice advance near 80 ka (Jyllandselv stade/MIS 5.2) on Greenland with those on Svalbard (Figure 6.48). After 65 ka, however, timing and regime of the glaciers on Greenland and Svalbard were very different. The latest phase (Flakkerhuk stade) is correlated with the entire period

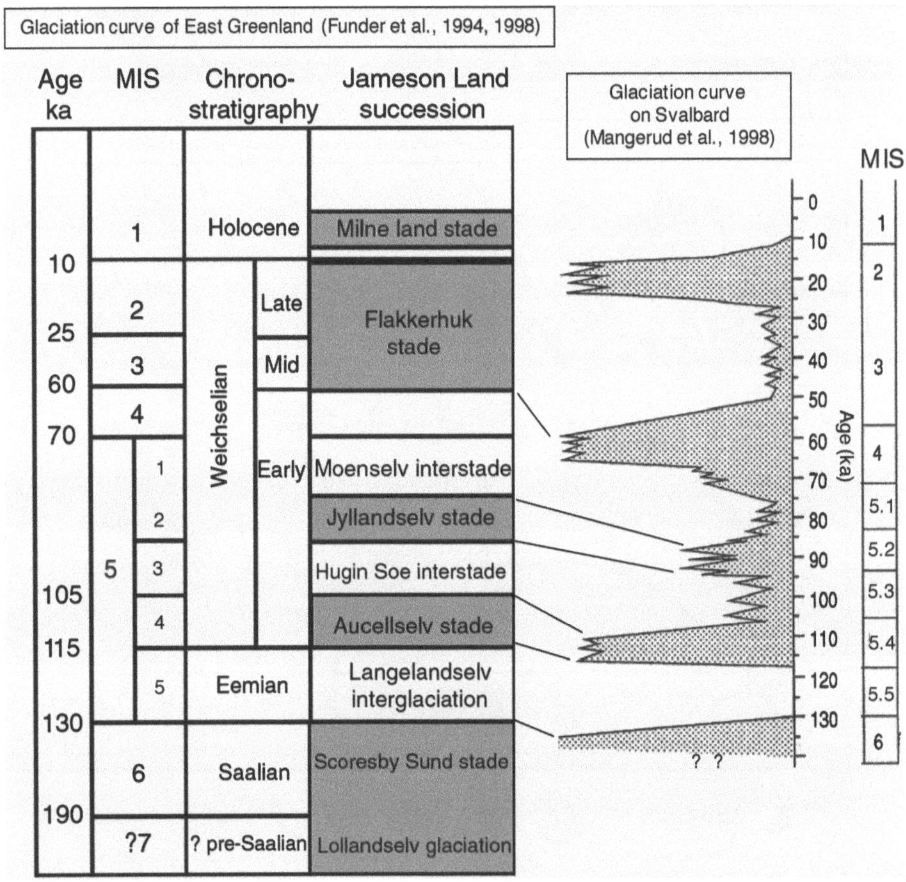

Figure 6.48 Glaciation history of East Greenland (according to Funder et al., 1998, 2004; taken from Wilkin & Mienert, 2006) and correlation with glaciations on Svalbard (according to Mangerud et al., 1998), based on terrestrial data.

from MIS 4 to MIS 2 (Figure 6.48). Even though it is the youngest phase of glaciation, extent and history are still discussed controversially (see Funder et al., 1998 and references therein; see also later discussions). In contrast to the 10 kyr long Aucellaelv glaciation which apparently left clear traces and deep impressions on landscapes, the more extensive, ~50 kyr long Flakkerhuk glaciation has left only thin and discontinuous till patches in the area (Elverhøi et al., 1998a). Thus, the terrestrial records of glaciation are often incomplete. According to Elverhøi et al. (1998a), the difference of these glaciations can be explained only by very different glacier regimes in the two periods. The Flakkerhuk stade glacier must be compared with present day glaciers in northernmost Greenland and Antarctica, with less melting and less calving. This more stable scenario for the Flakkerhuk glaciation seems to be valid at least for the East Greenland fjord zone north of Scoresby Sund between 68°N and 78°N (see Figures 3.7 and 3.10 for location). Mainly based on onshore data, glaciation around Scoresby Sund and northwards was largely restricted to outlet glaciers which filled the fjord troughs, and the inner shelf (Funder et al., 1998; Funder et al., 2004). To the south of Scoresby Sund, on the other hand, the inland ice of East Greenland expanded onto the continental shelf and, according to seismic stratigraphic evidence from the Kangerlussuaq area (65°N; see Figure 3.7 for location) in particular, spread across the 300 km-wide shelf to reach the shelf break (Andrews, Jennings, Cooper, Williams, & Mienert, 1996). This difference between the southern and northern glacial regimes can be explained by steep climatic gradient which exists over northern Denmark Strait, with mild sub-Arctic climate in the south and cold arctic, arid climate in the north (Elverhøi et al., 1998a).

Whereas stratigraphic onshore sections within glacial ice limits are often fragmentary because of glacial erosion or periods of non-deposition during subaerial exposure, marine records from the East Greenland continental margin may allow more continuous reconstructions of the glacial history (e.g., Marienfeld, 1991, 1992; Mienert, Andrews, & Milliman, 1992; Stein, Grobe, Hubberten, Marienfeld, & Nam, 1993; Stein et al., 1996; Hubberten et al., 1995; Nam et al., 1995; Andrews, Smith, Preston, Cooper, & Jennings, 1996, 1997; Nam & Stein, 1999; Evans et al., 2002). The amount of IRD found in the sediment cores off East Greenland suggests that several major pulses of glacial activity and supply of terrigenous material by glacio-marine processes occurred during the last ~200 ka, reflecting varying extensions of East Greenland glaciers (Figure 6.49; Nam et al., 1995; Stein et al., 1996; Nam, 1997). During MIS 2, 3, and 6/?7, a distinct increase in the amounts of IRD is observed. However, pulses of increased amounts of IRD also occur during warm isotope stages, suggesting that also during warm stages (such as MIS 5) East Greenland glaciers retreated and advanced, and supplied significant amounts of IRD into the East Greenland Current system. In general, the extent and frequency of the IRD-peaks along the East Greenland continental slope (Core PS1726-1) is much higher than that recorded from the deep sea (Core PS1730-2). This is explained by the proximity of glaciers on Greenland that led to more intensive discharge of icebergs and subsequent melting along the adjacent continental margin (Nam et al., 1995; Stein et al., 1996; Nam, 1997).

The maximum occurrence of IRD in the sedimentary records off Scoresby Sund (Core PS1726-1; Figure 6.49) is older than MIS 6 (i.e., pre-Saalian) and probably

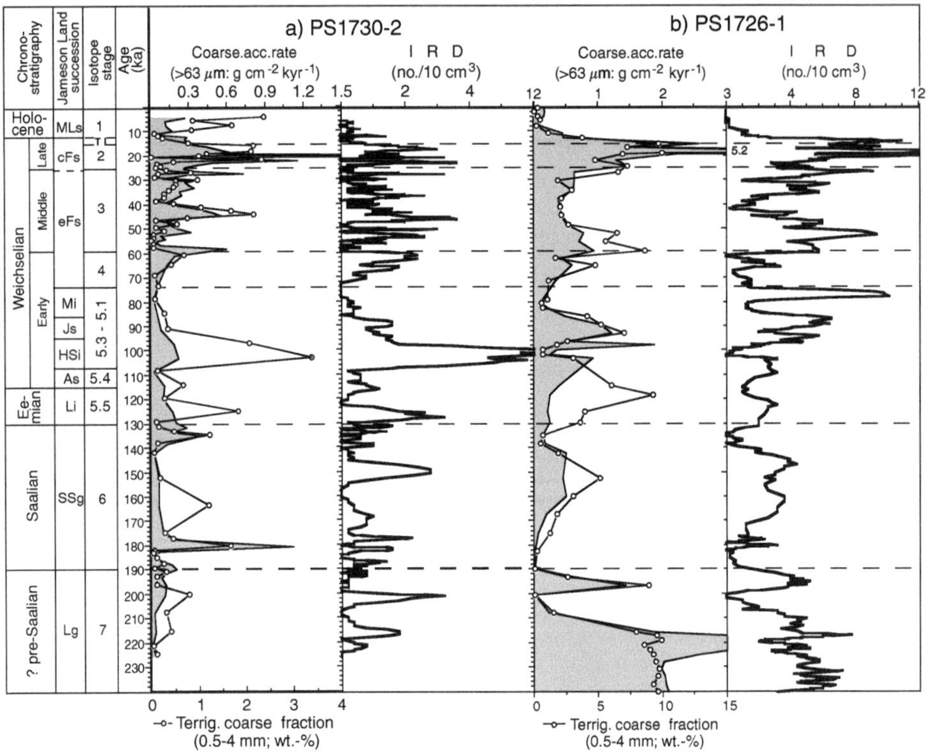

Figure 6.49 Comparison of the accumulation rate of coarse terrigenous matter (>63 μm; cAR), the estimated terrigenous coarse IRD content (0.5–4 mm; wt.%), and the IRD (>2 mm) content counted on X-radiographs from cores (a) PS1730-2 and (b) PS1726-1 with the Quaternary succession on Jameson Land, East Greenland (revised by Funder et al., 1998), and its proposed correlation with North European chronostratigraphy and marine oxygen isotope stratigraphy (from Nam & Stein, 1999). MLs, Milne Land stade; cFs, culmination of Flakkerhuk stade; eFs, early Flakkerhuk stade; Mi, Mønselv interstate; Js, Jyllandselv stade; HSi, Hugin Sø interstade; As, Aucellaelv stade; Li, Langelandselv interglaciation; SSg, Scoresby Sund glaciation; Lg, Lollandselv glaciation. TI, Termination I.

corresponds to the pre-Saalian "Lollandselv Glaciation" recorded on Jameson Land (Funder et al., 1994, 1998; Möller, Hjort, Adrielsson, & Salvigsen, 1994). During MIS 6, when a maximum extension of the East Greenland glaciers occurred (Saalian or "Scoresby-Sund Glaciation"; Hjort, 1981; Funder, 1984; Funder et al., 1994; Möller, Hjort, & Ingolfsson, 1991, 1994), the IRD abundances off Scoresby Sund are much lower than those described for the LGM (MIS 2) (Figure 6.49; Nam et al., 1995; Stein et al., 1996). This may suggest that more extensive continental ice masses on Greenland and a more extended sea-ice cover in the Greenland Sea during MIS 6 (in comparison to MIS 2) prevented the supply and accumulation of IRD at the continental slope off Scoresby Sund. The distinct IRD peak recorded at the well-dated Core PS1730-2 within MIS 5 (Figure 6.49) may coincide with the substage 5.4 or "Auccelaelv stage" glaciation described in the Scoresby Sund area (Funder et al., 1994; Israelson, Funder, & Kelly, 1994).

During MIS 4, the IRD content at the continental slope off Scoresby Sund was distinctly reduced (Figure 6.49). As during MIS 6, an extended sea-ice cover may have prevented iceberg drift and, therefore, IRD deposition in this area. Major pulses of IRD, however, are recorded in MIS 3. Significant transport of IRD onto the shelf can only take place when outlet glaciers filled the fjord basins off northern East Greenland, and it must therefore be concluded that in spite of the scarcity of traces of glaciation on land the fjord basins must have been filled by outlet glaciers for most of the time in MIS 4, MIS 3, and MIS 2. At the MIS 4/3 transition, a distinctly increased supply of IRD off Scoresby Sund was caused by enhanced iceberg drifting possibly produced by the rapid retreat of an ice sheet close to the shelf area. This early meltwater signal is also recorded in the eastern Arctic Ocean (e.g., Stein et al., 1994a; Nørgaard-Pedersen et al., 1998), the Fram Strait (Köhler, 1992), and the Greenland deep sea (Jünger, 1994).

In the well-dated cores PS1927, PS1726, PS1730, and PS1951, changes in IRD discharge are presented in more detail for MIS 2 and MIS 1, the last 30,000 years (Figure 6.50; Nam et al., 1995, Stein et al., 1996; Nam & Stein, 1999). Major IRD pulses occurred almost contemporaneously at all three cores near 27–26, 23–22, 21–20, 18–17, 16, and 15–14 ka. Thus, drastic events in iceberg discharge along the East Greenland continental margin recurred at very short intervals of 1,000–3,000 years, that is, on time scales of Dansgaard–Oeschger cycles (*cf.* Dansgaard et al., 1993; Bond & Lotti, 1995). At all four cores, maximum fluxes of coarse-grained terrigenous material were determined for the time interval between ~16 and 21 ka (Figure 6.50). At this time, the East Greenland fjord glaciers must have reached their maximum Late Weichselian extend and reached at least the inner shelf. Large amounts of terrigenous material deposited at the continental margin probably derived from the inner Scoresby Sund where almost all young unlithified sediments were eroded by glaciers (Dowdeswell, Whittington, & Marienfeld, 1994a; Dowdeswell, Uenzelmann-Neben, Whittington, & Marienfeld, 1994b; Uenzelmann-Neben, Jokat, & Vanneste, 1991; Marienfeld, 1992). This late Weichselian East Greenland Ice Sheet oscillation, with a first smaller ice advance at ~27 ka and the main advance near 21 ka, appears to be in phase with that in the northern Barents Sea area (Figure 6.35; Knies et al., 2000, 2001).

Stable oxygen isotope records from the Greenland ice cores (e.g., Dansgaard et al., 1982, 1993; Oeschger, 1992; Grootes et al., 1993; NGRIP Members, 2004) reveal repeated millennial-scale climate oscillations during the last glaciation (74 to 15 ka). A comparison of the IRD signal at Core PS1730-2 off Scoresby Sund and the isotope record from the GRIP Summit Ice Core (Dansgaard et al., 1993) suggests that most of the distinct IRD peaks coincide with intervals of very light isotope values, that is, times of colder air temperatures over Greenland (Figure 6.51; Nam et al., 1995; Stein et al., 1996; Nam, 1997; Nam & Stein, 1999). These IRD peaks seem to occur contemporaneously with the so-called "Heinrich Layers" or "Heinrich Events" (Heinrich, 1988), widespread features in the late Pleistocene North Atlantic and interpreted as short-lived massive discharge of icebergs (e.g., Broecker et al., 1992; Bond et al., 1992, 1993; Andrews & Tedesco, 1992; Bond & Lotti, 1995; Andrews, 1998; Hemming, 2004). Approximately 11 major IRD-events occurred during this period (Figure 6.51), indicating the repeated massive

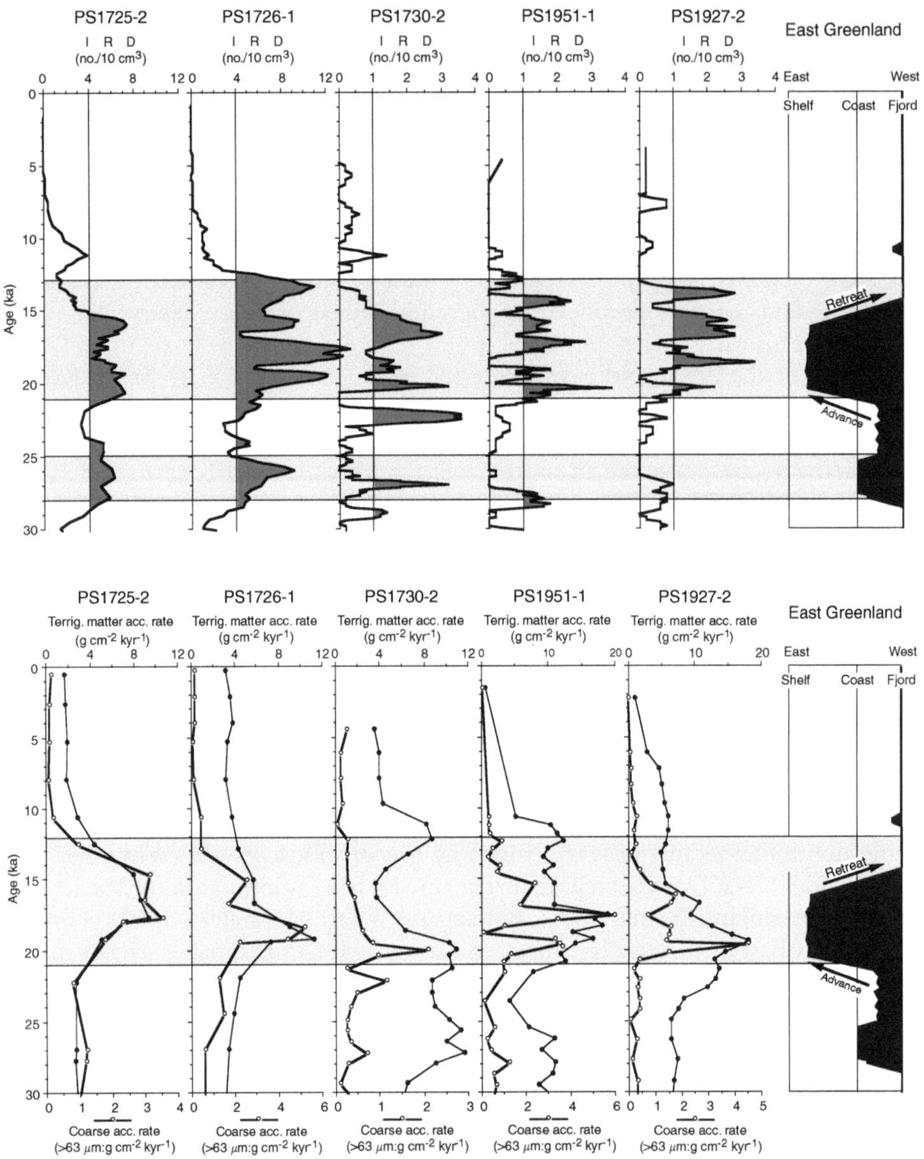

Figure 6.50 Comparison of IRD, terrigenous matter accumulation rate, accumulation rate of coarse-grained material > 63 μm (cAR), and postulated glacial advance and retreat of the Greenland Ice Sheet during the last 30 ka (Nam, 1997; Nam & Stein, 1999).

discharge of icebergs calved by glaciers that result from the instability of the Greenland Ice Sheet. In addition to the major IRD-events, several smaller IRD-cycles occurred between the major IRD-events. In particular, the smaller IRD-cycles are bundled into major IRD-events with asymmetric shapes and sharp

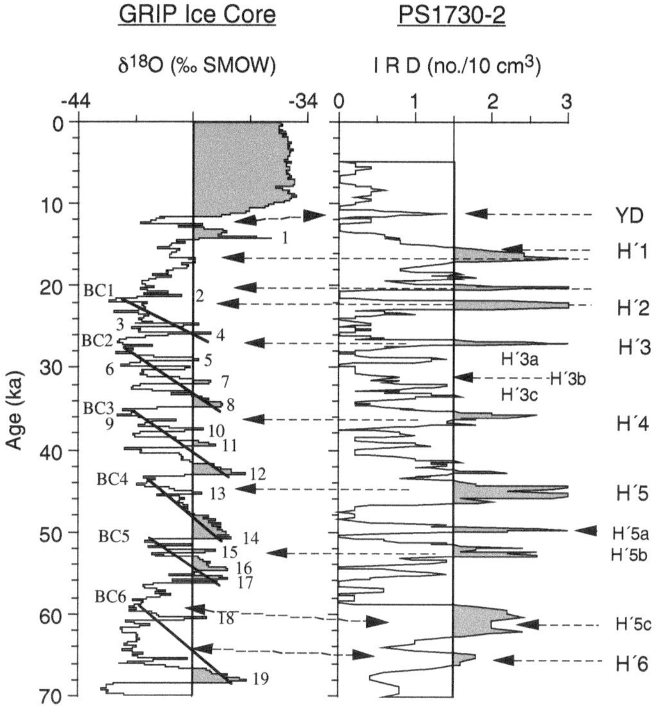

Figure 6.51 Correlation of the oxygen isotope record of the GRIP ice core with the IRD and oxygen isotope records of Core PS1730-2 (from Nam, 1997; Nam & Stein, 1999). Arrows indicate correlation between the parameters. Numbers in the GRIP ice core are interstadial numbers (Dansgaard et al., 1993). "Heinrich-type" IRD events H'6 to H'1 coincide with Heinrich Events H6 to H1 in the North Atlantic (e.g., Bond et al., 1992, 1993). BC1-6, "Bond Cycles" (Broecker, 1994).

boundaries (e.g., Bond et al., 1993), which closely correlate to the Dansgaard–Oeschger cycles in the GRIP record. The iceberg discharge in the Norwegian Sea coming from the Fennoscandian Ice Sheet correlates with the GRIP Greenland air temperature record (Fronval et al., 1995) in the same way as the IRD records off East Greenland. Thus, the data may support that coherent fluctuations in the Fennoscandian and Laurentide/Greenland ice sheets occurred on time scales of a few thousand of years (Stein et al., 1996).

A still ongoing debate is related to the question whether the LGM ice sheet has reached the open shelf north of Scoresby Sund (e.g., Funder et al., 1998; Evans et al., 2002; Ó Cofaigh et al., 2004). Some information can be obtained from shelf sediment cores. The transition from the LGM to the Holocene is well documented in the sedimentary sequence of Core PS2623 obtained from the shelf south of Shannon Island (Figure 6.52; Hubberten et al., 1995; Stein et al., 1995). The overconsolidated stiff diamicton/till recovered in Core PS2623 (unit D) suggest that the glaciers of the East Greenland continental ice sheet probably reached the shelf south of Shannon Island and extended to the east to at least 17°30′W. A similar overconsolidated diamicton was also recorded at the near-by cores PS1916 and PS2621 (see Figure 6.52 for location; Stein et al., 1993, 1995). According to terrestrial records on Hochstetter Forland, this glacial advance is suggested to be of early Weichselian age (Hjort, 1979, 1981). According to the facies succession at

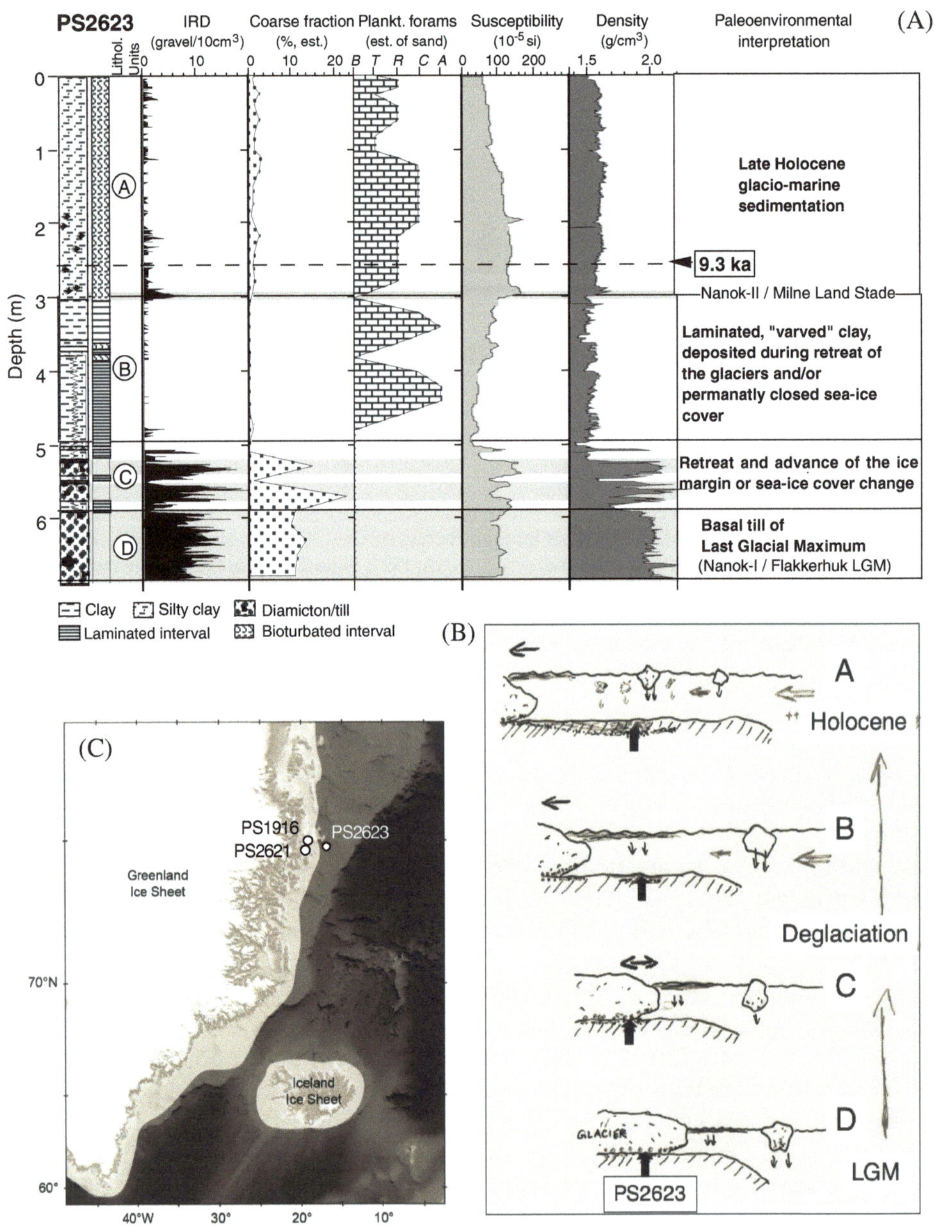

cores PS1916, PS2621, and PS2623, the occurrence of a similar diamicton (till?) with a ^{14}C age slightly older than 15 ka on the East Greenland shelf at 65°N (Mienert et al., 1992), together with the maximum occurrence of IRD recorded at the East Greenland continental slope between 69°N and 75°N during the LGM (Stein et al., 1993; Nam et al., 1995), a late Weichselian (MIS 2) age for this last major advance of glaciers reaching the continental shelf, appears to be more probable.

A LGM ice sheet reaching the mid to outer shelf is further supported by a prominent mid-shelf moraine which consists of unlithified sediments and marks the margin of a grounded palaeo-Greenland ice sheet on the shelf (Niessen & Whittington, 1995; Evans et al., 2002). This moraine is directly overlain by a thin ice-berg-rafted diamicton unit dated to ~13 ka (Core PS2630, Figure 6.53), indicating that the moraine is probably late Weichselian in age (Evans et al., 2002). The moraine therefore represents either the maximum ice-sheet advance during the LGM or marks a recessional position during late Weichselian deglaciation. Whereas Evans et al. (2002) favour the former interpretation, a LGM glacier advance (as ice streams?) onto the outer shelf or close to the shelf break may be supported by the occurrence of debris flows and turbidites that are documented at the East Greenland continental slope off Core PS2630 (Ó Cofaigh et al., 2004; see Chapter 3.3.3, Figure 3.27, and also Wilken & Mienert, 2006 for discussion). Part of the debris flow deposits recorded at cores PS2628 and PS2627 (Ó Cofaigh et al., 2004) are characterized by a very specific dark red colour not present in the other over- and underlying lithologies, which point to the mid-Devonian sand/siltstones of the hinterland being the source area of these diamictons (Figure 6.53; Stein et al., 1995). The sediments recovered at Core PS2640 taken in the inner fjord system, that means the area where these Devonian strata are outcropping, are entirely composed of red-coloured silt and sand lithologies (Figure 6.53). The well-defined,

Figure 6.52 (A) Lithology, IRD numbers, estimates of coarse-fraction content and planktonic foraminifers in the coarse fraction (B, barren; P, present; R, rare; C, common; A, abundant), magnetic susceptibility, and density at Core PS2623 (for location see Figure 6.53) and palaeoenvironmental interpretation (from Hubberten et al., 1995, based on Stein et al., 1995). According to an AMS^{14}C date of 9.3 ka at 258 cmbsf (J. Evans, unpublished data, 1996), the bioturbated glaciomarine sediments characterized by minor but significant amounts of IRD and foraminifers (unit A), probably represent the Holocene time interval. The IRD peak at ~300 cmbsf may represent the ice advance during the Milne Land stade (correlating with the YD cooling event). The overconsolidated stiff diamicton/till recovered in Core PS2623 (unit D) suggest that the glaciers of the East Greenland continental ice sheet probably reached the shelf at least to the core location. The deglaciation, that is, the retreat of the glaciers from the shelf at the end of the LGM, is recorded in the finely laminated clay-silty clay lithology (unit B) deposited in a distal proglacial environment (*cf.* Henrich, 1990). The retreat of the glaciers was gradual, interrupted by several re-advances as indicated in the intercalated diamictons in the lower part of the laminated sequence (unit C). The diamicton intervals are clearly indicated by maximum susceptibility and density values and maximum amounts of terrigenous coarse fraction (>63 μm); foraminifers are absent. (B) Schematic hand drawing (done by R. Stein onboard RV *Polarstern* during the Expedition of 1994) indicating advance and retreat of glaciers on the East Greenland shelf near the core location during LGM to Holocene times. (C) Map showing extension of LGM ice sheet (according to Funder et al., 1998) and core locations.

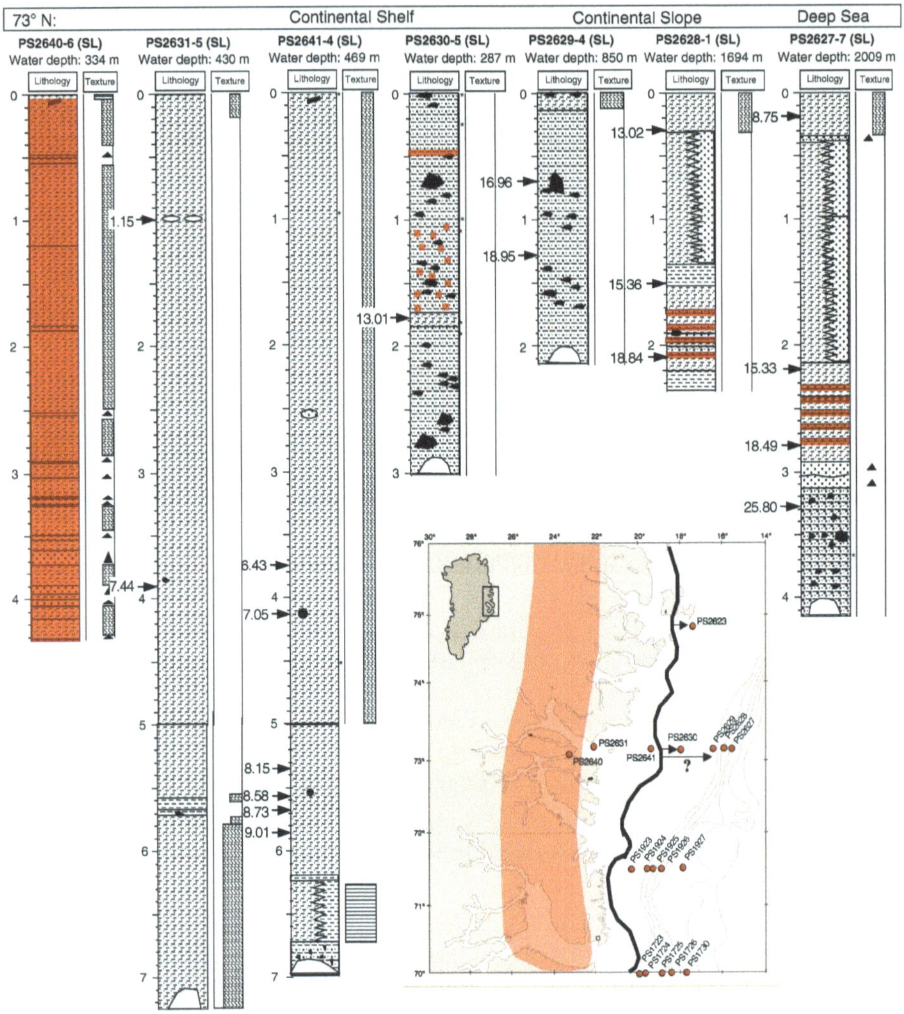

Figure 6.53 Lithological core description of cores from the East Greenland Continental Margin at 73°10′N (Stein et al., 1995). Prominent red-coloured lithologies/horizons are marked in red. In cores PS2628-1 and PS2627-7, these horizons are composed of debris-flow-type sediments (see Chapter 3.3.3, Figure 3.27). AMS ^{14}C ages in ka (reservoir correction of 550 yr) are from Evans et al. (2002). Map indicates core locations. The red area in this map indicates the occurrence of Phanerozoic red beds (e.g., Devonian silt/sand stones) in East Greenland (according to Bond & Lotti, 1995). Solid line marks extension of LGM ice sheet according to Funder et al. (1998), arrows indicate a further off shore location of ice margin as proposed here.

red-coloured lithology of the debris flows may favour rather a transport of sediment by the advanced glacier towards the shelf edge between ~16 and 19 ka than transport by single icebergs which would result probably in a more mixed type of sediment.

6.3. Circum-Arctic Glacial History, Sea-Ice Cover, and Surface-Water Characteristics: Quaternary Records from the Central Arctic Ocean

As outlined in Section 6.2, major parts of the land masses surrounding the Arctic Ocean were covered by ice sheets of variable size several times during the Quaternary. Once an ice sheet or glacier expands to the shoreline which may have extended north close to the shelf break during glacial (cold) intervals of lowered sea level, icebergs are released which can drift towards the open ocean, carrying terrestrial debris (IRD) of clay to boulder size adfrozen at the base, on their surfaces and within the ice. Especially near the end of glacial maxima and during times of sea-level rise (e.g., glacial terminations), icebergs from extensive shelf glaciations could easily escape to the open Arctic Ocean, whereas a decreasing flow velocity, a regional recession of an ice sheet from the shoreline, or a rapidly falling sea level would have diminished or even stopped the discharge of icebergs and deposition of coarse IRD (Spielhagen et al., 2004). Concerning the provenance of the IRD in central Arctic Ocean sediments, distinct spatial and temporal changes are obvious, as deduced, for example, from clay and heavy mineral composition, petrographic composition of lithic clasts, and/or the chemical composition of detrital iron oxide grains (e.g., Bischof et al., 1996; Spielhagen et al., 1997, 2004; Wahsner et al., 1999; Phillips & Grantz, 2001; Darby, 2003; Polyak et al., 2004; see Chapter 4.3 for use of proxies).

The history of continental glaciation, that is, the waning and waxing of ice sheets, can also be related to distinct changes in central Arctic Ocean palaeoenvironment, such as the variability of sea-ice cover and surface-water productivity. Whereas the amount and composition of coarse-grained IRD in deep-sea sediments may provide a direct link to the timing and regional extent of continental glaciations at sea level, the composition and abundance of marine microfossils, for example, planktonic and benthic foraminifer abundances, give information about the variability of sea-ice cover, that is, open-water environments and related increased surface-water productivity versus more closed sea ice and related reduced surface-water productivity (e.g., Polyak et al., 2004; Spielhagen et al., 2004).

Furthermore, the decay of the major circum-Arctic ice sheets during deglaciations must have resulted in a strong freshwater discharge into the central Arctic, causing a major decrease in surface-water salinity at those times. Such a decrease in salinity should be reflected in the isotope composition of planktonic foraminifers, as demonstrated in the oxygen isotope ($\delta^{18}O$) values of the planktonic foraminifer *N. pachyderma* (sin.) from Arctic Ocean surface sediments, which generally show a good correlation to the surface water salinity (Spielhagen & Erlenkeuser, 1994; Bauch et al., 1997, 2005; see Chapter 4.5 for details). Thus, the isotopic composition of planktonic foraminifers in marine sediment cores from the central Arctic Ocean, that is, outside the area near the Fram Strait that is directly affected by warm Atlantic water, can be used as a measure for surface-water salinity and its variation in the Arctic Ocean (e.g., Stein et al., 1994a; Nørgaard-Pedersen et al., 1998, 2003; Poore et al., 1999; Spielhagen et al., 2004). Although carbon isotope ($\delta^{13}C$) data from high-latitude planktonic foraminifers are more difficult to interpret than $\delta^{18}O$ records (e.g., Kohfeld, Anderson, & Lynch-Stieglitz, 2000), a prominent low $\delta^{13}C$ signal associated with meltwater spikes in the $\delta^{18}O$ records has been recorded in a large number of Arctic Ocean sediment cores representing the LGM-deglacial-Holocene time interval (Stein et al., 1994a; Poore et al., 1999; Nørgaard-Pedersen et al., 2003). This feature may be ascribed to reduced ventilation of surface waters caused by time-limited, extremely strong stratification from the surficial low-salinity freshwater lid (e.g., Spielhagen et al., 2004). Thus, strong negative excursions in both oxygen and carbon isotopes of planktonic foraminifers from Arctic sediment cores have been related to releases of freshwater from collapsing continental ice sheets during glaciations and especially at glacial terminations, resulting in major discharge of subglacial meltwater and icebergs to the ocean (e.g., Stein et al., 1994a, 1994b; Nørgaard-Pedersen et al., 1998, 2003; Poore et al., 1999; Spielhagen et al., 2004; see Section 6.3.5 for more detailed discussion). As outlined by Spielhagen et al. (2004), another important scenario causing strong freshwater discharge, could be the deglacial northward breakthrough and outburst of ice-dammed lakes which were fed by northbound rivers at times when ice sheets blocked the discharge to the Arctic Ocean (Houmark-Nielsen et al., 2001; Mangerud et al., 2001, 2004; Svendsen et al., 2004; see Section 6.2.1, Figure 6.25).

Thus, in summary, the history of continental glaciations and its palaeoenvironmental implications in the marine realm are also reflected in sedimentary records from the central Arctic Ocean. The number of reasonably well dated sediment cores from the central Arctic Ocean going beyond the range of AMS^{14}C dating and used for palaeoceanographic reconstruction, however, are still limited (e.g., Nørgaard-Pedersen et al., 1998, 2007a, 2007b; Jakobsson et al., 2001; Polyak et al., 2004; Spielhagen et al., 2004; Stein et al., 2004b). Although a major progress has been made during the last years concerning the Quaternary stratigraphy (e.g., Jakobsson et al., 2001; Backman et al., 2004 and references therein), a definitive judgement regarding an age model for these sediments has still to be done with caution, especially for sediments older than ~240 ka (see Section 6.1 for more details). Because age model choice strongly affects the interpretation of palaeoceanographic evolution, this uncertainty prompts an urgency in refining and verifying the chronostratigraphy for the Arctic Ocean sediments (Polyak et al., 2004).

In the following subchapters, some examples of palaeoenvironmental reconstructions using mainly sediment cores from the Lomonosov, Alpha-Mendeleev, and Northwind ridges and the Yermak Plateau (Figure 6.54) are presented, concentrating on the time interval MIS 6 to MIS 1, that is, the last ~200 ka.

6.3.1. Surface-Water Characteristics and Circum-Arctic Glaciations During MIS 6 to MIS 2

6.3.1.1. IRD variability and abundances of foraminifers

As shown in sediment cores from the Eurasian Basin as well as the Amerasian Basin (see later discussions) records of both IRD input and foraminifer abundances from the last 200 ka clearly demonstrate that sedimentary environments in the central

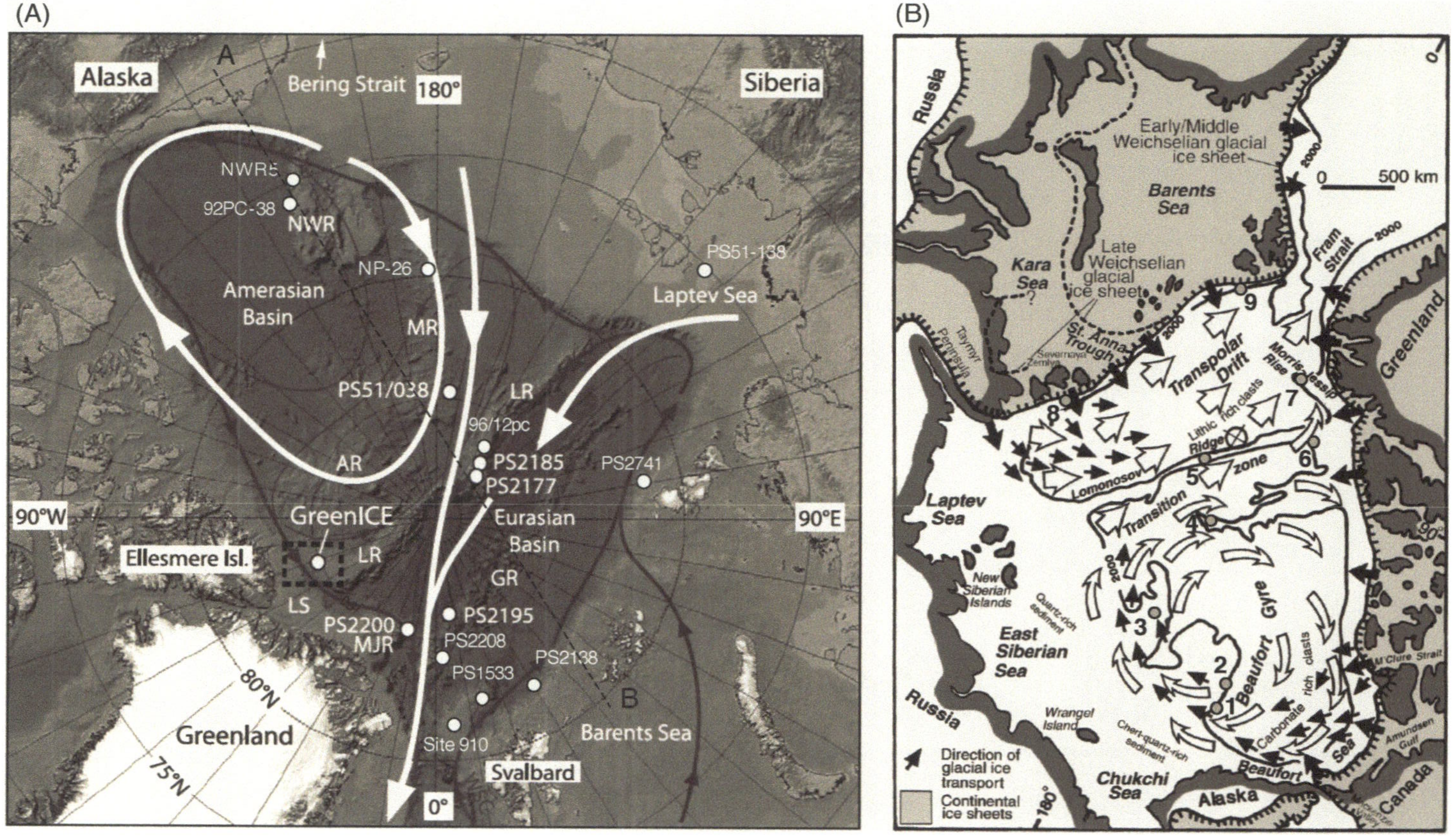

Figure 6.54 (A) Arctic Ocean bathymetry (based on IBCAO; Jakobsson, 2002; Jakobsson et al., 2003a), major surface currents (white), subsurface Atlantic water circulation system (dark grey), and locations of cores discussed in the text (from Nørgaard-Pedersen et al., 2007a, supplemented). LS, Lincoln Sea; LR, Lomonosov Ridge; MJR, Morris Jesup Rise; GR, Gakkel Ridge; AR, Alpha Ridge; MR, Mendeleev Ridge; NR, Northwind Ridge; WSC, West Spitsbergen Current. Stippled line (A–B) shows location of profile presented in Figure 6.65A. (B) Currents and glacial iceberg drift directions in the Arctic Ocean during late Pleistocene interglacial periods and boundaries of continental glacial ice sheets during Pleistocene glacial maxima (from Phillips & Grantz, 2001, supplemented). Pleistocene ice sheet boundaries for the Kara and Barents Seas during the Early/Middle and Late Weichselian from Svendsen et al. (1999). Contours in metres. Location of cores discussed in the text are shown: 1, NWR5; 2, 92PC-38; 3, NP26; 4, PS51/038; 5, PS2185; 6, GreenICE 10 and 11; 7, PS2200; 8, PS2741; 9, PS2138.

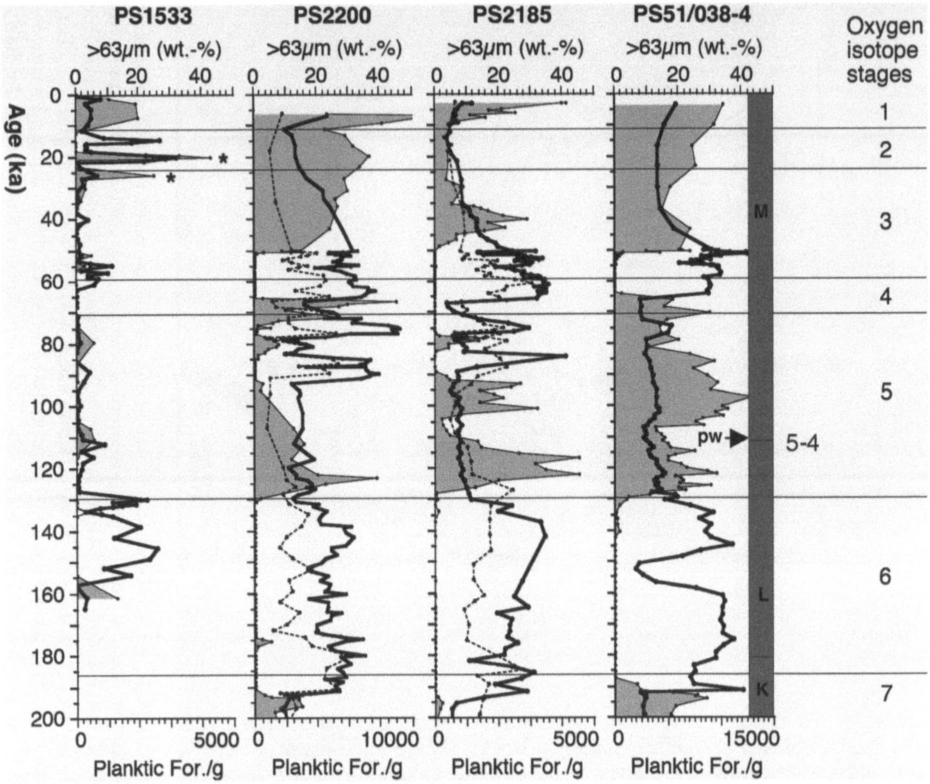

Figure 6.55 Records of abundance of planktonic foraminifers and coarse fraction in Arctic sediment cores during the last 200 ka (for location of cores see Figure 6.54) (from Spielhagen et al., 2004, supplemented). Asterisks in the record of Core PS1533 mark layers where high coarse-fraction content results from abundant planktonic foraminifers. Dotted lines represent smectite content in PS2185 and PS2200 (% of the <2 μm fraction; same scale as >63 μm). For Core PS51/038, lithological units K to M (standard units according to Clark et al., 1980) and occurrence of pink-white (pw) layer are shown (according to Stein et al., 1999b). Based on the age model proposed here (see Section 6.1 for details), the pw layer is of MIS 5.4 age.

Arctic Ocean were strongly variable (for stratigraphic framework see Section 6.1). In most sediment cores from the Arctic Ocean, that is, both the Eurasian and Amerasian basins, IRD content and abundances of planktonic foraminifers display a general anticorrelation (Figure 6.55; Pak et al., 1992; Darby et al., 1997; Spielhagen et al., 1997, 2004; Nørgaard-Pedersen et al., 1998, 2003, 2007a; Polyak et al., 2004). This anticorrelation can be interpreted by two major scenarios (Spielhagen et al., 2004). The foraminifer-rich intervals reflect times with an inflow of Atlantic water (*cf.* Hebbeln et al., 1994; Dokken & Hald, 1996; Nørgaard-Pedersen et al., 2003) of variable strength, temperature, and regional extension, at least seasonally open waters, and some increased surface-water productivity. In sediment cores from the Lomonosov Ridge (Core PS2185), the Yermak Plateau (Core PS1533), and the Morris Jesup Rise (Core PS2200) as well as the Amerasian Basin/Alpha Ridge

(Core PS51/038-4), peaks in planktonic foraminifer abundance are found from uppermost MIS 7 (7.1), substages MIS 5.5, 5.3, and 5.1 (best seen in cores PS2185 and PS51/038-4), the boundary between MIS 5 and 4 (PS2200 and PS51/038-4), and <50 ka (Figure 6.55). At Core PS2185, where the MIS 5/4 foraminifer peak was not found, a corresponding enhanced planktonic productivity can be inferred from a ^{10}Be double peak at 120–145 cm in this core (see Section 6.1.5, Figure 6.17). Foraminifers in this layer may be missing due to carbonate dissolution (Spielhagen et al., 2004).

The second major scenario is represented by the coarse-grained IRD layers which only contain very few or no foraminifers and which are related to iceberg transport of terrigenous material. Distinct maxima in IRD were recorded in uppermost MIS 7 to MIS 6 (190–130 ka), upper part of MIS 5 (substage 5.2, ~90–80 ka), near the MIS 5/4 boundary (~75 ka), and in the late MIS 4/early MIS 3 time interval (65–50 ka) (Figure 6.55; Spielhagen et al., 2004), indicating major continental glaciations during those times. Obviously, central Arctic sedimentation was not affected by the MIS 5.4 ice advance in the Svalbard/Barents Sea area (e.g., Mangerud et al., 1998; see discussion in Section 6.2.2, Figure 6.29). The central Arctic Ocean sediments from <50 ka show an upward decrease of IRD content and a minimum in the deposits from the LGM ~20 ka (Nørgaard-Pedersen et al., 1998, 2003; Spielhagen et al., 2004; see Section 6.3.3 for more details). In general, Holocene deep-sea sediments from the Eurasian Arctic Ocean have a low coarse-fraction content (<10 wt.%; Nørgaard-Pedersen et al., 1998, 2003) which reflects the scarcity of icebergs in the modern Arctic.

In general, a quite similar situation was described for Lomonosov Ridge Core 96/12-1pc as well as sediment cores from the Mendeleev Ridge (NP26) and the Northwind Ridge (NW5) (see Figure 6.54 for locations) where brown intervals with increased abundances of planktonic and benthic foraminifers are interpreted as interglacials (Figure 4.4: Core NP26; Figure 4.32: cores 96/-1pc and NW5). These interglacials in all three cores which can be correlated very well as based on lithology and microfossil assemblages, probably represent MIS 7.5, 7.1, 5.5, 5.3, 5.1, 3.3, 3.1 and 1, when using the "young" age model (Jakobsson et al., 2001; Polyak et al., 2004; Darby et al., 2006; see also Chapter 4.5.1, Figure 4.32, and Section 6.1 for more detailed discussion). The intercalated grey units representing glacial periods, on the other hand, are almost unfossiliferous and largely fine-grained, but some contain prominent IRD layers near the top and/or bottom of grey units that may extend into the adjacent brown interglacial unit (Figure 4.4; see also later discussions). As proposed by Polyak et al. (2004), the extremely low numbers or complete absence of biogenic remains in sediments from some glacial intervals may support the notion that the western Arctic Ocean was possibly covered by very thick pack ice or extensive ice shelves during Pleistocene glaciations.

Very recently, Nørgaard-Pedersen et al. (2007a, 2007b) published records from the hitherto unexplored and heavily ice-covered southernmost part of the Lomonosov Ridge off North Greenland/Ellesmere Island, at the western flank towards the Amerasian Basin, representing the MIS 7 to MIS 1 time interval (see Figure 6.54 for location). In these cores (GreenICE cores 10 and 11), two main lithologies can be distinguished: (1) brown beds composed of bioturbated silty clays rich in calcareous microfossils (i.e., planktonic and benthic foraminifers, ostracods and nannofossils) and (2) light-coloured beds of silty clays to sandy silts with occasional granule to small pebble-sized IRD and a sparse microfossil content (Figure 6.56). These alternating lithologies are very similar to those described from most other central Arctic Ocean sediment cores and related to glacial/interglacial (interstadial) variability (see earlier).

According to Nørgaard-Pedersen et al. (2007a), the association of the foraminifera peaks with dark brown, fine-grained mud poor in IRD was found in specific intervals of interglacial MIS 7, 5, and 1, as well as some interstadials of MIS 3 and 6 (Figure 6.56). Increased IRD deposition, on the other hand, took place during MIS 6 (early part and 6/5 transition), early-mid MIS 5 (5.4?), MIS 4, and late MIS 3 (Figure 6.56). As in other central Arctic Ocean records, there is no evidence of intensified ice-rafting during MIS 2 (including the LGM). The brown fine-grained, foraminifer-rich intervals support oxygenated bottom water conditions, dominant sea-ice sediment rafting, open-water leads during summer, and increased planktonic productivity (Poore et al., 1993; Phillips & Grantz, 1997; Nørgaard-Pedersen et al., 1998; Jakobsson et al., 2000a; Polyak et al., 2004; Spielhagen et al., 2004). In specific intervals, that is, substage MIS 5.5 (last interglacial) and MIS 5.1 (warm interstadial), peak abundances of small subpolar *Turborotalita quinqueloba* were found (Figure 6.56). At present, this subpolar species occurs abundantly in subsurface Atlantic waters close to the sea-ice margin north of Svalbard and in the Barents Sea Branch (see Chapter 4.5.1, Figure 4.26; Volkmann, 2000), and only a small percentage of subpolar specimens reach the interior Arctic Ocean as evidenced by living assemblages in the water column and late Holocene surface sediment samples from interior Arctic sites with excellent carbonate preservation (Nørgaard-Pedersen et al., 2007a). Even if the Atlantic water boundary current system was enhanced during the last interglacial period, it seems to be unlikely that abundant subpolar specimens were advected several thousands kilometres to the interior Arctic Ocean. Thus, Nørgaard-Pedersen et al. (2007a) conclude that sea ice conditions near the GreenICE site must have been reduced during the last interglacial.

6.3.1.2. Provenance of IRD and circum-Arctic glaciations

Concerning the provenance of the IRD and its variability through time in the Eurasian Arctic, data from Core PS2185 are shown as an example. In Core PS2185, bulk-, clay-, and heavy-mineral associations were used to identify source areas of the terrigenous (IRD) fractions (see Chapter 4.3 for background). Here, especially elevated smectite and kaolinite concentrations during MIS 6, upper MIS 5, and late MIS 4/early MIS 3, mostly coinciding with IRD maxima (Figure 6.57A), serve as tracer for an IRD origin from the western Laptev/southeastern Kara Sea/Franz Josef Land and central Barents Sea/Franz Josef Land, respectively (see Chapter 5.1.2, Figure 5.4B and D). The contemporaneous occurrence of elevated smectite and kaolinite concentrations in sediment cores from the Eurasian continental margin (see Section 6.2) and the central Arctic Ocean (Figure 6.57A; Spielhagen et al., 1997, 2004) may indicate the coincidence of glacigenic erosional processes in both

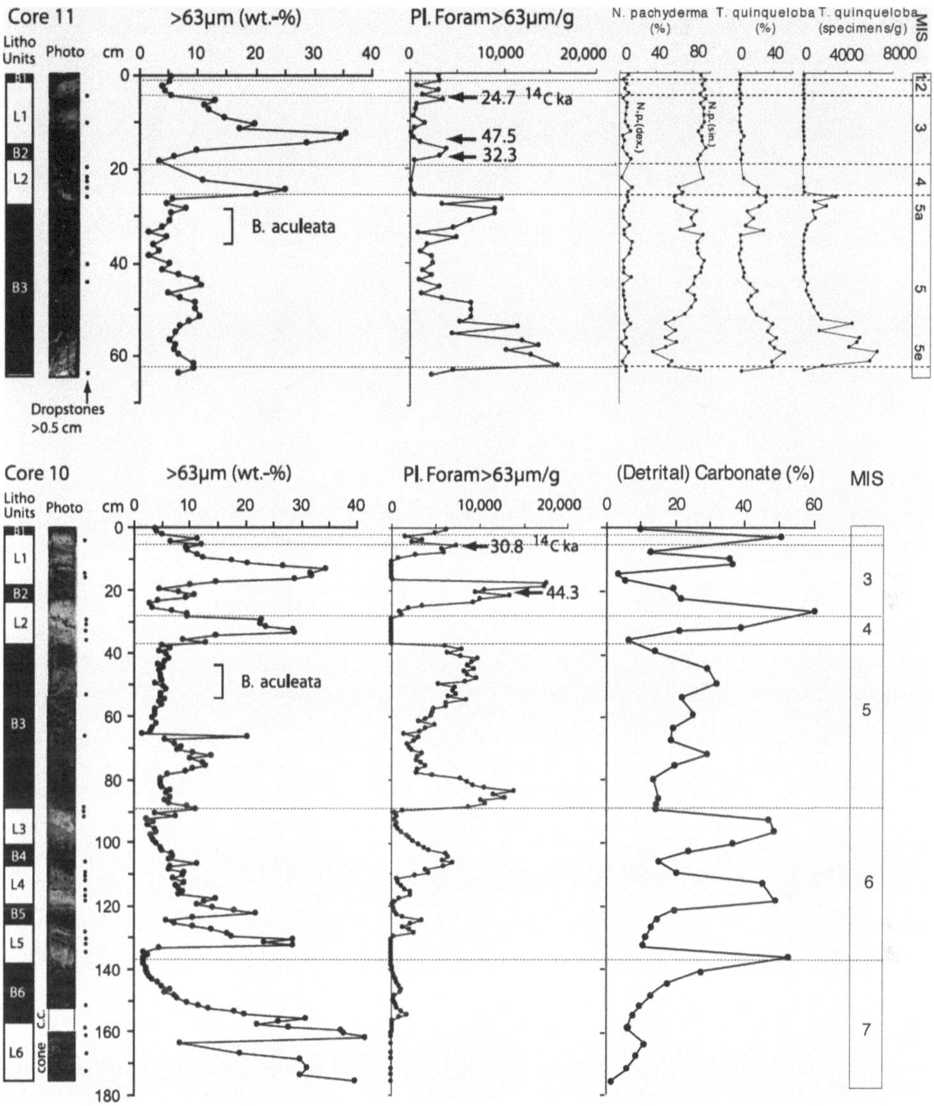

Figure 6.56 GreenICE cores 10 and 11, including core photos, lithogical units, coarse fraction content >63 μm (wt.%), occurrence of "dropstones" >0.5 cm, planktonic foraminifer abundance >63 μm, and age model (MIS 1–7) (Nørgaard-Pedersen et al., 2007a, modified and supplemented). The stratigraphic levels with AMS^{14}C dates (^{14}C ka) and the peak occurrence of benthic foraminifera *Bulimina aculeata* are indicated. For Core 11, also planktonic foraminifera assemblage records (specimens >63 μm) are shown. The two intervals assigned to substages 5e (5.5) and 5a (5.1) are characterized by abundant subpolar foraminifera *T. quinqueloba*, whereas polar species *N. pachyderma* (sin.) dominate totally in the other part of the record. For Core 10, carbonate concentrations are shown, mainly representing detrital carbonate (Nørgaard-Pedersen et al., 2007b).

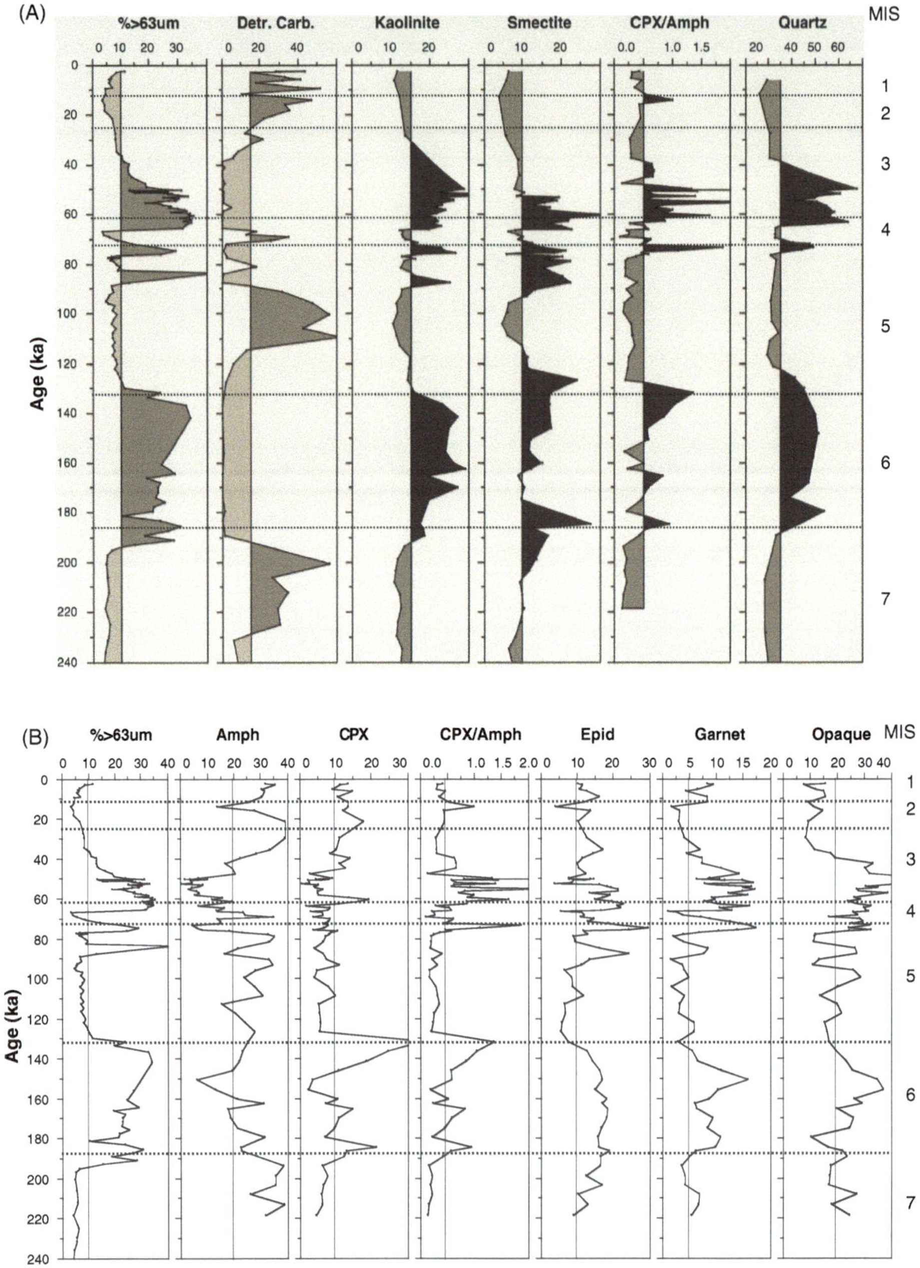

Figure 6.57 Summary plots showing coarse-fraction content >63 μm and XRD clay mineral assemblages (data from Spielhagen et al., 1997), quartz content (XRD data from Vogt, 1997, 2004), and heavy minerals (data from Behrends, 1999) at Core PS2185 for the last 240 ka (MIS 7 to MIS 1), using the age model of Spielhagen et al. (2004). Isotope stages MIS 1–7 are indicated. (A) Coarse-fraction content (wt.%), detrital carbonate (% of the coarse fraction >500 μm), kaolinite and smectite (% of clay minerals in the clay fraction <2 μm), clinopyroxene/amphibole (CPX/Amph) ratio, and quartz content (% of bulk fraction). (B) Coarse-fraction content (wt.%) and contents of amphibole (Amph), clinopyroxene (CPX), epidote (Epid), garnet and opaque minerals (% of heavy minerals), and clinopyroxene/amphibole (CPX/Amph) ratio.

source areas. High smectite concentrations not always coincide with high coarse fraction content. In some intervals, for example, in the uppermost and lowermost MIS 5 (5.1 and 5.5) smectite enrichments correlate with very low terrigenous coarse fraction which may suggest transport by sea ice (or currents) rather than icebergs.

Contemporaneously with the high smectite and kaolinite percentages, also specific heavy-minerals, such as epidote and garnet, were increased, indicating a Kara Sea/southern Barents Sea IRD source (Figure 6.57B; Behrends, 1999, using the age model of Spielhagen et al., 2004; see also Chapter 5.1.4, Figure 5.13C and D). A Kara Sea/western Laptev Sea source is supported by the significantly higher clinopyroxene/amphibole ratio in the IRD-rich intervals, although the absolute clinopyroxene values are surprisingly low in most of the samples (Figure 6.57, see also Figure 5.13B). Furthermore, the IRD-rich intervals are enriched in quartz, reaching values of >50% (Figure 6.57A). This is also consistent with a Eurasian source, although quartz alone is not specific enough for a source identification because it also occurs in major abundance in the Canadian Arctic (see later discussions).

During interglacials, that is, MIS 7 and middle part of MIS 5 as well as during the last ~20 ka, increased amount of detrital carbonate was determined in the record of Core PS2185 (Figure 6.57A; Spielhagen et al., 1997, using the stratigraphy of Spielhagen et al., 2004). Detrital carbonate is related to a sediment source in the Canadian Arctic (see Chapter 5.1 and later discussions), suggesting substantial sediment transport towards the Eurasia Basin by an extended Beaufort Gyre at those (mainly interglacial) time intervals (see Phillips & Grantz, 2001).

On the Mendeleev Ridge at NP26 cores, lithic fragments are dominated by quartz grains ranging between 40% and >80% (Figure 6.58A; Polyak et al., 2004). A major source area could be the Laptev/Kara Sea (see Figure 5.11) but also the Canadian Arctic, that is, the southwestern Canadian Archipelago, primarily Banks and Victoria Islands, plus parts of northern Canada from south of Amundsen Gulf to the Canadian/Alaskan border (Figures 4.16 and 5.2), where similar quartz contents were determined (e.g., Bischof et al., 1996; Vogt, 1997; Phillips & Grantz, 2001). Second most abundant IRD clasts are carbonate rock fragments, which peak at the pink-white layers (MIS 5.4, MIS 4/3, and MIS 2; Figure 6.58A) and indicate the Laurentide part of the Canadian Archipelago, that is, Banks and Victoria Islands, as the source area (see Figures 4.16 and 5.2) and a circulation pattern similar to that shown in Figure 6.59. This is also supported by distinct Fe oxide spikes from a Laurentide source (see Chapter 4.3.1 for background, Figures 4.13 and 4.14), characterizing the pink-white layers (Figure 6.58B; Polyak et al., 2004). During MIS 5.4, a pink-white layer was even found in the Alpha Ridge Core PS51/038-4 (Figure 6.55). In the lower part of the NP26 record (MIS 6; Figure 6.58B), the composition of Fe oxides points to elevated inputs from the Innuitian part of the Canadian Archipelago, that is, Queen Elizabeth Islands (see Figures 4.13 and 4.14), and a circulation pattern more similar to the modern one (Figure 6.54). These carbonate-rich IRD peaks are related to extended glaciations in the terrestrial North American Arctic during those times. The disintegration of ice shelves over the Amerasia Basin would have caused surges and outbursts of icebergs into the Arctic Ocean from the adjacent ice sheets (*cf.* Darby et al., 2002). Such probably

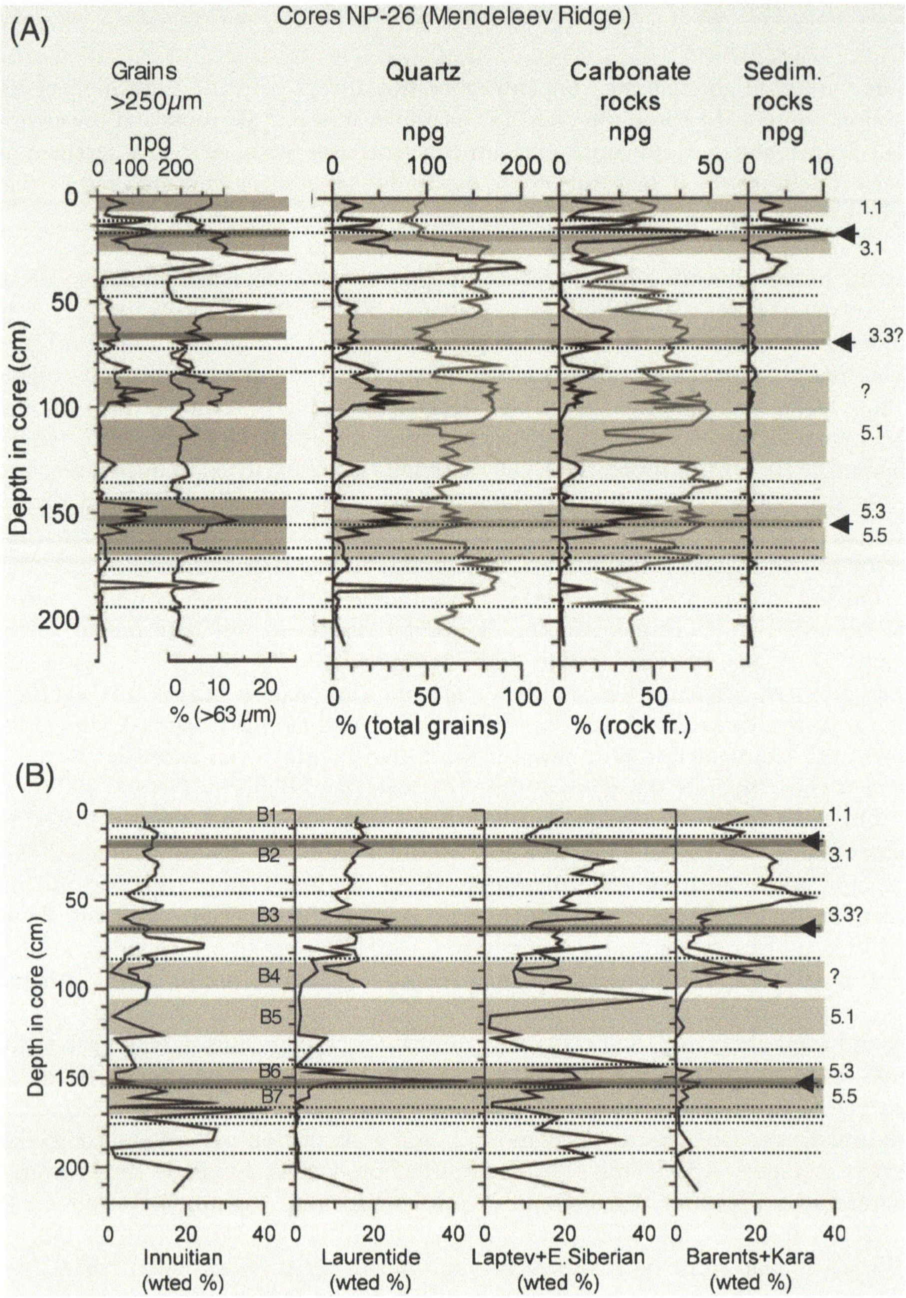

Figure 6.58 Amount and composition of coarse fraction of Mendeleev Ridge cores NP26 (from Polyak et al., 2004, supplemented). (A) Types of lithic clasts >250 μm in NP26 cores (see also Chapter 4.1, Figure 4.4, for lithology and total clast abundance). Black curves and upper scale bars show grain number per gram (note different scales), grey curves and lower scale bars show percent of quartz in total grains and percent of carbonates in rock fragments. (B) Iron oxide grain types (weighted %) in NP26 cores. Fe oxides are matched to source areas after Darby and Bischof (1996; see Chapter 4.3.1 for background). Stippled lines indicate periods of increased meltwater discharge as based on stable isotopes (see Figure 6.63). Light grey bars indicate interglacials with stage numbers MIS 1.1 to 5.5 as proposed by Darby et al. (2006), black arrows indicate pink-white layers. For location of cores see Figure 6.54A.

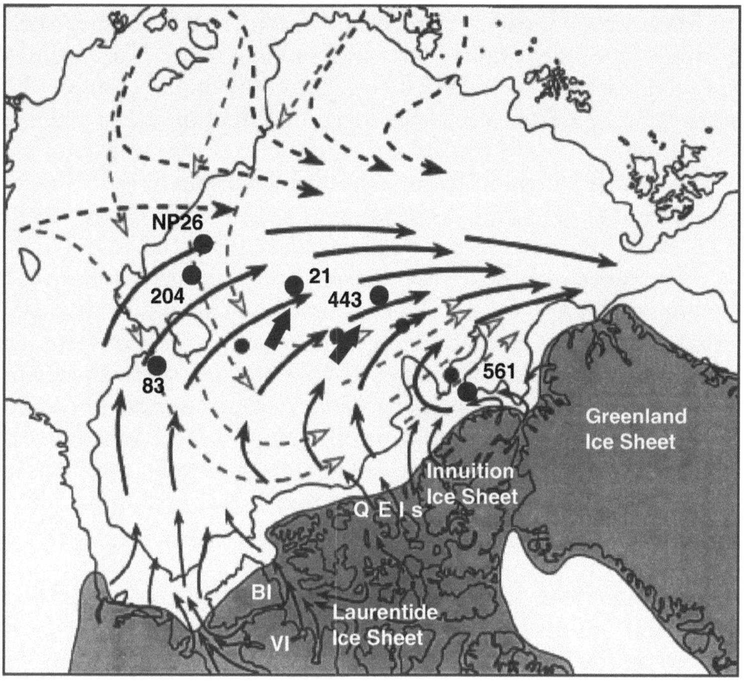

Figure 6.59 Map of the inferred surface current-driven iceberg drift directions from North American sources (solid arrows) and concurrent hypothesized drift of Siberian pack ice (broken arrows) during glacial intervals (from Bischof & Darby, 1997, supplemented). As proposed by these authors, the sea-ice drift in the western Arctic Ocean was mostly similar to the iceberg drift, but sea ice occasionally drifted from Siberian sources into the western Arctic Ocean along paths shown as shaded broken arrows. A similar sea-ice drift pattern was also described for the Holocene (see Figure 6.74). The extension of late Quaternary North American ice sheet is indicated, whereas the Eurasian ice sheets are not shown. Locations of sediment cores discussed in the text are shown.

short-lived discharge events are reflected in sedimentary records by distinct IRD spikes exemplified by pink-white carbonate layers and Fe grains indicative of material from the Laurentide part of the Canadian Archipelago (Bischof et al., 1996; Darby et al., 2002; Polyak et al., 2004), and synchronous low oxygen and carbon isotope values of *N. pachyderma*, indicative for meltwater input (see later discussions).

Throughout the NP26 cores, another major source of Fe oxides is the shelf of the Laptev and East Siberian Seas, whereas in the upper part of the record (< MIS 4), elevated contents of Fe oxides from the Kara and Barents seas (see Figures 4.13 and 4.14) were determined (Figure 6.58B; Polyak et al., 2004). Most of the Fe oxide peaks with a Siberian source coincide with low amount of terrigenous coarse fraction, indicating rather sea-ice than iceberg transport. Sediment input from the Siberian Arctic via icebergs and/or sea ice suggest a surface-water circulation different from that of today, that is, a distinct shift of the TPD system towards North America (Figure 6.59; Bischof & Darby, 1997; see Section 6.3.6).

In an earlier study, distinct changes in amount and composition of IRD $>250\,\mu m$ were also determined in low-sedimentation-rate sediment cores (compressed sections) from the Alpha Ridge, the Chukchi Cap, and the Northwind Ridge (Figure 6.60, see Figure 6.59 for locations; Bischof et al., 1996; Bischof & Darby, 1997). In these cores, a similar variability in the composition of lithic grains $>250\,\mu m$ as in the NP26 cores, that is, a dominance of quartz grains and carbonate grains, was determined (Figure 6.60; Bischof et al., 1996; Bischof & Darby, 1997). Bischof et al. (1996) interpreted the quasi-cyclic changes of IRD abundances and alternating changes between quartz-dominated and carbonate-dominated intervals as glacial/interglacial cycles. Often, IRD maxima coincide with carbonate maxima, whereas maxima in quartz coincide with IRD minima (most obvious at Core FL 21). The latter may point to sea-ice rather than iceberg transport. In general, highest quartz contents were found on the Alpha Ridge, highest carbonate contents on the Northwind Ridge. In Core 92PC-38 from the Northwind Ridge (see Figure 6.54 for location) characterized by higher sedimentation rates than the cores studied by Bischof et al. (1996), detrital carbonate minerals are also dominating the coarse (IRD) fraction (Phillips & Grantz, 2001).

On the basis of the distinct maxima of IRD content (probably related to late glacial maxima/early deglaciations; see Section 6.2) and the composition of the IRD, at least six major glaciations in the terrestrial North American Arctic were proposed by Bischof et al. (1996) for the time interval represented by lithological units I to M, that is, during the last ~ 1 Ma when using the "old stratigraphy model" (Figure 6.60). As outlined in Section 6.1, however, it is more probable that the age interval represented by the lithological units I to M is much younger. That means, for the four youngest glacials (lithological units L and M) an age of MIS 6 to MIS 2 is more probable. If this is correct, one can speculate that the IRD peak in unit L and the three IRD peaks in unit M (most clearly obvious in Core FL 21; Figure 6.60) may be related to extensive glaciations during MIS 6, MIS 5.4, MIS 4/3, and MIS 2. At least for the latter three, this tentative interpretation agrees well with the NP26 record (see earlier).

The IRD material in both GreenICE cores 10 and 11 from the Lomonosov Ridge off northern Greenland/Ellesmere Islands contains abundant detrital carbonates (Figure 6.56 for Core 10 record; Nørgaard-Pedersen et al., 2007b), supporting a dominant terrestrial source from North Canada/Greenland where Palaeozoic platform carbonates extensively crop out (see Chapter 5.1; Phillips & Grantz, 2001). In this respect, the GreenICE IRD record is more similar to Amerasian Basin records from the Northwind Ridge and the Mendeleev Ridge (e.g., Phillips & Grantz, 1997, 2001; Polyak et al., 2004) than to the records from the central Lomonosov Ridge (Spielhagen et al., 1997, 2004). In addition to coarse-grained, carbonate-rich intervals, also fine-grained, calcite-rich detrital carbonate layers occurred at glacial–interglacial transitions (Figure 6.56), which appears to be unique to this area. Nørgaard-Pedersen et al. (2007b) proposed that detrital carbonate was transported and deposited from nepheloid flows during deglaciations of Ellesmere Island, when the Atlantic water boundary current was gaining strength along the continental margin.

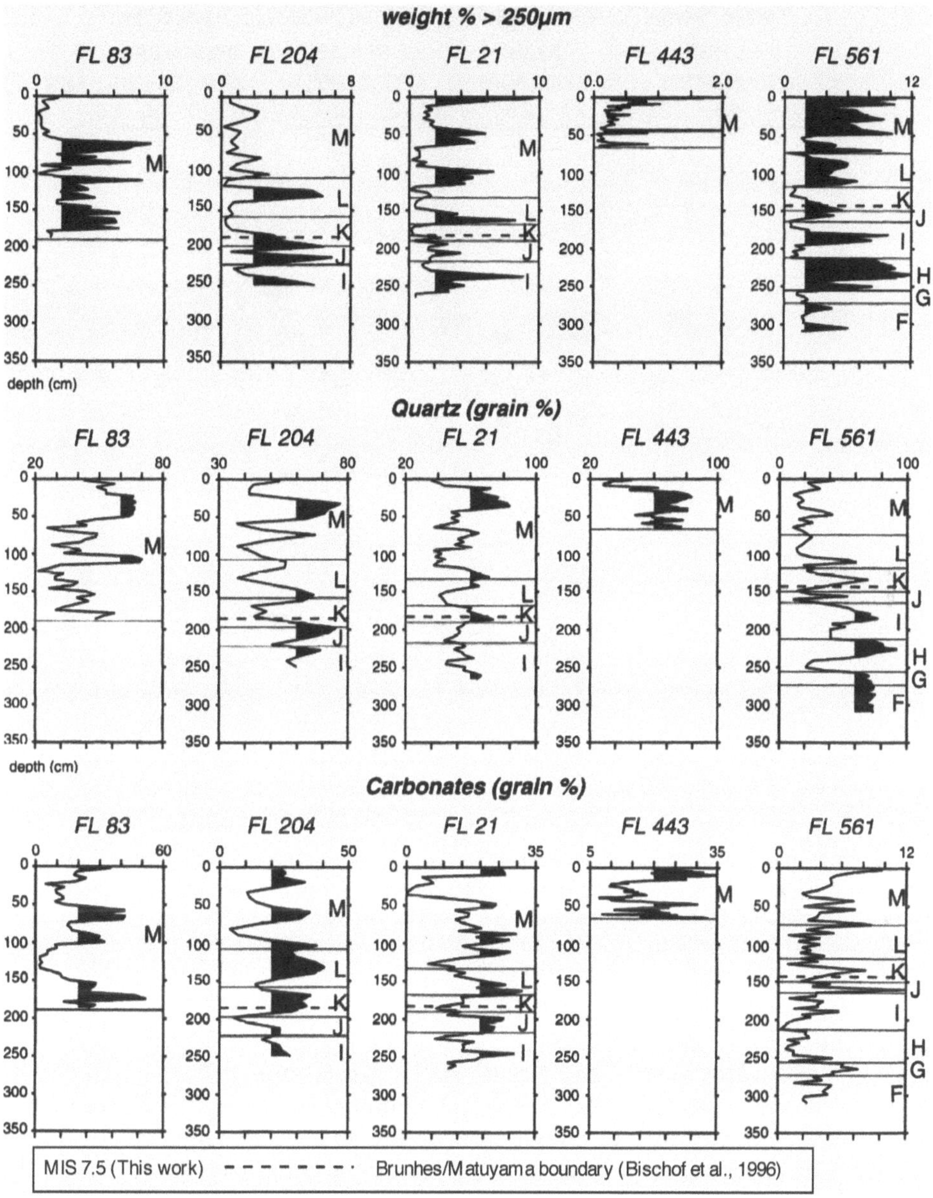

Figure 6.60 IRD, quartz, and carbonate contents in cores from the Amerasian Basin; for core locations see Figure 6.59 (from Bischof et al., 1996, supplemented). Standard litological units (F to M) are shown, stippled line indicate Brunhes/Matuyama boundary, based on the "old stratigraphy model" as used by Bischof et al. (1996). Based on the new age model described in Section 6.1, the stippled line is correlated with the Pringle Falls excursion of MIS 7.5 age (see Figure 6.20 and Table 6.2).

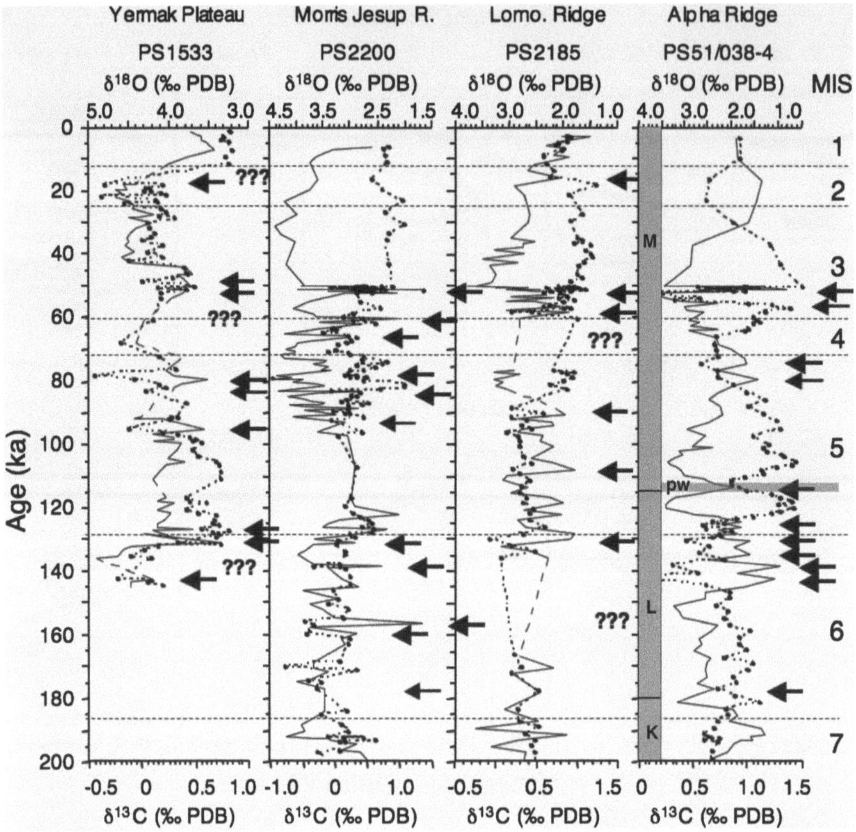

Figure 6.61 Planktonic oxygen and carbon isotope records of Arctic sediment cores for the last 200 ka (from Spielhagen et al., 2004, supplemented). Arrows mark identified freshwater events characterized by low values of both parameters. For Core PS51/038, lithological units K to M (standard units according to Clark et al., 1980) and occurrence of pink-white (pw) layer (see also Figure 6.55) are shown (Stein et al., 1999b).

6.3.1.3. Freshwater events and circum-Arctic glaciations

In the sediment cores from the Alpha and Lomonosov ridges, the Morris Jesup Rise, and the Yermak Plateau several events of freshwater discharge were recorded by both low oxygen isotope and low carbon isotope values during the last 200 ka (Figure 6.61). These freshwater events did not occur randomly at the various sites, but they were concentrated during defined time intervals at ~160 ka (MIS 6), 140–125 ka (late MIS 6/early MIS 5), 90–75 ka (upper MIS 5), 65–60 ka (MIS 4), and at ~50 ka (early MIS 3), and they were more or less synchronous with intervals of major IRD input, suggesting a common source (Spielhagen et al., 2004). Both freshwater supply and IRD input are related to the decay of major continental ice sheets in the Barents-Kara Sea area (Figure 6.62), causing major surges and outbursts of icebergs and meltwater plumes into the Arctic Ocean. One exception is a meltwater event recorded at Core PS51/038-4, probably correlating with MIS 5.4

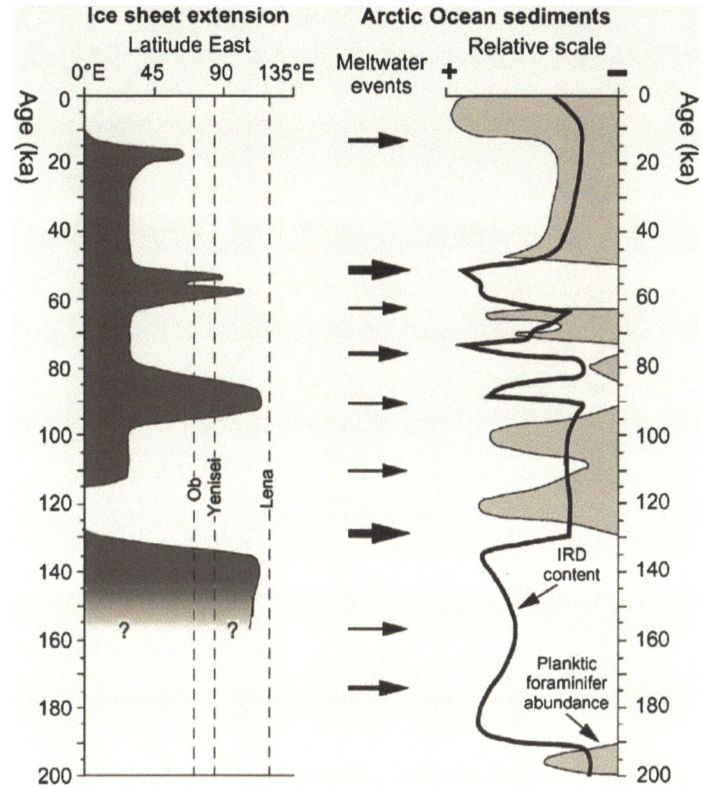

Figure 6.62 Comparison of reconstructions of the eastward extension of Eurasian ice sheets during the past 150 ka (based on Svendsen et al., 2004) with generalized records of IRD contents, abundance of planktonic foraminifers, and major meltwater peaks (black arrows) from the central Arctic Ocean (from Spielhagen et al., 2004). The geographical longitudes of present river mouths of Ob, Yenisei, and Lena are indicated by vertical lines.

and coinciding with a pink-white layer and the boundary of lithological units M and L (Figure 6.61). This event is related to a freshwater discharge due to the decay of North American ice sheets (see later discussions).

The most prominent freshwater events related to discharge from Siberia were most likely controlled by freshwater supply from major lakes formed during times of most extended glaciations. Based on the terrestrial reconstructions, the major Siberian rivers of Ob and Yenisei were dammed by major ice sheets at least three times during the past 200 ka, that is, at ~190–130 ka (Saalian), 90–80 ka (Early Weichselian), and 65–50 ka (Middle Weichselian) (see Section 6.2.2, Figures 6.24 and 6.25; Mangerud et al., 2001, 2004; Svendsen et al., 2004). The northward drainage of the rivers was blocked and led to the development of huge lake systems south of the ice sheet. During late glacial/deglaciation times, when a northern pathway for the water opened, the freshwater was released (probably as catastrophic event) into the Arctic Ocean, reflected in the prominent isotope spikes and IRD with a Siberian origin (Figures 6.61 and 6.62).

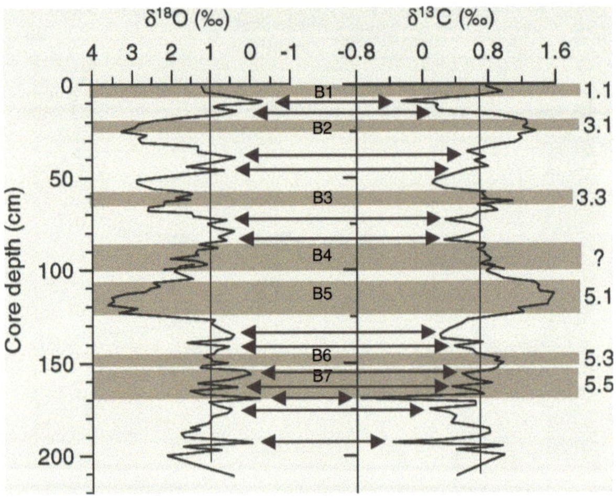

Figure 6.63 Stable-isotopic records from the Mendeleev Ridge, western Arctic, showing glacial–interglacial contrasts (Polyak et al., 2004; Darby et al., 2006, supplemented). Interglacial/interstadial intervals (B1 to B7) are shaded and tentatively correlated with standard Marine Isotope Stages (on the right) via correlation with stratigraphies in the eastern Arctic Ocean (Jakobsson et al., 2000a, 2001; Backman et al., 2004). Black arrows mark identified freshwater events characterized by low values of both parameters.

Polyak et al. (2004) also suggested that the down-core variations with distinct light values of both oxygen and carbon isotopes in the NP26 record from Mendeleev Ridge primarily reflect the history of freshwater budget rather than temperature changes. Using the stratigraphic model published in Darby et al. (2006) these minima representing freshwater pulses are concentrated in the late MIS 6/early to middle MIS 5, MIS 4, middle part of MIS 3, and the last deglaciation MIS 2/MIS 1 (Figure 6.63). Below B5 (i.e., probably older than MIS 5.1) light $\delta^{18}O$ values are more abundant, presumably indicating a particularly strong flux and storage of meltwater in the Amerasia Basin (Polyak et al., 2004). The youngest spike correlates to that in earlier investigated box core records from the Mendeleev Ridge (see Section 6.3.5, Figure 6.71; Poore et al., 1999) and represent a major meltwater event corresponding to the last deglaciation of the Arctic margins during Termination I (*cf.* Stein et al., 1994a; Nørgaard-Pedersen et al., 1998).

Concerning the source of the meltwater, distinct differences are obvious from the composition of contemporaneous IRD layers (Figure 6.58). Meltwater events at MIS 5.4, MIS 4/3, and MIS 2, for example, coincide with distinct IRD spikes characterized by pink-white carbonate layers and Fe oxide grains. This supports an origin of melting icebergs from the Laurentide part of the Canadian Archipelago. Meltwater events tentatively dated to MIS 6, upper MIS 5, and within MIS 3, on the other hand, correlate with IRD originated from the Siberian Arctic, that is, the East Siberian, Laptev, Kara, and/or Barents Sea. The meltwater signal following MIS 3.3 and characterized by a strong signal for a Barents/Kara Sea source (Figure 6.58) may correlate with the MIS 3 signal found in the records studied by Poore et al. (1999) (Section 6.3.5, Figure 6.71). If this correlation is correct, this would support Spielhagen et al.'s (2004) interpretation of a widespread meltwater event influencing the entire central Arctic and originated from Siberia, rather than Poore et al.'s (1999) interpretation who relate the MIS 3 meltwater signal on the Mendeleev Ridge to the deglaciation of the Innuitian Ice Sheet. On the other hand, minimum $\delta^{18}O$ values in the Mendeleev Ridge freshwater spikes are significantly lower than those in the cores from the Eurasian Basin/Lomonosov Ridge area, which suggests a closer proximity to the freshwater source. Thus, it remains open yet whether there were several (at least two) synchronous freshwater discharges from different sources into the Arctic Ocean near 50 ka or whether freshwater spread across much of the Arctic Ocean from a northern Eurasian source (Spielhagen et al., 2004).

Although, in general, periods of some of the major freshwater events seem to be contemporaneous in the Mendeleev Ridge core and in the cores from the Alpha and Lomonosov ridges, the Morris Jesup Rise, and the Yermak Plateau (i.e., MIS 6/MIS 5 and MIS 4/MIS 3) (Figures 6.61 and 6.63), a closer look at the correlation between the records is difficult due to the limited time control of the records for these time intervals.

In some interglacials and/or interstadial intervals in the western Arctic Ocean (i.e., especially in B5 and B2, probably corresponding to MIS 5.1 and 3.1; Figure 6.63) heavy spikes of planktonic $\delta^{18}O$ of $>3‰$ were found that are heavier than modern values by as much as 2‰. Because water could not have been much colder than modern temperatures that are just slightly above the freezing point (see Chapter 2.2), these heavy $\delta^{18}O$ spikes may reflect anomalous reductions in the amount of fresh water in the Arctic Ocean. The causes of these events of increased salinity are not well understood yet and require further investigation (Darby et al., 2006). Polyak et al. (2004) propose as one possible explanation that a large portion of the $\delta^{18}O$ increase was caused by a cessation of low-salinity Pacific surface waters, with a $\delta^{18}O$ value *ca.* 1.5‰ lower than Atlantic waters, when the Bering Strait was closed.

6.3.1.4. Freshwater events prior to MIS 6

Very recently, Knies et al. (2007b) showed in a detailed high-resolution study of sediments from ODP Site 910 (Yermak Plateau; see Figure 6.54 for location) that also in time intervals prior to MIS 6 at least 14 major Arctic Ocean freshwater events comparable to those described earlier, occurred (Figure 6.64). During most of these freshwater events, a contemporaneous decrease in NADW formation was observed, as recorded in a benthic $\delta^{13}C$ record from the North Atlantic (ODP Site 980, 55°29.1′N, 14°42.1′W; Flower et al., 2000), suggesting that the episodic freshwater discharge has contributed to fundamental perturbations of thermohaline circulation (THC) during the middle-late Pleistocene. Most prominent are freshwater pulses during glacial terminations such as the MIS 6/5, MIS 12/11, and MIS 16/15 transitions, as well as on top of the hiatus (MIS 20/19 transition) (Figure 6.64). Similar to the younger freshwater events, the disintegration of large ice sheets and the melting of icebergs are inferred to be the main sources for the freshwater pulses during these periods. As proposed by Knies et al. (2007b),

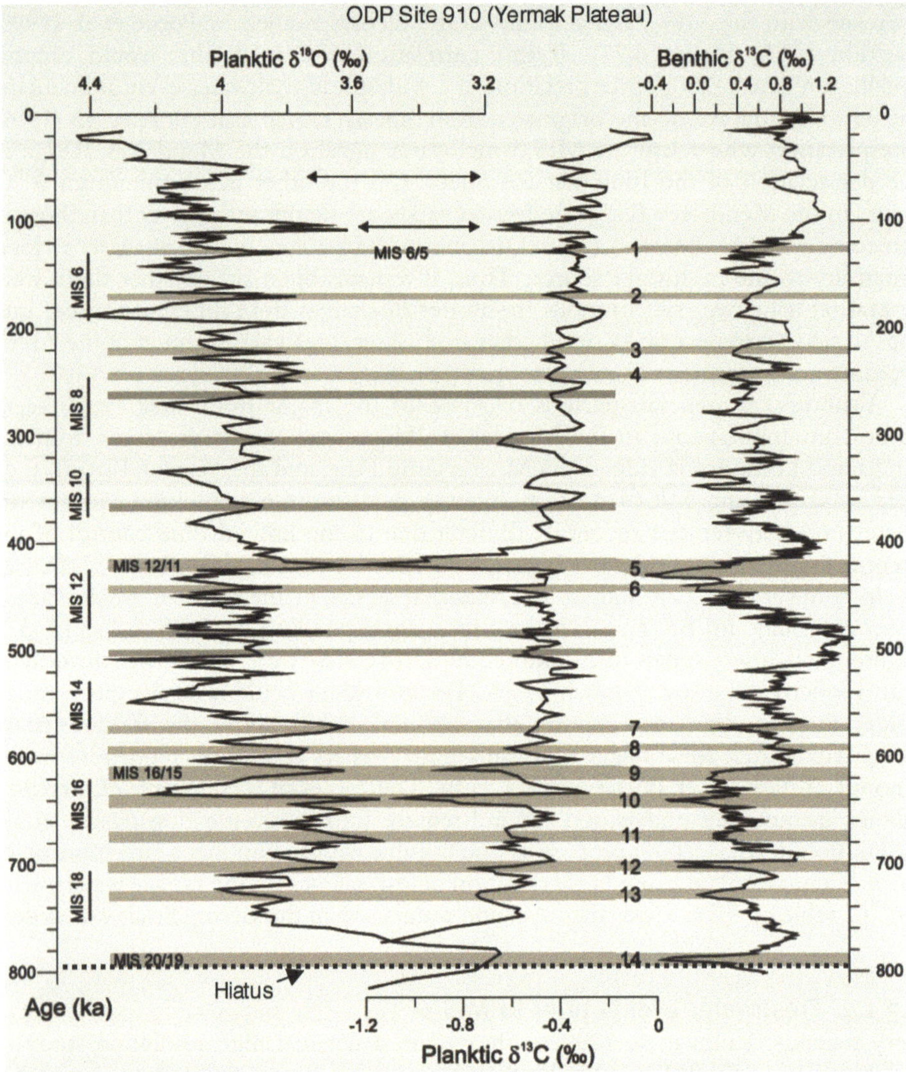

Figure 6.64 Climate proxy data from the Atlantic–Arctic gateway (from Knies et al., 2007b, supplemented). (Left) Three-point smoothed stable oxygen (δ^{18}O) and carbon (δ^{13}C) isotope values of planktonic foraminifer *N. pachyderma* (sin.) in Hole 910A versus age (Knies et al., 2007b). (Right) Smoothed benthic ^{13}C record from the North Atlantic (ODP Site 980, 55°29.1′N, 14°42.1′W; Flower et al., 2000). Grey bars mark identified freshwater events in the 910 record, characterized by low δ^{18}O and δ^{13}C values. Arrows indicate well-constrained Arctic freshwater pulses during early Weichselian (see Figure 6.62). Stippled line near 800 ka (MIS 20/19) marks proposed hiatus.

a prominent sequence including the hiatus (~790–950 ka) attributed to a pronounced ice sheet advance (Flower, 1997) or ploughing of deep-draft icebergs (Kristoffersen et al., 2004), and a subsequent distinct freshwater pulse are concurrent with the mid-Pleistocene transition. The latter is characterized by a significant increase in global ice volume from 942 to 892 ka relative to the background of low amplitude obliquity cycles (Mudelsee & Schulz, 1997). The timing of ice-collapse and the release of freshwater during the MIS 20/19 transition is presumably consistent with the oldest documented glaciation in northern Eurasia, older than 700 ka (Astakhov, 2004; see also Chapter 1.4, Figure 1.18), and can be traced into the Nordic Seas (Helmke, Bauch, & Erlenkeuser, 2003) suggesting a reorganization of ocean circulation triggered primarily by an Arctic freshwater source.

6.3.1.5. Summary

In summary, mineralogical and geochemical proxies determined in the central Arctic Ocean sediments on Lomonosov Ridge clearly point to several periods of strong IRD input via the TPD from the Eurasian continental margin. These data indicate extended glaciations in northern Eurasia at ~190–130 ka (MIS 6), 90–80 ka (MIS 5.2), ~75 ka (uppermost MIS 5), and 65–50 ka (MIS 4/3), when ice sheets/glaciers reached the shelf seas and discharged large quantities of icebergs into the Arctic Ocean during late glaciation maxima/early deglaciation (Figure 6.62). Almost contemporaneously with the IRD input, huge amounts of freshwater were released to the Arctic Ocean, as identified by $\delta^{18}O$-depleted meltwater events during ice-sheet decay at glacial terminations and/or collapses of ice-dammed lakes in the circum-Arctic. The large amounts of freshwater from the largest events at ~130 and 52 ka may even have reached the Nordic Seas and thus contributed to the contemporaneous low-salinity events in the northernmost North Atlantic (Spielhagen et al., 2004).

These results are in excellent correlation with those from recent field work in northern Scandinavia, European Russia, Siberia (see Section 6.1) as well as marine records from the Eurasian continental margin (see Section 6.2), also indicating extended glaciations at those time intervals — with one exception. Whereas in all of the sediment cores discussed earlier a very strong IRD signal with a Siberian source was determined ~75 ka (MIS 5.1), no terrestrial evidence for an extensive glaciation in northern Eurasia between Scandinavia and the Taymyr Peninsula at the MIS 5/4 boundary has been found yet. Because the stratigraphy of the deep-sea cores seems exceptionally well-constrained for the MIS 5.1 interval, Spielhagen et al. (2004) strongly believe that the arguments for some glaciation in the potential source area of smectite and kaolinite at the MIS 5/4 boundary are well substantiated. They speculate that possibly a number of smaller ice domes existed which were confined to the various archipelagos on the Barents and Kara seas shelves and supplied icebergs to the Arctic Ocean, a speculation, however, which has to be proven by further field work.

During the LGM (late Weichselian) time interval, relatively low amounts of IRD were found in central Arctic Ocean sediments which also agrees very well with the recent reconstruction of a very limited eastern ice sheet extension during this time interval (Spielhagen et al., 2004).

In general, the IRD-rich intervals related to major continental glaciations are alternating with periods of abundant planktonic foraminifers in the coarse fraction (Figure 6.62), indicating intensified Atlantic water inflow to the Arctic Ocean and related seasonally open waters (leads) and increased surface-water productivity in the central Arctic Ocean. The open-water conditions were an important moisture source which fostered the growth of ice sheets on northern Eurasia (*cf.* Hebbeln et al., 1994; Dokken & Hald, 1996; Knies et al., 2000).

The records from the Amerasian Basin indicate extended glaciations in the terrestrial North American Arctic during MIS 6, MIS 5.4, and MIS 4/3. These intervals are also alternating with interglacials/interstadials characterized by at least seasonally more open-water conditions with some increased productivity. That means, the evolution of glaciations in the Eurasian Arctic and in North America and the palaeoenvironment in the Amerasian and Eurasian basins seems to be similar and contemporaneous in general at least for the MIS 6 and MIS 4/3. Differences seem to be obvious for the MIS 5.2 (signal from Eurasia) and MIS 5.4 (signal from North America) glaciations. Limited age control (perhaps except for the MIS 2 to MIS 1 time interval; see Section 6.3.4), especially for the cores from the Amerasian Basin, however, makes a more detailed comparison of the glacial history in both areas difficult.

The appearance of sand-sized Fe grains from the Russian shelves, such as the Laptev Sea, found in the Amerasian Basin during the late Quaternary glacial/interglacial cycles suggests that sediment input was repeatedly significantly influenced by an extended TPD system towards North America, that is, a situation described for the Holocene (see Section 6.3.6; Polyak et al., 2004; Darby et al., 2006).

6.3.2. Erosional Event on Lomonosov Ridge: Indication for an Extended MIS 6 Ice Shelf?

Jakobsson (1999) detected erosional features on top of the Lomonosov Ridge (above *ca.* 1,000 m water depth) which were interpreted as evidence for ice grounding and the development of a floating ice shelf over part of the eastern Arctic Ocean in the Late Quaternary (Figure 6.65A; Polyak et al., 2001, 2007). This

Figure 6.65 (A) Schematic profile across the Arctic Ocean (from Spielhagen, 2001, supplemented; for location see Figure 6.54A), indicating the topography of the ocean basin and its subdivisions, and the possible extent of floating ice shelves as proposed by Polyak et al (2001). These ice shelves may have developed asynchronously, that is, MIS 2 and MIS 6. CP, Chukchi Plateau; LR, Lomonosov Ridge. (B) Swath sonar and (C) chirp images from Lomonosov Ridge (from Polyak et al., 2001, supplemented). The planed, eroded top surface of the crest is covered with ESE-trending low-relief flutes. The acoustically transparent debris lens drapes the lee (Amerasian) slope of the ridge. (D) Bathymetry of Yermak Plateau with depths shallower than 1,000 m colour-coded (light blue <1,000 m and light green <250 m) (from Kristoffersen et al., 2004a, supplemented). Areas of iceberg erosion (red) and deposition (brown) are outlined, and the deepest seabed impacts of deep draft ice are indicated by the deep blue line. Short light green line segments represent iceberg plowmarks in the southern part from Vogt, Crane, & Sundvor (1994) and in the northern part from new SCICEX data. Locations of Ocean Drilling Program Sites 910–912 are shown by circles. (E) Line drawings and seismic sections across Yermak Plateau (from Kristoffersen et al., 2004a, supplemented). Stratigraphic information on line AWI91128-31 from Myhre et al. (1995) and velocity information on line BU7917 from Austegard (1982).

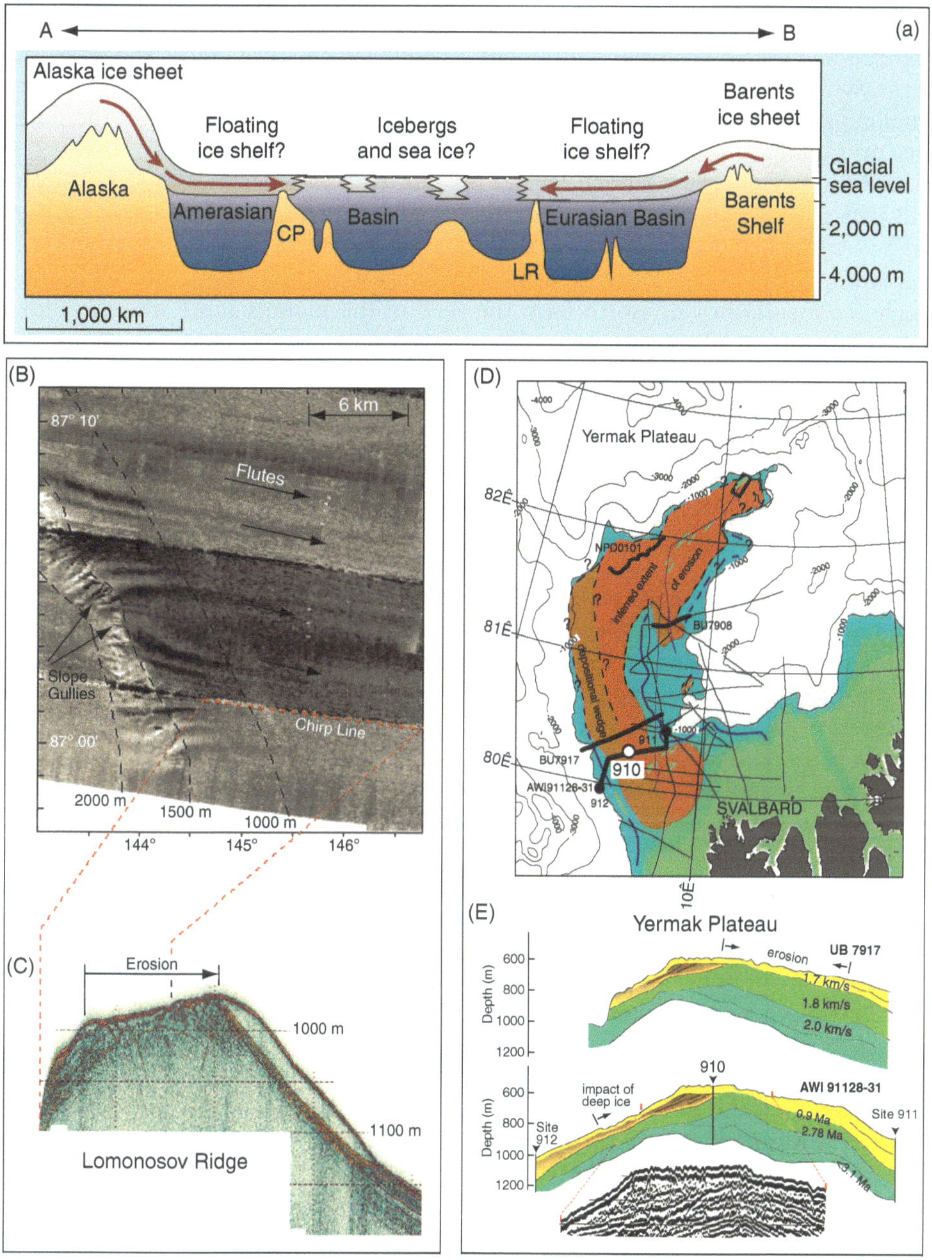
(a)
A
B
Alaska ice sheet
Barents ice sheet
Floating ice shelf?
Icebergs and sea ice?
Floating ice shelf?
Alaska
Amerasian
CP
Basin
LR
Eurasian Basin
Barents Shelf
Glacial sea level
2,000 m
4,000 m
1,000 km
(B)
87° 10'
6 km
Flutes
Slope Gullies
Chirp Line
87° 00'
2000 m
1500 m
1000 m
144°
145°
146°
(C)
Erosion
1000 m
1100 m
Lomonosov Ridge
(D)
Yermak Plateau
82E
81E
80E
NPD0101
inferred extent of erosion
depositional wedge
BU7908
BU7917
911
910
AWI91128-31
912
SVALBARD
10E
(E)
Yermak Plateau
erosion
UB 7917
1.7 km/s
1.8 km/s
2.0 km/s
Depth (m)
600
800
1000
1200
910
impact of deep ice
AWI 91128-31
Site 912
Site 911
0.9 Ma
2.78 Ma

interpretation is based on sea-floor images collected during the SCICEX-99 expedition using a submarine-mounted 12-kHz swath bathymetry and side-scan sonar, which display a variety of bedforms, including random or subparallel scours, parallel lineations, and transverse ridges. The direction of ice flow across the fluted area on Lomonosov Ridge is revealed in the chirp-sonar records, where the crest is planed by erosion over a stretch at least 50-km wide at a depth of almost 1,000 m, and a 20-m-thick lens of acoustically transparent sediment descends from the eroded surface >100 m down the Amerasian flank of the ridge (Figure 6.65B and C). According to Polyak et al. (2001), these features suggest erosion of the crest by a single ice massif, moving north from the shelf of the Barents and Kara seas, with a debris lobe of reworked material pushed downslope on the lee side. The dipping of the eroded surface towards the Eurasian side further shows that the ridge was a barrier to the passage of ice.

Removal of unconsolidated sediments along the Eurasia Basin side of the Lomonosov Ridge at water depths shallower than 1,000 m is also reflected in the multichannel seismic and chirp-sonar data from the same area published by Kristoffersen et al. (2004a). The erosion of the hemipelagic drape on top of the Lomonosov Ridge, however, seems to be incomplete. In contrast to Polyak et al. (2001) these authors propose that erosion was caused by deep-draft glacier ice reaching the central Arctic Ocean from the Eurasia Basin in the form of armadas of large icebergs embedded in sea ice rather than by a single, continuous floating ice shelf. New seismic data published by Jokat (2003), provide indications for the occurrence of iceberg scouring along the southeastern Lomonosov Ridge, however, the area of grounded ice is restricted to small areas between 88°N and 86°N, that is, to a maximum lateral extent on Lomonosov Ridge of 200 km. No evidence for a regional, massive glacial erosion is observed along the ridge north of 88°N and south of the 86°N. These data suggest that large icebergs may have affected Lomonosov Ridge sediments during glacial times, supporting the view published by Kristoffersen et al. (2004a) that there is little or no evidence for a large-scale ice sheet over the deep Arctic Ocean (Jokat, 2003). For a more detailed discussion about pros and cons of deep-draft icebergs versus grounded ice shelf causing erosion on Lomonosov Ridge, the reader is referred to the original papers by Polyak et al. (2001) and Kristoffersen et al. (2004a), and a review paper by Spielhagen (2001).

Concerning the timing of the erosional event on Lomonosov Ridge, ages of MIS 16 versus MIS 6 were discussed (Jakobsson et al., 2001; Polyak et al., 2001). Based on studies of several sediment cores from below and above the erosion level and using the "young stratigraphy model" (see Section 6.1), a MIS 6 age of the erosional event seems to be most probable (see Section 6.1.5, Figure 6.16; Jakobsson et al., 2001). Although the sediment cores studied by Spielhagen et al. (2004) are from water depths below the erosional unconformity and thus represent undisturbed records of sedimentation in MIS 6, the records neither support nor reject the ice shelf hypothesis of Polyak et al. (2001). Because there is no modern analogue for an ~1,000 m thick floating ice shelf which crossed a *ca.* 500 km wide deep-sea basin, Spielhagen et al. (2004) gave some speculation about the sediment type deposited under the floating ice. As outlined by these authors, such sediments must be free of planktonic organisms because those live in the uppermost euphotic zone of the water column. The high IRD contents in MIS 6 sediments are a feature which may result from sediment transport at the underside of the ice. Core PS51/038-4 from the Alpha Ridge, contains low amounts of planktic foraminifers throughout MIS 6 indicating low, but continuous bioproductivity at this site. Thus, the Alpha Ridge may not have been reached by the ice shelf. Cores PS2185 and PS2200, on the other hand, contain foraminifer-free intervals in the MIS 6 section, but none of these layers shows any special features or peaks in the grain-size distribution or clay mineral composition (Spielhagen et al., 2004).

Even a final judgement between the two hypotheses causing the erosion on Lomonosov Ridge seems to be not possible yet and more seismic and geological data along the Lomonosov Ridge are needed to determine the extent of erosion on Lomonosov Ridge, both hypotheses indicate a large-sized continental ice sheet in Eurasia during MIS 6, supporting records from the Eurasian continental margin (see Section 6.2.2) as well as terrestrial data (see Section 6.2.1, Figure 6.24).

Further erosional features related to glaciogenic processes were also found on the crest of the outer Morris Jesup Rise, on the Chukchi Plateau, and on the Yermak Plateau, probably developed during times of maximum glaciation of very different ages (see also Chapter 3.2). The features on Morris Jesup Rise were interpreted as the result of ice grounding and tentatively correlated in their age to the ice grounding event on the Lomonosov Ridge in MIS 6 (Spielhagen et al., 2004). On Chukchi Plateau, two stages of erosional events were dated to MIS 5.4 and the LGM, presumably caused by the grounding of ice shelves originating from the LIS (see Chapter 3.2, Figure 3.11; Polyak et al., 2007). The features related to iceberg erosion on the Yermak Plateau are iceberg plough marks present in the seabed down to 870 m water depth, and an erosional unconformity as well as overconsolidated sediments documented at ODP Site 910 (Kristoffersen et al., 2004a). Depositional wedges are consistently present along the western slope of Yermak Plateau and suggest erosion at the northern and eastern part of the plateau by ice exiting the Arctic Ocean (Figure 6.65D and E). Concerning the timing of the erosional event, however, erosion on the Yermak Plateau seems to be significantly older than the event on Lomonosov Ridge dated into MIS 6, as the unconformity at ODP Site 910 has probably an age of ~MIS 20/19 (see Figure 6.64; Knies et al., 2007b).

6.3.3. Sea-Ice Cover and Surface-Water Characteristics During the Last Glacial Maximum

Nørgaard-Pedersen et al. (2003) carried out a very detailed study on a large number of well-dated sediment cores from the Eurasian sector of the Arctic Ocean, representing the MIS 2/1 time interval (Figure 6.66). Based on sedimentological, micropalaeontological, and stable isotope data, these authors were able to characterize and map regions of different palaeoceanographic conditions for the LGM time slice (15–18 ^{14}C ka or 18.0–21.5 Cal. kyr BP according to GLAMAP; Sarnthein et al., 2003b). This reconstruction is based on (1) the spatial distribution of δ^{18}O values as proxy for the distribution of Atlantic and polar water masses (see Chapter 4.6), (2) flux records of planktonic foraminifers as productivity proxy

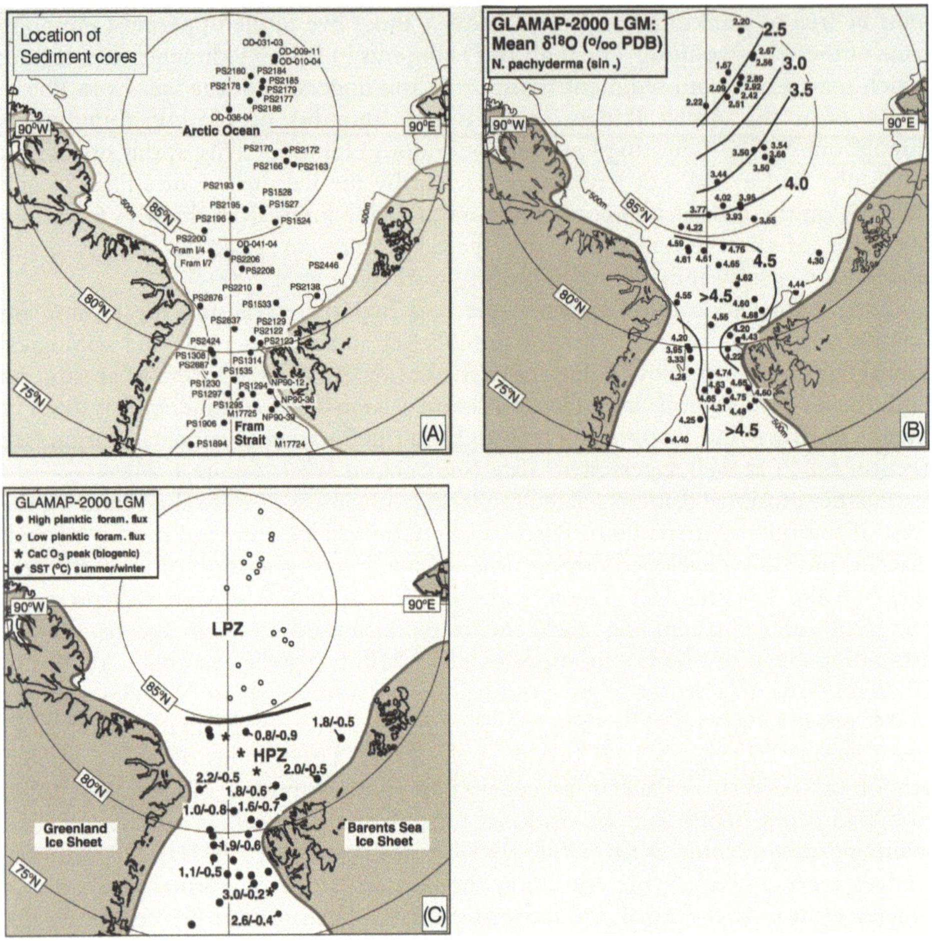

Figure 6.66 Stable oxygen isotopes, planktonic foraminifer fluxes, and SSTs during the LGM (from Nørgaard-Pedersen et al., 2003, supplemented). (A) Location of sediment cores used from the Fram Strait and the adjacent Arctic Ocean. The approximate extension of continental ice sheets during the LGM is indicated. (B) Distribution of mean $\delta^{18}O$ values of *N. pachyderma* (sin.) for the GLAMAP 2000 LGM time slice (21.5–18.0 Cal. kyr BP) in Fram Strait to central Arctic records. (C) Distribution of mean summer and winter SST estimated for the GLAMAP 2000 LGM time slice on selected records (according to Pflaumann et al., 2003). Also shown are LGM records with a high planktonic foraminiferal flux (solid circles) and/or a prominent biogenic $CaCO_3$ peak (HPZ, high productive zone), and the interior Arctic sites north of 85°N (open circles) characterized by an extremely low flux of planktonic foraminifers (LPZ, low productive zone).

reflecting nutrient supply and degree of ice cover (see Chapter 4.5.1; Hebbeln & Wefer, 1991; Hebbeln et al., 1994), and (3) the IRD content as proxy to estimate the input of terrigenous sediments transported by icebergs derived from continental ice sheets calving into the Arctic Ocean. As a result, Nørgaard-Pedersen et al. (2003) could separate three areas characterized by different sedimentation regimes and surface ocean properties during the LGM (Figure 6.67): (1) the eastern Fram

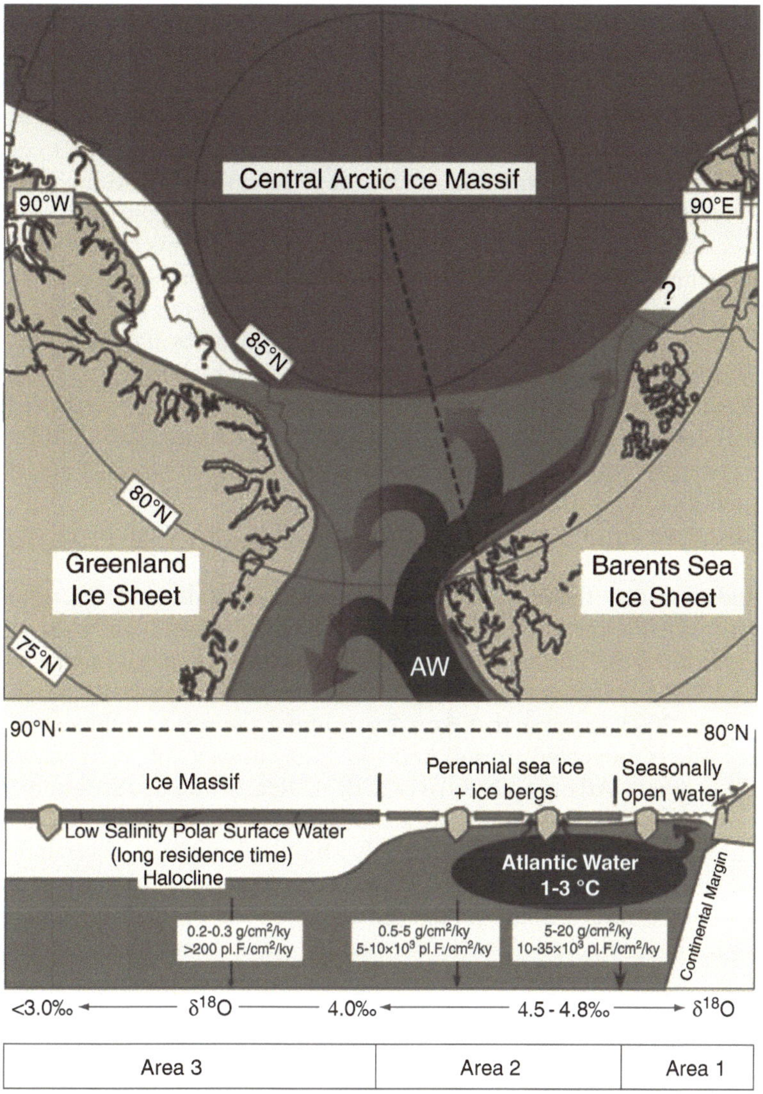

Figure 6.67 Simplified model of sea-ice and surface-ocean characteristics in the Fram Strait-central Arctic Ocean region during the LGM (from Nørgaard-Pedersen et al., 2003, supplemented). On the latitudinal transect from the northern Svalbard continental margin to the North Pole, related proxy data as bulk accumulation rates, planktonic foraminiferal flux, planktic $\delta^{18}O$ values, and mean summer SST values of the Atlantic water mass are shown. Advection and recirculation of relatively warm and saline Atlantic water in the Fram Strait-southwestern Eurasian Basin caused an open ice cover and a relatively high flux of biogenic and lithic material. Polynyas probably occurred at the Barents Sea continental margin. From $\sim$85°N and further into the interior Arctic, it is proposed that the sea-ice cover was permanently thick, resting upon cold, low-salinity halocline waters. This may explain the extreme low interior Arctic flux values. For definition and characteristics of areas 1–3 see text.

Strait and the northern Barents Sea margin area, (2) the western Fram Strait and the southwestern Eurasian Basin up to ~84–85°N, and (3) the central Arctic Ocean (north of 85°N in the Eurasian Basin). In summary, these areas have the following characteristics (for a more detailed discussion, see Nørgaard-Pedersen et al., 2003):

Area 1, the eastern Fram Strait and the northern Barents Sea margin area, is characterized by high sedimentation rates of 2–10 cm kyr^{-1} and high abundances of planktonic foraminifers (~4,000–6,000 specimens g^{-1} sediment). Today, such environments are found in areas of seasonally changing ice cover as the central Fram Strait (Hebbeln & Wefer, 1991). The high $\delta^{18}O$ values of *N. pachyderma* (sin.) of 4.5–4.8% (Figure 6.66B) and the estimates of summer SSTs of ~1.6–3.0°C (Figure 6.66C, according to Pflaumann et al., 2003) suggest a strong inflow of Atlantic water (Figure 6.67; see also Knies et al., 2001). The high abundances of planktonic foraminifers (and fluxes of 10–35 × 10^3 specimens cm^{-2} kyr^{-1}; Nørgaard-Pedersen et al., 2003) correspond to the glacial "high productive zones" (HPZ) first reported from the eastern Fram Strait and the Norwegian Sea by Hebbeln et al. (1994) and Dokken and Hald (1996).

Area 2, the western Fram Strait and the southwestern Eurasian Basin up to ~84–85°N, is characterized by lower sedimentation rates of 1–2 cm kyr^{-1} and moderately high abundances and fluxes of planktonic foraminifers, high $\delta^{18}O$ values of *N. pachyderma* (sin.) (Figure 6.66B), and summer SST estimates slightly lower than in Area 1 (Figure 6.66C). This region may have been characterized by an ice cover with some open leads in summer, similar to the present interior Arctic Ocean. Nørgaard-Pedersen et al. (2003) suppose that Area 2 was under the steady influence of Atlantic subsurface waters advected from Area 1. The impact of recirculating saline and relatively warm Atlantic water over a large area of the southwestern Eurasian Basin was the decisive factor causing a break-up of the ice cover, relatively high sedimentation rates and a comparatively high planktonic foraminiferal flux (~5–10 × 10^3 specimens cm^{-2} kyr^{-1}; Figure 6.67, Nørgaard-Pedersen et al., 2003).

Area 3, the central Arctic Ocean north of 85°N in the Eurasian Basin, is characterized by extremely low sedimentation rates (dominantly <1 cm kyr^{-1}; Table 6.1; see also Levitan & Stein, 2007), low abundances and fluxes of planktonic foraminifers (only a few hundred foraminifers g^{-1} sediment and <0.2 × 10^3 specimens cm^{-2} kyr^{-1}, respectively), and very low IRD (>500 μm) values <1% (Figure 6.67; Nørgaard-Pedersen et al., 2003). These data suggest the existence of an extensive and thick sea-ice cover with low seasonal variation, limiting planktonic productivity ("low productivity zone") (Figures 6.66C and 6.67) and minor IRD release in the eastern central Arctic during the LGM. Due to low temporal resolution, however, it has to be considered, that the data from the central Arctic Ocean are tentative and probably characterize average MIS 2 conditions rather than a specific LGM time slice (Nørgaard-Pedersen et al., 2003). Within Area 3, a persistent trend towards lower $\delta^{18}O$ values (>2%, similar to the present gradient, however, with a much steeper gradient) can be observed (Figure 6.66B; Chapter 4.6, Figure 4.46). Nørgaard-Pedersen et al. (2003) suggest that low-salinity halocline properties were maintained in the interior Arctic Ocean during the peak glacial. A very similar scenario has also already been proposed from an earlier study carried out in the central Arctic Ocean by Stein et al. (1994a, 1994b). Based on stable isotope data, these authors found that in the Lomonosov Ridge area the salinity of the last glacial was very similar to the modern one. Assuming a stable halocline, SSTs were most likely (as today) close to the freezing point (i.e., $-1.7°C$; see Chapter 2.2). Maintenance of the low-salinity layer may have been caused by yet unknown freshwater sources from the Siberian-North American margins (Nørgaard-Pedersen et al., 2003). Stein et al. (1994b) concluded that the TPD probably have also transported significant quantities of low-salinity waters from the Siberian shelf seas into the central Arctic Ocean during glacial times. This interpretation implies that the Siberian river discharge into the Arctic also continued during the LGM, a scenario supported by sediment and PARASOUND records from the southern Laptev Sea (Kleiber et al., 2001) as well as the southern Kara Sea (Dittmers et al., 2003, 2008).

MIS 2 (including the LGM) was already identified as a period of very limited bioproduction and extremely low sedimentation rates due to a massive sea-ice cover with limited seasonal variation from earlier studies of sediment cores from the eastern and western central Arctic Ocean (Darby et al., 1997; Nørgaard-Pedersen et al., 1998; Poore et al., 1999). Cores from the Northwind Ridge and the Chukchi Plateau (Amerasian Basin) are barren of foraminifers in the glacial interval (Darby et al., 1997; Phillips & Grantz, 1997), suggesting an even thicker and more coherent sea-ice cover in the western central Arctic Ocean, which had a sufficient thickness to reduce solar irradiance to levels that precluded photosynthesis. In some AMS^{14}C-dated sediment cores from the western Arctic Ocean even a non-depositional hiatus during the LGM was proposed; therefore, the possibility of an ice shelf floating over that part of the Arctic Ocean cannot be ruled out (Darby et al., 1997; Poore et al., 1999; Polyak et al., 2001, 2007).

6.3.4. Ice-Sheet Variability, Ice Export, and Sea-Ice versus Iceberg Transport During MIS 2–MIS 1

While there are very few icebergs calved into the Arctic Ocean, crossing the central basins, and exiting Fram Strait today (see Chapter 3.2, Figure 3.9), large quantities of glacial ice drifted across the central Arctic Ocean and finally through Fram Strait during the late Pleistocene. These pulses in IRD input during MIS 2 are mainly related to sudden, massive iceberg calving events in the North American Arctic ice sheets, as reconstructed by Darby et al. (2002) from sedimentary records from the western central Arctic Ocean to Fram Strait (Figure 6.68). In these cores, IRD composition $>250\,\mu m$ and Fe oxides were used to trace the IRD in the glacial marine sediments back to its sources (see Chapter 4.3.1 for background). Key core in their reconstruction is the AMS^{14}C-dated box Core PS1230 located in 1,235 m water depth at 78.9°N, 4.8°W in the central Fram Strait (Figure 6.68) where most of the Arctic sea ice is exported today (90% of the modern sea ice drifts through Fram Strait between 0°W and 10°W; Vinje et al., 1998).

The record of Core PS1230 representing the last 34 Cal. kyr BP (upper MIS 3 to MIS 1), show significant, rapid fluctuations in the percentages of Fe oxide grains from different sources (Figure 6.69; Darby et al., 2002). Whereas in the uppermost (Holocene) fine-grained part of the record, IRD input via sea ice is dominant (see

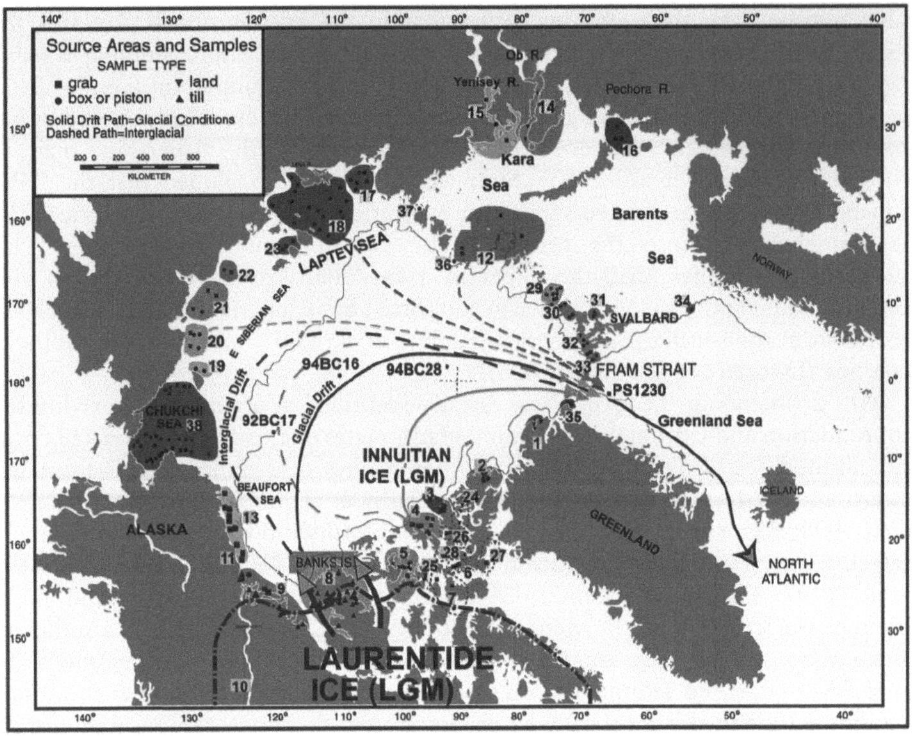

Figure 6.68 Map showing circum-Arctic source areas (1–38) defined by unique source compositions (see Chapter 4.3.1 for background) and drift paths of icebergs from Arctic LIS and Innuitian Ice Sheet to Fram Strait (solid drift paths), and the location of box Core PS1230 (from Darby et al., 2002). Dashed drift paths of sea ice in the Arctic show the influence of the Beaufort Gyre during warmer intervals like the Holocene by displacing North American ice drift paths westward.

Figure 6.69 Amount of terrigenous coarse fraction and Fe oxide grains from different circum-Arctic source areas determined in Fram Strait Core PS1230 (see Figure 6.68 for location) (from Darby et al., 2002, supplemented). (A) Arctic Laurentide (AL) and Innuitian (I) IRD events based on detrital Fe oxide mineral grain compositions matched to source area 8 on Figure 6.68 and source areas 2–7 plus 24–28, respectively, plotted versus core depth. Calendar years (kyr BP) and location of AMS^{14}C datings (crosses) as well as major lithologies are shown at the left-hand side of the figure. Numbers 14–34 in the white rectangle indicate ages of the main AL1–AL 5 events in Cal. kyr BP, the black bar the LGM. (B) IRD events originating from Siberian shelves, plotted versus calendar years. These events increase in frequency and amplitude during the last 13,000 years. Contemporaneously, amount of coarsed-grained IRD was decreased. In this figure, "ka" represents calendar years (Cal. kyr BP). Note that the Fe oxide peaks from the AL and I ice sheets rarely coincide with peaks in the overall abundance of coarse IRD, suggesting that the Fe oxide peaks are independent from fluctuations in the delivery of coarse IRD (Darby et al., 2002). SA, source area; for numbers see Figure 6.68.

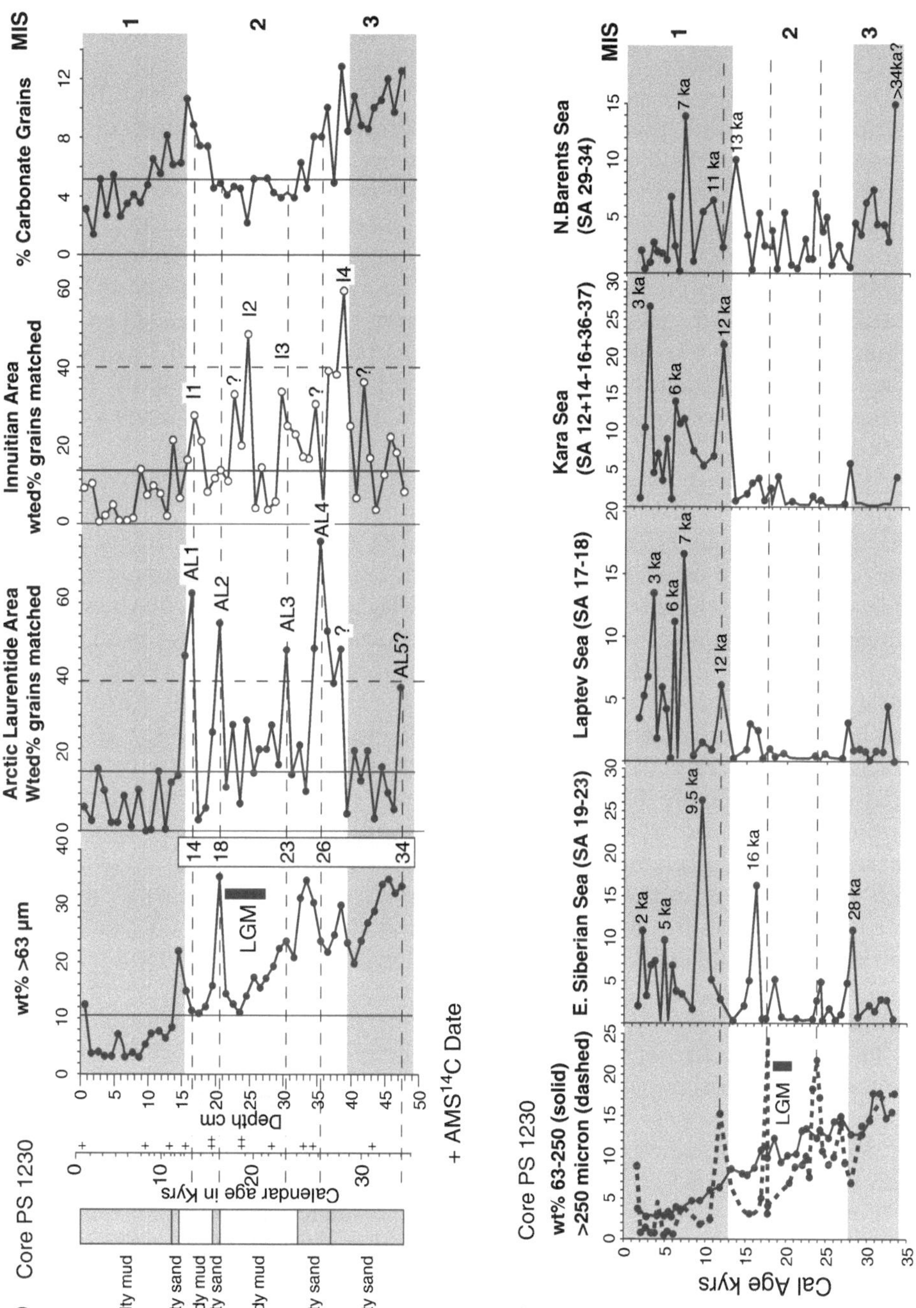
(A)
Core PS 1230
silty mud
silty sand
sandy mud
silty sand
sandy mud
silty sand
silty sand
Calendar age in Kyrs
Depth cm
+ AMS^{14}C Date
wt% >63 μm
LGM
Arctic Laurentide Area
Wted% grains matched
AL1
AL2
AL3
AL4
AL5?
Innuitian Area
wted% grains matched
I1
I2
I3
I4
% Carbonate Grains
MIS
(B)
Core PS 1230
wt% 63-250 (solid)
>250 micron (dashed)
Cal Age kyrs
LGM
E. Siberian Sea (SA 19-23)
Laptev Sea (SA 17-18)
Kara Sea
(SA 12+14-16+36-37)
N.Barents Sea
(SA 29-34)
>34ka?

later discussions), Fe oxide maxima in the pre-Holocene probably represent primarily iceberg transport on the basis of the abundance of coarse IRD in the Pleistocene section and on the similarity of the IRD grain size distribution to glacial tills (Darby et al., 2002).

During the late Pleistocene (MIS 3/MIS 2), a major source of IRD was the northwestern LIS that calved into the Arctic Ocean, identified by detrital Fe oxide grains in PS1230 that precisely matches those in the tills of Banks and Victoria Islands (source area 8 on Figure 6.68). A northwestern Laurentide source is also supported by the presence of light-coloured detrital carbonate, which is relatively abundant in tills and derived from the extensive Palaeozoic carbonates exposed on Victoria Island (e.g., Bischof et al., 1996). This is indicated by the coincidence of Fe oxide peaks in PS1230 with elevated detrital, light-coloured carbonate peaks of 4 to >8% (Figure 6.69A). Another important but secondary source of Fe oxide grains in Core PS1230 is the Innuitian ice sheet in the northern Canadian Islands (source areas 2–7 and 24–28 on Figure 6.68; Bischof & Darby, 1999; Darby et al., 2002).

As outlined and discussed in detail by Darby et al. (2002), the relative strength of IRD delivery to this location was rather controlled by the quantity of ice drifting overhead and differences in the amount of entrained IRD than by changing melt out rates or drift paths. The rapid onset and relatively short duration of the Fe oxide grain peaks suggest massive and fast deglaciation events in parts of the Laurentide and Innuitian ice sheets during MIS 2 (Figure 6.69A; at ~26, 23, 18, and 14 Cal. kyr BP), producing large armadas of icebergs, that is, events probably similar to those recorded during Heinrich Events in the North Atlantic. The most prominent event seems to be the AL 2 event near 18 Cal. kyr BP, which coincides with a distinct IRD maximum, probably representing the decay of the LGM ice sheet (Figure 6.69A). Fe oxide peaks related to IRD input from the northwestern LIS, occur at about the same time in cores from the Lomonosov, Mendeleev, and Northwind ridges (Figure 6.70; Darby et al., 2002). In general, Fe oxide grain peaks from the Laurentide and Innuitian ice sheets occur at about the same depth intervals in PS1230. Where the resolution allows (e.g., at events 2–4), however, the Innuitian IRD events seem to precede each Arctic Laurentide event by ~1–1.5 kyr (Figure 6.69) and possibly trigger a larger Laurentide event (Darby et al., 2002).

These major collapses of the LIS and iceberg discharge events through Fram Strait caused major pulses of fresh water export into the Greenland–Iceland–Norwegian (GIN) seas/North Atlantic where it may have been effective in arresting the deep-water formation/THC (*cf.* Peltier, 2007 and references therein). Darby et al. (2002) estimated an annual flux of fresh water bound in glacial icebergs of 350 $km^3 yr^{-1}$ through Fram Strait during a single deglaciation event. For comparison, the range of estimates for iceberg flux through Hudson Strait during a Heinrich event is 312–2,800 $km^3 yr^{-1}$ (Dowdeswell, Maslin, Andrews, & McCave, 1995; MacAyeal, 1993). In total, the annual freshwater flux by sea ice and icebergs through Fram Strait during a deglaciation event may have reached 3,870 km^3 with a range between 3,780 and 4,130 km^3 (Bischof, 2000). Based on these estimates, the amount of freshwater that was exported from the Arctic

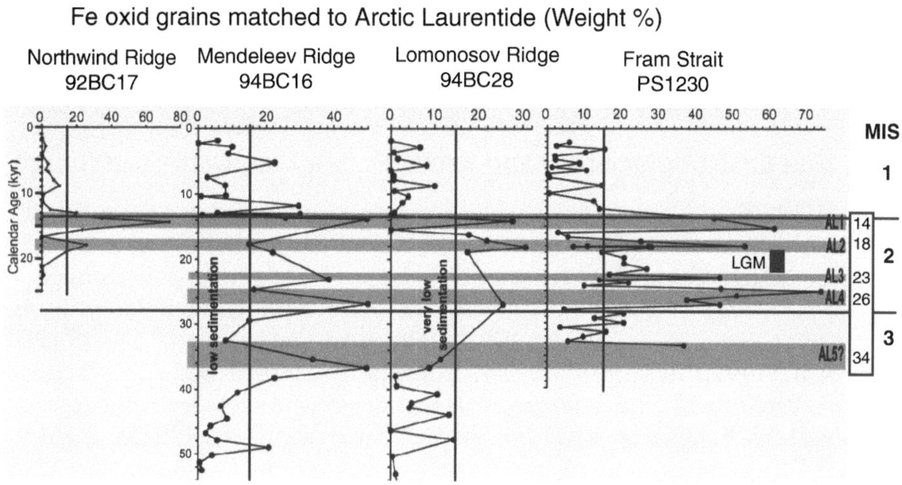

Figure 6.70 Correlation of Arctic Laurentide Fe oxide peaks in box cores across the Arctic Ocean (see Figure 6.68 for core locations) (from Darby et al., 2002, supplemented). Radiocarbon age models used for these correlations are from Darby et al. (1997) and Poore et al. (1999). Numbers 14–34 in the white rectangle indicate ages of the main AL1–AL 5 events in Cal. kyr BP, the black bar the LGM.

Ocean into the GIN seas by sea ice and glacial icebergs at that time was probably up to 65–80% larger than at present (Darby et al., 2002). In contrast, freshwater export rates through the Fram Strait region were probably much lower during the LGM than the present-day values (Nørgaard-Pedersen et al., 2003).

Darby et al. (2002) also compared the timing of the Arctic Laurentide and Innuitian IRD events with the timing of the Heinrich events. In general, the number of Arctic events and their occurrence intervals over the last 34 kyr are remarkable similar to those of Heinrich events, that is, the events 1–4 seems to correspond to H0–H3. The average age of each Arctic event, however, appears to precede the average age for each of the last four Heinrich events in the North Atlantic by ~1.5 kyr. Whether this lead–lag relationships which has important implications for ice sheet dynamics and the climate history, is real, however, has to be proven by further high-resolution studies of Arctic sediment cores. Yet, the time resolution of the Arctic Ocean is not good enough. Furthermore, it has to be considered, that larger reservoir age corrections of 0.8–1.5 kyr might have to be used for the Arctic cores as suggested by Waelbroeck et al. (2001) due to the change in circulation with climate changes in higher latitudes (>40°N). Such corrections would eliminate much, if not all, of the lead time between Arctic events and Heinrich events (Darby et al., 2002).

During MIS 1, the weight percentages of the >63 μm fraction decreased significantly, reaching <4% during the last 10 kyr (Figure 6.69). The predominance of silt and clay fraction suggest transport by sea ice rather than iceberg rafting (Pfirman et al., 1989; Reimnitz et al., 1998; Nürnberg et al., 1994). The contemporaneous increase in Fe oxide grains from unglaciated Eurasian shelves such as the Kara, Laptev, and East Siberian seas (Figure 6.69B) supports sea-ice rafting as a most important transport process during the Holocene (Darby et al., 2002; Darby, 2003; Darby & Bischof, 2004; see also Chapter 4.3.1, Figure 4.14).

6.3.5. The Last Deglaciation and Arctic Ocean Freshwater Discharge

At the MIS 1/2 boundary, a strong meltwater signal is recorded in a sharp depletion in $\delta^{18}O$ as well as $\delta^{13}C$ (Figure 6.71). This central Arctic Ocean meltwater event can be correlated from the Mendeleev Ridge and Makarov Basin through the Lomonosov Ridge and Amundsen Basin to the eastern Gakkel Ridge/Nansen Basin region (Stein et al., 1994a, 1994b; Nørgaard-Pedersen et al., 1998, 2003; Poore et al., 1999). In Core PS2170-4 from the Amundsen Basin this meltwater event was dated to 15.7 ka, interpreted as first deglacial meltwater event related to the early decay of the Barents Sea Ice Sheet (Stein et al., 1994a, 1994b). In most of the AMS^{14}C-dated cores, however, this meltwater event occurred between 15 and 13.5 ka, that is, during Termination Ia, suggesting a slightly younger age of the first extensive deglacial meltwater event in the Arctic Ocean (Nørgaard-Pedersen et al., 1998). A second meltwater event coincides with Termination Ib. The two deglacial meltwater episodes seem to culminate at 14–12 ka and 10–8 ka in the Arctic cores (Figure 6.71; Nørgaard-Pedersen et al., 1998) and thus more or less contemporaneously with the maximum rate of global sea-level rise as determined from the Barbados sea-level record (Fairbanks, 1989).

From the composition of IRD peaks related to these two meltwater events, Nørgaard-Pedersen et al. (1998) proposed that the first main meltwater pulse is primarily related to the early and rapid decay of the Barents Sea Ice Sheet (see also Polyak & Solheim, 1994; Elverhøi et al., 1995a; Lubinski et al., 1996; Knies et al., 2000), whereas the second younger one is related to the melting of icebergs derived from a Canadian/northern Greenland source. The latter is supported by terrestrial evidence pointing to a late retreat of the main ice sheet on Ellesmere Islands (i.e., the Innuitian Ice Sheet) and northern Greenland (Andrews, 1987; Funder, 1989; Hodgson, 1989). As outlined in Nørgaard-Pedersen et al. (1998), the northwestern part of the LIS, on the other hand, is thought to have disappeared earlier, that is, near 14–12 ka (Dyke & Prest, 1987; Hodgson, 1994; Peltier, 1994), and probably contributed to the meltwater event of Termination Ia recorded in the central Arctic Ocean. This interpretation is also supported by IRD-source proxies determined in further sediment cores from the Amerasian Basin as well as Fram Strait (see Section 6.3.4).

During the MIS 2/MIS 1 transition, a very strong freshwater event was also recorded in a sediment core from the Laptev Sea continental slope and dated to ~13 Cal. kyr BP or 11.2 ka (Figure 6.72C; Spielhagen et al., 2005). Based on the well-established chronology of Core PS2458 between 14.4 and 9 Cal. kyr BP, a correlation of the freshwater event at the Laptev Sea continental margin at 13 Cal. kyr BP with the global meltwater pulses (MWP-1A, MWP-1B) centred at 14.0 and 11.3 Cal. kyr BP (Fairbanks, 1989; Bard et al., 1996), can be excluded. This Laptev Sea freshwater event, however, seems to correlate with the beginning of the YD cooling interval lasting from 12.9 to 11.6 Cal. kyr BP or 10.9 to 10.1 ka (Figure 6.72C; Stuiver, Grootes, & Braziunas, 1995).

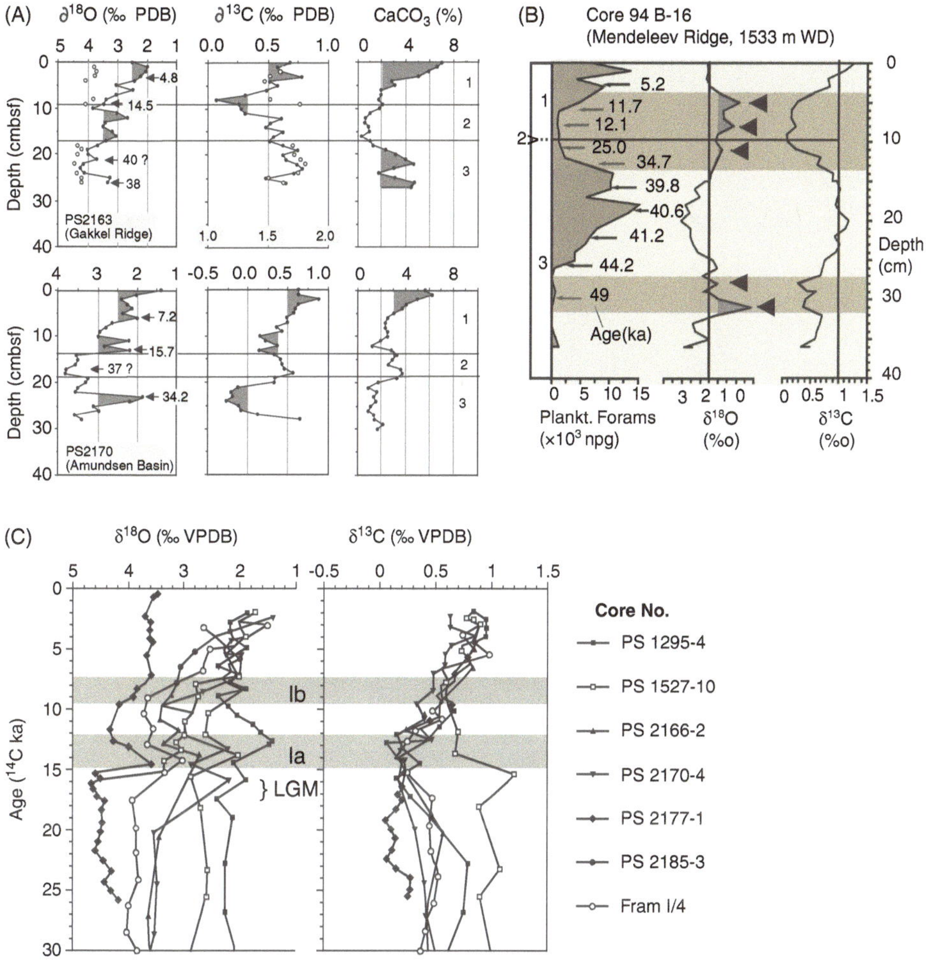

Figure 6.71 (A) Oxygen and carbon isotope records of *N. pachyderma* (sin.) and carbonate content versus depth in samples from cores PS2163 (Gakkel Ridge) and PS2170 (Amundsen Basin). Open circles (Core PS2163) indicate isotope data of benthic foraminifer *C. wuellerstorfi*. Numbers close to the black arrows are AMS^{14}C ages in ka; numbers at the right-hand side indicate isotope stages MIS 1 to MIS 3 (Stein et al., 1994a). (B) Abundance of planktonic foraminifers and oxygen and carbon isotopes of *N. pachyderma* (sin) versus depth in samples from cores 94 B-16, Mendeleev Ridge. AMS^{14}C ages are shown. Numbers at the left-hand side indicate isotope stages MIS 1 to MIS 3. Meltwater events are highlighted as black triangles and light grey intervals (modified from Poore et al., 1999). (C) AMS^{14}C-dated δ^{18}O and δ^{13}C records of *N. pachyderma* (sin) versus age from the Fram Strait to the central Arctic (Nørgaard-Pedersen et al., 1998, supplemented). Record of Core PS1295-4 from Jones and Keigwin (1988), PS1527-10 from Köhler (1992), PS2170-4 from Stein et al. (1994a), Fram I/4 from Markussen et al. (1985), and PS2166-2, PS2177-1, and PS2185-3 from Nørgaard-Pedersen et al. (1998). A standard reservoir correction of 400 years was used. For core locations see Figure 6.66A.

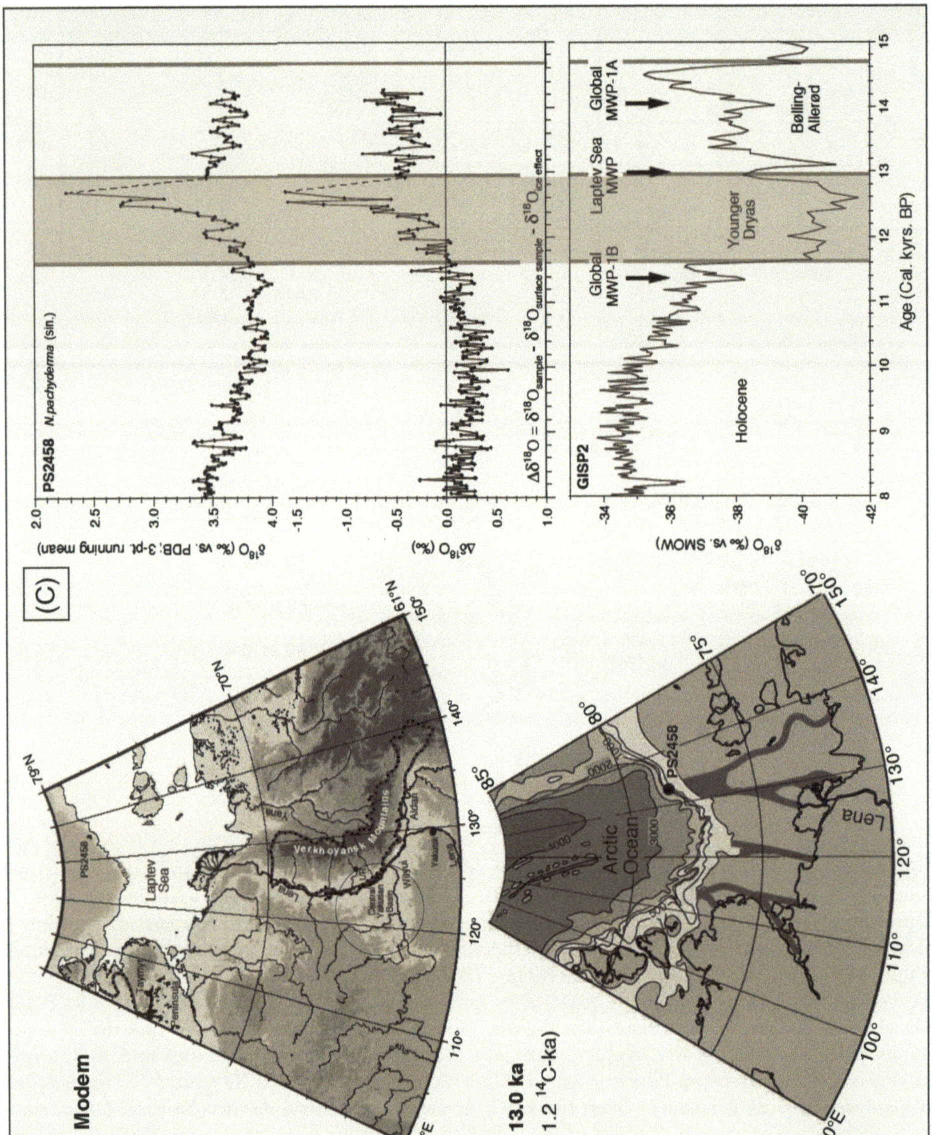
(A)
Modern
(B)
13.0 ka
(11.2 14C-ka)
(C)
PS2458
Laptev Sea
Verkhoyansk Mountains
Lena
Arctic Ocean
GISP2
Holocene
Younger Dryas
Bølling-Allerød
Global MWP-1A
Laptev Sea MWP
Global MWP-1B
Age (Cal. kyrs. BP)

What factors may have triggered this huge input of freshwater into the Laptev Sea near 13 Cal. kyr BP? For answering this question, first it is important to consider that Core PS2458 was located almost directly off the main Lena River mouth at that time of still lowered sea level (Figure 6.72B). Then, to explain the timing and short duration of the event, Spielhagen et al. (2005) favour — similar to the main Weichselian freshwater events (see Section 6.2.2) — a collapse of an ice- or morain-dammed lake in the hinterland and a rapid outburst of freshwater to the Arctic Ocean, which significantly reduced salinity and oxygen isotope ratio of the surface waters. For the duration of this event an estimate of only 200–250 years was given by these authors. How such an ice- or morain-dammed lake could have been developed during post-glacial times?

Several pre-Holocene moraine belts and glacio-lacustrine and fluvio-glacial sediments associated with the morainic deposits were distinguished by Kind, Kolpakov, and Sulerzhitsky (1971) and Kind (1975) in troughs and the foreland of the Verkhoyansk Mountains (Figure 6.72A). Glaciers originating from this mountain range must have crossed the present beds of the Lena and Aldan rivers and caused damming events (Kind et al., 1971; Kind, 1975). Concerning the chronology of these events, only a few ^{14}C-age data are available, suggesting that most of the studied far-reaching moraine belts are probably older than the LGM (Kind et al., 1971; Kind, 1974; Kolpakov & Belova, 1980). There are, however, also several large morainic arc systems that are not dated and loosely correlated to the deglacial time periods of 14–13 ka and *ca.* 11 ka (Kind et al., 1971; Kind, 1975). Furthermore, Kind et al. (1971) and Kind (1975) report on ^{14}C-dated glacio-lacustrine sequences (15.1 and 15.8 ka) from the Undyulyung River, a right tributary of the Lena River (Figure 6.72A), which are interpreted as deposits from ice and/or moraine-dammed lakes. As outlined by Spielhagen et al. (2005), these ^{14}C-datings were obtained from wood and peat extracted from fluvial and lake

Figure 6.72 Deglacial/Holocene meltwater event at the Laptev Sea continental margin and its palaeoenvironmental significance (from Spielhagen et al., 2005, supplemented). (A) Map of the Lena River area and the Verkhoyansk Mountains. Black dotted line indicates Taymyr Peninsula glaciation in the LGM (Svendsen et al., 2004) and maximum extent of Late Pleistocene ice sheets in the Verkhoyansk Mountains according to Isayeva (1984). Black dot indicates area with sites on Undyulyung River (UR) analysed by Kind et al. (1971) and Kind (1975), as mentioned in the text. (B) Palaeogeographic reconstruction of the Laptev Sea at ~13 Cal. kyr BP, based on Perry and Fleming (1986), Fairbanks (1989), and Bauch et al. (2001b). Dark grey areas mark palaeoriver channels as determined by Kleiber and Niessen (2000). At 13 Cal. kyr BP, Core PS2158 was located directly north of the palaeoriver channel and mouth. (C) Planktonic oxygen isotope record of PS2458 (three-point running mean; top) and regional isotopic changes (middle) (Spielhagen et al., 2005), and stable oxygen isotope record of the GISP2 ice core (Stuiver et al., 1995). The timing of the global meltwater pulses Ia and Ib, and of the Laptev Sea meltwater pulse are indicated. Grey bars mark the boundaries of the Holocene, YD, and Bølling/Allerød periods; the YD interval is light-grey shaded. To filter for the regional effects, the planktonic oxygen isotope values from Core PS2458 were normalized for the global ice effect and referred to modern conditions, that is, from each δ^{18}O result, the value of the surface sample (3.18‰) and the value of the ice effect for the individual sample age, as obtained from the continuous deglacial record of Fairbanks (1989) were subtracted. For more details of approach see Spielhagen et al. (2005).

sediments, suggesting that the dated material was redeposited and may in fact stem from older, eroded deposits. Thus, one should have in mind, that the reported radiocarbon ages have to be regarded as maximum ages, that is, the true depositional ages may be significantly younger.

A large LGM ice sheet has not been developed in this area due to extremely cold and dry conditions in northern and northeast Siberia at that time (see Sections 6.1 and 6.2 for details and references). For the early deglacial Bølling/Allerød period, however, palynological data suggest that summers in northernmost Siberia were warmer than today, but winters were probably colder, and precipitation was somewhat lower than today, but certainly higher than during the LGM (e.g., Andreev, Klimanov, & Sulerzhitsky, 1997; Andreev et al., 2002; Andreev et al., 2003; Klimanov, 1997; Velichko et al., 1997a, 2002; Hahne & Melles, 1999). Such conditions may have been more favourable to the development of glaciers on the western flank of the Verkhoyansk Mountains than the colder, but extremely dry conditions during the LGM, as stated by Spielhagen et al. (2005). These authors then speculate that some of the younger morains crossing the Lena valley which is quite narrow in this area, may have been formed during that time period and prohibited the river outflow towards the north. Under this scenario, an ice- or morain-dammed lake must have been developed in the central Yakutian lowland. If one of the damming events in fact occurred in late Allerød times, the development and subsequent flushing of a large lake in this area would certainly have supplied enough freshwater to the Laptev Sea continental margin to reduce the salinity significantly and thus explain the δ^{18}O spike recorded in the PS2458 record (Figure 6.72C). Spielhagen et al. (2005) concluded that in the lack of other plausible explanations for the low δ^{18}O values near 13 Cal. kyr BP in the PS2458 record, they favour the hypothesis of river damming by glacier ice or mixed snow/ice/sediment moraines, although they are aware of its speculative character. Further studies are needed to prove or reject this hypothesis.

Almost contemporaneously with the onset of the YD event, a huge outflow event of freshwater from North American glacial Lake Agassiz into the North Atlantic (i.e., the GIN seas) was also found, which may have weakened the THC during this interval (Broecker et al., 1988, 1989; Clark et al., 2002; Teller et al., 2002). For the Lake Agassiz even ten events of freshwater outbursts with partly different routing of runoff were described for the last deglaciation, when the LIS retreated and large volumes of water stored in this proglacial lake were episodically released into the oceans (Figure 6.73; Broecker et al., 1989; Barber et al., 1999; Teller et al., 2002). During this time interval, the three largest cooling events in the Northern Hemisphere closely followed four of the five largest outbursts from Lake Agassiz: (1) the YD, preceded by a release of $9.5 \times 10^3\,km^3$ at 10.9 ka (12.9 Cal. kyr BP); (2) the Preboreal Oscillation, preceded by releases of $9.3 \times 10^3\,km^3$ and $5.9 \times 10^3\,km^3$ at 10.1 ka (11.7 Cal. kyr BP) and 9.9 ka (11.2 Cal kyr BP), respectively; and (3) the "8.2 cold event" (7.7–7.4 ka or 8.2 Cal. kyr BP), preceded by a $163 \times 10^3\,km^3$ outburst; these are, respectively, fluxes of 0.30, 0.29, 0.19, and 5.2 Sv if assuming a release within one year (Teller et al., 2002). Based on modelling results, the freshwater budget in the North Atlantic has a strong influence on the strength of THC, and relatively small changes in the order of 0.1 Sv may result in

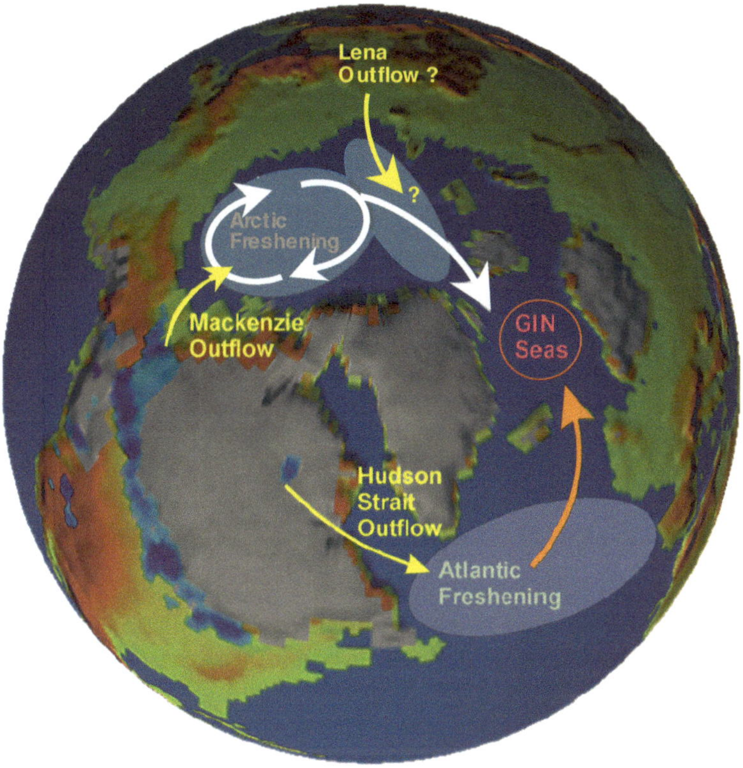

Figure 6.73 Areas in which intense freshwater forcing was applied to the surface of the ocean, and distribution of land ice (light grey) and pro-glacial lakes (blue) at the beginning of the YD (12.9 Cal. kyr BP) (modified from Peltier et al., 2006). As possible additional source for freshwater supply into the Arctic Ocean, Laptev Sea input was added (according to Spielhagen et al., 2005). With the beginning of the YD Event, the Barents Sea Ice Sheet was already absent as it probably disintegrated during the MWP-1a event. Furthermore, Bering Strait was dry land at that time implying that the only route whereby water added to the Arctic Basin could equilibrate with the global ocean was by outflow through Fram Strait into the GIN seas where deepwater forms today (see Peltier et al., 2006 and text for some more details).

significantly reduced deep-water formation (e.g., Stocker & Wright, 1991; Rahmsdorf, 1994; Manabe & Stouffer, 1995, 1997; Fanning & Weaver, 1997; Schmittner & Clement, 2002).

Thus, both data and models suggest that abrupt climate change during the last deglaciation may have originated through changes in the Atlantic THC in response to changes in the hydrological cycle, that is, a freshwater forcing due to direct rapid freshwater outburst into the North Atlantic and/or freshwater outburst into the western or eastern Arctic Basin with subsequent outflow through Fram Strait into the GIN seas (Figure 6.73; Clark et al., 2002; Tarasov & Peltier, 2005; Peltier, Vettoretti, & Stastna, 2006; Peltier, 2007). A final statement about the importance of the Laptev Sea freshwater event in respect to the changes in THC and related abrupt climate change during the YD, however, is difficult to make because it is not possible yet to determine the regional extent of the freshwater outburst and its influence on the salinity in a large basin like the Arctic Ocean (Spielhagen et al., 2005). High-resolution isotopic records from the central or eastern Arctic Ocean as well as the Fram Strait are needed to address this question.

6.3.6. Holocene Sediment Transport by Sea Ice and Its (Sub-) Millennial Variability

The drift of sea ice in the Arctic today is a direct response to surface atmospheric pressure gradients and resulting wind patterns, mainly related to the Arctic Oscillation (Thompson & Wallace, 1998; Rigor et al., 2002; see Chapter 2.3 for more details). During a positive Arctic Oscillation situation, the Arctic Oscillation induces the TPD to intensify, shifting farther eastward from Siberia towards North America, whereas during a negative Arctic Oscillation situation, the TPD is more restricted to the Russian Arctic, and the clockwise Beaufort Gyre dominates the Arctic Ocean near North America (Figure 6.74; Darby & Bischof, 2004; Darby et al., 2006). The appearance of Fe grains from Siberian sources in modern floes in the Beaufort Sea is direct evidence of the net effect of the TPD shifting towards North America and the capture of floes that originated in the Laptev Sea by the Beaufort Gyre (Darby, 2003; Darby & Bischof, 2004; see Chapter 4.3.1). Tremblay, Mysak, and Dyke (1997) also proposed the concept of an expanded TPD as the cause of the Russian driftwood found in northern Canada (see also Dyke, England, Reimnitz, & Jetté, 1997) Location of sediment cores discussed in the text, are shown.

The sources of Fe grains with origins from Siberia and North America display a distinct short-term variability in cores from the Chukchi Margin during Holocene times, suggesting changes in sea-ice drift and related atmospheric circulation patterns (Figure 6.75; Darby & Bischof, 2004). The high-resolution AMS^{14}C-dated Holocene record of Core P1-92-AR-P1 (4.1 m length) and its companion box Core P1-92-AR-B3, located near the confluence of the TPD and the Beaufort Gyre and thus in an ideal location (73° 42.4′N, 162° 44.6′W) to record ice-rafting from Siberia due to a TPD shift towards North America (Darby & Bischof, 2004), indicates that Siberian Fe grains are nearly always present throughout with variable contributions and average 38% (and up to 63%) of the matched Fe grains (Figure 6.75). The other major source is northern Canada, primarily the northern Queen Elizabeth Islands and the Banks and Victoria Islands area (Figure 6.75).

The concentration of smectite, a valuable proxy for the identification of source areas in Arctic Ocean sediments (see Chapter 5.1.2), in the $<2\,\mu m$ fraction of the same cores varies between ~20% and 40% in a quasiperiodic fashion throughout most of the Holocene. Possible source areas of the smectite could be either the Kara Sea/western Laptev Sea characterized by high smectite concentrations in surface sediments or the nearby Chukchi Sea where also smectite values of ~25% were determined (see Chapter 5.1.2, Figure 5.4B; Naidu & Mowatt, 1983; Wahsner

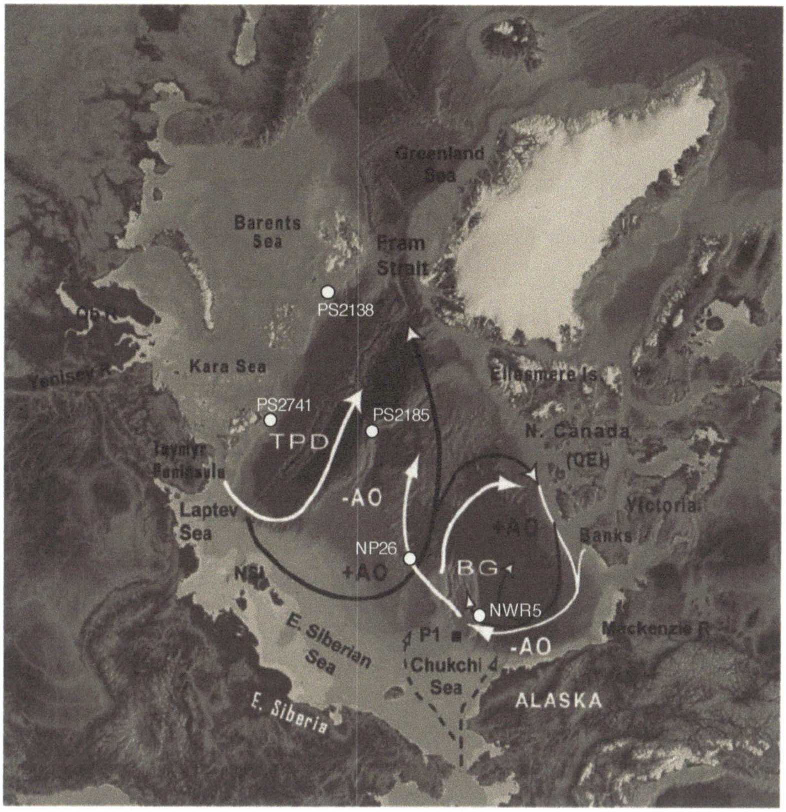

Figure 6.74 The modern Arctic ice drift pattern (white arrows) shows a fairly direct path of the TPD from the Laptev Sea to Fram Strait corresponding to a negative Arctic Oscillation (AO) (from Darby & Bischof, 2004, supplemented). When the AO is positive, the TPD shifts towards North America (black arrow) and the Beaufort Gyre (BG) diminishes. The Pacific ice drift through Bering Straits bifurcates across the Chukchi Sea (small dashed arrows), the eastern branch approximates the Alaskan Coastal Current and the western branch points towards Herald Canyon. NSI, New Siberian Islands; QEI, Queen Elizabeth Islands. Location of sediment cores discussed in the text, are shown.

et al., 1999; Viscosi-Shirley et al., 2003a; Viscosi-Shirley et al., 2003b; Stein et al., 2004a). Based on the results of Fe grain matches and assuming that most of the clay is transported by sea ice, the Laptev area seems to be a more probable source in most cases. The few instances where Siberian sources are low and smectite is high suggest that the smectite is derived from south of the core site, that is, the Chukchi Sea area. The rare occasions when Siberian Fe grain sources are high and smectite is low are due to Siberian sources other than the Laptev Sea, primarily the East Siberian Sea (Darby & Bischof, 2004).

Core intervals dominated by Fe grains from the Siberian shelves indicate that the TPD was shifted frequently towards North America, similar to what occurs during a more positive phase of the Arctic Oscillation (Darby & Bischof, 2004). Intervals with minor input from the Siberian shelves, on the other hand, suggest that the

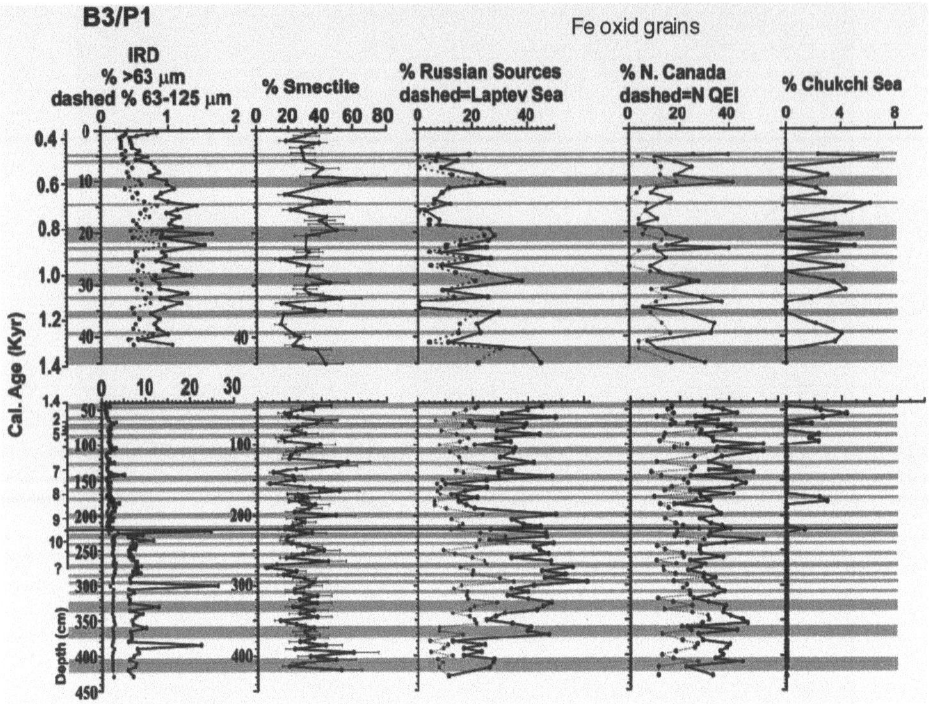

Figure 6.75 Sand and coarser fractions, smectite clay with ±25% confidence intervals, and Fe grains matched to Siberian (Kara Sea to East Siberian Sea) and northern Canadian source areas (from Darby & Bischof, 2004, supplemented). Shaded intervals are centred on the Siberian Fe grain peaks. The Chukchi grain matches indicate sources immediately south of the core location due to northward drift from the Bering Strait and show very little input from this area. The age scale changes at 1.4 ka.

TPD was restricted to the Russian half of the Arctic, far from the core site. Information about the timing of the centennial-scale oscillation of the TPD pattern can be obtained from the uppermost part of the Core P1-92-AR-P1/B3 which was sampled on a 1 cm resolution and represents the last 1,300 years (Figure 6.75). During this time interval, Siberian grains display rapid fluctuations of 20–30%, and maxima occur approximately every 50–150 years. Similar high-frequency oscillations might be present earlier in the record, that is, throughout the Holocene, but cannot be resolved with the sampling interval used below 50 cm core depth (Figure 6.75; Darby & Bischof, 2004). Thus, this fluctuation might represent the long-term oscillation of the Arctic Oscillation, which modulates the same TPD shift today.

Darby and Bischof (2004) proposed two possible drift paths for sea-ice delivery of Siberian grains to the core site, both involve a change in the TPD in which it moves closer to North America. This requires a dominant cyclonic flow regime in the Arctic as it occurs today during a positive Arctic Oscillation (Kwok, 2000; see also Chapter 1.1). First, sea-ice transport from the Siberian Arctic via the TPD deep into the western Arctic, where some of this sediment-laden ice is diverted south into a restricted Beaufort Gyre as the main TPD flows towards Fram Strait. A subsequent change to a more negative Arctic Oscillation pattern would strengthen the Beaufort Gyre and enhance the likelihood of captured Siberian ice delivery to the core site. This scenario would favour coincident peaks of northern Canadian and Siberian Fe grain sources which occurs for slightly more than half of the Russian grain peaks during the last 1,300 years at Core P1-92-AR-P1/B3 (Figure 6.75; Darby & Bischof, 2004). The second drift scenario which also involves the TPD expanding eastward even closer to North America and direct transport to the core site, requires a very weak Beaufort Gyre and would favour Russian grain peaks without input from North America.

6.4. Accumulation of Particulate Organic Carbon at the Arctic Continental Margin and Deep-Sea Areas during Late Quaternary Times

The palaeoclimatic evolution of the Arctic Ocean during Quaternary times is characterized by distinct changes in sea-ice cover, river discharge, and oceanic circulation patterns (e.g., Spielhagen et al., 1997, 2004; Nørgaard-Pedersen et al., 1998, 2003; Stein, 2000 and references therein; Darby et al., 2006 and references therein). As outlined in the previous chapters, the sediment input along the Eurasian continental margin is strongly influenced by the waxing and waning of continental ice sheets (e.g., Elverhøi et al., 1995a; Vorren and Laberg, 1997; Velitchko et al., 1997a; Mangerud et al., 1998; Svendsen et al., 1999, 2004). During times of lowered sea level, sediment transport towards the shelf edge and following down-slope transport by turbidity currents and slumps became another important factor controlling the sedimentation at the continental slope and adjacent deep sea (e.g., Laberg & Vorren, 1995; Kleiber et al., 2000). All these factors also influence the OC input and preservation at the continental margin and adjacent deep sea (see Stein & Macdonald, 2004a and references therein).

In the Subarctic North Atlantic (i.e., the NGS and Fram Strait) numerous OC records are available and discussed in relationship to palaeoceanographic and palaeoclimatic changes (e.g., Wagner & Henrich, 1994; Elverhøi et al., 1995a; Wagner & Hölemann, 1995; Nam, 1997; Notholt, 1998; Birgel & Hass, 2004; Birgel & Stein, 2004). Quaternary OC records in the Arctic Ocean are mainly restricted to the Eurasian continental margin (e.g., Knies & Stein, 1998; Fahl & Stein, 1999, 2007; Knies et al., 1999; Stein, Fahl, Niessen, & Siebold, 1999a, 2001, 2003a, 2004a; Müller-Lupp et al., 2000; Stein & Fahl, 2000, 2004a, 2004b; Bauch et al., 2001a; Schäfer, 2005), Morris Jesup Rise (Vogt, 1997), Northwind Ridge (Grantz et al., 1999), and the central Arctic Ocean (e.g., Stein et al., 1994a, 1994b, 2004b; Schubert, 1995; Schubert & Stein, 1996; Clough et al., 1997).

In the following subsections, examples of OC accumulation along the Eurasian continental margin and in the central Arctic Ocean will be presented. The locations of sediment cores discussed here are presented in Figure 6.76.

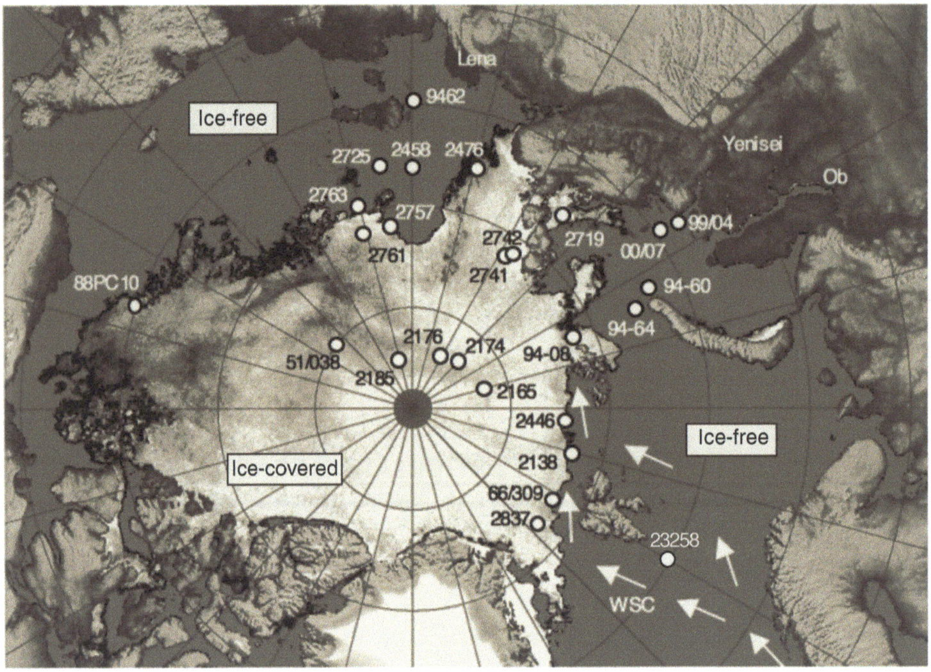

Figure 6.76 Map showing the distribution of sea ice in the Arctic Ocean on 6 August, 2007 (http://iup.physik.uni-bremen.de:8084/amsr/amsre.html; for background and technique see Spreen, Kaleschke, & Heygster, 2005) and locations of cores discussed in this chapter are shown.

6.4.1. Variability of Organic-Carbon Accumulation along the Eurasian Continental Margin

Along the Eurasian continental margin, the OC content varies between 0.2% and 2.5%, and distinct differences between late Quaternary glacial and interglacial intervals are obvious (Figure 6.77). In numerous cores from the area around Svalbard including the Yermak Plateau, distinct OC peaks were recorded which can be correlated over long distances very well (e.g., Elverhøi et al., 1995a; Vogt, 1997; Knies & Stein, 1998; Knies et al., 1999; Vogt et al., 2001; Schäfer, 2005; Winkelmann, 2007; Winkelmann et al., 2008b). These "OC events" highlighted in the OC record of Core PS66/309 (Figure 6.77), are mainly related to increased input of terrigenous OC during times of maximum glacier advances and early deglaciation in MIS 6, MIS 4, and MIS 2 (see later discussions).

6.4.1.1. Organic-carbon accumulation during MIS 6 to MIS 3

In most of the cores, MIS 6 is characterized by the occurrence of prominent dark grey intervals with maximum OC values and minimum hydrogen index values (Figures 6.77 and 6.78). The latter suggests that the organic matter is predominantly

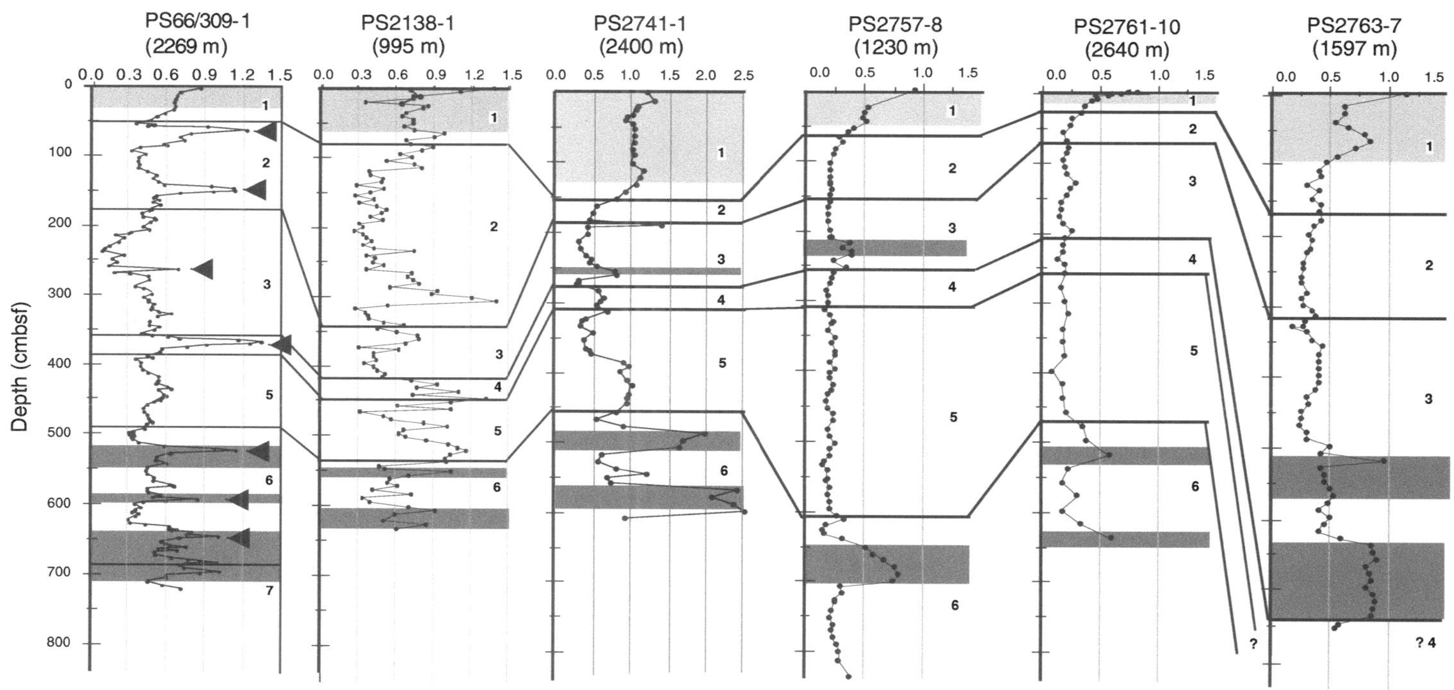

Figure 6.77 TOC records of long sediment cores (see Figure 6.76 for location), representing MIS 1 to 6: Core PS66/309 (Schäfer, 2005; stratigraphy according to Winkelmann, 2007); Core PS2138-1 (Knies, 1999; Knies et al., 2001); cores PS2741-1, PS2757-8, PS2763-7, and PS2761-10 (Stein et al., 2001). The Holocene time interval is marked by light grey shading. Dark grey, OC enriched lithologies are marked as dark grey bars. Black triangles in the PS66/309 record mark major OC events, contemporaneously occurring in several sediment cores around Svalbard and on Yermak Plateau and related to increased input of terrigenous OC by glacial erosion (see text and Winkelmann et al., 2008b for more detailed discussion). Water depths (m) are indicated.

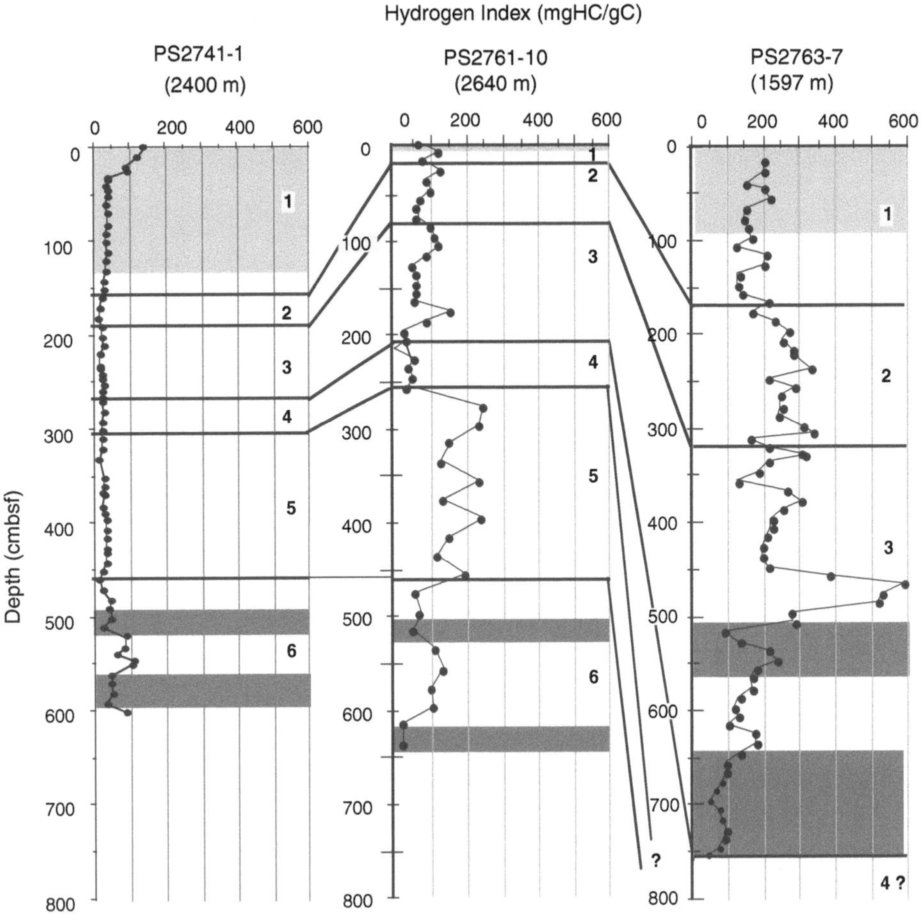

Figure 6.78 Hydrogen index values determined at long sediment cores PS2741-1, PS2763-7, and PS2761-10 (from Stein et al., 2001). The Holocene time interval is marked by shading. Dark grey, OC enriched lithologies are marked as dark grey bars. Water depths (m) are indicated. For location of cores see Figure 6.76.

of terrigenous origin. At Core PS2138-1 obtained from the northern Barents Sea continental margin, OC values are lower, but OC accumulation rates are increased during MIS 6 (Figure 6.79), indicating increased OC fluxes and dilution of OC percentages by significantly increased bulk (i.e., mainly siliciclastic) material at that time (Figure 6.31; Knies et al., 1999). Based on low hydrogen values <90 mgHC gC^{-1} (Knies et al., 1999) as well as low $\delta^{13}C$ values and high C/N ratios (Figure 6.79), terrigenous OC is also predominant at this core during MIS 6.

As described in Section 6.2, extended continental ice sheets existed in western (Barents-Kara Sea) as well as eastern (Taimyr Peninsula–Servernaya Semlya–New Siberian Islands) Eurasia (e.g., Velitchko et al., 1997a, 1997b; Mangerud et al., 1998, 2002; Svendsen et al., 1999, 2004). Probable source areas of the terrigenous organic matter deposited along the Eurasian continental margin at that time were the

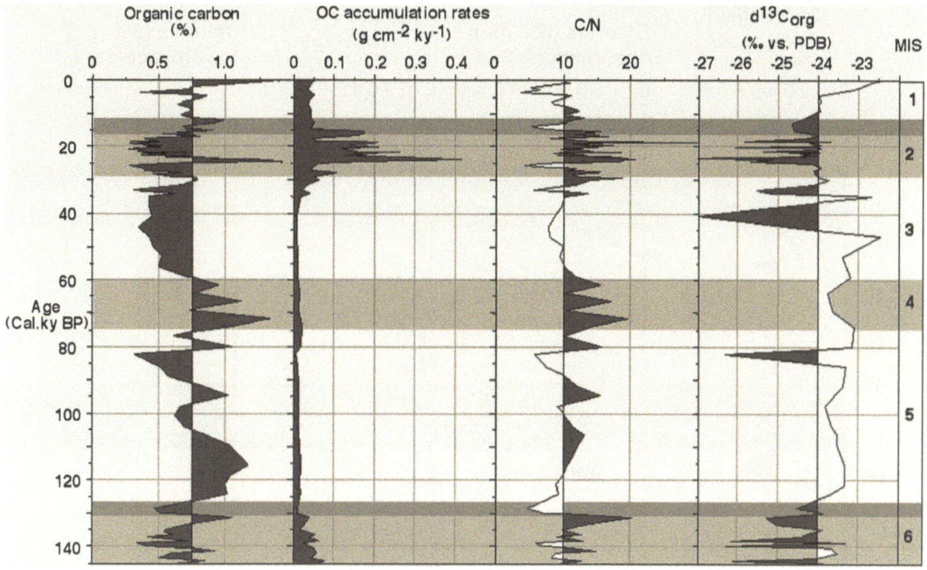

Figure 6.79 TOC, OC accumulation rates, C/N ratios, and stable carbon isotopes of the organic matter in Core PS2138-1 from the Barents Sea continental slope (data from Knies et al., 1999). MIS 1–6 are indicated. For location of core see Figure 6.76.

Mesozoic rocks in the Barents Sea and on Franz Josef Land (Elverhøi et al., 1989) and the coal deposits in the area between northern Taimyr Peninsula and the Lena River district (Anon, 1978). During ice advance the OC-rich bedrocks were eroded, and material was transported towards the adjacent continental margin into the influence of the eastward flowing Atlantic water (see Hebbeln & Wefer, 1997; Knies et al., 2001). The proposed source areas and transport processes are also supported by the occurrence of specific clay mineral (smectite–kaolinite) assemblages (Müller, 1999; see Section 6.2.2, Figure 6.32). Transport of OC via the Atlantic water boundary current towards the core locations may also explain the terrigenous OC-rich intervals in these cores during early MIS 3 (Figure 6.77; Stein et al., 2001) when, besides MIS 5 and the Holocene, strongest advection of Atlantic water has been proposed, contemporaneously with increased sediment input during deglaciation (Knies et al., 1999).

Besides the predominance of terrigenous OC along the Eurasian continental margin, also significant amount of marine OC seem to be occasionally preserved in the sediments, as suggested from marine OC proxies such as short-chain *n*-alkanes and chlorines (see Chapter 4.7.4). In general, the pattern of distribution of short-chain *n*-alkanes and chlorines along the northern Barents Sea margin seems to reflect glacial/interglacial variations with higher values during warmer and lower values during colder periods except event I and MIS 4 (Figure 6.80; Knies et al., 2000). Distinct peaks in marine OC input occurred during substages MIS 5.5 and MIS 5.1 as well as during the Early Holocene (Figure 6.80), suggesting an environment comparable to the recent situation with nutrient supply by ice-edge

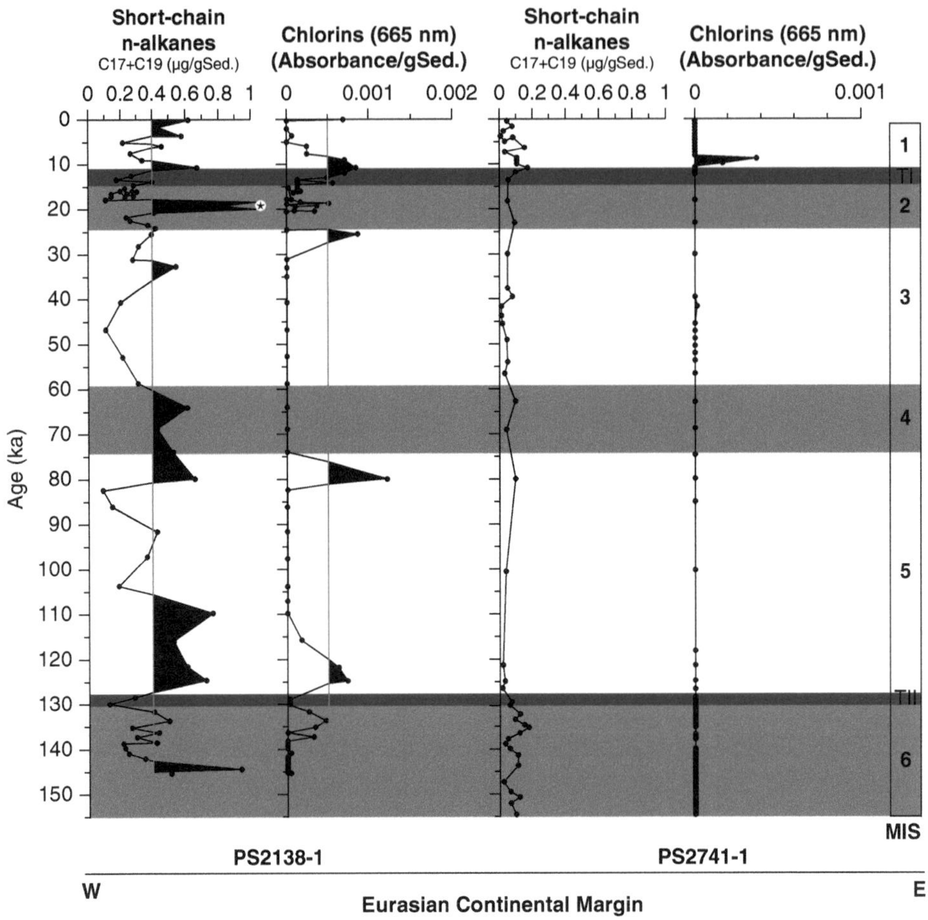

Figure 6.80 Short-chain *n*-alkane concentrations $\mu g/g_{Sed}$ and chlorine absorbance (665 nm) determined in sediment cores PS2138-1 and PS2741-1 versus age (from Knies et al., 2000). Asterisk marks high accumulation of marine organic matter during "Event I" as described in detail by Knies and Stein (1998). Marine isotope stages MIS 1–6 and Terminations I and II (TI and TII) are indicated. For location of cores see Figure 6.76.

upwelling, sea-ice melting, and Atlantic water inflow, at least close to the Franz Victoria Trough (Knies et al., 2000). Furthermore, fluctuations in those marine OC proxies show a distinct decrease in concentration from the western to the eastern Eurasian continental margin during the last 150 ka (Figure 6.80). Extremely low concentrations at the continental margin of Severnaya Semlya may reflect variations of significant environmental changes from an ice-edge upwelling regime with seasonally ice-free conditions during warm summers on the western margin (Core PS2138-1) to a permanent and stable sea-ice cover with very low surface-water productivity changes on the eastern margin (Core PS2741-1), as proposed by Knies et al. (2000).

In the easternmost cores, OC values were significantly reduced during MIS 5, MIS 4, and upper MIS 3 (Figure 6.77; Stein et al., 2001), probably due to a restricted extension of ice sheets and reduced terrigenous sediment input (see Section 6.2). In the two cores PS2763-7 and PS2761-10 from the western East Siberian Sea continental margin/Makarov Basin, that is, east of the Lomonosov Ridge, hydrogen index values increase during interglacials MIS 5 (Core PS2761-10) and mid-MIS 3 (PS2763-7) (Figure 6.78), suggesting some (higher) amounts of marine OC being preserved in the sediments. This increase in marine OC which is not observed in Core PS2741-1 (Figure 6.78), may indicate some enhanced primary productivity triggered by an at least seasonally reduced sea-ice cover on the southern Lomonosov Ridge area. Such a seasonally reduced sea-ice cover in this area has also been recorded in satellite photographs during summers of the early 1980s (Gloersen et al., 1992) as well as during summer 2007 (Figure 6.76). This interpretation is also supported by biomarker data from material sampled in 1995 from the water column by *in-situ* pumps across the Lomonosov Ridge, indicating significantly higher amounts of marine OC in the water column at locations east of the Lomonosov Ridge (Fahl, Nöthig, & Stein, 1997).

6.4.1.2. Organic-carbon accumulation during MIS 2

Whereas the Barents Sea area including Svalbard and Franz Josef Land was covered by a huge continental ice sheet during the last glacial MIS 2 (Figure 6.24; Svendsen et al., 2004), an extended continental ice sheet was absent in the eastern Kara Sea/Laptev Sea area (see Section 6.2 for more detailed discussion). This different climatic evolution in the western and eastern Eurasian continental margin areas is also documented in the OC records. In the Laptev Sea area, the supply of terrigenous (organic) matter was significantly reduced due to the lack of ice sheets and decreased river discharge (Sidorchuk & Panin, 1996), resulting in low OC values and low OC accumulation rates (Figure 6.77, Table 6.4; Knies et al., 1999, 2000; Stein et al., 2001). Along the Svalbard to Franz Josef Land continental margin, on the other hand, the flux of (terrigenous) OC increased markedly during MIS 2.

In Core PL94-8 located in the northern St. Anna Trough directly east of Franz Josef Land (see Figure 6.76 for location), maximum contents of terrigenous OC are found in a MIS 2 diamicton, indicated by maximum OC values, very low hydrogen index values and predominantly terrigenous macerals (Figure 6.81; Stein et al., 2001; Boucsein et al., 2002; Stein & Fahl, 2004a). The debris flows recorded in cores off the Franz Victoria Trough and representing times of LGM glacier advance (see Chapter 3.3.1, Figure 3.21), are also characterized by high OC contents and low hydrogen index values and high C/N ratios, indicating a terrigenous source of the organic matter (Figure 6.82; Kleiber et al., 2001). In Core PS2138-1, absolute maxima in accumulation rates of bulk sediment and OC of 38 g cm^{-2} kyr^{-1} and 0.2–0.4 gC cm^{-2} kyr^{-1}, respectively, were reached during MIS 2 (Figure 6.79, Table 6.4). This high input of terrigenous sediments was probably caused by increased glacial erosion of Mesozoic OC-rich bedrocks from Spitsbergenbanken and the central Barents Sea (Bjørlukke, Bue, & Elverhøi, 1978; Birkenmajer, 1989;

Table 6.4 Average Accumulation Rates of Bulk Sediment and Organic Carbon (Based on Stein et al., 2001, 2003a; Stein & Fahl, 2004b). For Core Locations See Figure 6.76 and Stein and Fahl (2004a, 2004b).

(A) Eurasian Continental margin and southern Lomonosov Ridge (MIS 1–6)

MIS	PS2138-1	PS2741-1	PS2474-3	PS2767-4	PS2763-7	PS2757-8	PS2761-10
Bulk accumulation rate ($g\,cm^{-2}\,kyr^{-1}$)							
1	9	9.02	12.15	17.39	10.59	4.8	1.98
2	38	1.65	8.54	8.76	10.42	5.35	3.02
3	6.7	2.14	10.06	12.36	13.22	3.14	4.43
4	3.2	2.31	nd	nd	nd	4.22	3.37
5	2.9	2.4	nd	nd	nd	5.14	3.83
6	10	3.6	nd	nd	nd	nd	2.55
Organic carbon accumulation rate ($gC\,cm^{-2}\,kyr^{-1}$)							
1	0.05	0.1	0.15	0.13	0.06	0.02	0.01
2	0.2	0.01	0.03	0.03	0.03	0.01	0.01
3	0.01	0.01	0.08	0.06	0.07	0.01	0.01
4	0.02	0.01	nd	nd	nd	0.01	0.01
5	0.02	0.02	nd	nd	nd	0.01	0.01
6	0.08	0.05	nd	nd	nd	nd	0.01

(B1) Northern (St. Anna Trough) and eastern Kara Sea, Holocene

	PL-60	PL-64	PL-67	PL-7	PL-8	PS2719-1
Bulk accumulation rate ($g\,cm^{-2}\,kyr^{-1}$)						
Holocene	28.1	32	20.8	3.3	4.3	28.3
Organic carbon accumulation rate ($gC\,cm^{-2}\,kyr^{-1}$)						
Holocene	0.45	0.48	0.25	0.02	0.03	0.34

(B2) Laptev Sea shelf and upper slope, Holocene

	PS51/080-13	PS51/092-12	PS51/135-4	PS51/159-10	PS2725-5	PS51/118-2	PS2458-4
Bulk accumulation rate ($g\,cm^{-2}\,kyr^{-1}$)							
Holocene	29	53	37	28	39	47	27
Organic carbon accumulation rate ($gC\,cm^{-2}\,kyr^{-1}$)							
Holocene	0.72	0.7	0.54	0.28	0.43	0.47	0.3

(C1) Laptev Sea shelf and upper slope, Holocene time slices

Cal. Yr BP	PS51/080-13	PS51/092-12	PS51/135-4	PS51/159-10	PS2725-5	PS51/118-2	PS2458-4
Bulk accumulation rate ($g\,cm^{-2}\,kyr^{-1}$)							
0–7,000	29	37	10	11	12	2	22
7,000–9,000	nd	153	84	15	12	2	22
9,000–11,000	nd	nd	112	102	239	349	71

Table 6.4 (*Continued*)

(C1) Laptev Sea shelf and upper slope, Holocene time slices							
Cal. Yr BP	PS51/ 080-13	PS51/ 092-12	PS51/ 135-4	PS51/ 159-10	PS2725-5	PS51/ 118-2	PS2458-4
Organic carbon accumulation rate ($gC\,cm^{-2}\,kyr^{-1}$)							
0–7,000	0.72	0.74	0.15	0.12	0.13	0.02	0.22
7,000–9,000	nd	2.07	1.21	0.15	0.13	0.02	0.25
9,000–11,000	nd	nd	1.62	1.02	2.63	3.49	0.82

(C2) Southern Kara Sea, Holocene time slices							
Cal. Yr BP	BP00-14/3	BP99-04/7	BP00-07/7	BP01-42/2	BP01-61/7	BP00-26/4	BP01-39/2
Bulk accumulation rate ($g\,cm^{-2}\,kyr^{-1}$)							
0–2,000	50–80	30–75	30	35	7	17	9
2,000–4,000	10	25	30	55	7	17	9
4,000–7,500	10	40–70	40	40	7	17	15
7,500–9,000	nd	100–125	240	28	37	25	32
9,000–11,000	nd	170	nd	nd	160	30–250	700
Organic carbon accumulation rate ($gC\,cm^{-2}\,kyr^{-1}$)							
0–2,000	0.8–1.2	0.5–1.2	0.5	0.4	0.05	0.2	0.05
2,000–4,000	0.2	0.3	0.5	0.6	0.05	0.2	0.05
4,000–7,500	0.2	0.5–0.7	0.5	0.4	0.05	0.2	0.5
7,500–9,000	nd	1.5	3	0.3	0.4	0.2	0.5
9,000–11,000	nd	3	nd	nd	1.5	1.5	10

Note: nd, no data.

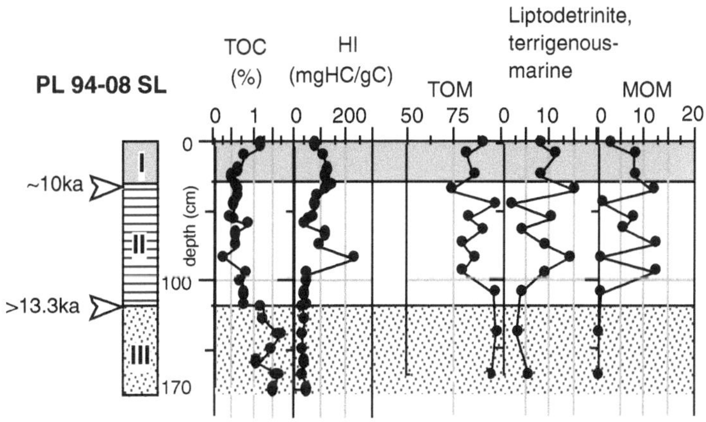

Figure 6.81 TOC and hydrogen index records and distribution of main maceral groups at Core PL94-08 obtained from the northern St. Anna Trough (from Stein et al., 2001; Boucsein et al., 2002). For the macerals, amounts of terrigenous organic matter (TOM) and marine organic matter (MOM) as well as liptodetrinite (marine and/or terrigenous origin) are presented. For location of core see Figure 6.76.

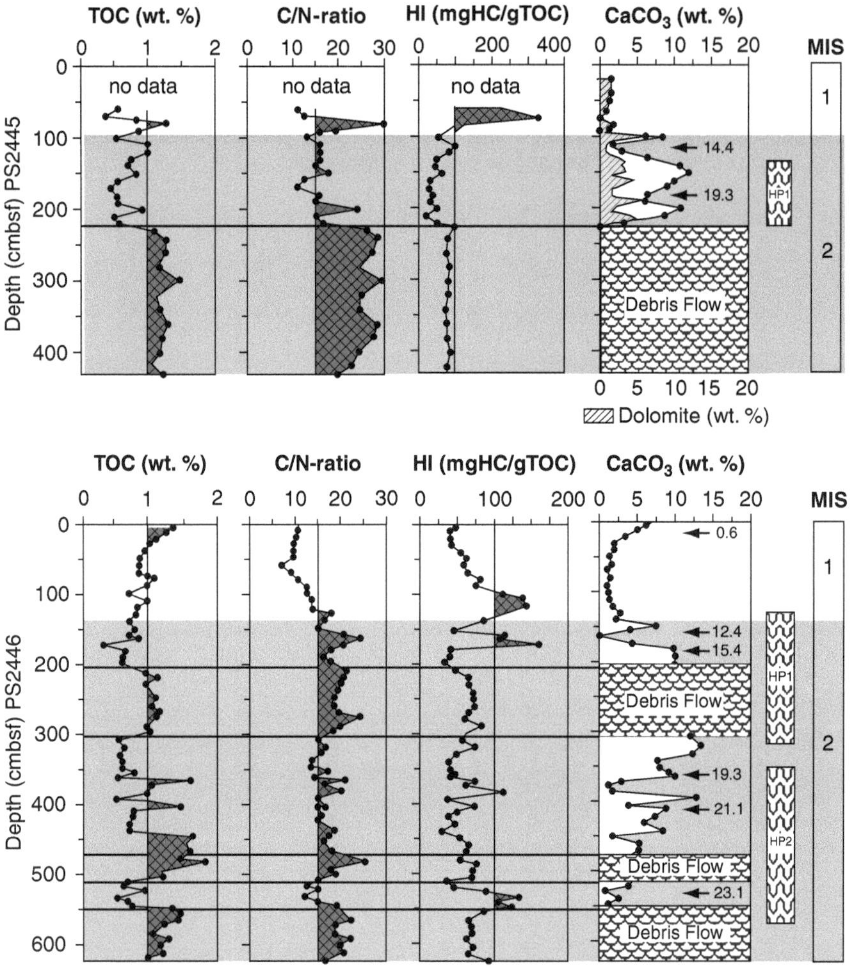

Figure 6.82 Compilation of the TOC, carbonate and dolomite content (all wt.%), C/N ratios and hydrogen index (HI) data versus core depth at cores PS2445 and PS2446 (from Kleiber et al., 2000). Dolomite content in Core PS2445-4 was published by Vogt (1997). Enrichment of carbonate rather than dolomite in Core PS2445-4 indicates high amounts of planktic and benthic foraminifers. AMS^{14}C datings and MIS are shown on the right. HP1 and 2 (high productive zones) are defined according to Dokken and Hald (1996). For location of cores see Figure 6.76.

Elverhøi et al., 1989, 1995; Polyak et al., 1997) and a continuous flux of turbid meltwater suspensions at the grounding line of the glaciers (Figure 6.83; Knies & Stein, 1998). By the latter process, also marine OC can be removed from the euphotic zone into the sediment due to incorporation of the labile marine OC into the fine-grained, OC-rich mineral matter of terrigenous origin. An enhanced preservation of marine OC in the PS2138-1 record at that time is indicated by the occurrence of short-chain *n*-alkanes and chlorine (Figure 6.80). That means, according to the model of Knies and Stein (1998) (Figure 6.83) the increased

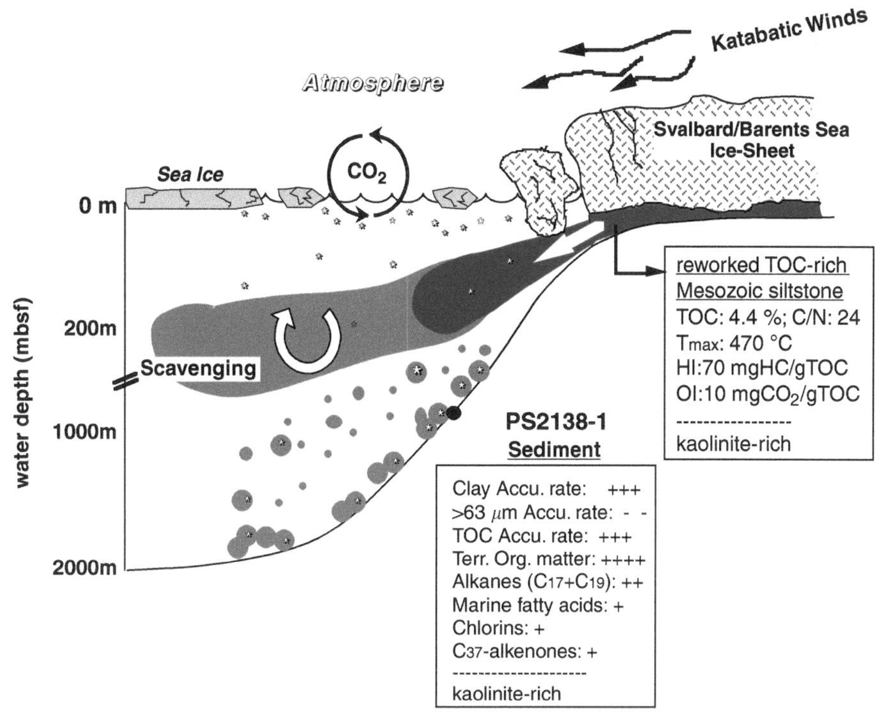

Figure 6.83 Model for OC accumulation, showing possible palaeoenvironment along the northern Barents Sea continental margin during "Event I" (from Knies & Stein, 1998). Reworked terrigenous OC-rich material transported by glaciers was picked out of the fraction $>63\,\mu m$. Climate-related proxies in the sediments of Core PS2138-1 are listed in shaded box.

amount of marine OC is explained by a combination of some higher surface-water productivity due to reduced sea-ice-cover and a more rapid transfer of marine OC into the deep-water sphere by forming large aggregates of marine and terrigenous particles. A similar palaeoenvironmental model has already been proposed for the OC accumulation in the Norwegian-Greenland Sea by Wagner and Hölemann (1995).

6.4.1.3. Postglacial to Holocene organic-carbon accumulation

During postglacial to Holocene times, all OC records from the Kara and Laptev Sea continental margin display distinct increases in OC contents; in most of the cores, these high OC values prevail throughout the Holocene (Figures 6.84A and 6.85A; Stein et al., 2001, 2003a). Furthermore, a terrigenous source of the OC is dominant in all cores from the shelf as well as the continental slope, as indicated by hydrogen index values dominantly $\ll 150\,mgHC\,gOC^{-1}$ (Figures 6.84B and 6.85B; Stein et al., 2001, 2003a). The dominance of terrigenous character of the OC is also supported by the dominance of long-chain *n*-alkanes and terrigenous macerals (Figure 6.86: Laptev Sea Core PS2458-4; for further data see Fahl & Stein, 1999;

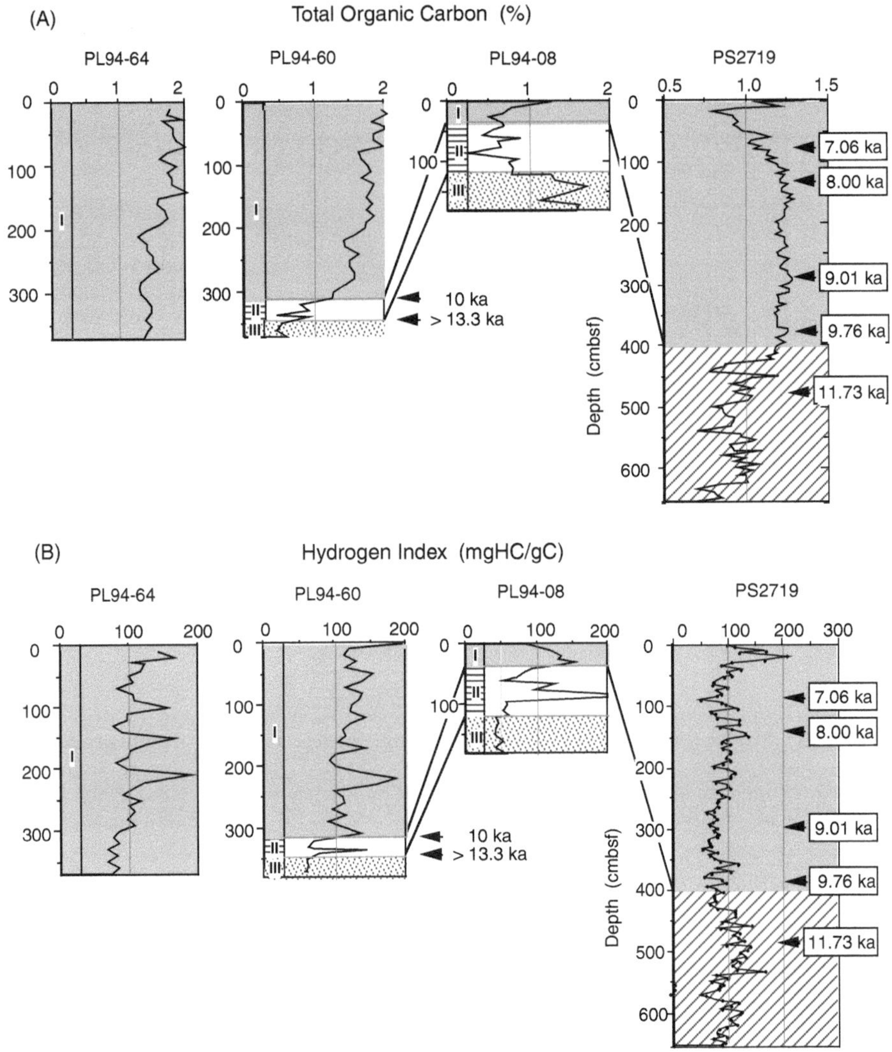

Figure 6.84 (A) TOC and (B) hydrogen index values at northern and eastern Kara Sea cores PL94-64, PL94-60, PL94-08, and PS2719-1 (from Stein et al., 2001). The Holocene time interval is marked by shading. At Core PS2719-1, arrows and numbers indicate reservoir-corrected AMS^{14}C ages. For discussion and ages of lithological units I to III see text. For location of cores see Figure 6.76.

Stein et al., 2001; Boucsein et al., 2002; Stein & Fahl, 2004a, 2004b). Based on long-chain *n*-alkane and maceral data, >80% of the OC is of terrigenous origin (Figures 6.86 and 6.87).

When looking into details, however, significant differences and variations in OC composition are obvious in most of the cores. Whereas the cores from the shelf areas of the southern and southeastern Kara Sea as well as the eastern Laptev Sea show a clear predominance of terrigenous matter with low variability throughout the

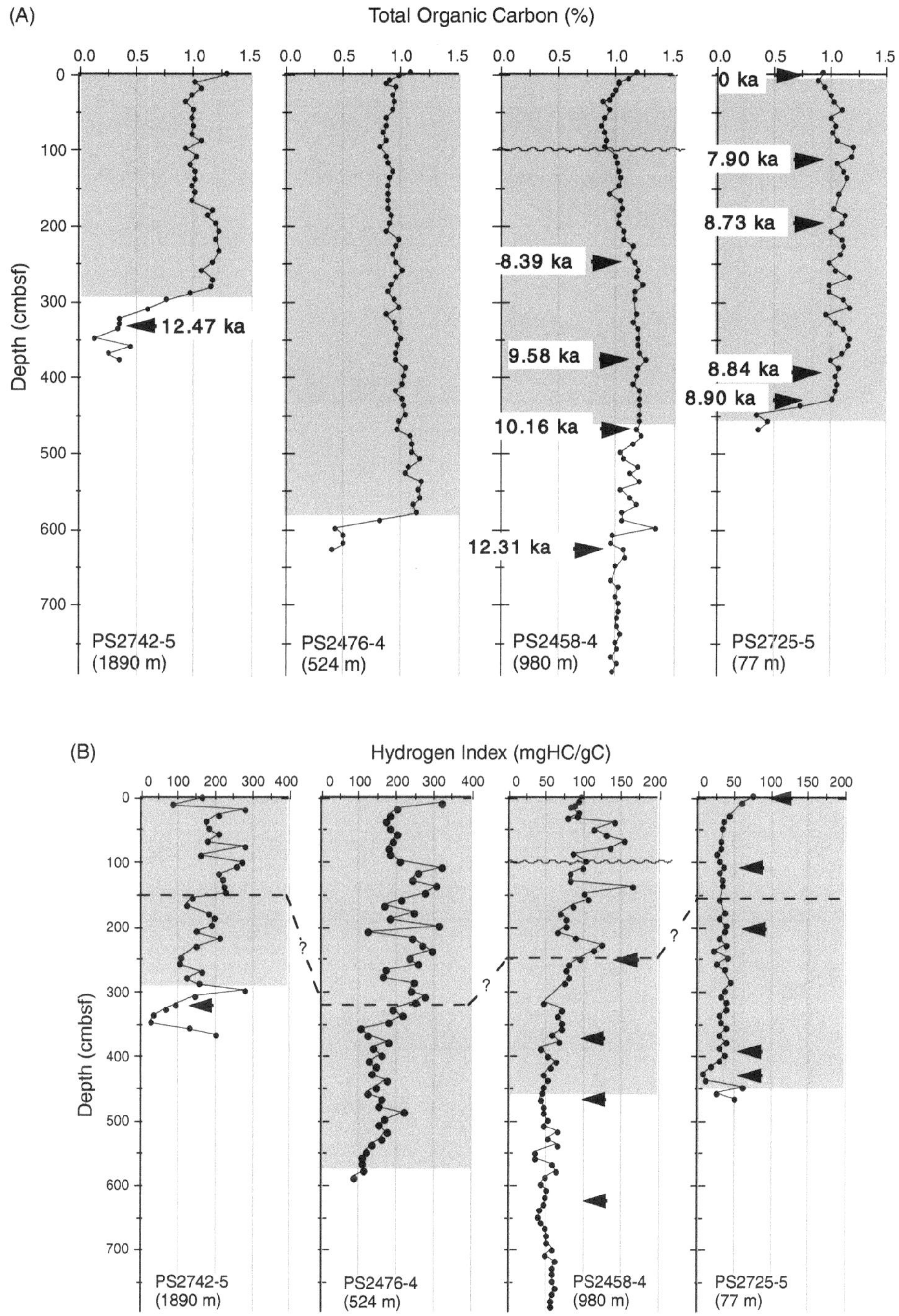

Figure 6.85 (A) TOC and (B) hydrogen index values at Laptev Sea cores PS2742-5, PS2476-4, PS2458-4, and PS2725-5 (from Stein & Fahl, 2000). The Holocene time interval is marked by shading. Arrows and numbers indicate reservoir-corrected AMS^{14}C ages. Wavy line indicates a major hiatus, and AMS^{14}C dates from Core PS2458-4 are according to Spielhagen et al. (2005). Water depths (m) are indicated. For location of cores see Figure 6.76.

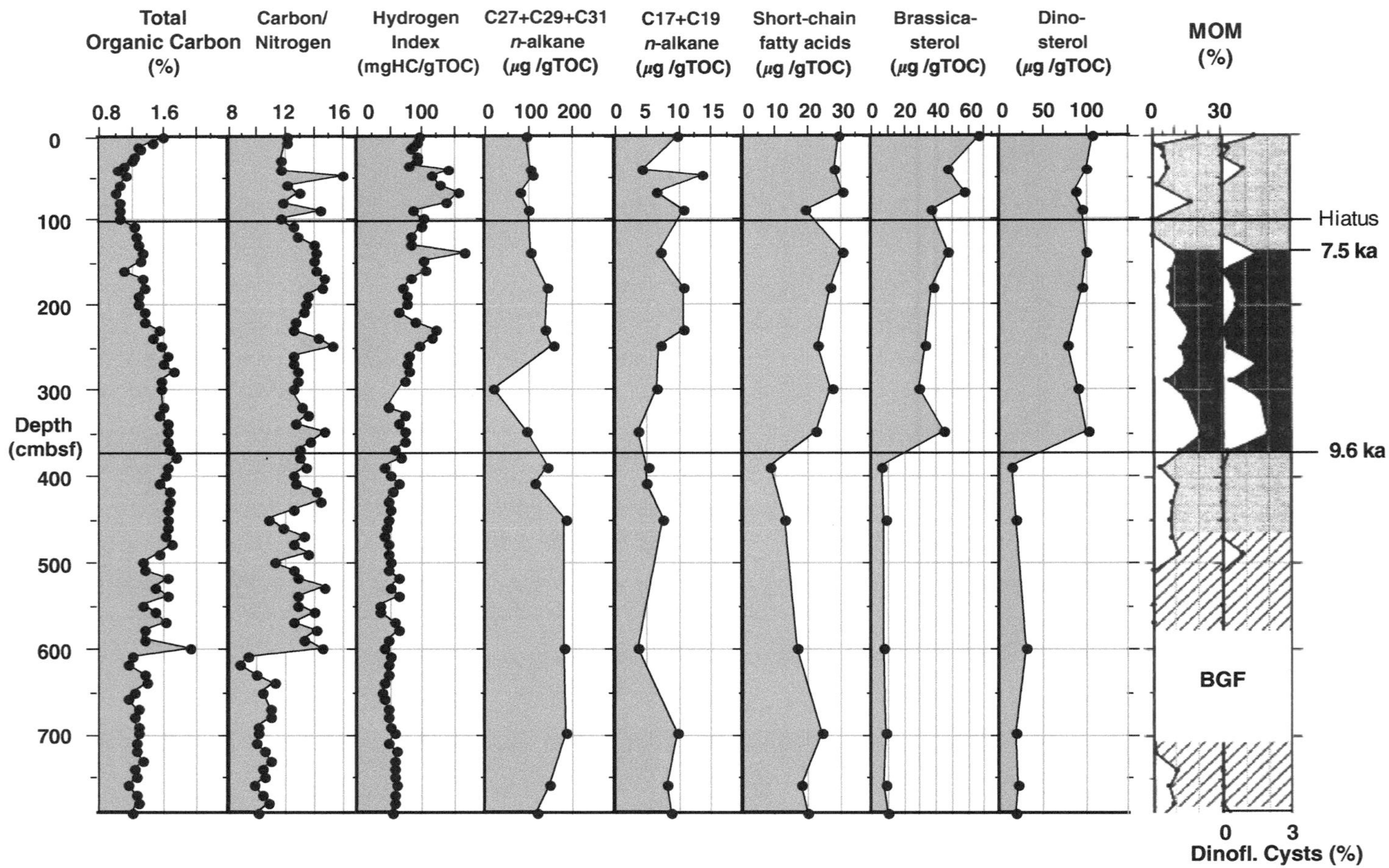

Figure 6.86 Organic-geochemical bulk parameters and specific biomarkers in the sedimentary sequence of Core PS2458-4 (from Fahl & Stein 1999, supplemented): TOC contents (%), TOC/total nitrogen (C/N) ratios, hydrogen index values (mgHC/gOC), long- and short-chain *n*-alkanes (µg/gOC), short-chain fatty acids (µg/gOC), brassicasterol (µg/gOC), and dinosterol (µg/gOC). In addition, the distribution of marine organic matter (MOM) and dinoflagellate cysts (grain %) is shown (Boucsein et al., 2002). BGF, Background fluorescence. AMS^{14}C ages according to Spielhagen et al. (2005). For location of core see Figure 6.76.

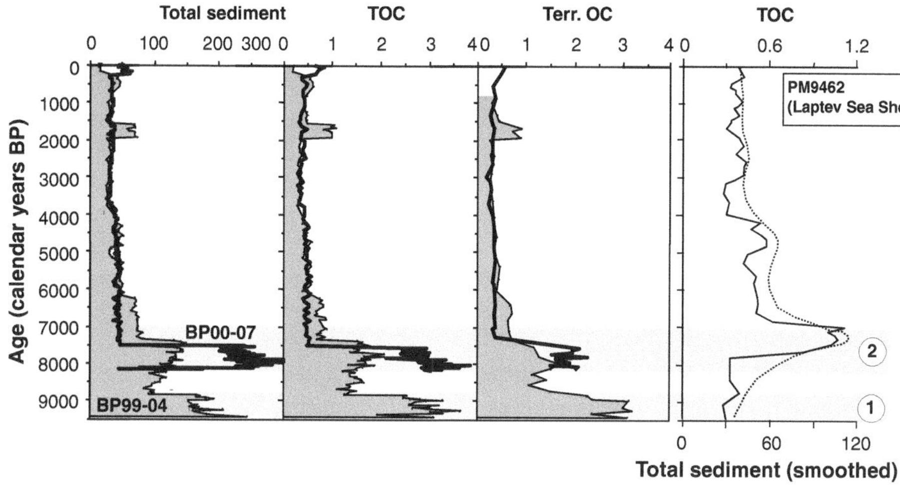

Figure 6.87 Accumulation rate records (in g.cm^{-2}. kyr^{-1}) of bulk (siliciclastic) sediments, TOC, and terrigenous OC from the AMS-^{14}C dated cores BP99-04/07 (grey-shaded) and BP00-07/07 (solid line) (Fahl & Stein, 2007, supplemented). The grey bars indicate the periods of final deglaciation of the Putoran Massif (1) and the mid-Holocene Climatic Optimum (2), respectively (see text for references). In addition, accumulation rates of bulk sediment and OC of Core PM9462 recovered from the Laptev Sea shelf are shown (Bauch et al., 1999).

Holocene (Stein et al., 2003a, 2004a; Stein & Fahl, 2004a, 2004b), in the other cores from the Kara and Laptev Sea continental slope terrigenous OC is also dominant, however, elevated hydrogen index values (Figure 6.85B) in the upper part of the sequence suggest some increased amount of marine OC being occasionally preserved in the Holocene sediments. This interpretation is also supported by an increase in algal biomarkers and increased abundances of marine macerals and dinoflagellate cysts during Holocene times, as shown for Core PS2458-4 from the Laptev Sea continental slope (Figure 6.86; Fahl & Stein, 1999; Boucsein et al., 2002). According to kerogen microscopy data, up to 15–20% of the OC appears to be of marine origin. The higher proportion of marine OC at the Kara and Laptev Sea continental slope is explained by the influence of Atlantic water masses flowing as a boundary current along the Eurasian continental margin (*cf.* Knies & Stein, 1998). Based on maceral data, pulses of Atlantic water inflow appear to be intensified between 9 and 7.5 ka (~10 and 8.3 Cal. kyr BP) and near 4 and 2.5 ka (~4.8 and 2.6 Cal. kyr BP) (Boucsein et al., 2002).

Average Holocene accumulation rates of OC reach maximum values of ~0.3–0.5 gC cm^{-2} kyr^{-1} in the central Kara Sea and at the Laptev Sea shelf and upper continental slope (Table 6.4). Close to the Lena river delta (Laptev Sea) and in the Yenesei and Ob estuaries (Kara Sea) even higher values of 1 to ~3 gC cm^{-2} kyr^{-1} are found (Stein & Fahl, 2000, 2004a, 2004b). Towards the open ocean, the mean accumulation rates decrease progressively down to 0.02 gC cm^{-2} kyr^{-1} (Table 6.4). The progressive decrease in accumulation rates from the inner shelf to the outer shelf/slope reflects the decrease in terrigenous organic matter supply with increasing offshore distance. Due to the high terrigenous (fluvial) input and relatively low primary productivity (see Chapters 2.4 and 2.5), the predominance of terrigenous organic matter in Arctic Ocean sediments from the continental margin as well as the deep sea accounts for the unique position of the Arctic Ocean in comparison to the other world oceans where the proportion of marine organic matter preserved in the sediments reaches much higher values.

When looking at the data of Holocene OC accumulation in the high-sedimentation-rate cores from the inner Kara and Laptev seas in more detail, a huge maximum in terrigenous OC supply is obvious in the early Holocene between ~8 and 10 ka or 9 and 11 Cal. kyr BP (Table 6.4). The correspondence between these cores means that between 9 and 11 Cal. kyr BP the supply of terrigenous (organic) matter sharply increased along the entire Kara Sea–Laptev Sea continental margin. This distinct maximum in terrigenous OC input occurred at times of a major postglacial sea-level rise, when the shallow inner Kara Sea and Laptev Sea shelf became widely flooded for the first time after the LGM, resulting in large-scale sea floor/coastal erosion and, thus, a distinctly increased supply of terrigenous (organic) matter (Polyak et al., 1997; Stein & Fahl, 2000; Bauch et al., 2001a, 2001b). The river discharge probably increased almost at the same time, transporting large amounts of riverine material towards the core locations, as shown in high-resolution records close to the Yenisei and Lena rivers.

At Core BP99-04/7 located directly north of the Yenisei Estuary, a maximum in siliciclastic as well as terrigenous OC accumulation was recorded between 7.3 and 8 Cal. kyr BP (Figure 6.87; Stein et al., 2003a; Fahl & Stein, 2007). Almost exactly at the same time (7.5–8.2 Cal. kyr BP), maximum accumulation rates were also recorded at Core BP00-07/7 (Figure 6.87). The absolute values at this core are even higher than those at Core BP99-04, caused by a location of Core BP00-07/7 within the centre of the "marginal filter" at the time of lowered sea level. This maximum in accumulation rates at both cores, which coincides with a distinct maximum in grain size of the siliciclastic material at Core BP99-04 (Stein et al., 2004a), is explained by distinctly increased fluvial discharge between 7.3 and 8.2 Cal. kyr BP. Higher contribution of river water between 7 and 8 Cal. kyr BP is also supported by the very low $\delta^{18}O$ values determined in ostracods at Core BP00-07/5 (Simstich et al., 2004). Between 7.5 and 8 Cal. kyr BP, a maximum in accumulation rates of both freshwater and marine diatoms were recorded at Core BP99-04/7, interpreted as an increase in marine productivity due to increased riverine nutrient supply (Polyakova & Stein, 2004). At 7.5 Cal. kyr BP, the diatom-specific fatty acids too reached a maximum (Fahl & Stein, 2007). Marine pollen spectra determined in the same core indicate the establishment of long-term favourable climate conditions in the coastal Kara Sea region between ~9 and 7.3 Cal. kyr BP (Kraus et al., 2003). Based on dendrochronological data, most favourable conditions for tree growth probably occurred between ~9.2 and 8 Cal. kyr BP at Yamal Peninsula (Hantemirov & Shiyatov, 2002). An early Holocene warming trend culminated in a temperature maximum between 7.8 and 6.8 Cal. kyr BP and characterized by increased Atlantic water inflow, has also been recorded in a sediment core from the northernmost Barents Sea (Duplessy et al., 2001). A very similar evolution has been described for the history of Lena river discharge in the Laptev Sea, as reflected in the record of

Core PM6492 (Figure 6.87; Bauch et al., 2001a, 2001b). This period of increased river discharge may coincide with the "Holocene Thermal Maximum (HTM)" (*cf.* Macdonald et al., 2000; Kaufman et al., 2004). The warmer and wetter climate in Siberia at that time may have resulted in high river runoff rates.

6.4.2. Variability of Organic-Carbon Accumulation in the Central Arctic Ocean

OC records representing older time intervals (MIS 4 and older) are still very rare for the central Arctic Ocean (e.g., Stein et al., 1994d, 2004b; Schubert, 1995; Schubert & Stein, 1996). Furthermore, the time control of these records is much weaker than for the AMS^{14}C-dated records from the last glacial and Holocene. In general, the OC content of the basin cores appears to be higher than those from ridge cores. On the Alpha and Lomonosov ridges (cores PS2185 and PS051/38), the OC contents vary between 0.05% and 0.3% (Figure 6.88), whereas in the Amundsen Basin cores PS2174-5 and PS2176-3 OC contents are significantly higher ranging between 0.4% and 1.5% (Figure 6.88). In Core PS2165 obtained from the Gakkel Ridge, surprisingly high OC contents of 0.2 to $\sim$1.2% were determined (Figure 6.88; see later discussions).

In the Amundsen Basin cores, maximum OC contents coincide with low hydrogen index values of $<100\,\mathrm{mgHC\,gOC^{-1}}$ and high C/N ratios of 10 to >20 (Figure 6.89), indicating high input of terrigenous organic matter. The clear dominance of long-chain *n*-alkanes over the short-chain *n*-alkanes in the Amundesen Basin Core PS2174-5 as well as the Lomonosov Ridge Core PS2185 (Figure 6.90) may also support a terrigenous origin of the organic matter (Schubert & Stein, 1996). The concentration of the long-chain *n*-alkanes $C_{27+29+31}$ in the Amundsen Core PS2174-5 reach high values of 200–600 $\mu\mathrm{g\,gOC^{-1}}$ (Schubert, 1995; Schubert & Stein, 1996) which point to the Eurasian shelf seas as a potential source of the organic matter (see Chapter 4.7.4, Figure 4.59; Stein et al., 2004b). Intervals with low OC contents, higher hydrogen index values and lower C/N ratios (Figure 6.89), probably represent reduced sea-ice cover and higher surface-water production during interglacials. Higher surface-water production for these intervals is also suggested from higher concentrations of short-chain *n*-alkanes and biogenic opal (Schubert & Stein, 1996).

Whereas on the Lomonosov and Alpha Mendeleev ridges OC accumulation probably derives from terrigenous OC input via sea ice and ocean currents, augmented by minor inputs from primary production, in the oceanic basins turbidity currents may become the prominent process for OC input and accumulation. Turbidites are very common in most (probably all) of the Arctic deep basins (e.g., Fütterer, 1992; Stein et al., 1994c; Grantz et al., 1996, 1999; Svindland & Vorren, 2002). In the Amundsen Basin Core PS2176-3, for example, turbidites are the dominant facies in the sedimentary sequence, especially in the lower half of the core (Figure 6.89; Svindland & Vorren, 2002; see also Chapter 3.4). In general, turbidite sedimentation probably became more important during glacial period of lowered sea level and glacigenic sediment transport close to the shelf edge (see later discussions). For the Amundsen Basin, Svindland and Vorren

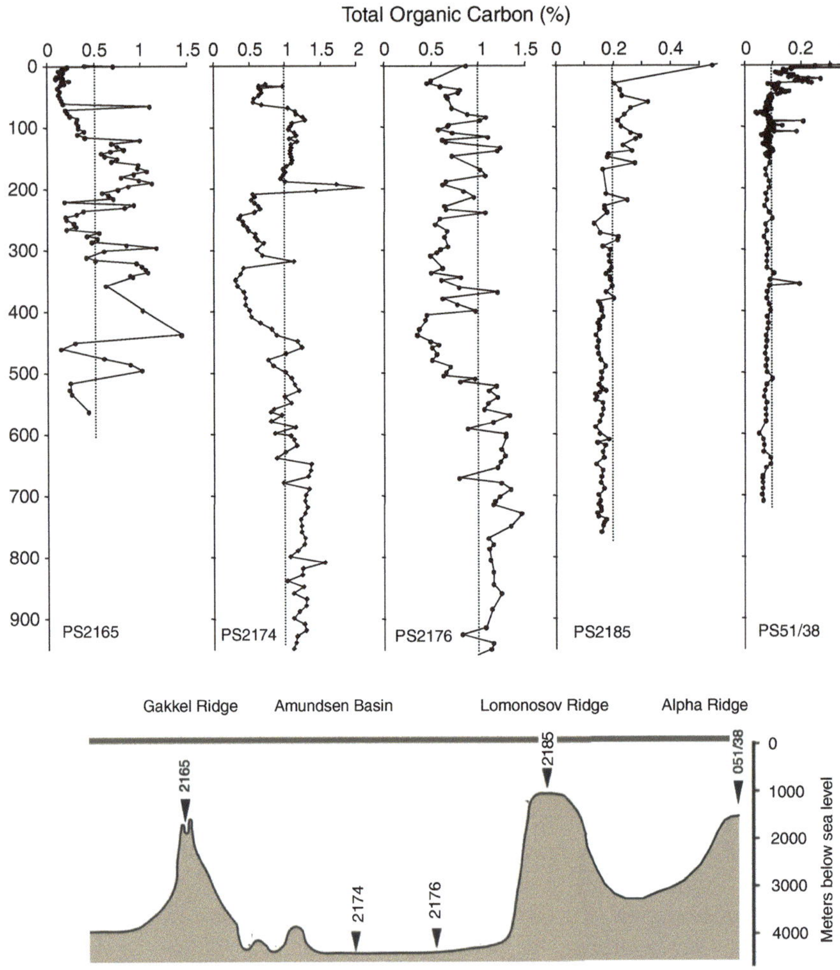

Figure 6.88 TOC contents of cores PS2165 (new data); PS2174-5, 2176-3, and PS2185-6 (Schubert, 1995); and PS51/038 (Stein et al., 2004b). The location of cores is shown in a schematic profile across the central Arctic Ocean as well as in Figure 6.76.

Figure 6.89 OC, hydrogen index, and C/N records (Schubert, 1995) and main lithologies of cores PS2174-5 and PS2176-3 (Amundsen Basin). The occurrence of turbidites is based on Svindland and Vorren (2002). The OC maximum at 200 cmbsf highlighted by a hatched bar, indicates a major turbidite (T). Light gray bars highlight brownish, more bioturbated, OC-poor intervals, probably representing interglacial (interstadial) periods. For location of cores see Figure 6.76.

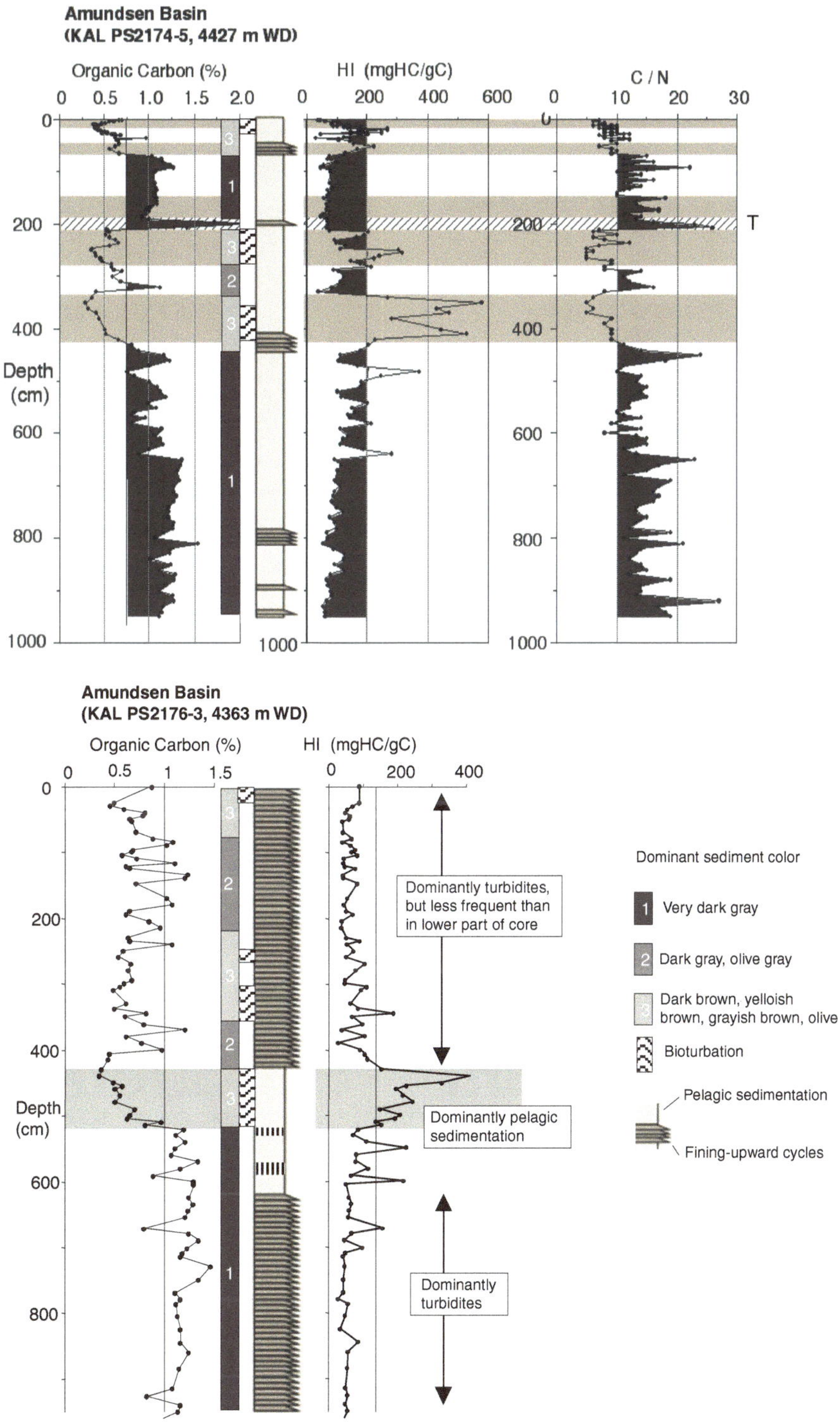
Amundsen Basin
(KAL PS2174-5, 4427 m WD)
Organic Carbon (%)
0 0.5 1.0 1.5 2.0
HI (mgHC/gC)
0 200 400 600
C / N
0 10 20 30
T
Depth
(cm)
0 200 400 600 800 1000
Amundsen Basin
(KAL PS2176-3, 4363 m WD)
Organic Carbon (%)
0 0.5 1 1.5
HI (mgHC/gC)
0 200 400
Dominantly turbidites, but less frequent than in lower part of core
Dominantly pelagic sedimentation
Dominantly turbidites
Dominant sediment color
Very dark gray
Dark gray, olive gray
Dark brown, yelloish brown, grayish brown, olive
Bioturbation
Pelagic sedimentation
Fining-upward cycles

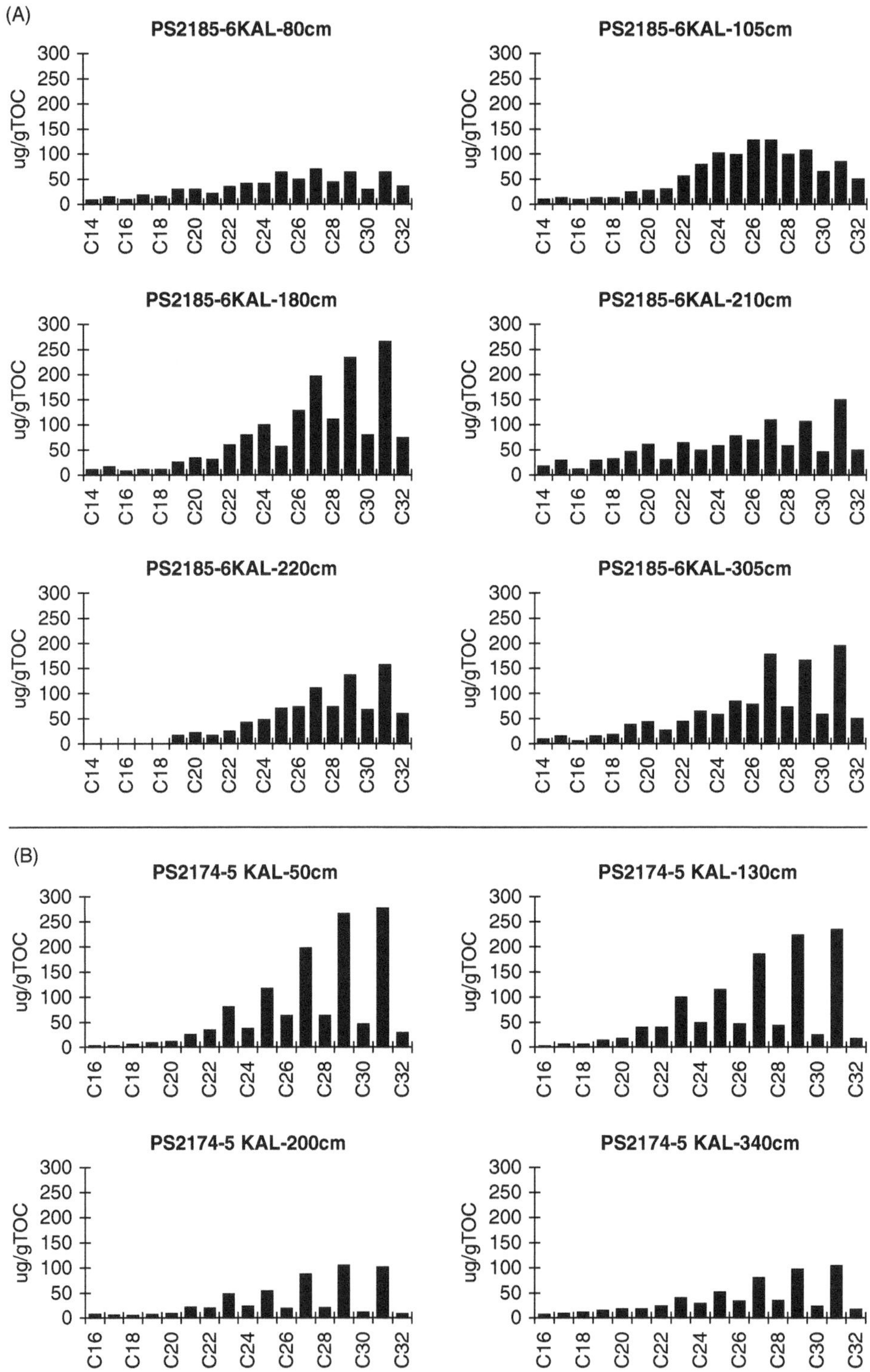
(A)
PS2185-6KAL-80cm
PS2185-6KAL-105cm
PS2185-6KAL-180cm
PS2185-6KAL-210cm
PS2185-6KAL-220cm
PS2185-6KAL-305cm
ug/gTOC
0 50 100 150 200 250 300
C14 C16 C18 C20 C22 C24 C26 C28 C30 C32
(B)
PS2174-5 KAL-50cm
PS2174-5 KAL-130cm
PS2174-5 KAL-200cm
PS2174-5 KAL-340cm
ug/gTOC
0 50 100 150 200 250 300
C16 C18 C20 C22 C24 C26 C28 C30 C32

(2002) determined high sedimentation rates of up to 25 cm kyr^{-1} in the turbidite sequences. That means, sedimentation rates are 20 to >50 times higher than those typical for normal pelagic sediments of the central Arctic (see Table 6.1) indicating the importance of turbidites for the OC budget in these areas.

The importance of turbidite sedimentation for the OC burial is also supported for the Holocene time interval recovered in Core 88PC10 from the Canada Abyssal Plain (see Figure 6.76 for location), where the OC of the turbidites reach values of 1–1.4%, whereas the intercalated pelagic/hemipelagic sediments show values of 0.5–1.1% (Grantz et al. 1999). AMS^{14}C measurements from the pelagic/ hemipelagic intervals allow the turbidite sequences to be dated and sedimentation rates to be calculated. For Core 88PC10, very high sedimentation rates of ~145 cm kyr^{-1} are obtained (Table 6.1), implying very high OC accumulation rates (i.e., ~1.5 gC cm^{-2} kyr^{-1}).

As outlined earlier, turbidite sedimentation is a common process for OC burial in the Arctic Ocean basins. The record of Core PS2165, however, suggests that turbidite sedimentation may also become (at least locally) important on the Gakkel Ridge (Figure 6.91). In the Nansen Basin between Franz Josef Land and the Gakkel Ridge, the occurrence of significant amounts of kaolinite in surface sediments has been related to turbidite sedimentation during (sub-) recent times (Figure 6.91A; Stein et al., 1994c; see Chapter 3.4). At Core PS2165, abundant thin distal turbidites were recorded between 325 and 515 cm and between 125 and 275 cm core depth (Figure 6.91B). These dark grey lithologies are distinctly enriched in OC, whereas the under- and overlying lithologies of brownish colours, and here especially the uppermost 125 cm, have significantly lower OC contents (Figure 6.91C). In the upper part of the sequence, OC values vary dominantly between 0.1% and 0.4%, that is, they are very similar to those determined for sediments from the Lomonosov and Alpha ridges (Figure 6.88). Most probably, turbidite sedimentation happened during glacial times of maximum glacier advances, and a Franz Josef Land/Barents Sea source area is assumed for these OC-rich turbidites (Figure 6.91D). According to the identification of MIS 5 based on the occurrence of coccoliths in ~25 cmbsf (Gard, 1993), the turbidite sequences should have an age of MIS 6 and/or older (Figure 6.91C).

Looking at a transect of short sediment cores from the Nansen Basin across the Gakkel and Lomonosov ridges towards the Makarov Basin representing the last glacial to Holocene time interval, distinct variations in OC content versus depth (and time) are obvious (Figure 6.92). In all records, OC contents increase in the Holocene sediments. Furthermore, OC contents are higher in the basin cores (especially the Nansen Basin) compared to the cores from the ridges. In sediments of last glacial age, OC contents appear to be significantly lower. The low OC values during glacial times are related to low input of both marine as well as terrigenous

Figure 6.90 *n*-alkane distributions of selected samples from (A) Lomonosov Ridge Core PS2185-6 and (B) Amundsen Basin Core PS2174-5 (from Schubert, 1995). In all samples (except one), the long-chain odd *n*-alkanes C_{27}, C_{29}, and C_{31} are dominant, indicating a terrigenous higher-plant OC source. For location of cores see Figure 6.76.

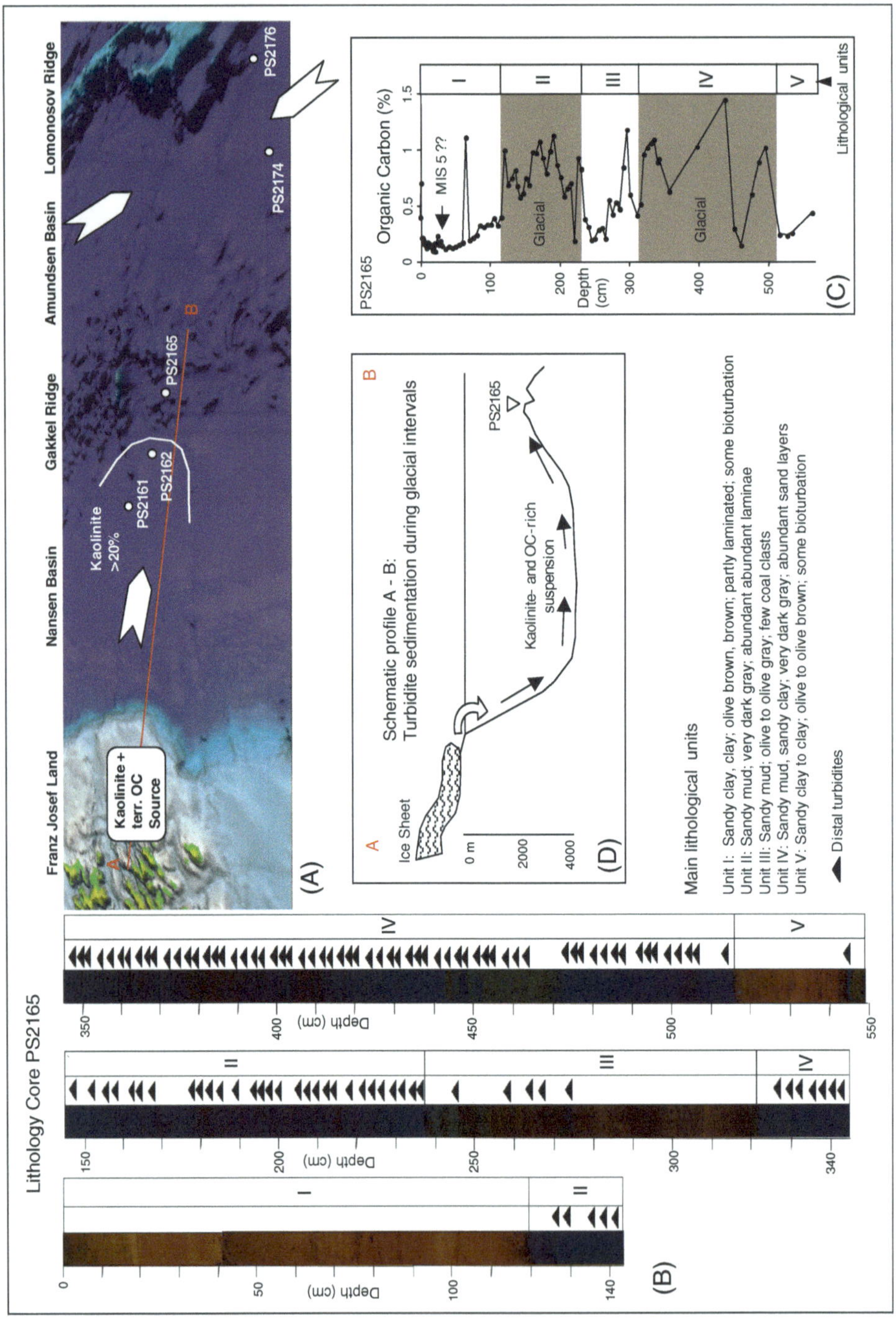
Lithology Core PS2165
Depth (cm)
(B)
Franz Josef Land
Nansen Basin
Gakkel Ridge
Amundsen Basin
Lomonosov Ridge
Kaolinite + terr. OC Source
Kaolinite >20%
PS2161
PS2162
PS2165
PS2174
PS2176
(A)
Schematic profile A - B:
Turbidite sedimentation during glacial intervals
Ice Sheet
Kaolinite- and OC-rich suspension
(D)
PS2165
Organic Carbon (%)
MIS 5 ??
Glacial
Depth (cm)
Lithological units
(C)
Main lithological units
Unit I: Sandy clay, clay; olive brown, brown; partly laminated; some bioturbation
Unit II: Sandy mud; very dark gray; abundant abundant laminae
Unit III: Sandy mud; olive to olive gray; few coal clasts
Unit IV: Sandy mud, sandy clay; very dark gray; abundant sand layers
Unit V: Sandy clay to clay; olive to olive brown; some bioturbation
Distal turbidites

OC due to a more closed sea-ice cover (Stein et al., 1994b, 2004b; Nørgaard-Pedersen et al., 1998).

During the Holocene following the Termination I, carbonate concentrations as well as $\delta^{13}C_{carb}$ values increased contemporaneously with the OC values (Stein et al., 1994a, 1994b). The Holocene increase in $\delta^{13}C_{carb}$ to modern values recorded in most of the cores suggests a well-oxygenated surface-water mass and more open-ice conditions prevailing in the Arctic Ocean at the end of Termination I (see Section 6.3.5). Because the carbonate material is mainly composed of planktonic foraminifera, coccoliths, and ostracods (Gard, 1993; Cronin et al., 1995; Nørgaard-Pedersen et al., 1998), the reduced sea-ice cover and at least seasonally open waters may have increased surface-water productivity. At the same time, carbon and nitrogen isotope values also increase steadily to modern times (Schubert et al., 2001; Stein et al., 2004b). In general, the increase in $\delta^{15}N$ values during the upper Holocene shows the higher nutrient utilization in surface waters due to more open ocean water conditions enabling the phytoplankton to receive enough light for reproduction. The OC levels, however, are not likely to depend entirely on marine productivity because major proportions of the organic matter are of terrigenous origin as indicated by biomarker compositions (Schubert & Stein, 1997), hydrogen index values, and C/N ratios (Stein et al., 1994a, 1994b, 2004b).

6.4.3. Organic-Carbon Accumulation in the Fram Strait

Whereas in the Arctic Ocean terrigenous OC accumulation is predominant (see previous chapters), in the northern central Fram Strait area influenced by strong Atlantic water inflow, marine OC related to surface-water productivity, becomes more significant. In this area, a strong glacial/interglacial variability in sea-ice cover and changes in productivity occurred, as reflected in the OC records of Core 2837-5 (Figure 6.93; Birgel & Hass, 2004; Birgel & Stein, 2004). During the LGM, OC accumulation and marine OC contents are very low due to a more dense sea-ice cover. During deglaciation, OC accumulation as well as the marine OC proportion significantly increased, reaching a very prominent maximum during the Bølling warm interval (Figure 6.93A). OC accumulation rates reach values as high as $\sim 1.5\,gC\,cm^{-2}\,kyr^{-1}$.

Figure 6.91 Lithology, OC record, and depositional model for Core PS2165. (A) Bathymetry of area between Franz Josef Land and Lomonosov Ridge based on the IBCAO Chart (Jakobsson et al., 2000b, 2003a). Core locations discussed in the text are indicated. In surface sediments of cores PS2161 and PS2162 increased amounts of kaolinite >20% were recorded, related to (sub-) recent turbidite sedimention with Franz Josef Land source (Stein et al., 1994c). Large white arrows indicate sediment supply by turbidity currents. Red line (A–B) indicate location of schematic profile shown in (C). (B) Core photograph and main lithologies of Core PS2165. (C) OC record of Core PS2165. Roman numbers indicate lithological units of (B). MIS 5 (?) is based on the occurrence of coccoliths (according to Gard, 1993). Units II and IV characterized by abundant fine-grained distal turbidites and elevated OC values, probably represent glacial intervals (MIS 6 and/or older?). (D) Schematic profile of turbidite sedimentation during glacial intervals. For location of profile see (A).

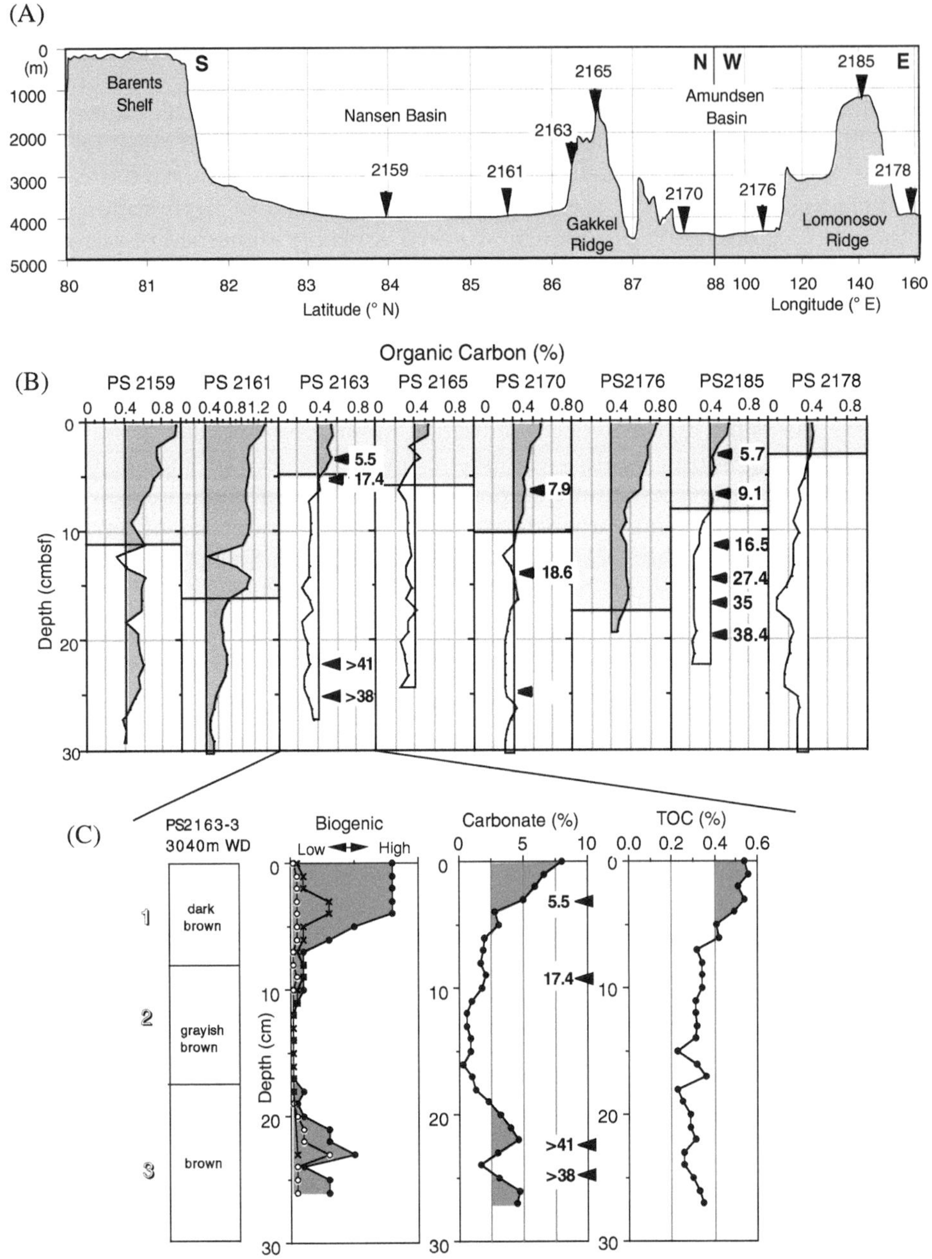

Figure 6.92 (A) and (B) Records of OC of selected sediment cores on a transect across the eastern central Arctic Ocean (from Stein et al., 2004b). (C) Data of Core PS2163-3 from the Gakkel Ridge: composition of biogenic coarse fraction (solid dots, planktonic foraminifera; open dots, benthic foraminifera; crosses, biogenic opal), as based on rough smear-slide estimates performed onboard *Polarstern*; and carbonate and TOC contents (from Stein et al., 1994d). Lithologic units 1, 2, and 3 approximately correspond to MIS 1, 2, and 3, respectively. The Holocene is marked as grey-shaded interval. Bold numbers indicate AMS^{14}C ages in ka (from Stein et al., 1994b; Nørgaard-Pedersen et al., 1998). For core locations see Figure 6.76.

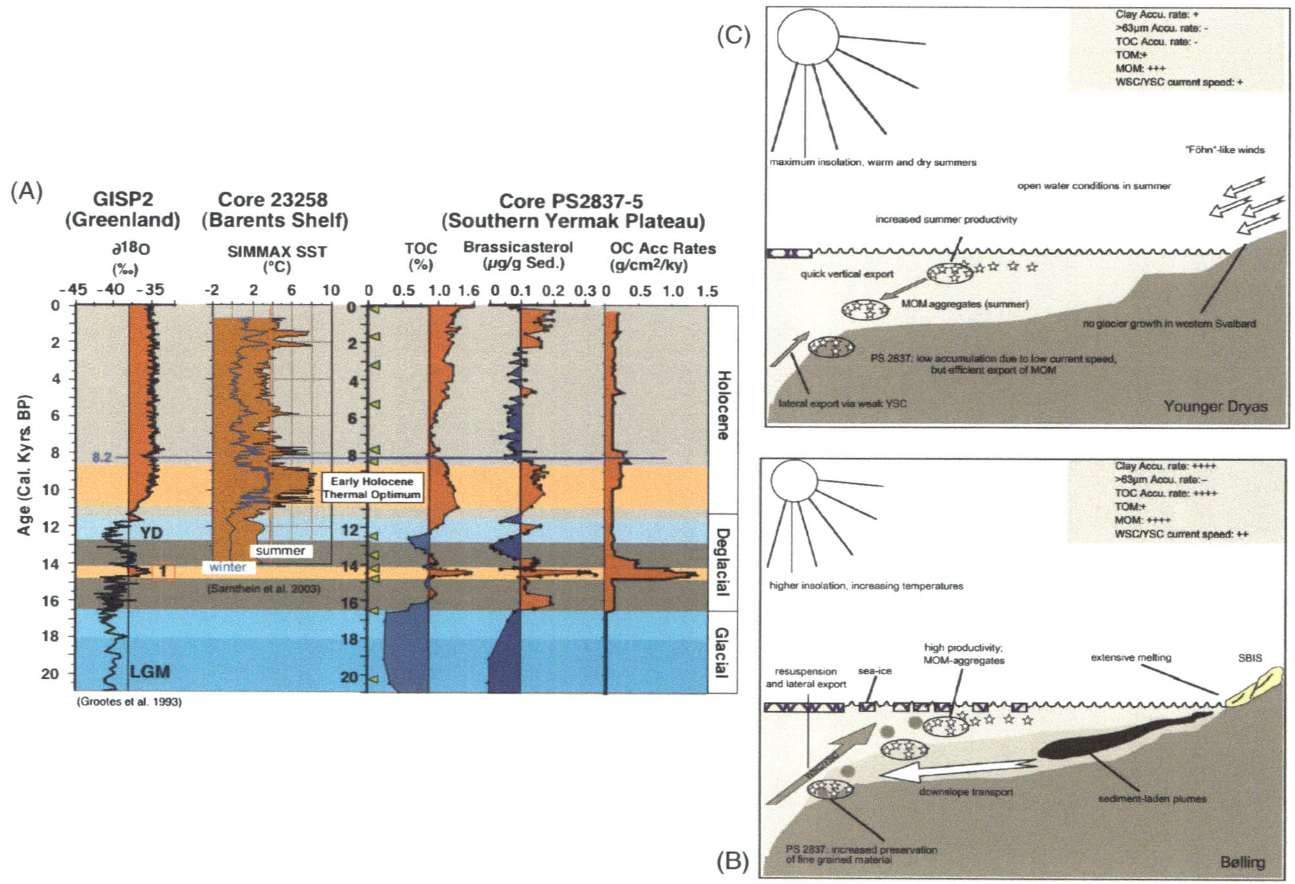

Figure 6.93 (A) TOC (%), brassicasterol (µg/gSed) indicative for diatom productivity, and OC accumulation rates ($gC\,cm^{-2}\,kyr^{-1}$) of the last 20 ka at Core PS2837-5; age scale in calendar years (data from Birgel & Hass, 2004; Birgel & Stein, 2004). Records are related to the GRIP Ice Core record (Grootes et al., 1993) and a SST record of a sediment core from the western Barents Sea (Sarnthein et al., 2003c). 1, Bølling interval. For location of core see Figure 6.76. Model showing the environment along the western Svalbard slope and the western Yermak slope during (B) the Bølling and (C) the YD (from Birgel & Hass, 2004). WSC/YSC, West Spitsbergen Current/Yermak Slope Current; SBIS, Svalbard-Barents Ice Sheet; MOM, Marine organic matter; TOM, Terrigenous organic matter. For further explanation see text.

To explain these maximum in OC accumulation, Birgel and Hass (2004) suggested that meltwater plumes carried large amounts of fine-grained terrigenous inorganic detritus as well as terrigenous organic matter (high C/N ratios, low hydrogen indices; Elverhøi et al., 1995a; Birgel & Hass, 2004; Birgel & Stein, 2004). The material released from decaying glaciers was transported by the WSC to northern Fram Strait (Andersen et al., 1996). Maximum bulk and organic matter accumulation rates were obtained on the Yermak slope by strong impact of sediment-laden meltwater plumes from the disintegrating SBIS (Figure 6.93B). High amounts of marine phytoplankton biomarkers (such as brassicasterol) were deposited and preserved in the matrix of large aggregates of marine and terrigenous organic matter (*cf.* Ittekkot, Haake, Bartsch, Nair, & Ramaswamy, 1992; Andreassen, Nöthig, & Wassmann, 1996; Knies & Stein, 1998). The export of fine-grained sediment particles and marine organic matter from the high productive marginal ice zones to locations under more dense sea ice via currents was also described from surface sediments along the Yermak slope (Soltwedel, Mokievsky, & Schewe, 2000; Rutgers van der Loeff et al., 2002). Marine organic compounds are protected by building aggregates of terrigenous and/or lithogenic fine sediments (Rutgers van der Loeff et al., 2002 and references therein).

OC contents and especially marine biomarkers increased steadily across the YD interval to reach a first maximum in concentration at the end of the YD (Figure 6.93A; Birgel & Hass, 2004). That means, there must have been open water periods at least during the summer seasons that allowed marine phytoplankton (e.g., diatoms, dinoflagellates) to grow. This interpretation is supported by phytoplankton data (diatoms) from the Greenland, Norwegian, and Iceland seas (Koç et al., 1993). Birgel and Hass (2004) proposed dryer and warmer summer conditions comparable with Holocene conditions with "Föhn" - like winds that enabled at least in summer open water conditions and higher productivity (Figure 6.93C).

During the early Holocene between ~10.5 and 8.5 Cal. kyr BP and in the late Holocene between ~2.5 and 0.2 Cal. kyr BP, OC contents as well as concentrations of marine OC increased (Figure 6.93A). The former maximum coincided with the early Holocene insolation maximum, and maximum SSTs in the western Barents Sea as reconstructed from planktonic foraminifers (Sarnthein et al., 2003c; see Chapter 4.5.1). During this early Holocene time span representing the Holocene Thermal Optimum/Maximum (e.g., Kaufman et al., 2004), significantly warmer surface waters resulted in a reduced sea-ice cover, triggering increased primary production and thus accumulation of marine OC at the sea floor. The distinct drop in marine OC concentration seems to be contemporaneous with the "8.2 Cal. kyr BP Cooling Event" (Figure 6.93A; e.g., Alley et al., 1997).

6.4.4. Organic Carbon Burial in Arctic Ocean Sediments During Holocene Times

On the basis of the sediment thickness obtained from acoustic profiles, AMS^{14}C-dated sediment cores and OC data, Stein and Macdonald (2004b) put together a first-order estimate of OC burial in the Arctic marginal seas and the central Arctic

Ocean for the Holocene time interval (for data sources and references as well as background and limitations of estimates, see Stein & Macdonald, 2004a, 2004b).

As shown in Figure 6.94 and Table 6.5, the average Holocene OC burial rates vary widely between the different marginal seas. In the Eurasian seas, OC burial rates reach values of $\sim 1\text{–}2.8 \times 10^6\,\text{t}\,\text{yr}^{-1}$. In the Beaufort Sea, $1.5 \times 10^6\,\text{t}\,\text{yr}^{-1}$ of OC are buried annually, whereas the lowest values for OC burial ($0.23 \times 10^6\,\text{t}\,\text{yr}^{-1}$), were determined for the Chukchi Sea. For the central Arctic Ocean (including continental slope and rise areas), an OC burial rates of

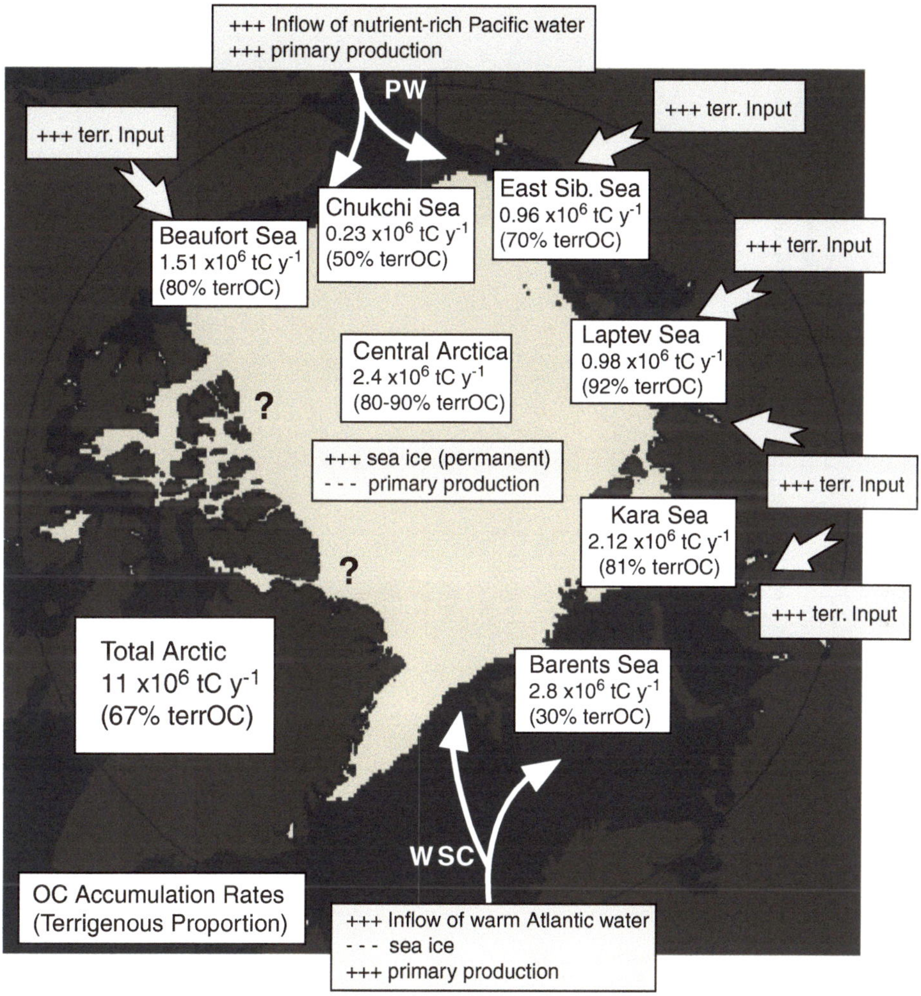

Figure 6.94 Average Holocene burial rates ($10^6\,\text{t}\,\text{yr}^{-1}$) of OC and proportions of terrigenous OC over the marginal seas and basins of the Arctic Ocean (based on Stein & Macdonald, 2004b). Base map with average SIC (September 1979–2004) according to Maurer (2007; http://nsidc.org/data/atlas/).

Table 6.5 Budget (10^6 t yr^{-1}) and Total Accumulation (0–11 Cal. kyr BP; 10^9 t) of Total Sediment and Organic Carbon Fractions (Holocene Average) (from Stein & Macdonald, 2004b) nd, no data. For Data Sources, Background of Calculation, and References See Stein and Macdonald (2004a, 2004b).

Area	Size (km^2)	Total sediment		Organic carbon (OC)		Terrigenous OC		Marine OC	
		(10^6 t yr^{-1})	(10^9 t)	(10^6 t yr^{-1})	(10^9 t)	(10^6 t yr^{-1})	(10^9 t)	(10^6 t yr^{-1})	(10^9 t)
Continental shelves									
Beaufort Sea	178	123	1353	1.51	16.6	1.21	13.3	0.3	3.3
Chukchi Sea	620	19	209	0.23	2.5	0.11	1.2	0.12	1.3
East Siberian Sea	987	109	1199	0.96	10.6	0.67	7.4	0.29	3.2
Laptev Sea	498	67	737	0.98	10.8	0.90	9.9	0.08	0.9
Kara Sea	926	194	2134	2.12	23.3	1.72	18.9	0.4	4.4
Barents Sea (incl. White Sea)	1597	259	2849	2.8	30.8	0.84	9.2	1.96	21.6
Candian Archipelago shelf	146	nd	nd	nd	nd	nd	nd	nd	nd
North Greenland Shelf	30	nd	nd	nd	nd	nd	nd	nd	nd
Lincoln Sea	64	nd	nd	nd	nd	nd	nd	nd	nd
Continental shelves	5052	771	8481	8.68	95.4	5.45	60	3.15	34.7
Continental Slopes	541	107	1177	1.29	14.2	1.03	11.4	0.26	2.9
Continental rises	1095	79	869	0.72	7.9	0.58	6.3	0.14	1.5
Abyssal plains	1367	30	330	0.25	2.8	0.2	2.2	0.05	0.6
Ridges	1506	21	231	0.11	1.2	0.09	1.0	0.02	0.2
Total Arctic	9555	1008	11088	11	121	7.4	80.9	3.6	39.9

$2.4 \times 10^6\,t\,yr^{-1}$ was calculated. In total, $\sim 11 \times 10^6\,t\,yr^{-1}$ of OC have been buried annually in the entire Arctic Ocean as a Holocene average (Figure 6.94). This corresponds to a total accumulation of $\sim 121 \times 10^9\,t$ (or 0.12 Gt) OC during the Holocene, that is, the last 11 Cal. kyr BP. In this calculation, the shelves of the Canadian Archipelago, North Greenland, and the Lincoln Sea have not been considered simply due to lack of data (Table 6.5).

About 79% of the total OC burial during Holocene times occurs as accumulation on the continental shelves ($8.68 \times 10^6\,t\,yr^{-1}$), the remaining 21% (central Arctic Ocean) can be separated into OC burial along the continental slopes and rises ($2.01 \times 10^6\,t\,yr^{-1}$ or 18% of the total burial) and on the central ridges and abyssal plains ($0.36 \times 10^6\,t\,yr^{-1}$ or 3% of the total burial) (Table 6.5).

On the basis of the origin of the OC, the OC burial rates can be divided into terrigenous and marine proportions. Whereas terrigenous OC is predominant in the sediments from the Beaufort, East Siberian, Laptev, and Kara seas as well as the central Arctic Ocean, marine OC is much more important in the Chukchi Sea and, especially, the Barents Sea. In the Arctic marginal seas, $\sim 5.5 \times 10^6\,t\,yr^{-1}$ of terrigenous and $3.2 \times 10^6\,t\,yr^{-1}$ of marine OC have been buried annually as a Holocene average. In the central Arctic Ocean $\sim 1.9 \times 10^6$ and $0.5 \times 10^6\,t\,yr^{-1}$ of terrigenous and marine OC, respectively, are buried (Table 6.5).

Taking the average global burial rate of OC of 100 to $160 \times 10^6\,t\,yr^{-1}$, that is, 0.1 to 0.16 Gt (Romankevich, 1984; Berner, 1989; Hedges & Keil, 1995; Hedges et al., 1997), the Arctic Ocean with $11 \times 10^6\,t\,yr^{-1}$ accounts for $\sim$7–11% of the global budget (Stein & Macdonald, 2004b). This number is disproportionately high considering that the Arctic Ocean accounts for only 2.5% of the global ocean area. Thus, the burial efficiency of OC is significantly higher in the Arctic Ocean than the global average, which is explained by the high proportion of (more refractive) terrigenous organic matter. When using these first-order estimates of OC burial rates in the Arctic Ocean, one should have in mind that an extrapolation of the numbers towards greater depth or longer time scales have to be seen with caution. During glacial times, probably much more sediment and OC were buried in the basins due to turbidite sedimentation and, thus, OC burial in the basins would become much more important for OC budget estimates than for the Holocene.

CHAPTER SEVEN

Mesozoic to Cenozoic Palaeoenvironmental Records of High Northern Latitudes

The long-term palaeoclimatic and palaeoceanographic history of the Arctic Ocean through late Mesozoic and/or Cenozoic times is still poorly known in comparison to other world ocean areas. Major information on the palaeoenvironment of the early Arctic is derived from petroleum exploration drill holes from the Arctic marginal seas (e.g., Mørk & Bjorøy, 1984; Dixon et al., 1992; Leith et al., 1992; Keller, Bird, & Evans, 1999; Montgomery, 2005) and DSDP and ODP drill cores from Subarctic regions (Talwani et al., 1976; Eldholm et al., 1989; Srivastava et al., 1987, 1989; Thiede et al., 1996; Larsen et al., 1994; Raymo, Jansen, Blum, & Herbert, 1999). Direct information from sediment cores derived from the central Arctic Ocean, however, are restricted to a very few short sections — at least prior to the IODP–ACEX drilling campaign in 2004 (Figure 7.1; Thiede, Clark, & Hermann, 1990; Backman et al., 2008; see Chapter 1.2 and below). From Figure 7.1 it is obvious that for the entire Miocene, Oligocene, Eocene (except one short core representing part of middle Eocene), and Palaeocene, no sediment-core data were available at all. For the Cretaceous, core data are limited to three short section representing isolated, discontinuous fragments of the late Campanian and Maastrichtian climate history and depositional environment.

In a number of cores from Northwind Ridge, a high-standing continental fragment in the Chukchi Borderland of the oceanic Amerasian Basin, Palaeozoic, Mesozoic, and Cenozoic sediments/sedimentary rocks were recovered and used — in combination with magnetic anomalies in the Canada Basin — to reconstruct the early tectonic evolution of the Amerasian Basin (Grantz et al., 1998).

In the following sections, some general information about the global palaeoenvironmental evolution through late Mesozoic–Cenozoic times, a review of existing Arctic Ocean and Subarctic records prior to the ACEX drilling campaign as well as some main results obtained from studies on the ACEX sections are presented and discussed.

7.1. Mesozoic High-Latitude Palaeoclimate and Arctic Ocean Palaeoenvironment

7.1.1. Mesozoic Palaeoclimate: General Trends and Variability

Up to the 1980s the palaeoclimate of the Cretaceous has been viewed as a period of great warmth over the globe (Frakes, 1979; Hallam, 1981, 1985). According to these authors tropical–subtropical conditions prevailed to at least 45°N and possibly

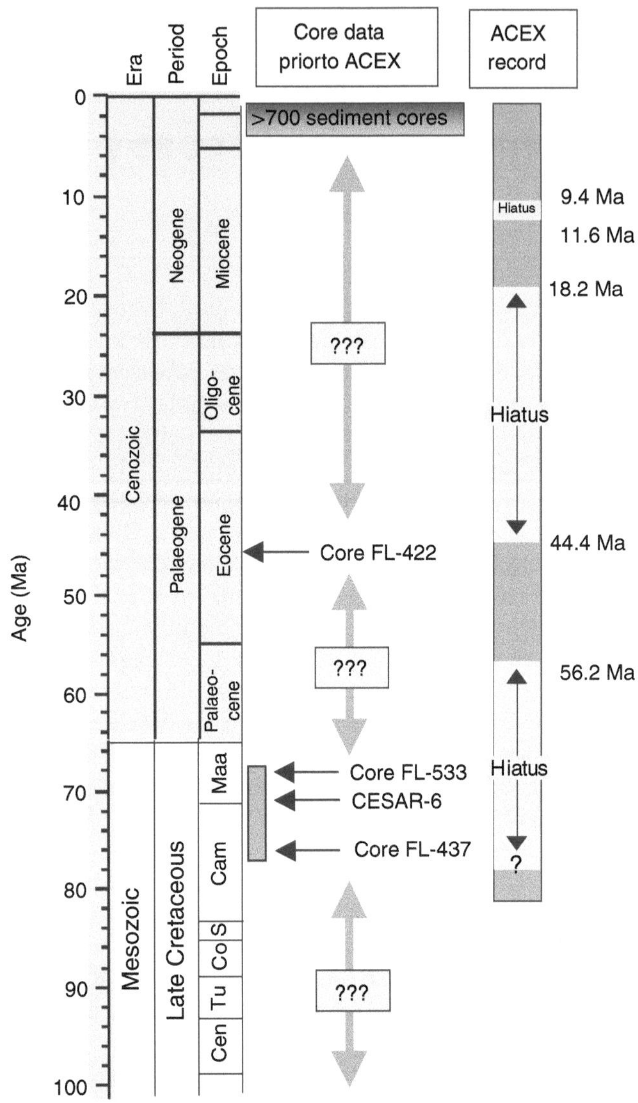

Figure 7.1 Stratigraphic coverage of existing cores in the central Arctic Ocean prior to ACEX (based on Thiede et al., 1990), and the section recovered during the ACEX drilling expedition (Backman et al., 2006, 2008). Well over 99% of all existing short cores were of Quaternary age prior to ACEX. The middle Eocene sediments recovered at Core FL-422 are arbitrarily placed at 45 Ma, the three Late Cretaceous (Maastrichtian/Campanian) cores are arbitrarily placed on the time axis as well. Each of these cores has a duration, at the most, of a few hundred kyr (Backman et al., 2008).

to 70°S, and warm to cool-temperate climates extended to the poles which were ice-free during those times. Mean annual temperatures were significantly higher and latitudinal gradients only about half of those of today. During the past two decades, however, this view has been questioned and a much more varied climatic history for the late Jurassic/Cretaceous has been suggested, replacing the somewhat antiquated view of constant equable warmth and sluggish oceanic circulation during the late Mesozoic (e.g., Kemper, 1987; Stoll & Schrag, 1996; Weissert & Lini, 1991; Hay et al., 1999; Mutterlose & Kessels, 2000; Price, Ruffell, Jones, Kalin, & Mutterlose, 2000; Mutterlose et al., 2003; Miller, Wright, & Browning, 2005).

In the mid-Cretaceous (124–83 Ma), magmatic superplumes are thought to have caused an episode of constant normal magnetic polarity, which resulted in high oceanic spreading rates, and generally high sea level and high palaeotemperatures (Figure 7.2; Larson, 1991a, 1991b). Along with this, increased rifting, oceanic plateau production, and island-arc volcanism subsequently caused an elevated

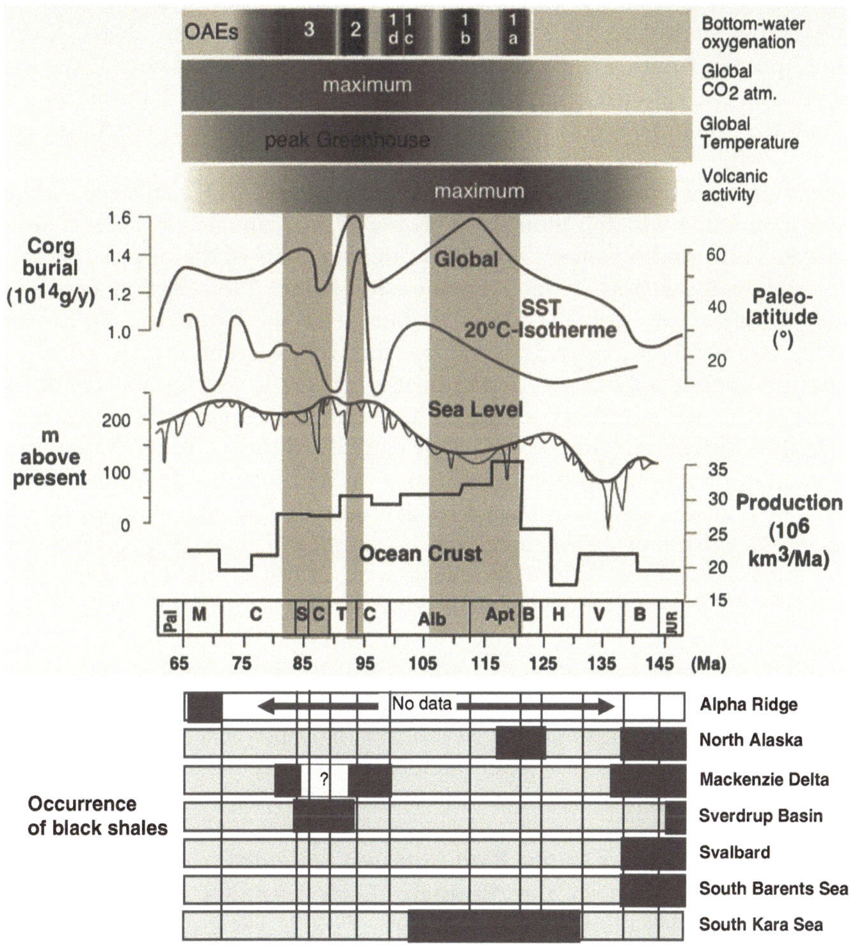

Figure 7.2 Timing of OAEs in relation to Cretaceous volcanic activity, palaeolatitude of SST 20°C isotherme, global C_{org} burial, sea-level change (Larson, 1991a; Wagner, 2002). Stratigraphic occurrence of OC-rich sediments/sedimentary rocks (black shales) drilled in the main circum-Arctic basins (according to Leith et al., 1992).

atmospheric CO_2 partial pressure, which was crucial for creating "greenhouse conditions" during these times. Even if warm high-latitude climates dominated much of the late Cretaceous (to early Eocene) and that warmth may have been caused by high CO_2, short intervals of more restricted glaciations may have occurred (at least) on Antarctica, as discussed by Miller et al. (2005). According to these authors, backstripped eustatic estimates from New Jersey and the Russian platform show large (>25 m) and rapid (<1 myr) sea-level changes during this time period of global greenhouse conditions, interpretated as indication of variable ice volume on Antarctica. Thus, they propose a vision of Earth's cryospheric evolution that reconciles warm, generally ice-free poles with cold snaps that resulted in glacioeustatic lowerings, probably driven by orbital forcing (Matthews & Frohlich, 2002; DeConto & Pollard, 2003). Based on the models of DeConto and Pollard (2003), late Cretaceous (to early Eocene) ice sheets may have reached maximum volumes of $8\text{–}12 \times 10^6\,km^3$ (20–30 m glacioeustatic equivalent). Such ice sheets — probably with the exception of an even larger Campanian/Masstrichtian event with an ice volume of $17 \times 10^6\,km^3$ (40 m glacioeustatic equivalent) — did not reach the Antarctic coast, which reconciles ice with relative warmth in coastal Antarctica (Miller et al., 2005).

The availability of nutrients and elevated atmospheric CO_2 may have stimulated primary production which is thought to be one of the main factors (besides bottom-water stagnation) causing anoxic conditions in large parts of the oceans (Figure 7.2; e.g., Schlanger & Jenkyns, 1976; Demaison & Moore, 1980; Arthur et al., 1984; Arthur, Schlanger, & Jenkyns, 1987; Arthur, Jenkyns, Brumsack, & Schlanger, 1988; Stein et al., 1986; Erbacher, Huber, Norris, & Markey, 2001). These major episodes are known as Oceanic Anoxic Events (OAEs) and have probably occurred on a regional to worldwide scale. Such extreme conditions favoured the accumulation and preservation of OC-rich marine sediments (black shales). Periods of widespread marine black-shale formation occurred repetitively in the Mesozoic. In the late Jurassic, widespread black shales occurred in the Kimmeridge (e.g., Cornford, Needham, & de Walque, 1986; Boussafir & Lallier-Verges, 1997; Sælen, Tyson, Telnæs, & Talbot, 2000; Riboulleau et al., 2003). In the Cretaceous, black-shale formation was concentrated in the Aptian–Albian (OAE 1a through 1d), the Cenomanian–Turonian (OAE 2), and the Coniacian–Santonian (OAE 3) (Figures 7.2 and 7.3; e.g., Arthur et al., 1987, 1988; Wagner et al., 2003). OAEs represent major perturbations in the global climate and ocean system defined by the massive OC burial in marine environments that probably have played a major role in the late Mesozoic climate evolution. Upper Jurassic to Cretaceous black shales are also wide-spread in the circum-Arctic area (see Section 7.1.2).

Palaeoclimatic conditions in the High Northern Latitudes. A dominantly warm climate probably existed in the Northern High Latitudes during mid to late Cretaceous times. Already since the discovery of the leaves and fruit of the tropical breadfruit tree in Cenomanian fluvio-deltaic sediments from west Greenland in 1883, it has been apparent that the mid Cretaceous climate of the Northern High Latitudes — at least during certain intervals — was far warmer than it is today

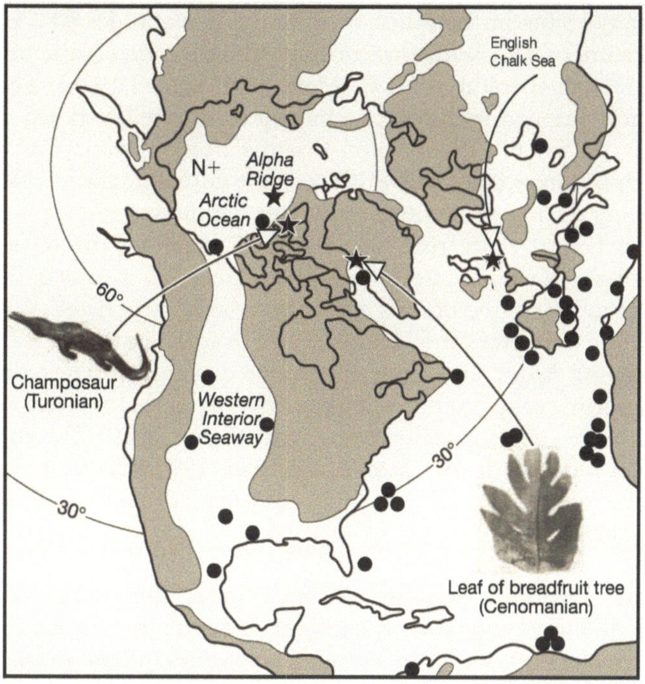

Figure 7.3 Northern Hemisphere view of the Arctic Ocean during the Late Cretaceous (Turonian), with likely connections of the Arctic Ocean to epicontinental seas, such as that of the western interior of North America and the north European shelf (based on Hay et al., 1993). Approximate location of English Chalk outcrops in southeast England (Jenkyns et al., 1994); position of fossil remains of breadfruit-tree leaves (image of the actual specimen recovered shown) after Nathorst (1890); position and image of champosaur (Huber, 1998; Tarduno et al., 1998); and approximate position of the Alpha ridge (Clark et al., 1986), are also shown. Black circles indicate locations of sites with Cenomanian/Turonian black shales (based on Arthur et al., 1987; Wagner et al., 2003 and further references therein). Figure from Jenkyns et al. (2004), supplemented.

(Nathorst, 1911). Other palaeontological data support a generally rather mild mid to late Cretaceous Arctic climate by the presence of deciduous trees and leaves with characteristic morphologies, at 80–85°N (Parrish & Spicer, 1988; Herman & Spicer, 1996), the presence of crocodiles beyond 60°N (Markwick, 1998) and, most specifically, the discovery of champosaurs (cold-blooded reptiles; closest living relatives are crocodiles) in the Turonian of the Sverdrup Basin (Axel Heiberg Island) at 72°N palaeolatitude (Tarduno et al., 1998; Huber, 1998) (Figure 7.3). The thermal limit for viable populations of crocodiles is marked by a coldest-month mean temperature of ~5.5°C. The preferred operating temperature of crocodiles is 25–35°C, and this temperature is maintained for sufficient duration in areas with a minimum mean annual temperature >14°C (Markwick, 1998). Fossil floras of Coniacian age from the North Slope of Alaska (palaeolatitude of 75°N) yield a mean annual temperature of 12.5°C, with a warm-month mean temperature of 20.0°C and a cold-month mean temperature of 5.7°C, and Turonian flora from Novaya Sibir in the Russian Arctic (palaeolatitude ~82°) yield a mean annual temperature of 9°C, with a warm-month mean temperature of 18.5°C and a cold-month mean temperature of 0°C (Herman & Spicer, 1996). These values are cooler than those implied by the Axel Heiberg fossil reptiles. Based on their floral data, Herman and Spicer (1996) stated that the nearby Arctic Ocean sea-surface temperature cannot have been below 0°C, or the adjacent land would have cooled to below freezing during the protracted winter darkness.

The palaeobotanical data from the Arctic Cretaceous, however, are not in agreement with global climatic trends established from other parts of the world from oxygen-isotope ratios of microfossils and bulk pelagic carbonates. Whereas Parrish and Spicer (1988) and Spicer and Parrish (1990) suggest a thermal maximum in the Coniacian based on Arctic flora, oxygen-isotope data from ODP cores in both the southern and northern hemispheres suggest that the global thermal maximum (or maxima) was/were developed between the Cenomanian/Turonian boundary and the late Turonian (Jenkyns, Gale, & Corfield, 1994; Clarke & Jenkyns, 1999; Huber, Hodell, & Hamilton, 1995; Huber, Norris, & MacLeod, 2002; Wilson, Norris, & Cooper, 2002).

Besides indication for a relatively warm (sub-) Arctic climate during Cretaceous times, there are also signals pointing to cold climatic conditions. In particular for parts of the early Cretaceous several hints for icehouse conditions or at least cool climates in the higher latitudes were found. Glendonites (pseudomorphs of the low-temperature hydrated form of calcium carbonate, ikaite) found in lower Valanginian and upper Aptian sediments from the Sverdrup Basin in Arctic Canada (70–80°N palaeolatitude), implies that early Cretaceous seawater temperatures were at times close to freezing (Kemper, 1987). Almost certainly these cooler temperatures record global changes because, at least in the case of the late Aptian, coeval glendonites are also known from the Southern Hemisphere, being found in the Eromanga Basin in Australia at a palaeolatitude of 65° (Frakes & Francis, 1988; De Lurio & Frakes, 1999). Furthermore, Frakes and Francis (1988) postulate at least seasonally cold ocean temperatures and limited polar ice caps for the early Cretaceous from the occurrence of ice-rafted deposits in Siberia, Australia, and Spitsbergen. Nannofossil assemblages from offshore mid-Norway and the Barents Sea also suggest cold conditions during the early Cretaceous (Mutterlose & Kessels, 2000; Mutterlose et al., 2003). Based on oxygen stable isotope palaeothermometry of well-preserved mid-Jurassic to lower Cretaceous belemnites from Kong Karls Land, Svalbard, cool high-latitude marine palaeotemperatures of <8°C were estimated for the early to middle Valanginian time interval, which may be compatible with the formation of high-latitude ice (Ditchfield, 1997).

7.1.2. Upper Jurassic/Lower Cretaceous Circum-Arctic Black Shales and Palaeoenvironmental Implications

Upper Jurassic to Cretaceous black shales were deposited in all the large circum-Arctic sedimentary basins (e.g., North Alaska Slope, Arctic Canada, the Barents Shelf, and western Siberia) that are highly productive in terms of gas and oil (Figure 7.4; e.g., Dixon et al., 1992; Leith et al., 1992; Bakke et al., 1998; USGS

World Petroleum Assessment, 2000; Montgomery, 2005; Vyssotski et al., 2006). Especially during the late Jurassic, black-shale-type sediments characterized by the preservation of labile organic matter, were found in all of the major basins (Figure 7.2), and anoxic environmental conditions were proposed for the Arctic Ocean at that time (Leith et al., 1992). Arctic Cretaceous black shales seem to occur not contemporaneously with the widespread Aptian–Albian and Cenomanian/Turonian black shales recorded in large parts of the Cretaceous world ocean and related to regional (to global) anoxia and/or increased primary production

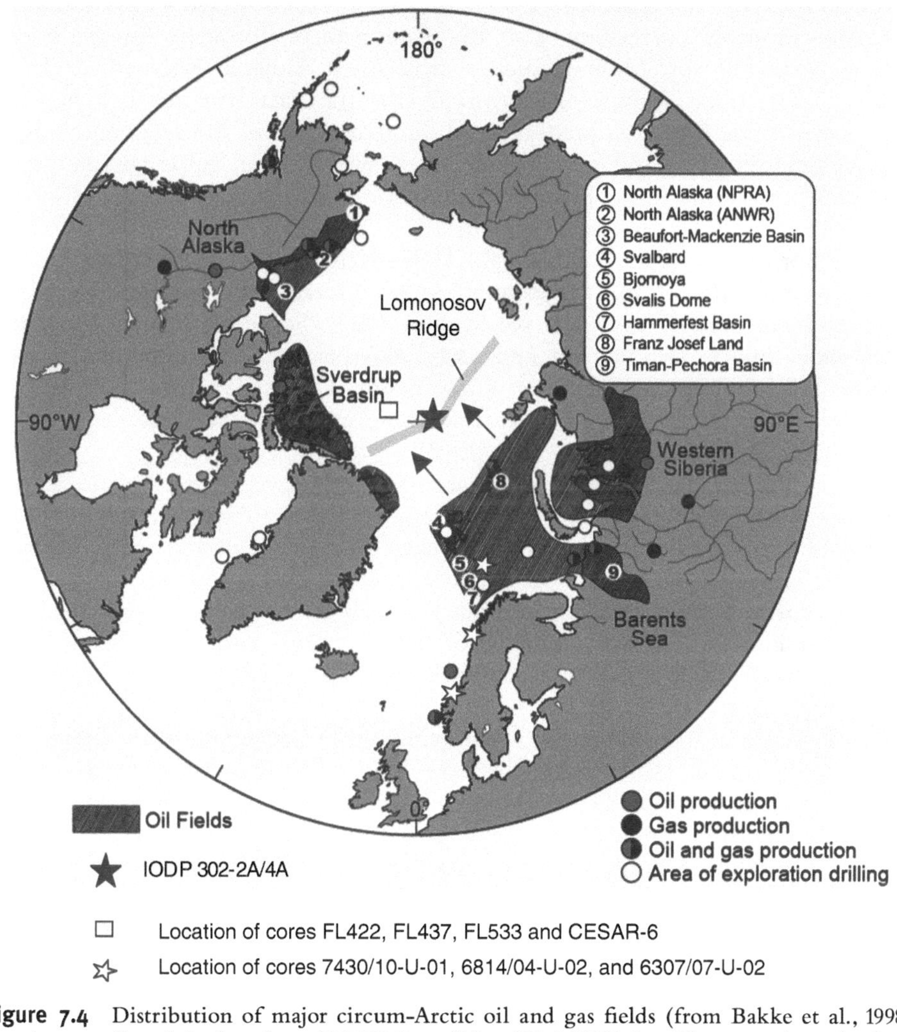

Figure 7.4 Distribution of major circum-Arctic oil and gas fields (from Bakke et al., 1998, supplemented) and the location of IODP Expedition 302 ACEX sites. The area where the four short sediment cores containing older pre-Neogene sediments, were recovered, are also shown. Black arrows indicate the drift of Lomonosov Ridge, separated from the Eurasian continental margin near 56 Ma (according to Kristoffersen, 1990).

(e.g., Schlanger & Jenkyns, 1976; Arthur et al., 1984; Stein et al., 1986). Early Cretaceous strata were mapped on Bennett Island, New Siberian Islands, which contain black fissile argillites (Kos'ko & Trufanov, 2002). In a drill hole on Sverdrup Island (southern Kara Sea, 74.65°N, 79.55°E), OC-rich sedimentary rocks of Aptian/Albian and Hauterivian/Barremian age were recovered, with typical OC values of 1–3%, but with maximum OC contents of 5–11% (absolute maximum of 41.7%) determined in coaly clay intervals (Shkola, 2007a). Black shales found in the North Alaskan Basin, Mackenzie Delta Basin, and Sverdrup Basin are of Aptian–Barremian, Cenomanian–Coniacian, and Turonian–Santonian ages, respectively (Figure 7.2; Leith et al., 1992). In the middle part of West Greenland along the coast towards Baffin Bay (Disco Island area), OC-rich sediments probably correlating with the Cenomanian/Turonian Anoxic Event, were deposited (Dam, Nøhr-Hansen, Christiansen, Bojesen-Koefoed, & Laier, 1998). It has to be considered, however, that the data base for late Jurassic/ Cretaceous Arctic Ocean sequences is still limited. Most of the sedimentological, micropalaeontological, and geochemical data available for palaeoenvironmental reconstructions of the Arctic realm are from petroleum exploration drill holes from the marginal seas (e.g., Mørk & Bjorøy, 1984; Worsley, Johansen, & Kristensen, 1988; Dixon et al., 1992; Leith et al., 1992; Keller et al., 1999; Montgomery, 2005), whereas data from the central Arctic Ocean are restricted to a few short sediment cores (Figure 7.1; see Section 7.1.3). Some of the data from drill holes taken in the Barents Sea and along the Norwegian continental margin (Figure 7.5), representing the late Jurassic/early Cretaceous time interval, are discussed later in more detail. At that time, these sites had palaeolatitudes between ~42°N and 67°N (Mutterlose et al., 2003).

During the late Jurassic to early Cretaceous, the Norwegian-Greenland Seaway (NGS) was one of three connections between the proto-Arctic Ocean with the Barents Sea at its southern end, and the Tethyan-Atlantic realm, the other two were across the Russian Platform and the Western Interior Seaway (Figures 7.3 and 7.5). Due to its long meridional extension the NGS may have played a crucial palaeoceanographic role as a conduit for heat transport and exchange of water masses between the low and high latitudes (e.g., Kazmin & Napatov, 1998; Hay et al., 1999; Mutterlose et al., 2003).

In the southern Barents Sea, OC-rich sediments of late Jurassic/early Cretaceous age were recovered, characterized by very high OC contents of up to ~35% (Figure 7.6; Langrock et al., 2003a). Hydrogen-index values are also high (Figure 7.6; 400–600 mgHC/gC), indicating a kerogen-type I/II with a very good potential for oil. The maceral composition of the sediments is dominated by lipid-rich organic matter derived from different marine/aquatic sources, supporting the type II kerogen characterization by Rock-Eval pyrolysis. Massive occurrences of well-preserved marine Prasinophycean algae (mainly *Leiosphaeridia*), minor abundance of juvenile *Botryococcus*-type algae (freshwater), a few *Tasmanites*-type algae (brackish), and *Reinschia*-type algae (freshwater puddle) as well as a few dinocysts (or acritarchs) were recorded at various depths (Langrock et al., 2003a).

The excellent preservation of labile, easily degradable organic matter, the abundance of freshwater algae, sporinites and bituminous fragments from boghead,

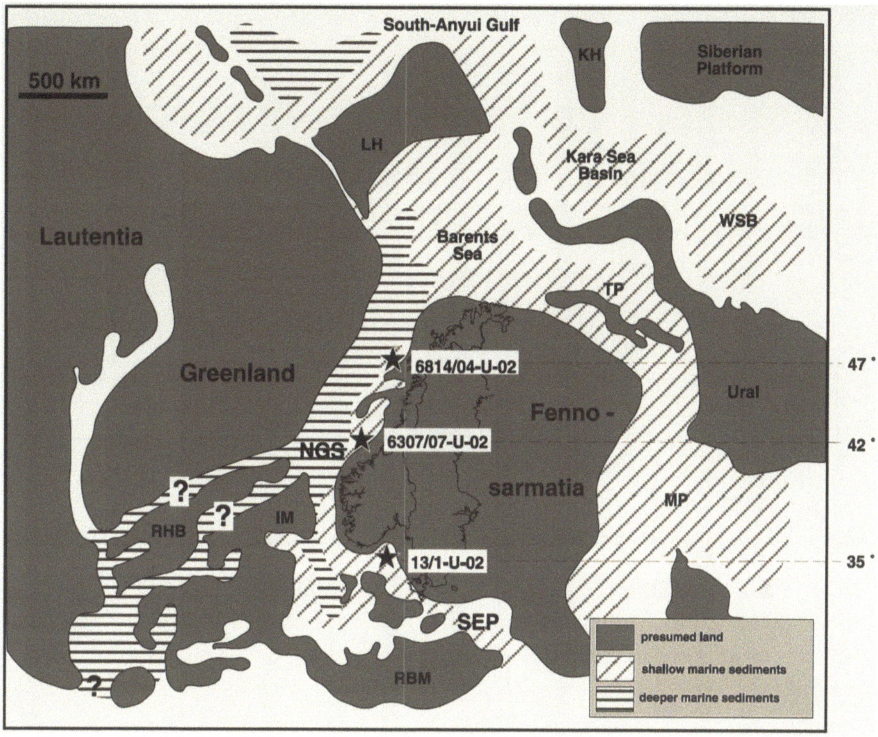

Figure 7.5 Palaeogeographic situation of the Norwegian-Greenland Seaway and surrounding continental masses during middle to late Berriasian times (from Langrock et al., 2003b, based on Ziegler, 1988 and Mutterlose et al., 2003). Locations of cores discussed in the text are shown relative to the modern coast line (outlines of Norway) and indicated by the asterisk. IM, Irish Massif; KH, Kara High; LH, Lomonossov High; MP, Moscow Platform; RBM, Rhenish Bohemian Massif; RHB, Rockall-Hatton Bank; TP, Timan Pechora Area; WSB, West Siberian Basin; NGS, Norwegian-Greenland Seaway; SEP, Southeastern Passage.

and cannel coals all argue for deposition in a restricted shallow basin very close to land rather than an upwelling regime (Figure 7.7; Arthur et al., 1987; Taylor et al., 1998). Strong anoxia is supported by high concentrations of Mo and high Re/Mo ratios (Figure 7.8; Langrock et al., 2003a; Lipinski et al., 2003; see also Chapter 4.4). Finally, very low sedimentation rates in the late Volgian of Core 7430/10-U-01 indicate that clastic dilution was strongly limited and that OM preservation was not controlled by rapid burial. In combination with very high (algae-type) OC contents, deposition under anoxic bottom-water conditions is the most reasonable mechanism for OC enrichment, rather than deposition in a high-productivity environment (see Chapter 4.7.2, Figure 4.55; *cf.* Stein, 1990, 1991a).

During the late Jurassic and early Cretaceous, such anoxic conditions proposed for large parts of the proto-Arctic Ocean, probably extended towards the south into the NGS as shown in geochemical and microscopical data sets from black shale sequences recovered in drill cores along the Norwegian shelf (Figure 7.5). These black shales contain high amounts of OC (mostly 3–7%) of dominantly aquatic

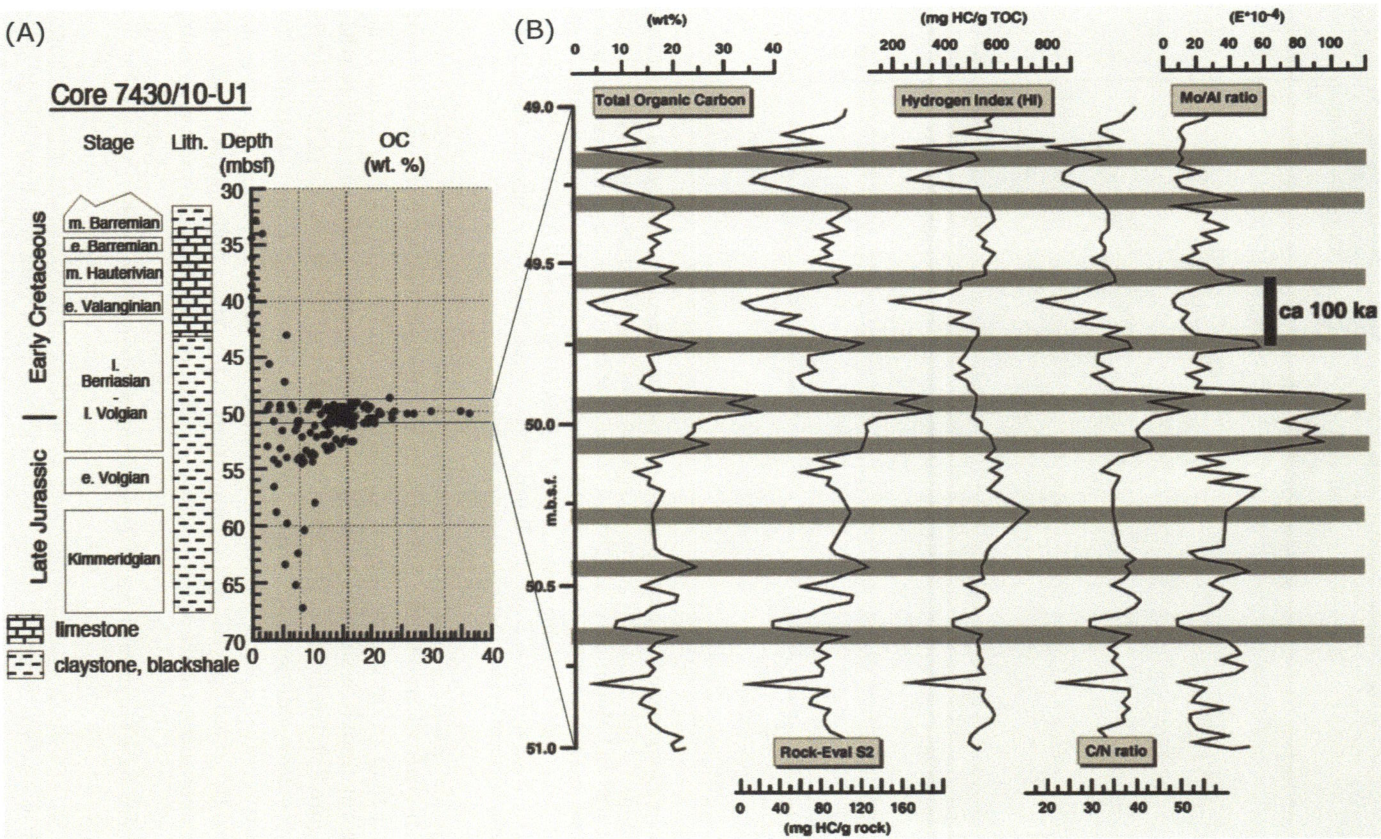

Figure 7.6 (A) Stratigraphy, main lithology, and TOC content of Core 7430/10-U-01 and (B) high-resolution section from 51 to 49 mbsf showing down-core variations in TOC content, Rock-Eval S2 and HI values, and C/N and Mo/Al ratios (from Langrock et al., 2003a). Maximum peaks are highlighted by shaded lines. Spacing between peaks may represent ~100 kyr, suggesting variable sedimentation rates.

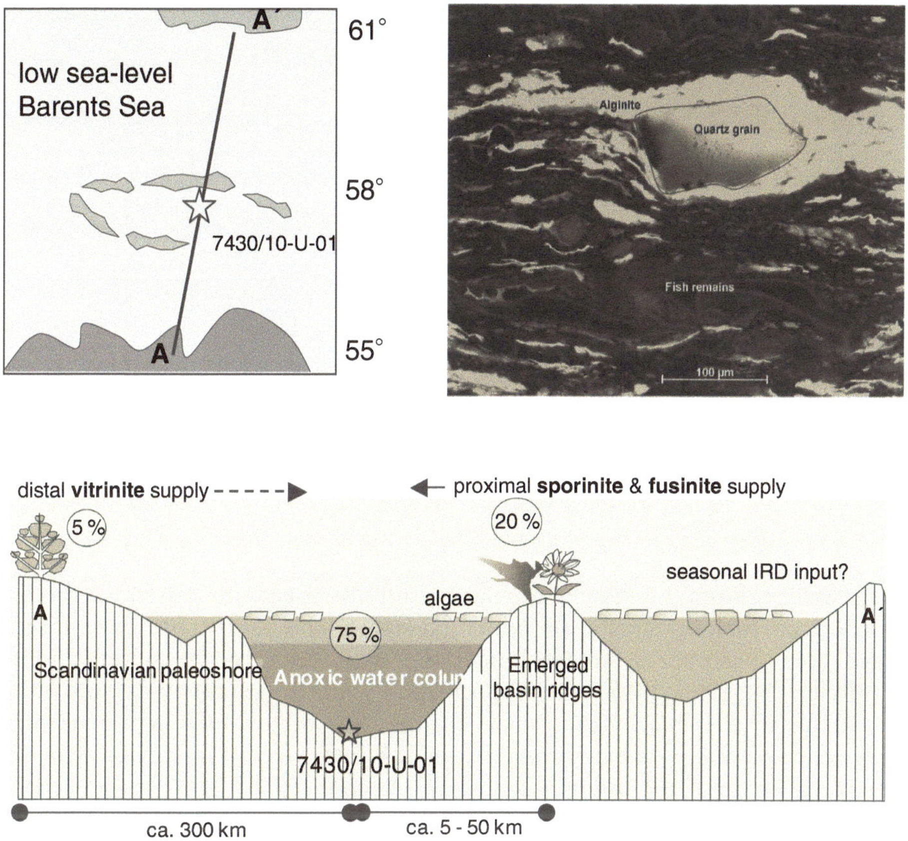

Figure 7.7 Model of depositional environment for black shale formation at Core 7430/10-U-01 (according to Langrock et al., 2003a). Black and white microscopic image showing composition of organic matter (reflected light microscopy with ultraviolet light, using a Zeiss Axiophot microcope equipped with a Zeiss Axiocam digital video camera). The light particles are mainly yellow-fluorescent alginites. A large-sized quartz grain is also shown. The patchy occurrence of such large isolated quartz grains of 100–200 µm in size may suggest some IRD input (seasonal sea-ice cover?).

origin and are characterized by well-developed lamination, suggesting suboxic to anoxic bottom-water conditions (Langrock et al., 2003b; Langrock & Stein, 2004). To assess the depositional conditions of black-shale formation in more detail, these authors used the pyrite size distribution (e.g., Wilkin, Barnes & Brantley, 1996; Wilkin, Arthur, & Dean, 1997), the TOC–Fe–S relationships (e.g., Brumsack, 1988; Dean & Arthur, 1989; Lückge et al., 1996; Hofmann et al., 2000), and the sedimentation rate/OC relationship (Stein, 1990, 1991a) (Figure 7.9). Autochthonous pyrite framboids were found in all studied cores and almost all depths, except for the Valanginian and Hauterivian of Core 6307/07-U-02 (Langrock et al., 2003b; Langrock & Stein, 2004). In general, autochthonous pyrite formation indicates an oxygen-deficient environment, but pyrite framboids that have formed within an

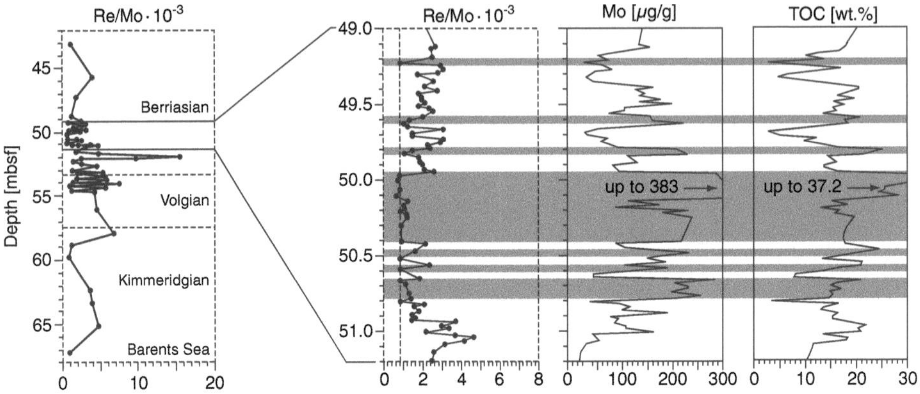

Figure 7.8 Re/Mo ratios, Mo, and TOC contents of all black shale samples analyzed by ICP-MS (Site 7430/10-U-01) (from Lipinski et al., 2003). Average Re/Mo ratio for the seawater (dotted line) is also given.

anoxic water column (i.e., under euxinic conditions), for example as in the modern Black Sea, are typically smaller (<7 µm) and less variable in size than those formed in the pore spaces of sediments underlying oxic and dysoxic water columns (*cf.* Wilkin et al., 1996, 1997). The Volgian and Ryazanian sediments of Core 6307/07-U-02 from the NGS basins show that ~98% of the total pyrite occur as framboids, and average diameters range from 6 to 7 µm with a low variability in size, supporting euxinic conditions (Langrock et al., 2003b). In the succeeding Valanginian and Hauterivian time period no pyrite was found in Core 6307/07-U-02, indicating the absence of H_2S.

Further support for euxinic conditions comes from the TOC–Fe–S relationships (Figure 7.9; Brumsack, 1988; Dean & Arthur, 1989; Lückge et al., 1996; Hofmann et al., 2000; Langrock et al., 2003b). The black shales of the Volgian and Ryazanian of Core 6307/07-U-02 plot below the "pyrite composition line" indicating limitations by reactive iron under oxygen-depleted conditions, followed by a change towards more oxic conditions, represented by the grey marl of the early Valanginian (right flank), and the red marl of the succeeding late Valanginian (upper corner). The same trend from anoxia in the Volgian and Ryazanian to oxic conditions in the Valanginian is reflected in the OC/sedimentation rate diagram (Figure 7.9; see Chapter 4.7.2, Figure 4.55 for background). Finally, the very low Re/Mo ratios (Figure 7.10) also point to a euxinic Black Sea type of environment.

On the basis of these data, the black shales in this part of the NGS were rather generated due to oxygen-depleted bottom waters ("stagnation or preservation model") than due to increased primary production ("productivity model"). Following the approach of Bralower and Thierstein (1984) for estimating palaeoproductivity for anoxic (stagnant) environments (see Chapter 4.7.2 for background), surface-water productivity may have reached 20–30 $gC\,m^{-2}\,yr^{-1}$ during Volgian to Ryazanian times (Figure 7.10). These values are much lower than average productivities estimated for modern coastal non-upwelling environments

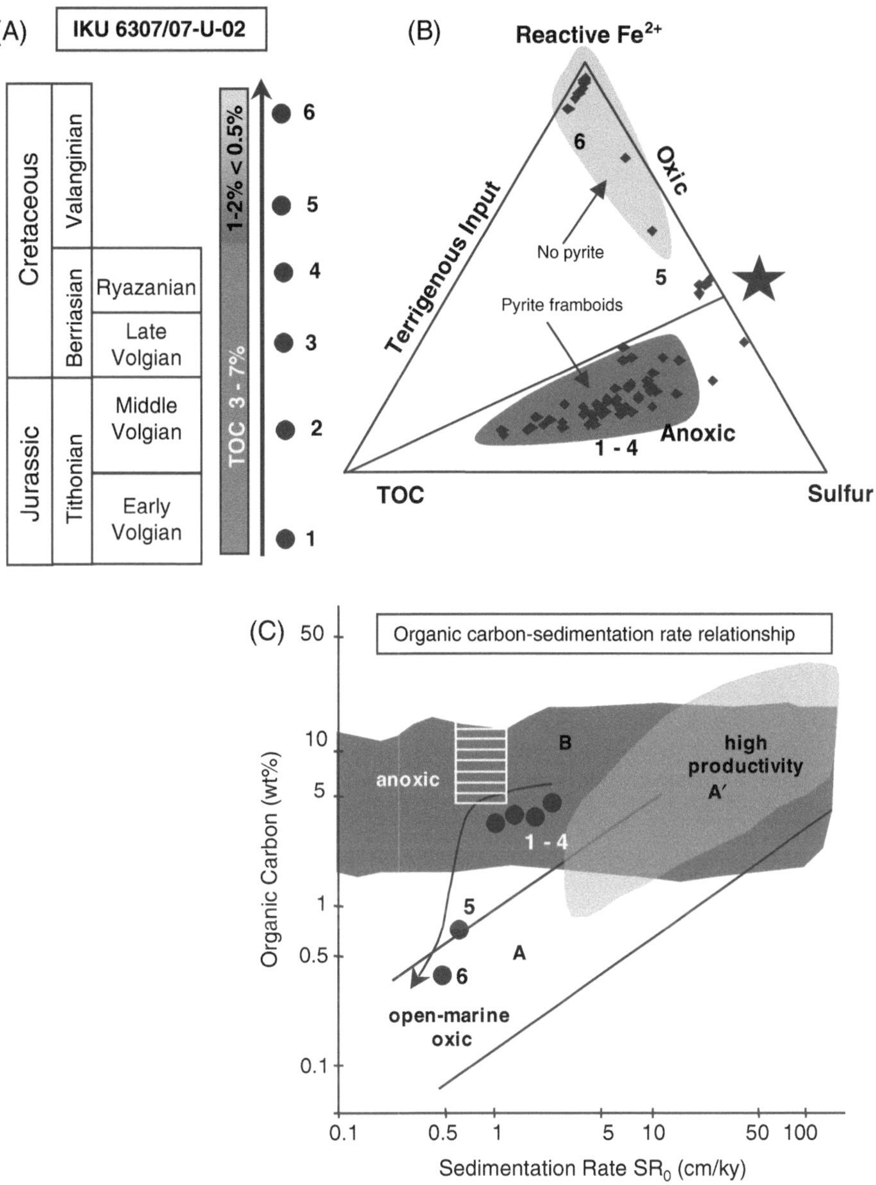

Figure 7.9 Lithology and geochemical data from Core 6307/07-U-02 (based on Langrock et al., 2003b; Langrock & Stein, 2004). (A) Lithology and stratigraphy. (B) TOC–Fe–S ternary diagram showing the limiting factors for pyrite formation (e.g., Brumsack, 1988; Dean & Arthur, 1989; Hofmann et al., 2000) of Core 6307/07-U-02. The amount of reactive iron was calculated using the empirical formula $Fe_{reactive} = Fe - 0.25 * Al$ (wt.%). Formation of pyrite (FeS_2) is provided along the solid line. Below this line pyrite formation is limited by reactive iron, probably reflecting anoxic conditions. Above the line pyrite formation is limited by the absence or composition of OM, for example, samples plot to the left if easily-to-metabolize compounds are absent. An evolutionary trend of changing depositional conditions is demonstrated by different numbers representing the early Volgian to late Volgian (1–4) and the early to late Valanginian (5–6). (C) Relationship between (marine) OC content and decompacted sedimentation rates for sediments from Core 6307/07-U-02. Each of the black circles represents average values for intervals 1–6; see (A). The white-hatched rectangle represents data from the Barents Sea Core 7430/10-U-01. The diagram is developed by Stein (1990, 1991a), where the different fields of deposition (A, open-marine oxic; A′, high productivity; B, anoxic) are based on results from Recent to Miocene sediments (see Chapter 4.7.2, Figure 4.55).

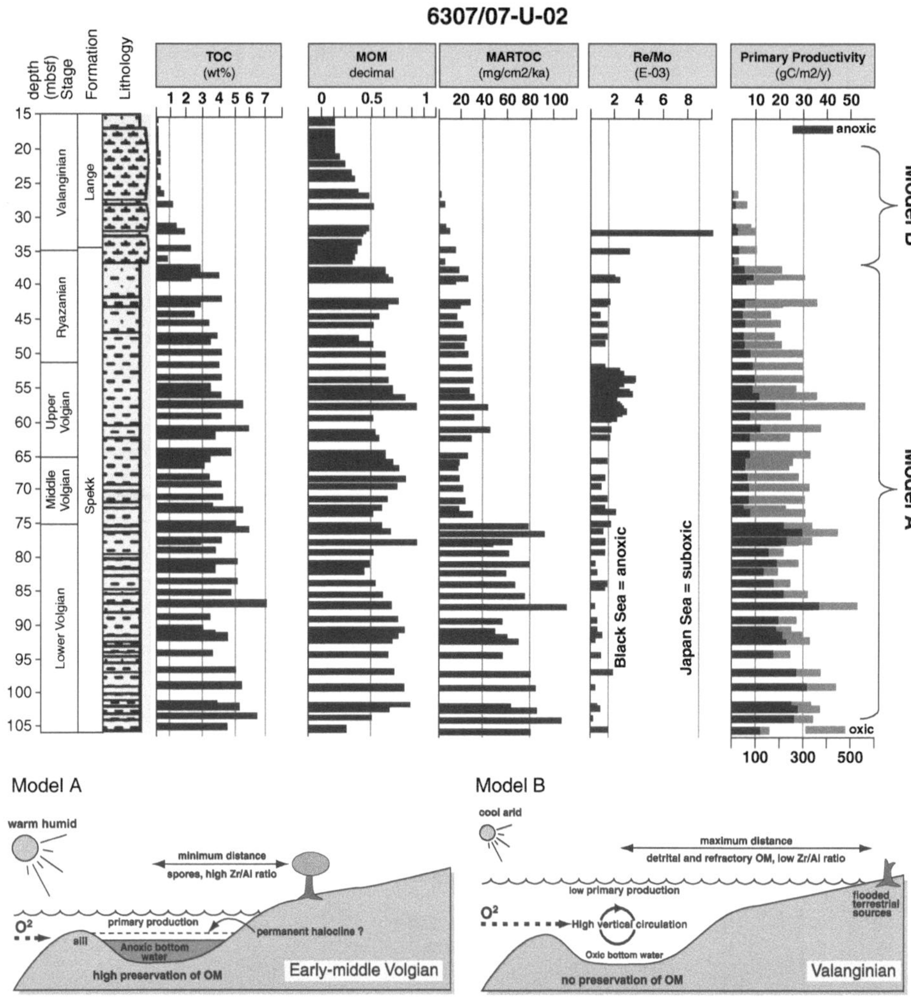

Figure 7.10 Stratigraphy, lithology, TOC content, proportion of marine organic matter (MOM), mass accumulation rates of total organic carbon (MARTOC), and Re/Mo ratios ($\times 10^{-3}$) to discriminate between anoxic and dysoxic-oxic environments (Lipinski et al., 2003), and palaeoproductivity estimates for both anoxic (black) and oxic (grey) conditions (see Chapter 4.7.2 for background and formulas used for palaeoproductivity calculation) for core 6307/07-U-02 (from Langrock & Stein, 2004). For the Volgian to Ryazanian interval characterized by euxinic conditions, productivity values shown by the black bars are more realistic (Model A), whereas for the Valanginian characterized by oxic conditions values shown by the grey bars (Model B) should be used (see text and Figure 7.9).

(100–150 $gC\,m^{-2}\,yr^{-1}$; e.g., Berger, Smetacek, & Wefer, 1989), suboxic sediments from the Quaternary Japan (120–150 $gC\,m^{-2}\,yr^{-1}$; e.g., Stein, 1991a), and for the OC-rich sediments from the modern Black Sea (50–90 $gC\,m^{-2}\,yr^{-1}$; e.g., Shimkus & Trimonis, 1974; Izdar et al., 1983; Calvert & Vogel, 1987).

However, the results are much more comparable to the general trend of mid-Cretaceous black shales from the Atlantic, which show generally low average palaeoproductivity values between 10 and 100 gC $m^{-2} yr^{-1}$ (Bralower & Thierstein, 1984; Stein, 1986; Stein et al., 1986).

Euxinic conditions were probably not the only factor controlling black-shale formation in the NGS; there may have been spatial and temporal differences. Whereas in the area of Core 6307/07-U-02, the stagnation model may best explain the enrichment of labile OC in the sediments, further to the north (Lofoten Basin, Core 6814/04-U-02; see Figure 7.5 for location) an oxygen minimum situation with some increased primary production may have been important too (Langrock et al., 2003b). This is supported by the higher Re/Mo ratios determined at the latter site (see Chapter 4.4; Lipinski et al., 2003).

With the continuous long-term change from (suboxic-) anoxic bottom-water conditions to more open-marine oxic conditions in the NGS OC contents significantly decreased (Figure 7.10), and the OC composition changed from more hydrogen-rich organic matter in the early and middle Volgian to more "woody" and refractory organic matter in the Valanginian (Langrock et al., 2003b; Langrock & Stein, 2004). These data are interpreted as indication for greater ventilation and more active flow through the seaway as the sea level rose. The foraminiferal assemblages also suggest low oxygen content bottom waters during the earlier Cretaceous, changing to more fully oxygenated conditions later (Mutterlose et al., 2003). The increasing amount of small-sized organic detritus derived from land-plants and the decreasing Zr/Al ratio (here used as proxy for the proximity to the land; Hinrichs, Schnetger, Schale, & Brumsack, 2001) suggest that the core positions moved farther from the palaeoshore, expanding the distance to the terrestrial sources, due to sea-level rise (Langrock et al., 2003b). Contemporaneously, a change from a more warm and humid climate in the Volgian to a more cool arid climate in the Valanginian/Hauterivian for the NGS was proposed from nannoplankton assemblages (Mutterlose et al., 2003).

Within the sequence of OC-rich sediments of Core 7430/10-U-01, patchily distributed, coarse-grained minerals, mainly quartz, were found which are 60–200 μm in diameter (Figure 7.7), however, they were not found at the more southern NGS sites. These minerals could not have been transported along with the claystone facies. A possible mechanism for the input of these large sand grains could be — although not yet proven — ice rafting, as proposed by Langrock et al. (2003a). Episodes of a relatively cool climate were already suggested for the earliest Cretaceous high-latitude realm (see earlier). Hence, episodic (or seasonal) IRD input cannot be excluded even for the late Volgian Barents Sea, which reveals a palaeolatitude of almost 55°N.

The long-term trend in climate evolution is superimposed by a short-term variability in the different palaeoenvironmental proxies (see Figures 7.6 and 7.7). Frequency analysis of high-resolution lightness logging data of all three sediments cores 7430/10-U-01, 6307/07-U-02, and 6814/04-U-02 suggests that this short-term variability is related to orbital control of climate, reflected in short-term fluctuations in the clastic influx and variations in OC production and/or oxygenation of bottom water (Mutterlose et al., 2003).

7.1.3. The Cretaceous Central Arctic Ocean and Its Palaeoenvironmental Characteristics

Cretaceous material has only been cored from three locations on the Alpha Ridge: cores Fl-437 and Fl-533 taken during the drift of the ice-island T-3 during the period 1963–1974 and Core 6 of the Canadian Expedition to Study the Alpha Ridge (CESAR) in 1983 (Figure 7.1, for core locations see Figure 7.11). More recently, an ~3 m thick interval of Campanian very dark grey clayey mud and silty sands with OC values of ~1–2% was recovered in the lowermost part of the ACEX record (Backman et al., 2006; Stein, 2007; see Section 7.2.2 for more details).

In Core Fl-533, the lowermost 67 cm of the 348 cm long sequence comprise OC-rich black mud (black shales), initially dated as late Campanian or late Campanian to Maastrichtian (Clark, Byers, & Pratt, 1986; Clark, 1988). Re-examination of the dinoflagellate cysts, together with acritarchs and prasinophytes, however, suggests an early Maastrichtian age (Firth & Clark, 1998). These black shales are characterized by very high OC contents of up to almost 16% and hydrogen index values of ~250–350 mgHC/gOC (Figure 7.11), and T_{max} values of ~420°C (Clark et al., 1986), indicating an immature, mixed terrigenous-marine type of organic matter. Whether these black shales recovered in Core FL533 resulted from anoxic conditions in an isolated local basin, a depositional environment under an oceanic water mass exhibiting an oxygen minimum, rapid burial and/or high terrigenous OC input remains an open question. Based on the mixed type of organic matter and the morphologically well-preserved nature of the organic matter

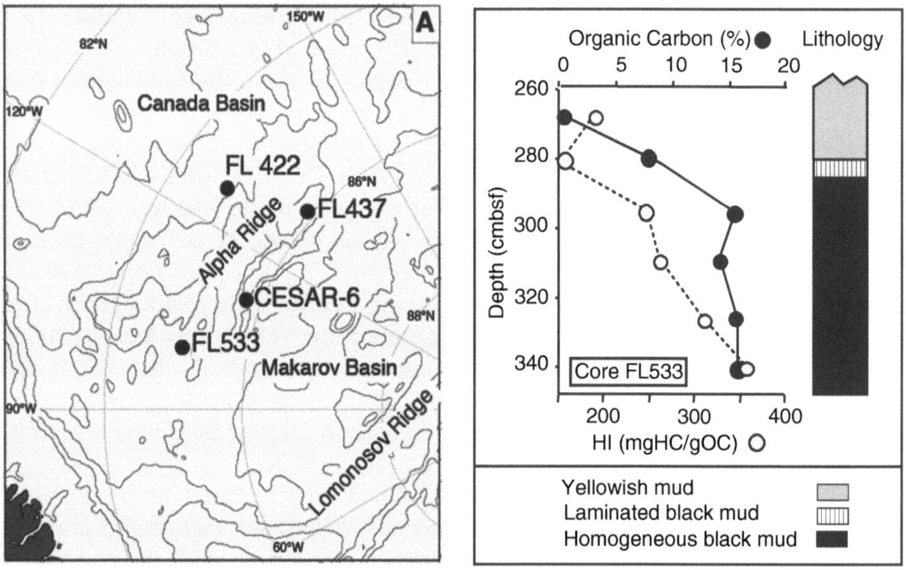

Figure 7.11 Location map of Cores FL-422, FL-437, FL-533, and CESAR-6, where upper Cretaceous and/or lower Cenozoic sediments were recovered from the Alpha Ridge, and lithology, OC, and hydrogen index values of the lower part of Core FL-533 (data from Clark et al., 1986; Firth & Clark, 1998).

in Core FL533, Clark et al. (1986) suggest a limited transport of organic detritus from the nearby Greenland-Svalbard-Scandinavia continental block and rapid burial being the most probable cause of black-shale formation rather than anoxia.

Core Fl-437 consists of yellowish laminated siliceous ooze rich in diatoms, ebrideans, silicoflagellates, and archaeomonads with sparse fish remains; the initial age attribution of mid- to late Maastrichtian (Clark, 1988) has later been revised to Campanian (Dell'Agnese & Clark, 1994). Very similar siliceous sediments with excellently preserved microfossils were retrieved in Core CESAR-6. These sediments are spectacularly laminated and contain variable amounts of iron and manganese that impart a characteristic dark pigment to some laminae (Stoffyn-Egli, 1987). OC contents are typically $<1\%$ (Mudie & Blasco, 1985). The age attribution of this core includes the interval Campanian–Maastrichtian, depending on whether diatoms, silicoflagellates or palynomorphs are taken as the prime biostratigraphic indicator (Barron, 1985; Bukry, 1985; Mudie, 1985).

In general, these data suggest a warm Arctic Ocean with strong seasonality and high palaeoproductivity, most likely associated with upwelling conditions (Clark et al., 1986). There are no dropstones in these sediments, so there is a complete lack of evidence for glacial activity. Differences in sediment composition between the cores may have been caused by lateral and temporal variations in nutrient conditions, oceanic currents, bottom-water oxygen levels, and basin topography (Kitchell & Clark, 1982; Dell'Agnese & Clark, 1994; Firth & Clark, 1998). Without doubt, the very different palaeogeographic boundary conditions during Cretaceous times have to be considered when interpreting these data. At that time, the Eurasian Basin did not exist, the Lomonosov Ridge fragment belonged to the Eurasian continental block, and the Alpha Ridge was probably located close to this block (Lawver & Scotese, 1990; see Chapter 1.3).

In order to estimate mid-Cretaceous central Arctic Ocean SST more quantitatively, Jenkyns et al (2004) analysed samples of the black shale from Core Fl-533 for membrane lipids of marine Crenarchaeota to determine the TEX_{86} (Schouten et al., 2002; see Chapter 4.7.5 for background). The TEX_{86} results indicate an average Arctic Ocean (80°N) SST of 15 ± 1°C for that part of the early Maastrichtian represented by the black muds (Figure 7.12). When comparing this high-latitude value with Maastrichtian SST values determined in the equatorial Pacific reaching 27–32°C (Wilson & Opdyke, 1996), this gives a Northern Hemisphere pole-equator temperature gradient of $\sim$15°C. A very similar gradient between low- and high-latitude SST was computed for the late Maastrichtian of the Southern Hemisphere, using oxygen–isotope ratios of well-preserved planktonic foraminifera (Huber et al., 1995).

Existing late Cretaceous SST records based on $\delta^{18}O$ data from several other localities show, independently of their environmental and diagenetic setting, a remarkable similarity, demonstrating that the climatic trends are probably global in nature (e.g., Jenkyns et al., 1994; Huber et al., 1995). There is a general warming trend during the Cenomanian reaching peak values at the Cenomanian/Turonian boundary, followed by a decline through the Coniacian to Campanian, as shown in the smoothed SST curve from the English Chalk (Figure 7.12; Jenkyns et al., 1994, 2004). Because this chalk SST curve extends into the lowest Maastrichtian, Jenkyns

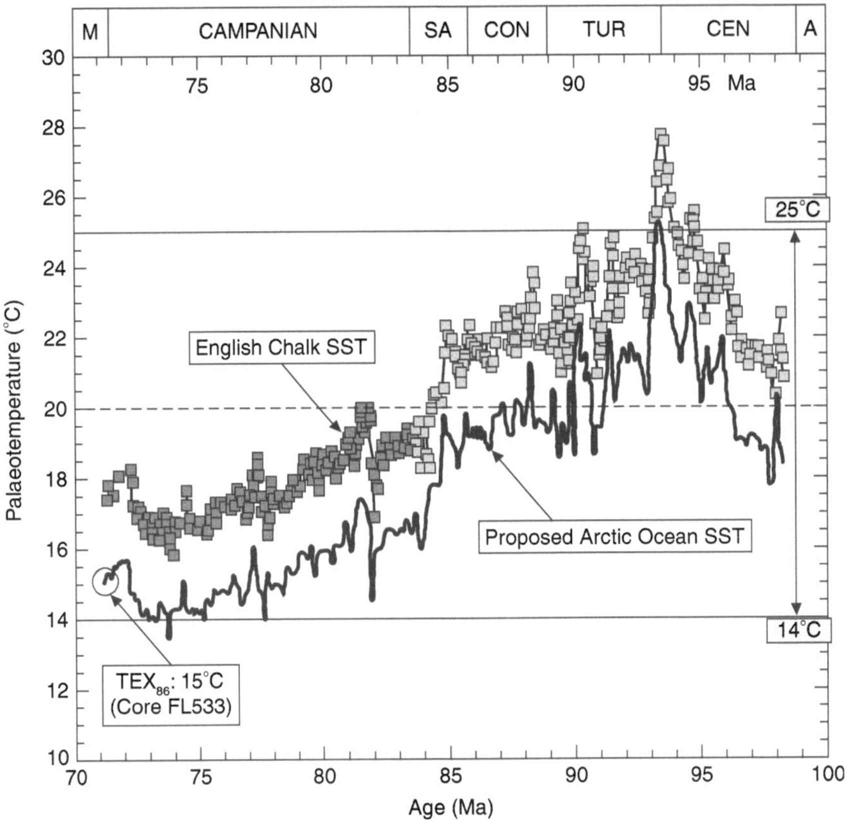

Figure 7.12 Proxy palaeotemperature curve for the Arctic Ocean (bold black line) based on oxygen–isotope ratios of bulk chalk from southern England (depositional palaeolatitude of ~40°N; Hay, Eicher, & DinerHay, 1993) and calibrated against a value of 15°C for the early Maastrichtian (open circle) (from Jenkyns et al., 2004, supplemented). Stratigraphic overlap in the lowest Campanian of the two curves (coastal outcrop in East Kent: unfilled squares) and borehole material in Norfolk (Trunch borehole: filled squares) permits adjustment of the $\delta^{18}O$ values, and hence calculated palaeotemperatures of core samples. Palaeotemperatures on a smoothed (three-point moving average) curve are calculated using a standard formula, assuming a value of −1.0‰ SMOW for an ice-free Cretaceous ocean (Jenkyns et al., 1994; see Jenkyns et al., 2004 for details).

et al. (2004) could generate a tentative mid to late Cretaceous palaeotemperature curve for the Arctic Ocean from the chalk record, taking the Maastrichtian SST value from Core FL-533 (15°C) as calibration point and assuming that the latitudinal gradient did not greatly change over the time period between ~70 and 100 Ma. Based on this (still tentative) approach, average Arctic Ocean SST probably reached values >20°C during much of the Cenomanian–Turonian time interval (Figure 7.12; Jenkyns et al., 2004).

On the basis of the still very weak data base on OC accumulation in the Arctic Ocean during the Cretaceous, Stein and Macdonald (2004d) did some speculation about the OC burial in the Arctic Ocean basin (excluding the shelf areas), which

Table 7.1 Calculation of Total Sediment (TS) and OC Budgets Using Different Linear Sedimentation Rates (LSR) and OC Contents. Typical Mean Values for Cretaceous North Atlantic Black Shales have been Used for Wet Bulk Density (WBD), Porosity, and Dry Density (DD) (See Stein et al., 1986 and Further References Therein). For the Size of the Cretaceous Arctic Ocean 50% of the Modern Size (Excluding the Shelf Areas) was Used Following Clark (1977).

	OC = 3%	OC = 5%	OC = 10%	OC = 3%	OC = 5%	OC = 10%
Size ($10^3\,km^2$)	2,250	2,250	2,250	2,250	2,250	2,250
LSR ($cm\,kyr^{-1}$)	2	2	2	5	5	5
WBD ($g\,cm^{-3}$)	2	2	2	2	2	2
Porosity (%)	40	40	40	40	40	40
DD ($g\,cm^{-3}$)	1.6	1.6	1.6	1.6	1.6	1.6
TS ($g\,cm^{-2}\,kyr^{-1}$)	3.2	3.2	3.2	7.9	7.9	7.9
OC ($g\,cm^{-2}\,kyr^{-1}$)	0.10	0.16	0.32	0.24	0.4	0.8
TS Budget ($10^6\,t\,yr^{-1}$)	71	71	71	178	178	178
OC Budget ($10^6\,t\,yr^{-1}$)	2.2	3.6	7.1	5.3	8.9	17.8
Maastrichtian (6 myr) OC (10^3 Gt = 10^{12} t)	13.2	21.6	42.6	31.8	53.4	106.8

was at that time probably ~50% smaller than today (Clark, 1977). Using a size of ~$2{,}250 \times 10^3\,km^2$, mean OC contents of 3–10%, and typical density ($1.6\,g\,cm^{-3}$), porosity (40%) and sedimentation rate ($2–5\,cm\,kyr^{-1}$) values determined for Cretaceous black shales from the North Atlantic (see Stein et al., 1986 and further references therein), OC burial rates between 2.2 and $17.8 \times 10^6\,t\,yr^{-1}$ were calculated (Table 7.1). These values are distinctly higher than those calculated for the modern (Holocene) central Arctic Ocean abyssal plains and ridges, reaching together ~$0.4 \times 10^6\,t\,yr^{-1}$ (Table 6.5; Stein & Macdonald, 2004b). For the Maastrichtian (6 million years) time interval, these estimates would result in total OC burial rates of $13.2–106.8 \times 10^3$ Gt (1 Gt = 10^9 t) (Table 7.1).

Furthermore, another very important difference between OC burial during the Quaternary and the Cretaceous is the OC composition. Whereas during the Quaternary the buried OC is mainly refractory terrigenous OC, the Cretaceous OC consists of major amount of labile algae-type OC. Although still very tentative, these estimates imply that the Cretaceous Arctic Ocean basin may have been an important sink for algae-type OC (and CO_2) and, thus, of importance for the global climate system of that time. To perform a more detailed reconstruction of the palaeoceanographic and palaeoenvironmental evolution of the early Arctic Ocean and its role for global climate change, however, long well-dated drill cores from different key areas are urgently needed (see Chapter 8).

7.2. Cenozoic High-Latitude Palaeoclimate and Arctic Ocean Palaeoenvironment

7.2.1. The Global Cenozoic Climate Record: From Greenhouse to Icehouse Conditions

A major element in the global climate evolution during Cenozoic times has been the transformation from warm Palaeogene oceans with low latitudinal and bathymetric thermal gradients into the more recent modes of circulation characterized by strong thermal gradients, oceanic fronts, cold deep oceans, and cold high-latitude surface waters (e.g., Miller et al., 1987; Zachos et al., 2001; Zachos, Dickens, & Zeebe, 2008). Throughout the course of the Cenozoic, the climate on Earth has changed from one extreme (Palaeogene greenhouse lacking major ice sheets) to another (Neogene icehouse with bipolar glaciation).

A strong greenhouse effect probably contributed to global warmth during the early Cenozoic. Pearson and Palmer (2000) estimated CO_2 concentrations of more than 2,000 ppm for the late Palaeocene and earliest Eocene periods (~60–52 Ma) (Figure 7.13; see also Lowenstein & Demicco, 2006). Bottom temperatures in the early Eocene, the time of maximum Cenozoic warmth that peaked with the early Eocene Climatic Optimum (~52–50 Ma), were of the order of 12–14°C, and large-scale continental ice sheets did not probably exist (Figure 7.13; Miller et al., 1987; Lear et al., 2000; Pearson & Palmer, 2000; Zachos et al., 2001, 2008). Based on stable isotope data of fossil mollusk shells from Ellesmere Island and Alaska,

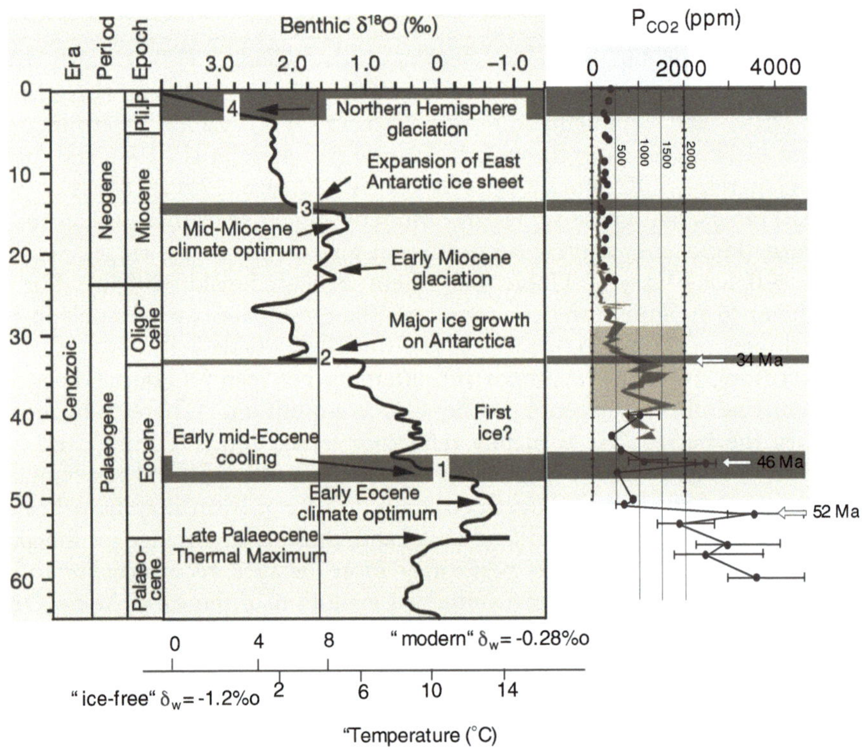

Figure 7.13 A smoothed global benthic foraminifer $\delta^{18}O$ time series of the Ceneozoic time interval (Pearson & Palmer, 2000, based on Shackleton, 1986; Miller et al., 1987; Zachos et al., 1994). Four major steps (numbered 1–4) in $\delta^{18}O$ and related events of climate change are indicated. The trend towards more positive $\delta^{18}O$ results from a combination of deep-sea cooling and global ice volume increases (Pearson & Palmer, 2000). In addition, CO_2 levels based on isotopic markers from marine algae (grey-shaded record; Pagani et al., 2005) and boron isotope composition of planktonic foraminifers (black circles; Pearson & Palmer, 2000) are shown.

temperatures of Arctic (80°N) coastal waters of ~10–22°C were reconstructed (Bice et al., 1996; Tripati, Zachos, Marincovich, & Bice, 2001). An early warm and equable Arctic climate was also inferred from Eocene terrestrial faunas and floras from Ellesmere Island (e.g., Marincovich, Brouwers, Hopkins, & McKenna, 1990; Kalgutkar & McIntyre, 1991; Greenwood & Wing, 1995).

The long-term history of Cenozoic high-latitude cooling starting at ~50 Ma is punctuated by four major steps in the early mid-Eocene (~48–45 Ma), at the Eocene/Oligocene boundary (near 34 Ma), in the mid-Miocene (at ~15 Ma), and in the mid-Pliocene (at ~3.5–2.6 Ma) (Figure 7.13; Miller et al., 1987; Zachos, Stott, & Lohmann, 1994; Zachos, Pagani, Sloan, Thomas, & Billups, 2001; Lear et al., 2000; Pearson & Palmer, 2000). Reconstructions of past atmospheric CO_2 concentrations based on isotopic markers from marine algae (Pagani, Zachos, Freeman, Tipple, & Bohaty, 2005) and boron isotope composition of planktonic foraminifers (Pearson & Palmer, 2000), show — although with obvious differences in absolute values — distinct drops in atmospheric CO_2 between ~50 and 25 Ma that generally correspond to the global cooling trend and development of major polar ice sheets, except for the interval around the early mid-Eocene cooling (Figure 7.13; see Pearson & Palmer, 2000 for discussion).

It is widely accepted that large ice sheets first appeared on Antarctica near the Eocene/Oligocene boundary at ~34 Ma ("Oi-1 glaciation"; e.g., Shackleton & Kennett, 1975; Kennett & Shackleton, 1976; Miller et al., 1987, 1991; Lear et al., 2000; Zachos et al., 2001), coincident with decreasing atmospheric carbon dioxide concentrations and a deepening of the CCD in the world's oceans (van Andel, 1975; Pearson & Palmer, 2000; Coxall, Wilson, Pälike, Lear, & Backman, 2005; Tripati, Backman, Elderfield, & Ferretti, 2005). For the Northern Hemisphere, on the other hand, it was indirectly inferred from Subarctic IRD records in the Norwegian-Greenland, Iceland, and Irminger seas and Fram Strait area that glaciations began much later, that is, in the middle Miocene as early as ~14 Ma (e.g., Wolf & Thiede, 1991; Fronval & Jansen, 1996; Wolf-Welling et al., 1996; Wright & Miller, 1996; Thiede et al., 1998; St. John & Krissek, 2002; Section 7.2.3). Based on more recent modelling results as well as new ODP/IODP sediment core data, this general picture has to be revised significantly — especially for the Northern Hemisphere (see Section 7.2.3).

The development of large ice sheets in Antarctica near the Eocene/Oligocene boundary is indicated by the occurrence of IRD at 58°S by Ocean Drilling Program Leg 120 (Zachos et al., 1992) as well as a prominent change in the clay–mineral assemblages at ODP Sites drilled on Maud Rise and Kerguelen Plateau from a smectite to an illite dominance (Ehrmann & Mackensen, 1992). The decrease of smectite content at these sites, however, already started in the late (/middle) Eocene (Ehrmann & Mackensen, 1992). These findings point to an earlier onset of significant Antarctic glaciations (see also summaries in Miller et al., 1991; Wise et al., 1991; Zachos et al., 1994).

Sediment records of calcium carbonate content as well as carbon and oxygen isotopic compositions from the tropical Pacific Ocean across the Eocene/Oligocene boundary indicate that the deepening of the CCD was more rapid than previously documented and occurred in two jumps of ~40,000 years each, synchronous with the stepwise onset of Antarctic ice-sheet growth (Coxall et al., 2005). Furthermore, there is evidence for synchronous deepening and subsequent oscillations in the CCD in the tropical Pacific and South Atlantic oceans already from 42 Ma, with a permanent deepening at 34 Ma (Tripati et al., 2005). These results demonstrate that the greenhouse–icehouse transition was closely coupled to changes in carbon cycling and indicate the presence of extensive ice accumulation before the Eocene–Oligocene boundary, with large fluctuations in ice volume during the late/middle Eocene.

The changes in oxygen–isotope composition across the Eocene/Oligocene boundary as well as in the late Eocene are too large to be explained by Antarctic ice-sheet growth alone and must therefore also indicate contemporaneous global cooling and/or NHG (Coxall et al., 2005; Tripati et al., 2005). Sedimentological evidence of such an early Cenozoic glaciation could recently be obtained from ODP and IODP sedimentary sequences (see Section 7.2.3).

7.2.2. The Palaeocene–Eocene Central Arctic Ocean and Its Palaeoenvironmental Characteristics

From the early Cenozoic "Greenhouse" time interval, continuous central Arctic Ocean sedimentary records, allowing a development of chronologic sequences of climate and environmental change through Cenozoic times and a comparison with global climate records were missing prior to the IODP Expedition 302 (Arctic Ocean Coring Expedition — ACEX; Backman et al., 2006; Moran et al., 2006; Backman, & Moran, 2008; see Chapter 1.2 for some general information about ACEX and its significance for Arctic Ocean research). The only direct information about palaeoenvironmental conditions in the Arctic Basin prior to 2004 was coming from a single short sediment core obtained from the drifting ice island T3 in 1969 (Core FL-422; see Figure 7.11 for location). The sediments of this core representing part of the middle Eocene time interval, are laminated, rich in siliceous microfossils, especially diatoms and silicoflagellates, and characterized by alternating layers of vegetative cells and resting spores, interpreted as evidence for strong seasonality, upwelling of nutrient-rich waters, and warm surface-water temperatures (Clark, 1974; Bukry, 1984; Dell'Agnese & Clark, 1994).

7.2.2.1. Main lithologies and stratigraphic framework of the ACEX sequence

With ACEX, an almost 430 m thick sedimentary sequence was recovered from Lomonosov Ridge (see Figures 1.8 and 7.4 for location) ranging in age from Quaternary to late Cretaceous (Backman et al., 2006; Moran et al., 2006). An update of the stratigraphic framework of the Cenozoic ACEX sequence was recently presented by Backman et al. (2008), which is based on biostratigraphic, cosmogenic isotope, magneto- and cyclostratigraphic data (Figure 7.14). Age/depth control points as well as resulting sedimentation rates and ages of lithological units are summarized in Table 7.2. On an average, Neogene and Palaeogene sedimentation rates reach values of $\sim$1.2 and 1.8 cm kyr^{-1}, respectively (Figure 7.14). Although the revised ACEX age model definitely confirms that the average sedimentation rate clearly is $>$1 cm kyr^{-1} (Backman et al., 2006), a highly resolved and robust age

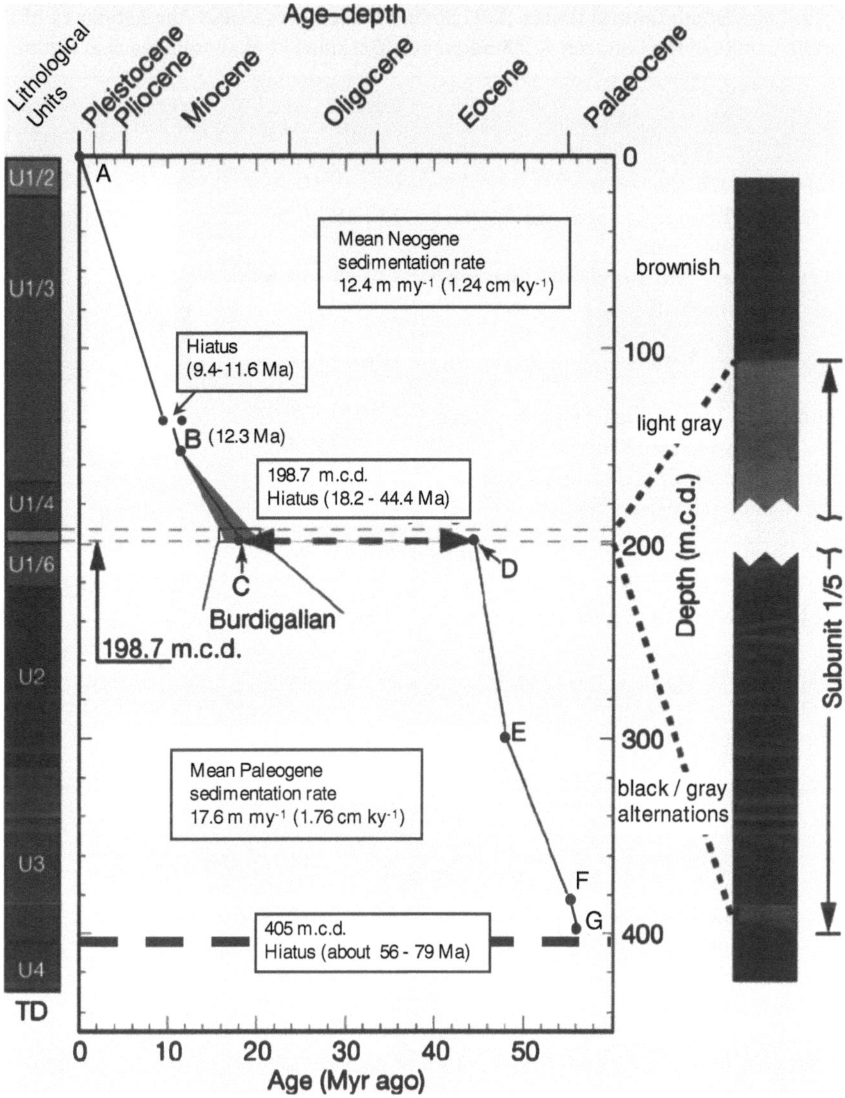

Figure 7.14 Age-depth diagram based on age control points A (0 Ma), B (12.3 Ma), C (18.2 Ma), D (44.6 Ma), E (48.6 Ma), F (55 Ma), and G (55.9 Ma) (from Jakobsson et al., 2007a, modified based on revised age model of Backman et al., 2008; see Table 7.2). Main lithological (sub-) units, average sedimentation rates, and the three major hiatuses are indicated. At the right-hand side, subunit I/5 ("Zebra Unit") is shown. The base of subunit I/5 at 198.7 mcd coincides with the hiatus 18.2–44.4 Ma. For a more detailed discussion of the age model and its limits see Jakobsson et al. (2007a) and Backman et al. (2008).

model for the ACEX cores is still challenging due to the poor core recovery (i.e., ~1/3 of the penetrated section was not recovered), the occurrence of an unexpected major hiatus, the limited availability of biostratigraphic indicators, and the enigmatic preservation of the geomagnetic polarity record (Backman et al., 2008).

Table 7.2 Age/Depth Control Points (CP), Sedimentation Rates, and Age Estimates of Lithostratigraphic Units of the Cenozoic ACEX Sequence (Backman et al., 2008; see also Figure 7.14).

TP	Event	Depth (mcd)	Age (Ma)	Interval	Sedimentation rate ($cm.kyr^{-1}$)
A	Top of ACEX sequence	0	0		
	Deepest ^{10}Be sample above middle-late Miocene hiatus	135.49	9.36		
	Next deeper ^{10}Be sample, middle-late Miocene hiatus	140.44	11.56		
B	Deepest ^{10}Be sample	151.28	12.31		
				B-C	0.8
C	Mid-point Burdigallian Stage = young end of Cenozoic hiatus	198.70	18.2		
	Extrapolate up from D, to old end of Cenozoic hiatus	198.70	44.4		
D	Last abundant occurrence of *P. clithridium*	202.95	44.6		
				D-E	2.43
E	Last occurrence of *Azolla* spp.	299.95	48.6		
				E-F	1.27
F	Last occurrence of *A. augustum*	381.42	55.0		
				F-G	2.01
G	Top Chron C25n	399.63	55.90		

Hole	Base of unit/subunit	Depth (mcd)	Age (Ma)
2A	Subunit 1/1	2.11	0.15
2A	Subunit 1/2	21.22	1.5
2A	Subunit 1/3	168.53	14.5
2A	Subunit 1/4	192.94	17.5
2A	Subunit 1/5	198.70	18.2
2A	Subunit 1/6	223.56	45.4
4A	Unit 2	313.61	49.7
4A	Unit 3	404.79	56.2

For a more detailed discussion of the most recent ACEX age model and its strengths and limits, the reader is referred to Backman et al. (2008), Frank et al. (2008), O'Regan et al. (2008a), and Pälike, Spofforth, O'Regan, and Gattacceca (2008).

On the basis of the visual core description and smear slide analysis as well as TOC content and X-ray diffraction (XRD) measured in core catcher samples, the ACEX sequence was divided into four main lithologic units (Backman et al., 2006; Stein, 2007; Figure 7.15, Table 7.2):

(1) Unit 1 (top to 223.6 mcd; Quaternary to Middle Eocene) is dominated by silty clay, ranging from light olive browns at the top, through olives and greys to very

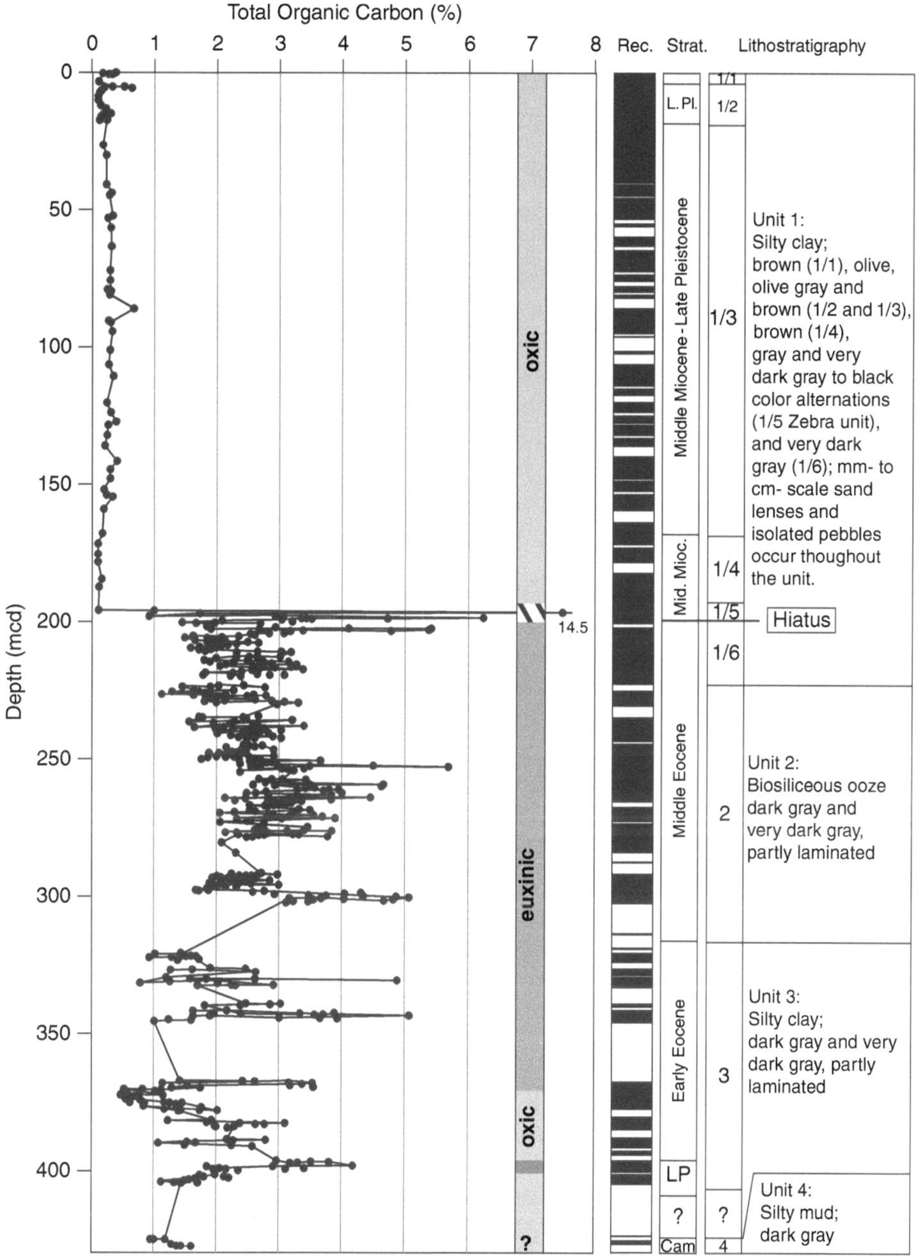

Figure 7.15 Record of TOC contents (from Stein, 2007) as determined in the composite ACEX sedimentary sequence (0–17.85 mcd: Hole MSP0004B; 18–265.23 mcd: Hole MSP0002A, and 265.34–427.57 mcd: Hole MSP0004A; see also Section 1.2.3). Data on recovery, stratigraphy, and lithological units (1–4) and subunits (1/1 to 1/6) from Backman et al. (2006). Dominantly oxic and euxinic are as based on C/S ratios (see Figures 7.16 and 7.18; Stein et al., 2006a). Cam, Campanian; LP, late Palaeocene; Mid Mioc., middle Miocene; L.Pl., late Pleistocene.

dark grey at the bottom; colour banding is strong. Millimetre-to-centimetre-scale sandy lenses and isolated pebbles also occur in unit 1. Unit 1 is subdivided into 6 subunits. A major hiatus was identified at ~198.7 mcd separating subunits 1/6 and 1/5 and spanning the time interval from ~44.4 to 18.2 Ma; another shorter hiatus lasting 2.2 my, occurs within subunit 1/3 in the late Miocene 9.4–11.6 Ma (Figure 7.14; Jakobsson et al., 2007a; Backman et al., 2008). Subunit 1/5 is outstanding due to its very prominent black and white colour banding ("Zebra Unit"; Figure 7.14).

(2) Unit 2 (223.6–313.6 mcd; Middle Eocene) is dominated by very dark grey mud-bearing biosiliceous ooze with submillimetre-scale laminations as well as isolated pebbles.

(3) Unit 3 (~313.6–404.8 mcd; Late Palaeocene to Early Eocene) is dominated by very dark grey clay with submillimetre-scale laminations. Units 3 and 4 are separated by a second major hiatus representing the time interval of ~56–79 Ma (Figure 7.14; Backman et al., 2008).

(4) Unit 4 (424.50–427.63 mbsf; Campanian) is dominated by very dark grey clayey mud and silty sands.

Whereas the upper (middle Miocene to Quaternary) part of the ACEX sequence (subunits 1/1 to 1/4) is composed of silty clay with very low OC contents of <0.5%, that is, values very similar to those known from upper Quaternary records determined in gravity cores from Lomonosov Ridge (Stein et al., 2004b), the Campanian and Palaeogene sediments of the ACEX sequence (units 2–4) are characterized by high TOC values of 1 to >5% (Figure 7.15). In subunit 1/5 (~193–199 mcd; late early Miocene; Jakobsson et al., 2007a; Backman et al., 2008) characterized by distinct grey/black colour bandings, even OC maxima of 7–14.5% were measured in samples from the black horizons (Figure 7.15).

7.2.2.2. Depositional environment, anoxia, and primary production

The data on amount and composition of the OC fraction in the Palaeogene part of the ACEX sequence give important information about the palaeoceanographic and palaeoenvironmental conditions of the early Cenozoic Arctic Ocean as will be outlined in the later section.

In the late Palaeocene interval, the organic matter is mainly of terrigenous origin (kerogen type III), indicated by very low HI values of 30–80 mgHC/gC (Figure 7.16; see also Chapter 4.7.1, Figure 4.52). The terrigenous source is also supported by biomarkers, palynomorphs as well as maceral data. Long-chain *n*-alkanes are clearly predominant over short-chain *n*-alkanes, phytane, and pristane (Figure 7.17; Weller & Stein, 2008). Values of the "BIT index", a measure for the amount of river-derived terrestrial organic matter relative to marine organic matter (see Chapter 4.7.4 for background and references), are relatively high with values of 0.5–0.7, and palynomorph assemblages from upper Palaeocene strata are dominated by terrestrial spores and pollen reaching 70 to >90% of the total palynomorphs (Sluijs et al., 2006; see PETM discussion later and Figure 7.21). Terrigenous macerals increased to 70–80% of total macerals (see Figure 4.53; Stein et al., 2006a; Boucsein

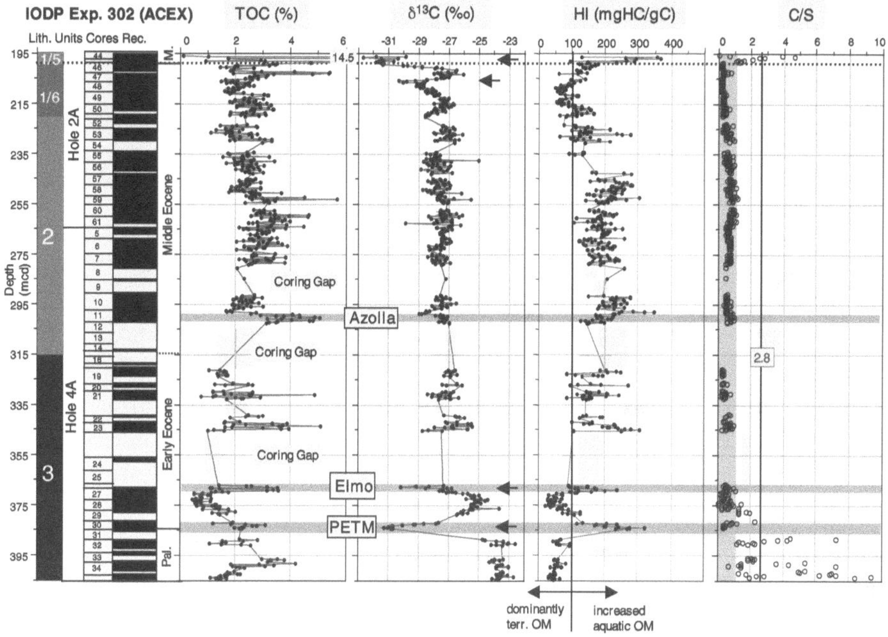

Figure 7.16 OC content (TOC), $\delta^{13}C_{org}$, hydrogen index (HI) values, and C/S ratios (from Stein et al., 2006a) as determined in the ACEX sequence between 195 and 405 mcd (Holes MSP0002A and MSP0004A). In addition lithological units, core numbers and recovery as well as the stratigraphy (M., Middle Miocene; Pal., Late Palaeogene) at the left-hand side (Backman et al., 2006). The PETM, Elmo, and Azolla events are marked as horizontal grey bars. The vertical line labelled with "2.8" indicates mean C/S ratio for oxic environments, vertical grey bar marks C/S values <1 typical for euxinic environments (for background and more details see Chapter 4.7.2 and Stein et al., 2006a). Dotted line at 198.7 med indicates major hiatus (see Figure 7.14).

& Stein, 2008). High C/S ratios in the range typical for recent oxic conditions (Figures 7.16 and 7.18) and the absence of small-sized pyrite framboids (Stein et al., 2006) point to an oxic depositional environment. Thus, significant amount of algae-type organic matter is not preserved in the sediments. Based on the Rock-Eval data, the composition of the organic matter of the upper Palaeocene sediments is similar to that of Quaternary sediments on Lomonosov Ridge (see Figure 4.52). The OC accumulation rates reaching up to 0.08 gC cm^{-2} kyr^{-1}, however, are by a factor of ~15 higher than those calculated for the Quaternary sediments (Figure 7.19).

In the middle Eocene sediments (unit 2) and the upper part of the early Eocene (220–345 mcd), elevated HI values of 100–350 mgHC/gC were determined (Figure 7.16). These values increase to 250–450 mgHC/gC when considering the dead OC present in the entire record with a background value of ~0.45% OC (see Figure 4.54), indicating that hydrogen-rich algae-type OC of aquatic (i.e., marine, brackish to freshwater) origin occurs in significant amounts (kerogen-types I/II and mixed II/III; see Figure 4.52). Based on the correlation between HI values and kerogen microscopy data determined in immature Mesozoic/Cenozoic marine sediments (Stein et al., 1986) and first maceral data obtained from ACEX sediment samples,

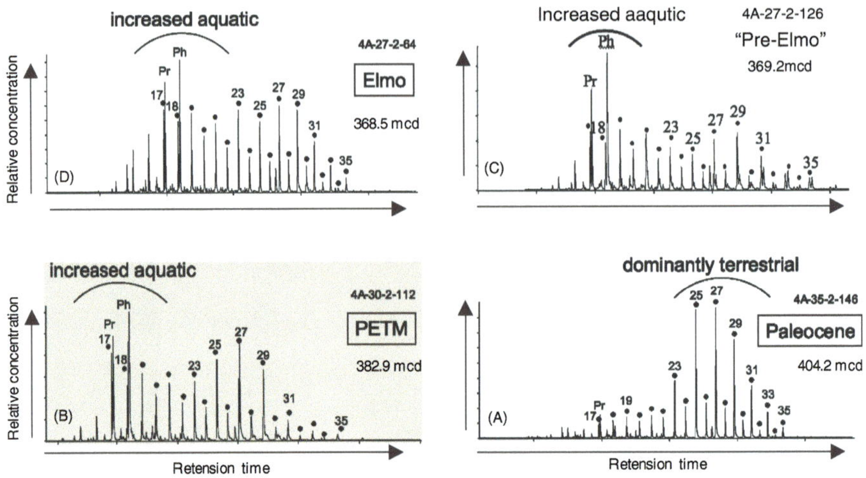

Figure 7.17 GC–MS trace (*m*/*z* 71) of three representative extractable hydrocarbon fractions in sediments from the ACEX sequence between 368 and 404 original mcd (from Weller & Stein, 2008). (A) Late Palaeocene; (B) PETM Event, (C) "Pre-Elmo", and (D) Elmo Event. ●, *n*-alkanes; Pr, pristane; Ph, phytane.

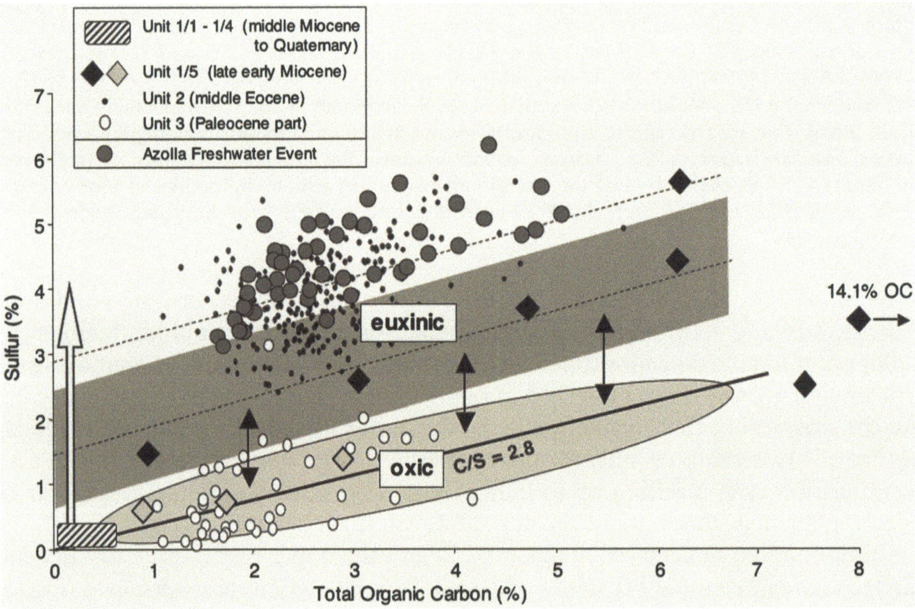

Figure 7.18 Plot of the TOC versus (pyritic) sulfur (C/S diagram) (from Stein et al., 2006a; see Chapter 4.7.2 for background). For Quaternary normal (oxic) marine fine-grained detrital sediments shown by the area underline in light grey, a positive correlation between pyritic sulfur and OC with a mean C/S ratio of ~2.8 exists (Goldhaber & Kaplan, 1974; Berner & Raiswell, 1983; Berner, 1984). In euxinic environments such as the modern Black Sea, however, H_2S already occurs in the water column, and framboidal pyrite can be initially formed, resulting in an excess of sulfur in the C/S diagram indicated by the open arrow and a positive sulfur intercept. ACEX data from the different time intervals/units are indicated by different symbols. Data from the modern euxinic Black Sea are shown by the grey bar (Leventhal, 1983).

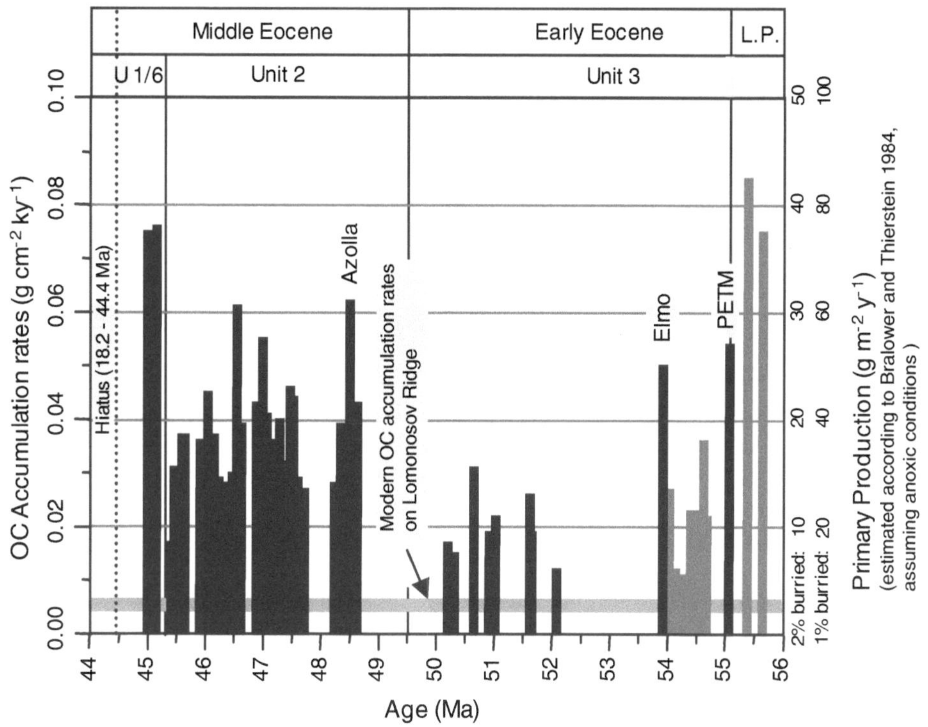

Figure 7.19 OC accumulation rates of the late Palaeocene to middle Eocene time interval (~56–44.4 Ma) of the ACEX record, including the PETM, Elmo, and Azolla events, based on sedimentation-rate data from Backman et al. (2008), physical-property data from Backman et al. (2006), and OC data from Stein et al. (2006a). Black bars represent samples with dominantly aquatic/marine OC, whereas grey bars represent samples dominated by terrigenous OC. The vertical dark grey bar indicates average modern (Holocene) OC accumulation on Lomonosov Ridge (Stein et al., 2001, 2004b). Following the approach of Bralower and Thierstein (1984), palaeoproductivity can be estimated from the OC accumulation rate data if OC has a mainly marine origin (see Chapter 4.7.2 for background). That means, this approach should not be used for data represented by the grey bars.

~40–60 (80)% of the OC seems to be of aquatic origin (see Figure 4.53). What are the dominant factors (e.g., anoxia and/or increased primary production) having controlled the enrichment of labile algae-type OC in the Eocene sediments?

On the basis of continously low C/S ratios (Figure 7.16), the distinct positive sulfur intercept in the C/S diagram (Figure 7.18), and the abundance of pyrite (Stein et al., 2006a) as well as the occurrence of fine lamination (Backman et al., 2006), euxinic conditions probably occurred throughout the early to middle Eocene Arctic Ocean, except for a short interval of oxic conditions in the lowermost Eocene, and were the predominant process for the preservation of labile algae-type OC. In the still limited set of Eocene samples used for kerogen microscopy, finely dispersed small-sized (Ø: 5 μm) pyrite framboids were found (Stein et al., 2006a; Boucsein & Stein, 2008), which supports microbial origin and *in-situ* formation of the framboids in an euxinic water column (Berner, 1970;

Wilkin et al., 1996, 1997). Low sulfur isotopes ($\delta^{34}S$) values with a mean of −37‰ were measured in units 2 and 3 (middle/late early Eocene), also indicating that the abundant sulfur in the ACEX sediments was related to microbial reduction from sea-water sulfate ion (Ogawa et al., 2008). In contrastto the PETM event (see later discussion), however, euxinic conditions probably did not reach the photic zone as suggested from the absence of isorenieratane and related isorenieratene derivates (see Chapter 4.7.5 for background) determined in a low-resolution biomarker study of the middle Eocene section (Weller & Stein, 2008). In general, the environment may have been similar to the modern Black Sea (Stein et al., 2006a). The extremely high sulfur values determined in the ACEX samples, with absolute maxima >10% in the middle Eocene subunit 1/6 (Stein et al., 2006a; Ogawa et al., 2008), that means, values significantly higher than those measured in Black Sea sediments, however, make the Arctic Ocean different from modern analogues.

The euxinic conditions at that time were most probably caused by widespread salinity stratification in the isolated Arctic Ocean at that time. A low surface-water salinity (brackish) environment during middle Eocene times is also suggested from the rare and sporadic occurrence of radiolarians (Backman et al., 2006). Runoff-related low salinity might be indicated as well by the abundance of terrestrial palynomorphs and green algae such as *Tasmanites* and *Botryococcus* (Backman et al., 2006).

Taken into account the high OC values of the Eocene interval of the ACEX section and the fact that sedimentation rates at that time were similar (or even twice) than during Holocene times (Stein et al., 2004b; Backman et al., 2006, 2008; Table 7.2), Eocene OC accumulation rates were 5 to >10 times higher than modern ones. Whereas very low OC accumulation rates of ~0.005 gC cm^{-2} kyr^{-1} are typical for the modern (Holocene) central Arctic Ocean on Lomonosov Ridge (Stein et al., 2004b), values of up to 0.07 gC cm^{-2} kyr^{-1} were reached during Eocene times (Figure 7.19). Based on the correlation between accumulation rates of (marine) OC and primary productivity in recent anoxic environments, ~1–2% of the primarily produced OC is preserved and accumulated in the sediments under anoxic conditions (Bralower & Thierstein, 1984; for background see Chapter 4.7.2). Using Bralower and Thierstein's approach for an Eocene anoxic Arctic Ocean, primary productivity may have reached values of 30–80 gC m^{-2} yr^{-1} (Figure 7.19; Stein, 2007). Knies et al. (2008) obtained quite similar values (40–100 gC m^{-2} yr^{-1}), using the Knies and Mann (2002) approach (see Chapter 4.7.2 for background). These values are not quite high (e.g., in comparison to high-productivity upwelling areas where >250 gC m^{-2} yr^{-1} are typical; Berger et al., 1989; Stein, 1991a and further references therein), but these values are at least ~3–8 times higher than modern ones (Sakshaug, 2004). For the middle Eocene biosiliceous ooze of unit 2 palaeoproductivity estimates are by a factor of 2 to 3 higher than those calculated for the underlying early Eocene silty clay of unit 3 — except for the PETM and Elmo events (see later discussion) which display productivity values similar to those of the biosiliceous ooze (Figure 7.19). Using a size for the Eocene Arctic Ocean Basin of ~3×10^6 km^2 (see Table 7.3 for calculation) and assuming similar environmental conditions for the whole Arctic Basin — a still speculative assumption which has to be proven by future drilling — this results in total mean burial rates of (mainly algae-type) OC of ~$0.9\text{–}1.2 \times 10^6$ t yr^{-1} (Table 7.3).

Table 7.3 Calculation of TS and OC Budgets for the Early and Middle Eocene as well as the PETM, Elmo, and Azolla events. LSR from Backman et al. (2008), WBD and DD from Backman et al. (2006), and OC Contents from Stein et al. (2006a). The Duration of the PETM and Elmo Events are Based on Sluijs et al., 2008 and Lourens et al. (2005), Respectively, the Duration of the Azolla Event is According to Brinkhuis et al. (2006). LSR and OC Numbers in Brackets in the PETM Column are from Sluijs et al. (2008). For the Size of the Early Palaeogene Arctic Ocean Basin an Estimate of 3×10^6 km^2 is Used. This Estimate Representing the Arctic Ocean Excluding the Shelf Areas was Calculated Using the Modern Value of 4.5×10^6 km^2 (See Chapter 2.1, Table 2.1) Minus the Size of the Eurasian Basin ($\sim 1.5 \times 10^6$ km^2; Jakobsson et al., 2003a) which was not Opened at 56 Ma (Kristoffersen, 1990).

Time interval	Early Eocene (48.6–55 Ma)	Middle Eocene (44.6–48.6 Ma)	PETM ($\sim$55 Ma)	Elmo ($\sim$53 Ma)	Azolla (near 49 Ma)
Size (10^3 km^2)	3,000	3,000	3,000	3,000	3,000
Average LSR (cm kyr^{-1})	1.27	2.43	2.01 (5.0)	1.27	(>)1.27
Wet bulk density (g cm^{-3})	1.6	1.4	1.8	1.74	1.42
Dry bulk density (g cm^{-3})	1.1	0.6	1.3	1.2	0.7
OC (%)	2	2.7	2.2 (2.8)	3.3	4.2
Bulk Sediment (g cm^{-2} kyr^{-1})	1.4	1.5	2.6 (6.5)	1.5	0.9
OC (g cm^{-2} kyr^{-1})	0.03	0.04	0.06 (0.18)	0.050	(>)0.04
OC Budget (10^6 t yr^{-1})	0.9	1.2	1.8 (5.4)	1.5	(>)1.2
Duration (myr)	6.4	4	0.17	0.1	0.8
Total burial					
(10^{12} t)	5.8	4.8	0.31 (0.92)	0.15	(>)0.96
(Gt)	5800	4800	310 (920)	150	(>)960

7.2.2.3. Sea-Surface temperatures of the Palaeocene/Eocene central Arctic Ocean

On the basis of the TEX_{86} index (see Chapter 4.7.5 for background information), Sluijs et al. (2006, 2008) estimated a SST of $\sim$18°C in the late Palaeocene/early Eocene, reaching maximum values as high as 23°C during the PETM (Figure 4.64 and Figure 7.20). The high SST values are much higher than those based on palaeoclimate models. As discussed in Sluijs et al. (2006), palaeoclimate models simulating the early Palaeogene world with 2,000 ppm of CO_2 in the atmosphere (Figure 7.13; Pearson & Palmer, 2000) underestimate Arctic Ocean summer SSTs by at least 15°C for the PETM and 10°C for the surrounding late Palaeocene and early Eocene (Huber et al., 2003; Shellito, Sloan, & Huber, 2003). The models consistently predict pole-to-equator temperature gradients of $\sim$30°C (Huber et al., 2003). Taking these SST estimates from the ACEX record, such gradients represent significant overestimates because they would imply unrealistically warm tropical SSTs (Sluijs et al., 2006). These high polar temperatures and

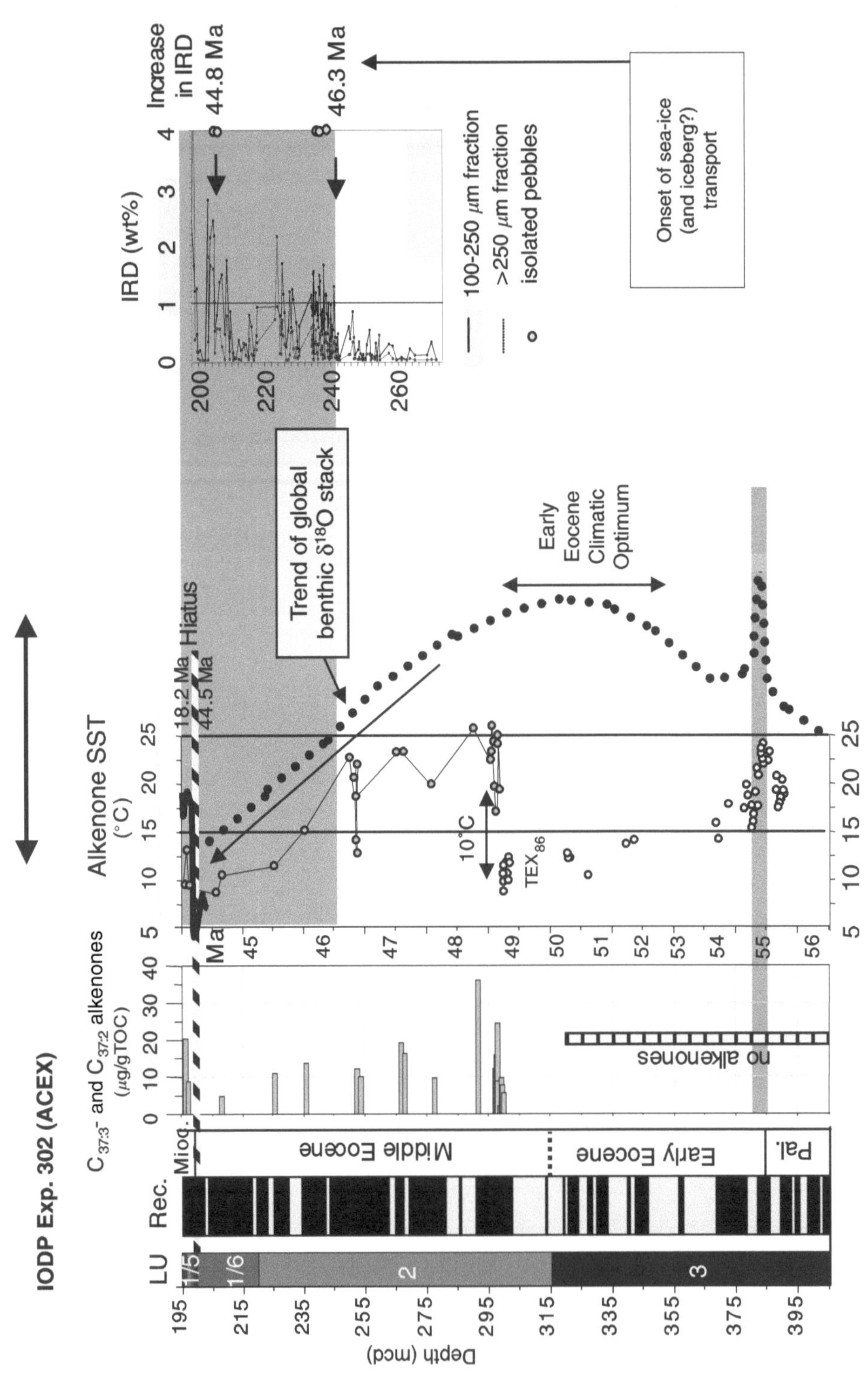

IODP Exp. 302 (ACEX)
Depth (mcd)
LU
Rec.
Mioc.
Middle Eocene
Early Eocene
Pal.
$C_{37:3}$- and $C_{37:2}$ alkenones
(μg/gTOC)
no alkenones
Ma
Alkenone SST
(°C)
10°C
TEX_{86}
18.2 Ma
44.5 Ma
Hiatus
Trend of global benthic $\delta^{18}O$ stack
Early Eocene Climatic Optimum
IRD (wt%)
100-250 μm fraction
>250 μm fraction
isolated pebbles
Increase in IRD
44.8 Ma
46.3 Ma
Onset of sea-ice (and iceberg?) transport

reduced pole-to-equator temperature gradients cannot be explained by invoking even greater greenhouse gas concentrations because this would elevate tropical SSTs, which in existing model predictions already match or exceed those determined from proxy records at low-latitude locations (Shellito et al., 2003). Sluijs et al. (2006) concluded that physical processes that are not yet incorporated in the models, operated in conjunction with high greenhouse gas concentrations to enhance polar warmth and reduce the pole-to-equator temperature gradient during the early Palaeogene. According to these authors, one of the processes could be high-latitude warming and tropical cooling through the enhancement of polar stratospheric clouds (Sloan & Pollard, 1998; see Sluijs et al., 2006, 2008 for more detailed discussion).

Weller and Stein (2008) used the $U^{K}_{37'}$ alkenone approach for estimating SST in the middle Eocene part of the ACEX section. In numerous alkenone studies on Quaternary to Eocene sediments from all the world oceans reliable SST records were obtained (e.g., Herbert & Schuffert, 1998; Peng, Yu, Jia, Song, & Zhang, 2004; McClymont, Rosell-Melé, Giraudeau, Pierre, & Loyd, 2005), suggesting that the calibration function developed from modern species (see Chapter 4.7.5 for background information) can be applied to ancient sediments. Mostly, long-chain alkenones are synthesized by marine phytoplankton (Volkman et al., 1980; Marlowe et al., 1984, 1990; Conte et al., 1992; Müller et al., 1998), however, they were also recorded in freshwater environments (Li, Philip, Pu, & Allen, 1995; Thiel et al., 1997). Therefore, for the interpretation of the ACEX alkenone data and — especially — the use of alkenone data for estimating SST, the marine origin of the alkenones has to be proven. Here, the distribution pattern of C_{37}- and C_{38} alkenones can be used for distinguishing between marine and lacustrine alkenone producers (Marlowe et al., 1984; Conte, Volkman, & Eglinton, 1994; Volkman et al., 1995; Müller et al., 1997; McClymont et al., 2005). Whereas marine alkenone-producing species (*E. huxleyi* and *G. oceanica*) have C_{37}/C_{38} alkenone ratios of 0.86–2.16 (mean 1.16) and ~0.6–0.8, respectively (Volkman et al., 1995; Conte et al., 1994), freshwater species (e.g., *Isochrysis galbana* and *Chrysotila lamellosa*) have much higher values of >8.5 (Marlowe et al., 1984). The C_{37}/C_{38} alkenone ratios of the ACEX samples vary between 0.8 and 1.6 with a mean value of 1.15 (Weller & Stein, 2008),

Figure 7.20 Low resolution record of occurrence of long-chain alkenones and $U^{K}_{37'}$ alkenone SST (light grey circles) for the interval between 195 and 300 original mcd representing the time interval of ~44.5–49 Ma (from Weller & Stein, 2008). The middle Eocene SST decrease coincided with the global cooling trend (thick dotted line) based on benthic oxygen isotopes (Zachos et al., 2001). For the late Palaeocene/early Eocene (below ~300 mcd), $TEX_{86'}$-SSTs (open circles) are shown (Sluijs et al., 2008). Note that across the Azolla Event the alkenone SSTs are ~10°C higher than the TEX_{86} SSTs (Brinkhuis et al., 2006; see text for further discussion). The uppermost three alkenone SST values are from the Zebra Unit (subunit 1/5), that is, above the hiatus, corresponding to a late early Miocene age. The hiatus (18.2–44.4 Ma) separating subunits 1/6 and 1/5, is marked as hatched bar. Across the hiatus, a TEX_{86}-derived SST record (thick black line) is also shown (Sangiorgi et al., 2008). In addition, the occurrence of IRD (wt.%) during the middle Eocene interval — with onset near 46.3 Ma — is presented (from St. John, 2008; see also Figure 7.31). *Note*: a higher-resolution alkenone-based SST record (Weller, unpubl. data 2008) support the low-resolution record presented here.

which is within the range of those determined for the marine species. Thus, the alkenones determined in the ACEX record were interpreted as a marine signal and used for SST calculation.

On the basis of a still limited set of samples from the middle Eocene, Weller and Stein (2008) present a first preliminary low-resolution SST record for the time interval between ~49 and 44.5 Ma. Within this interval, the SST record shows a distinct long-term decrease of ~15°C (Figure 7.20). This general temperature decrease follows very well the global cooling trend at the end of the Early Eocene Climate Optimum as deduced from the global benthic isotope stack (Figure 7.20; e.g., Miller et al., 1987; Pearson & Palmer, 2000; Zachos et al., 2001). Furthermore, this general long-term SST decrease seems to correlate quite well with the amount and flux of IRD >250 μm (St. John, 2008). At ~46.3 Ma, IRD first appeared, contemporaneously with a drop in SST to <15°C (Figure 7.20; see Section 7.2.3 for some more details).

The absolute values of the $U^{K}_{37'}$-based SST ranging between ~25°C and 10°C (Figure 7.20), seem to be surprisingly high. The maximum $U^{K}_{37'}$-based SST values between 49 and 47 Ma representing the final stage of the Early Eocene Climate Optimum are in the same range as those determined for the PETM using the TEX_{86} approach (Sluijs et al., 2006; Figure 7.20), that is, these are significantly higher than those predicted from climate models (see earlier). Furthermore, they are also distinctly higher than those calculated using the TEX_{86} index, at least in the Azolla interval. For the early Eocene time interval where TEX_{86} temperatures of ~10–20°C were determined (Figure 7.20; Sluijs et al., 2008), unfortunately no alkenone SSTs could be determined due to the absence of alkenones (see earlier). For the Azolla phase, the $U^{K}_{37'}$-based SST vary between ~20°C and 25°C, whereas the TEX_{86}-derived SST vary between 8°C and 13°C (Brinkhuis et al., 2006) (Figure 7.20). Assuming that both records are correct, how these differences may be explained?

Similar differences of ~10°C and more between alkenone- and TEX_{86}-based SSTs were also described from other studies (e.g., Menzel, Hopmans, Schouten, & Sinninghe Damsté, 2006; Huguet et al., 2007). As possible explanations different growth seasons (summer versus winter) and different water depths (surface waters versus subsurface waters) are discussed. While $U^{K}_{37'}$-based SST reflects the SST in the shallower euphotic zone (upper 10 m) (Müller et al., 1998), *Crenarchaeota* live deeper down (~100 m) (Wuchter et al., 2005). Furthermore, alkenone producing coccolithophores need sunlight for photosynthesis (Weaver et al., 1999; Conte, Weber, King, & Wakeham, 2001), while marine *Crenarchaeota* are not dependent on light for growth (Könnecke et al., 2005; Huguet et al., 2007). In addition, *Crenarchaeota* typically have their main phase of growth during the annual cycle outside the main period of phytoplankton blooms (Schouten et al., 2002; Menzel et al., 2006). Thus, $U^{K}_{37'}$-based SST probably reflects summer SST in the central Arctic Ocean (*cf.* Axelrod, 1984), whereas the TEX_{86}-derived values may represent more (annual mean) winter SST, that is, the difference between both data sets may represent the seasonal temperature variability. This interpretation is in agreement with reconstructions of a strong High Northern Latitudes seasonal temperature variability of >10°C during the early-middle Eocene, as estimated from morphological features of plant fossils (Basinger, Greenwood, & Sweda, 1994; Wolfe, 1994; Greenwood & Wing, 1995; see Weller & Stein, 2008, for more detailed discussion ACEX alkenone data and references).

Finally, it has to be mentioned that the low-resolution alkenone SST record (Figure 7.20; Weller & Stein, 2008) only allows preliminary interpretation and has to be verified by additional high-resolution data from the middle Eocene time interval. Furthermore, TEX_{86} data should be determined from the same time interval in a similar resolution in order to compare and interpret both data sets in more detail.

7.2.2.4. Palaeocene/Eocene palaeoceanographic events and their (global) significance

In the Palaeocene/Eocene ACEX black-shale section, three major events of global significance could be identified, which are characterized by very extreme palaeoclimatic and/or palaeoceanographic conditions: (1) the PETM Event (55 Ma; Pagani et al., 2006; Sluijs et al., 2006), (2) the Elmo Event ($\sim$53 Ma; Stein et al., 2006a; Sluijs et al., 2008), and (3) the Azolla Freshwater Event (near 49 Ma; Brinkhuis et al., 2006). At all three events, OC accumulation rate of $\sim 0.05\,gC\,cm^{-2}\,kyr^{-1}$ were reached (Figure 7.19). In total, 150 to $>$900 Gt of OC were buried in the sediments during these events (Table 7.3). Because most of this OC is derived from algae, by this process large amounts of CO_2 might have been extracted from the atmosphere and stored in the sediments, and thus — in analogy to the "Miocene Monterey Formation Hypothesis" (Vincent & Berger, 1985; Raymo, 1994) — may have contributed to the global decline in atmospheric CO_2 and cooling subsequently to the Early Eocene Climate Optimum (see Knies et al., 2008 and Sluijs et al., 2008 for further discussion). Here, however, more data are needed to prove these very preliminary and speculative results.

The Palaeocene/Eocene Thermal Maximum (PETM) Event. The PETM Event was a brief period of widespread, extreme climatic warming (e.g., Kennett & Stott, 1991; Röhl, Bralower, Norris, & Wefer, 2000; Zachos et al., 2001; Zachos, Pagani, Sloan, Thomas, & Billups, 2003; Tripati & Elderfield, 2005) that was probably associated with massive atmospheric greenhouse gas input (e.g., Dickens, O'Neil, Rea, & Owen, 1995). In the ACEX record, the PETM was identified from the top of Core 32X to within Core 29X (387–378.5 original mcd; Figure 7.21) by the occurrence of the dinocyst species *Apectodinium augustum*, which is diagnostic of the PETM (Bujak & Brinkhuis, 1998), as well as a distinct negative anomaly in $\delta^{13}C_{org}$ (Figures 7.16 and 7.22; Pagani et al., 2006; Sluijs et al., 2006, 2008; Stein et al., 2006a). During the PETM, a maximum TEX_{86} SST of $\sim$23°C was estimated (Figure 7.21; Sluijs et al., 2006; see earlier discussion).

With the PETM Event, a drastic change in OC composition is obvious. Whereas in the late Palaeocene terrigenous OC is predominant (see earlier), across the PETM the amount of labile OC significantly increased, as indicated by increased hydrogen index values (Figure 7.16) as well as increased preservation of algae-type biomarkers (Figure 7.17; Weller & Stein, 2008). Furthermore, palynomorph

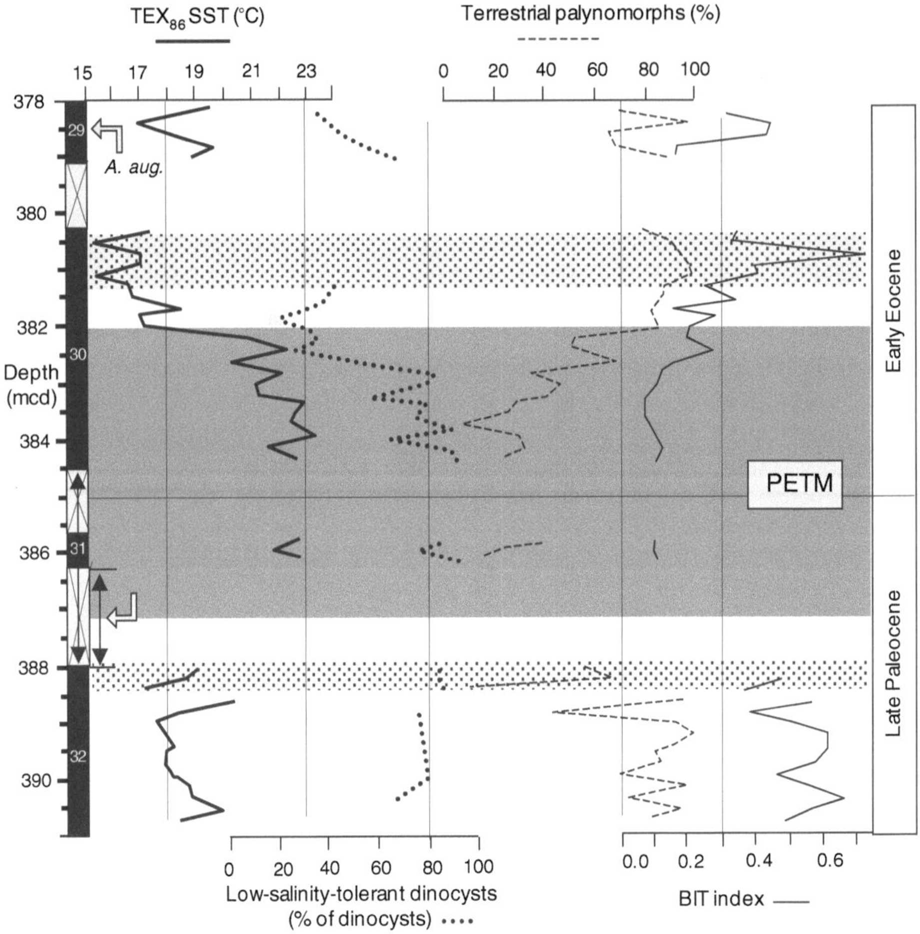

Figure 7.21 Core recovery, TEX_{86} sea-surface temperatures (see Chapter 4.7.5 for background), concentrations of low-salinity-tolerant dinocyst species and terrestrial palynomorphs, and "BIT index" values (see Chapter 4.7.4 for background) across the late Palaeocene to early Eocene time interval (including the PETM Event) of IODP Hole 302-4A (redrawn and modified from Sluijs et al., 2006). Black arrow connected to Core 31X in the recovery column indicates the uncertainty of its stratigraphic position. Dotted bars indicate intervals affected by drilling disturbance. Low-salinity-tolerant dinocysts include *Senegalinium* spp., *Cerodinium* spp., and *Polysphaeridium* spp., whereas *Membranosphaera* spp., *Spiniferites ramosus* complex, and *Areoligera-Glaphyrocysta* cpx. represent the typical normal marine species (Sluijs et al., 2005). Open arrows and *A. aug* indicate the first and last occurrences of dinocyst *A. augustum*. For further details see Sluijs et al. (2006, 2008).

assemblages from the PETM interval are characterized by abundant dinocysts (60%; Sluijs et al., 2006), a significant decrease in terrestrial palynomorphs, and substantially lower BIT indices of ~0.1 (Figure 7.21), indicating a relative decrease of riverine-derived OC (Sluijs et al., 2006). Sluijs et al. (2006) interpreted the enhanced influence of marine conditions during the PETM in terms of a sea-level rise, an interpretation which is consistent with evidence from other neritic

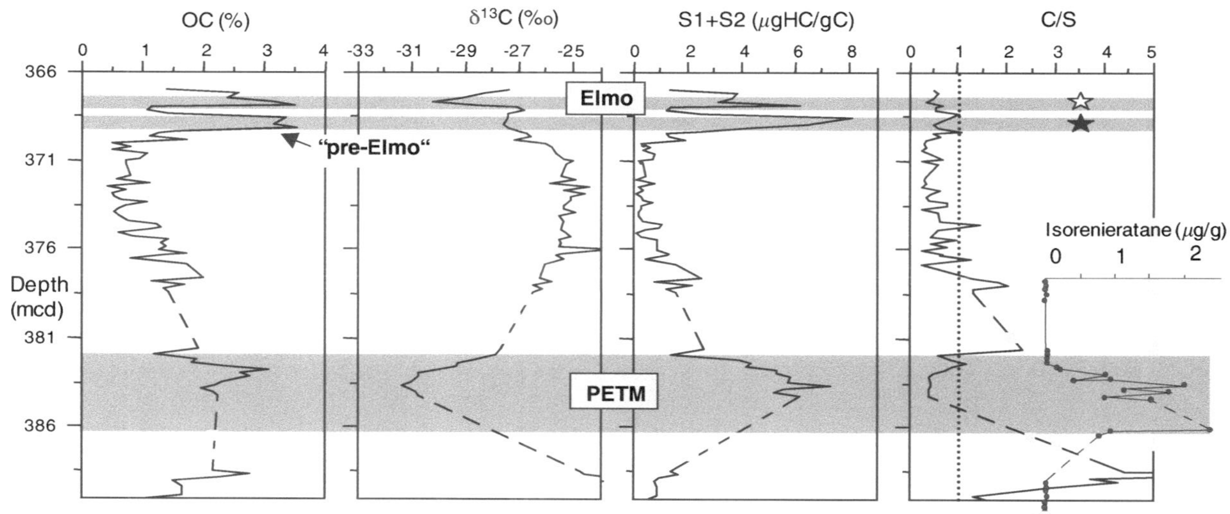

Figure 7.22 OC content (TOC), $\delta^{13}C_{org}$, S1 and S2 values, and C/S ratios of the depth interval 366–390 original mcd, including the PETM and Elmo events (Stein et al., 2006a; Stein, 2007). Open star indicates absence of isorenieratane (Elmo Event), solid star indicates presence of isorenieratane in the middle part of the "Pre-Elmo" Event (Weller & Stain, 2008). In addition, the concentration of isorenieratane across the PETM Event are shown at the right-hand side (data from Sluijs et al., 2006). For further explanation see text.

locations (Speijer & Morsi, 2002). Despite the transgression, however, low-salinity-tolerant dinocysts are dominant (Figure 7.21), suggesting that brackish surface waters persisted during the PETM.

The increased preservation of labile algae-type OC is related to a major change to euxinic conditions, as indicated by a drastic decrease in the C/S values (Figures 4.56 and 7.22), the occurrence of pyrite framboids (Stein et al., 2006a), the absence of benthic foraminiferal linings (Sluijs et al., 2006), and the occurrence of fine lamination (Backman et al., 2006). Fe/S ratios >1, however, point to incomplete pyritization probably due to a lack of seawater sulfate (Sluijs et al., 2008). During the PETM, euxinic conditions expanded even into the photic zone as suggested from the occurrence of the biomarker isorenieratene (Figure 7.22) related to photosynthetic green sulfur bacteria which requires euxinic conditions to thrive (Sinninghe-Damsté et al., 1993; Sluijs et al., 2006; see also Weller & Stein, 2008). Possible cause for the euxinic conditions was on one hand a salinity stratification suggested from the high abundance of low-salinity dinocysts (Figure 7.21; Sluijs et al., 2006). Because these dinocysts were already abundant prior to the PETM event, that is, during times of oxic water-mass conditions, increased flux of algae-type OC (indicated by increased S1, S2, and HI values; Figures 7.16 and 7.22) due to some enhanced primary production was probably needed as additional factor causing the change to euxinic conditions (Stein et al., 2006a). The increased primary production was probably related to increased fluvial nutrient supply at that time (Pagani et al., 2006).

Towards the end of the PETM Event, that is, in the lowermost Eocene, a gradual return to a more terrestrial influence is obvious (probably resulting from a subsequent regression; Sluijs et al., 2006), and oxic conditions re-occurred as clearly reflected in the C/S ratios (Figures 4.56 and 7.22) and the dominance of terrestrial OC (Figures 7.16 and 7.21). The termination of the euxinic conditions coincides with increasing surface-water salinities (Pagani et al., 2006) and cooling (Figure 7.21), suggesting an increased mixing water masses which may have caused this change (Sluijs et al., 2006).

Using a mean sedimentation of 2.01 cm kyr^{-1} and a mean OC value of 2.2%, $\sim 1.8 \times 10^6$ t yr^{-1} were buried in the Arctic Ocean Basin which gives a total burial rate of 310 Gt OC for the PETM Event (Table 7.3). This might be even a conservative estimate. According to Sluijs et al. (2008) who used a higher sedimentation rate of 5 cm kyr^{-1} for the PETM Event, the total OC burial may have been significantly higher, that is, $\sim 5.4 \times 10^6$ t yr^{-1} (or $\sim$920 Gt for the PETM Event) (Table 7.3). Comparing the background value of $\sim 0.9 \times 10^6$ t yr^{-1} and the PETM value of $\sim 5.4 \times 10^6$ t yr^{-1}, this gives an excess OC burial of $\sim 4.5 \times 10^6$ t yr^{-1} or a total of 765 Gt for the PETM Event. This high excess of OC burial may have been important for the global OC cycle and climatic cooling following the PETM Event (Sluijs et al., 2008 and further discussion therein).

The Elmo (or ETM2) Event. Within the early Eocene section of the ACEX sequence (at $\sim$368 original mcd), an event with similar characteristics as the PETM Event, has been identified. This event characterized by a prominent $\delta^{13}C$ minimum (Figures 7.16 and 7.22) and identified in the Arctic Ocean for the first time by Stein et al. (2006a), may correlate with the global "Elmo Event" described by Lourens et al. (2005) in a depth transect of ODP Leg 208 drill sites on Walvis Ridge. According to these authors, the Elmo horizon, a second pronounced early Eocene thermal maximum (ETM2), has similar geochemical and biotic characteristics as the PETM Event, but of smaller magnitude, and it is coincident with carbon isotope depletion events in other ocean basins, suggesting that it represents a second global thermal maximum.

In the ACEX record, this interval displays significantly increased OC contents mainly composed of labile algae-type organic matter, as reflected in the elevated Rock-Eval S1 and S2 (Figure 7.22) and hydrogen index values (Figure 7.16) as well as in the biomarker composition (Figure 7.17). Furthermore, very low C/S ratios <1 (Figure 7.22) point to an euxinic environment causing the preservation of the labile OC. Based on the OC and Rock-Eval data, this interval can be separated into two "sub-events", that is, the Elmo Event itself (coinciding with the $\delta^{13}C$ minimum) and the "Pre-Elmo Event" (Figure 7.22; Stein, Backman, & Moran, 2007). Using an average sedimentation rate of 1.3 cm kyr^{-1} (Backman et al., 2006; Jakobsson et al., 2007a), the two peaks are $\sim$100 kyr apart. Whereas C/S values are very low in both "sub-events", indicating bottom-water anoxia, the isorenieratane data are different (Figure 7.22; Weller & Stein, 2008). Isorenieratane and related isorenieratene derivates could not be detected in samples from the Elmo Event. Thus, euxinic conditions did probably not extend into the photic zone during this period. In the underlying "Pre-Elmo" event, on the other hand, the presence of isorenieratane and related isorenieratene derivates point to similar conditions as those of the PETM Event, that is, the water column was probably euxinic throughout. It has to be considered that these results are based on a still restricted number of samples; further studies on a larger number are needed to get a more detailed picture about the palaeoenvironment and its variability during the Elmo (and "Pre-Elmo") Event.

The Azolla Freshwater Event. The lowermost middle Eocene OC-rich section of the ACEX sequence ($\sim$299–305 original mcd) representing the time interval between $\sim$49 and 48.3 Ma are composed of microlaminated sediments with extraordinary abundances of microspore clusters (massulae) of the free-floating freshwater fern *Azolla* (Figure 7.23; Brinkhuis et al., 2006). Normally, *Azolla* known from modern freshwater bodies, such as ponds, canals and flooded rice fields in tropical, subtropical, and warm temperate regions, cannot tolerate salinities higher than 1–1.6‰ (Rai & Rai, 1998; Arora & Singh, 2003). Based on the presence of mature megaspores with and without attached massulae, single, small groups and large clusters of massulae and probable aborted megaspores of *Azolla*, and support by the relative scarcity of terrestrially derived palynomorphs and extremely low BIT index values of <0.1 indicating low river-derived terrestrial organic matter, Brinkhuis et al. (2006) favour that *Azolla* grew and reproduced in the Arctic Ocean rather than brought in by periodic mass transport from freshwater bodies on adjacent continents. That means, the *Azolla* Event probably represent a distinct episodic freshening of Arctic surface waters lasting less than $\sim$800 kyr. The freshening of surface waters supports stratification of water masses, causing the euxinic conditions reflected in the very low C/S ratios (Figure 7.18). Furthermore, the *Azolla* maxima coincide with density minima and OC maxima (Figure 7.23). Taken the average early Eocene sedimentation rates of ~1.3 cm kyr^{-1} (Backman et al., 2008), these oscillations may reflect orbital forcing of the 100-kyr eccentricity cycle, suggesting that the periodical freshening of the Arctic Ocean was astronomically modulated (Brinkhuis et al., 2006).

Abundant *Azolla* remains have previously been reported from the basal middle Eocene from all Nordic Seas (Figure 7.24; Boulter, 1986; Bujak & Mudge, 1994; Eldrett, Harding, Firth, & Roberts, 2004; Brinkhuis et al., 2006 and references therein). As *Azolla* fluxes to the sea floor were significantly higher in the Arctic Ocean than in the adjacent seas, Brinkhuis et al. (2006) propose a scenario of Arctic basin *Azolla* mats being transported through huge Arctic freshwater plumes towards the south as far as the North Sea. The termination of the Azolla Event in the Arctic coincides with a local SST rise from ~10°C to 13°C (Figure 7.23; Brinkhuis et al., 2006; Sluijs et al., 2008), which may suggest simultaneously increasing salt and heat supply due to the inflow of waters from adjacent oceans (Brinkhuis et al., 2006). This increased marine influence towards the end of the Azolla Event may also explain the increase in hydrogen index values (Figure 7.23; Stein et al., 2006a) as well as the occurrence of long-chain alkenones related to marine phytoplankton (Figure 7.20; Weller & Stein, 2008). Concerning the absolute TEX_{86} SST values, it

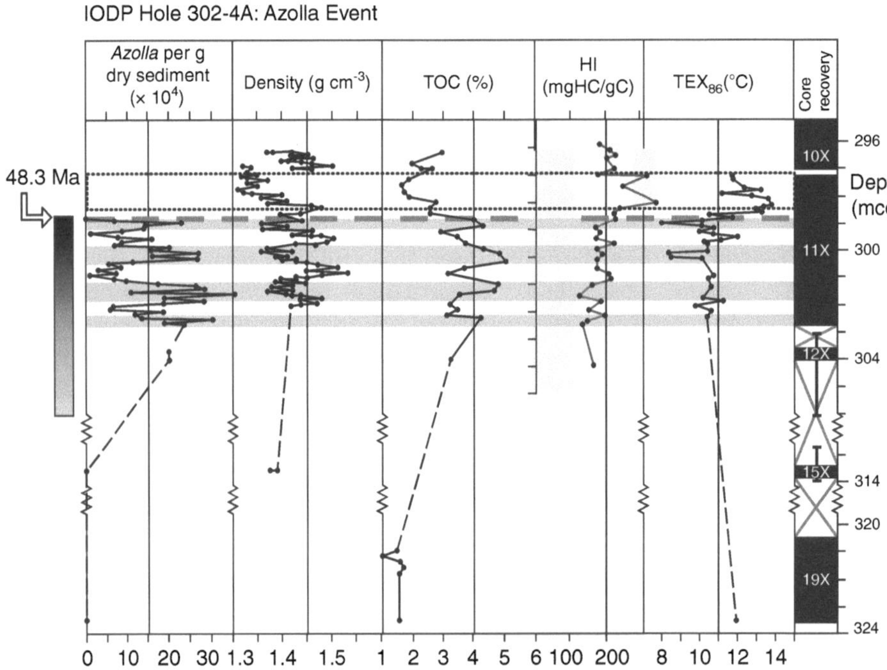

Figure 7.23 Abundance of *Azolla* species, density, TOC content, hydrogen index values, TEX_{86} sea-surface temperatures, and core recovery across the *Azolla* phase of IODP Hole 302-4A (from Brinkhuis et al., 2006, supplemented by hydrogen index values from Stein et al., 2006a). Numbers in core and recovery column refer to core numbers. The termination of the *Azolla* phase has been calibrated against mid-magnetochron C21r (~48.3 Ma) at three sites from the Greenland and Norwegian seas (Eldrett et al., 2004).

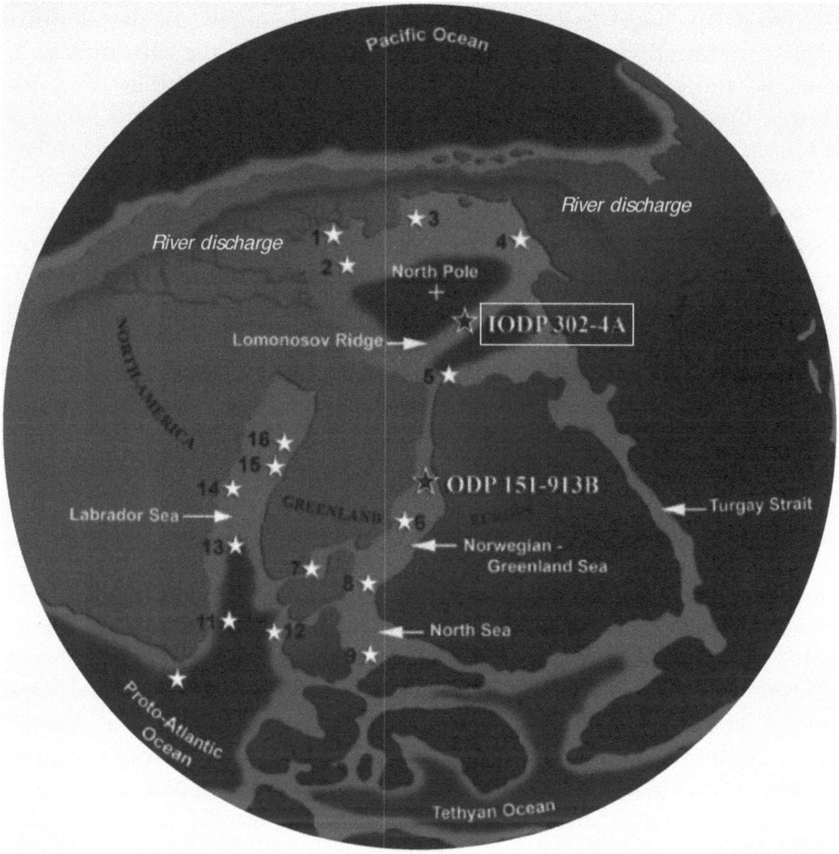

Figure 7.24 The early Eocene Arctic Basin, site locations, and geographic distribution of the *Azolla* pulse in adjacent basins (from Brinkhuis et al., 2006, supplemented). Black stars indicate locations of ODP Hole 151-913B and IODP Hole 302-4A where the Azolla Event was found and discussed in detail (Eldrett et al., 2004; Brinkhuis et al., 2006). White stars (1–16) show wells with earliest middle Eocene *Azolla* abundances in the various regions of the Nordic Seas. In this map, Brinkhuis et al. (2006) compiled notably industrial palynological information through numerous, mainly informal channels as most information is still confidential to oil and gas exploration companies.

has to be mentioned that the estimates based on alkenones are distinctly higher (Figure 7.20; Weller & Stein, 2008; see earlier discussion).

*7.2.2.5. **The final termination of anoxia in the early Cenozoic Arctic Ocean.*** On the basis of the data described earlier, the early Cenozoic Arctic Ocean was characterized by more or less euxinic conditions throughout most of this time interval. One prerequisite of this extreme palaeoenvironmental situation was the palaeogeographic boundary setting, that is, the early Arctic Ocean was isolated from the world ocean in terms of deep-water connection (Figure 1.16; Ziegler, 1988; Mutterlose et al., 2003; Backman et al., 2006; Jakobsson et al., 2007a).

Furthermore, the huge freshwater discharge has favoured the development of water-mass stratification resulting in a poor ventilation of the sub-surface, deeper water masses and causing the high OC preservation rate (see earlier). A remaining question is how and when the transition from poorly oxygenated to ventilated conditions in the Arctic Ocean did occur. Subunits 1/4 to 1/1 (17.5 Ma to Recent; Table 7.2) characterized by very low OC values of <0.5% (Figure 7.15; Stein, 2007) and C/S values falling into the "oxic" field (Figure 7.18), already represent palaeoenvironmental conditions similar to the modern ones, that is, fully ventilated water masses preventing preservation of high amount of labile algae-type OC. Thus, the transition from euxinic to oxic conditions in the central Arctic Ocean should have occurred within/around subunit 1/5 (Jakobsson et al., 2007a).

On the basis of a palaeogeographic and palaeobathymetric reconstruction of the Arctic Ocean, together with a physical oceanographic modelling of the evolving strait and sill conditions in the Fram Strait, Jakobsson et al. (2007a) suggest that across subunit 1/5 between 18.2 and 17.5 Ma the Arctic Ocean went from an oxygen-poor "lake stage", to a transitional "estuarine sea" phase with variable ventilation, and finally to the fully ventilated "ocean" phase at 17.5 Ma (Figure 7.25). As outlined by these authors, during Miocene times when the Fram Strait opened and deepened through sea-floor spreading (see also Figure 1.15), the water exchange between the Arctic and North Atlantic must have developed through a series of changes that also influenced the upstream basin circulation and ventilation conditions within the Arctic Ocean. During this evolution, the initial opening phase of the Fram Strait was restricted to a uni-directional hydraulically controlled fresh-water outflow and reduced ventilation of deep waters via seasonal convection. Later, when the Fram Strait widened and deepened, a compensating inflow of saline North Atlantic water became possible, resulting in a bi-directional, two-layer flow through the strait, similar to that of the modern Black Sea. An inflow of warm near-surface North Atlantics waters in the early Miocene (i.e., above the hiatus) may be supported by a distinct increase in TEX_{86}-derived SSTs to ~15–19°C (Figure 7.20; Sangiorgi et al., 2008). At this stage, alternations between euxinic and more oxic conditions (reflected in the black/light grey cycles of subunit 1/5) may have occurred due to oscillations in sea level (Jakobsson et al., 2007a). That means sea-level fluctuations could have acted as "on-off switch" for Arctic Ocean circulation, with reduced ventilation during times of lowered sea level and increased ventilation during times of high sea level. According to Miller et al. (2005), sea level may have varied between ~15 and 30 m during late early Miocene times. As the strait deepened further, the sea-level changes were no longer sufficient for a reversal to "Arctic lake" conditions, and oxic conditions similar to the modern ones prevailed (Jakobsson et al., 2007a).

According to the age-width estimation using new seismic reflection and aeromagnetic data (Jokat, Leinweber, Ehlers, Boebel, & Schenke, 2008), the Fram Strait began to open at great depths near 13.7 Ma, that is, 3.8 my after the change to a ventilated circulation stage in the Arctic Ocean, marked by the end of subunit 1/5 (Jakobsson et al., 2007a). As discussed by these authors, however, an initial corridor of immature seafloor spreading may have developed between Greenland and Svalbard before the full spreading extended through this gateway (Engen, 2005).

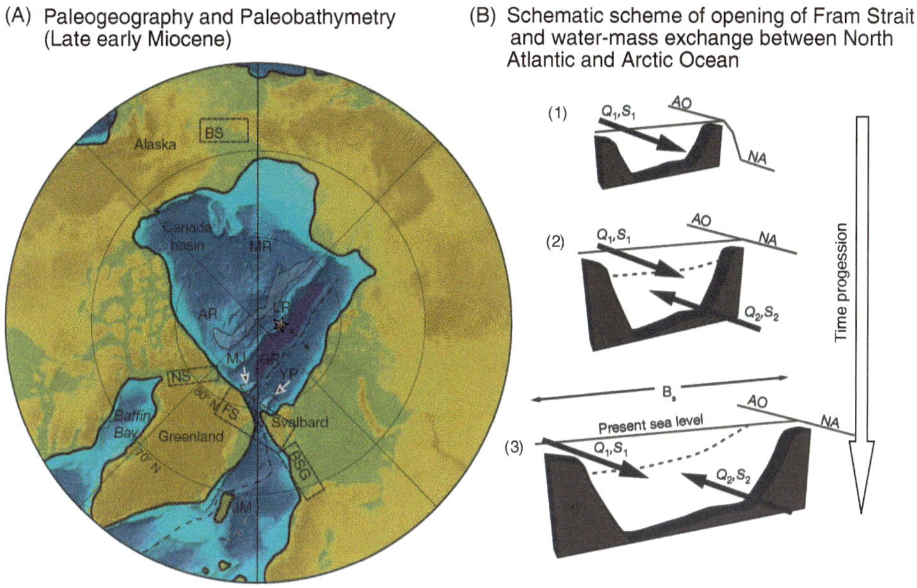

Figure 7.25 (A) Palaeogeographic/palaeobathymetric reconstruction for the late early Miocene, based on plate tectonic maps generated from the Ocean Drilling Stratigraphic Network (ODSN) tools available online at http://www.odsn.de (from Jakobsson et al., 2007a, supplemented). Physiographic features: AR, Alpha Ridge; BSG, Barents Sea gateway; BS, Bering Strait; FS, Fram Strait; GR, Gakkel Ridge; JM, Jan Mayen microcontinent; LR, Lomonosov Ridge; MR, Mendeleev Ridge; MJ, Morris Jessup rise; NS, Nares Strait; YP, Yermak Plateau. (B) Schematic illustration of the Fram Strait opening and hypothetical water exchange development between the Arctic Ocean and North Atlantic. AO, Arctic Ocean; NA, North Atlantic. (1) A narrow strait resulting in a uni-directional hydraulically controlled outflow from the Arctic. S1 is the salinity of the out-flowing flux of water Q1. (2) A wider and deeper strait allowing the establishment of a bi-directional, two-layer flow through the strait due to a compensating inflow (Q2) of saline (S2) North Atlantic water. This phase in the Arctic's palaeoceanographic development is analogous to the present Black Sea. (3) The Fram Strait becomes wide enough that the influence of the Earth's rotation changes the water flow through the strait to a rotationally controlled bi-directional two-layer flow. This opens the possibility of a barotropic current flow through the strait (modified from Jakobsson et al., 2007a).

In the lower part of subunit 1/5 characterized by black (OC-rich) and light grey (OC-poor) alternations (Figure 7.14), C/S ratios seem to support the alternation between more euxinic conditions (represented by the black intervals) and more oxic conditions (represented by the light gray horizons) (Figures 7.18 and 7.26), although this cyclicity is not fully reflected in the low-resolution record shown in Figure 7.26. These data support Jakobsson et al.'s (2007a) interpretation of euxinic/oxic alternations during the transitional stage of ventilation change. Surprisingly, the interval with absolute OC maxima of ~13–14%, maximum hydrogen index values (indicating the preservation of significant amount of algae-type OC), and very negative $\delta^{13}C_{org}$ values of $<-32‰$ at ~196.3–196.8 original mcd, has high C/S ratios suggesting a more oxic environment (Figure 7.26). At the same depth, also a prominent maximum in foram linings was found (Sangiorgi et al., 2008), supporting

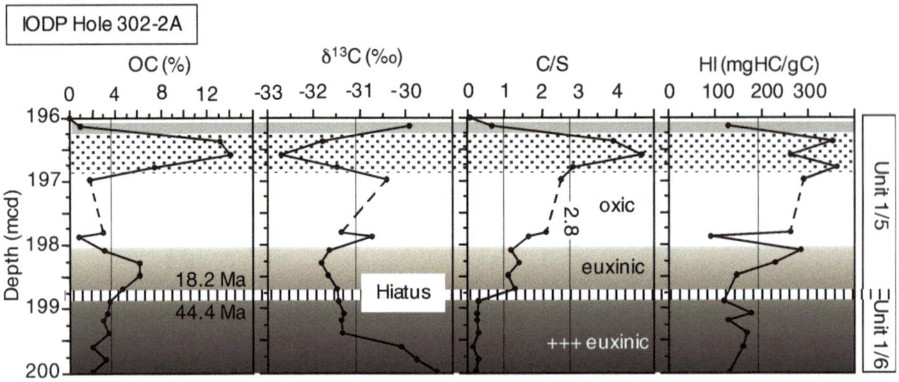

Figure 7.26 OC content (TOC), $\delta^{13}C_{org}$, hydrogen index values, and C/S ratios of the depth interval 196–200 original mcd, representing the interval across the major hiatus separating subunits 1/5 ("Zebra") and 1/6 (data from Stein et al., 2006a).

oxygenated bottom-water conditions. Furthermore, the maximum abundance of freshwater algae *Tasmanites* (Sangiorgi et al., 2008) coinciding with the high hydrogen index and light carbon isotope values, points to an increased input of terrestrial algae-type of organic matter causing the OC enrichment in these sediments.

Looking at the C/S ratios below and above the major hiatus separating subunits 1/5 and 1/6, there seems to be a prominent change across the hiatus (Figure 7.26). Whereas below the hiatus, that is, in the middle Eocene (subunit 1/6), C/S ratios are extremely low suggesting strong euxinic conditions, C/S ratios are distinctly higher above the hiatus, that is, in the late early Miocene (subunit 1/5), suggesting less strong euxinic to suboxic conditions. These data may suggest that the onset of change in ventilation of the Arctic Ocean from extreme euxinic to oxic conditions started earlier than proposed by Jakobsson et al. (2007a), that is, prior to 18.2 Ma, and, thus, is not preserved in the ACEX record.

7.2.2.6. Palaeogene black shales: Source-rock potential and hydrocarbon generation

As outlined in Chapter 4.7.1, Rock-Eval parameters give information about the source-rock potential and hydrocarbon generation. Most of the OC-rich middle Eocene sediments (black shales in a broader sense) have a (fair to) good source-rock potential, as indicated by the high S2 values of 5–10 mgHC/gSed (kgHC/tSed), whereas in the Campanian and upper Palaeocene sediments the source-rock potential is rather low (Figure 7.27; Stein, 2007). In the early Eocene the potential is also generally low, except for short intervals with higher potential similar to the middle Eocene. HI values $>$ 200 mgHC/gC and S2/S3 ratios $>$2.5 suggest that the intervals with a good source-rock potential are prone to generate a gas/oil mixture when the level of thermal maturity ("oil window") is reached. T_{max} values $<$435°C, however, indicate an immaturity of the organic matter, also supported by

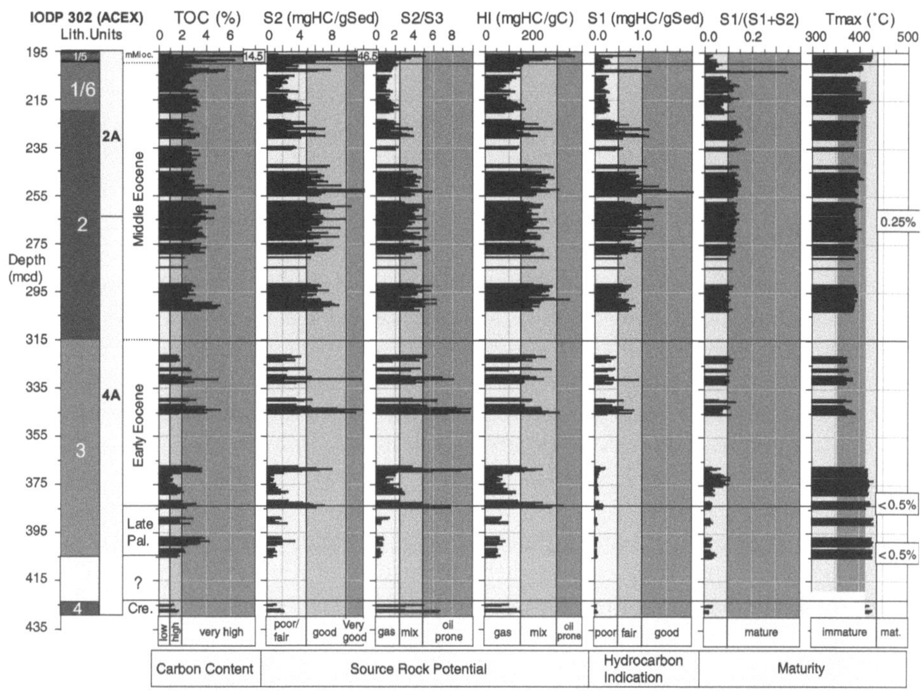

Figure 7.27 TOC content and Rock-Eval parameters S1, S2, S3, S2/S3, S1/(S1+S2), hydrogen index (HI), and T_{max} determined in lower 230 m of the ACEX drill site (Holes MSP0002A and MSP0004A), and interpretation in terms of source-rock potential, hydrocarbon indication, and maturity according to Peters (1986) (from Stein, 2007). Numbers in the T_{max} record are vitrinite reflectance values (from Stein et al., 2006a). Left, the stratigraphy and the lithological units are shown (Backman et al., 2006).

vitrinite reflectance (R_o) values <0.5% (Figure 7.27). Nevertheless, high S1 values may suggest free hydrocarbons being present in the middle Eocene interval (Figure 7.27). This interpretation is supported by biomarker data, showing a smooth short-chain *n*-alkane distribution, with no odd-over-even carbon number predominance (Weller & Stein, 2008).

Due to the low maturity, the presence of oil or gas already generated *in-situ* from the ACEX sediments at this part of Lomonosov Ridge can be ruled out. If these sediments are buried more deeply, however, *in-situ* hydrocarbon formation is possible. This situation might occur in the more southern part of Lomonosov Ridge closer to the Eurasian continental margin, where sedimentation rates are significantly higher. Sedimentation rates at the ACEX sites area (87°56′N, 140°E) are between ~1 and 3 cm kyr^{-1} (Backman et al., 2006), but the rates increase to 3.5–6 cm kyr^{-1} in Quaternary sediments recovered on Lomonosov Ridge at ~81°N (*Polarstern* Core PS2757-8, 81°09.8′N, 140°12.0′E, 1230 m water depth; Stein et al., 2001). The ACEX data do not also mean that in the underlying deeper (Mesozoic) sedimentary rocks from the Lomonosov Ridge belonging to the rifted continental crustal block of the Eurasian continental margin (Kristoffersen, 1990; Jokat et al., 1992), hydrocarbons could not have been generated. This is even probable because during pre-Tertiary times the Lomonosov Ridge was part of the Eurasian continental margin with its large sedimentary basins filled with Mesozoic OC-rich sediments (Figure 7.4). These sediments are the source rocks of the giant oil and gas fields explored in this area (e.g., Leith et al., 1992; Littke et al., 1999; Vyssotski et al., 2006). Thus, the presence of hydrocarbons in the ACEX sediments is most likely attributed to deeper buried equivalents of Eocene deposits and lateral migration, and/or vertical migration of hydrocarbons from underlying deeper Mesozoic OC-rich strata (Stein, 2007).

OC-rich deposits (source rocks) of Eocene age were also found in other circum Arctic areas, for example, in the Beaufort-Mackenzie Basin area (Snowdon, 1984; Brooks, 1986a, 1986b; Keller et al., 1999; Snowdon et al., 2004; Montgomery, 2005). Kos'ko and Trufanov (2002) reported marine shales of Eocene to Oligocene age at outcrop on the New Siberian Islands in the east of the Laptev Sea, which may have contributed to the area's oil potential (Cramer & Franke, 2005). Based on the widespread occurrence of Eocene black shales in the central Arctic Basin as well as in the surrounding marginal areas it may be speculated about the prevalence of a well-stratified, probably euxinic "Black Sea type" (Stein et al., 2006a) depositional environment that may have triggered the formation of highly prolific source rocks during the early Tertiary in the entire Arctic Ocean.

7.2.3. The Early Onset and Variability of Northern Hemisphere Glaciations During Cenozoic Times

7.2.3.1. Records from Subarctic areas: The knowledge prior to ACEX

Prior to ACEX, information on the long-term evolution of the palaeoenvironmental history of the Arctic Ocean, especially the onset and variability of NHGs, was mainly restricted to the Subarctic region. Since DSDP Leg 38 (Talwani et al., 1976) it is well known that (1) the Norwegian-Greenland Sea housed temperate to subtropical surface waters during the Palaeogene; (2) indications of early Cenozoic northern ice covers were not confined to the Quaternary, but also found in unspecified Mio-/Pliocene sediments (Warnke & Hansen, 1977); and (3) the history of NHG during the latest part of the Cenozoic was much more variable than indicated by the available terrestrial records from Europe and North America (Thiede et al., 1998). Following DSDP Leg 38, six ODP Legs were carried out in the North Atlantic between 1985 and 1995: ODP Leg 104 (Sites 642–644; Norwegian Sea, Vøring Plateau; Eldholm et al., 1989), ODP Leg 105 (Sites 645–647; Labrador Sea and Baffin Bay; Srivastava et al., 1987), ODP Leg 151(Sites 907–913; Yermak Plateau, Fram Strait, Greenland Basin, Iceland Plateau; Myhre et al., 1995; Thiede et al., 1996), ODP Leg 152 (Sites 914–919; East Greenland; Larsen et al., 1994), ODP Leg 162 (Sites 980–987; Norwegian-Greenland Sea; Jansen et al., 1996; Raymo et al., 1999), and ODP Leg 163 (Sites 988–990; Southeast Greenland margin; Larsen et al., 1999b). One of the main objectives of most of these legs was related to the onset and variability of High Northern Latitudes glaciations.

According to the results of studies from these legs, it is still uncertain when the Arctic Ocean became ice covered and when the first major continental ice sheets were built (Thiede et al., 1998). At the Vøring Plateau, the oldest pulses of glacial-derived IRD date back to ~12.6 Ma, indicating that glaciers reached sea level at this time (Fronval & Jansen, 1996). The history of the evolution and growth of the Barents Sea ice sheet, the Scandinavian ice sheet, and the Greenland ice sheet were examined at Sites 987 (East Greenland margin) and 986 (Svalbard margin), recovered during ODP Leg 162 (Jansen et al., 1996). Based on shipboard analyses, an ice sheet occurred on Greenland since the Miocene, whereas indications for a Barents Sea/Svalbard ice sheet were found in upper Pliocene sediments. The first occurrence of IRD recorded at DSDP-Site 408 (300 km SW of Iceland) is dated to 10.2 Ma, suggesting glacial deposits in the Iceland/Greenland area (Schaeffer & Spiegler, 1986). ODP Site 646 shows the onset and discontinuous existence of sea-ice cover in the Labrador Sea since the middle/late Miocene (Wolf & Thiede, 1991).

ODP Leg 151 with its drilling in the northern Fram Strait/Yermak Plateau area was the first ODP activity in the (marginal) Arctic Ocean, concentrating on the palaeoenvironmental history of the Arctic Gateway to the Norwegian-Greenland Sea (Myhre et al., 1995; Thiede et al., 1998). Based on shipboard results, at all Leg 151 sites except 907 dropstones appeared in the lower Pliocene with a marked increase at ~2.5 Ma (Figure 7.28; Myhre et al., 1995). More detailed shore-based analysis of the terrigenous coarse fraction indicates that the onset of glaciation certainly was much earlier. As an example, records from ODP Site 909 are shown in Figure 7.29. This site located on the Greenland Spitsbergen sill in a water depth of 2518 m is protected against sediment transport from the Spitsbergen continental slope due to its topographic setting. Thus, the terrigenous coarse fraction >63 μm found in the sediments was mainly controlled by melting of icebergs and/or sea ice. Changes in the terrigenous coarse fraction particle assemblage as well as the variation of grain-size distribution pattern at this site were used to document the sedimentation of IRD in the Fram Strait and its relationship to Northern Hemisphere cooling associated with regional restricted glaciations and sea-ice development (Wolf-Welling et al., 1996; Thiede et al., 1998; Winkler et al., 2002).

The results of this study give evidence that stepwise climatic changes related to Northern Hemisphere cooling probably started in the middle Miocene. The first major occurrence of IRD-related particles were found in the middle Miocene near 14 Ma at this site, also coinciding with an increase in accumulation rate of IRD and an increase in the percentages of angular to subrounded quartz (Figures 7.28 and 7.29). Single IRD spikes even occurred already at 15.2 and 16 Ma (Figure 7.29). IRD sediment input continued in pulses between 10 and 8.6 Ma, 7.5 and 6.8 Ma, and since ~2.7 Ma (Figure 7.29). It is still not clear, however, whether these IRD events represent local Svalbard and Greenland ice expansion events or whether the events can be correlated with processes in the central Arctic. A significant decrease of the smectite to illite and chlorite ratio at Site 909 support a middle Miocene cooling phase between 14.8 and 14.6 Ma, and a further cooling phase between 10 and 9 Ma (Winkler et al., 2002).

At ODP Leg 151 Site 911 (see Figure 7.28 for location), an abrupt major IRD pulse to the Yermak Plateau at ~2.7 Ma probably reflects distinct melting of sediment-laden icebergs derived from the SBIS (Knies, Matthiessen, Vogt, & Stein, 2002) and may indicate the protruding advance of the ice sheet onto the outer shelf at times of major intensification of NHG (e.g., Shackleton et al., 1984; Driscoll &

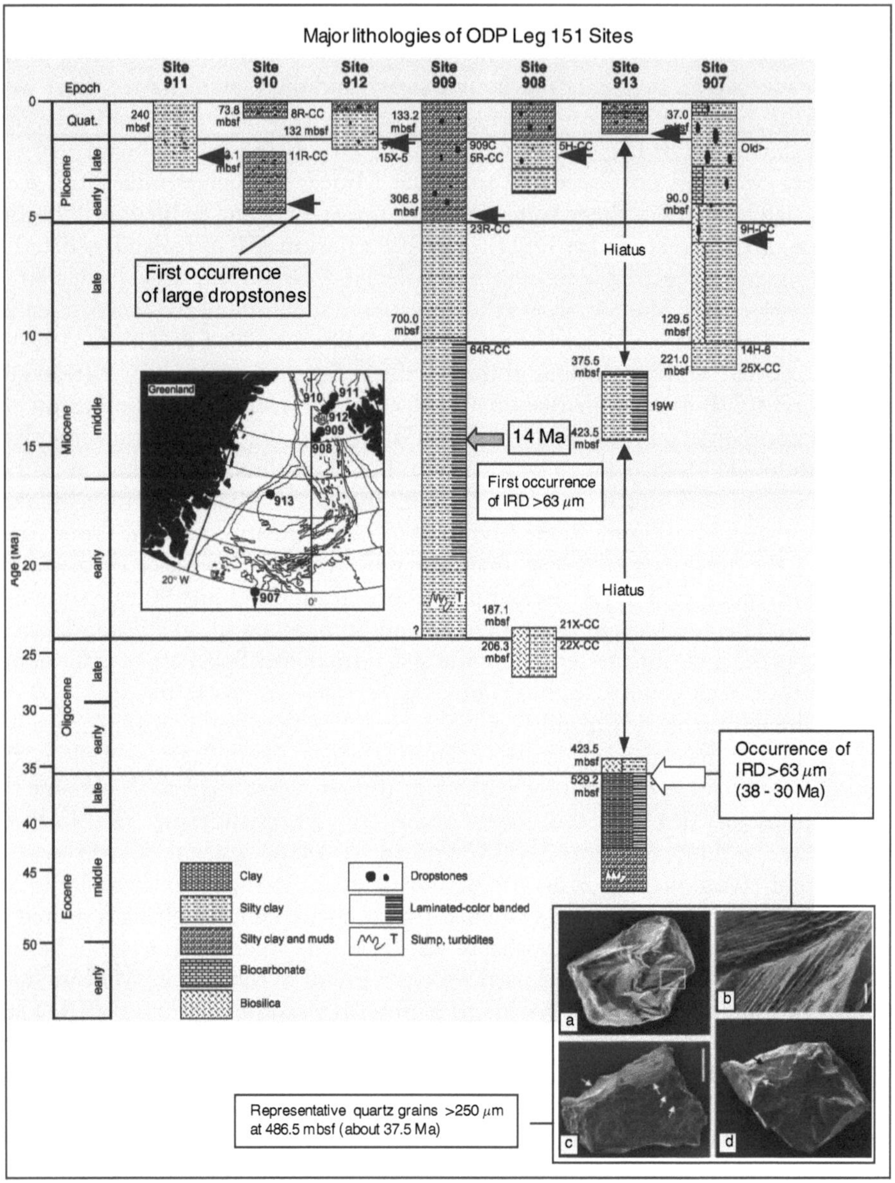

Figure 7.28 Lithostratigraphy of ODP Leg 151 sites versus age (Myhre et al., 1995; Thiede et al., 1996, supplemented). The first occurrence of large-sized dropstones is indicated. In addition, first occurrence of sand-sized terrigenous coarse fraction >63 μm at Site 909 is shown (Wolf-Welling et al., 1996; Thiede et al., 1998). A re-examination of the Oligocene/Eocene section of Site 913 gives evidence for the first occurrence of sand-sized IRD between 30 and 38 Ma. Representative scanning electron microscope images of >250 μm quartz grains from this interval, showing surface features indicative of glacial environments, are also presented (Eldrett et al., 2007). White arrows in c indicate conchoidal fractures. Scale bar in c (also valid for a and d) is 100 μm. b is a magnified view of the area inside the white box in a; scale bar in b is 10 μm.

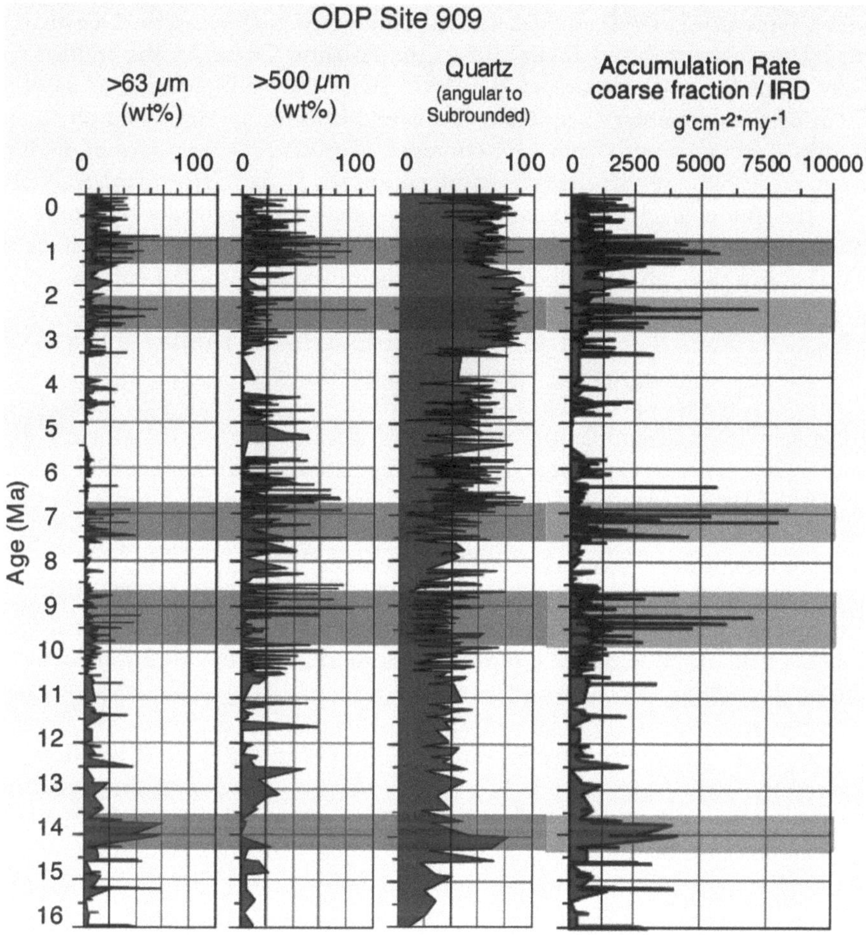

Figure 7.29 Percentage values of terrigenous coarse fractions >63 μm and >500 μm, percentage values of quartz, and accumulation rates of IRD >63 μm at Site 909 (Winkler et al., 2002). Intervals with increased input of terrigenous coarse fraction are highlighted by grey bars. Note that the IRD peaks in the >63 μm and >500 μm fractions did not often occur contemporaneously (see Chapter 4.2.2; Andrews & Principato, 2002).

Haug, 1998; Thiede et al., 1998). Furthermore, Knies et al. (2002) identified a warmer 300 kyr period with seasonally ice-free conditions in the marginal eastern Arctic Ocean in the Site 911 record, preceding the NHG intensification and coinciding with the "Mid-Pliocene (~3 Ma) Global Warmth (MPGW)" (e.g., Dowsett et al., 1992; Dowsett, Chandler, Cronin, & Dwyer, 2005; Haywood, Sellwood, & Valdes, 2000; Haywood, Dekens, Ravelo, & Williams, 2005). These ice-free conditions around Svalbard may have been an important regional moisture source and thus one decisive trigger for NHG intensification in the Svalbard/Barents Sea area near 2.7 Ma (Knies et al., 2002).

A very similar climatic record to that of Site 909 was obtained from ODP Leg 105 Site 645 recovered in Baffin Bay, a narrow intracontinental basin located between Greenland and Baffin Island and connected to the Arctic Ocean in the north by the narrow Nares Strait and to the Atlantic Ocean in the south by the shallow Davis Strait (Srivastava et al., 1987, 1989). Site 645 consists of an almost 1,150 m thick sedimentary sequence of lower Miocene to Holocene terrigenous clays, silts, and sands with high OC contents of up to 3% (Srivastava et al., 1989; Stein et al., 1989). At this site, clay-mineral and IRD data (Korstgärd & Nielsen, 1989; Thiebault et al., 1989) and accumulation rates of terrigenous OC (Stein et al., 1989) were used to reconstruct the long-term evolution of palaeoclimate in the Baffin Bay area through Miocene to Pleistocene times, with special emphasis on the High Northern Latitude cooling history (Stein, 1991a, 1991b, and further references therein). Based on these records, four major climatic intervals were distinguished, indicating a long-term cooling trend:

Interval I (early to middle Miocene, 20 to ~14.5 Ma) was characterized by relatively high accumulation rates of terrigenous OC ~0.1 gC cm^{-2} kyr^{-1} (Figure 7.30). This is explained by dense vegetation cover and fluvial sediment supply from Baffin Island and/or Greenland areas because of a more temperate palaeoclimate (Stein, 1991b). An increased fluvial supply is also supported by the large quantities of terrigenous clay (mainly smectites; Thiebault et al., 1989), at least for the upper part of this interval (Figure 7.30; Srivastava et al., 1987, 1989). The pollen and spores of Site 645 sediments indicate a climate in the surrounding continents that varied within a temperate regime during early to middle Miocene (Head et al., 1989).

During Interval II (middle Miocene, 14.5–9 Ma), accumulation rates of terrigenous OC fell significantly to <0.04 gC cm^{-2} kyr^{-1} (Figure 7.30). The stepwise decrease (with steps at 14.5, 12.5, and 9 Ma) coincided with the global mid-Miocene climatic cooling (e.g., Miller et al., 1987; Zachos et al., 2001), that resulted in the expansion of major Antarctic mid-Miocene ice caps (Kennett et al., 1975; Wright, Miller, & Fairbanks, 1992; Flower & Kennett, 1995) and the first (?) glacial deposits (tillites) on Iceland and in South Alaska ~10 Ma (Denton & Amstrong, 1969; Mudie & Helgason, 1983). Such a climatic deterioration may also have caused the widespread expansion of high-latitude steppe and tundra vegetation (Mudie & Helgason, 1983; Head et al., 1989) which could explain the decrease in terrigenous OC supply recorded at Site 645 (Stein, 1991b). The decrease in smectites which generally parallels the decrease in terrigenous OC supply (Figure 7.30) and the abrupt increase of chlorite near the end of this Interval II support this climatic change. The first occurrence of major amounts of chlorites followed by the first occurrence of dropstones in the sediments of Site 645 are strong arguments for the expansion of continental glaciation (on Greenland?) and the onset of ice rafting in Baffin Bay between 9 and 8 Ma (Figure 7.30; Korstgärd & Nielsen, 1989; Thiebault et al., 1989).

Interval III (late Miocene to early Pliocene, 9–3.4 Ma) is characterized by very low input of terrigenous OC (~0.04 gC cm^{-2} kyr^{-1}; Figure 7.30), interpreted as continued dominance of cold climatic conditions and reduced vegetation cover in the Baffin Island/Greenland area (Stein, 1991b). Isolated pebbles and granules found in sediments from this interval are indicators for the sporadic occurrence of ice-rafting in Baffin Bay during late Miocene to early Pliocene times (Figure 7.30; Korstgärd & Nielsen, 1989; Srivastava et al., 1989).

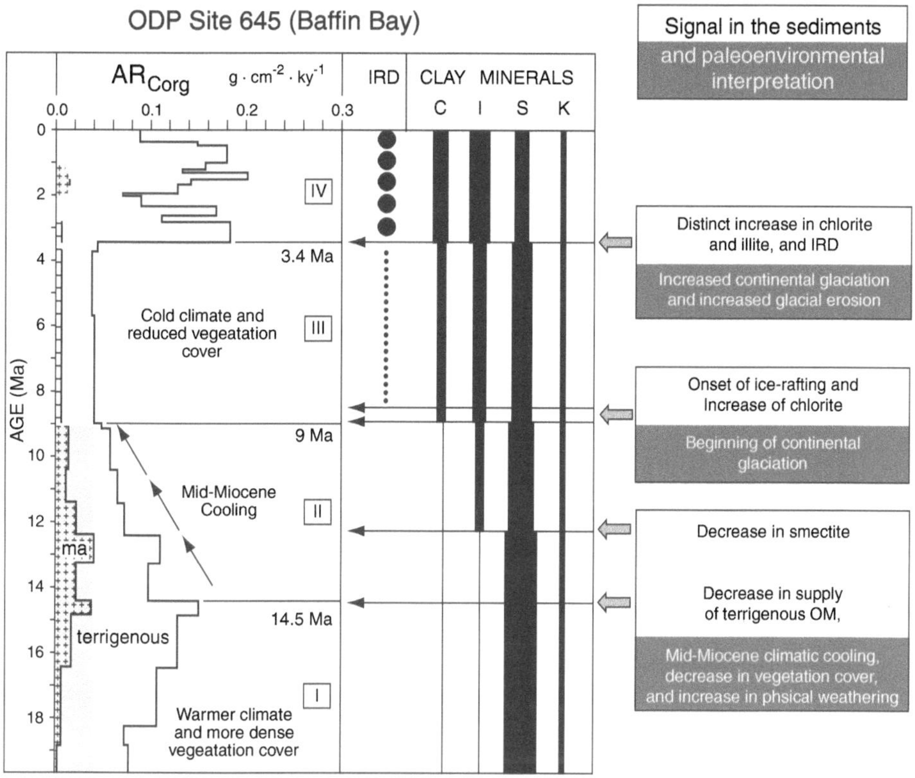

Figure 7.30 Summary plot of accumulation rates of marine and terrigenous OC, occurrence of IRD (according to Srivastava et al., 1987), and clay mineral composition (according to Thiebault et al., 1989) at Site 645, and the palaeoenvironmental interpretation (from Stein, 1991b). C, chlorite; I, illite; S, smectite; K, kaolinite; ma, marine.

Near 3.4 Ma, the supply of terrigenous OC distinctly increased by a factor of ~5 (Figure 7.30). Furthermore, Interval IV (late Pliocene to recent, 3.4–0 Ma) is characterized by high-amplitude short-term variations in climatically modulated terrigenous OC input between 0.07 and 0.20 gC cm^{-2} kyr^{-1} (Stein, 1991b). The prominent increase in supply of terrigenous OC near 3.4 Ma is paralleled by an even more distinct increase in accumulation rates of terrigenous inorganic (siliciclastic and detricarbonate) material and the distinct increase in IRD (Figure 7.30; Srivastava et al., 1987, 1989). This may suggest that glacial erosion and fluvial (meltwater) outwash of the adjacent islands caused the synchronous changes in organic and inorganic matter supply. This eroded material is partly derived from pre-Quaternary sedimentary formations cropping out around Baffin Bay, as indicated by the abundance of reworked (Cretaceous and Tertiary) palynomorphs found in Site 645 sediments (De Vernal & Mudie, 1989; Hillaire-Marcel et al., 1989).

On the basis of the sediment core data available prior to ACEX it was generally accepted that the onset of major NHG was ~14 Ma. This picture has changed drastically with the new results of the ACEX studies.

7.2.3.2. The ACEX record from Lomonosov Ridge: Evidence for early Cenozoic onset and variability of Northern Hemisphere glaciations

On the basis of the visual core description, isolated pebbles and granules interpreted as IRD or "dropstones", occur together with sand lenses throughout subunits 1/1 to 1/5, that is, the upper ~195 m of Miocene–Pleistocene siliclastic silty clays of the ACEX sequence, but unexpectedly also in ~50 m of the underlying middle Eocene biosiliceous silty clays and oozes (subunit 1/6 and top of unit 2) (Figure 7.31; Backman et al., 2005, 2006). The deepest dropstone, a gneiss of 1 cm

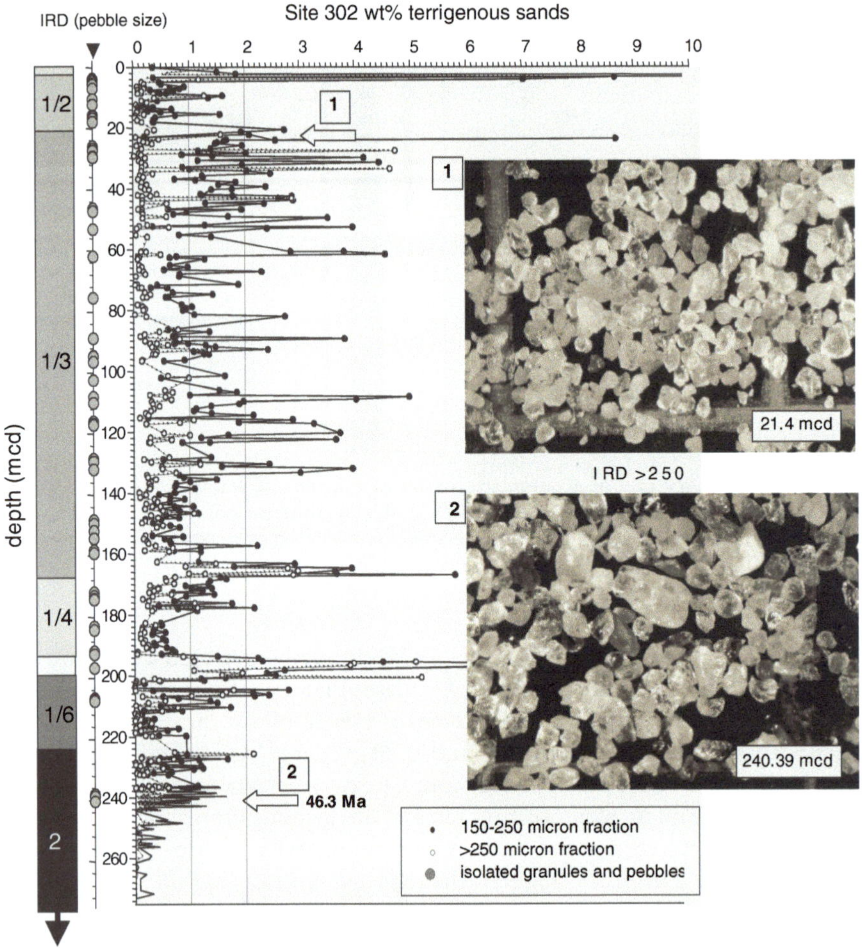

Figure 7.31 Weight percent abundance of terrigenous sands in the >250 μm (dotted line and open circles), and 150–250 μm (black line and solid circles) size fractions, and isolated granules and pebbles (large grey circles) (Backman et al., 2006) versus depth (mcd) (modified from St. John, 2008). Generalized lithologic column is shown at the right-hand side (based on Backman et al., 2006). Representative photographs from the >250 micron fraction; 1, sample 302-2A-5X-1-106 cm (21.4 mcd; 1.6 Ma); 2, sample 302-2A-55X-4–120 cm (240.39 mcd, ~46 Ma). The grid scale in each photomicrograph is 0.5 cm.

in diameter, was found in unit 2 at ~240 mcd (Sample M0002A-55X-4-122 cm). This dropstone was recovered from an undisturbed section of core and could not have been reworked or moved downward from higher in the sedimentary section. The location of the ACEX site during this time period, although probably in shallow water (~200 m), was distal from the Siberian continental coast and isolated from it by the Gakkel Ridge, suggesting a delivery of the dropstone by ice as most probable process (Moran et al., 2006). Thus, this dropstone was interpreted as a hint for an onset of NHG as early as ~46 Ma, that is, sea-ice formation and/or iceberg transport took place ~30 my earlier than previously thought (Moran et al., 2006, using the revised stratigraphic age model of Backman et al., 2008).

The ACEX dropstones, while strong direct evidence of the presence of ice, are stratigraphically discontinuous, and only ~60 granules and pebbles of ~0.2–3.0 cm in diameter were visually identified between 0 and 240 mcd (Backman et al., 2005, 2006). In order to get a more continuous record of the ice-rafting history of the central Arctic Ocean, St. John (2008) studied the terrigenous coarse sand (IRD) fraction throughout the entire depth interval 0–240 mcd of the ACEX sequence in much more detail (Figure 7.31). Along with information on grain size, composition, and mass accumulation rates of IRD, SEM imaging of representative quartz grains was used to distinguish between sea-ice and iceberg transport. As outlined by St. John (2008), surface features of iceberg-transported grains are dominated by those produced by mechanical breakage (e.g., angular edges, high relief, and step-fractures), whereas surface features of sea-ice transported grains show more rounded edges and chemical features, such as silica-dissolution and precipitation (Krinsley & Doornkamp, 1973; Helland & Holmes, 1997; Dunhill, 1998).

The results from St. John's (2008) study of IRD percentages (Figure 7.31) and derived IRD mass accumulation rates (Figure 7.32) confirm the pebble-based interpretation made by the IODP 302 Scientific Party (Backman et al., 2006; Moran et al., 2006) that ice initiated in the Arctic in the middle Eocene near 46.3 Ma. At that time, IRD 150–250 μm percentages and accumulation rates reached values >1% and 0.02 g $cm^{-2} kyr^{-1}$, respectively. Qualitatively, the IRD coarse fraction >250 μm mainly composed of quartz, looks very similar in samples from the middle Eocene and from the Pleistocene (Figure 7.31).

Contemporaneously with the onset of IRD near 46.3 Ma, both the abundance of needle-like diatoms interpreted as indicator for diatom concentration in stratified waters (Stickley et al., 2008; see Kemp et al., 2006, and references therein), and salinity reconstructed from oxygen isotope analysis of fish bone carbonate (Waddell & Moore, 2008) increased. These sedimentological, micropalaeontological, and geochemical data are interpreted by St. John (2008) as strong indication for sea-ice formation in the middle Eocene. That means, Arctic sea ice could have transported the debris, helped to stratify the underlying surface waters, and increased their salinity. Iceberg transport, however, was probably also already present in the middle Eocene, as indicated by mechanical surface-texture features on quartz grains from this interval (St. John, 2008). A further three-steps increase in IRD input occurred near 44.9, 44.7–44.8, and 44.5 Ma (Figure 7.32).

With the first occurrence of significant amount of IRD near 46.3 Ma the alkenone SST in the ACEX record dropped by ~7.5°C, and temperatures of

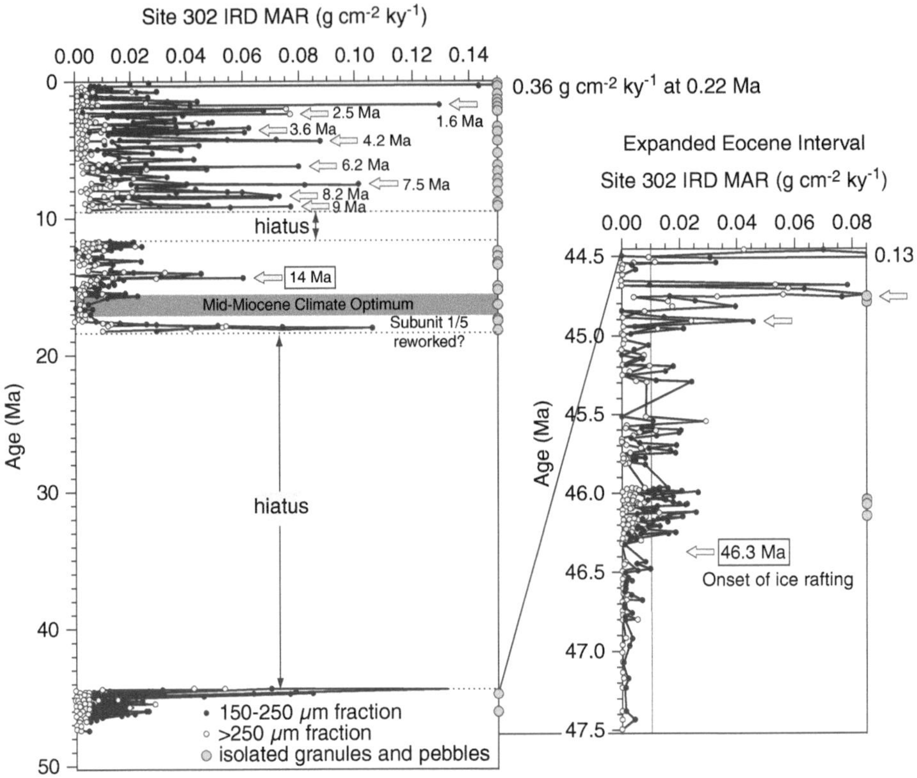

Figure 7.32 IRD mass accumulation rates ($g\,cm^{-2}\,kyr^{-1}$) in the $>250\,\mu m$ (dotted line and open circles) and 150–250 μm (black line and solid circles) size fractions (St. John, 2008, supplemented), along with isolated granules and pebbles (large grey circles) (from Backman et al., 2006) versus age (Ma). Open arrows indicate major pulses of IRD input. On the right is an enlargement of the middle Eocene interval (44.5–47.5 Ma) of this dataset. The mid-Miocene Climate Optimum (e.g., Flower & Kennet, 1995; Zachos et al., 2001) is marked as grey bar.

10–17°C were determined for the time interval 46.3–44.8 Ma (Figure 7.20; Weller & Stein, 2008). Such SSTs seem to be not unrealistic. Assuming that the alkenone SST represents rather the summer SST and considering the strong seasonal temperature variability of >10°C during the early middle Eocene (see Section 7.2.2), favourable conditions for sea-ice formation may have occurred during winter time. This could have been a situation similar to that observed in the modern Baltic Sea where summer SSTs of >15°C and winter SSTs <1°C with sea-ice formation are typical (Wüst & Brogmus, 1955; Krause, 1969).

Following the discovery of Eocene dropstones in the ACEX record (Backman et al., 2005, 2006; Moran et al., 2006), Eldrett et al. (2007) and Tripati et al. (2008) re-examined the late Eocene–Oligocene sediment record from ODP Site 913 in the Greenland Basin (Figure 7.28). At this site, initial shipboard investigations identified an isolated, macroscopic lithic clast at 453.16 mbsf but concluded that it was the result of down-hole contamination from the overlying Miocene glacial sequence, rather than an *in situ* glacially derived clast (Myhre et al., 1995). Based on microscopic examination of the surface features of clasts (Figure 7.28) and detailed core observations giving evidence for impact-induced deformation of laminae underlying clasts, however, Eldrett et al. (2007) re-interpret the clasts as being *in situ* and of dropstone origin. Together with data on mass accumulation rates of grain-size fractions >63, >125, and >250 μm as well as data on the composition of the coarse fractions, these authors report an interval of extensive but highly variable ice rafting in the Greenland Sea between 30 and 38 Ma, related to iceberg transport (rather than sea ice) with East Greenland as the likely source. Tripati et al. (2008) extended the occurrence of IRD at ODP Site 913 even to 44 Ma. Furthermore, they determined distinct peaks in IRD at 42–40 Ma, 39–37 Ma, and 34–30 Ma, interpreted as phases of extended glaciation on Greenland. These results indicate the existence of (at least) isolated glaciers on Greenland at times close to the onset of major glaciations in Antarctica (Figure 7.33).

The records from ACEX (Backman et al., 2006; Moran et al., 2006; St. John, 2008) and ODP Site 913 (Eldrett et al., 2007; Tripati et al., 2008) prove an early

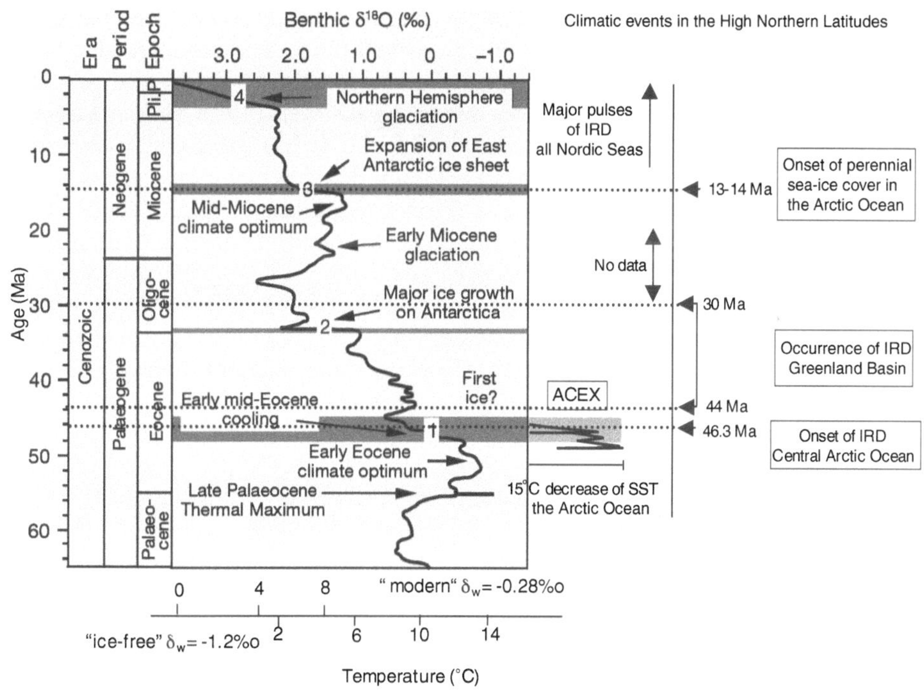

Figure 7.33 A smoothed global benthic foraminifer $\delta^{18}O$ time series of the Ceneozoic time interval (Pearson & Palmer, 2000, based on Shackleton, 1986; Miller et al., 1987; Zachos et al., 1994). Four major steps (numbered 1–4) in $\delta^{18}O$ and related events of climate change are indicated. The trend towards more positive $\delta^{18}O$ results from a combination of deep-sea cooling and global ice volume increases (Pearson & Palmer, 2000). In addition, major events of IRD input related to NHGs (see text for details and references) and major drop in alkenone SST ~46.3 Ma coinciding with the onset of IRD in the ACEX record (Weller & Stein, 2008), are shown.

onset/intensification of NHGs during Eocene times, as already proposed from changes in oxygen–isotope composition across the Eocene/Oligocene boundary and in the late Eocene in records from the tropical Pacific and South Atlantic (Coxall et al., 2005; Tripati et al., 2005). Furthermore, the increases in IRD in the ACEX and ODP Site 913 records coincided with major decreases in atmospheric CO_2 concentrations (Figure 7.13; Pearson & Palmer, 2000; Pagani et al., 2006; Lowenstein & Demicco, 2006). These data suggest that the Arctic and Antarctic Cenozoic climate evolutions are more closely timed (Figure 7.33), that is, the Earth's transition from the "greenhouse" to the "icehouse" world was bipolar, which points to greater control of global cooling linked to changes in greenhouse gases in contrast to tectonic forcing (Moran et al., 2006). The decline of atmospheric concentrations of CO_2 in the middle Eocene may have driven both poles across the temperature threshold that enabled the nucleation of glaciers on land and partial freezing of the surface Arctic Ocean, especially during times of low insolation (St. John, 2008).

The ACEX record also gives information about the variability of Northern Hemisphere climate during Neogene times. Directly above the main hiatus recorded at 198.7 mcd in the ACEX sequence, mass accumulation rates reached very high values between 17.5 and 18.2 Ma (Figure 7.32). However, because this is a condensed sediment interval containing reworked Oligocene dinocysts (Backman et al., 2008), it is possible that the terrigenous coarse fraction is also reworked and thus does not represent true input from overlying ice melt in the middle Miocene surface waters (St. John, 2008). Between ~17.5 and 16 Ma a distinct minimum in ice-rafting in the central Arctic was found (Figure 7.32), which may correspond to the late mid-Miocene climate optimum (Flower & Kennet, 1995; Zachos et al., 2001; Moran et al., 2006), suggesting minimal ice conditions in the Arctic during this period (St. John, 2008). Between ~15 and 14 Ma, IRD accumulation in the ACEX record distinctly increased, contemporaneously with similar IRD maxima in the Greenland Sea (Figure 7.29; Wolf-Welling et al., 1996) and the onset of cooling in Baffin Bay (Figure 7.30; Stein, 1991b). This may suggest a shift to larger-scale or permanent ice in the Northern Hemisphere high latitudes, which is consistent in timing to the establishment of more extensive ice sheets in Antarctica (Figure 7.33) and greater Antarctic Bottom Water formation (Wright et al., 1992; Flower & Kennet, 1995), and a corresponding eustatic sea level regression (Miller et al., 1998). Most of the late Cenozoic maxima in IRD mass accumulation rates found in the ACEX record and interpreted as evidence of Arctic ice (i.e., sea ice and/or icebergs) intensification (Figure 7.32; St. John, 2008), co-occurred with either initial or intensified ice-rafting events at Subarctic sites (Figures 7.29 and 7.30; Wolf & Thiede, 1991; Stein, 1991b; Fronval & Jansen, 1996; Wolf-Welling et al., 1996; Thiede et al., 1998; St. John & Krissek, 2002). That means, St. John's (2008) data provide a long-term pattern of Arctic ice expansion and decay, on the order of those determined for the Subarctic oceans.

Concerning the IRD composition of the ACEX sequence, there seems to be a major break at ~13–14 Ma, when looking at the heavy-mineral assemblages (Figure 7.34, Table 7.4; Krylov et al., 2008a). The low-resolution record shows high clinopyroxene and low hornblende (amphibole) values and resulting high Cpx/Hbl

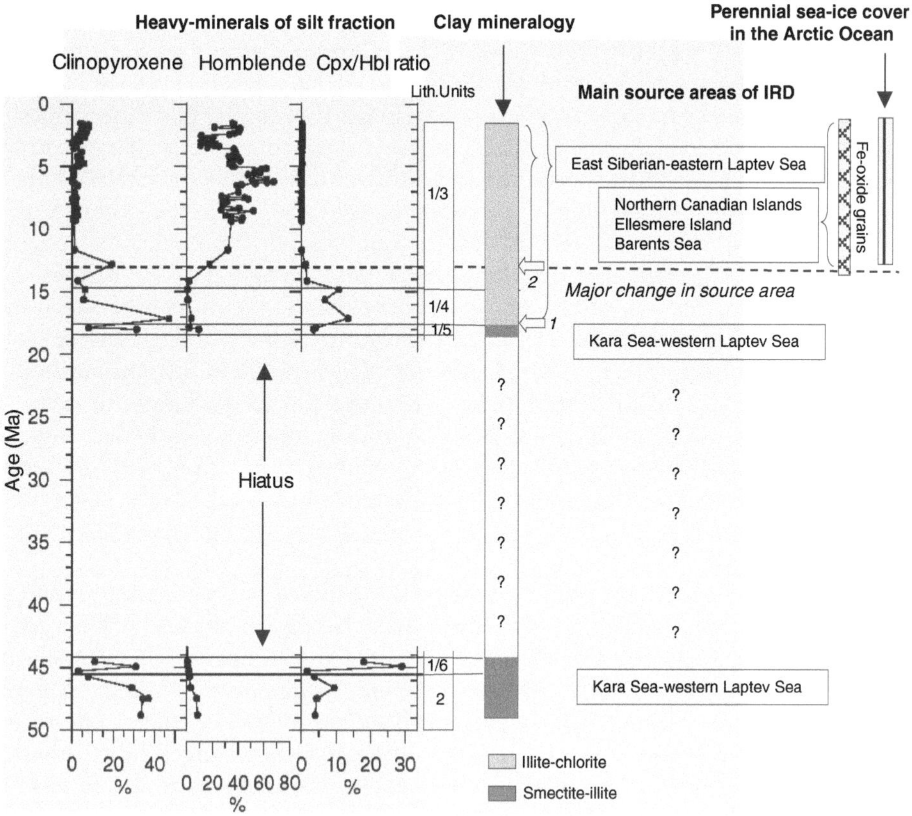

Figure 7.34 Percentage values of clinopyroxene and hornblende (amphibole) in the heavy mineral silt fraction as well as the clinopyroxene/hornblende (cpx/hbl) ratio, determined in the ACEX sequence (Krylov et al., 2008a), and interpretation in terms of source areas of the terrigenous (IRD) fraction (see Chapter 5.1.4 for background information) and sea-ice cover. In addition, source-area identification based on Fe-oxide grains (see Chapter 4.3.1 for background information) (Darby, 2008) and clay–mineral assemblages (Krylov et al., 2008a) are shown at the right-hand side.

Table 7.4 Mean Values of Relative Abundance of Heavy Minerals (Clinopyroxene and Hornblende/Amphibole in Percentage of Total Heavy Minerals of the 50–100 μm Fraction) and Clay Minerals (in Percentage of Sum of Illite, Chlorite, Kaolinite, and Smectite in the <2 μm Fraction) for Units 1/3, 1/4, 1/5, 1/6, and 2 of the ACEX Sequence (from Krylov et al., 2008a).

	Samples	Clinopyroxene	Hornblende	CPX/Hbl	Illite	Chlorite	Kaolinite	Smectite
Unit 1/3	68	2.6	34.0	0.1	58	20	14	8
Unit 1/4	3	19.6	1.6	10.3	61	17	18	4
Unit 1/5	2	19.5	5.5	3.8	38	8	19	35
Unit 1/6	3	15.0	1.2	16.3	54	7	16	23
Unit 2	5	28.3	5.9	5.3	30	6	13	52

ratios below 14 Ma, and low clinopyroxene and high hornblende (amphibole) values and resulting low Cpx/Hbl ratios above 14 Ma. Considering that sea ice is the main transport agent for the heavy minerals, this points to the western Laptev/Kara Sea as major IRD source prior to ~14 Ma and the eastern Laptev Sea/East Siberian Sea as major IRD source area after 14 Ma (see Chapter 5.1 for source-area characteristics), assuming a Transpolar Drift system similar to that of today (St. John, 2008). Due to the fact that the distance between the amphibole source region in the eastern Laptev Sea/East Siberian Sea and the ACEX drill site requires a drift time that exceeds one year if assuming present-day drift trajectories and velocities (see Chapter 2.3, Figure 2.14), Krylov et al. (2008a) proposed that the sea ice must have survived a melting season. This strongly points to the development of a perennial sea-ice cover in the Arctic Ocean at ~13–14 Ma. Darby (2008) who studied the composition of Fe oxide grains in the ACEX section (see Chapter 4.3.1 for background), came to the conclusion that during the last 14 Ma a significant amount of IRD derived from the northern Canadian islands and Ellesmere Island (Figure 7.34), with additional sources in the Eurasian Arctic, and also proposed a perennial sea-ice cover since ~14 Ma.

The dominance of smectite in the lower part of the ACEX record (Figure 7.34, Table 7.4) supports the heavy-mineral data, that is, indicates a terrigenous sediment input from the western Laptev/Kara Sea. The changes in the clay- and heavy-mineral assemblages, however, do not occur exactly in phase. The decrease in smectite (and increase in illite) already occurred within unit 1/5, that is, a few million years earlier (Figure 7.34). As outlined by Krylov et al. (2008a), the reason for this discrepancy could be related to climate-driven changes in weathering conditions in the source regions, that is important for formation of clay in soils, or to different mechanisms of clay- and heavy-minerals transportation.

Because of the strong short-term variability of IRD sources known from studies of Quaternary Arctic sediments (see Chapter 6.3, Figure 6.57), however, it is difficult to interpret these low-resolution records of IRD provenance available from the ACEX sequence so far, in more detail. Here, more high-resolution studies (on the ACEX material) are needed to get information about the variability in sea-ice cover during Neogene times.

CHAPTER EIGHT

Open Questions and Future Geoscientific Arctic Ocean Research

Although major progress in Arctic Ocean research has been made during the past decades (see previous chapters), the knowledge of its short- and long-term palaeoceanographic and palaeoclimatic history as well as its plate-tectonic evolution is much behind that from the other world's oceans. That means — despite the importance of the Arctic in the climate system (see Chapter 1.1) — the data base we have from this area is still very weak, and large parts of the climate history have not been recovered at all in sedimentary sections (Figure 8.1). Furthermore, there is an uneven data base for the various marginal seas and the central Arctic Ocean. Much more data, for example, are available from the Beaufort, Barents, Laptev, and Kara seas than for the East Siberian Sea and the central Arctic Ocean. A larger number of well-dated (AMS^{14}C) sediment cores are available only from the Barents, Laptev, and Kara seas. For the heavily ice-covered shelf area of the Canadian Archipelago as well as large parts of the central Arctic, no data at all are available (see Chapter 1.2, Figure 1.5). This lack of knowledge is mainly caused by the major technological/logistic problems in reaching this permanently ice-covered region with normal research vessels and in retrieving long and undisturbed sediment cores.

Concerning future Arctic geoscientific research, the climate history must be studied at different temporal scales from 10 to 10^6 years, as palaeoclimate research and climate models demonstrate that processes and varying conditions in polar regions play a key role in driving and amplifying global climate variability and sea-level change on time scales of decades to millions of years. Furthermore, to reconstruct past climate variability and the interrelation of different processes for times prior to direct measurements, one has to rely on indirect evidence, that is, on information provided by "proxies" determined in sediment (and ice) cores. These proxies may be used for reconstruction of past temperature, salinity, sea ice cover, global ice volume, nutrients, marine biological productivity, etc. (Fischer & Wefer, 1999; Weinelt et al., 2001; Hillaire-Marcel & de Vernal, 2007 for most recent overview). Whereas a large number of these proxies are routinely used, others are new and still have to be fully established in Arctic Ocean environmental research (see Chapter 4). In this context, one key issue of overall importance in Arctic research is the establishment of a stratigraphic/chronological framework at highest accuracy possible, to be achieved with a combined effort of different stratigraphic methods including radiometric dating, microfossils, stable isotopes, cosmogenic isotopes, and geomagnetic records. Such a framework is the prerequisite for all palaeoenvironmental studies and for the correlation of Arctic sedimentary records with other sedimentary records from the low, mid, and southern high latitudes, as well as climate records from land and ice cores.

In the following, some of the key topics of future research are presented. The topics mentioned here are certainly not complete and have to be seen as examples.

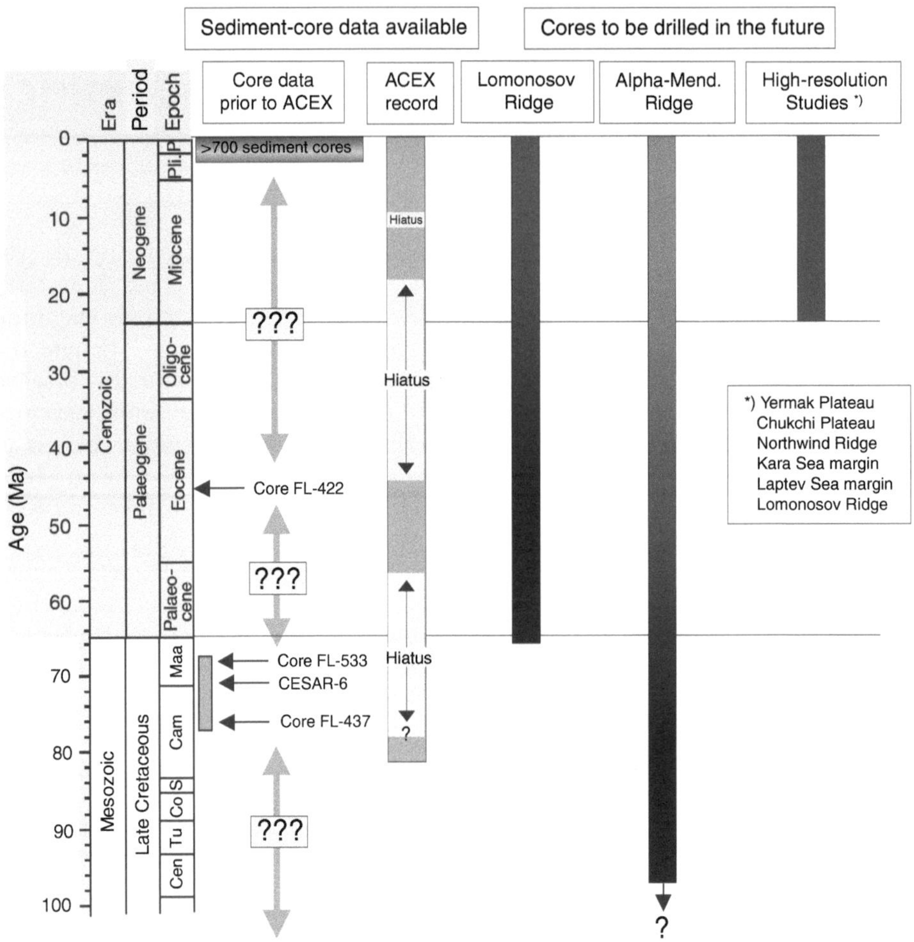

Figure 8.1 Stratigraphic coverage of existing cores in the central Arctic Ocean prior to ACEX (based on Thiede et al., 1990) and the section recovered during the ACEX drilling expedition (Backman et al., 2006; Backman et al., 2008), and stratigraphic coverage and key locations of sites to be drilled in the future.

Furthermore, in this short outline it has been concentrated especially on key questions related to the long-term evolution of the Arctic Ocean where our lack in knowledge is most extensive.

8.1. Quaternary and Neogene Climate Variability on Sub-Millennial to Milankovich Time Scales

Arctic high-resolution (sub-millennial) climate records are still rare and known from only a few locations in the Nordic Seas. Risebrobakken, Jansen, Andersson,

Melde, and Hevrøy (2003) and Sarnthein et al. (2003c), for example, analysed in decadal resolution the abundances and isotope values of foraminifers in sediment cores from the Norwegian continental margin and the western Barents Sea, respectively. These authors found evidence for a strong variability of northward heat advection, but statistical evaluations gave different results concerning dominant frequencies in the variability. Darby and Bischof (2004) presented a Holocene record of sub-millennial changes in sea-ice drift and its relationship to the Arctic Oscillation (see Chapter 6.3.6). Some high-resolution records are also known from the Kara Sea indicating Holocene centennial- to millennial-scale variability of Siberian river discharge (Stein et al., 2003a, 2004a; see Chapter 4.3.5, Figure 4.23). This short-term variability in Siberian climate and river discharge seems to correlate with the Greenland ice core record, changes in sea-ice cover in the North Atlantic and advance/retreat of Norwegian glaciers, and may reflect natural cyclic climate variations related to changes of the NAO/AO pattern as described for the past decades (see Chapter 1.1). Considering the importance of the Arctic Ocean for the global ocean circulation and the climate system, more high-resolution palaeoceanographic data sets from key areas are needed to reconstruct the climatically important parameters (e.g., sea-surface temperature and salinity, inflow of Atlantic Water, sea-ice cover, river discharge). This kind of data sets going back beyond the time scale of direct discharge measurements can be used to determine the natural variability of these parameters as a background for an assessment of anthropogenically influenced changes in the past 100–150 years.

In this context, the strategic approach is to merge atmosphere and ocean signals from ice, marine, and land (lake/permafrost) climate records at orbital, millennial, and up to decadal time scales to decipher their complex inter-relationship and impact, their relation with processes in low and middle latitudes, and the effect of external forcing. Of overall interest is the understanding of the polar mechanisms and thresholds in triggering rapid (10^2–10^3 yr) climate changes during warm and cold climate conditions (ice–permafrost–ocean–atmosphere mechanisms) and the comparison of the spatial and temporal evolution of such changes in Northern and Southern High Latitudes. How does the teleconnection between climate change on the Southern and Northern Hemisphere ("bipolar seesaw") as proposed for the last deglaciation in the Atlantic Ocean (Knorr & Lohmann, 2003; Stocker, 2003) function on different time scales?

Key areas for high-resolution records are the Fram Strait/Yermak Plateau area (sites to study the variability of North Atlantic water inflow, sea-ice cover, primary production, etc.), the continental margins surrounding the Arctic Ocean (sites to study the variability of circum-Arctic river discharge, marginal-ice-zone processes, Atlantic- and Pacific-water influence, etc.), and the central Arctic Ocean (sites to study the variability of sea-ice cover and drift, water-mass characteristics, primary production, etc.). Coring sites characterized by high sedimentation rates, have to be carefully selected by bathymetric (Hydrosweep) and acoustic/seismic (PARASOUND and multi-channel seismic) surveys. For studying the Holocene climate variability, cores may be retrieved by conventional piston and kastenlot coring. For studying the (sub-) millennial climate variability on longer time scales, that is, under different boundary conditions during Quaternary/Neogene times, drill cores are needed (Figure 8.1; see below for some more details).

In addition to and, especially, in combination with high-resolution records from marine sediments cores as well as lake and ice core archives, examination of potential impacts of climate changes in the Arctic region has to be done with the aid of numerical models, using global coupled circulation models, regional models for the Arctic realm, as well as data–model comparison (e.g., Prange & Lohmann, 2003; Lorenz & Lohmann, 2004; Steiner et al., 2004; Lohmann, Lorenz, & Prange, 2005).

8.2. The Mesozoic–Cenozoic History of the Arctic Ocean

Despite the success of IODP Expedition 302 — ACEX — (see Chapter 7.2), major key questions related to the climate history of the Arctic Ocean and its long- and short-term variability during Mesozoic–Cenozoic times, cannot be answered from the ACEX record due to the poor core recovery and, especially, a major mid-Cenozoic hiatus (Figure 8.1). This hiatus just spans the critical time when prominent changes in global climate took place during the transition from the early Cenozoic Greenhouse world to the late Cenozoic Icehouse world (Figure 8.2; e.g., Miller et al., 1987, 1991; Lear et al., 2000; Pearson & Palmer, 2000; Zachos et al., 2001). The success of ACEX has certainly opened the door for further scientific drilling in the Arctic Ocean. The ACEX results will frame the next round of questions to be answered from new drill holes to be taken during a series of drilling legs. Some of the main scientific topics are listed below.

(1) The Cenozoic transition of the Earth's climate from one extreme (Palaeogene greenhouse lacking ice) to another (Neogene icehouse with bipolar glaciation characterized by an Antarctic continental ice-cap and seasonally variable but persistent sea-ice cover in the Arctic) is linked to increased latitudinal gradients and oceanographic changes that connected surface and deep-sea circulation between high and low latitude oceans. The general Cenozoic cooling trend, however, is interrupted by warming intervals as well as short-term extreme cooling transients, such as the late Oligocene warming and the mid-Miocene Climatic Optimum, and the Oi-1 Glaciation and the Mi-1 Glaciation (Figure 8.2; Miller et al., 1987, 1991; Lear et al., 2000; Zachos et al., 2001; Coxall et al., 2005; Tripati et al., 2005). Some of the related key questions are: Did the Arctic Ocean climate follow the global trend shown in Figure 8.2? Are the Early Eocene Climate Optimum (poor recovery in the ACEX record) and the Oligocene and mid-Miocene warmings also reflected in Arctic Ocean records? Were extensive glaciations developed synchronously in both the Northern and Southern Hemispheres? Are the Oi-1 and Mi-1 glaciations bipolar, that is, are there indications for major Northern Hemisphere glaciations at that

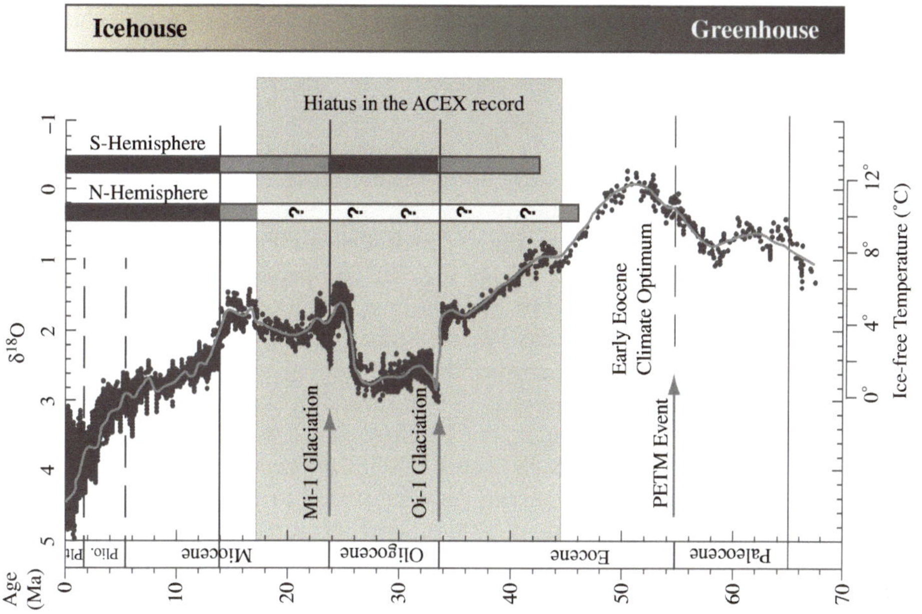

Figure 8.2 A smoothed global benthic foraminifer $\delta^{18}O$ time series showing the long-term cooling and the Greenhouse/Icehouse transition through Cenozoic times (Zachos et al., 2001, supplemented). The occurrence of Cenozoic glaciations on the Northern and Southern Hemisphere are shown as black (strong glaciations) and dark grey (less intense glaciations) vertical bars, major glaciations Oi-1 and Mi-1 are indicated (based on Miller et al., 1987, 1991; Lear et al., 2000; Zachos et al., 2001; Backman et al., 2006). The grey interval represents the hiatus in the ACEX record. For further explanations and references see text.

time? What was the variability of sea ice in terms of frequency, extent, and magnitude?

(2) Black biosiliceous silty clays and clayey silts rich in organic carbon were found throughout the upper early to middle Eocene of the ACEX record (Chapter 7.2.2; Backman et al., 2006; Moran et al., 2006; Stein et al., 2006a), but also in a short sediment core from the Alpha Ridge, representing the late Cretaceous (Chapter 7.1.3; Clark et al., 1986). These data indicate a palaeoenvironment characterized by poorly ventilated bottom waters and elevated but variable primary production. When and how did these extreme conditions develop during Mesozoic and Cenozoic times? When and how did the change to oxygenated bottom waters typical for the Neogene and Quaternary Arctic Ocean occur? Did anoxia occur down to the deep basins?

(3) Drill sites more proximal to the Siberian margin allow a detailed study of the history of river discharge and its palaeoenvironmental significance. In this context, the Miocene uplift of the Himalayan-Tibetan region is of particular interest as it may have triggered enhanced flow of Siberian rivers and changed the fresh-water balance of the Arctic's surface waters, considered to be

a key factor for the formation of Arctic sea-ice and onset of major glaciations (Driscoll & Haug, 1998), a hypothesis to be tested by drilling along the Kara and Laptev Sea continental margin influenced by discharge of the large rivers Lena, Ob, and Yenisei.

(4) How critical is the exchange of water masses between the Arctic Ocean and the Atlantic and Pacific oceans for the long-term climate evolution as well as rapid climate change? High-resolution records of Neogene/Quaternary Arctic Climate to be obtained from the Yermak Plateau/Fram Strait area and the Chukchi Plateau/Northwind Ridge area in comparison with similar records from the North Atlantic (IODP Expeditions 303/306; Channell et al., 2006a, 2006b; Stein et al., 2006b) and Bering Strait (Takahashi, Ravelo, & Alvarez-Zarikian, 2007) may help to answer this question.

(5) During the Pliocene warm period, SST in several ocean basins was substantially warmer (Marlow, Lange, Wefer, & Rosell-Melé, 2000; Dowsett et al., 2005; Haywood, Dekens, Ravelo, & Williams, 2005; Lawrence, Liu, & Herbert, 2006) and global mean surface temperature was estimated to be at least ~3°C higher than today (Haywood & Valdes, 2004). Furthermore, cooling in the surface ocean seems to have started at least 1 myr before the intensification of the NHG as shown for example for the Eastern Equatorial Pacific (EEP), implying that while the growth of Northern Hemisphere ice sheets undoubtedly played a major role as a climatic feedback during the Plio-Pleistocene transition, it did not force or initiate EEP cooling (Lawrence et al., 2006). How did the Arctic Ocean evolve during the Pliocene warm period and succeeding cooling? Based on limited ostracode data from the Canadian Arctic, for example, there seems to be some indication for elevated Pliocene temperatures in the Arctic Ocean (Cronin et al., 1993).

(6) What is the cause of the major hiatus recovered in the ACEX record? Is it related to the subsidence history of Lomonosov Ridge? Was there a phase of uplift and exposure of the ridge in the Oligocene, tentatively linked to a transpressional/compressional episode in the formation of the Amundsen Basin caused in part by the northward motion of Greenland in the Palaeogene (Brozena et al., 2003)? Was the hiatus a response to increased ventilation during the opening of surface- and deep-water connections via the Fram Strait (Moore & the Expedition 302 Scientists, 2006)? Recovery of the missing mid-Cenozoic interval as well as more complete basement recovery from additional drill sites can be used to test these hypotheses.

In addition to these objectives related to the palaeoceanographic and palaeoclimatic history, the tectonic evolution will be a major focus of future research. Tectonic objectives, for example, may focus on the evolution of the Alpha Ridge and its consequences on the palaeooceanography in the Mesozoic Arctic. In the case the Alpha-Mendeleev Ridge is a large igneous province, its formation might have caused profound disturbances in the Arctic climate.

In order to study the long-term Mesozoic–Cenozoic climate evolution, we need to obtain undisturbed and complete sedimentary sequences to be drilled on depth transects across the major ocean ridge systems, that is, the Lomonosov Ridge, the

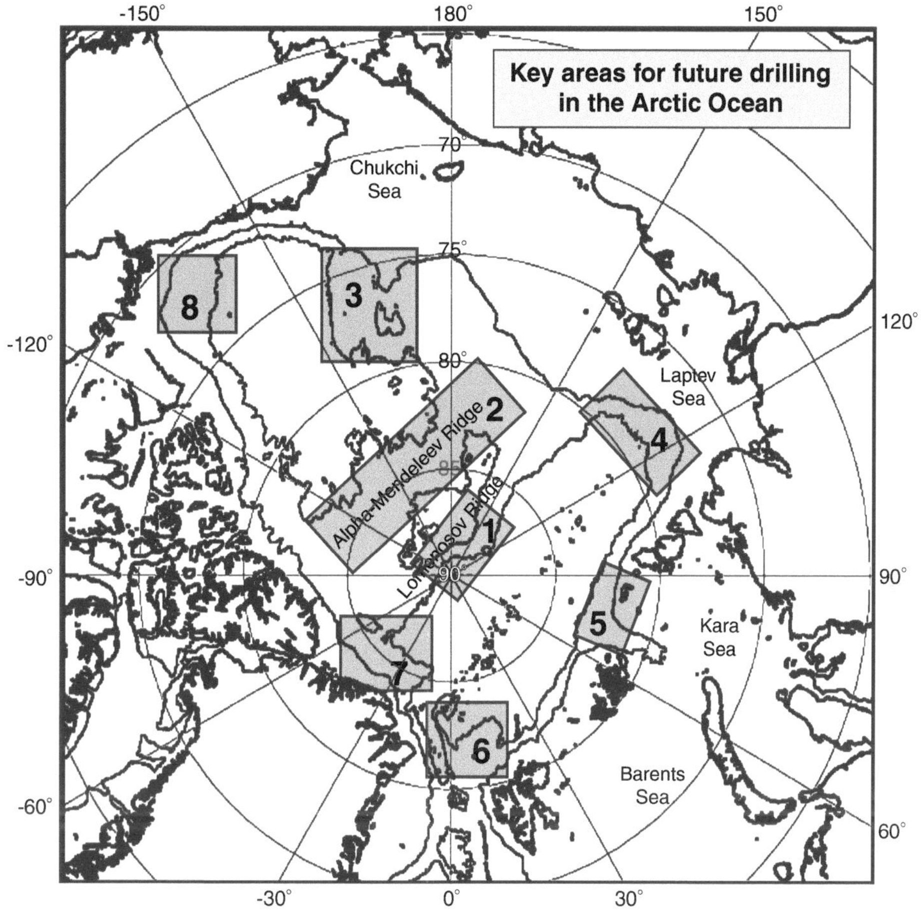

Figure 8.3 Key areas for future drilling areas in the Arctic Ocean. 1, Lomonosov Ridge; 2, Alpha-Mendeleev Ridge; 3, Chukchi Plateau/Northwind Ridge; 4, Laptev Sea continental margin; 5, Kara Sea continental margin; 6, Fram Strait/Yermak Plateau; 7, Northern Greenland continental margin/Morris Jesup Rise; 8, Beaufort Sea (Mackenzie) continental margin. For further explanations see text.

Alpha-Mendeleev Ridge, the Chukchi Plateau/Northwind Ridge, and Morris Jesup Rise (Figures 8.1 and 8.3, key areas 1–3 and 7). For studying the earliest Arctic Ocean climate history, the Alpha Ridge where Cretaceous sediments are outcropping (see Chapter 7.1.3, Figure 7.11), is of special interest. High-resolution records allowing to study climate variability on Milankovich and millennial to sub-millennial time scales can be drilled along the continental margins characterized by high sedimentation rates. Here, key areas are the Beaufort, Kara and Laptev seas characterized by their large river discharge (Figure 8.3, key areas 4, 5 and 8). Key location for studying the history of exchange of the Arctic Ocean with the world's oceans are the Fram Strait/Yermak Plateau and Chukchi Plateau areas (Figure 8.3, areas 3 and 6).

For the precise planning of future drilling campaigns including site selection, evaluation of proposed drill sites for safety and environmental protection aspects, etc., comprehensive site survey data are needed. For some of the potential study areas, the site survey data base is already quite good. For example from the Lomonosov Ridge, a large number of deep-penetration reflection seismic profiles were acquired on icebreaker-based expeditions in 1991, 1996, 1998, and 2005 (Jokat et al., 1992, 1995, 1999; Kristoffersen, Buravtsev, Jokat, & Poselov, 1997; Darby et al., 2005; Jokat, 2003), an intensive PARASOUND survey (in combination with coring) was carried out in 1995 and 1998 (Rachor, 1997; Jokat et al., 1999), and the first high-resolution chirp profiles were collected in 1996 (Jakobsson, 1999). In 1999, the SCICEX program collected high-resolution chirp sub-bottom profiler data, swath bathymetry and side-scan sonar backscatter data on Lomonosov Ridge from an US nuclear submarine (Edwards & Coakley, 2003), contributing significantly to the much improved bathymetric chart of the Arctic Ocean (Jakobsson, 2002; Jakobsson et al., 2004). During the 1995, 1996, and 1998 expeditions, a large number of sediment cores were taken by piston, gravity, and kastenlot coring in the Lomonosov Ridge area (Backman et al., 1997; Rachor, 1997; Jokat et al., 1999; Stein et al., 2001). That means, in combination with the results from the ACEX drilling campaign (Backman et al., 2006), future drill areas/sites on Lomonosov Ridge can be identified more accurately (Figure 8.4). In other key areas for future drilling, for example, the Alpha-Mendeleev Ridge, on the other hand, site survey expeditions still have to be carried out before a detailed planning and drill site selection can start.

Concerning the short- and long-term evolution of the Arctic Ocean and its importance for the understanding of the global climate history, most of the key questions mentioned above as well as the key areas for scientific drilling in the Arctic Ocean needed to answer these questions, were already identified on several workshops during the past two decades and published in outcoming reports. Here, especially the Science Plan (Thiede and the NAD Science Committee, 1992) and the Implementation Plan of the "Nansen Arctic Drilling Program" (NAD, 1997) as well as the final report of the "Arctic's Role in Global Climate Change Program Planning Group (APPG)" (Hovland, 2001) and the most recent report of the "2nd International Conference on Arctic Research Planning — ICARP II" (Bowden, Corell, Hassol, & Symon, 2007) have to be mentioned. Over the years, however, scientific drilling in the ice-covered Arctic Ocean remained a dream. The ACEX drilling in 2004 (Backman et al., 2006; Backman & Moran, 2008) was the first major step that part of this dream became reality. Now, further drilling campaigns are needed to follow-up in the future. Here, the construction of a new large icebreaker with deep-water drilling capability will certainly be the next milestone in Arctic Ocean research. Such a vessel will guarantee a commitment to Arctic deep drilling as well as a continuous drilling program, and could be a potential contribution to the IODP and succeeding programmes, as already outlined in the APPG Report (Hovland, 2001). Plans for the development of *Aurora borealis*, an icebreaker with deep-water drilling capability, are pushed forward over the past years, and it seems to be not unrealistic that within the next decade such a ship will become available for the international research community and open a new dimension in multi-disciplinary Arctic Ocean research (Thiede & Egerton, 2004).

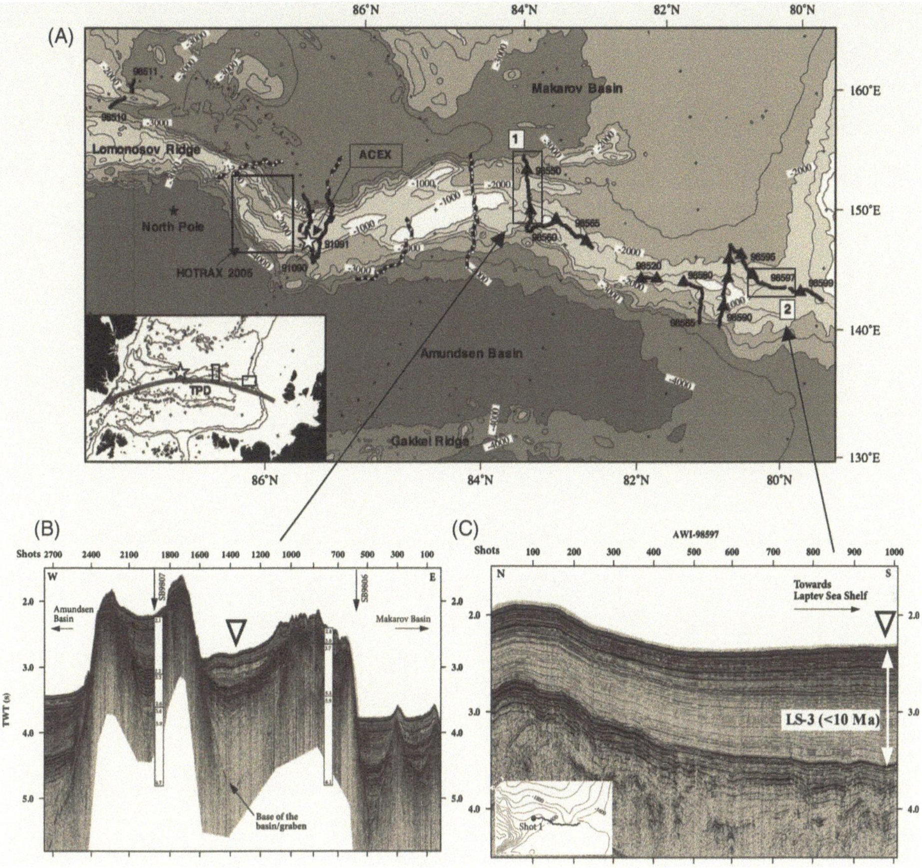

Figure 8.4 (A) Location map of seismic profiles across the Lomonosov Ridge (Jokat, 2003), study area of the HOTRAX 2005 Expedition (Darby et al., 2005), and ACEX drill site (open star; Backman et al., 2006) (from Jokat, 2003, supplemented). The bathymetry is taken from the IBCAO grid (Jakobsson et al., 2000b, 2003a). The contour interval is 500 m and labelled every 1,000 m. Solid line, Arctic-91 seismic profiles; dashed, Oden'96 seismic profiles; bold line, Arctic-98 seismic profiles; triangles, position of deployed sonobuoys. Stacked seismic sections of lines (B) AWI-98550 and (C) AWI-98597 are shown. In line 98597, the acoustic basement is visible at ~4–4.5 s TWT. The sedimentary unit above is slightly affected by faulting. The age of the base of Unit LS-3 is estimated to be ~10 Ma (Jokat, 2003). On Lomonosov Ridge, basins surrounded by topographic highs, were identified (Line AWI-98550). These topographic highs (water depth >1,000 m) might have prevented strong erosion of the basin in-between and thus contain complete Cenozoic sections to study the long-term climate history. Similar sequences can probably be recovered in the HOTRAX area (Darby et al., 2005). Towards the Laptev Sea, thick sequences of Quaternary/Neogene sediments can be recovered (Line AWI-98550) to be used for high-resolution studies of the younger climate history. Possible drill sites are marked as open triangles.

References

Aagaard, K., Barrie, L. A., & Carmack, E. C. (1996). U.S. and Canadian Researchers Explore Arctic Ocean. *EOS Trans. Amer. Geophys. Un.*, 77(22), 209–210.

Aagaard, K., & Carmack, E. C. (1989). The role of sea ice and other fresh water in the Arctic circulation. *Journal of Geophysical Research, 94*(C10), 14485–14498.

Aagaard, K., & Carmack, E. C. (1994). The Arctic Ocean and climate: A perspective. In: O. M. Johannessen, R. D. Muench & J. E. Overland (Eds), *The Polar Oceans and their role in shaping the global environment: The Nansen Centennial Volume* (Vol. 85, pp. 5–20). Washington, DC: Geophysical Monograph, American Geophysical Union.

Aagaard, K., Swift, J. H., & Carmack, E. C. (1985). Thermohaline circulation in the Arctic Mediterranean seas. *Journal of Geophysical Research, 90*(C3), 4833–4846.

Abelmann, A. (1992). Diatom assemblages in Arctic Sea ice — indicator for ice drift pathways. *Deep-Sea Research, 39*(Suppl. 2), S525–S538.

Abelmann, A., Brathauer, U., Gersonde, R., Sieger, R., & Zielinski, U. (1999). A radiolarian-based transfer function for the estimation of sea-surface temperatures in the Southern Ocean (Atlantic Sector). *Paleoceanography, 14*, 410–421.

Abelmann, A., & Gowing, M. M. (1997). Spatial distribution of living polycystine radiolarian taxa-baseline study for paleoenvironmental reconstructions in the Southern Ocean (Antarctic sector). *Marine Micropaleontology, 30*, 3–28.

Abelmann, A., & Nimmergut, A. (2005). Radiolarians in the Sea of Okhotsk and their ecological implication for paleoenvironmental reconstructions. *Deep-Sea Research Part II, 52*, 2302–2331.

ACIA, (2004). *Impacts of a warming Arctic: Arctic climate impact assessment.* Cambridge: Cambridge University Press, 139pp. (http://www.acia.uaf.edu).

ACIA, (2005). *Arctic climate impact assessment.* Cambridge: Cambridge University Press, 1042p.

Aksu, A.E. (1985). Paleomagnetic stratigraphy of CESAR cores. In: H. R. Jackson, P. J. Mudie & S. M. Blasco (Eds), *Initial geological report on CESAR — the Canadian expedition to study the Alpha Ridge* (pp. 101–114), Arctic Ocean. Geollogical Survey of Canadian Paper 84–22.

Aksu, A. E., de Vernal, A., & Mudie, P. E. (1989). High-resolution foraminifer, palynologic, and stable isotopic records of upper Pleistocene sediments from the Labrador Sea: Paleoclimatic and paleoceanographic trends. In: S. P. Srivastava, M. A. Arthur, B. Clement, et al. (Eds). *Proceedings of ODP Science Results* (Vol. 105, pp 617–652), College Station, Texas (Ocean Drilling Program).

Aksu, A. E., & Mudie, P. J. (1985). Magnetostratigraphy and palynology demonstrate at least 4 million years of Arctic sedimentation. *Nature, 318*, 280–283.

Aksu, A. E., Mudie, P. J., Macko, S. A., & de Vernal, A. (1988). Upper Cenozoic history of the Labrador Sea, Baffin Bay, and the Arctic Ocean, a paleoclimatic and paleoceanographic summary. *Paleoceanography, 5*, 519–538.

Alabyan, A. M., Chalov, R. S., Korotaev, V. N., Sidorchuk, A. Y., & Zaitsev, A. A. (1995). Natural and technogenic water and sediment supply to the Laptev Sea. *Report on Polar Research, 176*, 265–271.

Aldahan, A. A., Ning, S., Possnert, G., Backman, J., & Bostrom, K. (1997). ^{10}Be records from sediments of the Arctic Ocean covering the past 350 ka. *Marine Geology, 144*, 147–162.

Aldahan, A. A., Possnert, G., Scherer, R., Shi, N., Backman, J., & Bostroem, K. (2000). Trace-element and major-element stratigraphy in Quaternary sediments from the Arctic Ocean and implications for glacial termination. *Journal of sedimentary research. Section A, Sedimentary Petrology and Processes: An International Journal of SEPM (Society for Sedimentary Geology), 70*, 1095–1106.

Alekin, O. A., & Brazhnikova, L. V. (1964). *Discharge of dissolved substances from the USSR territory.* Moscow: Nauka, 143pp. (in Russian).

Aleksandrova, V. D. (1980). *The Arctic and Antarctic: Their division into geobotanical areas.* Cambridge: Cambridge University Press, 247pp.

Aleksandrova, V. D. (1988). *Vegetation of the Soviet Polar deserts. Studies in Polar research.* Cambridge: Cambridge University Press, 228pp.

Alexanderson, H., Adrielsson, L., Hjort, C., Möller, P., Antonov, O., Eriksson, S., & Pavlov, M. (2002). Depositional history of the North Taymyr ice-marginal zone, Siberia — a landsystem approach. *Journal of Quaternary Science, 17*(4), 361–382.

Alldredge, A. L., & Silver, M. W. (1988). Characteristics, dynamics and significance of marine snow. *Progress in Oceanography, 20*, 41–82.

Alley, R., Anandakrishnan, S., & Jung, P. (2001). Stochastic resonance in the North Atlantic. *Paleoceanography, 16*(2), 190–198.

Alley, R. B., Mayewski, P. A., Sowers, T., Stuiver, M., Taylor, K. C., & Clark, P. U. (1997). Holocene climatic instability: A prominent, widespread event 8200 yr ago. *Geology, 25*, 483–486.

Altabet, M. A., & Francois, R. (1994). Sedimentary nitrogen isotopic ratio as a recorder for surface ocean nitrate utilization. *Global Biogeochemical Cycles, 8*, 103–116.

AMAP (1998). *AMAP Assessment report: Arctic pollution issues. Arctic monitoring and assessment program (AMAP).* Oslo, Norway, 859pp.

Amiel, D., Cochran, J. K., & Hirschberg, D. J. (2002). ^{234}Th/^{238}U disequilibrium as an indicator of the seasonal export flux of particulate organic carbon in the North Water Polynya. *Deep-Sea Research, 49*, 5191–5209.

Andersen, E. S., Dokken, T. M., Elverhøi, A., Solheim, A., & Fossen, I. (1996). Late Quaternary sedimentation and glacial history of the western Svalbard margin. *Marine Geology, 133*, 123–156.

Andersen, N., Paul, H. A., Bernasconi, S. M., McKenzie, J. A., Behrens, A., Schaeffer, P., & Albrecht, P. (2001). Large and rapid climate variability during the Messinian salinity crisis: Evidence from deuterium concentrations of individual biomarkers. *Geology, 29*(9), 799–802.

Anderson, J. B., & Andrews, J. T. (1999). Radiocarbon constraints on ice sheet advance and retreat in the Weddell Sea, Antarctica. *Geology, 27*, 179–182.

Anderson, J. B., & Molnia, B. F. (1989). *Glacial-marine sedimentation. Short course in geology* (Vol. 9). Washington, DC: American Geophysical Union, 127pp.

Anderson, L. G., Björk, G., Holby, O., Jones, E. P., Kattner, G., Koltermann, K. P., Liljeblad, B., Lindegren, R., Rudels, B., & Swift, J. (1994). Water masses and circulation in the Eurasian Basin: Results from the Oden 91 North pole expedition. *Journal of Geophysical Research, 99*(C2), 3273–3283.

Anderson, L. G., & Carlsson, M. L. (Eds). (1991). International Arctic Ocean Expedition 1991, *Icebreaker ODEN — Cruise Report*, Swedish Polar Research Secretariat, 128pp.

Andreassen, I., Nöthig, E. M., & Wassmann, P. (1996). Vertical particle flux on the shelf off northern Spitsbergen. *Marine Ecology Progress Series, 137*, 215–218.

Andreassen, K., Nilssen, L. C., Rafaelsen, B., & Kuilman, L. (2004). Three-dimensional seismic data from the Barents Sea margin reveal evidence of past ice streams and their dynamics. *Geology, 32*, 729–732.

Andreev, A. A., & Klimanov, V. A. (2000). Quantitative Holocene climatic reconstruction from Arctic Russia. *Journal of Paleolimnology, 24*, 81–91.

Andreev, A. A., Klimanov, V. A., & Sulerzhitsky, L. D. (1997). Younger Dryas Pollen records from central and southern Yakutia. *Quaternary International, 41/42*, 111–117.

Andreev, A. A., Siegert, C., Klimanov, V. A., Derevyagin, A. Y., Shilova, G. N., & Melles, M. (2002). Late Pleistocene and Holocene vegetation and climate on the Taymyr lowland, northern Siberia. *Quaternary Research, 57*, 138–150.

Andreev, A. A., Tarasov, P. E., Siegert, C., Ebel, T., Klimanov, V. A., Melles, M., Bobrov, A. A., Dereviagin, A. Y., Lubinski, D. J., & Hubberten, H.-W. (2003). Late Pleistocene and Holocene vegetation and climate on the northern Taymyr Peninsula, Arctic Russia. *Boreas, 32*, 484–505.

Andrews, J. T. (1987). The late Wisconsian glaciation and deglaciation of the Laurentide Ice Sheet. In: W. F. Ruddiman & H. E. Wright, Jr. (Eds), *The Geology of North America, Vol. K3, North America and Adlacent Oceans During the Last Deglaciation* (pp. 13–38). Boulder: Geological Society of America.

Andrews, J. T. (1998). Abrupt changes (Heinrich events) in late Quaternary North Atlantic marine environments: A history and review of data and concepts. *Journal of Quaternary Science, 13*(1), 3–16.

Andrews, J. T. (2000). Icebergs and iceberg rafted detritus (IRD) in the North Atlantic: Facts and assumptions. *Oceanography, 13*(3), 100–108.

Andrews, J. T., Hardadottir, J., Stoner, S. D., Mann, M. E., Greta B. Kristjansdottir, G. B., & Koc, N. (2003). Decadal to millennial-scale periodicities in North Iceland shelf sediments over the last 12 000 cal yr: Long-term North Atlantic oceanographic variability and solar forcing. *Earth and Planetary Science Letters, 210*, 453–465.

Andrews, J. T., Helgadottir, G., Geirsdottir, A., & Jennings, A. E. (2001). Multicentury-scale records of carbonate (hydrographic?) variability on the northern Iceland margin over the last 5000 years. *Quaternary Research, 56*, 199–206.

Andrews, J. T., Jennings, A. E., Cooper, T., Williams K. M., & Mienert, J. (1996). Late Quaternary sedimentation along a fjord to shelf (trough) transect, East Greenland (c. 68°N). In: J. Andrews, W.E.N. Austin, H. Bergsten & A. E. Jennings (Eds), *Late Quaternary Paleoceanography of the North Atlantic margins* (pp. 153–166). Geological Society of London (Special Publication 111).

Andrews, J. T., & Principato, S. M. (2002). Grain-size characteristics and provenance of ice proximal glacial marine sediments. In: J. A. Dowdeswell & C. Cofaigh (Eds), *Glacier-influenced sedimentation on high-latitude continental margins* (pp. 305–324). Geological Society of London (Special Publication 203).

Andrews, J. T., Smith, L. M., Preston, R., Cooper, T., & Jennings, A. E. (1997). Spatial and temporal patterns of iceberg rafting (IRD) along the East Greenland margin, ca. 68°N, over the last 14 cal. ka. *Journal of Quaternary Science, 12*, 1–13.

Andrews, J. T., & Tedesco, K. (1992). Detrital carbonate-rich sediments, northwestern Labrador Sea: Implications for ice-sheet dynamics and iceberg rafting (Heinrich) events in the North Atlantic. *Geology, 20*, 1087–1090.

Anisimov, O. A. (2007). Potential feedback of thawing permafrost to the global climate system through methane emission. *Environmental Research Letter., 2*. doi: 10.1088/1748-9326/2/4/045016.

Anisimov, O. A., Shiklomanov, N. I., & Nelson, F. E. (1997). Effects of global warming on permafrost and active-layer thickness: Results from transient general circulation models. *Global and Planetary Change, 15*, 61–77.

Anon (1978). *Polar Regions Atlas*. Produced by the Foreign Assessment Center, Central Intelligence Agency.

Antonow, M., Goldschmidt, P. M., & Erlenkeuser, H. (1997). The climatesensitive Vesterisbanken area (central Greenland Sea): Depositional environment and paleoceanography during the past 250,000 years. In: H. C. Hass & M. A. Kaminski (Eds), *Contributions to micropaleontology and paleoceanography of the North Atlantic* (pp. 101–118). Krakow: Grzybowski Foundation Special Publication.

Are, F. E. (1999). The role of coastal retreat for sedimentation in the Laptev Sea. In: H. Kassens, H. Bauch, I. Dmitrenko, H. Eicken, H.-W. Hubberten, M. Melles, J. Thiede & L. Timokhov (Eds), *Land-Ocean systems in the Siberian Arctic: Dynamics and history* (pp. 287–299). Heidelberg: Springer-Verlag.

Arkhipov, S. A., Bespaly, V. G., Faustova, M. A., Glushkova, O., Isayeva, L. L., & Velichko, A. A. (1986b). Ice-sheet reconstructions. *Quaternary Science Review, 5*, 475–483.

Arkhipov, S. A., Ehlers, J., Johnson, R. G., & Wright, H. E. (1995). Glacial drainage towards the Mediterrean during the Middle and Late Pleistocene. *Boreas, 24*, 196–206.

Arkhipov, S. A., Isayeva, L. L., Bespaly, V. G., & Glushkova, O. Y. (1986a). Glaciation of Siberia and northeast USSR. *Quaternary Science Review, 5*, 463–474.

Armand, L. K., & Leventer, A. (2003). Paleosea ice distribution — reconstruction and paleoclimatic significance. In: D. N. Thomas & G. S. Diekmann (Eds), *Sea ice: An introduction to its physics, chemistry, biology and geology* (pp. 333–372). Oxford: Blackwell.

Arora, A., & Singh, P. K. (2003). Comparison of biomass productivity and nitrogen fixing potential of Azolla SPP. *Biomass Bioenergy, 24*, 175–178.

Artemiev, V. E. (1997). *Geochemistry of organic matter in river-sea system*. Dordrecht: Kluwer, 190pp.

Arthur, M. A., Dean, W .E., & Stow, D. A. V. (1984). Models for the deposition of Mesozoic — Cenozoic fine-grained organic-carbon-rich sediments in the deep sea. In: D.A.V. Stow &

D. J. W. Piper (Eds), *Fine-grained sediments: Deep-water processes and facies* (pp. 527–560). Geological Society London (Special Publication 15).

Arthur, M. A., Jenkyns, H., Brumsack, H., & Schlanger, S. (1988). Stratigraphy, geochemistry, and paleoceanography of organic-carbon-rich mid-Cretaceous sequences. In: B. Beaudoin & R. Ginsburg (Eds), *Cretaceous resources, events, and rhythms* (pp. 25–70). Report of NATO Advanced Research Workshop, Digne/France, September 16-23.

Arthur, M. A., Schlanger, S. O., & Jenkyns, H. C. (1987). The Cenomanian-Turonian Oceanic anoxic event, ii. Palaeoceanographic controls on organic-matter production and Preservation, In: J. Brooks & A. Fleet (Eds), *Marine petroleum source rocks* (Vol. 26, pp. 401–420). *Geological Society (Special Publication).*

Astakhov, V. (2004). Middle Pleistocene glaciations of the Russian North: Quat. *Science Review, 23,* 1285–1311. doi: 10.1016/j.quascirev.2003.12.011.

Astakhov, V. I., Svendsen, J. I., Matiouchkov, A., Mangerud, J., Maslenikova, O., & Tveranger, J. (1999). Marginal formations of the last Kara and Barents ice sheets in northern European Russia. *Boreas, 28,* 23–45.

Atakan, K., & Ojeda, A. (2005). Stress transfer in the Storegga area, offshore mid-Norway. *Marine and Petroleum Geology, 22,* 161–170.

Atlas Arktiki (Atlas of the Arctic) (1985). G.U.G.K., Moscow (in Russian).

Austegard, A. (1982). Velocity analysis of sonobuoy data from the northern Svalbard margin. Sci. Rep. 9, Seismol. Obs., University of Bergen, 26pp.

Axelrod, D. I. (1984). An interpretation of Cretaceous and tertiary biota in Polar regions. *Palaeogeography, Palaeoclimatology, Palaeoecology, 45,* 105–147.

Backman, J., Jakobsson, M., Frank, M., Sangiorgi, F., Brinkhuis, H., Stickley, C., O'Regan, M., Løvlie, R., Pälike, H., Spofforth, D., Gattacecca, J., Moran, K., King, J., & Heil, C. (2008). Age model and core-seismic integration for the Cenozoic ACEX sediments from the Lomonosov Ridge. *Paleoceanography, 23,* PA1S03, doi: 10.1029/2007PA001476.

Backman, J., Jacobson, M., Knies, J., Knudsen, J., Kristoffersen, Y., Lif, A., Musatov, E., & Stein, R. (1997). *Geological coring and high resolution chirp sonar profiling.* Cruise report, Polarforskningssekretaariatets arsbok 1995/96, Stockholm, pp. 64–66.

Backman, J., Jakobsson, M., Løvlie, R., Polyak, L., & Febo, L. A. (2004). Is the central Arctic Ocean a sediment starved basin? *Quaternary Science Review, 23,* 1435–1454.

Backman, J., & Moran, K. (2008). Introduction to special section on Cenozoic Paleoceanography of the Central Arctic Ocean. *Paleoceanography, 23,* PA1S01, doi:10.1029/2007PA001516.

Backman, J., Moran, K., McInroy, D., et al. (2005). IODP expedition 302, Arctic coring expedition (ACEX): A first look at the Cenozoic paleoceanography of the central Arctic Ocean. *Science Drilling, 1,* 12–17.

Backman, J., Moran, K., McInroy, D. B., Mayer, L. A., & the Expedition 302 Scientists (2006). *Proceedings IODP, 302,* Edinburgh (Integrated Ocean Drilling Program Management International, Inc.). doi: 10.2204/iodp.proc.302.2006.

Backman, J., & Raffi, I. (1997). Calibration of Miocene nannofossil events to orbitally tuned cyclostratigraphies from Ceara Rise. *Proceedings of ODP Science Results, 154,* 83–99.

Bakke, T., Hameedi, J., Kimstach, V., Macdonald, R., Melnikov, S., Robertson, A., et al. (1998). Chapter 10, Petroleum hydrocarbons. In: A. Robertson (Ed.), *AMAP assessment report: Arctic pollution issues* (pp. 662–716). AMAP, Oslo.

Barber, D., Dyke, A., Hillaire-Marcel, C., Jennings, A., Andrews, J., Kerwin, M. W., Bilodeau, G., McNeely, R., Southon, J., Morehead, M. D., & Gagnon, J.-M. (1999). Forcing of the cold event of 8200 years ago by catastrophic drainage of Laurentide lakes. *Nature, 400,* 344–348.

Bard, E., Hamelin, B., Arnold, M., Montaggioni, L., Cabioch, G., Faure, G., & Rougerie, F. (1996). Deglacial sea-level record from Tahiti corals and the timing of global meltwater discharge. *Nature, 382,* 241–244.

Barnes, P. W., & Lien, R. (1988). Iceberg rework shelf sediments to 500 m off Antarctica. *Geology, 16,* 1130–1133.

Barron, J. A. (1985). Diatom biostratigraphy of the CESAR 6 Core, Alpha Ridge. In: H. R. Jackson, P. J. Mudie & S. M. Blasko (Eds). *Initial geological report on CESAR: The Canadian expedition to study the Alpha* (pp. 137–148). Geological Survey of Canadian Paper 84–22.

Barry, R. G. (1989). The present climate of the Arctic Ocean and possible past and future states. In: Y. Herman (Ed.), *The Arctic Seas* (pp. 1–46). New York: Van Nostrand Reinhold Co.

Barry, R. G. (1996). The parameterization of surface albedo for sea ice and its snow cover. *Progress in Physical Geography, 20*, 63–79.

Barry, R. G., Serreze, M. C., & Maslanik, J. A. (1993). The Arctic sea ice-climate system: Observations and modeling. *Reviews of Geophysics, 31*, 397–422.

Bart, P. J., & Anderson, J. B. (1995). Seismic record of glacial events affecting the Pacific margin of the northwestern Antarctic Peninsula. In: A. K. Cooper, P. F. Barker & G. Brancolini (Eds), *Geology and seismic stratigraphy of the Antarctic margin. Antarctic research series* (Vol. 68, pp. 74–95). Washington, DC: American Geophysical Union.

Bartek, L. R., Andersen, J., & Oneacre, T. (1997). Ice stream troughs and variety of Cenozoic seismic stratigraphic architecture from a high southern latitude section: Ross Sea, Antarctica. In: T. A. Davies, T. Bell, A. K. Cooper, H. Josenhans, L. Polyak, A. Solheim, M. S. Stoker & J. A. Stravers (Eds), *Glaciated continental margins: An atlas of acoustic images* (pp. 250–253). London: Chapman & Hall.

Bartels, J., Laurson, V., Brooks, C. E. P., Paton, J., Benyon, W. J. G., Vestine, E. H., & Nagata, T. (1959). The second international Polar year. *Annual International Geophysics Year, 1*, 205–382.

Basinger, J. F., Greenwood, D. G., & Sweda, T. (1994). Early tertiary vegetation of Arctic Canada and its relevance to paleoclimatic interpretation. In: M. C. Boulter & H. C. Fischer (Eds), *Cenozoic plants and climates of the high Arctic NATO ASI series 1* (Vol. 68, pp. 175–198). Heidelberg: Springer-Verlag.

Baskaran, M., Swarzenski, P. W., & Porcelli, D. (2003). Role of colloidal material in the removal of ^{234}Th in the Canada Basin of the Arctic Ocean. *Deep-Sea Research, 50*, 1353–1373.

Bathmann, U. V., Peinert, R., Noji, T. T., & Bodungen, B. V. (1990). Pelagic origin and fate of sedimenting particles in the Norwegian Sea. *Progress in Oceanography, 24*, 117–125.

Bauch, D., Carstens, J., & Wefer, G. (1997). Oxygen isotope composition of living *Neogloboquadrina pachyderma* (sin.) in the Arctic Ocean. *Earth and Planetary Science Letters, 146*, 47–58.

Bauch, D., Erlenkeuser, H., & Andersen, N. (2005). Water mass processes on Arctic shelves as revealed from δ^{18}O of H_2O. *Global and Planetary Change, 28*, 165–174.

Bauch, H. A., Erlenkeuser, H., Fahl, K., Spielhagen, R. F., Weinelt, M. S., Andruleit, H., & Henrich, R. (1999). Evidence for a steeper Eemian than Holocene sea surface temperature gradient between Arctic and sub-Arctic regions. *Palaeogeography, Palaeoclimatology, Palaeoecology, 145*, 95–117.

Bauch, H. A., Erlenkeuser, H., Jung, S. J. A., & Thiede, J. (2000). Surface and deep water changes in the subpolar North Atlantic during Termination II and the last interglaciation. *Paleoceanography, 15*, 76–84.

Bauch, H., & Kassens, H. (2005). Arctic Siberian shelf environments — an introduction. *Global and Planetary Change, 48*, 1–8.

Bauch, H. A., Kassens, H., Naidina, O. D., Kunz-Pirrung, M., & Thiede, J. (2001a). Composition and flux of Holocene sediments on the Eastern Laptev Sea shelf, Arctic Siberia. *Quarternary Research, 55*, 344–351.

Bauch, H. A., Mueller-Lupp, T., Taldenkova, E., Spielhagen, R. F., Kassens, H., Grootes, P. M., Thiede, J., Heinemeier, J., & Petryashov, V. V. (2001b). Chronology of the Holocene transgression at the North Siberian margin. *Global and Planetary Change, 31*, 125–141.

Bauch, H. A., & Polyakova, Y. I. (2000). Late Holocene variations in Arctic shelf hydrology and sea-ice regime: Evidence from north of the Lena Delta. *International Journal of Earth Science, 89*, 569–577.

Bauch, H. A., & Polyakova, Ye. I. (2003). Diatom-inferred salinity records from the Arctic Siberian margin: implications for fluvial runoff patterns during the Holocene. *Palaeoceanography, 18*(2), 5.1–5.10.

Bauch, D., Schlosser, P., & Fairbanks, R. G. (1995). Freshwater balance and the sources of deep and bottom waters in the Arctic Ocean inferred from the distribution of $H_2^{18}O$. *Progress in Oceanography, 35*, 53–80.

Bauerfeind, E. (2004). North East Water Polynya. In: R. Stein & R. W. Macdonald (Eds), *The Organic Carbon Cycle in the Arctic Ocean* (pp. 106–109). Heidelberg: Springer-Verlag.

Baumann, K.-H., Andruleit, H., & Samtleben, C. (2000). Coccolithophores in the Nordic Seas: comparison of living communities with surface sediment assemblages. *Deep-Sea Research Part II, 47*, 1743–1772.

Baumann, K.-H., Lackschewitz, K. S., Mangerud, J., Spielhagen, R. F., Wolf-Welling, T. C. W., Henrich, R., & Kassens, H. (1995). Reflection of Scandinavian ice-sheet fluctuations in Norwegian Sea sediments during the past 150000 years. *Quat. Res., 43*, 185–197.

Baumann, M. (1990). Coccoliths in sediments of the eastern Arctic Basin. In: U. Bleil & J. Thiede (Eds), *Geological History of the Polar Oceans: Arctic Versus Antarctic, NATO ASI Series C* (Vol. 108, pp. 437–445). Dordrecht: Kluwer.

Bé, A. W. H., & Tolderlund, D. S. (1971). Distribution and ecology of living planktonic foraminifera in surface waters of the Atlantic and Indian Oceans. In: B. F. Funnel (Ed.), *The Micropaleontology of the Oceans* (pp. 105–149). New York: Cambridge University Press.

Behrends, M. (1999). Reconstruction of sea-ice drift and terrigenous sediment supply in the Late Quaternary: Heavy-mineral associations in sediments of the Laptev-Sea continental margin and the central Arctic Ocean. *Reps. Polar Research, 310*, 167pp.

Behrends, M., Hoops, E., & Peregovich, B. (1999). Distribution Patterns of Heavy Minerals in Siberian Rivers, the Laptev Sea and the eastern Arctic Ocean: An Approach to Identify Sources, Transport and Pathways of Terrigenous Matter. In: H. Kassens, H. Bauch, I. Dmitrenko, H. Eicken, H. W. Hubberten, M. Melles, J. Thiede & L. Timokhov (Eds), *Land-Ocean Systems in the Siberian: Dynamics and History* (pp. 265–286). Heidelberg: Springer-Verlag.

Belicka, L. L., Macdonald, R. W., & Harvey, H. R. (2002). Sources and transport of organic carbon to shelf, slope, and basin surface sediments of the Arctic Ocean. *Deep-Sea Research Part I, 49*, 1463–1483.

Belicka, L. L., Macdonald, R. W., Yunker, M. B., & Harvey, H. R. (2004). The role of depositional regime on carbon transport and preservation in Arctic Ocean sediments. *Marine Chemistry, 86*, 65–88.

Belov N.A., & Lapina N.N. (1961). *Bottom deposits of the Arctic Basin.* Publ. H. ìMorskoy transportî, L., 152pp.

Belt, S. T., Allard, W. G., Masse', G., Robert, J.-M., & Rowland, S. J. (2000). Highly branched isoprenoids (HBIs): identification of the most common and abundant sedimentary isomers. *Geochimica et Cosmochimica Acta, 64*, 3839–3851.

Belt, S. T., Allard, W. G., Masse', G., Robert, J.-M., & Rowland, S. J. (2001). Structural characterisation of C_{30} highly branched isoprenoid alkenes (rhizenes) in the marine diatom *Rhizosolenia setigera. Tetrahedron Lett, 42*, 5583–5585.

Belt, S. T., Massé, G., Rowland, S. J., Poulin, M., Michel, C., & LeBlanc, B. (2007). A novel chemical fossil of palaeo sea ice: IP25. *Organic Geochemistry, 38*, 16–27.

Bendle, J., & Rosell-Melé, A. (2004). Distribution of UK37 and in the surface waters and sediments of the Nordic Seas: Implications for paleoceanography. *Geochemistry Geophysics Geosystems, 5*, Q11013. doi: 10.1029/2004GC000741.

Berger, W. H., Smetacek, V., & Wefer, G. (1989). *Productivity of the Ocean: Past and Present. Life Sciences Research Report* (Vol. 44). New York: Wiley & Sons, 471pp.

Berggren, W. A., Kent, D. V., Swisher, C. C., III, & Aubry, M.-P. (1995). A revised Cenozoic geochronology and chronostratigraphy. *SEPM Special Publication, 54*, 129–212.

Berggren, W. A., Kent, D. V., & Van Couvering, J. (1985). The Neogene: Part 2. Neogene geochronology and chronostratigraphy. In: N.J. Snelling (Ed.), *The Chronology of the Geological Record. Geol. Soc. Lond. Mem.* (Vol. 10, pp. 211–260).

Bergsten, H. (1994). Recent benthic foraminifera of a transect from the North Pole to the Yermak Plateau, eastern central Arctic Ocean. *Marine Geology, 119*, 251–267.

Berner, H. (1991). Mechanismen der Sedimentbildung in der Framstraße, im Arktischen Ozean und in der Norwegischen See. Ber. Fachbereich Geowissenschaften, Bremen University, Vol. 20, 167pp.

Berner, H., & Wefer, G. (1990). Physiographic and biologic factors controlling surface sediment distribution in the Fram Strait. In: U. Bleil, J. Thiede (Eds), *Geological History of the Polar Oceans: Arctic versus Antarctic. NATO ASI Series C, 308*, 317–335.

Berner, R. A. (1970). Sedimentary pyrite formation. *American Journal of Science, 268*, 1–23.

Berner, R. A. (1984). Sedimentary pyrite formation: An update. *Geochimica et Cosmochimica Acta, 48*, 605–615.

Berner, R. A. (1989). Biogeochemical cycles of carbon and sulfur and their effect on atmospheric oxygen over Phanerozoic time. *Palaeogeogr. Palaeoclim. Palaeoecol., 75*, 97–122.

Berner, R. A., & Raiswell, R. (1983). Burial of organic carbon and pyrite sulfur in sediments over Phanerozoic time: A new theory. *Geochimica et Cosmochimica Acta, 47*, 855–862.

Bernstein, T. (1932). Über einige arktische Radiolarien. *Archive für Protistenkunde, 76*, 217–227.

Betzer, P. R., Showers, W. J., Laws, E. A., Winn, C. D., Ditullo, G. R., & Kroopnick, P. M. (1984). Primary productivity and particle fluxes on a transect of the equator at 153 W in the Pacific Ocean. *Deep-Sea Research, 31*, 1–11.

Bice, K. L., Arthur, M. A., & Marincovich, L., Jr. (1996). Late Paleocene Arctic Ocean shallow-marine temperatures from mollusc stable isotopes. *Paleoceanography, 11*(3), 241–249.

Birgel, D., & Hass, C. (2004). Oceanic and atmospheric variations during the last deglaciation in the Fram Strait (Arctic Ocean): a coupled high-resolution organic-geochemical and sedimentological study. *Quaternary Science Review, 23*, 29–47.

Birgel, D., & Stein, R. (2004). Northern Fram Strait und Yermak Plateau: Distribution, Variability and Burial of Organic Carbon and Paleoenvironmental Implications. In: R. Stein & R. W. Macdonald (Eds), *The Organic Carbon Cycle in the Arctic Ocean* (pp. 279–295). Heidelberg: Springer Verlag.

Birgel, D., Stein, R., & Hefter, J. (2004). Aliphatic lipids in recent sediments of the Fram Strait/ Yermak Plateau (Arctic Ocean): Composition, sources and transport processes. *Marine Chemistry, 88*, 127–160.

Birkenmajer, K. (1989). The geology of Svalbard, the western part of the Barents Sea, and the continental margin of Scandinavia. In: A. E. M. Nairn, M. Churkinund & F. G. Stehli (Eds), *The Arctic Ocean. The Ocean Basins and Margins* (Vol. 5, pp. 265–330). New York: Plenum.

Birks, C. J. A., & Koç, N. (2002). A high-resolution diatom record of late-Quaternary sea-surface temperatures and oceanographic conditions from the eastern Norwegian Sea. *Boreas, 31*(4), 323–344.

Biryukov, V. Yu., Faustova, M. A., Kaplin, P. A., Pavlidis, Yu. A., Romanova, E. A., & Velichko, A. A. (1988). The paleogeography of Arctic shelf and coastal zone of Eurasia at the time of the last glaciation (18,000 yrs B.P.). *Palaeogeography, Palaeoclimatology, Palaeoecology, 68*, 117–125.

Bischof, J. F. (2000). *Ice Drift, Ocean Circulation, and Climate Change.* Heidelberg: Springer Verlag, 215pp.

Bischof, J. F., Clark, D. L., & Vincent, J. S. (1996). Pleistocene paleoceanography of the central Arctic Ocean: the sources of ice rafted debris and the compressed sedimentary record. *Paleoceanography, 11*, 743–756.

Bischof, J. F., & Darby, D. A. (1997). Mid to Late Pleistocene ice drift in the western Arctic Ocean: evidence for a different circulation in the past. *Science, 277*, 74–78.

Bischof, J. F., & Darby, D. A. (1999). Quaternary ice transport in the Canadian Arctic and extent of late Wisconsinan glaciation in the Queen Elizabeth Islands. *Canadian Journal of Earth Science, 36*, 2007–2022.

Bischof, J., Koch, J., Kubisch, M., Spielhagen, R.F., & Thiede, J. (1990). Nordic seas surface ice drift reconstructions: Evidence from ice-rafted coal fragments during oxygen isotope stage 6. In: J. A. Dowdeswell & J. D. Scourse (Eds), *Glacimarine Environments: Processes and Sediments* (Vol. 53, pp. 235–251). *Geological Society of Special Publication.*

Bishop, J. K. B. (1988). The barite-opal-organic carbon association in oceanic particulate matter. *Nature, 332*, 341–343.

Bjørklund, K. R., Cortese, G., Swanberg, N., & Schrader, H. J. (1998). Radiolarian faunal provinces in surface sediments of the Greenland, Iceland and Norwegian (GIN) Seas. *Marine Micropaleontology, 35*(1–2), 105–140.

Bjørklund, K. R., & Kruglikova, S. B. (2003). Polycystine radiolarians in surface sediments in the Arctic Ocean basins and marginal seas. *Marine Micropaleontology, 49*, 231–273.

Bjørlukke, K., Bue, B., & Elverhøi, A. (1978). Quaternary sediments in the northwestern part of the Barents Sea and their relation to the underlying Mesozoic bedrocks. *Sedimentology, 25*, 227–246.

Blackburn, T. H. (1987). Microbial food webs in sediments. In: M. A. Sleigh (Ed.), *Microbes in the sea Chapter 2* (pp. 39–58). New York, NY: Ellis Horwood/John Wiley.

Bleil, U., & Thiede, J. (Eds). (1990). *Geological History of the Polar Oceans: Arctic versus Antarctic.* NATO ASI Series C 308, 823pp.

Blumer, M., Guillard, R. R. L., & Chase, T. (1971). Hydrocarbons of marine phytoplankton. *Marine Biology, 8*, 183–189.

Bobrovitskaya, N. N., Kokorev, A. V., & Lemeshko, N. A. (2003). Regional patterns in recent trends in sediment yields of Eurasian and Siberian rivers. *Global and Planetary Change, 39*, 127–146.

Böggild, O. B. (1906). On the bottom deposits of the north Polar seas. *Scientific results of the Norwegian North Polar Expedition 1893–1896, 5*, 1–62.

Bond, G., Broecker, W., Johnsen, S., McManus, J., Labeyrie, L., Jouzel, J., & Bonani, G. (1993). Correlations between climate records from North Atlantic sediments and Greenland ice. *Nature, 365*, 143–147.

Bond, G., Heinrich, H., Broecker, W., Labeyrie, L., McManus, J., Andrews, J., Huon, S., Jantschik, R., Clasen, S., Simet, C., Tedesco, K., Klas, M., Bonani, G., & Ivy, S. (1992). Evidence for massive discharges of icebergs into the North Atlantic ocean during the last glacial period. *Nature, 360*, 245–249.

Bond, G. C., & Lotti, R. (1995). Iceberg discharges into the North Atlantic on millennial time scales during the last glaciation. *Science, 267*, 1005–1010.

Bondarev, V.N., & Kosheleva, V.A. (2007). *Grain size composition of sediments from Hole AMIGE-1993-157 drilled in the Kara Sea.* PANGAEA, doi: 10.1594/PANGAEA.587087.

Boon, J. J., Rijpstra, W. I. C., de Lange, F., de Leeuw, J. W., Yoshioka, M., & Shimizu, Y. (1979). The Black Sea sterols - A molecular fossil for dinoflagellate blooms. *Nature, 277*, 125–127.

Bordowskiy, O. K. (1965a). Sources of organic matter in marine basins. *Marine Geology, 3*, 5–31.

Bordowskiy, O. K. (1965b). Accumulation of organic matter in bottom sediments. *Marine Geology, 3*, 33–82.

Boström, K., & Thiede, J. (1984). YMER-80, Swedish Arctic Expedition - Cruise Report for Marine Geology and Geophysics, Sediment Core Descriptions. *Medd. Stockholms Univ. Geol. Inst., 260*, 123pp.

Boucsein, B., Fahl, K., Siebold, M., & Stein, R. (1999). Quantity and quality of organic carbon in suspended matter and surface sediments of the Ob and Yenisei estuaries and adjacent coastal areas. *Report on Polar Research, 300*, 116–126.

Boucsein, B., Fahl, K., & Stein, R. (2000). Late Quaternary organic matter records from the Laptev Sea continental margin: Evidence for the variability of river discharge during the last 15000 years BP. *International Journal of Earth Science, 89*, 578–591.

Boucsein, B., Knies, J., & Stein, R. (2002). Organic matter deposition along the Kara and Laptev Sea contineneta1 margin (eastern Arctic Ocean) during last deglaciation and Holocene: Evidence from organic-geochemical and petrographical data. *Marine Geology, 183*, 67–88.

Boucsein, B., & Stein, R. (2000). Particulate organic matter in surface sediments of the Laptev Sea (Artic Ocean): Application of maceral analysis as organic-carbon-source indicator. *Marine Geology, 162*, 573–586.

Boucsein, B., & Stein, R. (2008). Black shale formation in the late Paleocene/early Eocene Arctic Ocean and paleoenvironmental conditions: New results from a detailed organic petrological study. *Marine and Petroleum Geology*, in press, doi:10.1016/j.marpetgeo.2008.04.001.

Boulter, M. C. (1986). Pollen and spore events from the marine Tertiary of North Europe. *Journal Micropalaeontology, 5*, 75–84.

Bourke, R. H., & Garret, R. P. (1980). Sea-ice thickness distribution in the Arctic Ocean. *Cold Regions Science and Technology, 13*, 259–280.

Bourke, R. H., & Garrett, R. P. (1987). Sea ice thickness distribution in the Arctic Ocean. *Cold Regions Science and Technology, 13*, 259–280.

Boussafir, M., & Lallier-Verges, E. (1997). Accumulation of organic matter in the Kimmeridge Clay Formation (KCF): An update fossilisation model for marine petroleum source-rocks. *Marine and Petroleum Geology, 14*, 75–83.

Bowden, S., Corell, R. W., Hassol, S. J., & Symon, C. (Eds). (2007). Arctic rsearch: A global responsibility. *Report of ICARP-II (2nd International Conference on Arctic Research Planning)*, Copenhagen, Denmark, 10–12 November 2005, McCallum Printing Group Inc., Canada (36pp. and CD-Rom).

Bralower, T. J., & Thierstein, H. R. (1984). Low productivity and slow deep-water circulation in mid-Cretaceous Oceans. *Geology, 12*, 614–618.

Brassell, S. C., & Dumitrescu, ODP Leg 198 Shipboard Scientific Party, (2004). Recognition of alkenones in a lower Aptian porcellanite from the west-central Pacific. *Organic Geochemistry, 35*, 181–188.

Brassell, S. C., Eglinton, G., Marlowe, I. T., Pflaumann, U., & Sarntheim, M. (1986). Molecular stratigraphy: A new tool for climate assessment. *Nature, 320*, 129–133.

Brett, C. P., & Zarudzki, E. F. K. (1979). Project Westmar: A shallow marine geophysical survey on the West Greenland continental shelf. *Gronlands Geol. Unders. Rapp., 87*, 27pp.

Brinkhuis, H., Schouten, S., Collinson, M. E., Sluijs, A., Damsté, J. S. S., Dickens, G. R., Huber, M., Cronin, T. M., Onodera, J., Takahashi, K., Bujak, J. P., Stein, R., van der Burgh, J., Eldrett, J. S., Harding, I. C., Lotter, A. F., Sangiorgi, F., Cittert, H. v. K.-v., de Leeuw, J. W., Matthiessen, J., Backman, J., & Moran, K.the Expedition 302 Scientists, (2006). Episodic fresh surface waters in the Eocene Arctic Ocean. *Nature, 441*, 606–609.

Bro, E. G. (2007). Chemical characteristics of hydrocarbons from sedimentary rocks of Heiss-1 Hole drilled in the Heiss Island, Franz Josef Land Archipelago. *All-Russian Research Inst. for Geology and Mineral Resources of the World Ocean*, PANGAEA, doi: 10.1594/PANGAEA.615254.

Broecker, W. S. (1994). Massive iceberg discharges as triggers for global climate change. *Nature, 372*, 421–424.

Broecker, W. S. (1997). Thermohaline circulation, the Achilles Heel of our climate system: Will man-made CO_2 upset the current balance? *Science, 278*, 1582–1588.

Broecker, W. S., Andree, M., Wolfli, W., Oeschger, H., Bonani, G., Kennett, J., & Peteet, D. (1988). The chronology of the last deglaciation: Implications to the case of the Younger Dryas event. *Paleoceanography, 3*, 1–19.

Broecker, W. S., Bond, G., Klass, M., Clark, E., & McManus, J. (1992). Origin of the northern Atlantic's Heinrich events. *Climate Dynamics, 6*, 265–273.

Broecker, W. S., & Denton, G. H. (1989). The role of ocean-atmosphere reorganizations in glacial cycles. *Geochimica et Cosmochimica Acta, 53*, 2465–2501.

Broecker, W. S., Kennett, J. T., Flower, B. P., Teller, J. T., Trumbore, S., Bonani, G., & Wolfli, W. (1989). Routing of meltwater from the Laurentide ice sheet during the Younger Dryas cold episode. *Nature, 341*, 318–320.

Brooks, P. W. (1986a). Biological marker geochemistry of crude oils and condensates from the Beaufort-Mackenzie Basin. *Bulletin of Canadian Petroleum Geology, 34*, 490–505.

Brooks, P. W. (1986b). Unusual biological marker geochemistry of oils and possible source rocks, offshore Beaufort-Mackenzie Delta, Canada. *Organic Geochemistry, 10*, 401–406.

Brown, J., Ferrians, O. J., Heginbottom, J. A., & Melnikov, E. S. (1998). Digital circum-arctic map of permafrost and groundice conditions. *Circumpolar Active-Layer Permafrost System (CAPS) CD-ROM, version 1.0.* National Snow and Ice Data Center, University of Colorado at Boulder.

Brown, J., Ferrians, O. J. J., Heginbottom, J. A., & Melnikov, E. S. (1997). *International Permafrost Association Circum-Arctic Map of Permafrost and Ground Ice Conditions*, Scale 1:10,000,000. U.S. Geological Survey.

Brown, J., Jorgenson, M. T., Smith, O. P., & Lee, W. (2003). Long-term rates of coastal erosion and carbon input, Elson Lagoon, Barrow, Alaska. In: M. Phillips, S. M. Springman & L. U. Arenson (Eds). *Proceeding of the Eighth International Conference on Permafrost* (pp. 101–106). A.A. Balkema, Lisse.

Brozena, J. M., Childers, V. A., Lawver, L. A., Gahagan, L. M., Forsberg, R., Fileide, J.-I., & Eldholm, O. (2003). New aerogeophysical study of the Eurasia basin and Lomonosov ridge: Implications for basin development. *Geology, 31*, 825–828.

Brumsack, H.-J. (1980). Geochemistry of Cretaceous black shales from the Atlantic Ocean (DSDP Legs 11, 14, 36 and 41). *Chemical Geology, 31*, 1–25.

Brumsack, H.-J. (1986). The inorganic geochemistry of Cretaceous black shales (DSDP Leg 41) in comparison to modern upwelling sediments from the Gulf of California. In: N. J. Shackleton & C. P. Summerhayes (Eds). *North Atlantic Paleoceanography* (pp. 447–462). Geological Society (Special Publication 21).

Brumsack, H.-J. (1988). Rezente Corg-reiche Sedimente als Schlüssel zum Verständnis fossiler Schwarzschiefer, Habilitationsschrift, Universität Göttingen.

Brumsack, H.-J. (1989). Geochemistry of recent TOC-rich sediments from the Gulf of California and the Black Sea. *International Journal of Earth Science, 78*, 851–882.

Bryn, P., Berg, K., Forsberg, C. F., Solheim, A., & Kvalstad, T. J. (2005). Explaining the Storegga Slide. *Marine and Petroleum Geology, 22*, 11–19.

Bugge, T., Befring, S., Belderson, R. H., Eidvin, T., Jansen, E., Kenyon, N. H., Holtedahl, H., & Sejrup, H. P. (1987). A giant three-stage submarine slide off Norway. *Geo-Marine Letters*, 7, 191–198.

Bugge, T., Belderson, R. H., & Kenyon, N. H. (1988). The Storegga Slide. *Philosophical transactions of the Royal Society of London. Series A, 325*, 357–388.

Bujak, J. P., & Brinkhuis, H. (1998). In: M.-P. Aubry, S. G. Lucas & W. A. Berggren (Eds), *Late Paleocene-early Eocene climatic and biotic events in the marine and terrestrial records* (pp. 277–295). New York: Columbia University Press.

Bujak, J. P., & Mudge, D. C. (1994). A high-resolution North Sea Eocene dinocyst zonation. *Journal of the Geological Society, 151*, 449–462.

Bukry, D. (1984). Paleogene paleoceanography of the Arctic Ocean is constrained by the middle or late Eocene age of USGS Core FI-422: Evidence from silicoflagellates. *Geology, 12*, 199–201.

Bukry, D. (1985). Correlation of Late Cretaceous Arctic silicoflagellates from Alpha Ridge. In: H. R. Jackson, P. J. Mudie & S. M. Blasko (Eds), *Initial geological report on CESAR: The Canadian expedition to study the Alpha Ridge* (pp. 125–135), Geological Survey of Canadian Paper 84–22.

Burenkov, V. I., & Vasilkov, A. P. (1995). The influence of runoff from land on the distribution of hydrologic characteristics of the Kara Sea. *Oceanology, 34*(5), 591–599 (English Translation).

Calvert, S. E. (2004). Beware intercepts: Interpreting compositional ratios in multi-component sediments and sedimentary rocks. *Organic Geochemistry, 35*, 981–987.

Calvert, S. E., & Pedersen, T. F. (1992). Organic carbon accumulation and preservation in marine sediments: How important is anoxia? In: J. K. Whelan & J. W. Farrington (Eds), *Organic matter: Productivity, accumulation, and preservation in recent and ancient sediments* (p. 533pp). New York: Columbia University Press.

Calvert, S. E., & Vogel, J. S. (1987). Carbon accumulation rates and the origin of the Holocene sapropel in the Black Sea. *Geology, 15*, 918–921.

Campbell, J. S., & Clark, K. D. (1977). Pleistocene turbidites of the Canada Abyssal Plain of the Arctic Ocean. *Journal of Sedimentary Petrology, 47*, 657–670.

Cande, S. C., & Kent, D. V. (1992). A new geomagnetic polarity time scale for the Late Cretaceous and Cenozoic. *Journal of Geophysical Research, 97*, 13917–13951.

Cande, S. C., & Kent, D. V. (1995). Revised calibration of the geomagnetic polarity timescale for the Late Cretaceous and Cenozic. *Journal of Geophysical Research, 100*, 6093–6095.

Carmack, E. (1990). Large-scale physical oceanography of Polar Oceans. In: W. O. Smith (Ed.), *Polar pceanography, part A, physical science* (pp. 171–222). San Diego: Academic Press Inc.

Carmack, E., Barber, D., Christensen, J., Macdonald, R. W., Rudels, B., & Sakshaug, E. (2006). Climate variability and physical forcing of the food webs and the carbon budget on panarctic shelves. *Progress in Oceanography, 71*, 145–181.

Carstens, J., Hebbeln, D., & Wefer, G. (1997). Distrubution of planktic foraminifera at the ice margin in the Arctic (Fram Strait). *Marine Micropaleontology, 29*, 257–269.

Carstens, J., & Wefer, G. (1992). Recent distribution of planktonic foraminifera in the Nansen Basin, Arctic Ocean. *Deep-Sea Research, 30*, 507–524.

Cauwet, G., & Sidorov, I. S. (1996). The biogeochemistry of Lena River: Organic carbon and nutrients distribution. *Marine Chemistry, 53*, 211–227.

Cavalieri, D. J., Gloersen, P., Parkinson, C. L., Comiso, J. C., & Zwally, H. J. (1997). Observed hemispheric asymmetry in global sea ice changes. *Science, 278*, 1104–1106.

Chakrapani, G. J. (2005). Factors controlling variations in river sediment loads. *Current Science, 88*(4), 569–575.

Chamley, H. (1989). *Clay sedimentology.* Heidelberg: Springer-Verlag, 623pp.

Champbell, J. S., & Clark, K. D. (1977). Pleistocene turbidites of the Canada Abyssal Plain of the Arctic Ocean. *Journal of Sedimentary Petrology, 40,* 657–670.

Champion, D. E., Lanphere, M. A., & Kuntz, M. A. (1988). Evidence for a new geomagnetic reversal from lava flows in Idaho: Discussion of short polarity reversals in the Brunhes and late Matuyama chrons. *Journal of Geophysical Research, 93*(11), 611–667, 680.

Channell, J. E. T., Curtis, J. H., & Flower, B. P. (2004). The Matuyama-Brunhes boundary interval (500–900 ka) in North Atlantic drift sediments. *Geophysics Journal International, 158,* 489–505.

Channell, J. E. T., Hodell, D. A., & Lehman, B. (1997). Relative geomagnetic paleointensity and ^{18}O at ODP Site 983 (Gardar Drift, North Atlantic) since 350 ka. *Earth and Planetary Science Letters, 153,* 103–118.

Channell, J. E. T., Kanamatsu, T., Sato, T., Stein, R., Alvarez Zarikian, C. A., Malone, M. J., & the Expedition 303/306 Scientists (2006a). *Proceedings of the Intergrated Ocean Drilling Program, Vol 303/306 Expeditions Report, North Atlantic Climate,* doi: 10.2204/iodp.proc.303306.2006.

Channell, J. E. T., & Kleiven, H. F. (2000). Geomagnetic palaeointensities and astrochronological ages for the Matuyama-Brunhes boundary and the boundaries of the Jaramillo Subchron: Palaeomagnetic and oxygen isotope records from ODP Site 983 Phil. *Transactions of the Royal Society of London. Series A, 358,* 1027–1047.

Channell, J. E. T., Mauzaud, A., Sullivan, P., Turner, S., & Raymo, M. E. (2002). Geomagnetic excursions and paleointensities in the Matuyama Chron at ocean drilling sites 983 and 984 (Iceland Basin). *Journal of Geophysical Research, 107.* doi: 10.1029/2001JB000491.

Channell, J. E. T., Sato, T., Kanamatsu, T., Stein, R., Malone, M., & Alvarez-Zarikian, C. the Expedition 303/306 Scientists, (2006b). IODP Expeditions 303 and 306 monitor Miocene-Quaternary climate in the North Atlantic. *Scientific Drilling, 2,* 4–10.

Channell, J. E. T., Stoner, J. S., Hodell, D. A., & Charles, C. D. (2000). Geomagnetic paleointensity from late Brunhes-age piston cores from the subantarctic South Atlantic. *Earth and Planetary Science Letters, 175,* 145–160.

Chapman, M. R., & Shackleton, N. J. (1999). Global ice-volume fluctuations, North Atlantic ice-rafting events, and deep-ocean circulation changes between 130 and 70 ka. *Geology, 27,* 795–798.

Chappell, J., Omura, A., Esat, T., McCulloch, M., Pandolfi, J. Y. O., & Pillans, B. (1996). Reconciliation of late Quaternary sea levels derived from coral terraces at Huon Peninsula with deep sea oxygen isotope records. *Earth and Planetary Science Letters, 141,* 227–236.

Chappell, J., & Shackleton, N. J. (1986). Oxygen isotopes and sea level. *Nature, 324,* 137–140.

Charles, C. D., & Fairbanks, R. G. (1990). Glacial to interglacia l changes in the isotopic gradient s of the Southern Ocean surface waters. In: U. Bleil & J. Thiede (Eds), *Geologic History of the Polar Oceans* (pp. 519–538). Dordrecht: Kluwer Academics Publishers.

Cherkis, N. Z., Max, M. D., Vogt, P. R., Crane, K., Midthassel, A., & Sundvor, E. (1999). Large-scale mass wasting on the north Spitsbergen continental margin, Arctic Ocean. *Geo-Marine Letters, 19,* 131–142.

Choubert, G., & Faure-Muret, A. (1976). *Geological world atlas 1/10,000,000.* commission for the geological map of the world-UNESCO, sheets 10–12.

Clark, D. L. (1970). Magnetic reversals and sedimentation rates in the Arctic Basin. *Geological Society of America Bulletin, 81,* 3129–3134.

Clark, D. L. (1974). Late Mesozoic and early Cenozoic sediment cores from the Arctic Ocean. *Geology, 2,* 41–44.

Clark, D. L. (1977). Paleontologic response to post-Jurassic crustal plate movements in the Arctic Ocean. *Biology Geology, 2,* 55–76.

Clark, D. L. (1988). Early history of the Arctic Ocean. *Paleoceanography, 3,* 539–550.

Clark, D. L. (1990). Arctic Ocean ice cover; geologic history and climatic signifcance. In: A. Grantz, L. Johnson & J. F. Sweeney (Eds), *The Arctic Ocean region. (The Geology of North America)* (pp. 53–62). Geological Society of America (Special Paper).

Clark, D. L. (1996). The Pliocene record in the central Arctic Ocean. *Marine Micropaleontology, 27*, 157–164.

Clark, D. L., Byers, C. W., & Pratt, L. M. (1986). Cretaceous black mud from the central Arctic Ocean. *Paleoceanography, 1*, 265–271.

Clark, D. L., Chern, L. A., Hogler, J. A., Mennicke, C. M., & Atkins, E. D. (1990). Late Neogene climate evolution of the central Arctic Ocean. *Marine Geology, 93*, 69–94.

Clark, D. L., & Hanson, A. (1983). Central Arctic Ocean sediment texture: A key to ice transport mechanism. In: B. F. Molnia (Ed.), *Glacial-marine sedimentation* (pp. 301–330). New York: Plenum Press.

Clark, D. L., Vincent, J.-S., Jones, G. A., & Morris, W. A. (1984). Correlation of marine and continental glacial and interglacial events, Arctic Ocean and Banks Island. *Nature, 311*, 147–149.

Clark, D. L., Whitman, R. R., Morgan, K. A., & Mackey, S. D. (1980). Stratigraphy and glacialmarine sediments of the Amerasian Basin, central Arctic Ocean. *Geological Society of America, Special Paper, 181*, 57.

Clarke, L. J., & Jenkyns, H. C. (1999). New oxygen-isotope evidence for long-term Cretaceous climate change in the Southern Hemisphere. *Geology, 27*, 699–702.

Clark, P. U., & Mix, A. C. (2002). Ice sheets and sea level of the Last Glacial Maximum. *Quaternary Science Review, 21*, 1–7.

Clark, P. U., Pisias, N. G., Stocker, T. F., & Weaver, A. J. (2002). The role of the thermohaline circulation in abrupt climate change. *Nature, 415*, 863–869.

CLIMAP (1981). *Seasonal reconstructions of the earth's surface at the last glacial maximum*. Geological Society of America, Map and Chart Ser., MC-36.

Clough, L. M., Ambrose, W. G., Cochran, J. K., Barnes, C., Renaud, P. E., & Aller, R. C. (1997). Infaunal density, biomass, and bioturbation in the sediments of the Arctic Ocean. *Deep-Sea Research, 44*, 1683–1704.

Cochran, J. K., Barnes, C., Achman, D., & Hirschberg, D. J. (1995). Thorium-234/Uranium-238 disequilibrium as an indicator of scavenging rates and particulate organic carbon fluxes in the Northeast Water Polynya, Greenland. *Journal of Geophysical Research (C), 100*, 4399–4410.

Colodner, D., Edmond, J., & Boyle, E. (1995). Rhenium in the Black Sea: Comparison with molybdenum and uranium. *Earth and Planetary Science Letters, 131*, 1–15.

Colodner, D., Sachs, J., Ravizza, G., Turekian, K., Edmont, J., & Boyle, E. (1993). The geochemical cycle of rhenium: A reconnaissance. *Earth and Planetary Science Letters, 117*, 205–221.

Colony, R., & Thorndike, A. S. (1985). Sea ice motion as a drunkard's walk. *Journal of Geophysical Research, 90*, 965–974.

Conlon, D. M., & Curtin, T. B. (2004). The ONR high latitude dynamics program. *Arctic Research of the United States, 18*, 2–5.

Conte, M. H., Eglinton, G., & Madureira, L. A. S. (1992). Long-chain alkenones and alkyl alkenoates as palaeotemperature indicators: Their production, flux and early sedimentary diagenesis in the eastern North Atlantic. *Organic Geochemistry, 19*, 287–298.

Conte, M. H., Volkman, J. K., & Eglinton, G. (1994). Lipid biomarkers of Haptophyta. In: B. Leadbeater & J. C. Green (Eds), *The Haptophyte algae* (pp. 351–377). Oxford: Clarendon.

Conte, M. H., Weber, J. C., King, L. L., & Wakeham, S. G. (2001). The alkenone temperature signal in western North Atlantic surface waters. *Geochimica et Cosmochimica Acta, 65*(23), 4275–4287.

Coppola, L., Roy-Barman, M., Wassmann, P., Mulsow, S., & Jeandel, C. (2002). Calibration of sediment traps and particulate organic carbon export using ^{234}Th in the Barents Sea. *Marine Chemistry, 80*, 11–26.

Cornford, C., Gardner, P., & Burgess, C. (1998). Geochemical truths in large data sets, I: Geochemical screening data. *Organic Geochemistry, 29*(1–3), 519–530.

Cornford, C., Needham, C. E. J., & de Walque, L. (1986). Geochemical habitat of North Sea oils and gases. In: A. M. Spencer (Ed.), *Habitat of Hydrocarbons on the Norwegian Continental Shelf* (pp. 38–54). London: Graham & Trotman.

Cortijo, E., Labeyrie, L., Vidal, L., Vautravers, M., Chapman, M., Duplessy, J.-C., Elliot, M., Arnold, M., Turon, J.-L., & Auffret, G. (1997). Changes in sea surface hydrology associated with Heinrich event 4 in the North Atlantic Ocean between 40 and 60 N. Earth Planet. *Science Letters, 146*, 29–45.
Coxall, H. K., Wilson, P. A., Pälike, H., Lear, C. H., & Backman, J. (2005). Rapid stepwise onset of Antarctic glaciation and deeper calcite compensation in the Pacific Ocean. *Nature, 433*, 53–57.
Cramer, B., & Franke, D. (2005). Indications for an active petroleum system in the Laptev Sea, NE Siberia. *Journal Petroleum Geology, 28*, 369–384.
Crane, K., Galasso, J., Brown, C., Cherkashov, G., Ivanov, G., Petrova, V., & Vanstayn, B. (2001). Northern ocean inventories of organochlorine and heavy metal contamination. *Marine Pollution Bulletin, 43*, 28–60.
Cranston, R. E. (1997). Organic carbon burial rates across the Arctic Ocean from the 1994 Arctic Ocean Section expedition. *Deep-Sea Research, 44*, 1705–1724.
Cremer, H. (1999). Distribution patterns of diatom surface sediment assemblages in the Laptev Sea (Arctic Ocean). *Marine Micropaleontology, 38*, 39–67.
Cronin, T. M. (1977). Champlain sea Foraminifera and Ostracoda: A systematic and paleoecological synthesis. *Géographie Physique et Quaternaire, 31*, 107–122.
Cronin, T. M. (1981). Paleoclimatic implications of late Pleistocene marine ostracodes from the St. Lawrence Lowlands. *Micropalaeontology, 27*(4), 384–418.
Cronin, T. M. (1989). Paleozoogeography of postglacial Ostracoda from Northeastern North America. In: N. R. Gadd (Ed.), *The late Quaternary development of the Champlain Sea Basin* (pp. 125–144). Geological Association of Canada (Special Paper 35).
Cronin, T. M. (1996). Distribution of deep-sea ostracoda in the Arctic Ocean. In: R. Stein, G. Ivanov, M. Levitan & K. Fahl (Eds), *Surface-sediment composition and sedimentary processes in the central Arctic Ocean and adjacent Eurasian continental margin. Report on Polar Research, 212*, 269–282.
Cronin, T. M., Holtz, T. R., Stein, R., Spielhagen, R., Fütterer, D., & Wollenburg, J. (1995). Late Quaternary deep water and mid-depth history of the Eurasian Basin, Arctic Ocean. *Paleoceanography, 10*, 259–281.
Cronin, T. M., Holtz, T. R., & Whatley, R. C. (1994). Quaternary paleoceanography of the deep Arctic Ocean based on quantitative analysis of Ostracoda. *Marine Geology, 119*, 305–332.
Cronin, T. M., Smith, S. A., Eynaud, F., O'Regan, M., & King, J. (2008). Quaternary paleoceanography of the Central Arctic based on IODP ACEX 302 foraminiferal assemblages. *Paleoceanography, 23*, PA1S18, doi:10.1029/2007PA001484.
Crosta, X., Pichon, J.-J., & Burckle, L. H. (1998). Application of the modern analog technique to marine Antarctic diatoms: Reconstruction of maximum sea-ice extent at the Last Glacial Maximum. *Paleoceanography, 13*(3), 284–297.
Crusius, J., Calvert, S., Pedersen, T., & Sage, D. (1996). Rhenium and molybdenum enrichments in sediments as indicators of oxic, suboxic and sulfidic conditions of deposition. Earth Planet. *Science Letters, 145*, 65–78.
Crusius, J., & Thomson, J. (2000). Comparative behavior of authigenic Re, U, and Mo during reoxidation and subsequent long-term burial in marine sediments. *Geochimica et Cosmochimica Acta, 64*, 2233–2242.
Dale, B. (1996). Dinoflagellate cyst ecology: Modeling and geological applications. In: J. Jansonius & D. C. McGregor (Eds), *Palynology: Principles and applications* (Vol. 3, pp. 1249–1275). Dallas, TX: American Association of Stratigraphic Palynologists Foundation.
Dallimore, S. R. (1991). Geological, geotechnical and geophysical studies along an onshore-offshore transect of the Beaufort Shelf. *Geological Survey of Canadian Open File, 2408*, 325pp.
Dallimore, S. R., Wolfe, S. A., & Solomon, S. M. (1996). Influence of ground ice and permafrost on coastal evolution, Richards Island, Beaufort Sea coast, N.W.T. Can. *Journal of Earth Science, 33*, 664–675.
Dalrymple, R. W., & Maass, O. C. (1987). Clay mineralogy of late Cenozoic sediments in the CESAR cores, Alpha Ridge, central Arctic ocean. *Canadian Journal of Earth Science, 24*, 1562–1569.
Dam, G. D., Nøhr-Hansen, H., Christiansen, F. G., Bojesen-Koefoed, J. A., & Laier, T. (1998). The oldest marine Cretaceous sediments in West Greenland (Umiivik-1 borehole) -record of the Cenomanian-Turonian Anoxic Event? *Geological Greenland Survey Bulletin, 180*, 128–137.

Danilov, I. D., Komarov, I. A., & Vlasenko, A. Y. (1998). Pleistocene-Holocene permafrost of the east Siberian Eurasian arctic shelf. In: A. G. Lewkowicz & M. Allard (Eds), *Proceedings of the Seventh International Conference on Permafrost* (pp. 207–212). Centre d'Etudes Nordiques, Université Laval, Québec.

Dansgaard, W., Clausen, H. B., Grundestrup, N., Hammer, C. U., Johnsen, S. J., Kristinsdorttir, P. M., & Reeh, N. (1982). A new Greenland deep ice core. *Science, 218*, 1273–1277.

Dansgaard, W., Johnen, S. J., Clausen, H. B., Dahl-Jensen, D., Grundestrup, N. S., Hammer, C. U., Hvidberg, C. S., Steffensen, J. P., Sveinbjörnsdottir, A. E., Jouzel, J., & Bond, G. (1993). Evidence for general instability of past climate from a 250-kyr ice-core record. *Nature, 364*, 218–220.

Darby, D. A. (1975). Kaolinite and other clay minerals in Arctic Ocean sediments. *Journal of Sedimentary Petrology, 45*, 272–279.

Darby, D. A. (2003). Sources of sediment found in sea ice from the western Arctic Ocean, new insights into processes of entrainment and drift patterns. *Journal of Geophysical Research, 108*, C8, 3257. doi: 10.1029/2002JC001350.

Darby, D. A. (2008). The Arctic perennial ice cover over the last 14 million years. *Paleoceanography*, 23, PA1S07, doi: 10.1029/2007PA001479.

Darby, D. A., & Bischof, J. F. (1996). A statistical approach to source determination of lithic and Fe oxide grains: An example from the Alpha Ridge, Arctic Ocean. *Journal of Sediment Research, 66*, 599–607.

Darby, D. A., & Bischof, J. F. (2004). A Holocene record of changing Arctic Ocean ice drift, analogous to the effects of the Arctic Oscillation. *Paleoceanography*, 19, PA1027, doi: 10.1029/2003PA000961.

Darby, D. A., Bischof, J. F., & Jones, G. A. (1997). Radiocarbon chronology of depositional regimes in the western Arctic Ocean. *Deep Sea Research Part II, 44*(8), 1745–1757.

Darby, D. A., Bischof, J. F., Spielhagen, R. F., Marshall, S. A., & Herman, S. W. (2002). Arctic ice export events and their potential impact on global climate during the late Pleistocene. *Paleoceanography, 17*(2), doi: 10.1029/2001PA000639.

Darby, D. A., Burckle, L. H., & Clark, D. L. (1974). Airborne dust on the Arctic pack ice: Its composition and fallout rate. *Earth and Planetary Science Letters, 24*, 166–172.

Darby, D. A., Jakobsson, M., & Polyak, L. (2005). Icebreaker expedition collects key Arctic seafloor and ice data. *Eos, 86*(52), 549–552.

Darby, D. A., Naidu, A. S., Mowatt, T. C., & Jones, G. (1989). Sediment composition and sedimentary processes in the Arctic Ocean. In: Y. Herman (Ed.), *The Arctic Seas — Climatology, Oceanography, Geology, and Biology* (pp. 657–720). New York: Van Nostrand Reinhold.

Darby, D. A., Polyak, L., & Bauch, H. (2006). Past glacial and interglacial conditions in the Arctic Ocean and marginal seas - a review. *Progress in Oceanography, 71*, 129–144.

Davies, T. A., Bell, T., Cooper, A. K., Josenhans, H., Polyak, L., Solheim, A., Stoker, M. S. & Stravers, J. A. (Eds). (1997). *Glaciated continental margins: An atlas of acoustic images*. London: Chapman & Hall. 315pp.

De Leeuw, J. W., Rijpstra, W. I. C., Schenck, P. A., & Volkman, J. K. (1983). Free, esterified, and residual bound sterols in Black Sea Unit I sediments. *Geochimica et Cosmochimica Acta, 47*, 455–465.

De Lurio, J. L., & Frakes, L. A. (1999). Glendonites as a paleoenvironmental tool: Implications for early Cretaceous high latitude climates in Australia. *Geochimica et Cosmochimica Acta, 63*, 1039–1048.

de Vernal, A., Henry, M., Matthiessen, J., Mudie, P. J., Rochon, A., Boessenkool, K. P., Eynaud, F., Grøsfjeld, K., Guiot, J., Hamel, D., Harland, R., Head, M. J., Kunz-Pirrung, M., Levac, E., Loucheur, V., Peyron, O., Pospelova, V., Radi, T., Turon, J.-L., & Voronina, E. (2001). Dinoflagellate cyst assemblages as tracers of sea-surface conditions in the northern North Atlantic, Arctic and sub-Arctic seas: The new n = 677 data base and its application for quantitative palaeoceanographic reconstruction. *Journal of Quaternary Science, 16*(7), 681–698.

de Vernal, A., & Mudie, P. J. (1989). Late Pliocene to Holocene palynostratigraphy at ODP-Site 645, Baffin Bay. In: S. P. Srivastava, M. A. Arthur, et al. (Eds), *Proceeding of ODP, Science Results* (pp. 387–400), 105, College Station, Texas (Ocean Drilling Program).

de Vernal, A., Turon, J.-L., & Guiot, J. (1994). Dinoflagellate cyst distribution in high-latitude marine environments and quantitative reconstruction of sea-surface salinity, temperature, and seasonality. *Canadian Journal of Earth Science, 31*, 48–62.

De'ry, S. J., & Wood, E. F. (2005). Decreasing river discharge in northern Canada. *Geophysical Research Letters, 32*, L10401. doi: 10.1029/2005GL022845.

Dean, W. E., & Arthur, M. A. (1989). Iron-sulfur-carbon relationships in organic-carbon-rich sequences: Cretaceous western interior seaway. *American Journal of Science, 289*, 708–743.

Dean, W. E., Arthur, M. A., & Claypool, G. E. (1986). Depletion of ^{13}C in Cretaceous marine organic matter: Source, diagenetic, or environmental signal?. *Marine Geology, 70*, 119–157.

DeConto, R. M., & Pollard, D. (2003). Rapid Cenozoic glaciation of Antarctica induced by declining atmospheric CO_2. *Nature, 421*, 245–249.

Dehairs, F., Chesselet, R., & Jedwab, J. (1980). Discrete suspended particles of barite and the barium cycle in the open ocean. *Earth and Planetary Science Letters, 49*, 528–550.

Dell'Agnese, D. J., & Clark, D. L. (1994). Siliceous microfossils from the warm late Cretaceous and early Cenozoic Arctic Ocean. *Journal of Paleontology, 68*, 31–47.

Demaison, G. J., & Moore, G. T. (1980). Anoxic environments and oil source bed genesis. *Organic Geochemistry, 2*, 9–31.

Denton, G. H., & Amstrong, R. L. (1969). Miocene-Pliocene glaciations in southern Alaska. *American Journal of Science, 267*, 1121–1142.

Denton, G. H., & Hughes, T. J. (1981). *The last great ice sheets*. New York: Wiley, 484pp.

Denton, G. H., & Hughes, T. J. (2002). Reconstructing the Antarctic Ice Sheet at the Last Glacial Maximum. *Quaternary Science Review, 21*, 193–202.

DeRosa, M., & Gambacorta, A. (1988). The lipids of archaebacteria. *Progress in lipid research, 27*, 153–175.

Dethleff, D. (2005). Entrainment and export of Laptev Sea ice sediments, Siberian Arctic. *Journal of Geophysical Research, 110*(C07009), doi: 10.1029/2004JC002740.

Dethleff, D., Loewe, P., Weiel, D., Nies, H., Kuhlmann, G., Bahe, C., & Tarasov, G. (1998). Winter expedition to the Southwestern Kara Sea — investigations on formation and transport of turbid sea-ice. *Report on Polar Research, 271*, 40pp.

Dethleff, D., Nürnberg, D., Reimnitz, E., Saarso, M., & Savchenko, Y. P. (1993). East Siberian Arctic region expedition '92: The Laptev Sea — its significance for Arctic Sea ice formation and transpolar sediment flux. *Report on Polar Research, 120*, 3–37.

Dethleff, D., Rachold, V., Tintelnot, T., & Antonow, M. (2000). Sea-ice transport of riverine particles from the Laptev Sea to Fram Strait based on clay mineral studies. *International Journal of Earth Science, 89*, 496–502.

Dibb, J. E. (1996). Overview of field data on the deposition of aerosol-associated species to the surface snow of Polar glaciers, particularly recent work in Greenland. In: E. W. Wolff & R. C. Bales (Eds), *Chemical exchange between the atmosphere and Polar snow* (pp. 249–274). Heidelberg: Springer-Verlag.

Dickens, G. R., O'Neil, J. R., Rea, D. K., & Owen, R. M. (1995). Dissociation of oceanic methane hydrate as a cause of the carbon isotope excursion at the end of the Paleocene. *Paleoceanography, 10*, 965–971.

Dickson, R. R., Osborn, T. J., Hurrell, J. W., Meincke, J., Blindheim, J., Adlandsvik, B., Vinje, T., Alekseev, G., & Maslowski, W. (2000). The Arctic Ocean response to the North Atlantic Oscillation. *Journal of Climate, 13*(15), 2671–2696.

Ditchfield, P. W. (1997). High northern palaeolatitude Jurassic-Cretaceous palaeotemperature variation: New data from Kong Karls Land, Svalbard. *Palaeogeography, Palaeoclimatology, Palaeoecology, 130*, 163–175.

Dittmers, K., Niessen, F., & Stein, R. (2003). Holocene sediment budget and sedimentary history of the Ob and Yenisei estuaries. In: R. Stein, K. Fahl, D. K. Fütterer & E. M. Galimov (Eds), *Siberian River run-off in the Kara Sea: Characterisation, quantification, variability, and environmental significance. Proceeding of Marine Science* (Vol. 6, pp. 457–484). Amsterdam: Elsevier.

Dittmers, K., Niessen, F., & Stein, R. (2008). Late Weichselian fluvial evolution on the southern Kara Sea Shelf, North Siberia. *Global and Planetary Change*, in press.

Dixon, J., Dietrich, J., Snowdon, L. R., Morrell, G., & McNeil, D. H. (1992). Geology and petroleum potential of upper Cretaceous and tertiary strata, Beaufort Mackenzie area, Northwest Canada. *Bull. Am. Assoc. Petrol. Geol., 76*, 927–947.

Dmitrenko, I. A., Gribanov, V. A., Volkov, D. L., & Kassens, H. (1999). Impact of river discharge on the sea land fast ice extention in the Russian arctic shelf area. In: J. Tuhkuri & K. Riska (Eds), *Proceeding of 15th International Conference on Port and Ocean Enginering under Arctic Conditions, Espoo* (Vol. 1, 311–321), Finland, August 23–27.

Dokken, T. M., & Hald, M. (1996). Rapid climatic shifts during isotope stages 2-4 in the Polar North Atlantic. *Geology, 24*(7), 599–602.

Dolginow, J., & Kropatschjow, S. (1994). *Abriss der Geologie Russlands und angrenzender Staaten.* Schweizerbart'sche Verlagsbuchhandlung Stuttgart, 163pp.

Dolven, J. K., & Bjørklund, K. R. (2001). An early Holocene peak occurrence and recent distribution of Rhizoplegma boreale (Radiolaria): A biomarker in the Norwegian Sea. *Marine Micropaleontology, 42*, 25–44.

Domack, E. W., & Domack, C. R. (1992). *Cenozoic glaciation — the marine record established by ocean drilling.* Joint Oceanographic Instituttions, U.S. Science Sopprot Program (JOI/USSSP), Washington DC, 49pp.

Dowdeswell, J. A., Elverhøi, A., & Spielhagen, R. (1998). Glacimarine sedimentary processes and facies on the Polar North Atlantic margins. *Quaternary Science Review, 17*, 243–272.

Dowdeswell, J. A., Kenyon, N. H., & Laberg, J. S. (1997). The glacierinfluenced Scoresby Sund Fan, East Greenland continental margin: Evidence from GLORIA and 3.5 kHz records. *Marine Geology, 143*, 207–221.

Dowdeswell, J. A., Kenyon, N. H., Elverhøi, A., Laberg, J. S., Hollender, F.-J., Mienert, J., & Siegert, M. (1996). Large-scale sedimentation on the glacier-influenced Polar North Atlantic margins: Long-range side-scan sonar evidence. *Geophysical Research Letters, 23*, 3535–3538.

Dowdeswell, J. A., Maslin, M. A., Andrews, J. T., & McCave, I. N. (1995). Iceberg production, debris rafting, and the extent and thickness of Heinrich layers (H-1, H-2) in North Atlantic sediments. *Geology, 23*, 301–304.

Dowdeswell, J. A. &, Ó Cofaigh, C. (Eds). (2002). Glacier-influenced sedimentation on high-latitude continental margins. *Geological Society of London (Special Publication), 203*, 378pp.

Dowdeswell, J. A., Ó Cofaigh, C., Taylor, J., Kenyon, N. H., Mienert, J., & Wilken, M. (2002). On the architecture of high-latitude continental margins: The influence of ice-sheet and sea-ice processes in the Polar North Atlantic. In: J. A. Dowdeswell, & C. Ó Cofaigh (Eds), *Glacier influenced sedimentation on high-latitude continental margins, Geological Society of London* (Special Publication), *203*, 33–54.

Dowdeswell, J. A., & Scourse, J. D. (Eds). (1990). *Glacimarine environments: Processes and sediments.* Geological Society (Special Publication), *53*.

Dowdeswell, J. A., & Siegert, M. J. (1999). Ice-sheet numerical modelling and marine geophysical measurements of glacier-derived sedimentation on the Eurasian Arctic continental margins. *Geological Society of America Bulletin, 111*, 1080–1097.

Dowdeswell, J. A., Uenzelmann-Neben, G., Whittington, R. J., & Marienfeld, P. (1994b). The late Quaternary sedimentary record in Scoresby Sund, East Greenland. *Boreas, 23*, 294–310.

Dowdeswell, J. A., Villinger, H., Whittington, R. J., & Marienfeld, P. (1993). Iceberg scouring in Scoresby Sund and on the East Greenland continental shelf. *Marine Geology, 111*, 37–53.

Dowdeswell, J. A., Whittington, R. J., & Hodgkins, R. (1992). The sizes, frequencies and freeboards of East Greenland icebergs observed using ship radar and sextant. *Journal of Geophysical Research, 97*, 3515–3528.

Dowdeswell, J. A., Whittington, R. J., & Marienfeld, P. (1994a). The origin of massive diamicton facies by iceberg rafting and scouring, Scoresby Sund, East Greenland. *Sedimentology, 41*, 21–35.

Dowsett, H. J., Chandler, M. A., Cronin, T. M., & Dwyer, G. S. (2005). Middle Pliocene sea surface temperature variability. *Paleoceanography, 20*, PA2014. doi: 10.1029/2005PA001133.

Dowsett, H. J., Cronin, T. M., Poore, R. Z., Thompson, R. S., Whatley, R. C., & Wood, A. M. (1992). Micro-paleontological evidence for increased meridional heat transport in the North Atlantic Ocean during the Pliocene. *Science, 258*, 1133–1155.

Driscoll, N. W., & Haug, G. H. (1998). A short circuit in thermohaline circulation: A cause for Northern Hemisphere glaciation? *Science, 282*, 436–438.

Duce, R. A., Liss, P. S., & Merrill, J. T. (1991). The atmospheric input of trace species to the World Ocean. *Global Biogeochemical Cycles, 5*(3), 193–259.

Dunayev, N. N., & Pavlidis, J. A. (1988). A model of the late Pleistocene glaciation of Eurasiatic Arctic shelf. In: V.M. Kotlyakov, & V. E. Sokolov (Eds), *Arctic research — advances and prospects proceedings of the conference of Arctic and Nordic countries on coordination of research in the Arctic* (Vol. 2, pp. 70–72), Leningrad (Academy of Sciences of the USSR.

Dunhill, G. (1998). *Comparison of sea-ice and glacial-ice rafted debris: grain size, surface features, and grain shape.* Open-File Report — U. S. Geological Survey, Report OF 98-0367, 74pp.

Duplessy, J.-C., Ivanova, E., Murdmaa, I., Paterne, M., & Labeyrie, L. (2001). Holocene paleoceanography of the northern Barents Sea and variations of the northward heat transport by the Atlantic Ocean. *Boreas, 30,* 2–16.

Duzhikov, O. A., & Strunin, B. M. (Eds). (1992). *Geology and metallogeny of sulfide deposits.* Norilisk region: Moscow: USSR SEG Special Publication, 60pp.

Dyke, A. S., Andrews, J. T., Clark, P. U., England, J. H., Miller, G. H., Shaw, J., & Veillette, J. J. (2002). The Laurentide and Innuitian ice sheets during the Last Glacial Maximum. *Quaternary Science Review, 21,* 9–31.

Dyke, A. S., England, J., Reimnitz, E., & Jetté, H. (1997). Changes in driftwood delivery to the Canadian Arctic Archipelago: The hypothesis of postglacial oscillations of the transpolar drift. *Arctic, 50,* 1–16.

Dyke, A. S., Morris, T. F., Green, D. E. C., & England, J. (1992). Quaternary geology of Prince of Wales Island, Arctic Canada. *Geolgical Survey of Canada,* Memoir 433.

Dyke, A. S., & Prest, V. K. (1987). LateWisconsinan and Holocene history of the Larentide ice sheet. *Geogr. Phys. Quat., 41,* 237–263.

Dymond, J., Suess, E., & Lyle, M. (1992). Barium in deep sea sediment: A geochemical indicator of paleoproductivity. *Paleoceanograpgy, 7,* 163–181.

Dzvonik, J. P. (1996). *Alkenones as records of oceanic paleotemperatures: Studies of Eocene and Oligocene sediments from the North, South, and Equatorial Atlantic.* Monograph, 21pp.

Edwards, M. H., & Coakley, B. J. (2003). SCICEX investigations of the Arctic Ocean system. *Chemie der Erde, 63,* 281–392.

Edwards, M. H., & Coakley, B. J. (2004). The SCICEX program-Arctic Ocean investigations from a U.S. Navy Nuclear-Powered Submarine. *Arctic Research of the United States, 18,* 14–20.

Edwards, M. H., Kurras, G. J., Tolstoy, M., Bohnenstiehl, D. R., Coakley, B. J., & Cochran, J. R. (2001). Evidence of recent volcanic activity on the ultraslow-spreading Gakkel Ridge. *Nature, 409,* 808–812.

Eglinton, G., & Hamilton, R. J. (1963). The distribution of alkanes. In: T. Swan (Ed.), *Chemical Plant Taxonomy* (pp. 187–208). Academic Press: London.

Ehlers, J., & Gibbard, P. L. (2003). Extent and chronology of glaciations. *Quaternary Science Review, 22,* 1561–1568.

Ehlers, J., & Gibbard, P. L. (Eds). (2004a). *Quaternary glaciations-extent and chronology, Part I: Europe. developments in Quaternary science* (Vol. 2a). Amsterdam: Elsevier.

Ehlers, J., & Gibbard, P. L. (Eds). (2004b). *Quaternary glaciations-extent and chronology, Part II: North America. Developments in Quaternary science* (Vol. 2b). Amsterdam: Elsevier.

Ehlers, J., & Gibbard, P. L. (Eds). (2004c). *Quaternary glaciations-extent and chronology, Part III: South America, Asia, Africa, Australasia, Antarctica. Developments in Quaternary science* (Vol. 2c). Amsterdam: Elsevier.

Ehlers, J., & Gibbard, P. L. (2007). The extent and chronology of Cenozoic global glaciation. *Quaternary International, 164–165,* 6–20.

Ehrmann, W. U., & Mackensen, A. (1992). Sedimentological evidence for the formation of an East Antarctic ice sheet in Eocene/Oligocene time. *Palaeogeography, palaeoclimatology, palaeoecology, 93,* 85–112.

Ehrmann, W. U., Melles, M., Kuhn, G., & Grobe, H. (1992). Significance of clay mineral assemblages in the Antarctic Ocean. *Marine Geology, 107,* 249–273.

Eicken, H. (2004). The role of Arctic sea ice in transporting and cycling terrigenous organic matter. In: R. Stein & R. W. Macdonald (Eds), *The organic carbon cycle in the Arctic Ocean* (pp. 45–53). Heidelberg: Springer-Verlag.

Eicken, H., Gradinger, R., Graves, A., Mahoney, A., & Rigor, I. (2005). Sediment transport by sea ice in the Chukchi and Beaufort Seas: Increasing importance due to changing ice conditions? *Deep-Sea Research Part II, 52*, 3281–3302.

Eicken, H., Kolatschek, J., Freitag, J., Lindemann, F., Kassens, K., & Dmitrenko, I. (2000). A key source area and constraints on entrainment for basin scale sediment transport by Arctic sea ice. *Geophysical Research Letters, 27*(13), 1919–1922.

Eicken, H., Lensu, M., Leppäranta, M., Tucker, W. B. I., Gow, A. J., & Salmela, O. (1995). Thickness, structure, and properties of level summer multiyear ice in the Eurasian sector of the Arctic Ocean. *Journal of Geophysical Research (C), 11*, 22,697–22,710.

Eicken, H., Reimnitz, E., Alexandrov, V., Martin, T., Kassens, H., & Viehoff, T. (1997). Sea-ice processes in the Laptev Sea and their importance for sediment export. *Continental Shelf Research, 17*(2), 205–233.

Eide, E.A. (Coord.), (2002). BATLAS, Mid Norway plate reconstruction atlas with global and Atlantic perspectives. *Geological Survey of Norway*, Trondheim, 75pp.

Eidvin, T., Goll, R. M., Grogan, P., Smelror, M., & Ulleberg, K. (1994). En stratigrafisk undersøkelse av øvre del av brønn 7316/5-1 (Bjørnøya Vest). *Nor. Pet. Director. Contrib., 38*, 81pp.

Eidvin, T., Jansen, E., & Riis, F. (1993). Chronology of tertiary fan deposits off western Barents Sea: implications for the uplift and erosion history of the Barents Sea Shelf. *Marine Geology, 112*, 109–131.

Eidvin, T., & Riis, F. (1989). Nye dateringer av de tre vestligste borhullene i Barentshavet. *Resultater og konsekvenser for den tertiaere hevingen. Nor. Pet. Director. Contrib., 27*, 44pp.

Eiken, O. (Ed.) (1994). *Seismic Atlas of Western Svalbard: A selection of regional seismic transects, Meddelelser* (vol. 130), Oslo, Seiten: Norsk Polarinstitutt.

Eisenhauer, A., Meyer, H., Rachold, R., Tütken, T., Wiegand, B., Hansen, B. T., Spielhagen, R. F., Lindemann, F., & Kassens, H. (1999). Grain size separation and sediment mixing in Arctic Ocean sediments: Evidence from the strontium isotope systematic. *Chemical Geology, 158*, 173–188.

Eisenhauer, A., Spielhagen, R. F., Frank, M., Hentzschel, G., Mangini, A., Kubik, P. W., Dittrich-Hannen, B., & Billen, T. (1994). ^{10}Be records of sediment cores from high northern latitudes-Implications for environmental and climatic changes. Earth Planet. *Science Letters, 124*, 171–184.

Eldholm, O., Thiede, J., & Taylor, E., et al. (Eds). (1989). *Proceedings of ODP Science Results, 104*, College Station, Texas (Ocean Drilling Program), doi:10.2973/odp.proc.sr.104.1989.

Eldrett, J. S., Harding, I. C., Firth, J. V., & Roberts, A. P. (2004). Magnetostratigraphic calibration of Eocene-Oligocene dinoflagellate cyst biostratigraphy from the Norwegian-Greenland Sea. *Marine Geology, 204*, 91–127.

Eldrett, J. S., Harding, I. C., Wilson, P. A., Butler, E., & Roberts, A. P. (2007). Continental ice in Greenland during the Eocene and Oligocene. *Nature*, doi: 10.1038/nature05591.

Elverhøi, A., Andersen, A. E., Dokken, T., Hebbeln, D., Spielhagen, R. F., Svendsen, J. I., Sørflaten, M., Rørnes, A., Hald, M., & Forsberg, C. F. (1995a). The growth and decay of the Late Weichselian ice sheet in western Svalbard and adjacent areas based on provenance studies of marine sediments. *Quaternary Research, 44*, 303–316.

Elverhøi, A., de Blasio, F. V., Butt, F. A., Issler, D., Harbitz, C., Engvik, L., Solheim, A., & Marr, J. (2002). Submarine mass-wasting on glacially-influenced continental slopes: Processes and dynamics. In: J. A. Dowdeswell & C. Ó Cofaigh (Eds), *Glacier influenced sedimentation on high-latitude continental margins* (Vol. 203, pp. 73–88). *Geological Society of London (Special Publication).*

Elverhøi, A., Dowdeswell, J. A., Funder, S., Mangerud, J., & Stein, R. (Eds). (1998a). Glacial and oceanic history of the Polar North Atlantic margins. *Quaternary Science Review, 17*, 302pp.

Elverhøi, A., Hooke, R. L. B., & Solheim, A. (1998b). Late Cenozoic Erosion and Sediment Yield from the Svalbard-Barents Sea Region: Implications for understanding erosion of glacierized basins. *Quaternary Science Review, 17*, 209–241.

Elverhøi, A., Pfirman, S. L., Solheim, A., & Larsen, B. B. (1989). Glaciomarine sedimentation in epicontinental seas exemplified by the Northern Barents Sea. *Marine Geology, 85*, 225–250.

Elverhøi, A., Svensen, J. I., Solheim, A., Andersen, E. S., Milliman, J., Mangerud, J., & Hooke, R. L. (1995b). Late Quaternary sediment yield from the high Arctic Svalbard Area. *Journal of Geology, 103*, 1–17.

Emery, K. O. (1963). Organic transportation of marine sediments. In: M. N. Hill (Ed.), *The Sea* (pp. 776–793). New York: Wiley Interscience.
Engen, Ø. (2005). *Evolution of high Arctic Ocean Basins and continental margins*. Oslo: University of Oslo, 181pp.
England, J. (1999). Coalescent Greenland and Innutian ice during the last Glacial Maximum: Revising the Quaternary of the Canadian High Arctic. *Quaternary Science Review, 18*, 421–456.
Englebrecht, A. C., & Sachs, J. P. (2005). Determination of sediment provenance at drift sites using hydrogen isotopes and unsaturation ratios in alkenones. *Geochimica et Cosmochimica Acta, 69*(17), 4253–4265.
English, T. S. (1961). Some biological observations in the central North Polar Sea. Drift Station Alpha 1957–1958. *Arctic Inst. N. Am. Res. Pap., 13*, 8–80.
Environmental Working Group (EWG). (1998). Oceanography atlas for the summer period, in Joint U.S.-Russian Atlas of the Arctic Ocean [CD-ROM]. University of Colorado, Boulder, Colorado.
Erbacher, J., Huber, B. T., Norris, R. D., & Markey, M. (2001). Increased thermohaline stratification as a possible cause for an ocean anoxic event in the Cretaceous period. *Nature, 409*, 325–327.
Espitalié, J., Laporte, J. L., Madec, M., Marquis, F., Leplat, P., Paulet, J., & Boutefeu, A. (1977). Méthode rapide de caractérisation des roches mères, de leur potentiel pétrolier et de leur degre d'évolution. *Rev. Inst. Franc. Pétrol., 32*, 23–42.
Espitalié, J., Makadi, K. S., & Trichet, J. (1984). Role of mineral matrix during kerogen pyrolysis. *Organic Geochemistry, 6*, 365–382.
Evans, J., Dowdeswell, J. A., Grobe, H., Hubberten, H. -W., Niessen, F., Stein, R., & Whittington, R. J. (2002). Late Quaternary sedimentation in Kejser Franz Josephs Fjord and the continental margin of East Greenland. In: J. A. Dowdeswell & C. Ó Cofaigh (Eds), *Glacier-influenced sedimentation on high-latitude continental margins* (pp. 149–179). Geological Society of London (Special Publication 203).
Fahl, K., Cremer, H., Erlenkeuser, H., Hanssen, H., Hölemann, J., Kassens, H., Knickmeier, K., Kosobokova, K., Kunz-Pirrung, M., Lindemann, F., Markhaseva, E., Lischka, S., Petryashov, V., Piepenburg, D., Schmid, M., Spindler, M., Stein, R., & Tuschling, K. (2001). Sources and pathways of organic carbon in the modern Laptev Sea (Arctic Ocean): Implications from biological, geochemical and geological data. *Polarforschung, 69*(1999), 193–205.
Fahl, K., & Nöthig, E.-M. (2007). Lithogenic and biogenic particle fluxes on the Lomonosov Ridge (central Arctic Ocean) and their relevance for sediment accumulation: Vertical vs. lateral transport. *Deep-Sea Research Part I, 54*(8), 1256–1272.
Fahl, K., Nöthig, E., & Stein, R. (1997). Organischer Kohlenstoff in der Laptewsee und auf dem Lomonossowrücken. Zweijahresbericht 1996/97, Alfred-Wegener-Institut Bremerhaven, pp. 45–46.
Fahl, K., & Stein, R. (1997). Modern organic-carbon-deposition in the Laptev Sea and the adjacent continental slope: Surface-water productivity vs. terrigenous input. *Organic Geochemistry, 26*(5/6), 379–390.
Fahl, K., & Stein, R. (1999). Biomarkers as organic-carbon-source and environmental indicators in the Late Quaternary Arctic Ocean: "Problems and perspectives". *Marine Chemistry, 63*, 293–309.
Fahl, K., & Stein, R. (2007). Biomarker records, organic carbon accumulation, and river discharge in the Holocene southern Kara Sea (Arctic Ocean). *Geo-Marine Letters, 27*, 13–25.
Fahl, K., Stein, R., Gaye-Haake, B., Gebhardt, C., Kodina, L. A., Unger, D., & Ittekkot, V. (2003). Biomarkers in surface sediments from Ob and Yenisei estuaries and southern Kara Sea: Evidence for particulate organic carbon sources, pathways, and degradation. In: R. Stein, K. Fahl, D. K. Fütterer, E. M. Galimov & O. V. Stepanets (Eds), *Siberian River run-off in the Kara Sea: Characterisation, Quantification, variability, and environmental significance, Proceeding of Marine Science* (Vol. 6, pp. 329–348). Amsterdam: Elsevier.
Fahrbach, E. (Ed.). (2002). Die Expedition ARKTIS XVII/1 des Forschungsschiffes POLARSTERN 2001. *Report on Polar Marine Research, 433*, 60pp.
Fahrbach, E., Meincke, J., Østerhus, S., Rohardt, G., Schauer, U., Tverberg, V., & Verduin, J. (2001). Direct measurements of volume transports through Fram Strait. *Polar Research, 20*, 217–224.
Fairbanks, R. G. (1989). A 17,000-year glacio-eustatic sea level record: Influence of glacial melting rates on the Younger Dryas event and deep-ocean circulation. *Nature, 342*, 637–642.

Faleide, J. l., Solheim, A., Fiedler, A., Hjelstuen, B. O., Andersen, E. S., & Vanneste, K. (1996). Late Cenozoic evolution of the western Barents Sea-Svalbard continental margin. *Global and Planetary Change, 12*, 53–74.

Fanning, A. F., & Weaver, A. J. (1997). Temporal-geographical meltwater influences on the North Atlantic conveyor: Implications for the Younger Dryas. *Paleoceanography, 12*, 307–320.

Farrell, J. W., Pedersen, T. F., Calvert, S. E., & Nielsen, B. (1995). Glacial-interglacial changes in nutrient utilization in the equatorial Pacific Ocean. *Nature, 377*, 514–517.

Farrimond, P., Eglinton, G., & Brassell, S. C. (1986). Alkenones in Cretaceous black shales, Blake Bahama Basin, western North Atlantic. In: D. Leythaeuser & J. Rullkötter (Eds), *Advances in organic geochemistry, 1985. Organic Geochemistry, 10*, 897–903.

Fernandes, M. B., & Sicre, M. A. (2000). The importance of terrestrial organic carbon inputs on Kara Sea shelves as revealed by *n*-alkanes, OC and $\delta^{13}C$ values. *Organic Geochemistry, 31*, 363–374.

Fiedler, A., & Faleide, J. l. (1996). Cenozoic sedimentation along the southwestern Barents Sea margin in relation to uplift and erosion of the shelf. *Global and Planetary Change, 12*, 75–93.

Fietzke, J., Liebetreu, V., Eisenhauer, T., & Dullo, C. (2006). Determination of uranium isotope ratios by multi-static MIC-ICP-MS: Method and implementation for precise U- and Th-series isotope measurements. *Journal of Analytical Atomic Spectrometry, 20*, 395-401.

Firth, J. V., & Clark, D. L. (1998). An early Maastrichtian organic-walled phytoplankton cyst assemblage from an organic-walled black mud in Core F1-533, Alpha Ridge: Evidence for upwelling conditions in the Cretaceous Arctic Ocean. *Marine Micropaleontology, 34*, 1–27.

Fischer, G., & Wefer, G. (Eds). (1999). *Use of proxies in paleoceanography: Examples from the South Atlantic*. Heidelberg: Springer-Verlag. 735pp.

Flower, B., & Kennett, J. (1995). Middle Miocene deepwater paleoceanography in the Southwest Pacific; relations with East Antarctic ice sheet development. *Paleoceanography, 10*, 1095–1112.

Flower, B. P. (1997). Overconsolidated section on the Yermak Plateau, Arctic Ocean: Ice sheet grounding prior to ca. 660 ka? *Geology, 25*, 147–150.

Flower, B. P., Oppo, D. W., McManus, J. F., Venz, K. A., Hodell, D. A., & Cullen, J. L. (2000). North Atlantic intermediate to deep water circulation and chemical stratification during the past 1 Myr. *Paleoceanography, 15*, 388–403. doi: 10.1029/1999PA000430.

Forman, S. L., Ingolfsson, O., Gataullin, V., Manley, W. F., & Lokrantz, H. (1999). Late Quaternary stratigraphy of western Yamal Peninsula, Russia: New constraints on the configuration of the Eurasian ice sheet. *Geology, 27*, 807–810.

Forman, S. L., Lubinski, D. J., Ingolfsson, O., Zeeberg, J. J., Snyder, J. A., & Matishov, G. G. (2004). A review of postglacial emergence on Svalbard, Franz Josef Land and Novaya Zemlya, northern Eurasia. *Quaternary Science Review, 23*(11–13), 1391–1434.

Frahm, A. (2003). Qualitative Untersuchungen einer ausgeprägten Rinnenstruktur am ostgrönländischen Kontinentalhang. Unveröff. Diplomarbeit (Master thesis), Hochschule Hamburg, 153pp.

Frahm, A., Hohmann, & Matthiessen, J. (2003). Bathymetrical survey. In: P. Lemke (Ed.), *The expedition ARKTIS XIII/1 a, b of the research vessel Polarstern in 2002* (Vol. 446, pp. 37–38), *Report on Polar Marine Research*. Bremerhaven: Alfred Wegene Institute.

Frakes, L. A. (1979). *Climates through geological time*. Amsterdam: Elsevier, 310pp.

Frakes, L. A., & Francis, J. E. (1988). A guide to Phanerozoic cold Polar climates from high-latitude ice-rafting in the Cretaceous. *Nature, 333*, 547–549.

Frakes, L. A., Francis, J. E., & Syktus, J. I. (1992). *Climate modes of the Phanerozoic*. Cambridge: Cambridge University Press, 274pp.

Francis, J. A., Hunter, E., Key, J. R., & Wang, X. (2005). Clues to variability in Arctic minimum sea ice extent. *Geophysical Research Letters, 32*, L21501. doi: 10.1029/2005GL024376.

Francois, R., & Altabet, M. A. (1992). Glacial to interglacial changes in surface nitrate utilization in the Indian sector of the Southern Ocean as recorded by sediment $\delta^{15}N$. *Paleoceanography, 7*, 589–603.

Frank, M., Backman, J., Jakobsson, M., Moran, K., O'Regan, M., King, J., Haley, B. A., Kubik, P. W., & Garbe-Schönberg, D. (2008). Beryllium isotopes in central Arctic Ocean sediments over the past 12.3 million years: Stratigraphic and paleoclimatic implications. *Paleoceanography, 23*, PA1S02, doi: 10.1029/2007PA001478.

Frank, M., Schwarz, B., Baumann, S., Kubik, P. W., Suter, M., & Mangini, A. (1997). A 200 kyr record of cosmogenic radionuclide production rate and geomagnetic field intensity from ^{10}Be in globally stacked deep-sea sediments. *Earth and Planetary Science Letters, 149*, 121–129.

Frederichs, T. (1995). Regional and temporal variations of rock magnetic parameters in Arctic marine sediments. *Report on Polar Research, 164*, 212p.

Fredskild, B. (1973). Studies in the vegetational history of Greenland. *Palaeobotanical investigations of some Holocene lake and bog deposits. Meddelelser om Grønland, 198*(4), 245.

Frey, K. E., McClelland, J. W., Holmes, R. M., & Smith, L. C. (2007). Impacts of climate warming and permafrost thaw on the riverine transport of nitrogen and phosphorus to the Kara Sea. *Journal of Geophysical Research, 112*, G04S58. doi: 10.1029/2006JG000369.

Fronval, T., & Jansen, E. (1996). Late Neogene paleoclimates and paleoceanography in the Iceland Norweigian Sea: Evidence from the Iceland and Vøring Plateaus. In: J. Thiede, A. M. Myhre, J. V. Firth, G. L. Johnson & W. F. Ruddiman (Eds). *Proceedings of ODP, Science Results* (Vol. 151, pp. 455–468). College Station, Texas (Ocean Drilling Program).

Fronval, T., & Jansen, E. (1997). Eemian and early Weichselian 140-60 ka paleoceanography and paleoclimate in the Nordic seas with comparisons to Holocene conditions. *Paleoceanography, 12*, 443–462.

Fronval, T., Jansen, E., Bloemendal, J., & Johnsen, S. (1995). Oceanic evidence for coherent fluctuations in Fennoscandian and Laurentide ice sheets on millennium timescales. *Nature, 374*, 443–446.

Funder, S. (1984). Chronology of the last interglacial/glacial cycle in Greenland: First approximation. In: W. C. Mahaney (Ed.), *Correlation of Quaternary chronologies* (pp. 261–279). Geo Books: Norwich.

Funder, S. (1989). Quaternary geology of the ice-free areas and adjacent shelves of Greenland. In: R. J. Fulton (Ed.), *Quaternary geology of Canada and Greenland* (pp. 743–792), Geological Survey of Canada, Geology of Canada, no. 1 (Geol. Soc. Am., The Geology of the North America, v. K-1).

Funder, S., Hjort, C., & Landvik, J. Y. (1994). The last glacial cycles in East Greenland, an overview. *Boreas, 23*, 283–293.

Funder, S., Hjort, C., Landvik, J. Y., Nam, S., Reeh, N., & Stein, R. (1998). History of a stable ice sheet margin — East Greenland during the Middle and Upper Pleistocene. In: A. Elverhøi, J. A. Dowdeswell, S. Funder, J. Mangerud & R. Stein (Eds), *Glacial and oceanic history of the Polar North Atlantic Margins, Quaternary Science Review, 17*, 77–124.

Funder, S., Jennings, A., & Kelly, M. (2004). Middle and late Quaternary glacial limits in Greenland. In: J. Ehlers & P. L. Gibbard (Eds), *Quaternary glaciations — extent and chronology, Part II* (pp. 425–430). Amsterdam: Elsevier.

Fütterer, D. K. (Ed.). (1992). ARCTIC'91: The expedition ARK-VIII/3 of RV “"Polarstern” in 1991. *Report on Polar Research, 107*, 267pp.

Fütterer, D.K. (Ed.). (1994). The Expedition ARCTIC'93 Leg ARK IX/4 of RV “Polarstern” 1993. *Report on Polar Research, 149*, 244pp.

Ganopolski, A., & Rahmstorf, S. (2002). Abrupt glacial climate changes due to stochastic resonance. *Physical Review Letters, 88*(3), 038501. doi: 10.1103/PhysRevLett.88.038501.

Gard, G. (1988). Late Quaternary calcareous nannofossil biozonation, chronology and palaeo-oceanography in areas north ofthe Faeroe-Iceland Ridge. *Quaternary Science Review, 7*, 65–78.

Gard, G. (1993). Late Quaternary coccoliths at the North Pole: Evidence of ice-free conditions and rapid sedimentation in the central Arctic Ocean. *Geology, 21*, 227–230.

Gard, G., & Backman, J. (1990). Synthesis ofArctic and Sub-Arctic coccolith biochronology and history ofNorth Atlantic Drift water influx during the last 500,000 years. In: U. Bleil, & J. Thiede (Eds). *Geological history of the Polar Oceans: Arctic versus Antarctic* (Vol. 308, pp. 417–436), NATO ASI Series C, Kluwer Academic Publishers, Dordrecht.

Gard, G., & Crux, J. A. (1994). Reworked Jurassic-Neogene calcareous nannofossils in the central Arctic Ocean. *Marine Geology, 119*, 287–300.

Gataullin, V., Mangerud, J., & Svendsen, J.-I. (2002). The extent of the Late Weichselian ice sheet in the southeastern Barents Sea. *Global and Planetary Change, 31*, 453–474.

Gaye, B., Fahl, K., Kodina, L. A., Lahajnar, N., Nagel, B., Unger, D., & Gebhardt, A. C. (2007). Particulate matter fluxes in the southern and central Kara Sea compared to sediments: Bulk fluxes, amino acids, stable carbon and nitrogen isotopes, sterols and fatty acids. *Continental Shelf Research, 27*, 2570–2594.

Gebhardt, A. C., Gaye-Haake, B., Unger, D., Lahajnar, N., & Ittekkot, V. (2004). Recent particulate organic carbon and total suspended matter fluxes from the Ob and Yenisei Rivers into the Kara Sea (Siberia). *Marine Geology, 207*, 225–245.

Geissler, W., & Jokat, W. (2004). A geophysical study of the northern Svalbard continental margin. *Geophysical Journal International, 158*, 50–66. doi: 10.1111/j.1365-246X.2004.02315.x.

Gersonde, R., Abelmann, A., Brathauer, U., Becquey, S., Bianchi, C., Cortese, G., Grobe, H., Kuhn, G., Niebler, H.-S., Segl, M., Sieger, R., Zielinski, U., & Fütterer, D. K. (2003). Last glacial sea surface temperatures and sea-ice extent in the Southern Ocean (Atlantic-Indian sector): A multiproxy approach. *Paleoceanography, 18*(3), 1061. doi: 10.1029/2002PA000809.

Gersonde, R., Crosta, X., Abelmann, A., & Armand, L. (2005). Sea-surface temperature and sea ice distribution of the Southern Ocean at the EPILOG Last Glacial Maximum a circum-Antarctic view based on siliceous microfossil records. *Quaternary Science Review, 24*, 869–896. doi: 10.1016/j.quascirev.2004.07.015.

Gersonde, R., & Zielinski, U. (2000). The reconstruction of late Quaternary Antarctic sea-ice distribution — the use of diatoms as a sea-ice proxy. *Palaeogeography, Palaeoclimatology, Palaeoecology, 162*(3/4), 263–286. doi: 10.1016/S0031-0182(00)00131-0.

Gibson, J. A. E., Trull, T., Nichols, P. D., Summons, R. E., & McMinn, A. (1999). Sedimentation of ^{13}C-rich organic matter from Antarctic sea-ice algae: A potential indicator of past sea-ice extent. *Geology, 27*(4), 331–334.

Gilbert, M. W., & Clark, D. L. (1983). Central Arctic Ocean paleoceanographyinterpretations based on late Cenozoic calcareous dinoflagellates. *Marine Micropaleontology, 7*, 385–401.

Gilbert, R. (1984). The movement of gravel by the algae *Fucus vesiculosus* (L.) on an arctic intertidal flat. *Journal of Sedimentary Petrology, 54*, 463–468.

Gilbert, R. (1990). Rafting in glacimarine environments. In: J. A. Dowdeswell, & J. D. Scourse (Eds), *Glacimarine environments: Processes and sediments* (pp. 105–120), Geological Society (Special Publication 53).

Gingele, F., & Dahmke, A. (1994). Discrete barite particles and barium as tracers of paleoproductivity in South Atlantic sediments. *Paleoceanography, 9*, 151–168.

Gloersen, P., Campbell, W. J., Cavalieri, D. J., Comiso, J. C., Parkinson, C. L., & Zwally, H. J. (1992). *Arctic and Antarctic sea ice, 1978–1987: Satellite passive-microwave observations and analysis*, NASA SP-511, 290pp.

Gobeil, C., Macdonald, R. W., & Sundby, B. (1997). Diagenetic separation of cadmium and manganese in suboxic continental margin sediments. *Geochimica et Cosmochimica Acta, 61*, 4647–4654.

Gobeil, C., Sundby, B., Macdonald, R. W., & Smith, J. N. (2001). Recent change in organic carbon flux to Arctic Ocean deep basins: Evidence from acid volatile sulfide, manganese and rhenium discord in sediments. *Geophysical Research Letters, 28*, 1743–1746.

Goericke, R., & Fry, B. (1994). Variations of marine plankton $\delta^{13}C$ with latitude, temperature, and dissolved CO_2 in the world ocean. *Global Biogeochemical Cycles, 8*, 85–90.

Goldberg, R. H. (1983). Stratigraphy and sediemntology of ice-rafted and turbidite sediment, Canada Basin, Arctic Ocean. In: B. F. Molnia (Ed.), *Glacial-marine sedimentation* (pp. 367–400). New York: Plenum Publishing Corp.

Goldhaber, M. B., & Kaplan, I. R. (1974). The sulfur cycle. In: E. D. Goldberg (Ed.), *The Sea* (Vol. 5, pp. 569–655). New York: Wiley.

Goldschmidt, P., Pfirman, S. L., Wolllenburg, I., & Henrich, R. (1992). Origin of sediment pellets from the Arctic seafloor: Sea ice or icebergs. *Deep-Sea Research, 39*(Suppl. 2), S539–S565.

Goldstein, S. J., & Jacobsen, S. B. (1988). Nd and Sr isotopic systematics of river water suspended material: Implications for crustal evolution. *Earth and Planetary Science Letters, 87*, 249–265.

Goldstein, S. L., O'Nions, R. K., & Hamilton, J. (1984). A Sm-Nd isotopic study of atmospheric dusts and particulates from major river systems. *Earth and Planetary Science Letters, 70*, 221–236.

Goñi, M. A., & Hedges, J. I. (1992). Lignin dimers: Structures, distribution, and potential geochemical applications. *Geochimica et Cosmochimica Acta, 56*, 4025–4043.
Goñi, M. A., Ruttenberg, K. C., & Eglinton, T. I. (1997). Sources and contribution of terrigenous organic carbon to surface sediments in the Gulf of Mexico. *Nature, 389*, 275–278.
Goñi, M. A., Yunker, M. B., Macdonald, R. W., & Eglinton, T. I. (2000). Distribution and sources of organic biomarkers in arctic sediments from the Mackenzie River and Beaufort Shelf. *Marine Chemistry, 71*, 23–51.
Goñi, M. A., Yunker, M. B., Robie, W., Macdonald, R. W., Timothy, I., & Eglinton, T. I. (2005). The supply and preservation of ancient and modern components of organic carbon in the Canadian Beaufort Shelf of the Arctic Ocean. *Marine Chemistry, 93*, 53–73.
Gordeev, V. V. (2000). River input of water, sediment, major ions, nutrients and trace metals from Russian territory to the Arctic Ocean. In: E. L. Lewis (Ed.), *Freshwater budget of the Arctic Ocean* (pp. 297–322). Dordrecht: Kluwer.
Gordeev, V. V., Martin, J. M., Sidorov, I. S., & Sidorova, M. V. (1996). A Reassessment of the Eurasian River input of water, sediment, major elements, and nutrients to the Arctic Ocean. *American Journal of Science, 296*, 664–691.
Gordeev, V. V., & Tsirkunov, V. V. (1998). River fluxes of dissolved and suspended substances. In: V. Kimstach, M. Meybeck & E. Baroudy (Eds), *A water quality assessment of the former Soviet Union* (pp. 311–350). London and New York: E & FN Spon.
Gordillo, S., & Aitken, A. E. (2001). Postglacial succession and palaeoecology of Late Quaternary macrofaunal assemblages from the Canadian Arctic Archipelago. *Boreas, 30*, 61–72.
Gorshkova, T. J. (1931). Chemical-mineralogical investigations of Barents and White Seas sediments. *Trudi GOIN, 1*(2–3), Seiten.
Gosselin, M., Levasseur, M., Wheeler, P. A., Horner, R. A., & Booth, B. C. (1997). New measurements of phytoplankton and ice algal production in the Arctic Ocean. *Deep-Sea Research Part II, 44*(8), 1623–1644.
Goulden, M. L., Wofsy, S. C., Harden, J. W., Trumbore, S. E., Crill, P. M., Gower, S. T., Fries, T., Daube, B. C., Fan, S. M., Sutton, D. J., Bazzaz, A., & Munger, J. W. (1998). Sensitivity of boreal forest carbon balance to soil thaw. *Science, 279*, 214–217.
Grantz, A., Clark, D. L., Phillips, R. L., & Srivastava, S. P. (1998). Phanerozoic stratigraphy of Northwind Ridge, magnetic anomalies in the Canada Basin, and the geometry and timing of rifting in the Amerasia Basin. *Arctic Ocean. Geological Society of American Bulletin, 110*, 801–820.
Grantz, A., Johnson, L., & Sweeney, J. F. (1990). The Arctic Ocean Region, Geological Society of America. *Geology of North America, L*, 644pp.
Grantz, A., Phillips, R. L., & Jones, G. A. (1999). Holocene pelagic and turbidite sedimentation rates in the Amerasia Basin, Arctic Ocean from radiocarbon age-depth profiles in cores. *GeoResearch Forum, 5*, 209–222.
Grantz, A., Phillips, R. L., Mullen, M. W., Starratt, S. W., Jones, G. A., Naidu, S. A., & Finney, B. P. (1996). Character, paleoenvironment, rate of accumulation, and evidence for seismic triggering of Holocene turbidites, Canada Abyssal Plain, Arctic Ocean. *Marine Geology, 133*, 51–73.
Grebmeier, J. M., & Harvey, H. R. (2005). The Western Arctic Shelf-Basin Interactions (SBI) project: An overview. *Deep Sea Research Part II (Topical Studies in Oceanography), 52*, 3109–3115.
Greenwood, D. G., & Wing, S. L. (1995). Eocene continental climates and latitudial temperatures gradients. *Geology, 23*, 1044–1048.
Grigoriev, M. N., Rachold, R., Hubberten, H.-W., & Schirrmeister, (2004). Organic carbon input to the Arctic Seas through coastal erosion. In: R. Stein & R. W. Macdonald (Eds), *The organic carbon cycle in the Arctic Ocean* (pp. 41–45). Heidelberg: Springer-Verlag.
Grobe, H. (1987). A simple method for the determination of ice-rafted debris in sediments cores. *Polarforschung, 57*, 123–126.
Grönlund, E. (Ed.). (2001). Swedarctic 2001. *Polarforskningssekretariatets Årsbok 2001*, pp. 44–76.
Grootes, P. M., Stuiver, M., White, J. W. C., Johnsen, S., & Jouzel, J. (1993). Comparison of oxygen isotope records from the GISP2 and GRIP Greenland ice cores. *Nature, 366*, 552–554.
Grosswald, M. G. (1980). Late-Weichselian ice sheet of northern Eurasia. *Quaternary Research, 13*, 1–32.

Grosswald, M. G. (1993). Extent and melting history of the Late Weichselian ice sheet, the Barents-Kara continental margin. In: R. W. Peltier (Ed.), *Ice in the climate system, NATO ASI series 12* (pp. 1–20). Heidelberg: Springer-Verlag.

Grosswald, M. G. (1998). Late-Weichselian ice sheets in Arctic and Pacific Siberia. *Quaternary International, 45/46,* 3–18.

Grosswald, M. G. (1999). *Cataclysmic Megafloods in Eurasia and the Polar ice sheets.* Moscow: Scientific World, 120pp (in Russian).

Grosswald, M. G., & Hughes, T. J. (2002). The Russian component of an Arctic ice sheet duringthe Last Glacial Maximum. *Quaternary Science Review, 21,* 121–146.

Grousset, F. E., Biscaye, P. E., Zindler, A., Prospero, J., & Chester, R. (1988). Nd isotopes as tracers in marine sediments and aerosols: North Atlantic. *Earth and Planetary Science Letters, 87,* 367–378.

Grousset, F. E., Pujol, C., Labeyrie, L., Auffret, G., & Boelaert, A. (2000). Were the North Atlantic Heinrich events triggered by the behavior of the European ice-sheets? *Geology, 28,* 123–126.

Guay, C. K., & Falkner, K. K. (1997). Barium as a tracer of Arctic halocline and river waters. *Deep-Sea Research, 44*(8), 1543–1569.

Gunn, J. T., & Muench, R. D. (2001). Observed changes in Arctic Ocean temperature structure over the past half decade. *Geophysical Research Letters, 28/6,* 1035–1038.

Guo, L., Semiletov, I., Gustafsson, O., Ingri, J., Andersson, P., Dudarev, O., & White, D. (2004). Characterization of Siberian Arctic coastal sediments: Implications for terrestrial organic carbon export. *Global Biogeochemical Cycles, 18,* GB1036.

Gurevich, V. I. (1995). Recent sedimentogenesis and environment of the Arctic shelf of western Eurasia. *Norsk Polarinstitut Meddelelser, 131,* 92pp.

Guyodo, Y., & Valet, J.-P. (1996). Relative variations in geomagnetic intensity from sedimentary records: The past 200 000 years. *Earth and Planetary Science Letters, 143,* 23–26.

Haake, F.-W., & Pflaumann, U. (1989). Late Pleistocene Foraminiferal stratigraphy on the Vøring Plateau, Norwegian Sea. *Boreas, 18,* 343–356.

Haflidason, H., Lien, R., Sejrup, H. P., Forsberg, C. F., & Bryn, P. (2005). The dating and morphometry of the Storegga Slide. *Marine and Petroleum Geology, 22,* 123–136.

Haflidason, H., Sejrup, H. P., Nygard, A., Mienert, J., Bryn, P., Lien, R., Forsberg, C. F., Berg, K., & Masson, D. (2004). The Storegga slide: Architecture, geometry and slide development. *Marine Geology, 213,* 201–234.

Hahne, J., & Melles, M. (1997). Late- and post-glacial vegetation and climate history of the southwestern Taymyr Peninsula, central Siberia, as revealed by pollen analysis of a core from Lake Lama. *Veget Hist Archaeobot, 6,* 1–8.

Hahne, J., & Melles, M. (1999). Climate and vegetation history on the Taymyr Peninsula since Middle Weichselian time — palynological evidence from lake sediments. In: H. Kassens, H. A. Bauch, I. A. Dmitrenko, H. Eicken, H. W. Hubberten, M. Melles, J. Theide & L. A. Timokhov (Eds), *Land-ocean systems in the Siberian Arctic: Dynamics and history* (pp. 407–423). Heidelberg: Springer-Verlag.

Hald, M., Danielsen, T. K., & Lorentzen, S. (1989). Late Pleistocene and Holocene benthic foraminiferal distribution in the southwestern Barents Sea: Paleoenvironmental implications. *Boreas, 18,* 367–388.

Hald, M., Dokken, T., & Mikalsen, G. (2001). Abrupt climatic change during the last interglacial-glacial cycle in the polar North Atlantic. *Marine Geology, 176,* 121–137.

Hald, M., Kolstad, V., Polyak, L., Forman, S. L., Herlihy, F. A., Ivanov, G., & Nescheretov, A. (1999). Late-glacial and Holocene paleoceanography and sedimentary environments in the St. Anna Trough, Eurasian Arctic Ocean margin. *Palaeogeography, Palaeoclimatology, Palaeoecology, 146,* 229–249.

Hald, M., & Steinsund, P. I. (1992). Distribution of surface sediment benthic foraminifera in the southwestern Barents Sea. *Journal of Foraminiferal Research, 22,* 347–362.

Haley, B., Frank, M., Spielhagen, R. F., & Fietzke, J. (2008). Radiogenic isotope record of Arctic Ocean circulation and weathering inputs of the past 15 million years. *Paleoceanography, 23,* PA1S13, doi:10.1029/2007PA001486.

Hall, J. K. (1973). Geophysical evidence for ancient sea-floor spreading from Alpha Cordidillera and Mendeleev Ridge. In: M. G. Pilcher (Ed.), *Arctic geology,* Am. Ass. Petr. Geol. Mem, *19,* 542–561.

Hall, J. K. (1979). Sediment waves and other evidence of paleo-bottom currents at two locations in the deep Arctic Ocean. *Marine Geology, 23*, 269–299.

Hallam, A. (1981). *Facies interpretation and the stratigraphic record.* Oxford, SanFranzisco: Freeman & Company, 291pp.

Hallam, A. (1985). A review of Mesozoic climates. *Journal of the Geological Society, 142*, 433–445.

Hamel, D., de Vernal, A., Gosselin, M., & Hillaire-Marcel, C. (2002). Organic-walled microfossils and geochemical tracers: Sedimentary indicators of productivity changes in the North Water and northern Baffin Bay during the last centuries. *Deep-Sea Research Part II, 49*, 5277–5295.

Hantemirov, R. M., & Shiyatov, S. G. (2002). A continuous multimillennial ring-with chronology in Yamal, southwestern Siberia. *The Holocene, 12*(6), 717–726.

Harada, N., Shin, K. H., Murata, A., Uchida, M., & Nakatani, T. (2003). Characteristics of alkenones synthesized by a bloom of Emiliania huxleyi in the Bering Sea. *Geochimica et Cosmochimica Acta, 67*(8), 1507–1519.

Hargrave, B. T. (2004). North water Polynya. In: R. Stein & R. W. Macdonald (Eds), *The organic carbon cycle in the Arctic Ocean* (pp. 103–106). Heidelberg: Springer-Verlag.

Hargrave, B. T., von Bodungen, B., Conover, R. J., Fraser, A. J., Phillips, G., & Vass, W. P. (1989). Seasonal changes in sedimentation of particulate matter and lipid content of zooplankton collected by sediment trap in the Arctic Ocean off Axel Heiberg Island. *Polar Biology, 9*, 467–475.

Hargrave, B. T., von Bodungen, B., Stoffyn-Egli, P., & Mudi, P. J. (1994). Seasonal variability in particle sedimentation under permanent ice cover in the Arctic Ocean. *Continental Shelf Research, 14*, 279–293.

Hargrave, B. T., Walsh, I. D., & Murray, D. W. (2002). Seasonal and spatial patterns in mass and organic matter sedimentation in the North Water Polynya. *Deep-Sea Research Part II, 49*(22–23), 5227–5244.

Hass, H. C. (2002). A method to reduce the influence of ice-rafted debris on a grain-size record from the northern Fram Strait. *Arctic Ocean Polar Research, 21*, 299–306.

Hastings, D. W., Russell, A. D., & Emerson, S. R. (1998). Foraminiferal magnesium in Globeriginoides sacculifer as a paleotemperature proxy. *Paleoceanography, 13*(2), 161–169. doi: 10.1029/97PA03147.

Hatch, J. R., & Leventhal, J. S. (1992). Relationship between inferred redox potential of the depositional environment and geochemistry of the upper Pennsylvanian (Missourian) Stark Shale member of the Dennis Limestone, Wabaunsee County, Kansas, USA. *Chemical Geology, 99*, 65–82.

Hay, W. W., De Conto, R. M., Wold, C. N., Wilson, K. M., Vogt, S., Schultz, M., Wold-Rossby, A., Dullo, W. -C., Ronov, A. B., Balukhovsky, A. N., & Söding, E. (1999). Alternative global Cretaceous paleogeography. In: E. Barrera, & C. C. Johnson, (Eds), *Evolution of the Cretaceous ocean-climate system* (Vol. 332, pp. 1–47). Geological Society of America.

Hay, W. W., Eicher, D. L., & Diner, R. (1993). Physical oceanography and water masses in the Cretaceous Western Interior Seaway. In: W. G. E Caldwell, E. G. Kauffman (Eds), *Evolution of the Western Interior Basin*, Geological Association of Canada (Special Paper), *39*, 297–318.

Hayashida, A., Verosub, L., Heider, F., & Leonhardt, R. (1999). Magnetostratigraphy and relative paleointensity of late Neogene sediments at ODP Leg 167 Site 1010 off Baja California. *Geophysical Journal International, 139*, 829–840.

Hays, J. D., Imbrie, J., & Shackleton, N. J. (1976). Variations in the Earth's Orbit: Pacemaker of the ice ages. *Science, 194*, 1121–1132.

Hays, J. D., & Morley, J. J. (2003). The Sea of Okhotsk: A window on the ice age ocean. *Deep-Sea Research Part I, 50*, 1481–1506.

Haywood, A. M., Dekens, P., Ravelo, A. C., & Williams, M. (2005). Warmer tropics during the mid-Pliocene? Evidence from alkenone paleothermometry and a fully coupled ocean-atmosphere GCM. *Geochemistry Geophysics Geosystems, 6*, Q03010.

Haywood, A. M., Sellwood, B. W., & Valdes, P. J. (2000). Regional warming: Pliocene (3 Ma) paleoclimate of Europe and the Mediterranean. *Geology, 28*, 1063–1066.

Haywood, A. M., & Valdes, P. J. (2004). Modelling Pliocene warmth: Contribution of atmosphere, oceans and cryosphere. *Earth and Planetary Science Letters, 218*, 363–377.
Head, M. J. (1996). Modern dinoflagellate cysts and their biological affinities. In: J. Jansonius & D. C. McGregor (Eds), *Palynology: Principles and applications* (Vol. 3, pp. 1197–1248). Dallas, TX: American Association of Stratigraphic Palynologists Foundation.
Head, M. J., Norris, G., & Mudie, P. J. (1989). Palynology and dinocyst stratigraphy of the Miocene in ODP Leg 105, Hole 645E, Baffin Bay. In: S. P. Srivastava, M. A. Arthur, B. Clement, et al. (Eds), *Proceeding of ODP, Science Results* (Vol. 105, pp. 467–514). College Station: Texas (Ocean Drilling Program).
Heath, G. R., Moore, T. C., & Dauphin, J. P. (1977). Organic carbon in deep-sea sediments. In: N. R. Anderson & A. Malahoff (Eds), *The fate of fossil fuel CO2 in the oceans* (pp. 605–625). New York: Plenum Press.
Hebbeln, D. (1992). Weichselian glacial history of the Svalbard area: Correlating the marine and terrestrial records. *Boreas, 21*, 295–304.
Hebbeln, D. (2000). Flux of ice-rafted detritus from sea ice in the Fram Strait. *Deep-Sea Research Part II, 47*, 1773–1790.
Hebbeln, D., & Berner, H. (1993). Surface sediment distribution in the Fram Strait. *Deep-Sea Research Part I, 40*(9), 1731–1745.
Hebbeln, D., Dokken, T., Andersen, E. S., Hald, M., & Elverhøi, A. (1994). Moisture supply for northern ice-sheet growth during the Last Glacial Maximum. *Nature, 370*, 357–360.
Hebbeln, D., Henrich, R., & Baumann, K.-H. (1998). Paleoceanography of the last interglacial/glacial cycle in the Polar North Atlantic. *Quaternary Science Review, 17*, 125–153.
Hebbeln, D., & Wefer, G. (1991). Effects of ice coverage and ice-rafted material on sedimentation in the Fram Strait. *Nature, 350*, 409–411.
Hebbeln, D., & Wefer, G. (1997). Late Quaternary palaeoceanography in the Fram Strait. *Palaeoceanography, 12*, 65–78.
Hedges, J. I., Clark, W. A., & Cowie, G. l. (1988). Organic matter sources to the water column and surficial sediments of a marine bay. *Limnology and Oceanography, 33*, 1116–1136.
Hedges, J. I., Clark, W. A., Quay, P. D., Richey, J. E., Devol, A. H., & Santos, U. D. M. (1986). Composition and fluxes of particulate organic material in the Amazon River. *Limnology and Oceanography, 31*, 717–738.
Hedges, J. I., & Keil, R. (1995). Sedimentary organic matter preservation: An assessment and speculative synthesis. *Marine Chemistry, 49*, 81–115.
Hedges, J. I., Keil, R., & Benner, R. (1997). What happens to terrestrial organic matter in the ocean? *Organic Geochemistry, 27*, 195–212.
Hedges, J. I., & Mann, D. C. (1979). The lignin geochemistry of marine sediments from the southern Washington coast. *Geochimica et Cosmochimica Acta, 43*, 1809–1818.
Heezen, B. C., & Ewing, M. (1961). The Mid-Oceanic Ridge and its extension through the Arctic Basin. In: G. O. Raasch (Ed.), *Geology of the Arctic* (pp. 622–642). Toronto: University of Toronto Press.
Heinrich, H. (1988). Origin and consequences of cyclic ice rafting in the Northeast Atlantic Ocean during the past 130,000 years. *Quaternary Research, 29*, 142–152.
Heiskanen, A. S., & Keck, A. (1996). Distribution and sinking rates of phytoplankton, detritus, and particulate biogenic silicia in the Laptev Sea and Lena River (Arctic Siberia). *Marine Chemistry, 53*, 229–245.
Helland, P. E., & Holmes, M. A. (1997). Surface textural analysis of quartz sand grains from ODP Site 918 off the southeast coast of Greenland suggests glaciation of southern Greenland at 11Ma. *Palaeogeography, Palaeoclimatology, Palaeoecology, 135*, 109–121.
Helmke, J. P., Bauch, H. A., & Erlenkeuser, H. (2003). Development of glacial and interglacial conditions in the Nordic seas between 1.5 and 0.35 Ma. *Quaternary Science Review, 22*, 1717–1728. doi: 10.1016/S0277-3791(03)00126-4.
Hemleben, C., Spindler, M., & Anderson, O. R. (1989). *Modern planktonic foraminifera*. Heidelberg: Springer-Verlag, 363pp.
Hemming, S. R. (2004). Heinrich events: Massive Pleistocene detritus layers of the North Atlantic and their global climate imprint. *Reviews of Geophysics, 42*, RG1005. doi: 10.1029/2003RG000128.

Hemming, S. R., Broecker, W. S., Sharp, W. D., Bond, G. C., Gwiazda, R. H., McManus, J. F., Klas, M., & Hajdas, I. (1998). Provenance of Heinrich layers in core V28-82, northeastern Atlantic: 40Ar/39Ar ages of ice-rafted hornblende, Pb isotopes in feldspar grains, and Nd-Sr-Pb isotopes in the fine sediment fraction. *Earth and Planetary Science Letters, 164*, 317–333.

Henrich, R. (1986). A calcite dissolution pulse in the Norwegian-Greenland Sea during the last deglaciation. *International Journal of Earth Science, 75*(3), 805–827.

Henrich, R. (1989). Glacial/interglacial cycles in the Norwegian Sea: Sedimentology, paleoceanography and evolution of Late Pliocene to Quaternary Northern Hemisphere climate. In: O. Eldholm, J. Thiede, E. Taylor, et al., *Proceedings of ODP Science Results, 104* (pp. 189–232), College Station, TX (Ocean Drilling Program).

Henrich, R. (1990). Cycles, rhythms, and events in Quaternary Arctic and Antarctic glaciomarine deposits. In: U. Bleil & J. Thiede (Eds), *Geological history of the Polar Oceans: Arctic versus Antarctic* (pp. 213–244). Dordrecht, Kluwer Academic Publishers, Nato ASI Series C308.

Henrich, R., Kassens, H., Vogelsang, E., & Thiede, J. (1989). Sedimentary facies of glacial-interglacial cycles in the Norwegian Sea during the last 350 ka. *Marine Geology, 86*, 283–319.

Henrich, R., Wagner, T., Goldschmidt, P., & Michels, K. (1995). Depositional regimes in the Norwegian-Greenland Sea: The last two glacial to interglacial transitions. *International Journal of Earth Science, 84*, 28–48.

Herbert T. D., & Schuffert J. D. (1998). Alkenone unsaturation estimates of late Miocene through late Pliocene sea-surface temperatures at Site 958. In: J. V. Firth (Ed.), *Proceedings of ODP, Science Research 159T* (pp. 17–21), Texas A & M University, Ocean Drilling Program.

Herfort, L., Schouten, S., Boon, J. P., & Sinnighe Damsté, J. S. (2006). Application of the TEX86 temperature proxy to the southern North Sea. *Organic Geochemistry, 37*, 1715–1726.

Herman, A. B., & Spicer, R. A. (1996). Palaeobotanical evidence for a warm Cretaceous Arctic Ocean. *Nature, 380*, 330–333.

Herman, Y. (1974). Arctic Ocean sediments, microfauna, and the climatic recoord in late Cenozoic time. In: Y. Herman (Ed.), *Marine geology and oceanography of the Arctic Seas* (pp. 283–348). Heidelberg: Springer-Verlag.

Herman, Y. (1989). *The Arctic Seas — climatology, oceanography, geology, and biology.* New York: Van Nostrand Reinhold Company, 888pp.

Hibler, W. D. (1989). Arctic ice-ocean dynamics. In: Y. Herman (Ed.), *The Arctic Seas: Climatology, oceanography, geology, and biology* (pp. 47–92). New York: Van Nostrand Reinhold Company.

Hilgen, F. J. (1991). Extension of the astronomically calibrated (polarity) time scale to the Miocene/Pliocene boundary. *Earth and Planetary Science Letters, 107*, 349–368.

Hill, P., Mudie, P. J., Moran, K., & Blasco, S. M. (1985). A sea-level curve for the Canadian Beaufort Shelf. *Canadian Journal of Earth Science, 22*, 1383–1393.

Hill, P. R., Blasco, S. M., Harper, J. R., & Fissel, D. B. (1991). Sedimentation on the Canadian Beaufort Shelf. *Continental Shelf Research, 11*, 821–842.

Hillaire-Marcel, C., & de Vernal, A. (Eds). (2007). Proxies in late Cenozoic paleoceanpgraphy. *Developments in Marine Geology* (Vol. 1, 843pp), Elsevier, Amsterdam.

Hillaire-Marcel, C., de Vernal, A., Aksu, A. E., & Macko, S. (1989). High-resolution isotopic micropaleontological studies of upper Pleistocene sediments at ODP-Site 645, Baffin Bay. In: S. P. Srivastava, M. A. Arthur, et al. (Eds). *Proceedings ODP, Science Results* (Vol. 105, pp. 599–616), College Station, Texas (Ocean Drilling Program).

Hillaire-Marcel, C., de Vernal, A., Bilodeau, G., & Wu, G. (1994). Isotope stratigraphy, sedimentation rates, deep circulation, and carbonate events in the Labrador Sea during the last 200 ka. *Canadian Journal of Earth Science, 31*, 68–89.

Hilmer, M., & Jung, T. (2000). Evidence for a recent change in the link between the NAO and Arctic sea ice export. *Geophysical Research Letters, 27*(7), 989–992.

Hinrichs, J., Schnetger, B., Schale, H., & Brumsack, H. J. (2001). A high resolution study of NE Atlantic sediments at station Bengal: Geochemistry and early diagenesis of Heinrich layers. *Marine Geology, 177*, 79–92.

Hjelstuen, B. O., Eldholm, O., & Faleide, J. I. (2007). Recurrent Pleistocene mega-failure on the SW Barents Sea margin. *Earth and Planetary Science Letters, 258*, 605–618.

Hjelstuen, B. O., Elverhøi, A., & Faleide, J. I. (1996). Cenozoic erosion and sediment yield in the drainage area of the Storfjorden Fan. *Global and Planetary Change, 12*, 95–117.

Hjort, C. (1979). Glaciation in northern East Greenland during the Late Weichselian and Early Flandrian. *Boreas, 8*, 281–296.

Hjort, C. (1981). A glacial chronology for the northern East Greenland. *Boreas, 10*, 259–274.

Hodgson, D. A. (1989). Quaternary geology of the Queen Elisabeth Islands. In: R. J. Fulton (Ed.), *The geology of North America* (Vol. K1, pp. 443–477), *Quaternary Geology of Canada and Greenland, Geology Society of America*, Boulder.

Hodgson, D. A. (1994). Episodic ice streams and ice shelves during retreat of the northwesternmost sector of the Wisconsian Laurentide ice sheet over the central Canadian arctic archipelago. *Boreas, 23*, 14–28.

Hofmann, P., Ricken, W., Schwark, L., & Leythaeuser, D. (2000). Carbon-sulfur-iron relationships and $\delta^{13}C$ of organic matter for late Albian sedimentary rocks from the North Atlantic Ocean: Paleoceanographic implications. *Palaeogeography, Palaeoclimatology, Palaeoecology, 163*, 97–113.

Holloway, G., & Sou, T. (2001). Has Arctic Sea ice rapidly thinned? *Journal of Climate, 15*, 1691–1701.

Holmes, M. L. (1967). *Late Pleistocene and Holocene history of the Laptev Sea.* Master thesis, University of Washington, 99pp.

Holmes, R. M., McClelland, J. W., Peterson, B. J., Shiklomanov, A. I., Zhulidov, A. V., Gordeev, V. V., & Bobrovitskaya, N., (2002). A circumpolar perspective on fluvial sediment flux to the Arctic Ocean. *Global Biogeochemical Cycles, 16/4*, Seiten, doi: 10.1029/2002GB001920.

Holmes, R. M., Peterson, B. J., Gordeev, V. V., Zhulidov, A. V., Meybeck, M., Lammers, R. B., & Vörösmarty, C. J. (2000). Flux of nutrients from Russian rivers to the Arctic Ocean: Can we establish a baseline against which to judge future changes? *Water Resources Research, 36*(8), 2309–2320.

Holmes, R. M., Peterson, B. J., Zhulidov, A. V., Gordeev, V. V., Makkaveev, P. N., Stunzas, P. A., Kosmenko, L. S., Kohler, G. H., & Shiklomanov, A. I. (2001). Nutrient chemistry of the Ob' and Yenisey rivers, Siberia: Results from June 2000 expedition and evaluation of long-term data sets. *Marine Chemistry, 75*, 219–227.

Honjo, S., Manganini, S., Karowe, A., & Wooward, B. L. (1987). *Particle fluxes, northeastern Nordic Seas (1983–1986).* Technical Report WHOI-87-17, 84pp.

Honjo, S., Manganini, S., & Wefer, G. (1988). Annual particle flux and a winter outburst of sedimentation in the northern Norwegian Sea. *Deep-Sea Research, 35*, 1223–1234.

Hoops, E. (1999). *Die Charakterisierung von Sedimenten der in die Laptev-See mündenden Flüsse anhand von Schwermineralverteilungen.* Dissertation, Uni Potsdam, 95pp.

Hopmans, E. C., Weijers, J. W. H., Schefuß, E., Herfort, L., Sinninghe Damsté, J. S., & Schouten, S. (2004). A novel proxy for terrestrial organic matter in sediments based on branched and isoprenoid tetraether lipids. *Earth and Planetary Science Letters, 224*, 107–116.

Houghton, J. T., Meira Filho, L. J., Callander, B. A., Harris, N., Kattenberg, A., & Maskell, K. (1996). *Climate change 1995. The science of climate change.* Cambridge: Cambridge University Press, 584pp.

Houmark-Nielsen, M., Demidov, I., Funder, S., Grosfjeld, K., Kjær, K. H., Larsen, E., Lavrova, N., Lysa, A., & Nielsen, J. K. (2001). Early and Middle Valdaian glaciations, ice-dammed lakes and periglacial interstadials in northwest Russia: New evidence from the Pyoza River area. *Global and Planetary Change, 31*, 215–237.

Hovland, M. (Ed.). (2001). *The high-Arctic drilling challenge.* Final Report of the Arctic's Role in Global Change Program Planning Group (APPG), Ocean Drilling Program, 38pp.

Huang, Y., Bryan, S., Yi, W., & Thompson, W. I. (2002). Hydrogen isotope ratios of palmitic acid in lacustrine sediments record late Quaternary climate variations. *Geology, 30*(12), 1103–1106.

Huang, W. Y., & Meinschein, (1976). Sterols as source indicators of organic material in sediments. *Geochimica et Cosmochimica Acta, 40*, 323–330.

Hubberten, H.-W., Andreev, A., Kuhnke, T., Melles, M., & Siegert, C. (2001). The Late Quaternary climatic and environmental history of the Taymyr Peninsula, northern Central Siberia with special emphasis on the last 10,000 years. *Terra Nostra, 3*, 100–104.

Hubberten, H. W., Grobe, H., Jokat, W., Melles, M., Niessen, F., & Stein, R. (1995). Glacial history of East Greenland explored. EOS, Transactions. *Am. Geophys. Union, 76*(36), 353–356.

Hubberten, H. -W., & Romanovskii, N. N. (2003). The main features of permafrost in the Laptev Sea region, Russia: A review. *Permafrost: Proceedings of the 8th International Conference on Permafrost* (pp. 431–436), A. A. Balkema, Brookfield, Vt.

Huber, B. T. (1998). Tropical paradise at the Cretaceous poles? *Science, 282*, 2199–2200.

Huber, B. T., Hodell, D. A., & Hamilton, C. P. (1995). Mid-to Late Cretaceous climate of the southern high latitudes. Stable isotopic evidence for minimal equator-to-pole thermal gradients. *Geological Society of America Bulletin, 107*, 1164–1191.

Huber, B. T., Norris, R. D., & MacLeod, K. G. (2002). Deep-sea paleotemperature record of extreme warmth during the Cretaceous. *Geology, 30*, 123–126.

Huber, M., Sloan, L. C., & Shellito, C. J. (2003). Early Paleogene oceans and climate: A fully coupled modelling approach using NCAR's CSM. In: S. L. Wing, P. D. Gingerich, B. Schmitz & E. Thomas (Eds), *Causes and consequences of globally warm climates in the early Palaeogene* (pp. 25–47). Geological Society of America (Special Paper 369), Boulder.

Hughes, T. (1996). Can ice sheets trigger abrupt climatic change? *Arctic and Alpine Research, 28*(4), 448–465.

Hughes, T., Denton, G. H., & Grosswald, M. G. (1977). Was there a late Würm Arctic ice sheet? *Nature, 266*, 596–602.

Huguet, C., Schimmelmann, A., Thunell, R., Lourens, L. J., Sinninghe Damsté, J. S., & Schouten, S. (2007). A study of TEX 86 paleothermometer in the water column and sediments of the Santa Barbara Basin, California. *Paleoceanography, 22*, PA3203 1029/2006PA001310.

Hülsemann, K. (1963). Radiolaria in plankton from the Arctic Drift in Station T-3. Including the description of three new species. *Arctic Institute of North America Technical Papers, 13*, 1–52.

Hunkins, K., Thorndike, E. M., & Mathieu, G. (1969). Nepheloid layers and bottom currents in the Arctic Ocean. *Journal of Geophysical Research, 74*, 6995–7008.

Hunkins, K. L. (1961). Seismic studies of the Arctic Ocean floor, Geol. *Arctic, 1*, 645–665.

Hunkins, K. L., Herron, T., Kutschale, H. W., & Peter, G. (1962). Geophysical studies of the Chukchi Cap, Arctic Ocean. *Journal of Geophysical Research, 67*, 235–247.

Hurrell, J. W. (1995). Decadal trends in the North Atlantic oscillation: Regional temperatures and precipitation. *Science, 269*, 676–679.

Huybrechts, P. (2002). Sea-level changes at the LGM from ice-dynamic reconstructions of the Greenland and Antarctic ice sheets during glacial cycles. *Quaternary Science Review, 21*, 203–231.

Huybrechts, P., Steinhage, D., Wilhelms, F., & Bamber, J. L. (2000). Balance velocities and measured properties of the Antarctic ice sheet from a new compilation of gridded datasets for modeling. *Annual Glaciology, 30*, 52–60.

Huybrechts, P., & T'siobbel, S. (1997). A three-dimensional climate-icesheet model applied to the Last Glacial Maximum. *Annual Glaciology, 25*, 333–339.

Imbrie, J., Hays, J. D., Martinson, D. G., McIntyre, A., Mix, A. C., Morley, J. J., Pisias, N. G., Prell, W. L., & Shackleton, N. J. (1984). The orbital theory of Pleistocene climate: Support from a revised chronology of the marine $\delta^{18}O$ record. In: A. L. Berger, J. Imbrie, G. Hays, G. Kukla & B. Saltzman (Eds), *Milankovitch and climate* (1, pp. 269–305). Dordrecht: D. Reidel Publishing Company.

Imbrie, J., & Kipp, N. G. (1971). A new micropaleontological method for quantitative paleoclimatology: Application to a late Pleistocene Caribbean core. In: K. K. Turekian (Ed.), *The Late Cenozoic Glacial Ages* (pp. 71–181). New Haven, CT: Yale University Press.

Intergovernmental Panel on Climate Change (IPCC), (IPCC) (2001). *Climate change 2000-third assessment report.* Cambridge: Cambridge University Press.

Intergovernmental Panel on Climate Change (IPCC), (IPCC) (2007). *Climate change 2007-fourth assessment report.* Cambridge: Cambridge University Press.

Isayeva, L. L. (1984). Late Pleistocene glaciation of North-Central Siberia. In: A. A. Velichko (Ed.), *Late Quaternary environments of the Soviet Union* (pp. 21–30). Minneapolis: University of Minnesota Press.

Ishman, S. E., & Foley, K. M. (1996). Modern benthic foraminifer distribution in the Amerasian Basin, Arctic Ocean. *Micropaleontol, 42*, 206–220.

Ishman, S. E., Polyak, L., & Poore, R. Z. (1996). An expanded record of Pleistocene deep Arctic change: Canada Basin, western Arctic Ocean. *Geology, 24*, 139–142.

Israelson, C., Funder, S., & Kelly, M. (1994). The Aucellaelv stade at Aucellaelv, the first Weichselian glacier advance in Scoresby Sund, East Greenland. *Boreas, 23*, 424–431.

Itaki, T., Ito, M., Narita, M., Ahagon, N., & Sakai, H. (2003). Depth distribution of radiolarians from the Chukchi and Beaufort Seas, western Arctic. *Deep-Sea Research Part I, 50*, 1507–1522.

Ittekkot, V. (1988). Global trends in the nature of organic matter in river suspensions. *Nature, 332*, 436–438.

Ittekkot, V., Haake, B., Bartsch, M., Nair, R. R., & Ramaswamy, V. (1992). Organic carbon removalin the sea: The continental connection. In: C. P. Summerhayes, W. L. Prell & K. C. Emeis (Eds), *Upwelling systems: Evolution since the early Miocene* (pp. 167–176). Geological Society of London (Special Publication 64).

Izdar, E., Konuk, T., Ittekkot, V., Kempe, S., & Degens, E. T. (1987). Particle flux in the Black Sea: Nature of the organic matter. In: E. T. Degens, E. Izdar & S. Honjo (Eds), *Particle flux in the ocean* (Vol. 62, pp. 1–18), *Mitt. Geol. Pal. Inst.*, Hamburg University.

Jackson, H. R., Johnson, G. L., Sundvor, E., & Myhre, A. M. (1984). The Yermak Plateau: Formed at a triple junction. *Journal of Geophysical Research, 89*(B5), 3223–3232.

Jackson, H. R., Mudie, P. J., & Blasco, S. M. (1985). Initial geological report on CESAR — The Canadian expedition to study the Alpha Ridge. *Arctic Ocean Geological Survey, Canadian Paper, 84–22*, 177pp.

Jackson, H. R., & Oakey, G. N. (1990). Sedimentary thickness map of the Arctic Ocean. In: A. Grantz, L. Johnson & J. F. Sweeney (Eds), *The Arctic Ocean Region* (Vol. L, Plate 5). Boulder: Geological Society of America, The Geology of North America.

Jacobs, L., Emerson, S., & Huested, S. S. (1987). Trace metal geochemistry in the Cariaco Trench. *Deep-Sea Research, 34*, 965–981.

Jacobs, L., Emerson, S., & Skei, J. (1985). Partitioning and transport of metals across the O_2/H_2S interface in a permanently anoxic basin: Framvaren Fjord, Norway. *Geochimica et Cosmochimica Acta, 49*, 1433–1444.

Jakobsson, M. (1999). First high-resolution chirp sonar profiles for the central Arctic Ocean reveal erosion of Lomonosov Ridge sediments. *Marine Geology, 158*, 111–123.

Jakobsson, M. (2002). Hypsometry and volume of the Arctic Ocean and its constituent seas. *Geochemistry Geophysics Geosystems, 3*, 1–18.

Jakobsson, M., Backman, J., Murray, A., & Løvlie, R. (2003b). Optically stimulated luminescence dating supports central Arctic Ocean cm-scale sedimentation rates. *Geochemistry Geophysics Geosystems, 4*, 1–11.

Jakobsson, M., Backman, J., Rudels, B., Nycander, J., Frank, M., Mayer, L., Jokat, W., Sangiorgi, F., O'Regan, M., Brinkhuis, H., King, J., & Moran, K. (2007a). The early Miocene onset of a ventilated circulation regime in the Arctic Ocean. *Nature, 447*, 986–990.

Jakobsson, M., Cherkis, N., Woodward, J., Coakley, B., & Macnab, R. (2000b). A new grid of Arctic bathymetry: A significant resource for scientists and mapmakers. *EOS (Trans. Am. Geophys. Un.), 81*(9), 89, 93, 96.

Jakobsson, M., Gardner, J. V., Vogt, P., Mayer, L. A., Armstrong, A., Backman, J., Brennan, R., Calder, B., Hall, J. K., & Kraft, B. (2005). Multibeam bathymetric and sediment profiler evidence for ice grounding on the Chukchi Borderland, Arctic Ocean. *Quaternary Research, 63*, 150–160.

Jakobsson, M., Grantz, A., Kristoffersen, Y., & Macnab, R. (2003a). Physiographic provinces of the Arctic Ocean seafloor. *GSA Bulletin, 115*(12), 1443–1455.

Jakobsson, M., Grantz, A., Kristoffersen, Y., & Macnab, R. (2004). Physiography and bathymetry of the Arctic Ocean. In: R. Stein & R. W. Macdonald (Eds), *The organic carbon cycle in the Arctic Ocean* (pp. 1–6). Heidelberg: Springer-Verlag.

Jakobsson, M., Løvlie, R., Al-Hanbali, H., Arnold, E., Backman, J., & Mörth, M. (2000a). Manganese and color cycles in Arctic Ocean sediments constrain Pleistocene chronology. *Geology, 28*, 23–26.

Jakobsson, M., Løvlie, R., Arnold, E. M., Backman, J., Polyak, L., Knutsen, J.-O., & Musatov, E. (2001). Pleistocene stratigraphy and paleoenvironmental variation from Lomonsov Ridge sediments, central Arctic Ocean. *Global and Planetary Change, 31*, 1–22.

Jakobsson, M., Macnab, R., Mayer, L., Anderson, R., Edwards, M., Hatzky, J., Schenke, H. W., & Johnson, P. (2008). An improved bathymetric portrayal of the Arctic Ocean: Implications for ocean modeling and geological, geophysical and oceanographic analyses. *Geophysical Research Letters, 35*, L07602, doi:10.1029/2008GL033520.

Jakobsson, M., Polyak, L., & Darby, D. A. (2007b). Arctic Ocean: Glacial history from multibeam mapping and coring during HOTRAX (2005) and LOMROG (2007). *EOS Trans. AGU, 88*(52), Fall Meet. Suppl. Abstract PP42B-06.

Janecek, T. R., & Rea, D. K. (1983). Eolian deposition in the northeast Pacific Ocean: Cenozoic history of atmospheric circulation. *Geological Society of America Bulletin, 94*, 730–738.

Jansen, E., Befring, S., Bugge, T., Eidvin, T., Holtedahl, H., & Sejrup, H. P. (1987). Large submarine slides on the continental margin: Sediments, transport and timing. *Marine Geology, 78*, 77–107.

Jansen, E., Bleil, U., Henrich, R., & Slettemark, B. (1988). Paleoenvironmental changes in the Norwegian sea and the Northeast Atlantic during the last 2.8 Ma: DSDP/ODP Sites 610, 642, 643 and 644. *Paleoceanography, 3*, 563–581.

Jansen, E., Raymo, M. E., Blum, P., et al. (1996). *Proceedings of ODP, Init. Results, 162*, College Station, TX (Ocean Drilling Program).

Jansen, E., Sjølholm, J., Bleil, U., & Erichsen, J. A. (1990). Neogene and Pleistocene glaciationsin the Northern Hemisphere and late Miocene-Pliocene global ice volume fluctuations: Evidence from the Norwegian Sea. In: U. Bleil & J. Thiede (Eds), *Geological history ofthe Polar Oceans: Arctic versus Antarctic, Vol. 308. NATO ASI Series C* (pp. 677–706). Dordrecht: Kluwer Academic Publishers.

Jasper, J. P., & Gagosian, R. B. (1990). The sources and deposition of organic matter in the Late Quaternary Pigmy Basin, Gulf of Mexico. *Geochimica et Cosmochimica Acta, 54*, 1117–1132.

Jenkyns, H. C., Forster, A., Schouten, S., & Sinninghe Damsté, J. S. (2004). High temperatures in the late Cretaceous Arctic Ocean. *Nature, 432*, 888–892.

Jenkyns, H. C., Gale, A. S., & Corfield, R. M. (1994). Carbon-and oxygen-isotope stratigraphy of the English Chalk and Italian Scaglia and its palaeoclimatic significance. *Geological Magazine, 131*, 1–34.

Jiang, H., Eiriksson, J., Schulz, M., Knudsen, K.-L., & Seidenkrantz, M.-S. (2005). Evidence for solar forcing of sea-surface temperature on the North Icelandic Shelf during the late Holocene. *Geology, 33*, 73–76.

Johannessen, O. M., Bengtsson, L., Miles, M. W., Kuzmina, S. I., Semenov, V. A., Alekseev, G. V., Nagurnyi, A. P., Zakharov, V. F., Bobylev, L. P., Pettersson, L. H., Hasselmann, K., & Cattle, H. P. (2004). Arctic climate change, observed and modelled temperature and sea-ice variability. *Tellus, 56A*(4), 328–341.

Johannessen, O. M., Shalina, E., Kuzmina, S., Miles, M. W., & Nagurnyi, A. (2001). Shrinking of the Arctic cover over the last decades. In: W. L. Smith, Y. M. Timofeyev, (Eds), *Proceeding of International Radiation Symposium* (pp. 1007–1011), St Petersburg, Russia, 24–29 July 2000, Deepak Publishing, Hampton, USA.

Johannessen, O. M., Shalina, E. V., & Miles, M. W. (1999). Satellite Evidence for an Arctic Sea Ice Cover in Transformation. *Science, 286*(5446), 1937–1939.

Johansson, A., & Gee, D. G. (1999). The late Palaeoproterozoic Eskolabreen granitoids of southern Ny Friesland, Svalbard Caledonides — geochemistry, age and origin. *GFF, 121*, 113–126.

Johansson, A., Gee, D. G., Bjørklund, L., & Witt-Nilsson, P. (1995). Isotope studies of granitoids from Bangenhuk Formation, Ny Friesland Caledonides. *Svalbard Geological Magazine, 132*, 303–320.

Johansson, A., Larionov, A. N., Tebenkov, A. M., Gee, D. G., Whitehouse, M. J., & Vestin, J. (2000). Grenvillian magmatism of western and central Nordaustlandet, northeastern Svalbard. *Transactions of the Royal Society of Edinburgh: Earth Sciences, 90*, 221–254.

Johns, L., Wraige, E. J., Belt, S. T., Lewis, C. A., Masse', G., Robert, J.-M., & Rowland, S. J. (1999). Identification of C25 highly branched isoprenoid (HBI) dienes in Antarctic sediments, sea-ice diatoms and laboratory cultures of diatoms. *Organic Geochemistry, 30*, 1471–1475.

Johnsen, S. J., Clausen, H. B., Dansgaard, W., Gundestrup, N. S., Hammer, C. U., Andersen, U., Andersen, K. K., Hvidberg, C. S., Dahl-Jensen, D., Steffensen, J. P., Shoji, H., Sveinbjornsdottir, A. E., White, J., Jouzel, J., & Fisher, D. (1997). The delta O-18 record along the Greenland Ice Core Project deep ice core and the problem of possible Eemian climatic instability. *Journal of Geophysical Research, 102*, 26397–26410.

Johnson, G. L. (1969). Morphology of the Eurasian Arctic Basin. *The Polar Record, 14*, 619–628.

Johnson, G. L. (1983). The FRAM expeditions: Arctic Ocean studies from floating ice 1979–1982. *The Polar Record, 21*, 583–589.

Jokat, W. (Ed.). (1999). Arctic'98: The expedition ARK-XIV/1a of RV Polarstern in 1998, *Report Polar Research, 308*, 159pp.

Jokat, W. (2003). The sedimentary structure of the Lomonosov Ridge between 88°N and 80°N. *Geophysical Journal International, 163*, 698–726.

Jokat, W., & Micksch, U. (2004). The tectonic evolution of the Arctic Ocean: Overview and perspectives. In: R. Stein & R. W. Macdonald (Eds), *The organic carbon cycle in the Arctic Ocean* (pp. 21–24). Heidelberg: Springer-Verlag.

Jokat, W., Leinweber, V., Ehlers, B. M., Boebel, T., & Schenke, H. W. (2008). Timing and geometry of the Fram Strait opening. *Geophysical Journal International*, in press.

Jokat, W., & Micksch, U. (2004). Sedimentary structure of the Nansen and Amundsen basins. *Geophysical Research Letters, 31*. doi: 10.1029/2003GL018352.

Jokat, W., Stein, R., Rachor, E., & Schewe, I. (1999). Expedition gives fresh view of central Arctic geology. *EOS Trans, 80*, 465, 472–473.

Jokat, W., Uenzelmann-Neben, G., Kristoffersen, Y., & Rasmussen, T. M. (1992). Lomonosov Ridge: A double-sided continental margin. *Geology, 20*, 887–890.

Jokat, W., Weigelt, E., Kristoffersen, Y., & Rasmussen, T. M. (1995). New insights into the evolution of Lomonosov Ridge and the Eurasia Basin. *Geophysical Journal International, 122*, 378–392.

Jones, G. A. (1987b). The central Arctic Ocean sediment record: current progress in moving from a litho-to a chronostratigraphy. *Polar Research, 5*, 309–311.

Jones, G. A., & Keigwin, L. D. (1988). Evidence from Fram Strait (78°N) for early deglaciation. *Nature, 336*, 56–59.

Jones, E. P. (2001). Circulation in the Arctic Ocean. *Polar Research, 20*, 139–146.

Jones, E. P., Anderson, L. G., & Swift, J. H. (1998). Distribution of Atlantic and Pacific waters in the upper Arctic Ocean: Implications for circulation. *Geophysical Research Letters, 25*(6), 765–768.

Jones, R. W. (1987a). Organic facies. *Advances in Petroleum Geochemisrty, 2*, 1–80.

Jorgenson, M. T., Racine, C. H., Walters, J. C., & Osterkamp, T. E. (2001). Permafrost degradation and ecological changes associated with a warming climate in central Alaska. *Climatic Change, 48*, 551–571.

Josenhans, H. W., Zevenhuizen, J., & Klassen, R. A. (1986). The Quaternary geology of the Labrador Shelf. *Canadian Journal of Earth Science, 23*, 1190–1213.

Jünger, B. (1994). Tiefenwassererneuerung in der Grönlandsee während der letzten 340000 Jahre. *GEOMAR Rep, 35*, 109pp.

Kaban'kov, V., Andreeva, I., Ivanov, V., & Petrova, V. (2004). *The geotectonic nature of the Central Arctic Morphostructures and geological implications of bottom sediments for its interpretation geotectonics, 6*, 33–48.

Kalgutkar, R. M., & McIntyre, D. J. (1991). Helicosporous fungi and early Eocene pollen, Eureka Sound Group, Axel Heiberg Island, Northwest Territories. *Canadian Journal of Earth Science, 28*, 364–371.

Kalinenko, V. V., Shelekhova, E. S., & Wahsner, M. (1996). Clay minerals in surface sediments of the East Siberian and Laptev Seas. In: R. Stein, G. I. Ivanov, M. A. Levitan & K. Fahl (Eds). *Surface-sediment composition and sedimentary processes in the central Arctic Ocean and along the Eurasian Continental Margin. Report on Polar Research, 212*, 43–50.

Kaneda, T. (1991). Iso- and anteiso-fatty acids in bacteria: biosynthesis, function, and taxonomic significance. *Microbiological Review, 55*(2), 288–302.

Karasik, A. M. (1968). Magnetic anomalies of Hakkel Ridge and origin of the Eurasia subbasin of the Arctic Ocean. *Geophysical methods of prospecting in the Arctic* (Vol. 5, pp. 8–19). Leningrad.

Karasik, A. M. (1973). Anomalous magnetic field of the Eurasian Basin of the Arctic Ocean. *Akademiya Nauk SSSR Doklady, Moscow, 211*, 86–89.

Kassens, H., Bauch, H., Dmitrenko, I., Eicken, H., Hubberten, H. W., Melles, M., Thiede, J. & Timokhov, L. (Eds). (1999). *Land-ocean systems in the Siberian: Dynamics and history.* Heidelberg: Springer-Verlag. 711pp.

Kassens, H., & Thiede, J. (1994). Climatological significance of Arctic sea ice at present and in the past. *Report on Polar Research, 144,* 81–85.

Katsuki, K., & Takahashi, K. (2005). Diatoms as paleoenvironmental proxies for seasonal productivity, sea-ice and surface circulation in the Bering Sea during the late Quaternary. *Deep-Sea Research Part II, 52,* 2110–2130.

Katsuki, K., Takahashi, K., & Okada, M. (2003). Diatom assemblage and productivity changes during the last 340,000 years in the subarctic Pacific. *Journal of Oceanography, 59,* 695–707.

Kattner, G., Lobbes, J. M., Fitznar, H. P., Engbrodt, R., Nöthig, E.-M., & Lara, R. J. (1999). Tracing dissolved organic substances and nutrients from the Lena River through Laptev Sea (Arctic). *Marine Chemistry, 65,* 25–39.

Kaufman, D. S., Ager, T. A., Anderson, N. J., Anderson, P. M., Andrews, J. T., Bartlein, P. J., Brubaker, L. B., Coats, L. L., Cwynar, L. C., Duvall, M. L., Dyke, A. S., Edwards, M. E., Eisner, W. R., Gajewski, K., Geirsdóttir, A., Hu, F. S., Jennings, A. E., Kaplan, M. R., Kerwin, M. W., Lozhkin, A. V., MacDonald, G. M., Miller, G. H., Mock, C. J., Oswald, W. W., Otto-Bliesner, B. L., Porinchu, D. F., Rühland, K., Smol, J. P., Steig, E. J., & Wolfe, B. B. (2004). Holocene thermal maximum in the western Arctic (0–180°W). *Quaternary Science Review, 23,* 529–560.

Kawai, N. (1984). Paleomagnetic study of the Lake Biwa sediments. In: S. W. Horie (Ed.), *Lake Biwa* (pp. 399–416). Dordrecht: W. Junk.

Kawai, N., Yaskawa, K., Nakajima, T., Torii, M., & Horie, S. (1972). Oscillating geomagnetic field with a recurring reversal discovered from Lake Biwa. *Proceedings of the Japan Academy, 48,* 186–190.

Kazmin, V. G., & Napatov, L. M. (Eds). (1998). *The paleogeographic Atlas of Northern Eurasia, Institute Tecton.* Lithospheric Plates, Russian Academy Natural Science, Moscow, Russia, 28pp.

Keigwin, L. D., & Lehman, S. J. (1994). Deep circulation change linked to heinrich event I and younger dryas in a mid-depth North Atlantic core. *Paleoceanography, 9,* 185–194.

Keil, R. G., Tsamakis, E., Fuh, C. B., Giddings, C., & Hedges, J. I. (1994). Mineralogical and textural controls on organic composition of coastal marine sediments: Hydrographic separation using SPLITT fractionation. *Geochimica et Cosmochimica Acta, 57,* 879–893.

Keller, M. A., Bird, K. J., & Evans, K. R. (1999). Petroleum source rock evaluation based on sonic and resistivity logs. In the oil and gas resource potential of the Arctic National Wildlife Refuge, 1002 area, Alaska: U.S. Geological Survey Open-File Report 98-34, CD-ROM, chapter SR, 62pp.

Kellogg, T. B. (1980). Paleoclimatology and paleo-oceanography of the Norwegian and Greenland seas: Glacial-interglacial contrasts. *Boreas, 9,* 115–137.

Kellogg, T. B., & Kellogg, D. E. (1988). Antarctic cryogenic sediments: Biotics and inorganic facies of ice shelf and marine-based ice sheet environments. *Palaeogeography, Palaeoclimatology, Palaeoecology, 67,* 51–74.

Kelly, P. M., Jones, P. D., Sear, C. B., Cherry, B. S. G., & Tavakol, R. K. (1982). Variations in surface air temperature: Part 2. Arctic regions, 1881–1980. *Monthly Weather Review, 110,* 71–82.

Kemp, A. E. S., Pearce, R. B., Grigorov, I., Rance, J., Lange, C. B., Quilty, P., & Salter, I. (2006). Production of giant diatoms and their export at oceanic frontal zones: Implications for Si and C flux from stratified waters. *Global Biogeochemical Cycles, 20,* GB4S04, doi: 10.1029/2006GB002698.

Kempema, E. W., Reimnitz, E., & Barnes, P. W. (1989). Sea ice sediment entrainment and rafting in the Arctic. *Journal Sedimentary Petrology, 59,* 308–317.

Kemper, E. (1987). Das Klima der Kreidezeit. *Geol. Jahrb. A, 96,* 5–185.

Kennett, J. P., Houtz, R. E., Andrews, P. B., Edwards, A. R., Gostin, V. A., Hajos, M., Hampton, M., Jenkins, D. G., Margolis, S. V., Ovenshine, A. T., & Perch-Nielsen, K. (1975). Cenozoic paleoceanography in the southwest Pacific Ocean, Antarctic glaciation, and the development of the Circum-Antarctic Current. In: J. P. Kennett, R. E. Houtz, et al., *Init. Reps. DSDP 29,* Washington (U.S. Govt. Printing Office) (pp. 1155–1169).

Kennett, J. P., & Shackleton, N. J. (1976). Oxygen isotopic evidence for the development of the psychrosphere 38 Myr ago. *Nature, 260,* 513–515.

Kennett, J. P., & Srinivasan, M. S. (1983). *Neogene planktonic foraminifera. A phylogenetic Atlas.* Hutchinson Ross Publishing Company, 265pp.

Kennett, J. P., & Stott, L. D. (1991). Abrupt deep-sea warming, palaeoceanographic changes and benthic extinctions at the end of the Palaeocene. *Nature, 353,* 225–229.

Kenyon, N. H. (1987). Mass wasting features on the continental slope of north west Europe. *Marine Geology, 74*, 57–77.

Kerr, R. A. (2007). Is battered Arctic sea ice down for the count? *Science, 318*, 33–34.

Kierdorf, C. (2006). *Variability of organic carbon along the ice-covered Polar continental margin of East Greenland*. Ph.d. thesis, Bremen University, 146pp.

Kim, J. H., Schouten, S., Hopmans, E. C., & Donner, B. Sinninghe Damsté, (2008). Global sediment core-top calibration of the TEX86 paleothermometer in the ocean. *Geochimica et Cosmochimica Acta, 72*, 1154–1173.

Kind, N. V. (1974). Late quaternary geochronology according to isotope data (Geokhronologia pozdniego antropogena po izotopnym dannym). *Transactions of Geological Institute of Academic Science, 257*, 255pp. (in Russian).

Kind, N. V. (1975). Glaciation in the Verkhoyansk mountains and their place in the radiocarbon geochronology of the Siberian Late Anthropocene. *Biuletyn Peryglac, 24*, 41–54.

Kind, N. V., Kolpakov, V. V., & Sulerzhitsky, L. D. (1971). On the age of glaciations in the Verkhoyansk highlands. Izvestiya academii nauk SSSR. *Serial Geology, 10*, 135–144 (in Russian).

Kipp, N. G. (1976). New transfer function for estimating past sea-surface conditions from sea-bed distribution of planktonic foraminiferal assemblages in the North Atlantic. In: R. M. Cline, & J. D. Hays (Eds), *Investigation of Southern Ocean paleoceanography and paleoclimatology, Memories* (Vol. 45, pp. 3–42), Geological Society of America.

Kisakürek, B., Eisenhauer, A., Böhm, F., Garbe-Schönberg, D., & Erez, J. (2008). Controls on shell Mg/Ca and Sr/Ca in cultured planktonic foraminiferan, *Globigerinoides ruber* (white). *Earth and Planetary Science Letters*, in press.

Kiselev, Y. G. (Ed.) (1986). *Crustal geology of the Arctic Basin*. Moscow: Nedra Publishers. 224pp.

Kitchell, J. A., & Clark, D. L. (1982). Late cretaceous Paleogene paleogeography and paleocirculation: Evidence of north polar upwelling. *Palaeogeography, Palaeoclimatology, Palaeoecology, 40*, 135–165.

Kleiber, H. P. (1999). *Late quaternary paleoclimatic reconstructions along the Eurasian continental margin*. Ph.D. thesis, University of Bremen, 79pp.

Kleiber, H.-P., Knies, J., & Niessen, F. (2000). The Late Weichselian glaciation of the Franz Victoria Trough, northern Barents Sea: Ice sheet extent and timing. *Marine Geology, 168*, 25–44.

Kleiber, H. P., & Niessen, F. (2000). Variations of continental discharge pattern in space and time-implications from the Laptev Sea continental margin, Arctic Siberia. *International Journal of Earth Science, 89*(3), 605–616.

Kleiber, H. P., Niessen, F., & Weiel, D. (2001). The Late Quaternary evolution of the western Laptev Sea continental margin, Arctic Siberia — implications from sub-bottom profiling. *Global and Planetary Change, 31*, 105–124.

Klein, B., LeBlanc, B., Mei, Z.-P., Beret, R., Michaud, J., Mundy, C.-J., von Quillfeldt, C. H., Garneau, M.-É., Roy, S., Gratton, Y., Cochran, J. K., Bélanger, S., Larouche, P., Pakulski, J. D., Rivkin, R. B., & Legendre, L. (2002). Phytoplankton biomass, production and potential export in the North Water. *Deep-Sea Research Part II, 49*, 4983–5002.

Klenova, M. V. (1960). Geology of the Barents Sea. Problem House AN SSSR, M., 368pp.

Klimanov, V. A. (1997). Late glacial climate in northern Eurasia: The last climatic cycle. *Quaternary International, 41/42*, 141–152.

Kling, S. A., & Boltovskoy, D. (1995). Radiolarian vertical distribution patterns across the southern California current. *Deep-Sea Research Part I, 42*, 191–231.

Knies, J. (1999). Late quaternary paleoenvironment along the northern Barents and Kara seas continental margin: A multi-parameter analysis. *Report on Polar Reasearch, 304*, 159pp.

Knies, J., Brookes, S., & Schubert, C. J. (2007a). Re-assessing the nitrogen signal in continental margin sediments: New insights from the high northern latitudes. *Earth and Planetary Science Letters, 253*, 471–484.

Knies, J., Hald, M., Ebbesen, H., Mann, U., & Vogt, C. (2003). A deglacial-middle Holocene record of biogenic sedimentation and paleoproductivity changes from the northern Norwegian continental shelf. *Paleoceanography, 18*, 1096. doi: 10.1029/2002PA000872.

Knies, J., Kleiber, H. P., Matthiessen, , Müller, C., & Nowaczyk, N. (2001). Marine ice-rafted debris records constrain maximum extent of Saalian and Weichselian ice-sheets along the northern Eurasian margin. *Global and Planetary Change, 31*, 45–64.

Knies, J., & Mann, U. (2002). Depositional environmentand source rock potential of Miocene strata from the central Fram Strait: Introduction of a new computing tool for simulating organic facies variations. *Marine and Petroleum Geology, 19*, 811–828.
Knies, J., Mann, U., Popp, B., Stein, R., & Brumsack, H. (2008). Surface-water productivity and paleoceanographic implications in the Cenozoic Arctic Ocean. *Paleoceanography, 23*, PA1S16, doi: 10.1029/2007PA001455.
Knies, J., Matthiessen, J., Mackensen, A., Stein, R., Vogt, C., Frederichs, T., & Nam, S. (2007b). Effects of Arctic freshwater forcing on thermohaline circulation during the Pleistocene. *Geology, 35*, 1075–1078.
Knies, J., Matthiessen, J., Vogt, C., & Stein, R. (2002). Evidence of mid-pliocene (~3 Ma) global warmth. *Boreas, 31*(1), 82–93.
Knies, J., Müller, C., Nowaczyk, N., Vogt, C., & Stein, R. (2000). A multiproxy approach to reconstruct the environmental changes along the Eurasian continental margin over the last 150 kyr. *Marine Geology, 163*, 317–344.
Knies, J., & Stein, R. (1998). New aspects of organic carbon deposition and its paleoceanographic implications along the northern Barents Sea margin during the last 30,000 years. *Paleoceanography, 13*, 384–394.
Knies, J., & Vogt, C. (2003). Freshwater pulses in the eastern Arctic Ocean during Saalian and early Weichselian ice-sheet collapse. *Quaternary Research, 60*, 243–251.
Knies, J., Vogt, C., & Stein, R. (1999). Late quaternary growth and decay of the Svalbard-Barents-Sea ice sheet and paleoceanographic evolution in the adjacent Arctic Ocean. *Geo-Marine Letters, 18*, 195–202.
Knorr, G., & Lohmann, G. (2003). Southern Ocean origin for the resumption of Atlantic thermohaline circulation during deglaciation. *Nature, 424*, 532–536.
Koç, N., & Jansen, E. (1992). A high-resolution diatom record of the last deglaciation from the SE Norwegian Sea: Documentation of rapid climatic changes. *Paleoceanography*, 7(4), 499–520.
Koç, N., Jansen, E., & Haflidason, H. (1993). Paleoceanographic reconstructions of surface ocean conditions in the Greenland, Iceland and Norwegian seas through the last 14 ka based on diatoms. *Quaternary Science Review, 12*, 115–140.
Koç, N., Jansen, E., Hald, M., & Labeyrie, L. (1996). Late Glacial-Holocene sea surface temperatures and gradients between the North Atlantic and the Norwegian Sea: Implications for the Nordic heat pump. In: J. T. Andrews, W. E. N. Austin, H. Bergsten & A. E. Jennings (Eds), *Late Quaternary palaeoceanography of the North Atlantic Margins* (pp. 177–185), Geological Society (Special Publications 111).
Koç, N., Kristensen, D. K., Hasle, K., Forsberg, C. F., & Solheim, A. (2002). Late glacial paleoceanography of Hinlopen Strait, northern Svalbard. *Polish Research, 21*(2), 307–314.
Kohfeld, K. E., Anderson, R. F., & Lynch-Stieglitz, J. (2000). Carbon isotopic disequilibrium in Polar planktonic foraminifera and its impact on modern and Last Glacial Maximum reconstructions. *Paleoceanography, 15*(1), 53–64.
Kohfeld, K. E., Fairbanks, R. G., Smith, S. L., & Walsh, L. D. (1996). *Neogloboquadrina pachyderma* (sinistral coiling) as paleoceanographic tracer in Polar oceans: Evidence from Northeast Water Polynya plankton tows, sediment traps, and surface sediments. *Paleoceanography, 11*, 679–699.
Köhler, S. E. I. (1992). Spätquartäre paläo-ozeanographische Entwicklung des Nordpolarmeeres und europäischen Nordmeeres anhand von Sauerstoff-und Kohlenstoffisotopenverhältnissen der planktischen Foraminifere *Neogloboquadrina pachyderma* (sin.). GEOMAR Rep. 13, 104pp.
Köhler, S. E. I., & Spielhagen, R. F. (1990). The enigma of oxygen isotope stage 5 in the central Fram Strait. In: U. Bleil & J. Thiede (Eds), *Geological history of the Polar oceans: Arctic versus Antarctic NATO ASI series C* (Vol. 308, pp. 489–497). Dordrecht: Kluwer Academic Publishers.
Koide, M., Hodge, V. F., Yang, J. S., Stallard, M., Goldberg, E. G., Calhoun, J., & Bertine, K. K. (1986). Some comparative marine chemistries of rhenium, gold, silver and molybdenum. *Applied Geochemistry, 1*, 705–714.
Koizumi, I., Shiga, K., Irino, T., & Ikehara, M. (2003). Diatom record of the late Holocene in the Okhotsk Sea. *Marine Micropaleontology, 49*, 139–156.

Kolatschek, J., Eicken, H., Alexandrov, V. Y., & Kreyscher, M. (1996). The sea-ice cover of the Arctic Ocean and the Eurasian marginal seas: A brief overview of present day patterns and variability. In: R. Stein, G. Ivanov, M., Levitan, & K. Fahl (Eds), *Surface-sediment composition and sedimentary processes in the central Arctic Ocean and adjacent Eurasian continental margin* (Vol. 212, pp. 2–18). Report on Polar Research.

Kollatukudy, P. E. (1976). *Chemistry and biochemistry of natural waxes*. Amsterdam: Elsevier, 459pp.

Kolpakov, V. V., & Belova, A. P. (1980). *Radiocarbon dating in the glacial area of Verkhoyansk highlands and its surroundings. Geokhronologiya chetvertichnogo perioda*. Nauka: Moscow, pp. 230–235 (in Russian).

Könnecke, M., Bernhard, A. E., de la Tore, J. R., Walker, C. B., Waterbury, J. B., & Stahl, D. A. (2005). Isolation of an autotropic ammonia-oxidizing marine Archaeon. *Nature*, *437*, 543–546.

Koopmans, M. P., Köster, J., van Kaam-Peters, H. M. E., Kenig, F., Schouten, S., Hartgers, W. A., de Leeuw, J. W., & Sinninghe Damsté, J. S. (1996). Diagenetic products of isorenieratene: Molecular indicators of photic zone anoxia. *Geochimica et Cosmochimica Acta*, *60*, 4467–4496.

Kordikov, A. A. (1953). Sediments of the Kara Sea. *Transactions of Niiga*, *56*, 142pp.

Korotkevich, E. S., Egizarov, B. H., & Faleev, V. I. (Eds) (1981). Atlas of grounds of the Arctic Ocean. Problem House GUNIO Min. of defense of USSR, 104pp.

Korstgärd, J. A., & Nielsen, O. B. (1989). Provenance of dropstones in Baffin Bay and Labrador Sea, Leg 105. In: S. P. Srivastava, M. A. Arthur, B. Clement, et al. (Eds), *Proceedongs of ODP, Science Results* (pp. 65–70), 105, College Station, Texas (Ocean Drilling Program).

Kos'ko, M. K., & Trufanov, G. V. (2002). Middle Cretaceous to eopleistocene sequences on the New Siberian Islands: An approach to interpret offshore seismic. *Marine and Petroleum Geology*, *19*, 901–919.

Kovacs, L. C. Bernero, C., Johnson, G. L., Pilger, R. H., Srivastava, S. P., Vink, G. E., & Vogt, P. R. (1985). Residual magnetic anomaly chart of the Arctic Ocean region, U.S. Naval Research Laboratory and Naval Ocean Research and Development Activity, scale 1:6,000,000. Geological Society of America Map and Chart Series MC-53.

Kraus, M., Matthiessen, J., & Stein, R. (2003). A Holocene marine pollen record from the northern Yenisei Estuary (southeastern Kara Sea, Siberia). In: R. Stein, K. Fahl, D. K. Fütterer, E. M. Galimov & O. V. Stepanets (Eds), *Siberian river run-off in the Kara Sea: Characterisation, quantification, variability, environmental significance, Proceedings in Marine Sciences* (Vol. 6, pp. 435–456). Amsterdam: Elsevier.

Krause, G. (1969). Ein Beitrag zum Problem der Erneuerung des Tiefenwassers im Arkona-Becken. *Kieler Meeresforschung*, *25*, 268–271.

Krause, G., & Schauer, U. (Eds). (2001). The expeditions ARKTIS XVI/1 and ARKTIS XVI/2 of the Research Vessel "Polarstern" in 2000. Report on Polar Marine Research 389, 107pp.

Krinsley, D. H., & Doornkamp, H. C. (1973). *Atlas of quarts sand surface textures*. London: Cambridge University Press, 91pp.

Krishnamurthy, R. V., Machavaram, M., Baskaran, M., Brooks, J. M., & Champs, M. A. (2001). Organic carbon flow in the Ob, Yenisei rivers and Kara Sea of the Arctic region. *Marine Pollution Bulletin*, *42*, 726–732.

Kristoffersen, Y. (1990). Eurasian Basin. In: A. Grantz, L. Johnson, & J. F. Sweeny (Eds), *The Arctic Ocean region, geology of North America* (Vol. L., pp. 365–378), Geological Society of America Bulletin, Boulder, Colorado.

Kristoffersen, Y., Buravtsev, V., Jokat, W., & Poselov, V. (1997). Seismic reflection surveys during Arctic Ocean-96. Cruise report, Polarforskningssekretaariatets arsbok 1995/96, Stockholm, pp. 75–77.

Kristoffersen, Y., Coakley, B., Jokat, W., Edwards, M., Brekke, H., & Gjengedal, J. (2004a). Seabed erosion on the Lomonosov Ridge, central Arctic Ocean: A tale of deep draft icebergs in the Eurasia Basin and the influence of Atlantic water inflow on iceberg motion? *Paleoceanography*, *19*, PA3006. doi: 10.1029/2003PA000985.

Kristoffersen, Y., Sorokin, M. Y., Jokat, W., & Svendsen, O. (2004b). A submarine fan in the Amundsen Basin. *Arctic Ocean Marine Geology*, *204*, 317–324.

Kruglikova, S. B. (1989). Arctic Ocean radiolarians. In: Y. Herman (Ed.), *The Arctic Seas-climatology, oceanography, geology, and biology* (pp. 461–480). New York: Van Nostrand Reinhold Company.

Kruglikova, S. B. (1999). Distribution of polycystine radiolarians from recent and pleistocene sediments of the Arctic-Boreal zone. *Report on Polar Research, 306*, 120–131.

Krylov, A. A., Andreeva, I. A., Vogt, C., Backman, J., Krupskaya, V. V., Grikurov, G. E., Moran, K., & Shoji, H. (2008a). A Shift in heavy and clay mineral provenance indicates a middle miocene onset of a Perennial Sea-Ice cover in the Arctic Ocean. *Paleoceanography, 23*, PA1S06, doi: 10.1029/2007PA001497.

Krylov, A., Weiel, D., Sapega, V., Ivanov, G., Stein, R., Vogt, C., & Ryskova, E. (2008b). Clay minerals as indicator of deposition of upper quaternary sediments within the St. *Anna Trough (Kara Sea). Oceanology, 48*, 85–93.

Kubisch, M. (1992). Die Eisdrift im Arktischen Ozean während der letzten 250.000 Jahre. Geomar Report, Geomar, Kiel, 100pp.

Kulikov, N. N. (1961). The sedimentation in the Kara Sea. Symp. Sovr. osadki morei i okeanov (Modern sediments of Seas and Oceans). Trudi sovesh. on May 24-27 1960, Problem House AN SSSR, M, Seiten.

Kulikov, N. N., & Khitrova, R. M. (1982). Palynological study of Kara Sea floor sediment cores. In: A. I. Tolmachev (Ed.), *The Arctic Ocean and its coast in the Cenozoic era* (pp. 34–38). New Delhi: Amerind Publishing Co. PVT. Ltd.

Kulikov, N. N., Lapina, N. N., & Ivanov, G. I. (1999). The history of investigation of bottom sediments of the Western Arctic shelf: the Barents and Kara seas. In: R. Stein, K. Fahl, G. I. Ivanov, M. A. Levitan & G. Tarasov (Eds), *Modern and Late Quaternary depositional environment of the St. Anna Trough area, Northern Kara Sea* (Vol. 342, pp. 3–9), Report on Polar Research.

Kunz-Pirrung, M. (1998). Rekonstruktion der Oberflächenmassen der östlichen Laptevsee im Holozän anhand aquatischen Palynomorphen. *Report on Polar Research, 281*, 116pp.

Kunz-Pirrung, M. (1999). Distribution of aquatic palynomorphs in surface sediments of the Laptev Sea, eastern Arctic Ocean. In: H. Kassens, H. A. Bauch, I. Dmitrenko, H. Eicken, H.-W. Hubberten, ThiedeJ. Melles. M. & L. Timokhov (Eds), *Land-Ocean systems in the Siberian Arctic: Dynamics and history* (pp. 561–576). Heidelberg: Springer-Verlag.

Kuptsov, V. M., & Lisitzin, A. P. (1996). Radiocarbon of Quaternary along shore and bottom deposits of the Lena and the Laptev Sea sediments. *Marine Chemistry, 53*, 301–312.

Kusakabe, M., Ku, T. L., Southon, J. R., Vogel, J. S., Nelson, D. E., Measures, C. I., & Nozaki, Y. (1987). Distribution of ^{10}Be and ^{9}Be in the Pacific Ocean: Earth planet. *Science Letters, 82*, 231–240.

Kutzbach, J. E., & Guetter, P. J. (1986). The influence of changing orbital parameters and surface boundary conditions on climate simulations for the past 18 000 years. *Journal Atmospheric Science, 43*(16), 1726–1759.

Kuvaas, B., & Kristoffersen, Y. (1996). Mass movements in glaciomarine sediments on the Barents Sea continental slope. *Global and Planetary Change, 12*, 287–307.

Kuypers, M. M. M., Pancost, R. D., Nijenhuis, I. A., & Sinninghe Damsté, J. S. (2002). Enhanced productivity led to increased organic carbon burial in the euxinic North Atlantic basin during the late Cenomanian oceanic anoxic event. *Paleoceanography, 17*. doi: 10.1029/2000PA000569.

Kwok, R. (2000). Recent changes in Arctic Ocean sea ice motion associated with the NAO. *Geophysical Research Letters, 27*(6), 775–778.

Kwok, R., & Rothrock, D. A. (1999). Variability of Fram Strait ice flux and North Atlantic oscillation. *Journal of Geophysical Research, 104*(C3), 5177–5189.

Laberg, J. S., & Vorren, T. O. (1993). A Late Pleistocene submarine slide on the Bear Island trough mouth fan. *Geo-Marine Letters, 13*, 227–234.

Laberg, J. S., & Vorren, T. O. (1995). Late Weichselian debris flow deposits on the Bear Island Trough Mouth Fan. *Marine Geology, 127*, 45–72.

Laberg, J. S., & Vorren, T. O. (1996a). The glacier-fed fan at the mouth of Storfjorden trough, western Barents Sea: A comparative study. *Geologische Rundschau, 85*, 338–349.

Laberg, J. S., & Vorren, T. O. (1996b). Late Pleistocene evolution of the Bear Island Trough Mouth Fan. *Global and Planetary Change, 12*, 309–330.

Laberg, J. S., & Vorren, T. O. (2000). The Traenadjupet Slide, offshore Norway-morphology, evacuation and triggering mechanisms. *Marine Geology, 171*, 95–114.

Laberg, J. S., Vorren, T. O., Dowdeswell, J. A., Kenyon, N. H., & Taylor, J. (2000). The Andøya Slide and the Andøya Canyon, northeastern Norwegian-Greenland Sea. *Marine Geology, 162*, 259–275.

Labeyrie, L., Leclaire, H., Waelbroeck, C., Cortijo, E., Duplessy, J. -C., Vidal, L., Elliot, M., Le Coat, B., & Auffret, G. (1999). Temporal variability of the surface and deep waters of the North West Atlantic Ocean at orbital and millennial scales. In: P. U. Clark, R. S. Webb, & L. D. Keigwin (Eds), *Mechanisms of global climate change at millennial time scales* (Vol. 112, pp. 77–98), Geophysical Monograph. Series, AGU, Washington, DC.

Lachenbruch, A. H., & Sass, J. H. (1982). Permafrost, heat flow, and the geothermal regimes of Prudhoe Bay, Alaska. *Journal of Geophysical Research, 87*(B11), 9301–9316.

Lagoe, M. B. (1977). Recent benthic foraminifera from the central Arctic Ocean. *Journal Foraminiferal Research*, 7, 106–129.

Laj, C., Kissel, C., Mazaud, A., Channell, J. E. T., & Beer, J. (2000). North Atlantic paleointensity stack since 75 ka (NAPIS-75) and the duration of the Laschamp event. *Philosophical Transactions of the Royal Society of London, 358*, 1009–1026.

Lambeck, K., Yokoyama, Y., & Purcell, T. (2002). Into and out of the Last Glacial Maximum sea-level change during oxygen isotope stages 3 and 2. *Quaternary Science Review, 21*, 343–360.

Lammers, R. B., Shiklomanov, A. I., Vörösmarty, C. J., & Peterson, B. J. (2001). Assessment of contemporary arctic river runoff based on observational discharge records. *Journal of Geophysical Research, 106*, 3321–3334.

Landa, E. R., Reimnitz, E., Beals, D. M., Pochkowski, J. M., Winn, W. G., & Rigor, I. (1998). Transport of ^{137}Cs and ^{239,240}Pu with ice-rafted debris in the Arctic Ocean. *Arctic, 51*, 27–39.

Landvik, J. Y., Bolstad, M., Lycke, A. K., Mangerud, J., & Sejrup, H. P. (1992). Weichselian stratigraphy and palaeoenvironments at Bellsund, western Svalbard. *Boreas, 21*, 335–358.

Landvik, J. Y., Bondevik, S., Elverhøi, A., Fjeldskaar, W., Mangerud, J., Salvigsen, O., Siegert, M. J., Svendsen, J. I., & Vorren, T. O. (1998). The last glacial maximum of Svalbard and the Barents Sea area: Ice sheet extent and configuration. *Quaternary Science Review, 17*, 43–75.

Langereis, C. G., Dekkers, MJ., de Lange, G. J., Paterne, M., & van Santvoort, P. J. M. (1997). Magnetostratigraphy and astronomical calibration of the last 1.1 Ma from an eastern Mediterranean piston core and dating of short events in the Brunhes. *Geophysical Journal International, 129*, 75–94.

Langford, F. F., & Blanc-Valleron, M. M. (1990). Interpreting rock-eval-pyrolysis data using graphs of pyrolysable hydrocarbons vs. total organic carbon. *AAPG Bulletin, 74*, 799–804.

Langrock, U., & Stein, R. (2004). Origin of marine petroleum source rocks from the Late Jurassic/ Early Cretaceous Norwegian Greenland Seaway - evidence for stagnation and upwelling. *Marine Petroleum Geology, 21*, 157–176.

Langrock, U., Stein, R., Lipinski, M., & Brumsack, H. (2003a). Late Jurassic to early Cretaceous blak shale formation and paleoenvironment in high northern latitudes — examples from the Norwegian-Greenland-Seaway. *Paleoceanography*, doi: 10.1029/2002PA000867.

Langrock, U., Stein, R., Lipinski, M., & Brumsack, H. (2003b). Paleoenvironment and sea-level change in the early Cretaceous Barents Sea -implications from near-shore marine sapropels. *Geo-Marine Letters, 23*, 34–42.

Lapina, N. N. (1959). Mineralogical provinces in the modern bottom sediments of the Arctic Ocean. Leningrad, Proceedings of NIIGA 107.

Lapina, N. N. (1965). The determination of distribution paths of sediments, based on mineralogical investigations of marine deposits (example Laptev Sea). Uchennye Zapiski NIIGA. *Ser. Region. Geol.*, 7, 139–157 (in Russian).

Lapina, N. N., Belov, N. A., Kulikov, N. N., Semenov, Yn. P., & Spiridonov, M. A. (1970). Bottom sediments of the Arctic seas. In: B. V. Tkachenko, & B. H. Egizarov (Eds), *Geology of the USSR* (Vol. XXYI, pp. 485–530), Ostrova Sovetskoi Arctiki, Nedra, M.

Lara, R. J., Rachold, V., Kattner, G., Hubberten, H.-W., Guggenberger, G., Skoog, A., & Thomas, D. N. (1998). Dissolved organic matter and nutrietnts in the Lena River, Siberian Arctic: Characteristics and distribution. *Marine Chemistry, 59*, 301–309.

Larsen, E., Lysa, A., Demidov, I., Funder, S., Houmark-Nielsen, M., Kjær, K. H., & Murray, A. S. (1999a). Age and extent of the Scandinavian ice sheet in northwest Russia. *Boreas, 28*, 115–132.
Larsen, H. C., Duncan, R. A., Allan, J. F., & Brooks, K. (Eds). (1999b). Proceeding ODP, Science Results, 163, College Station, Texas (Ocean Drilling Program), doi:10.2973/odp.proc.sr.163.1999.
Larsen, H. C., Saunders, A. D., Clift, P. D., Beget, J., Wei, W., & Spezzaferri, S.ODP Leg 152 Scientific Party, (1994). Seven million years of glaciation in Greenland. *Science, 264*, 952–955.
Larson, R. L. (1991a). Latest pulse of earth: Evidence for a mid-Cretaceous superplume. *Geology, 19*, 547–550.
Larson, R. L. (1991b). Geological consequences of superplumes. *Geology, 19*, 963–966.
Larssen, B. B., Elverhøi, A., & Aagaard, P. (1987). Study of particulate material in sea ice in the Fram Strait: A contribution to paleoclimate research. *Polar Research, 5*, 313–315.
Lauterjung, J. (1994). Quantitative phase analysis with X-ray powder diffraction. (Anleitung zum Softwareprogramm QUAX), Potsdam (GeoForschungsZentrum Potsdam).
Lauterjung, J., Will, G., & Hinze, E. (1985). A fully automatic peak-search program for the evaluation of Gauss-shaped diffraction pattern. *Nuclear Instruments and Methods in Physics Research, A239*, 281–287.
Lawrence, K. T., Liu, Z., & Herbert, T. D. (2006). Evolution of the eastern tropical Pacific through Plio-Pleistocene glaciation. *Science, 312*, 79–83.
Lawver, L. A., Grantz, A., & Gahagan, L. M. (2002). Plate kinematic evolution of the present Arctic region since the Ordovician. In: E. L. Miller, A. Grantz, & S. L. Klemperer (Eds), *Tectonic evolution of the Bering Shelf — Chukchi Sea — Arctic Margin and adjacent landmasses* (Vol. 360, pp. 333–358), Boulder, Colorado, Geological Society of America Special Paper.
Lawver, L. A., & Müller, R. D. (1994). Iceland hotspot track. *Geology, 22*, 311–314.
Lawver, L. A., Müller, R. D., Srivastava, S. P., & Roest, W. (1990). The opening of the Arctic Ocean. In: U. Bleil, & J. Thiede (Eds), *Geological history of the Polaroceans: Arctic versus Antarctic, NATO ASI Series C* 308, pp. 29–62.
Lawver, L. A., & Scotese, C. R. (1990). A review of tectonic models for the evolution of the Canadian Basin. In: A. Grantz, L. Johnson & J. F. Sweeney (Eds), *The Arctic Ocean region, The Geology of North America* (Vol. L, pp. 593–618).
Lear, C. H., Elderfield, P. A., & Wilson, P. A. (2000). Cenozoic deep-sea temperatures and global ice volumes from Mg/Ca in benthic foraminiferal calcite. *Science, 287*, 269–272.
Leith, T. L., Weiss, H. M., Mørk, A., Arhus, N., Elvebakk, G., Embry, A. F., Brooks, P. W., Stewart, K. R., Pchelina, T. M., Bro, E. G., Verba, M. L., Danyushevskaya, A., & Borisov, A. V. (1992). Mesozoic hydrocarbon source-rocks of the Arctic region. In: T. O. Vorren et al. (Eds), *Arctic geology and Petroleum Potential* (Vol. 2, pp. 1–25), NPF (Special Publication).
Lemke, P. (Ed.) (2003). The expedition ARKTIS XIII/1 a, b of the research vessel polarstern in 2002. *Report Polar Marine Research, 446*, 120pp.
Leo, R. F., & Parker, P. L. (1966). Branched-chain fatty acids in sediments. *Science, 152*, 649–650.
Letréguilly, A., Huybrechts, P., & Reeh, N. (1991). Steady-state characteristics of the Greenland ice sheet under different climates. *Journal of Glaciology, 37*, 149–157.
Letzig, T. (1995). Meereistransportiertes lithogenes Feinmaterial in spätquartären Tiefseesedimenten des zentralen östlichen Arktischen Ozeans und der Framstraße. *Report on Polar Research, 162*, 98pp.
Levac, E., de Vernal, A., & Blake, W., Jr. (2001). Holocene paleoceanography of the north water Polynya: Palynological evidence. *Journal of Quaternary Science, 16*, 353–363.
Leventhal, J. S. (1983). An interpretation of carbon and sulfur relationships in Black Sea sediments as indicators of environments of deposition. *Geochimica et Cosmochimica Acta, 47*, 133–137.
Levi, B. G. (2000). The decreasing Arctic ice cover. *Physics Today January, 2000*, 19–20.
Levitan, M. A., & Stein, R. (2007). History of sedimentation rates in the Arctic Ocean during the last 130 ka. In: I. A. Chistyakova & I. M. Khoreva (Eds), *Bulletin of Commision for study of the Quaternary 67* (pp. 33–43). Moscow: GEOS.
Levitan, M. A., Arnold, M., Bourtman, M. V., Ivanova, E. V., & Marina, M. M. (2000). History of Holocene sedimentation in the eastern Kara Sea. *Oceanology, 40*, 577–582.
Levitan, M. A., Bourtman, M. V., Demina, L. L., Chudetsky, M. Y., & Schoster, F. (2005). Facies variability of surface sediments from the Ob-Yenisei Shoal and the Ob and Yenisei Estuaries. *Lithology and Mineral Resources, 40*, 408–419.

Levitan, M. A., Dekov, V. M., Gorbunova, Z. N., Gurvich, E. G., Muyashin, S. I., Nürnberg, D., Pavlidis, M. A., Rusakova, N. P., Shelekhova, E. S., Vasilkov, A. V., & Wahsner, M. (1996). The Kara Sea: A reflection of modern environment in grain size, mineralogy, and chemical composition of the surface layer of bottom sediments. *Report on Polar Research, 212*, 58–80.

Levitan, M. A., Lavrushin, Y. A., & Stein, R. (2007). Outlines of sedimentation history of the Arctic Ocean and Subarctic Seas fort he last 130 ka. GEOS, ISBN 5-89118-378-8, 404pp. (in Russian).

Levitan, M. A., Nürnberg, D., Pavlidis, J. A., & Shelekova, E. S. (1994). Distribution of clay minerals in surface-layer sediments from the eastern Barents and southwestern Kara Seas, in Preparations.

Levitan, M. A., Tarasov, G. A., Bourtman, M. V., & Kukina, N. A. (1999). Mineral composition of the St. Anna Trough surface sediments. *Oceanology, 39*, 821–829.

Levitus, S., & Boyer, T. P. (1994). World ocean Atlas 1994, vol. 4, temperature, NOAA Atlas NESDIS, 129pp.

Lewis, E. L. (2000). *The freshwater budget of the Arctic Ocean.* Dordrecht, Seiten: Kluwer Academic Publishers.

Li, J., Philip, R. P., Pu, F., & Allen, J. (1995). Long-chain alkenones on Qinghai lake sediments. *Geochimica et Cosmochimica Acta, 60*(2), 235–241.

Lien, R., Solheim, A., Elverhøi, A., & Rokoengen, K. (1989). Iceberg scouring and sea bed morphology on the eastern Weddell Sea shelf, Antarctica. *Polar Research*, 7, 43–57.

Lindberg, B., Laberg, J. S., & Vorren, T. O. (2004). The Nyk Slide — morphology, progression and age of a partly buried submarine slide offshore Northern Norway. *Marine Geology, 213*, 277–289.

Lipinski, M., Warning, B., & Brumsack, H.-J. (2003). Trace metal signatures of Jurassic/Cretaceous black shales from the Norwegian Shelf and the Barents Sea. *Palaeogeography, Palaeoclimatology, Palaeoecology, 190*, 459–475.

Lisiecki, L. E., & Raymo, M. E. (2005). A Pliocene-Pleistocene stack of 57 globally distributed benthic $\delta^{18}O$ records. *Paleoceanography, 20*, PA1003. doi: 10.1029/2004PA001071.

Lisitzin, A. P. (1972). Sedimentation in the world ocean. SEPM Special Publication 17, Tulsa, Oklahoma, 218pp.

Lisitzin, A. P. (1995). The marginal filter of the ocean. *Oceanology, 34*(5), 671–682.

Lisitzin, A. P. (1996). Oceanic sedimentation. American Geophysical Union, Washington, DC, 400pp.

Lisitzin, A. P. (2002). *Sea-ice and iceberg sedimentation in the ocean. Recent and past.* Heidelberg, Berlin: Springer-Verlag, 564pp.

Lisitzin, A. P., & Vinogradov, M. E. (1995). International high-latitude expedition in the Kara Sea (the 49th cruise of the R/V Dmitriy Mendeleev). *Oceanology, 34*(5), 583–590.

Littke, R. (1993). Deposition, diagenesis, and weathering of organic matter-rich sediments. Lectures Notes Earth Science (Vol. 47), Springer-Verlag, Heildelberg, 216pp.

Littke, R., Cramer, B., Gerling, P., Lopatin, N. V., Poelchau, H. S., Schaefer, R. G., & Welte, D. H. (1999). Gas generation and accumulation in the West Siberian basin. *American Association of Petroleum Geologists Bulletin, 83*(10), 1642–1665.

Lloyd, J. M., Kroon, D., Boulton, G. S., Laban, C., & Fallick, A. (1996). Ice rafting history from the Spitsbergen ice cap over the last 200 kyr. *Marine Geology, 131*, 103–121.

Lobbes, J. M., Fitznar, H. P., & Kattner, G. (2000). Biogeochemical characteristics of dissolved and particulate organic matter in Russian rivers entering the Arctic Ocean. *Geochimica et Cosmochimica Acta, 64*, 2973–2983.

Loeng, H. (1991). Features of the physical oceanographic conditions of the Barents Sea. *Polar Research, 10*, 5–18.

Loeng, H., Ozhigin, V., Adlandsvik, B., & Sagen, H. (1993). Current measurements in the northeastern Barents Sea. ICES Statutory Meeting.

Lohmann, G., Lorenz, S. J., & Prange, M. (2005). Northern high-latitude climate changes during the Holocene as simulated by circulation models. In: H. Drange, T. Dokken, T. Furevik, R. Gerdes, & W. Berger (Eds), *The Nordic Seas: An integrated perspective* (Vol. 158, pp. 273–288), *Geophysical Monographs*, American Geophysical Union.

Lorenz, S., & Lohmann, G. (2004). Acceleration technique for Milankovitch type forcing in a coupled atmosphere-ocean circulation model: method and application for the Holocene. *Climate Dynamics, 23*, 727–743.

Loubere, P. (1994). Quantitative estimation of surface ocean productivity and bottom water oxygen concentration using benthic foraminifera. *Paleoceanography, 9*, 723–737.

Loubere, P. (2000). Marine control of biological production in the eastern equatorial Pacific Ocean. *Nature, 406*, 497–500.

Loubere, P., & Fariduddin, M. (1999). Quantitative estimation of global patterns of surface ocean biological productivity and its seasonal variation on timescales from centuries to millennia. *Global Biogeochemical Cycles, 13*, 115–133.

Lourens, L. J., Sluijs, A., Kroon, D., Zachos, J. C., Thomas, E., Röhl, U., Bowles, J., & Raffi, I. (2005). Astronomical pacing of late Palaeocene to early Eocene global warming events. *Nature, 435*, 1083–1087.

Løvlie, R., Markussen, B., Sejrup, H-P., & Thiede, J. (1986). Magnetostratigraphy in three Arctic Ocean cores; arguments for geomagnetic excursions within oxygen-isotope stage 2–3. *Physics of the Earth and Planetary Interiors, 43*, 173–184.

Lowenstein, T. K., & Demicco, R. V. (2006). Elevated Eocene atmospheric CO_2 and its subsequent decline. *Science, 313*, 1928pp.

Lozano, J. A., & J. D. Hays (1976). Relationship of radiolarian assemblages to sediment types and physical oceanography in the Atlantic and western Indian oceanic sectors of the Antarctic Ocean. In: R. M. Cline, & J. D. Hays, (Eds), *Investigations of Late Quaternary paleoceanography and paleoclimatology, Geological society of American Memories, 145*, 303–336.

Lubinski, D. J., Korsun, S., Polyak, L., Forman, S. L., Lehman, S. J., Herlihy, F. A., & Miller, G. H. (1996). The last deglaciation of the Franz Victoria Trough, northern Barents Sea. *Boreas, 25*, 89–100.

Lubinski, D. J., Polyak, L., & Forman, S. L. (2001). Freshwater and Atlantic water inflows to the deep northern Barents and Kara seas since ca.13 ^{14}C ka: Foraminifera and stable isotopes. *Quaternary Science Review, 20*, 1851–1879.

Lückge, A. B., Lallier-Vergès, M., & Littke, R. (1996). Comparative study of organic matter preservation in immature sediments along the continental margins of Peru and Oman, part I: Results of petrographical and bulk geochemical data. *Organic Geochemistry, 24*(4), 437–451.

Luepke, G., & Escowitz, E. C. (1989). Grain-size, heavy-mineral, and geochemical analyses of sediments from the Chukchi Sea, Alaska. *U.S.G.S. Bulletin, 1896*, 1–12.

Lund, S., Stoner, J. S., Channell, J. E. T., & Acton, G. (2006). A summary of Brunhes paleomagnetic field variability recorded in Ocean Drilling Program cores. *Physics of the Earth and Planetary Interiors, 156*, 194–204.

Lund, S. P., Acton, G., Clement, B., Hastedt, M., Okada, M., & Williams, T. (1998). ODP Leg 172 Scientific party, geomagnetic field excursions occurred often during the last million years. *Transactions, American Geophysical Union (EOS), 79*(14), 178–179.

MacAyeal, D. R. (1993). Purge oscillations of the Laurentide Ice sheet as a cause of the North Atlantic Heinrich events. *Paleoceanography, 8*(6), 775–784.

Macdonald, G. M., Velichko, A. A., Kremenetski, C. V., Borisova, O. K., Goleva, A. A., Andreev, A. A., Cwynar, L. C., Riding, R. T., Forman, S. L., Edwards, T. W. D., Aravena, R., Hammarlund, D., Szeicz, J. M., & Gattaulin, V. N. (2000). Holocene treeline history and climate change across northern Eurasia. *Quaternary Research, 53*, 302–311.

Macdonald, R. W. (1996). Awakenings in the Arctic. *Nature, 380*, 286–287.

Macdonald, R. W., Anderson, L. G., Christensen, J. P., Miller, L. A., Semiletov, I. P., & Stein, R. (2008). The Arctic Ocean: Budgets and fluxes. In: K.-K. Liu, L. Atkinson, R. Quinones & L. Talaue-McManus (Eds), *Carbon and nutrient fluxes in continental margins: A global synthesis.* Springer-Verlag. in press.

Macdonald, R. W., & Bewers, J. M. (1996). Contaminants in the arctic marine environment: Priorities for protection. *ICES Journal of Marine Science, 53*, 537–563.

Macdonald, R. W., Harner, T., Fyfe, J., Loeng, H., & Weingartner, T. (2003). AMAP Assessment 2002: The influence of global change on contaminant pathways to, within, and from the Arctic. Arctic Monitoring and Assessment Programme (AMAP), Oslo, Norway. xii+65pp.

Macdonald, R. W., Naidu, A. S., Yunker, M. B., & Gobail, C. (2004b). The Beaufort Sea: Distribution, sources, fluxes, and burial of organic carbon. In: R. Stein & R. W. Macdonald (Eds), *The organic carbon cycle in the Arctic Ocean* (pp. 6–21). Heidelberg: Springer-Verlag.

Macdonald, R. W., Paton, D., Carmack, E., & Omstedt, A. (1995a). The freshwater budget and under-ice spreading of Mackenzie River water in the Canadian Beaufort Sea based on salinity and $^{18}O/^{16}O$ measurements in water and ice. *Journal of Geophysical Research, 100*, 895–919.

Macdonald, R. W., Paton, D., Carmack, E., & Omstedt, A. (1995b). The freshwater budget and under-ice spreading of Mackenzie River water in the Canadian Beaufort Sea based on salinity and $^{18}O/^{16}O$ measurements in water and ice. *Journal of Geophysical Research, 100*, 895–919.

Macdonald, R. W., Sakshaug, E., & Stein, R. (2004a). The Arctic Ocean: Modern status and recent climate change. In: R. Stein & R. W. Macdonald (Eds), *The organic carbon cycle in the Arctic Ocean* (pp. 6–21). Heidelberg: Springer-Verlag.

Macdonald, R. W., Solomon, S. M., Cranston, R. E., Welch, H. E., Yunker, M. B., & Gobeil, C. (1998). A sediment and organic carbon budget for the Canadian Beaufort Shelf. *Marine Geology, 144*, 255–273.

Macdonald, R. W., Wong, C. S., & Erickson, P. E. (1987). The distribution of nutrients in the southeastern Beaufort Sea: Implications for water circulation and primary production. *Journal of Geophysical Research, 92*, 2939–2952.

Machida, H., Arai, F., & Yokoyama, T. (1991). Re-examination of marker-tephra layers in the 200 m Lake Biwa core. *The Quaternary Research (Daiyonki-kenkyuu), 30*, 439–442.

Macko, S. A., & Aksu, A. E. (1986). Amino acid epimerization in planktonic foraminifera suggests slow sedimentation rates for Alpha Ridge, Arctic Ocean. *Nature, 322*, 730–732.

Manabe, S., & Broccoli, A. J. (1985). The influence of continental ice sheets on the climate of an ice age. *Journal of Geophysical Research, 90*, 2167–2190.

Manabe, S., & Stouffer, R. J. (1995). Simulation of abrupt climate change induced by freshwater input to the North Atlantic Ocean. *Nature, 378*, 165–167.

Manabe, S., & Stouffer, R. J. (1997). Coupled ocean-atmosphere model response to freshwater input: Comparison to Younger Dryas event. *Paleoceanography, 12*, 321–336.

Mangerud, J., Astakhov, V., Jakobsson, M., & Svendsen, J. I. (2001). Huge ice-age lakes in Russia. *Journal Quaternary Science, 16*(8), 773–777.

Mangerud, J., Astakhov, V. I., & Svendsen, J. I. (2002). The extent of the Barents-Kara ice sheet during the Last Glacial Maximum. *Quaternary Science Review, 21*, 111–119.

Mangerud, J., Dokken, T., Hebbeln, D., Heggen, B., Ingolfsson, O., Landvik, J. Y., Meydahl, V., Svendsen, J. I., & Vorren, T. O. (1998). Fluctuations of the Svalbard-Barents Sea ice sheet during the last 150.000 years. *Quaternary Science Review, 17*, 11–42.

Mangerud, J., & Gulliksen, S. (1975). Apparent radiocarbon ages of recent marine shells from Norway, Spitsbergen, and Arctic Canada. *Quaternary Research, 5*, 273–296.

Mangerud, J., Jakobsson, M., Alexanderson, H., Astakhov, V., Clarke, G. K. C., Henriksen, M., Hjort, C., Krinner, G., Lunkka, J. P., Möller, P., Murray, A., Nikolskaya, O., Saarnisto, M., & Svendsen, J. I. (2004). Ice-dammed lakes, rerouting of the drainage of Northern Eurasia during the last glaciation. *Quaternary Science Review, 23*(11–13), 1313–1332.

Mangerud, J., Jansen, E., & Landvik, J. Y. (1996). Late Cenozoic history of the Scandinavian and Barents Sea ice sheets. In: A. Solheim, F. Riis, A. Elverhøi, J. J. Faleide, L. N. Jensen, & S. Cloetingh (Eds), *Impact of glaciations on basin evolution: Data and models from the Norwegian Margins and Adjacent Basins* (pp. 11–26). Global and Planetary Change (Special Issue 12).

Mangerud, J., & Svendsen, J. I. (1992). The last interglacial-glacial period on Spitsbergen, Svalbard. *Quaternary Science Review, 11*, 633–664.

Mangerud, J., Svendsen, J. I., & Astakhov, V. I. (1999). Age and extent of the Barents and Kara ice sheets in Northern Russia. *Boreas, 28*, 46–80.

Manighetti, B., & McCave, I. N. (1995). Late glacial and Holocene palaeocurrents through South Rockall Gap, NE Atlantic Ocean. *Paleoceanography, 10*, 611–626.

Mann, U., Knies, J., Chand, S., Jokat, W., Stein, R., & Zweigel, J. (2008). Evaluation and modelling of Tertiary source rocks in the central Arctic Ocean. *Marine and Petroleum Geology*, in review.

Marienfeld, P. (1991). Holozäne Sedimentationsentwicklung im Scoresby Sund, Ost-Grönland. *Report on Polar Research, 96*, 166pp.

Marienfeld, P. (1992). Postglacial sedimentary history of Scoresby Sund, East Greenland. *Polarforschung, 60*(3), 181–195.

Marincovich, L., Jr., Brouwers, E. M., Hopkins, D. M., & McKenna, M. C. (1990). Late Mesozoic and Cenozoic paleogeographic and paleoclimatic history of the Arctic Ocean basin, based on shallow-water marine faunas and terrestrial vertebrates. *Geological Society of America*, (1), 403–426.

Markussen, B., Zahn, R., & Thiede, J. (1985). Late Quaternary sedimentation in the eastern Arctic Basin: Stratigraphy and depositional environment. *Palaeogeography, Palaeoclimatology, Palaeoecology, 50*, 271–284.

Markwick, P. J. (1998). Fossil crocodilians as indicators of Late Cretaceous and Cenozoic climates: Implications for using palaeontological data in reconstructing palaeoclimate. *Palaeogeography, Palaeoclimatology, Palaeoecology, 137*, 205–271.

Marlow, J. R., Lange, C. B., Wefer, G., & Rosell-Melé, A. (2000). Upwelling intensification as part of the Pliocene-Pleistocene climate transition. *Science, 290*, 2288–2291.

Marlowe, I. T., Brassell, S. C., Eglinton, G., & Green, J. C. (1984). Longchain unsaturated ketones and esters in living algae and marine sediments. *Organic Geochemistry, 6*, 135–141.

Marlowe, I. T., Brassell, S. C., Eglinton, G., & Green, J. C. (1990). Long-chain alkenones and alkyl alkenoates and the fossil coccolith record of marine sediments. *Chemical Geology, 88*, 349–375.

Martin, J. M., Guan, D. M., Elbaz-Poulichet, F., Thomas, A. J., & Gordeev, V. V. (1993). Preliminary assessment of the distributions of some trace elements (As, Cd, Cu, Fe, Ni, Pb and Zn) in a pristine aquatic environment: The Lena River estuary (Russia). *Marine Chemistry, 43*, 185–199.

Martinson, D. G., Pisias, N. G., Hays, J. D., Imbrie, J., Moore, T. C., & Shackleton, N. J. (1987). Age dating and the orbital theory of ice ages: Development of high-resolution 0 to 300,000-year chronostratigraphy. *Quaternary Research, 27*, 1–29.

Martrat, B., Grimalt, J. O., Villanueva, J., van Krefeld, S., & Sarnthein, M. (2003). Climatic dependence of the organic matter contributions in the north eastern Norwegian Sea over the last 15,000 years. *Organic Geochemistry, 34*, 1057–1070.

Maslanik, J. A., Serreze, M. C., & Barry, R. G. (1996). Recent decreases in Arctic summer ice cover and linkages to atmospheric circulation anomalies. *Geophysical Research Letters, 23*, 1677–1680.

Massé, G., Belt, S. T., Rowland, S. J., & Rohmer, M. (2004). Isoprenoid biosynthesis in the diatoms *Rhizosolenia setigera* (Brightwell) and *Haslea ostrearia* (Simonsen). *Proceedings of the National Academy of Sciences of the United States of America, 101*, 4413–4418.

Massé, G., Rowland, S. J., Sicre, M.-A., Jacob, J., Jansen, E., & Belt, S. T. (2008). Abrupt climate changes for Iceland during the last millennium: Evidence from high resolution sea ice reconstructions. *Earth and Planetary Science Letters*, 269, 564–568.

Matthews, R. K., & Frohlich, C. (2002). Maximum flooding surfaces and sequence boundaries: Comparisons between observations and orbital forcing in the Cretaceous and Jurassic (65–190 Ma). *GeoArabia*, 7, 503–538.

Matthiessen, J. (1995). Distribution patterns of dinoflagellate cysts and other organic-walled microfossils in recent Norwegian–Greenland Sea sediments. *Marine Micropaleontology, 24*, 307–334.

Matthiessen, J. (1999). Distribution of palynomorphs in surface sediments from the estuaries of rivers Ob and Yenisei, Kara Sea. *Report on Polar Research, 300*, 222–235.

Matthiessen, J., & de Vernal, A. (2001). Dinoflagellate cysts and paleoceanography of high latitude marine environments. *Journal Quaternary Science, 16*(7), 595–751.

Matthiessen, J., & Knies, J. (2001). Dinoflagellate cyst evidence for warm interglacial conditions at the northern Barents Sea margin during marine oxygen isotope stage 5. *Journal Quaternary Science, 16*, 727–738.

Matthiessen, J., Knies, J., Nowaczyk, N. R., & Stein, R. (2001). Late Quaternary dinoflagellate cyst stratigraphy at the Eurasian continental margin, Arctic Ocean: Indications for Atlantic water inflow in the past 150,000 years. *Global and Planetary Change, 31*, 65–86.

Matthiessen, J., Kunz-Pirrung, M., & Mudie, P. J. (2000). Freshwater chlorophycean algae in recent marine sediments of the Beaufort, Laptev and Kara Seas (Arctic Ocean) as indicators or river runoff. *International Journal of Earth Science, 89*, 470–485.

Matthiessen, J., Rogenhagen, J., Usbeck, R., Clausing, H., Kierdorf, C., & von Seggern, B. (2003). Sediment echosounding. In: P. Lemke (Ed.) *The expedition ARKTIS XIII/1 a, b of the research vessel polarstern in 2002, Report on Polar Marine Research, 446*, 39–40.

Matul, A., Abelmann, A., Tiedemann, R., Kaiser, A., & Nürnberg, D. (2002). Late Quaternary polycystine radiolarian datum events in the Sea of Okhotsk. *Geo-Marine Letters, 22*, 25–32.

Matul, A. G. (1995). Late Quaternary paleoceanology of the North Atlantic based on radiolaria analysis data. *Oceanology, 31*, 550–555.

Matul, A. G., & Abelmann, A. (2005). Pleistocene and Holocene distribution of the radiolarian Amphimelissa setosa Cleve in the north Pacific and North Atlantic: Evidence for water mass movement. *Deep-Sea Research Part II, 52*, 2351–2364.

Maurer, J. (2007). *Atlas of the Cryosphere*. Boulder, Colorado, U.S.A., National Snow and Ice Data Center, digital media; Retrieved from http://nsidc.org/data/atlas/.

Mayer, L. M. (1994). Surface area control of organic carbon accumulation in continental shelf sediments. *Geochimica et Cosmochimica Acta, 58*, 1271–1284.

Mazaud, A., Laj, C., & Bender, M. (1994). A geomagnetic chronology for antarctic ice accumulation. *Geophysical Research Letters, 21*, 337–340.

McCabe, A. M., & Clark, P. U. (1998). Ice-sheet variability around the North Atlantic Ocean during the last glaciation. *Nature, 392*, 373–377.

McCave, I. N., & Hall, I. R. (2006). Size sorting in marine muds: Processes, pitfalls, and prospects for paleoflow-speed proxies. *Geochemical Geophysical Geosystem*, 7, Q10N05. doi: 10.1029/2006GC001284.

McCave, I. N., Manighetti, B., & Beveridge, N. A. S. (1995b). Changes in circulation of the North Atlantic during the last 25,000 years inferred from grain size measurements. *Nature, 374*, 149–152.

McCave, I. N., Manighetti, B., & Robinson, S. G. (1995a). Sortable silt and fine sediment size/composition slicing: Parameters for paleocurrent speed and paleoceanography. *Paleoceanography, 10*, 593–610.

McClelland, J. W., De'ry, S. J., Peterson, B. J., Holmes, R. M., & Wood, E. F. (2006). A pan-arctic evaluation of changes in river discharge during the latter half of the 20th century. *Geophysical Research Letters, 33*, L06715. doi: 10.1029/2006GL025753.

McClymont, E. L., Rosell-Melé, A., Giraudeau, J., Pierre, C., & Loyd, J. M. (2005). Alkenone and coccolith records of the mid-Pleistocene in the south-east Atlantic. Implications for U^{K}_{37} index and South African climate. *Quaternary Science Review, 24*, 1559–1572.

McCulloch, M. T., & Wasserburg, G. J. (1978). Sm-Nd and Rb-Sr chronology of continental crust formation. *Science, 200*, 1003–1011.

McIntyre, A., Kipp, N. G., Be, A. W. H., Crowley, T., Kellogg, T. B., Gardner, J. V., Prell, W., & Ruddiman, W. F. (1976). Glacial North Atlantic 18,000 years ago: A CLIMAP reconstruction. In: R. M. Cline, & J. D. Hays (Eds), *Investigation of Southern Ocean paleoceanography and paleoclimatology, Memories* (Vol. 145, pp. 43–76), Geological Society of America.

McLaughlin, F. A., Carmack, E. C., Macdonald, R. W., & Bishop, J. K. B. (1996). Physical and geochemical properties across the Atlantic/Pacific water mass boundary in the southern Canadian Basin. *Journal of Geophysical Research, 101*(C1), 1183–1197.

McLennan, S. M. (1993). Weathering and global denudation. *The Journal of Geology, 101*, 295–303.

McManus, J. F., Bond, G. C., Broecker, W. S., Johnsen, S., Labeyrie, L., & Higgins, S. (1994). High-resolution climate records from the North Atlantic during the last interglacial. *Nature, 371*, 326–329.

McMillen, K. J., & Casey, R. E. (1978). Distribution of living polycystine radiolarians in the Gulf of Mexico and Caribbean Sea, and comparison with the sediment record. *Marine Micropaleontology., 3*, 121–145.

Meade, R. H., Bobrovitskaya, N. N., & Babkin, V. I. (2000). Suspended-sediment and freshwater discharges in the Ob and Yenisei rivers, 1960–1988. *International Journal of Earth Sciences, 89*, 461–469.

Melling, H. (2000). Exchanges of freshwater through the shallow straits of the North American Arctic. In: E. L. Lewis, E. P. Jones, P. Lemke, T. D. Prowse & P. Wadhams (Eds), *The freshwater budget of the Arctic Ocean* (pp. 479–502). Netherlands: Kluwer Academic Publishers.
Menard, H. W., & Smith, S. M. (1966). Hypsometry of ocean basin provinces. *Journal of Geophysical Research, 71*, 4305–4325.
Menzel, D., Hopmans, E. C., Schouten, S., & Sinninghe Damsté, J. S. (2006). Membrane teraether lipids of planctonic Crenachaeota in pliocene sapropels of the eastern Mediterranean Sea. *Palaeogeography, Palaeoclimatology, Palaeoecology, 239*, 1–15.
Meybeck, M. (1982). Carbon, nitrogen, and phosphorous transport by world rivers. *American Journal of Science, 282*, 401–450.
Meyers, P. A. (1994). Preservation of elemental and isotopic source identification of sedimentary organic matter. *Chemical Geology, 144*, 289–302.
Meyers, P. A. (1997). Organic geochemical proxies of paleoceanographic, paleolimnologic, and paleoclimatic processes. *Organic Geochemistry, 27*, 213–250.
Meynadier, L., Valet, J. P., Weeks, R., Shackleton, N. J., & Hagee, V. L. (1992). Relative geomagnetic intensity of the field during the last 140 ka. *Earth and Planetary Science Letters, 114*, 39–57.
Michaelson, G. J., Ping, C. L., & Kimble, J. M. (1996). Carbon storage and distribution in tundra soils of Arctic Alaska, U.S.A. *Arctic and Alpine Research, 28*, 414–424.
Mienert, J. (2004). COSTA — continental slope stability: Major aims and topics. *Marine Geology, 213*, 1–8.
Mienert, J., Andrews, J. T., & Milliman, J. D. (1992). The East Greenland continental margin (65° N) since the last deglaciation: Changes in seafloor properties and ocean circulation. *Marine Geology, 106*, 217–238.
Mienert, J., Hollender, F. -J., & Kenyon, N. H. (1995). GLORIA survey of the East Greenland margin: 70°N to 80°N. In: K. Crane, & A. Solheim (Eds), *Seafloor Atlas of the Northern Norwegian-Greenland Sea. Nor. Polarinst. Medd., 137*, 150–151.
Mienert, J., Kenyon, N. H., Thiede, J., & Hollender, F.-J. (1993). Polar continental margins: Studies off East Greenland. *EOS. Transactions American Geophysics Union, 74*, 225, 234, 236.
Mienert, J., Posewang, J., & Lukas, D. (2001). Changes in the hydrate stability zone on the Norwegian margin and their consequences for methane and carbon releases into the oceanosphere. In: P. Schäfer, W. Ritzrau, M. Schlüter & J. Thiede (Eds), *The northern North Atlantic; a changing environment* (pp. 259–280). Heidelberg: Springer-Verlag.
Mienert, J., Vanneste, M., Bünz, S., Andreassen, K., Haflidason, H., & Sejrup, H.-P. (2005). Ocean warming and gas hydrate stability on the mid-Norwegian margin at the Storegga Slide. *Marine Petroleum Geology, 22*, 233–244.
Mienert, J., & Weaver, P. (2003). *European margin sediment dynamics: Side-scan sonar and deismic images.* Berlin: Springer, 309pp.
Miller, K. G., Fairbanks, R. G., & Mountain, G. S. (1987). Tertiary oxygen isotope synthesis, sea level history, and continental margin erosion. *Paleoceanography, 2*, 1–19.
Miller, K. G., Mountain, G. S., Browning, J. V., Kominz, M., Sugarman, P. J., Christie-Blick, N., Katz, M. E., & Wright, J. D. (1998). Cenozoic global sea-level, sequences, and the New Jersey transect: Results from coastal plain and slope drilling. *Review Geophysiology, 36*, 569–601.
Miller, K. G., Wright, J. D., & Browning, J. V. (2005). Visions of ice sheets in a greenhouse world. *Marine Geology, 217*, 215–231.
Miller, K. G., Wright, J. D., & Fairbanks, R. G. (1991). Unlocking the ice house: Oligocene-Miocene oxygen isotopes, eustasy, and margin erosion. *Journal of Geophysical Research, 96*(B4), 6829–6849.
Miller, R. G., O'Nions, R. K., Hamilton, J., & Welin, E. (1986). Crustal residence ages of clastic sediments, orogeny and continental evolution. *Chemical Geology, 57*, 87–99.
Milliman, J. D., & Meade, R. H. (1983). World-wide delivery of river sediment to the oceans. *Journal of Geology, 91*, 1–21.
Milliman, J. D., & Syvitski, J. P. M. (1992). Geomorphic/tectonic control of sediment discharge to the ocean: The importance of small mountainous rivers. *Journal of Geology, 100*(5), 524–544.

Milne, G. A., Mitrovica, J. X., & Schrag, D. P. (2002). Estimating past continental ice volume from sea-level data. *Quaternary Science Review, 21*, 361–376.

Minicucci, D. A., & Clark, D. L. (1983). A late Cenozoic stratigraphy for glacial-marine sediments of the eastern Alpha Cordillera, central Arctic Ocean. In: B. F. Molnia (Ed.), *Glacial-marine sedimentation* (pp. 331–365). New York: Plenum Press.

Miyaka, Y. E., & Wada, E. (1967). The abundance ratio of $^{15}N/^{14}N$ in marine environments. *Records of Oceanographic Works in Japan, 9*, 32–53.

Möller, P., Hjort, C., Adrielsson, L., & Salvigsen, O. (1994). Glacial history of interior Jameson Land, East Greenland. *Boreas, 23*, 320–348.

Möller, P., Hjort, C., & Ingolfsson, O. (1991). The last interglacial/glacial cycle: Preliminary report on the PONAM fieldwork in Jameson Land and Scoresby Sund, East Greenland. *Lundqua Report, 33*, 181.

Montgomery, S. L. (2005). Petroleum geology and resource assessment: 1002 area, Arctic National Wildlife Refuge. *AAPG Bulletin, 89*, 291–310.

Moore, T. C., & the Expedition 302 Scientists (2006). Sedimentation and subsidence history of the Lomonosov Ridge. In: J. Backman, K. Moran, D. B. McInroy, & L. A. Mayer (Eds), *The expedition 302 scientists, Proceeding IODP, 302: Edinburgh (Integrated Ocean Drilling Program Management International, Inc.).* doi: 10.2204/iodp.proc.302.105.2006.

Moran, K., Backman, J., Brinkhuis, H., Clemens, S. C., Cronin, T., Dickens, G. R., Eynaud, F., Gattacceca, J., Jakobsson, M., Jordan, R. W., Kaminski, M., King, J., Koc, N., Krylov, A., Martinez, N., Matthiessen, J., McInroy, D., Moore, T. C., Onodera, J., O'Regan, A. M., Pälike, H., Rea, B., Rio, D., Sakamoto, T., Smith, D. C., Stein, R., St. John, K., Suto, I., Suzuki, N., Takahashi, K., Watanabe, M., Yamamoto, M., Frank, M., Jokat, W., & Kristoffersen, Y. (2006). The Cenozoic palaeoenvironment of the Arctic Ocean. *Nature, 441*, 601–605.

Moran, S. B., Ellis, K. M., & Smith, J. N. (1997). $^{234}Th/^{238}U$ disequilibria in the central Arctic Ocean: Implications for particulate organic carbon export. *Deep-Sea Research Part II, 44*, 1593–1606.

Moran, S. B., & Smith, J. N. (2000). ^{234}Th as a tracer of scavenging and particle export in the Beaufort Sea. *Continental Shelf Research, 20*, 153–167.

Moran, S. B., Weinstein, S. E., Edmonds, H. N., Smith, J. N., Kelly, R. P., Pilson, M. E. Q., & Harrison, W. G. (2003). Does $^{234}Th/^{238}U$ disequilibrium provide an accurate record of the export flux of particulate organic carbon from the upper ocean?. *Limnology and Oceanography, 48*, 1018–1029.

Morison, J., Aagaard, K., & Steele, M. (2000). Recent environmental changes in the Arctic: A review. *Arctic, 53*, 359–371.

Morison, J., Alexander, V., Codispoti, L., Delworth, T., Dickson, B., Eicken, H., Grebmeier, J., Kruse, J., Overland, J., Overpeck, J., Schlosser, P., Serreze, M., & Walsh, J. (2001). SEARCH: Study of environmental Arctic change, science plan. *Polar Science Center, Applied Physics Laboratory* (89pp). University of Washington, Seattle.

Moritz, R. E., Bitz, C. M., & Steig, E. J. (2002). Dynamics of recent climate change in the Arctic. *Science, 297*, 1497–1502.

Mørk, A., & Bjorøy, M. (1984). Mesozoic source-rocks on Svalbard. In: A. M. Spencer et al. (Eds), *Petroleum geology of the North European margin* (pp. 371–382). London: Graham & Trotman.

Mørk, M. B. E., & Duncan, R. A. (1993). Late pliocene basaltic volcanism on the Western Barents Shelf margin: Implications from petrology and ^{40}Ar-^{39}Ar dating of volcaniclastic debris from a shallow drill core. *Norsk Geology Tidsskrift, 73*, 209–225.

Morley, J. J., & Hays, J. D. (1983). Oceanographic conditions associated with high abundances of the radiolarian Cycladophora davisiana. *Earth and Planetary Science Letters, 66*, 63–72.

Moros, M., Emeis, K., Risebrobakken, B., Snowball, I., Kuijpers, A., McManus, J., & Jansen, E. (2004). Sea surface temperatures and ice rafting in the Holocene North Atlantic: Climate influences on northern Europe and Greenland. *Quaternary Science Review, 23*, 2113–2126.

Morris, T. H., Clark, D. L., & Blasco, S. M. (1985). Sediments of the Lomonosov Ridge and Makarov Basin: A pleistocene stratigraphy for the North Pole. *Geological Society of America Bulletin, 96*, 901–910.

Mostafawi, N. (1990). Ostracods in Late Pleistocene and Holocene sediments from the Fram Strait, eastern Arctic. In: R. Whatley & C. Maybuty (Eds), *Ostracoda and global events* (pp. 489–494). London: Chapman & Hall.

Mowatt, T. C., & Naidu, A. S. (1987). A brief overview of the clay mineral assemblages in sediments of the major rivers of Alaska and adjacent Arctic Canada. *Mitteilungen Geology-Pal. Institute University Hamburg, 64*, 269–277.

Mudelsee, M., & Schulz, M. (1997). The Middle Pleistocene climate transition: Onset of 100kyr cycle lags ice volume buildup by 280 ka. *Earth and Planetary Science Letters, 151*, 117–123. doi: 10.1016/S0012-821X(97)00114-3.

Mudie, P. J. (1982). Pollen distribution in recent marine sediments, eastern Canada. *Canadian Journal of Earth Science, 19*(4), 729–747.

Mudie, P. J. (1985). Palynology of the CESAR cores, Alpha Ridge. In: H. R. Jackson, P. J. Mudie, & S. M. Blasko (Eds), *Initial seological report on CESAR: The Canadian expedition to study the Alpha Ridge* (Vol. 84–22, pp. 148–174), Geographical Survey Canadian Papers.

Mudie, P. J. (1989). Palynology and dinocyst biostratigraphy of the Late Miocene to Pleistocene, Norwegian Sea: ODP Leg 104, Sites 642 to 644. In: O. Eldholm, J. Thiede, & E. Taylor (Eds), *Proceedings of the Ocean Drilling Program, Scientific Results* (pp. 587–610). Texas: Ocean Drilling Program, College Station.

Mudie, P. J. (1992). Circum-Arctic Quaternary and Neogene marine palynofloras: Paleoecology and statistical analysis. In: M. J. Head, & J. H. Wrenn (Eds), *Neogene and Quaternary dinoflagellate cysts and acritarchs* (pp. 347–390), American Association of Stratigraphy Palynology Foundations, Dallas.

Mudie, P. J., & Blasco, S. M. (1985). Lithostratigraphy of the CESAR cores. In: H. R. Jackson, P. J. Mudie, & S. M. Blasco (Eds), *Initial Geological report on CESAR: The Canadian expedition to study the Alpha Ridge* (Vol. 84–22, pp. 59–99), Geological Survey Canadian Papers.

Mudie, P. J., de Vernal, A., & Head, M. J. (1990). Neogene to recent palynostratigraphy of circum-Artic basins: Results of ODP Leg 104, Norwegian Sea, Leg 105, Baffin Bay, and DSDP Site 611, Irminger Sea. In: U. Bleil & J. Thiede (Eds), *Geological history of the Polar Oceans: Arctic versus Antarctic. NATO ASI Series C — mathematical and physical sciences* (pp. 609–646). Dordrecht: Kluwer Academic Publishers.

Mudie, P. J., & Helgason, J. (1983). Palynological evidence for Miocene climatic cooling in eastern Iceland about 9.8 Myr ago. *Nature, 303*, 689–692.

Mudie, P. J., & McCarthy, F. M. G. (1994). Late Quaternary pollen transport processes, western North Atlantic: Data from box models, cross-margin and N-S transects. *Marine Geology, 118*, 79–105.

Mudie, P. J., & Rochon, A. (2001). Distribution of dinoflagellate cysts in the Canadian Arctic marine region. *Journal Quaternary Science, 16*(7), 603–620.

Mudie, P. J., & Short, S. K. (1985). Marine palynology of Baffin Bay. In: J. T. Andrews (Ed.), *Quaternary environments — eastern Canadian Arctic, Baffin Bay and Western Greenland* (pp. 263–308). Boston: Allen & Unwin.

Mulholland, P. J., & Watts, J. A. (1982). Transport of organic carbon to the oceans by rivers of North America: A synthesis of existing data. *Tellus, 34*, 176–186.

Mullen, R. E., Darby, D. A., & Clark, D. L. (1972). Significance of atmospheric dust and ice rafting for Arctic Ocean sediment. *Bulletin of Geological Society of America, 83*, 205–212.

Müller, C. (1999). Rekonstruktion der Paläo-Umweltbedingungen am Laptev-See-Kontinentalrand während der beiden letzten Glazial-/Interglazial-Zyklen anhand sedimentologischer und mineralogischer Untersuchungen. *Report on Polar Research, 328*, 146pp.

Müller, C., & Stein, R. (2000). Variability of fluvial sediment supply to the Laptev Sea continental margin during Late Weichselian to Holocene times: Implications from clay-mineral records. *International Journal of Earth Science, 89*, 592–604.

Müller, P. J. (1977). C/N ratios in Pacific deep-sea sediments: Effect of inorganic ammonium and organic nitrogen compounds sorbed by clays. *Geochimica et Cosmochimica Acta, 41*, 765–776.

Müller, P. J., Cepek, M., Ruhland, G., & Schneider, R. R. (1997). Alkenone and coccolithophorid species changes in Late Quaternary sediments from the walvis ridge: Iimplications for the alkenone paleotemperature method. *Palaeogeography, Palaeoclimatology, Palaeoecology, 135*, 71–96.

Müller, P. J., Kirst, G., Ruhland, G., von Storch, I., & Rosell-Melé, A. (1998). Calibration of the alkenone paleotemperature index Uk37based on core-tops from the eastern South Atlantic and the global ocean (60 degrees N -60 degrees S). *Geochimica et Cosmochimica Acta, 62*(10), 1757–1772.

Müller, P. J., & Suess, E. (1979). Productivity, sedimentation rate, and sedimentary organic matter in the oceans. I.-Organic matter preservation. *Deep-Sea Research, 26*, 1347–1362.

Müller-Lupp, T., Bauch, H. A., Erlenkeuser, H., Hefter, J., Kassens, H., & Thiede, J. (2000). Changes in the deposition of terrestrial organic matter on the Laptev Sea shelf during the Holocene: Evidence from stable carbon isotopes. *International Journal of Earth Science, 89*, 563–568.

Mutterlose, J., Brumsack, H., Flögel, S., Hay, W., Klein, C., Langrock, U., Lipinski, M., Ricken, W., Söding, E., Stein, R., & Swientek, O. (2003). The Greenland-Norwegian Seaway: A key for understanding Late Jurassic to Early Cretaceous paleoenvironments. *Paleoceanography, 18*, 1010. doi: 10.1029/2001PA000625.

Mutterlose, J., & Kessels, K. (2000). Early Cretaceous calcareous nannofossils from high latitudes: Implications for palaeobiogeography and palaeoclimate. *Palaeogeography, Palaeoclimatology, Palaeoecology, 160*, 347–372.

Myhre, A. M., Thiede, J., Firth, J. V., et al. (Eds). (1995). *Proceedings of ODP, Init. Results, 151*, College Station, Texas (Ocean Drilling Program).

Mysak, L. A. (2001). Patterns of Arctic circulation. *Science, 293*, 1269–1270.

Mysak, L. A., & Venegas, S. A. (1998). Decadal climate oscillations in the Arctic: A new feedback loop for atmospheric-ice-ocean interactions. *Geophysical Research Letters, 25*(19), 3607–3610.

NAD (1997). Nansen Arctic drilling implementation plan. *Report of "Nansen Arctic Drilling Program Implementation Plan Workshop*", Arctic and Antarctic Research Institute, St. Petersburg, Russia, 14–15 October 1996, Joint Oceanographic Institutions, Washington, DC, 50pp.

Nagurnyi, A. P., Korostelev, V. G., & Abaza, P. A. (1994). Wave method for evaluating effective ice thickness of sea ice in climate monitoring. *Bull. Russian Acad. Sci. Phys. Suppl. Phys. Vib., 58*, 168–174.

Nagurnyi, A. P., Korostelev, V. G., & Ivanov, V. V. (1999). Multiyear variability of sea ice thickness in the arctic basin measured by elasticgravity waves on the ice surface. *Meteorology and Hydrology, 3*, 72–78 (In Russian).

Naidina, O. D. (1995). Holocene climatic, vegetation, and pollen data of Siberia adjacent to the Laptev Sea, In: H. Kassens, D. Pipenburg, J. Thiede, L. Timokhov, H. W. Hubberten, & S. Priamikov, (Eds). *Russian-German cooperation: The Laptev Sea System. Reps. Polar Research* (Vol. 176, pp. 235–253).

Naidina, O. D., & Bauch, H. A. (1999). Distribution of pollen and spores in surface sediments of the Laptev Sea. In: H. Kassens, H. Bauch, I. Dmitrenko, H. Eicken, H. W. Hubberten, M. Melles, J. Thiede & L. Timokhov (Eds), *Land-ocean systems in the Siberian: Dynamics and history* (pp. 577–585). Heidelberg: Springer-Verlag.

Naidina, O. D., & Bauch, H. A. (2001). A Holocene pollen record from the Laptev Sea shelf, Northern Yakutia. *Global and Planetary Change, 31*, 141–153.

Naidu, A. S. (1985). Organic carbon, nitrogen, and C/N ratios of deltaic sedimens, North Arctic Alaska, In: E. T. Degens, S. Kempe, & R. Herrera (Eds). *Transport of carbon and minerals in major world rivers* (pp. 311–321), SCOPE 58, Mitt. Geol.-Palaeontol. Institute Univ. Hamburg.

Naidu, A. S., Cooper, L. W., Finney, B. P., Macdonald, R. W., Alexander, C., & Semiletov, I. P. (2000). Organic carbon isotope ratios ($\delta^{13}C$) of Amerasian continental shelf sediments. *International Journal of Earth Science, 89*, 522–532.

Naidu, A. S., Cooper, L. W., Grebmeier, J. M., Whitledge, T. E., & Hameedi, M. J. (2004). The continental margin of the North Bering-Chukchi Sea: Distribution, sources, fluxes, and burial rates of organic carbon. In: R. Stein & R. W. Macdonald (Eds), *The organic carbon cycle in the Arctic Ocean* (pp. 193–203). Heidelberg: Springer-Verlag.

Naidu, A. S., Creager, J. S., & Mowatt, T. C. (1982). Clay mineral dispersal patterns in the North Bering and Chukchi seas. *Marine Geology, 47*, 1–15.

Naidu, A. S., & Mowatt, T. C. (1983). Sources and dispersial patterns of clay minerals in surface sediments from the continental-shelf areas off Alaska. *Geological Society of America Bulletin, 94*, 841–854.

Naidu, A. S., Mowatt, T. C., Hawkins, D. B., & Hood, D. W. (1975). Clay minerals and geochemistry of some Arctic Ocean sediments: Significance of paleoclimate interpretation. In: G. Weller & S. A.

Bowling (Eds), *Climate of the Arctic, Geophysics* (pp. 59–67). Fairbanks: Institute, University of Alaska.
Naidu, A. S., Scalan, R. S., Feder, H. M., Goering, J. J., Hameedi, M. J., Parker, P. L., Behrens, E. W., Caughey, M. E., & Jewett, S. C. (1993). Stable organic carbon isotopes in sediments of the North Bering-south Chukchi seas, Alaskan-Soviet Arctic shelf. *Continental Shelf Research, 13*, 669–691.
Nam, S., & Stein, R. (1999). Late Quaternary variations in sediment accumulation rates and their paleoenvironmental implications: A case study from the East Greenland Continental Margin. In: P. Bruns, & H. C. Hass (Eds). *On the determination of sediment accumulation rates* (pp. 223–240), GeoResearch Forum 5, Trans. Tech. Publications.
Nam, S.-I. (1997). *Late Quaternary glacial history and paleoceanographic reconstructions along the East Greenland continental margin: Evidence from high-resolution records of stable isotopes and ice-rafted debris.* Ph.D. Dissertation, University of Bremen, 251pp.
Nam, S.-I., Stein, R., Grobe, H., & Hubberten, H. (1995). Late quaternary glacial-interglacial changes in sediment composition at the East Greenland continental margin and their paleoceanographic implications. *Marine Geology, 122*, 243–262.
Nansen, F. (1897). *Farthest north.* Whitehall Gardens: Archibald Constabel & Co, 510pp.
Nansen, F. (1902). *The Norwegian North Polar Expedition 1893–1896: Scientific results, III.* Toronto: Longmans, 427pp.
Nansen, F. (1904). The bathymetrical features of the North Polar Seas, with a discussion of continental shelves and previous oscillations of the shoreline, Norwegian North Polar expedition, 1893–1896. *Science Research, 4*, 232pp.
Nathorst, A. G. (1890). Ueber die Reste eines Brotfruchtbaums ARTOCARPUS DICKSONI N. SP., aus den cenomanen Kreideablagerungen Grönlands. *Kongl. Svenska Vetenskaps-Akad. Hand, 24*, 2–9.
Nathorst, A. G. (1911). On the value of fossil floras of the Arctic regions as evidence of geological climates. *Geological Magazine, 8*, 217–225.
Naugler, F. P. (1967). Recent sediments of the East Siberian Sea. Unpublished Master thesis, University of Washington.
Naugler, F. P., Silverberg, N., & Creager, J. S. (1974). Recent sediments of the East Siberian Sea. In: Y. Herman (Ed.), *Oceanography of the Arctic Seas, Marine Geology* (pp. 191–210). Heidelberg: Springer-Verlag.
Nelson, F. E., Lachenbruch, A. H., Woo, M. K., Koster, E. A., Osterkamp, T. E., Gavrilova, M. K., & Cheng, G. D. (1993). Permafrost and changing climate. In: *Proceedings of the Sixth International Conference on Permafrost* (pp. 987–1005). Wushan, Guangzhou, China: South China University of Technology Press.
Nelson, F. E., Shiklomanov, N. I., & Mueller, G. R. (1999). Variability of active-layer thickness at multiple spatial scales, northcentral Alaska, U.S.A. *Arctic, Antarctic, and Alpine Research, 31*, 158–165.
Nesje, A., Matthews, J. A., Dahl, S. O., Berrisford, M. S., & Andersson, C. (2001). Holocene glacier fluctuations of Flatebreen and winter-precipitation changes in the Jostedalsbreen region, western Norvay, based on glaciolacustrine sediment records. *The Holocene, 11*, 267–280.
Nghiem, S. V., Rigor, I. G., Perovich, D. K., Clemente-Colón, P., Weatherly, J. W., & Neumann, G. (2007). Rapid reduction of Arctic perennial sea ice. *Geophysical Research Letters, 34*, L19504. doi: 10.1029/2007GL031138.
NGRIP Members, (2004). High-resolution record of Northern Hemisphere climate extending into the last interglacial period. *Nature, 431*, 147–151.
Nichols, P. D., Volkman, J. K., Palmisano, A. C., Smith, G. A., & White, D. C. (1988). Occurrence of an isoprenoid C25 diunsaturated alkene and high neutral lipid content in Antarctic sea ice diatom communities. *Journal of Phycology, 24*, 90–96.
Niebauer, H. J., Bond, N. A., Yakunin, L. P., Plotnikov, V. V. (1999). An update of the climatology and sea ice of the Bering Sea. In: T. R. Loughlin & K. Ohtani (Eds), *The physical oceanography of the Bering Sea* (pp. 29–59). Fairbanks: University of Alaska Seagrant AK-SG-03, University of Alaska Fairbanks.
Nies, H., Harms, H., Karcher, M. J., Dethleff, D., Bahe, C., Kuhlmann, G., Oberhuber, J. M., Backhaus, J. O., Kleine, E., Loewe, P., Matishov, D., Stepanov, A., & Vasiliev, O. F. (1998).

Anthropogenic radioactivity in the Nordic Seas and the Arctic Ocean - results of a joint project. *Deutsche Hydrographische Zeitschrift (German Journal of Hydrography), 50*, 313–343.

Niessen, F., Ebel, T., Kopsch, C., & Federov, G. B. (1999). High-resolution seismic stratigraphy of lake sediments on the Taymyr Peninsula, Central Siberia. In: H. Kassens, H. A. Bauch, I. Dmitrenko, H. Eicken, H.-W. Hubberten, M. Melles, J. Thiede & L. Timokhov (Eds), *Land–Ocean systems in the Siberian Arctic: Dynamics and history* (pp. 437–456). Berlin: Springer-Verlag.

Niessen, F., Jarrard, R. D., & Bücker, C. (1998). Log-based physical properties of the CRP-1 core, Ross Sea, Antarctica. *Terra Antarctica, 5*, 299–310.

Niessen, F., & Weiel, D. (1996). Distribution of magnetic susceptibility on the eurasian shelf and continental slope -Implications for source area of magnetic minerals. *Report on Polar Research, 212*, 81–88.

Niessen, F., & Whittington, R. (1995). Marine sediment echosounding using PARASOUND. In: H. -W. Hubberten (Ed), *Die expedition ARKTIS-X/2 mit FS "Polarstern" 1994, Report on Polar Research, 174*, 62–68.

Nijenhuis, I. A., Bosch, H.-J., Sinninghe Damsté, J. S., Brumsack, H.-J., & de Lange, G. J. (1999). Organic matter and trace element rich sapropels and black shales: A geochemical comparison. *Earth and Planetary Science Letters, 169*, 277–290.

Nijenhuis, I. A., Brumsack, H. -J., de Lange, G. J. (1998). The trace element composition of organic carbon-rich sapropels versus black shales: A comparison. In: A. H. F. Robertson, K. -C. Emeis, C. Richter, & A. Camerlenghi (Eds), *Proceedings of ODP Science Research* (Vol. 160, pp. 199–206), College Station, Texas.

Nimmergut, A., & Abelmann, A. (2002). Spatial and seasonal changes of radiolarian standing stocks in the Sea of Okhotsk. *Deep-Sea Research Part I, 49*, 463–493.

Nordenskjold, A. E. (1882). *The voyage of the Vega around Aisa and Europe with a historical review of previous journeys along the north coast of the old world.* New York: Macmillan & Company, 756pp.

Nørgaard-Pedersen, N., Mikkelsen, N., Lassen, S. J., & Kristoffersen, Y. (2007b). Arctic Ocean record of last two glacial-interglacial cycles off North Greenland/Ellesmere Island — implications for glacial history. *Marine Geology, 244*, 93–108.

Nørgaard-Pedersen, N., Mikkelsen, N., Lassen, S. J., Kristoffersen, Y., & Sheldon, E. (2007a). Reduced sea ice concentrations in the Arctic Ocean during the last interglacial period revealed by sediment cores off Northern Greenland. *Paleoceanography, 22*, PA1218. doi: 10.1029/2006PA001283.

Nørgaard-Pedersen, N., Spielhagen, R. F., Erlenkeuser, H., Grootes, P. M., Heinemeier, J., & Knies, J. (2003). The Arctic Ocean during the Last Glacial Maximum: Atlantic and Polar domains of surface water mass distribution and ice cover. *Paleoceanography, 18*, 1–19.

Nørgaard-Pedersen, N., Spielhagen, R. F., Thiede, J., & Kassens, H. (1998). Central Arctic surface ocean environment during the past 80.000 years. *Paleoceanography, 13*, 193–204.

Notholt, H. (1998). The implication of the "NorthEastWater"-Polynya on the sedimentation by NE-Greenland and Late Quaternary Paleo-oceanic investigations. *Report on Polar Research, 275*, 183pp.

Nowaczyk, N. R. (1991). High-resolution magnetostratigraphy of late Quaternary Arctic marine sediments. *Report on Polar Research, 78*, 187pp.

Nowaczyk, N. R. (1997). High-resolution magnetostratigrahy of four sediment cores from the Greenland Sea—II. Rock magnetic and palaeointensity data. *Geophysical Journal International, 131*, 325–334.

Nowaczyk, N. R., & Baumann, M. (1992). Combined high-resolution magnetostratigraphy and nannofossil biostratigraphy for late Quaternary Arctic Ocean sediments. *Deep-Sea Research, 39*, 567–701.

Nowaczyk, N. R., & Frederichs, T. (1999). Geomagnetic events and relative paleointensity variations during the last 300 ka as recorded in Kolbeinsey Ridge sediments, Iceland Sea, indication for a strongly variable geomagnetic field. *International Journal of Earth Science, 88*, 116–131.

Nowaczyk, N. R., Frederichs, T. W., Eisenhauer, A., & Gard, G. (1994). Magnetostratigraphic data from late quaternary sediments from the Yermak Plateau, Arctic Ocean. Evidence for four geomagnetic polarity events within the last 170 ka of the Brunhes Chron. *Geophysical Journal International, 117*, 453–471.

Nowaczyk, N. R., Frederichs, T. W., Kassens, H., Norgaard-Pedersen, N., Spielhagen, R., Stein, R., & Weiel, D. (2001). Sedimentation rates in the Makarov Basin, central Arctic Ocean: A paleomagnetic and rock magnetic approach. *Paleoceanography, 16*, 368–389.

Nowaczyk, N. R., Frederichs, T. W., Kassens, H., Nørgaard-Petersen, N., Spielhagen, R., Stein, R., & Weiel, D. (2000). Sedimentation rates in the Makarov Basin, Central Arctic Ocean — a paleo- and rock magnetic approach. *Paleoceanography, 16*, 368–389.

Nowaczyk, N. R., & Knies, J. (2000). Magnetostratigraphic results from the eastern Arctic Ocean, AMS ^{14}C ages and relative palaeointensity data of the Mono Lake and Laschamp geomagnetic reversal excursions. *International Journal of Geophysical, 140*, 185–197.

Nürnberg, D. (1996). Biogenic barium and opal in shallow eurasian shelf sediments in relation to the pelagic arctic ocean environment. In: R. Stein, G. I. Ivanov, M. A. Levitan, & K. Fahl (Eds), *Surface sediment composition and sedimentary processes in the central Arctic Ocean and adjacent Eurasian continental margin*, Rep. Pol. Res. 212, pp. 96–118.

Nürnberg, D., Levitan, M. A., Pavlidis, J. A., & Shelekhova, E. S. (1995). Distribution of clay minerals in surface sediments from the Eastern Barents and Southwestern Kara seas. *Geological Rundschau, 84*, 665–682.

Nürnberg, D., Wollenburg, I., Dethleff, D., Eicken, H., Kassens, H., Letzig, T., Reimnitz, E., & Thiede, J. (1994). Sediments in Arctic sea ice: Implications for entrainment, transport and release. In: J. Thiede, T. O. Vorren & R. F. Spielhagen (Eds), *Marine Geology, 119*, 185–214.

Ó Cofaigh, C., Dowdeswell, J. A., Evans, J., Kenyon, N. H., Mienert, J., & Wilken, M. (2004). Timing and significance of glacially-influenced mass-wasting in the submarine channels of the Greenland Basin. *Marine Geology, 207*, 39–54.

Ó Cofaigh, C., Dowdeswell, J. A., & Kenyon, N. H. (2006). Geophysical investigations of a high-latitude submarine channel system and associated channel-mouth lobe in the Lofoten Basin, Polar North Atlantic. *Marine Geology, 226*, 41–50.

Ó Cofaigh, C., Pudsey, C. J., Dowdeswell, J. A., & Morris, P. (2003a). Evolution of subglacial bedforms along a paleo-ice stream, Antarctic Peninsula continental shelf. *Geophysical Research Letters, 29*, 41-1–41-4.

Ó Cofaigh, C., Taylor, J., Dowdeswell, J. A., & Pudsey, C. J. (2003b). Palaeo-ice streams, trough mouth fans and high-latitude continental slope sedimentation. *Boreas, 32*, 37–55.

Ó Cofaigh, C., Taylor, J., Dowdeswell, J. A., Rosell-Mele', A., Kenyon, N. H., Evans, J., & Mienert, J. (2002). Sediment reworking on high-latitude continental margins and its implications for palaeoceanographic studies: Insights from the Norwegian-Greenland Sea, In: J. A. Dowdeswell, & C. O' Cofaigh (Eds). *Glacier-influenced sedimentation on high-latitude continental margins* (pp. 325–348), Geological Society of London (Special Publication 203).

O'Brien, M. C., Macdonald, R. W., Melling, H., & Iseki, K. (2006). Particle fluxes and geochemistry on the Canadian Beaufort Shelf: Implications for sediment transport and deposition. *Continental Shelf Research, 26*, 41–81.

O'Grady, D. B., & Syvitski, J. P. M. (2002). Large-scale morphology of Arctic continental slopes: The influence of sediment delivery on slope form. In: J. A. Dowdeswell & C. Ó Cofaigh (Eds), *Glacier influenced sedimentation on high-latitude continental margins* (pp. 11–32). Geological Society of London (Special Publication 203).

O'Leary, M. H. (1988). Carbon isotopes in photosynthesis. *Bioscience, 38*, 328–336.

O'Neill, B. J. (1981). Pliocene and Pleistocene benthic foraminifera from the central Arctic Ocean. *Journal of Paleontology, 55*, 1141–1170.

O'Regan, M., King, J., Backman, J., Jakobsson, M., Pälike, H., Moran, K., Heil, C., Sakamoto, T., Cronin, T. M., & Jordan, R. W. (2008a). Constraints on the Pleistocene Chronology of Sediments from the Lomonosov Ridge. *Paleoceanography, 23*, PA1S19, doi:10.1029/2007PA001551.

O'Regan, M., Moran, K., Backman, J., Jakobsson, M., Sangiorgi, F., Brinkhuis, H., Pockalny, R., Skelton, A., Stickley, C., Koç, N., Brumsack, H., & Willard, D. (2008b). Mid-Cenozoic tectonic and paleoenvironmental setting of the central Arctic Ocean. *Paleoceanography, 23*, PA1S20, doi:10.1029/2007PA001559.

Oeschger, H. (1992). Working hypotheses for glaciation/deglaciation mechanisms. In: G. J. Kulka & E. Went (Eds), *Start of a Glacial* (pp. 273–289). NATO ASI Series I 3.

Ogawa, Y., Takahashi, K., & Yamanaka, T. (2008). Paleoceanography of the middle Eocene Arctic Ocean based on geochemical measurements of biogenic matter. *Memoirs of the Faculty of Science Kyushu University Series D Earth and Planetary Sciences, 32*(1), 31–48.

Okazaki, Y., Takahashi, K., Yoshitani, H., Nakatsuka, T., Ikehara, M., & Wakatsuchi, M. (2003). Radiolarians under seaice covered conditions in the Sea of Okhotsk: Flux and their implications for paleoceanography. *Marine Micropaleontology, 49*, 195–230.

Okulitch, A. V. (1991). Geology of the Canadian Archipelago and North Greenland. In: H. P. Trettiln (Ed). *Innuitian orogen and Arctic platform: Canada and Greenland. The geology of North America*. The Geological Society of America, Boulder, Colorado, E, 1:200,000.

Okulitch, A. V., Lopatin. B. G., & Jackson, H. R. (1989). Circumpolar geological map of the Arctic. Geological Survey of Canada, 1:6,000,000. Ottawa (Geological Survey of Canada), Map 1765A.

Opsahl, S., Benner, R., & Amon, R. M. W. (1999). Major flux of terrigenous dissoved organic matter through the Arctic Ocean. *Limnology and Oceanography, 44*, 2017–2023.

Osterkamp, T. E., Baker, G. C., Harrison, W. D., & Matava, T. (1989). Characteristics of the active layer and shallow subsea permafrost. *Journal of Geophysical Research, 94*(C11), 16,227–16,236.

Osterkamp, T. E., & Gosink, J. P. (1984). Observations and analyses of sediment-laden sea ice. In: P. W. Barnes, D. M. Schell & E. Reimnitz (Eds), *The Alaskan Beaufort Sea: Ecosystems and environments* (pp. 73–93). Orlando: Academic Press.

Ottesen, D., Dowdeswell, J.A., Rise, L., Rokoengen, K., & Henriksen, S. (2002). Large-scale morphological evidence for past ice-stream flow on the mid-Norwegian continental margin. In: J. A. Dowdeswell, & C. Ó Cofaigh (Eds). *Glacier-influenced sedimentation on high-latitude continental margins* (pp. 245–258), Geological Society of London (Special Publication 203).

Pace, M. L., Knauer, G. A., Karl, D. M., & Martin, J. H. (1987). Primary production, new production and vertical flux in the eastern Pacifc Ocean. *Nature, 325*, 803–804.

Pagani, M., Pedentchouk, N., Huber, M., Sluijs, A., Schouten, S., Brinkhuis, H., Sinninghe Damsté, J. S., Dickens, G. R., & the IODP Expedition 302 Scientists (2006). The Arctic's hydrologic response to global warming during the Palaeocene-Eocene thermal maximum. *Nature, 442*, 671–675.

Pagani, M., Zachos, J. C., Freeman, K. H., Tipple, B., & Bohaty, S. (2005). Marked decline in atmospheric carbon dioxide concentrations during the Paleogene. *Science, 309*, 600–603.

Pagels, U. (1991). Sedimentologische Untersuchungen und Bestimmungen der Karbonatlösung in spätquartären Sedimenten des östlichen Arktischen Ozeans. *GEOMAR Report, 10*, 106.

Pak, D. K., Clark, D. L., & Blasco, S. M. (1992). Late Pleistocene stratigraphy and micropaleontology of a part of the Eurasian Basin (Fram Basin), central Arctic Ocean. *Marine Micropaleontology, 20*, 1–22.

Pälike, H., Spofforth, D. J. A., O'Regan, M., & Gattacceca, J. (2008). Orbital scale variations and timescales from the Arctic Ocean. *Paleoceanography, 23*, PA1S10, doi:10.1029/2007PA001490.

Pancost, R. D., & Boot, C. S. (2004). The palaeoclimatic utility of terrestrial biomarkers in marine sediments. *Marine Chemistry, 92*, 239–261.

Parkinson, C. L., Cavalieri, D. J., Gloersen, P., Zwally, J. H., & Comiso, J. C. (1999). Arctic sea ice extents, areas, and trends, 1978–1996. *Journal of Geophysical Research, 104*, 20837–20856.

Parrish, J. T., & Spicer, R. A. (1988). Late Cretaceous terrestrial vegetation: A nearpolar temperature curve. *Geology, 16*, 22–25.

Paul, H. A. (2002). Application of novel stable isotope methods to reconstruct paleoenvironments: Compound-specific hydrogen isotopes and pore-water oxygen isotopes. Diss. ETH Nr. 14593, Swiss Federal Institute of Technology Zürich, Zürich, 140pp.

Paull, C. K., Ussler, W., III, Holbrook, W. S., Hill, T. M., Keaten, R., Mienert, J., Haflidason, H., Johnson, J. E., Winters, W. J., & Lorenson, T. D. (2008). Origin of pockmarks and chimney structures on the flanks of the Storegga Slide, offshore Norway. *Geo-Marine Letters, 28*, 43–51.

Pavlidis, Yu. A., Dunayev, N. N., & Shecherbakov, F. A. (1997). The Late Pleistocene palaeogeography of Arctic Eurasian shelves. In: A. A. Velitchko, P. M. Dolukhanov, N. W. Rutter, & N. R. Catto (Eds). *Quaternary of Northern Eurasia: Late Pleistocene and Holocene landscapes, stratigraphy and environments. Quaternary International, 41/42*, 3–9.

Pchelina, T. M., & Komarnitsky, V. M. (2007). Chemical characteristics of bitumen from sedimentary rocks of VNIIO-1984-80 Hole drilled in the Barents Sea. All-Russian Research Institute for Geology and Mineral Resources of the World Ocean, PANGAEA, doi: 10.1594/PANGAEA.626692.

Pearson, P. N., & Palmer, M. R. (2000). Atmospheric carbon dioxide concentrations over the past 60 million years. *Nature, 406*, 695–699.

Pedersen, T. F., Waters, R. D., & Macdonald, R. W. (1989). On the natural enrichment of cadmium and molybdenum in the sediments of ucluelet inlet, British Columbia. *The Science of the Total Environment, 79*, 125–139.

Peinert, R., Antia, A., Bauerfeind, E., v. Bodungen, B., Haupt, O., Krumbholz, M., Peeken, I., Ramseier, R. O., Voss, M., & Zeitschel, B. (2001). Particle flux variability in the Polar and Atlantic biogeochemical provinces of the Nordic Seas. In: P. Schäfer, W. Ritzrau, M. Schlüter & J. Thiede (Eds), *The Northern North Atlantic: A changing environment* (pp. 53–68). Heidelberg: Springer-Verlag.

Peinert, R., & Noji, T. T. (2004). Greenland Sea. In: R. Stein & R. W. Macdonald (Eds), *The organic carbon cycle in the Arctic Ocean* (pp. 109–112). Heidelberg: Springer-Verlag.

Peltier, W. R. (1994). Ice age paleotopography. *Science, 265*, 195–201.

Peltier, W. R. (2002). On eustatic sea level history: Last Glacial Maximum to Holocene. *Quaternary Science Review, 21*, 377–396.

Peltier, W. R. (2007). Rapid climate change and Arctic Ocean freshening. *Geology, 35*, 1147–1148.

Peltier, W. R., Vettoretti, G., & Stastna, M. (2006). Atlantic meridional overturning and climate response to Arctic Ocean freshening. *Geophysical Research Letters, 33*. doi: 10.1029/2005GL025251.

Peng, P., Yu, C., Jia, G., Song, J., & Zhang, G. (2004). Data report: Marine and terrigenous lipids in the sediments from the South China Sea, Site 1148, Leg 184. In: W. L. Prell, P. Wang, P. Blum, D. K. Rea, & S. C. Clemens (Eds). *Proceedings of ODP, Science Research, 184*, 1–16.

Peregovich, B. (1999). Postglacial depositional history of the Laptev Sea: Mineralogy and sedimentology. *Reps. Polar Research, 316*, 85pp.

Peregovich, B., Hoops, E., & Rachold, V. (1999). Sediment transport to the Laptev Sea (Siberian Arctic) during the Holocene - evidence from the heavy mineral composition of fluvial and marine sediments. *Boreas, 28*, 205–214.

Perry, G. J., Volkman, J. K., Johns, R. B., & Bavor, H. J., Jr. (1979). Fatty acids of bacterial origin in contemporary marine sediments. *Geochimica et Cosmochimica Acta, 43*(11), 1715–1725.

Perry, R .K., & Fleming, H. S. (1986). Bathymetry of the Arctic Ocean. The Geological Society of America Map and Chart Series, Boulder, MC-56.

Peters, K. E. (1986). Guidelines for evaluating petroleum source rocks using programmed pyrolysis. *American Association of Petroleum Geologists Bulletin, 70*, 318–329.

Peters, K. E., & Moldowan, J. M. (1993). *The biomarker guide: Interpreting molecular fossils in petroleum and ancient sediments*. Prentice Hall, 363pp.

Peterson, B. J., Holmes, R. M., McClelland, J. W., Vorosmarty, C. J., Lammers, R. B., Shiklomanov, A. I., Shiklomanov, I. A., & Rahmstorf, S. (2002). Increasing river discharge to the Arctic Ocean. *Science, 298*, 2171–2173.

Petrova, V. I., Batova, G. I., Zinchenko, A. G., Kursheva, A. V., & Narkevskiy, E. V. (2004). The East Siberian Sea: Distribution, sources, and burial of organic carbon. In: R. Stein & R. W. Macdonald (Eds), *The organic carbon cycle in the Arctic Ocean* (pp. 204–212). Heidelberg: Springer-Verlag.

Pfirman, S. L., Bauch, D., & Gammelsrød, T. (1994). The northern Barents Sea: Water mass distribution and modification. In: O. M. Johannessen, R. D. Meunch & J. E. Overland (Eds), *The Polar Oceans and their role in shaping the global environment* (pp. 77–94). Washington, DC: American Geophysical Union.

Pfirman, S. L., Colony, R., Nürnberg, D., Eicken, H., & Rigor, I. (1997). Reconstructing the origin and trajectory of drifting Arctic sea ice. *Journal of Geophysical Research, 102*(C6), 12575–12586.

Pfirman, S. L., Gascard, J.-C., Wollenburg, I., Mudie, P., & Abelmann, A. (1989). Particle-laden Eurasian Arctic sea ice: Observations from July and August 1987. *Polar Research, 7*, 59–66.

Pfirman, S. L., & Kassens, H. (1995). Seafloor echo character of the northern Norwegian-Greenland Sea. In: K. Crane & A. Solheim (Eds), *Seafloor atlas of the northern Norwgian-Greenland Sea. Norsk Polarinstitutt Meddelelser* (pp. 14–16). Oslo: Norsk Polarinstitutt.

Pfirman, S. L., Lange, M. A., Wollenburg, I., & Schlosser, P. (1990). Sea ice characteristics and the role of sediment inclusions in deep-sea deposition: Arctic-Antarctic comparisons. In: U. Bleil & J. Thiede

(Eds), *Geological history of the Polar Oceans: Arctic versus Antarctic* (pp. 187–211). Netherlands: Kluwer Academic Publishers.

Pfirman, S. L., & Solheim, A. (1989). Subglacial meltwater discharge in the open marine tidewater glacier environment: Observations from Nordaustlandet. *Marine Geology, 86*, 283–319.

Pflaumann, U., Duprat, J., Pujol, C., & Labeyrie, L. D. (1996). SIMMAX: A modern analog technique to deduce Atlantic sea surface temperatures from planktonic foraminifera in deep-sea sediments. *Paleoceanography, 11*, 15–35.

Pflaumann, U., Sarnthein, M., Chapman, M., d, A. L., Funnell, B., Huels, M., Kiefer, T., Maslin, M., Schulz, H., Swallow, J., van, K. S., Vautravers, M., Vogelsang, E., & Weinelt, M. (2003). Glacial North Atlantic sea-surface conditions reconstructed by GLAMAP 2000. *Paleoceanography, 18*, 1065. doi: 10.1029/2002PA000774.

Phillips, R. L., & Grantz, A. (1997). Quaternary history of sea ice and paleoclimate in the Amerasia Basin, Arctic Ocean, as recorded in the cyclical strata or Northwind Ridge. *Geological Society of America Bulletin, 109*, 1101–1115.

Phillips, R. L., & Grantz, A. (2001). Regional variations in provenance and abundance of ice-rafted clasts in Arctic Ocean sediments: Implications for the configuration of Late Quaternary oceanic and atmospheric circulation in the Arctic. *Marine Geology, 172*, 91–115.

Phillips, R. L., Grantz, A., Mullen, M. W., Rieck, H. J., McLaughlin, M. W., & Selkirk, T. L., (1992). Summary of lithostratigraphy and stratigraphic correlations in piston cores from Northwind Ridge, Arctic Ocean, from USCGC POLAR STAR, 1988. U.S. Geological Survey Open-File Report 92–426, 110pp.

Piper, D. J. W. (1978). Turbidites, muds and silts on deep-sea fans and abyssal plains. In: D. J. Stanley & G. Kelling (Eds), *Sedimentation in submarine Canyons, Fans and Trenches* (pp. 163–175). Stroudsburg, PA: Dowden, Hutchinson & Ross.

Piper, D. Z. (1994). Seawater as the source of minor elements in black shales, phosphorites and other sedimentary rocks. *Chemical Geology, 114*, 95–114.

Pogrebitskii, Y. E. (1983). Tectonic map of the Arctic Ocean and of contiguous territories. Leningrad (in Russian).

Polyak, L. V. (1986). New data on microfauna and stratigraphy of bottom sediments of the Mendeleev Ridge, Arctic Ocean. In: S. I. Andreev (Ed.), *Sedimentogenesis and nodule-formation in the Ocean* (pp. 40–50). Leningrad: Sevmorgeologia. (in Russian).

Polyak, L. V., Curry, W. B., Darby, D. A., Bischof, J., & Cronin, T. M. (2004). Contrasting glacial/interglacial regimes in the Western Arctic Ocean as exemplied by a sedimentary record from the Mendeleev Ridge. *Palaeogeography, Palaeoclimatology, Palaeoecology, 203*, 73–93.

Polyak, L. V., Darby, D. A., Bischoff, J. F., & Jakobsson, M. (2007). Stratigraphic constraints on late Pleistocene glacial erosion and deglaciation of the Chukchi margin, Arctic Ocean. *Quaternary Research, 67*, 234–245.

Polyak, L. V., Edwards, M. H., Coakley, B. J., & Jakobsson, M. (2001). Ice shelves in the Pleistocene Arctic Ocean inferred from glaciogenic deep-sea bedforms. *Nature, 410*, 453–457.

Polyak, L. V., Forman, S. L., Herlihy, F. A., Ivanov, G., & Krinitsky, P. (1997). Late Weichselian deglacial history of the Svyataya (Saint) Anna Trough, northern Kara Sea, Arctic Russia. *Marine Geology, 143*, 169–188.

Polyak, L. V., Gataullin, V., Epshtein, O., Okuneva, O., & Stelle, V. (2000a). New constraints on the limits of the Barents-Kara ice sheet during the Last Glacial Maximum based on borehole stratigraphy from the Pechora Sea. *Geology, 28*, 611–614.

Polyak, L. V., Gataullin, V., Gainanov, V., Gladysh, V., & Goremykin, Yu. (2002a). Kara Sea expedition yields insight into extent of LGM ice sheet. *Eos, 83*, 525–529.

Polyak, L. V., Korsun, S., Febo, L., Stanovoy, V., Khusid, T., Hald, M., Paulsen, B. E., & Lubinski, D. A. (2002b). Benthic foraminiferal assemblages from the southern Kara Sea, a river-influenced arctic marine environment. *Journal of Foraminiferal Research, 32*(3), 252–273.

Polyak, L. V., Levitan, M., Gataullin, V., Khusid, T., Mikhailov, V., & Mukhina, V. (2000b). The impact of glaciation, river-discharge, and sea-level change on Late Quaternary environments in the Southwestern Kara Sea. *International Journal of Earth Science, 89*, 550–562.

Polyak, L. V., & Mikhailov, V. (1996). Post-glacial environments of the southeastern Barents Sea: Foraminiferal evidence. In: J. T. Andrews, W. E. N.Austin, H. Bergsten, & A. E. Jennings (Eds). *Late Quaternary palaeoceanography of the North Atlantic margins* (pp. 323–339), Geological Society (Special Publication III).

Polyak, L. V., & Solheim, A. (1994). Late-and postglacial environments in the northern Barents Sea West of Franz Josef land. *Polar Research, 13*, 197–207.

Polyakov, I. V., & Johnson, M. A. (2000). Arctic decadal and interdecadal variability. *Geophysical Research Letters, 27*, 4097–4100.

Polyakov, I. V., Johnson, M. A., Colony, R. L., Bhatt, U., & Alekseev, G. V. (2002). Observationally based assessment of Polar amplification of global warming. *Geophysical Research Letters, 29*, 1878pp. doi: 1029/2002GL011111.

Polyakova, Y. I. (2003). Diatom assemblages in the surface sediments of the Kara Sea (Siberian Arctic) and their relationship to oceanological conditions. In: R. Stein, K. Fahl, D. K. Fütterer, E. M. Galimov & O. V. Stepanets (Eds), *Siberian River run-off in the Kara Sea: Characterisation, quantification, variability, environmental significance, Proceedings in Marine Sciences* (Vol. 6, pp. 375–399). Amsterdam: Elsevier.

Polyakova, Y. I., & Stein, R. (2004). Holocene paleoenvironmental implications of diatom and organic carbon records from the Southeastern Kara Sea (Siberian Margin). *Quaternary Research, 62*, 256–266.

Poore, R. Z., Ishman, S. E., Phillips, L., & McNeil, D. (1994). Quaternary stratigraphy and paleoceanography of the Canada Basin, Western Arctic Ocean. *US Geological Survey Bullettin, 2080*, 1–32.

Poore, R. Z., Osterman, L., Curry, W. B., & Phillips, R. L. (1999). Late Pleistocene and Holocene meltwater events in the Western Arctic Ocean. *Geology, 27*(8), 759–762.

Poore, R. Z., Phillips, R. L., & Rieck, H. J. (1993). Paloeclimate record for Northwind Ridge, Western Arctic Ocean. *Paleoceanography, 8*, 149–159.

Powers, L. A., Werne, J. P., Johnson, T. C., Hopmans, E. C., Sinninghe Damste', J. S., & Schouten, S. (2004). Crenarchaeotal membrane lipids in lake sediments: A new paleotemperature proxy for continental paleoclimate reconstruction? *Geology, 32*, 613–616.

Praeg, D., MacLean, B., Piper, D. J. W., & Shor, A. N. (1987). Study of iceberg scours across the continental shelf and slope off Southeast Baffin Island using the SeaMARC I midrang sidescan sonar. *Geological Survey of Canada, Current Research, 87-1A*, 847–857.

Prahl, F. G., & Carpenter, R. (1984). Hydrocarbons in Washington coastal sediments. *Esturine Coastal Shelf and Science, 18*, 703–720.

Prahl, F. G., Ertel, J. R., Goñi, M. A., Sparrow, M. A., & Eversmeyer, B. (1994). Terrestrial organic-carbon contributions to sediments on the Washington margin. *Geochimica et Cosmochimica Acta, 58*(14), 3035–3048.

Prahl, F. G., & Muehlhausen, L. A. (1989). Lipid biomarkers as geo-chemical tools for paleoceanographic study. In: W. H. Berger, V. S. Smetacek & G. Wefer (Eds), *Productivity of the oceans: Present and past* (pp. 271–289). New York: Wiley.

Prahl, F. G., & Wakeham, S. G. (1987). Calibration of unsaturation pattern in long-chain ketone compositions for paleotemperature assessment. *Nature, 330*, 367–369.

Prange, M., & Lohmann, G. (2003). Effects of mid-Holocene river runoff on the Arctic ocean-sea ice system: A numerical study. *The Holocene, 13*(3), 335–342.

Premke-Kraus, M. (2008). *Holocene environmental history of the Kara Sea (Siberian Arctic, Russia) inferred from marine palynological records.* Unpublished Ph.D thesis, Bremen University, 149pp.

Premuzic, E. T., Benkovitz, C. M., Gaffney, J. S., & Walsh, J. J. (1982). The nature and distribution of organic matter in the surface sediments of world oceans and seas. *Organic Geochemistry, 4*, 53–77.

Prest, V. K. (1984). Late Wisconsin glacier complex. Map 1257A. *Geological Survey of Canada, und weiter?*

Price, G. D., Ruffell, A. H., Jones, C. E., Kalin, R. M., & Mutterlose, J. (2000). Isotopic evidence for temperature variation during the early Cretaceous (late Ryazanian-mid Hauterivian). *Journal of the Geological Society, 155*, 335–343.

Proshutinsky, A. Y., & Johnson, M. A. (1997). Two circulation regimes of the wind-driven Arctic Ocean. *Journal of Geophysical Research, 102*, 12493–12514.
Rachold, V. (1999). Major, trace, and rare earth element geochemistry of suspended particulate material of East Siberian rivers draining to the Arctic Ocean. In: H. Kassens, H. A. Bauch, I. Dmitrenko, H. Eicken, H.-W. Hubberten, M. Melles, J. Thiede & L. A. Timokhov (Eds), *Land-ocean systems in the Siberian Arctic: Dynamics and history* (pp. 199–222). Heidelberg: Springer-Verlag.
Rachold, V., Are, F. E., Atkinson, D. E., Cherkashov, G., & Solomon, S. M. (2004b). Arctic Coastal Dynamics (ACD): An introduction. *Geo-Marine Letters, 25*, 63–68.
Rachold, V., Bolshiyanov, D. Y., Grigoriev, M. N., Hubberten, H. W., Junker, R., Kunitsky, V. V., Merker, F., Overduin, P., & Schneider, W. (2007). Nearshore Arctic subsea permafrost in transition. *EOS, 88*(13), 149–156.
Rachold, V., Eicken, H., Gordeev, V. V., Grigoriev, M. N., Hubberten, H. W., Lisitzin, A. P., Shevchenko, V. P., & Schirmeister, L. (2004a). Modern terrigenous organic carbon input to the Arctic Ocean. In: R. Stein & R. W. Macdonald (Eds), *The organic carbon cycle in the Arctic Ocean* (pp. 33–56). Heidelberg: Springer-Verlag.
Rachold, V., Eisenhauer, A., Hubberten, H.-W., Hansen, B. T., & Meyer, H. (1998). Sr Isotope composition of suspended particulate material (SPM) of East Siberian rivers: Sediment transport to the Arctic Ocean. *Arctic, Antarctic, and Alpine Research, 4*, 422–429.
Rachold, V., Grigoriev, M. N., Are, F. E., Solomon, S., Reimnitz, E., Kassens, H., & Antonow, M. (2000). Coastal erosion vs riverine sediment discharge in the Arctic shelf seas. *International Journal of Earth Science, 89*, 450–460.
Rachold, V., & Hubberten, H.-W. (1999). Carbon isotope composition of particulate organic material in East Siberian rivers. In: H. Kassens, H. Bauch, I. Dmitrenko, H. Eicken, H. W. Hubberten, M. Melles, J. Thiede & L. Timokhov (Eds), *Land-ocean systems in the Siberian: Dynamics and history* (pp. 223–238). Heidelberg: Springer-Verlag.
Rachor, E. (Ed.). (1997). Scientific cruise report of the Arctic Expedition ARK-XI/1 of RV "Polarstern" in 1995, *Report on Polar Research, 226*, 157pp.
Rahmstorf, S. (2002). Ocean circulation and climate during the past 120,000 years. *Nature, 419*, 207–214.
Rahmstorf, S., & Ganopolski, A. (1999). Long-term global warming scenarios computed with an efficient coupled climate model. *Climatic Change, 43*, 353–367.
Rai, V., & Rai, A. K. (1998). Growth behaviour of Azolla pinnata at various salinity levels and induction of high salt tolerance. *Plant Soil, 206*, 79–84.
Rau, G. H., Riebesell, U., & Wolf-Gladrow, D. (1997). $CO_{2(aq)}$-dependent photosynthetic ^{13}C fractionation in the ocean: A model versus measurements. *Global Biogeochemical Cycles, 11*, 267–278.
Rau, G. H., Sullivan, C. W., & Gordon, L. I. (1991). $\delta^{13}C$ and $\delta^{15}N$ variations in Wedell Sea particulate organic matter. *Marine Chemistry, 35*, 355–369.
Rau, G. H., Sweeney, R. E., & Kaplan, I. R. (1982). Plankton ^{13}C:^{12}C ratio changes with latitude: Differences between northern and Southern Oceans. *Deep-Sea Research, 29*, 1035–1039.
Rau, G. H., Takahashi, T., & Des Marais, D. J. (1989). Latitudinal variations in plankton $\delta^{13}C$: Implications for CO_2 and productivity in the past oceans. *Nature, 341*, 516–518.
Raymo, M. E. (1994). The Himalayas, organic carbon burial, and climate in the Miocene. *Paleoceanography, 9*, 399–404.
Raymo, M. E., Jansen, E., Blum, P., & Herbert, T. D. (Eds). (1999). *Proceedings of ODP, Science Results*, 162, College Station, Texas (Ocean Drilling Program).
Raymo, M. E., Rind, D., & Ruddiman, W. F. (1990). Climatic effects of reduced Arctic sea ice limits in the GISS II general circulation model. *Paleoceanography, 5*, 367–382.
Reimnitz, E., Barnes, P. W., & Weber, W. S. (1993a). Particulate matter in pack ice of the Beaufort Gyre. *Journal of Glaciology, 39*, 186–198.
Reimnitz, E., Clayton, J. R., Kempema, E. W., Payne, J. R., & Weber, W. S. (1993c). Interaction of rising frazil with suspended particles, Tank experiments with applications to nature. *Cold Regions Science and Technology, 21*, 117–135.

Reimnitz, E., Eicken, H., & Martin, T. (1995). Multiyear fast ice along the Taymyr Peninsula. *Siberia. Arctic, 48*, 359–367.

Reimnitz, E., Graves, S. M., & Barnes, P. W. (1988). Beaufort Sea coastal erosion, sediment flux, shoreline evolutionand the erosional shelf profile. U.S. Geological Survey. To accompany Map I-1182-G, 22pp.

Reimnitz, E., Marincovich, L., McCormick, M., & Briggs, W. M. (1992). Suspension freezing of bottom sediment and biota in the Northwest passage and implications for Arctic Ocean sedimentation. *The Canadian Journal of Earth Science, 29*, 693–703.

Reimnitz, E., McCormick, M., Bischof, J., & Darby, D. (1998). Comparing sea-ice sediment load with Beaufort Sea shelf deposits: Is entrainment selective? *Journal of Sedimentary Research, 68*, 777–787.

Reimnitz, E., McCormick, M., McDougall, K., & Brouwers, E. (1993b). Sediment export by ice rafting from a coastal polynya, Arctic Alaska, U.S.A. *Arctic Alpine Research, 25*, 83–98.

Rekant, P., Cherkashev, G., Vanstein, B., & Krinitsky, P. (2005). Submarine permafrost in the nearshore zone of the Southwestern Kara Sea. *Geological Marine Letters, 25*, 167–182.

Revel, M., Sinko, J. A., & Grousset, F. E. (1996). Sr and Nd isotopes as tracers of North Atlantic lithic particles: Palaeoclimatic implications. *Palaeogeography, Palaeoclimatology, Palaeoecology, 11*, 95–113.

Riboulleau, A., Baudin, F., Deconick, J.-F., Derenne, S., Largeau, C., & Tribovillard, N. (2003). Depositional conditions and organic matter preservation pathways in an epicontinental environment: The upper Jurassic Kashpir oil shales (Volga Basin, Russia). *Palaeogeography, Palaeoclimatology, Palaeoecology, 197*, 171–197.

Rigor, I. G., & Wallace, J. M. (2004). Variations in the age of Arctic sea-ice and summer sea-ice extent. *Geophysical Research Letters, 31*, L09401. doi: 10.1029/2004GL019492.

Rigor, I. G., Wallace, J. M., & Colony, R. (2002). Response of sea ice to the Arctic oscillation. *Journal of Climate, 15*(18), 2648–2663.

Risebrobakken, B., Jansen, E., Andersson, C., Melde, E., & Hevrøy, K. (2003). A high-resolution study of Holocene paleoclimatic and paleoceanographic changes in the Nordic Seas. *Paleoceanography, 18*(1), doi: 10.1029/2002PA000764.

Rivkin, F. M. (1998). Release of methane from permafrost as a result of global warming and other disturbances. *Polar Geography, 22*, 105–118.

Roach, A. T., Aagaard, K., Pease, C. H., Salo, S. A., Weingartner, T., Pavlov, V., et al. (1995). Direct measurements of transport and water properties through the Bering Strait. *Journal of Geophysical Research, 100*, 18,443–18,457.

Robinson, S. D., & Moore, T. R. (1999). Carbon and peat accumulation over the past 1200 years in a landscape with discontinuous permafrost, northwestern Canada. *Global Biogeochemical Cycles, 13*(2), 591–601.

Robe, R. Q. (1980). In: S. C. Colbeck (Ed.), *Dynamics of snow and ice masses* (pp. 211–259). New York: Academic.

Robert, C., & Kennett, J. P. (1992). Paleocene and Eocene kaolinite distribution in the South Atlantic and Southern Ocean: Antarctic climate and paleoceanographic implications. *Marine Geology, 103*, 99–110.

Robert, C., & Kennett, J. P. (1994). Antarctic subtropical humid episode at the Paleocene–Eocene boundary: Clay mineral evidence. *Geology, 22*, 211–214.

Robinson, N., Eglinton, G., Brassell, S. C., & Cranwell, P. A. (1984). Dinoflagellate origin for sedimentary 4a-methlysteroids and 5a (H)-stanols. *Nature, 308*, 439–442.

Robinson, S. D., Turetsky, M. R., Kettles, I. M., & Wieder, R. K. (2003). Permafrost and peatland carbon sink capacity with increasing latitude. In: M. Phillips, S. M. Springman, & L. U. Arenson (Eds). *Proceedings of the Eighth International Conference on Permafrost* (pp. 965–970), A.A. Balkema, Lisse.

Rochon, A., de Vernal, A., Turon, J.-L., Matthiessen, J., & Head, M. J. (1999). Distribution of dino£agellate cysts in surface sediments from the North Atlantic Ocean and adjacent basins and quantitative reconstruction of sea-surface parameters. *American Association of Stratigraphic Palynologists, Control Series, 35*, 150pp.

Roeckner, E., Bengtsson, L., Feichter, J., Lelieveld, J., & Rodhe, H. (1999). Transient climate change simulations with a coupled atmosphere-ocean GCM including the tropospheric sulfur cycle. *Journal of Climate, 12*, 3004–3032.

Röhl, U., Bralower, T. J., Norris, G., & Wefer, G. (2000). A new chronology for the late Paleocene thermal maximum and its environmental implications. *Geology, 28*, 927–930.

Rohmer, M., Bouvier-Nave, P., & Ourisson, G. (1984). Distribution of hopanoid triterpenes in prokaryotes. *Journal of General Microbiology, 130*, 1137–1150.

Romankevich, E. A. (1984). *Geochemistry of organic matter in the Ocean*. Heidelberg: Springer-Verlag, 334pp.

Romankevich, E. A., & Vetrov, A. A. (2001). *Carbon cycle in the Russian Arctic seas*. Moscow: Nauka, 302pp. (in Russian).

Romanov, I. P. (1993). *Atlas morphometric characteristics of ice and snow in the Arctic Basin*. Russia: St. Petersburg, 152pp.

Romanovskii, N. N., Hubberten, H. W., Gavrilov, A. V., Eliseeva, A. A., & Tipenko, G. S. (2005). Offshore permafrost and gas hydrate stability zone on the shelf of the East Siberian Seas. *Geological Marine Letters, 25*, 167–182.

Romanovskii, N. N., Hubberten, H.-W., Gavrilov, A. V., Tumskoy, V. E., & Kholodov, A. L. (2004). Permafrost of the east Siberian Arctic shelf and coastal lowlands. *Quaternary Science Reviews, 23*, 1359–1369.

Romanovskii, N. N., Hubberten, H.-W., Gavrilov, A. V., Tumskoy, V. E., Tipenko, G. S., & Grigoriev, M. N. (2000). Thermokarst and land-ocean interactions, Laptev Sea region, Russia. *Permafrost and Periglacial Process, 11*, 137–152.

Romanov, I. P. (1993). *Atlas of ice and snow of the Arctic Basin and Siberian shelf seas*. Backbone Publishing Company, 277pp.

Rosell-Melé, A. (1998). Interhemispheric appraisal of the value of alkenone indices as temperature and salinity proxies in high-latitude locations. *Paleoceanography, 13*(6), 694–703.

Rosell-Melé, A. (2001). Examination of the use of biomarker proxies for the reconstruction of paleoceanographic conditions in the northern North Atlantic. In: P. Schäfer, W. Ritzrau, M. Schlüter & J. Thiede (Eds), *The northern North Atlantic; a changing environment* (pp. 353–363). Heidelberg: Springer-Verlag.

Rosell-Melé, A., Bard, E., Emeis, K.-C., Grimalt, J. O., Müller, P., Schneider, R., Bouloubassi, I., Epstein, B., Fahl, K., Fluegger, A., Freeman, K., & Goni, M. (2001). Precision of the current methods to measure the alkenone proxy UK. *Geochemical Geophysical Geosystem, 2*, 2000GC000141.

Rosell-Melé, A., Carter, J., & Eglinton, G. (1994). Distribution of longchain alkenones and alkyl alkenoates in marine surface sediments from the North East Atlantic. *Organic Geochemistry, 22*, 501–509.

Rosell-Melé, A., Jansen, E., & Weinelt, M. (2002). Appraisal of a molecular approach to infer variations in surface ocean freshwater inputs into the North Atlantic during the last glacial. *Global and Planetary Change, 34*(3-4), 143–152.

Rosell-Melé, A., Weinelt, M., Sarnthein, M., Koc, N., & Jansen, E. (1998). Variability of the Arctic front during the last climatic cycle; application of a novel molecular proxy. *Terra Nova, Eur. J. Geosci., 10*(2), 86–89.

Ross, D. A., Degens, E. T., & MacIlvaine, J. (1970). Black Sea: Recent sedimentary history. *Science, 170*, 163–165.

Rossak, B. T., Kassens, H., Lange, H., & Thiede, J. (1999). Clay mineral distribution in surface sediments of the Laptev Sea: Indicator for sediments provinces, dynamics and sources. In: H. Kassens, H. Bauch, I. Dmitrenko, H. Eicken, H. W. Hubberten, M. Melles, J. Thiede & L. Timokhov (Eds), *Land-Ocean systems in the Siberian: Dynamics and history* (pp. 587–600). Heidelberg: Springer-Verlag.

Rothrock, D. A., Yu, Y., & Maykut, G. A. (1999). Thinning of the Arctic sea-ice cover. *Geophysical Research Letters, 26*(23), 3469–3472.

Rothrock, D. A., Zhang, J., & Yu, Y. (2003). The arctic ice thickness anomaly of the 1990s: A consistent view from observations and models. *Journal of Geophysical Research, 108*. doi: 10.1029/2001JC001208.

Rothwell, R.G. (Ed). (2006). *New techniques in sediment core analysis*. Geological Society of London (Special Publication 267), 266pp.

Rowland, S. J., Belt, S. T., Wraige, E. J., Masse', G., Roussakis, C., & Robert, J.-M. (2001). Effects of temperature on polyunsaturation in cytostatic lipids of *Haslea ostrearia*. *Phytochemistry*, *56*, 597–602.

Rowland, S. J., & Robson, J. N. (1990). The widespread occurrence of highly branched acyclic C_{20}, C_{25} and C_{30} hydrocarbons in recent sediments and biota -a review. *Marine EnvironMental Research*, *30*, 191–216.

Ruddiman, W. F. (2002). *Earth's climate: Past and future*. New York: W.H. Freeman & Company, 465pp.

Ruddiman, W. F., & McIntyre, A. (1981). Oceanic mechanisms for amplification of the 23,000-year ice-volume cycle. *Science*, *212*, 617–627.

Rudels, B., Jones, E. P., Anderson, L. G., & Kattner, G. (1994). On the intermediate depth waters of the Arctic Ocean. *AGU Geophysics Monograph*, *85*, 33–46.

Rudels, B., Meyer, R., Fahrbach, E., Ivanov, V. V., Østerhus, S., Quadfasel, D., Schauer, U., Tverberg, V., & Woodgate, R. A. (2000). Water mass distribution in fram strait and over the Yermak Plateau in summer 1997. *Annales Geophysicae*, *18*, 687–705.

Rullkötter, J. (2000). Organic matter: The driving force for early diagenesis. In: H. D. Schulz & M. Zabel (Eds), *Marine geochemistry* (pp. 129–172). Heidelberg: Springer-Verlag.

Rutgers van der Loeff, M., Meyer, R., Rudels, B., & Rachor, E. (2002). Resuspension and particle transport in the benthic nepheloid layer in and near Fram Strait in relation to faunal abundance and ^{234}Th depletion. *Deep- Sea Research Part I*, *49*, 1941–1958.

Sachs, J. P., Pahnke, K., Smittenberg, R., & Zhang, Z. (2008). Biomarker indicators of past climate. In: S. Elias (Ed.), *Encyclopedia of quaternary science*. Amsterdam: Elsevier, in press.

Sachse, D., Radke, J., & Gleixner, G. (2004). Hydrogen isotope ratios of recent lacustrine sedimentary *n*-alkanes record modern climate variability. *Geochimica et Cosmochimica Acta*, *68*(23), 4877–4889.

Sælen, G., Tyson, R. V., Telnæs, N., & Talbot, M. R. (2000). Contrasting watermass conditions during deposition of the Whitby Mudstone (Lower Jurassic) and Kimmeridge Clay (Upper Jurassic) formations, UK. *Palaeogeography, Palaeoclimatology, Palaeoecology*, *163*, 163–196.

Saettem, J., Bugge, T., Fanavoll, S., Goll, R. M., Mork, A., Mork, M. B. E., Smelror, M., & Verdenius, J. G. (1994). Cenozoic margin development and erosion of the Barents Sea: Core evidence from Southwest of Bjørnøya. *Marine Geology*, *118*, 257–281.

Saks V.N. (1952). Conditions of bottom sediments formation in the Arctic seas of the USSR. Glavsevmorput. L.-M.

Sakshaug, E. (2004). Primary and secondary production in the Arctic Seas. In: R. Stein & R. W. Macdonald (Eds), *The organic carbon cycle in the Arctic Ocean* (pp. 57–82). Heidelberg: Springer-Verlag.

Samoilov, Y. V., & Gorshkova, M. V. (1924). The sediments of the Barents and Kara seas. *Trudi. PLAVMORNINS*, *1*(14),

Samoilov, Ya. V., & Klenova, M. V. (1927). To lithology of the Barents Sea. *Tr. PLAVMORNINA*, *II*(3), Seiten.

Sancetta, C., & Robinson, S. W. (1983). Diatom evidence on Wisconsin and Holocene events in the Bering Sea. *Quaternary Research*, *20*, 232–245.

Sancetta, C., & Silvestri, S. M. (1986). Pliocene-Pleistocene evolution of the North Pacific ocean-atmosphere system, interpreted from fossil diatoms. *Paleoceanography*, *1*, 163–180.

Sangiorgi, F., Brumsack, H., Willard, D. A., Brinkhuis, H., Schouten, S., Stickley, C., O'Regan, M., Reichart, G. J., & Sinninghe Damste, J. S. (2008). A 26 million year gap in the central Arctic record at the Greenhouse-Icehouse transition: Looking for clues. *Paleoceanography*, 23, PA1S02, doi: 10.1029/2007PA001477.

Sarnthein, M., Gersonde, R., Niebler, S., Pflaumann, U., Spielhagen, R., Thiede, J., Wefer, G., & Weinelt, M. (2003b). Overview of glacial Atlantic Ocean mapping (GLAMAP 2000). *Paleoceanography*, *18*(2), 8-1-8-6.

Sarnthein, M., Grootes, P. M., Kennett, J. P., & Nadeau, M. J. (2007). ^{14}C reservoir ages show deglacial changes in ocean currents and carbon cycle. In: A. Schmittner, J. Chiang, & S. Hemmings (Eds). *Ocean circulation: Mechamisms and impacts. Geophysics Monograph Serier*, *173*, 175–196.

Sarnthein M., Pflaumann U., Ross R., Tiedemann R., & Winn K. (1992). Transfer functions to reconstruct ocean paleoproductivity: A comparison. In: C. P. Summerhayes, W. L. Prell, & K. C. Emeis (Eds). *Upwelling systems. Evolution since the early Miocene* (pp. 411–427), Geological Society (Special Publication 64).
Sarnthein, M., Pflaumann, U., & Weinelt, M. (2003a). Past extent of sea ice in the northern North Atlantic inferred from foraminiferal paleotemperature estimates. *Paleoceanography, 18*(2), 25-1-25-8.
Sarnthein, M., Stattegger, K., Dreger, D., Erlenkeuser, H., Grootes, P., Haupt, B. J., Jung, S., Kiefer, T., Kuhnt, W., Pflaumann, U., Schaefer, N. C., Schultz, H., Schultz, M., Seidov, D., Simstich, J., van, K. S., Vogelsang, E., Voelker, A., & Weinelt, M. (2001). Fundamental modes and abrupt changes in Northern Atlantic circulation and climate over the last 60 ky; concepts, reconstruction and numerical modeling. In: P. Schäfer, W. Ritzrau, M. Schlüter & J. Thiede (Eds), *The northern North Atlantic; a changing environment* (pp. 365–410). Heidelberg: Springer-Verlag.
Sarnthein, M., van Kreveld, S., Erlenkeuser, H., Grootes, P. M., Kucera, M., Pflaumann, U., & Schulz, M. (2003c). Centennial-to-millennial-scale periodicities of Holocene climate and sediment injections off the Barents shelf, 75°N. *Boreas, 32*, 447–461.
Sarnthein, M., Winn, K., & Zahn, R. (1987). Paleoproductivity of oceanic upwelling and the effect on atmospheric CO_2 and climatic change during deglaciation times. In: W. H. Berger & L. Labeyrie (Eds), *Abrupt climatic change* (pp. 311–337). Dordrecht: Riedel Publication.
Sasaki, H., & Fukuchi, M. (2004). Northern Bering Sea. In: R. Stein & R. W. Macdonald (Eds), *The organic carbon cycle in the Arctic Ocean* (pp. 118–120). Heidelberg: Springer-Verlag.
Sauer, P. E., Eglinton, T. I., Hayes, J. M., Schimmelmann, A., & Sessions, A. L. (2001). Compound-specific D/H ratios of lipid biomarkers from sediments as a proxy for environmental and climate conditions. *Geochimica et Cosmochimica Acta, 65*(2), 213–222.
Saukel, C. (2006). *Tonmineral- und Korngrößenverteilung in oberflächennahen Sedimenten aus dem Weißen Meer als Indikatoren für rezente Transportprozesse und Quellen der terrigenen Fraktion.* Unpublished Master thesis, Osnabrück University, Germany, 91pp.
Schaeffer, R., & Spiegler, D. (1986). Neogene Kälteeinbrüche und Vereisungsphasen im Nordatlantik. *Z. Deutsch. Geol. Ges., 137*, 537–552.
Schäfer, C. (2005). *Untersuchungen zu Menge und Zusammensetzung des organischen Kohlenstoffs in spätquartären Sedimenten des Yermak-Plateaus (Arktischer Ozean) und Umweltbedingungen.* Unpublished Dipl. Arbeit, Trier University.
Schauer, U. (Ed). (2008). The expedition ARKXXII-2 of the Research Vessel "Polarstern" in 2007 — a contribution to the International Polar Year 2007/08. *Report on Polar Marine Research*, in press.
Schauer, U., Loeng, H., Rudels, B., Ozhigin, V. K., & Dieck, W. (2002). Atlantic water flow through the Barents and Kara Seas. *Deep-Sea Research, 49*, 2281–2298.
Schauer, U., Muench, R. D., Rudels, B., & Timokhov, L. (1997a). Impact of eastern Arctic shelf waters on the Nansen Basin intermediate layers. *Journal of Geophysical Research, 102*, 3371–3382.
Schauer, U., Rudels, B., Loeng, H., Jones, P., Muench, R., Swift, J., & Bjoerk, G. (1997b). Barents Sea Water input to the Eurasian Basin through St. Anna Trough. In: K. Aagaard, D. Hartmann, V. Kattsov, D. Martinson, R. Stewart & A. Weaver (Eds). *Conference on Polar processes and global climate. Arctic climate system study (ACSYS)* (pp. 236–238). Orcas Island, Washington, USA.
Scheffer, F., & Schachtschabel, P. (1984). *Lehrbuch der Bodenkunde.* Stuttgart: Enke-Verlag, 442pp.
Schefuß, E., Ratmeyer, V., Stuut, J.-B. W., Jansen, J. H. F., & Sinninghe Damsté, J. S. (2003). Carbon isotope analysis of *n*-alkanes in dust from the lower atmosphere over the central Eastern Atlantic. *Geochimica et Cosmochimica Acta, 67*, 1757–1767.
Schlanger, S. O., & Jenkyns, H. C. (1976). Cretaceous oceanic anoxic events - causes and consequences. *Geologie en Mijnbouw, 55*, 179–184.
Schlitzer, R. (2001). *Ocean Data View,* Retrieved from http://www.awi-bremerhaven.de/GEO/ODV.
Schlosser, P., Swift, J. H., Lewis, D., & Pfirman, S. L. (1995). The role of large-scale Arctic Ocean circulation in the transport of contaminants. *Deep Sea Research, 42*, 1341–1367.
Schlüter, M., Sauter, E. J., Schäfer, A., & Ritzrau, W. (2000). Spatial budget of organic carbon £ux to the sea£oor of the Northern North Atlantic (60°N-80°N). *Global Biogeochemical Cycles, 14*, 329–340.

Schmidt, G. A., Bigg, G. R., Rohling, E. J. (1999). *Global seawater oxygen-18 database*, Retrieved from http://www.giss.nasa.gov/data/o18data/

Schmittner, A., & Clement, A. (2002). Sensitivity of the thermohaline circulation to tropical and high-latitude freshwater forcing during the last glacial-interglacial cycle. *Paleoceanography, 17*, 7.1-7.14.

Schneider, D. A., Backman, J., Curry, W. B., & Possnert, G. (1996). Paleomagnetic constraints on sedimentation rates in Eastern Arctic Ocean. *Quaternary Research, 46*, 62–71.

Schnetger, B. (1997). Trace element analysis of sediments by HR-ICP-MS using low and medium resolution and di¡erent acid digestions. *Fresenius Journal of Analytical Chemistry, 359*, 468–472.

Schoning, K., & Wastegard, S. (1999). Ostracod assemblages in late Quaternary varved glaciomarine clay of the Baltic Sea Yoldia stage in eastern middle Sweden. *Marine Micropaleontology, 37*, 313–325.

Schoster, F. (2005). Terrigenous sediment supply and paleoenvironment in the Arctic Ocean during the late Quaternary: Reconstructions from major and trace elements. *Report on Polar Marine Research, 498*, 149pp.

Schoster, F., Behrends, M., Müller, C., Stein, R., & Wahsner, M. (2000). Modern river discharge in the Eurasian Arctic Ocean: Evidence from mineral assemblages and major and minor element distributions. *International Journal of Earth Science, 89*, 486–495.

Schouten, J., Hopmans, E. C., & Sinninghe Damsté, J. S. (2004). The effect of maturity and depositional redox conditions on archaeal tetraether lipid palaeothermometry. *Organic Geochemistry, 35*, 567–571.

Schouten, S., Hopmans, E., Forster, A., van Breugel, Y., Kuypers, M. M. M., & Sinninghe Damste', J. S. (2003). Extremely high sea-surface temperatures at low latitudes during the middle Cretaceous as revealed by archaeal membrane lipids. *Geology, 31*, 1069–1072.

Schouten, S., Hopmans, E. C., Pancost, R. D., & Sinninghe Damste', J. S. (2000). Widespread occurrence of structurally diverse tetraether membrane lipids: Evidence for the ubiquitous presence of low-temperature relatives of hyperthermophiles. *Proceedings of National Academy of Sciences, 97*, 14,421–14,426.

Schouten, S., Hopmans, E. C., Schefuß, E., & Sinninghe Damste', J. S. (2002). Distributional variations in marine crenarchaeotal membrane lipids: A new tool for reconstructing ancient sea water temperatures? *Earth and Planetary Science Letters, 204*, 265–274.

Schubert, C. J. (1995). Organic carbon in late Quaternary Arctic Ocean sediments: Terrigenous supply and marine productivity. *Report on Polar Research, 177*, 178pp.

Schubert, C. J., & Calvert, S. E. (2001). Nitrogen and carbon isotopic composition of marine and terrestrial organic matter in Arctic Ocean sediments: Implications for nutrient utilization and organic matter composition. *Deep-Sea Research, 48*, 789–810.

Schubert, C. J., & Stein, R. (1996). Deposition of organic carbon in Arctic Ocean sediments: Terrigenous supply vs marine productivity. *Organic Geochemistry, 24*, 421–436.

Schubert, C. J., & Stein, R. (1997). Lipid distribution in surface sediments from the eastern central Arctic Ocean. *Marine Geology, 138*, 11–25.

Schubert, C. J., Stein, R., & Calvert, S. E. (2001). Tracking nutrient and productivity variations over the last deglaciation in the Arctic Ocean. *Paleoceanography, 16*, 199–211.

Schubert, C. J., Villanueva, J., Calvert, S. E., Cowie, G. L., von Rad, U., Schulz, H., & Berner, U. (1998). Stable phytoplankton community in the Arabian Sea over the last 200,000 years. *Nature, 394*, 563–566.

Schultheiss, P. J., & McPhail, S. D. (1989). An automated p-wave logger for recording fine scale compressional wave velocity in sediments. In: W. Ruddiman, & M. Sarntheim, et al. (Eds). *Procedings of ODP Init. Report, 108*, pp. 407–413.

Schultheiss, P. J., Mienert, J., & Shipboard Scientific Party. (1987). Whole-core p-wave velocity and gamma-ray attenuation logs from Leg 108 (sites 657 through 668). In: M. Ruddiman et al. (Eds), *Proceedings of ODP Init. Report, 108*, pp. 1015–1017.

Schwarzbach, M. (1974). *Das Klima der Vorzeit*. Stuttgart: Enke-Verlag, 380pp.

Schytt, V., Boström, K., & Hjort, C. (1981). Geoscience during the Ymer-80 expedition to the Arctic. *Geol. Fören. Stockholm Förhandl., 103*, 109–119.

Scott, D. B., Mudie, P. J., Baki, V., MacKinnon, K. D., & Cole, F. E. (1989). Biostratigraphy and Late Cenozoic paleoceanography of the Arctic Ocean: Foraminiferal, lithostratigraphic, and isotopic evidence. *Geological Society of America Bulletin, 101*, 260–277.

Scott, D. B., & Vilks, G. (1991). Benthic foraminifera in the surface sediments of the deep-sea Arctic Ocean. *Journal Foraminiferal Research, 21*(1), 20–38.

Sejrup, H. P., Gifford, H. M., Brigham-Grette, J., Løvlie, R., & Hopkins, D. (1984). Amino acid epimerization implies rapid sedimentation rates in Arctic Ocean cores. *Nature, 310*, 772–775.

Seki, O. (2005). Decreased surface salinity in the Sea of Okhotsk during the last glacial period estimated from alkenones. *Geophysical Research Letters, 32*(8), doi: 10.1029/2004GL022177.

Semiletov, I. P., Savelieva, N. I., Weller, G. E., Pipko, I. I., Pugach, S. P., Gukov, A. Y., & Vasilevskaya, L. N. (2000). The dispersion of Siberian river flows into coastal waters: Meteorological, hydrological and hydrochemical aspects. In: E. L. Lewis (Ed.), *The freshwater budget of the Arctic Ocean* (pp. 323–366). Dordrecht: Kluwer Academic Publishers.

Serreze, M. C., Holland, M. M., & Stroeve, J. (2007). Perspectives on the Arctic's shrinking Sea-ice cover. *Science, 315*, 1533–1536.

Serreze, M. C., Walsh, J. E., Chapin, F. S., Osterkamp, T., Dyurgerov, M., Romanovsky, V., Oechel, W. C., Morison, J., Zhang, T., & Barry, R. G. (2000). Observational evidence of recent change in the northern high-latitude environment. *Climatic Change, 46*, 159–207.

Sessions, A. L., Burgoyne, T. W., Schimmelmann, A., & Hayes, J. M. (1999). Fractionation of hydrogen isotopes in lipid biosynthesis. *Organic Geochemistry, 30*(9), 1193–1200.

Shackleton, N. J. (1967). Oxygen isotope analyses and Pleistocene temperatures re-assessed. *Nature, 215*, 15–17.

Shackleton, N. J. (1986). Palaeogene stable isotope events. *Palaeogeography, Palaeoclimatology, Palaeoecology, 57*, 91–102.

Shackleton, N. J. (1987). Oxygen isotopes, ice volume and sea level. *Quaternary Science Review, 6*, 183–190.

Shackleton, N. J., Backman, J., Zimmerman, H., Kent, D. V., Hall, M. A., Roberts, D. G., Schnitker, D., Baldauf, J. G., Desprairies, A., Homrighausen, R., Huddlestun, P., Keene, J. B., Kaltenback, A. J., Krumsiek, K. A. O., Morton, A. C., Murray, J. W., & Westberg-Smith, J. (1984). Oxygen isotope calibration of the onset of ice-rafting and history of glaciation in North Atlantic region. *Nature, 307*, 620–623.

Shackleton, N. J., Berger, A., & Peltier, W. R. (1990). An alternative astronomical calibration of the lower Pleistocene timescale based on ODP site 677. *Transaction of Royal Society of Edinburgh, 81*, 251–261.

Shackleton, N. J., Crowhurst, S., Hagelberg, T., Pisias, N. G., & Schneider, D. A. (1995). A new late Neogene time scale: Application to leg 138 sites. *Proceedings of ODP Science Results, 138*, 73–101.

Shackleton, N. J., Crowhurst, S. J., Weedon, G. P., & Laskar, J. (1999). Astronomical calibration of oligocene-miocene time. *Philosophical Transaction of Royal Society, 357*, 1907–1929.

Shackleton, N. J., & Kennett, J. P. (1975). Paleotemperature history of the Cenozoic and the initiation of Antarctic glaciation: Oxygen and carbon isotope analysis in DSDP Sites 277, 279, and 281. *Initial Report on Deep Sea Drilling Project, 29*, 743–755.

Sharma, M., Basu-Asish, R., & Nesterenko, G. V. (1992). Temporal Sr-, Nd-, and Pb-isotopic variations in the Siberian flood basalts, implications for the plume-source characteristics. *Earth and Planetary Science Letters, 113*, 365–381.

Sharma, P., Mahannah, R., Moore, W. S., Ku, T. L., & Southon, J. R. (1987). Transport of ^{10}Be and ^{9}Be in the ocean: Earth planet. *Science Letters, 86*, 69–76.

Shellito, C. J., Sloan, L. C., & Huber, M. (2003). Climate model sensitivity to atmospheric CO_2 levels in the Early-Middle. *Paleogene. Palaeogeography, Palaeoclimatology, Palaeoecology, 193*, 113–123.

Sherwood, C. R. (2000). Numerical model of frazil ice and suspended sediment concentrations and formation of sediment-laden ice in the Kara Sea. *Journal of Geophysical Research, 105*(C6), 14061–14080.

Shevchenko, V. (2003). The influence of aerosols on the oceanic sedimentation and environmental conditions in the Arctic. *Report on Polar Marine Research, 464*, 149.

Shevchenko, V., Lisitzin, A., Vinogradova, A., & Stein, R. (2003). Heavy metals in aerosols over the seas of the Russian Arctic. *The Science of the Total Environment, 306*, 11–25.

Shevchenko, V. P., & Lisitzin, A. P. (2004). Aeolian input. In: R. Stein & R. W. Macdonald (Eds), *The organic carbon cycle in the Arctic Ocean* (pp. 53–54). Heidelberg: Springer-Verlag.

Shevchenko, V. P., Lisitzin, A. P., Kuptsov, V. M., Van Malderen, H., Martin, J.-M., Van Grieken, R., & Huang, W. W. (1999). Composition of aerosols in the surface boundary layer of the atmosphere over the seas of the Western Russian Arctic. *Oceanology, 39/1*, 128–136.

Shevchenko, V. P., Lisitzin, A. P., Stein, R., Serova, V. V., Isaeva, A. B., & Politova, N. V. (1998). The composition of the coarse fraction of aerosols in the marine boundary layer over the Laptev, Kara and Barents seas. In: H. Kassens, H. A. Bauch, I. Dmitrenko, H. Eicken, H.-W. Hubberten, M. Melles, J. Thiede & L. Timokhov (Eds), *Land-ocean systems in the Siberian Arctic: Dynamics and history* (pp. 53–58). Heidelberg: Springer-Verlag.

Shevchenko, V. P., Lisitzin, A. P., Vinogradova, A. A., Smirnov, V. V., Serova, V. V., & Stein, R. (2000). Arctic aerosols. Results of ten-year investigations. *Atmospheric and Oceanic Optics (English translation), 13*, 510–533.

Shevchenko, V. P., Lisitzin, A. P., Vinogradova, A. A., Vasil'ev, L. Yu., Ivanov, G. I., Klyuvitkin, A. A., Kriews, M., Novigatsky, A. N., Nöthig, E.-M., Politova, N. V., Seleznev, P. V., Serova, V. V., Smirnov, V. V., Sokolov, V. T., Haas, C., & Stein, R. (2004). *New view on the influence of aeolian transport on the modern sedimentation and environment in the Arctic — results of studies of aerosols and snow. New Ideas in the Oceanology* (Vol. 2, pp. 168–214). Moscow: Nauka.

Shevchenko, V. P., Stein, R., Vinogradova, A. A., Bergholter, U., Eicken, H., Kolatschek, J., Lisitzin, A. P., Smirnov, V. V., & Ivanov, G. I. (2001). Elemental composition of aerosold in the marine boundary layer over the Laptev Sea in July-September 1995. *Journal of Aerosol Science, 32*, S471–S472.

Shiga, K., & Koizumi, I. (2000). Latest Quaternary oceanographic changes in the Okhotsk Sea based on diatom records. *Marine Micropaleontology, 38*, 91–117.

Shiklomanov, I. A., & Skakalsky, B. G. (1994). Studying water, sediment and contaminant run off of Siberian Rivers. Modern status and prospects. *Arctic Research, 8*, 295–306.

Shimkus, K. M., & Trimonis, E. S. (1974). Modern sedimentation in the Black Sea. In: E. T. Degens, & D. A. Ross (Eds), *The Black Sea — geology, chemistry, and biology, AAPG Memories*, 20, 249–278.

Shipp, S. S., Anderson, J., & Domack, E. (1999). Late Pleistocene-Holocene retreat of the West Antarctic ice-sheet system in the Ross Sea: Part 1-Geophysical results. *Geological Society of America Bulletin, 111*, 1486–1516.

Shipp, S. S., Wellner, J. S., & Anderson, J. B. (2002). Retreat signature of a polar ice stream: Sub-glacial geomorphic features and sediments from the Ross Sea, Antarctica. In: J. A. Dowdeswell & C. Ó Cofaigh (Eds), *Glacier influenced sedimentation on high-latitude continental margins* (pp. 277–304). Geological Society of London (Special Publication 203).

Shkola, O. V. (2007a). Chemical characteristics of bitumen from sedimentary rocks of Hole Sverdrup-1 drilled in the Sverdrup Island, Kara Sea. All-Russian Research Inst. for Geology and Mineral Resources of the World Ocean, PANGAEA, doi: 10.1594/PANGAEA.615781.

Shkola, O. V. (2007b). Clay mineral composition of <0.001 mm grain size fraction of sediments and sedimentary rocks from Hole Nagursk-1 drilled in the Franz Josef Land Archipelago. All-Russian Research Institute for Geology and Mineral Resources of the World Ocean, PANGAEA, doi: 10.1594/PANGAEA.614565.

Sicre, M. A., Bard, E., Ezat, U., & Rostek, F. (2002). Alkenone distributions in the North Atlantic and Nordic sea surface waters. *Geochemical Geophysical Geosystem, 3*(2), doi: 10.1029/2001GC000159.

Sicre, M. A., Ternois, Y., Paterne, M., Boireau, A., Beaufort, L., Martinez, P., & Bertrand, P. (2000). Biomarker stratigraphic records over the last 150 k years off the NW African coast at 25° N. *Organic Geochemistry, 31*, 577–588.

Sidorchuk, A. Y., & Panin, A. V. (1996). Water supply from the Yana River basin since Late Pliocene. *Terra Nostra, 9*, 97.

Siegert, M. J., & Dowdeswell, J. A. (2002). Late Weichselian iceberg, meltwater and sediment production from the Eurasian high Arctic ice sheet: Results from numerical ice-sheet modelling. *Marine Geology, 188*, 109–127.

Siegert, M. J., & Dowdeswell, J. A. (2004). Numerical reconstructions of the Eurasian ice sheet and climate during the late Weichselian. *Quaternary Science Review, 23*, 1273–1283.

Siegert, M. J., Dowdeswell, J. A., Hald, M., & Svendsen, J. I. (2001). Modeling the Eurasian ice sheet through a full (Weichselian) glacial cycle. *Global and Planetary Change, 31*, 367–385.

Siegert, M. J., Dowdeswell, J. A., & Melles, M. (1999). Late Weichselian glaciation of the Eurasian high Arctic. *Quaternary Research, 52*, 273–285.

Siegert, M. J., & Marsiat, I. (2001). Numerical reconstructions of LGM climate across the Eurasian high Arctic. *Quaternary Science Review, 20*(15), 1595–1605.

Sikes, E. L., Farrington, J. W., & Keigwin, L. D. (1991). Use of the alkenone unsaturation ratio U^{k}_{37} to determine past sea surface temperatures: Core-top SST calibrations and methodology considerations. *Earth and Planetary Science Letters, 104*, 36–47.

Sikes, E. L., Howard, W. R., Neil, H. L., & Volkman, J. K. (2002). Glacial-interglacial sea surface temperature changes across the subtropical front east of New Zealand based on alkenone unsaturation ratios and foraminiferal assemblages. *Paleoceanography, 17*. doi: 10.1029/2001PA000640.

Sikes, E. L., & Sicre, M. A. (2002). Relationship of the tetra-unsaturated C_{37} alkenone to salinity and temperature: Implications for paleoproxy applications. *Geochemistry Geophysics Geosystems*, 3 art. no.-1063.

Sikes, E. L., & Volkman, J. K. (1993). Calibration of alkenone unsaturation ratios ($U^{K}_{37\prime}$) for paleotemperatureestimation. *Geochimica et Cosmochimica Acta, 57*, 1883–1889.

Silverberg, N. (1972). *Sedimentology of the surface sediments of the east Siberian and Laptev Seas*. Ph.D. thesis, University of Washington, 184pp.

Simstich, J., Stanovoy, V., Bauch, D., Erlenkeuser, H., & Spielhagen, R. (2004). Holocene variability of bottom water hydrography on the Kara Sea shelf (Siberia) depicted in multiple single-valve analyses of stable isotopes in ostracods. *Marine Geology, 206*, 147–164.

Sinninghe Damsté, J. S., Hopmans, E. C., Pancost, R. D., Schouten, S., & Geenevasen, J. A. J. (2000). Newly discovered nonisoprenoid glycerol dialkyl glycerol tetraether lipids in sediments. *Chemical Communications*, (1), 1683–1684.

Sinninghe Damsté, J. S., & Köster, J. (1998). An euxinic southern North Atlantic Ocean during the Cenomanian/Turonian oceanic anoxic event. *Earth and Planetary Science Letters, 158*, 165–173.

Sinninghe Damsté, J. S., Muyzer, G., Abbas, B., Rampen, S. W., Massé, G., Allard, W. G., Belt, S. T., Robert, J.-M., Rowland, S. J., Moldowan, J. M., Barbanti, S. M., Fago, F. J., Denisevich, P., Dahl, J., Trindade, L. A. F., & Schouten, S. (2004). The rise of the rhizosolenid diatoms. *Science, 304*, 584–587.

Sinninghe Damsté, J. S., Rijpstra, W. I. C., Hopmans, E. C., Prahl, F. G., Wakeham, S. G., & Schouten, S. (2002a). Distribution of membrane lipids of planktonic Crenarchaeota in the Arabian Sea. *Applied and Environmental Microbiology, 68*, 2997–3002.

Sinninghe Damsté, J. S., Schouten, S., Hopmans, E. C., van Duin, A. C. T., & Geenevasen, J. A. J. (2002b). Crenarchaeol: The characteristic core glycerol dibiphytanyl glycerol tetraether membrane lipid of cosmopolitan pelagic crenarchaeota. *Journal of Lipid Research, 43*, 1641–1651.

Sinninghe Damsté, J. S., Wakeham, S. G., Kohnen, M. E. L., Hayes, J. M., & de Leeuw, J. W. (1993). A 6,000-year sedimentary molecular record of chemocline excursions in the Black Sea. *Nature, 362*, 827–829.

Skopintsev, B. A., & Krylova, L. P. (1955). Organic substances input by the biggest rivers of the Soviet Union. *Reports of the USSR Academy of Sciences (Doklady Akademii Nauk SSSR), 105*(4), 770–773 (in Russian).

Sloan, L. C., & Pollard, D. (1998). Polar stratospheric clouds: A high latitude warming mechanism in an ancient greenhouse world. *Geophysical Research Letters, 25*, 3517–3520.

Sluijs, A., Pross, J., & Brinkhuis, H. (2005). From greenhouse to icehouse; organic-walled dinoflagellate cysts as paleoenvironmental indicators in the Paleogene. *Annual Review of Earth and Planetary Sciences, 68*, 281–315.

Sluijs, A., Röhl, U., Schouten, S., Sangiorgi, F., Brumsack, H., Sinninghe Damste, J.S., & Brinkhuis, H. (2008). Arctic late Paleocene-Early Eocene paleoenvironments with special emphasis on the Paleocene-Eocene thermal maximum (Lomonosov Ridge, IODP Expedition 302), *Paleoceanography,* 23, PA1S11, doi: 10.1029/2007PA001495.

Sluijs, A., Schouten, S., Pagani, M., Woltering, M., Brinkhuis, H., Damsté, J. S. S., Dickens, G. R., Huber, M., Reichart, G.-J., Stein, R., Matthiessen, J., Lourens, L. J., Pedentchouk, N.,

Backman, J., & Moran, K.the Expedition 302 Scientists, (2006). Subtropical Arctic Ocean temperatures during the Palaeocene/Eocene thermal maximum. *Nature, 441*, 610–613.

Smethie, W. M., Jr., Schlosser, P., Bonisch, G., & Hopkins, T. S. (2000). Renewal and circulation of intermediate waters in the Canadian Basin observed on the SCICEX 96 cruise. *Journal of Geophysical Research, 105*(C1), 1105–1121.

Smith, D. (1998). Recent increase in the length of the melt season of perennial Arctic sea ice. *Geophysical Research Letters, 25*, 655–658.

Smith, J. N., Ellis, K. M., & Boyd, T. (1999). Circulation feature in the central Arctic Ocean revealed by nuclear fuel reprocessing tracer from Scientific Ice Expeditions 1995 and 1996. *Journal of Geophysical Research, 104/C12*, 29,663–29,677.

Smith, S. D., Muench, R. D., & Pease, C. H. (1990). Polynyas and leads: An overview of physical processes and environment. *Journal of Geophysical Research, 95*, 9461–9479.

Smith, W. O. (1995). Primary productivity and new production in the Northeast Water (Greenland) Polynya during summer 1992. *Journal of Geophysical Research, 100*(C3), 4341–4356.

Smith, W. O., Jr., Gosselin, M., Legendre, L., Wallace, D., Daly, K., & Kattner, G. (1997). New production in the Northeast Water Polynya. *Journal of Marine System, 10*, 199–209.

Snowdon, L. R. (1984). Organic geochemical data for the Eocene Richards Formation, Beaufort-Mackenzie Basin. *Geological Survey of Canada Open File Report, 1007*, 129p.

Snowdon, L. R., Stasiuk, L. D., Robinson, R., Dixon, J., Dietrich, J., & McNeil, D. H. (2004). Organic geochemistry and organic petrology of a potential source rock of early Eocene age in the Beaufort–Mackenzie Basin. *Organic Geochemistry, 35*, 1039–1052.

Solheim, A., Andersen, E. S., Elverhoi, A., & Fiedler, A. (1996). Late Cenozoic depositional history of the western Svalbard continental shelf, controlled by subsidence and climate. *Global and Planetary Change, 12*, 135–148.

Solheim, A., Bryn, P., Sejrup, H. P., Mienert, J., & Berg, K. (2005). Ormen Lange – an integrated study for the safe development of a deepwater gas field within the Storegga Complex, NE Atlantic continental margin: Executive summary. *Marine and Petroleum Geology, 22*, 1–9.

Solheim, A., Faleide, J. I., Andersen, E., Elverhøi, A., Forsberg, C. F., Vanneste, K., Uenzelmann-Neben, G., & Channell, J. E. T. (1998). Late Cenozoic seismic stratigraphy and glacial geological development of the East Greenland and Svalbard-Barents Sea continental margin. *Quaternary Science Review, 17*, 155–184.

Solheim, A., Milliman, J. D., & Elverhøi, A. (1988). Sediment distribution and sea-floor morphology of Storbanken: Implications for the glacial history of the northern Barents Sea. *Canadian Journal of Earth Science, 25*, 547–556.

Solheim, A., Russwurm, L., Elverhøi, A., & Nyland-Berg, M. (1990). Glacial geomorphic features in the northern Barents Sea: Direct evidence for grounded ice and implications for the pattern of deglaciation and late glacial sedimentation. *Geological Society, Special Publication, 53*, 253–268.

Soltwedel, T., Mokievsky, V., & Schewe, I. (2000). Benthic activity and biomass on the Yermak Plateau and in adjacent deep-sea regions northwest of Svalbard. *Deep-Sea Research Part II, 47*, 1761–1785.

Speijer, R. P., & Morsi, A.-M. M. (2002). Ostracode turnover and sea-level changes associated with the Paleocene-Eocene thermal maximum. *Geology, 30*, 23–26.

Spicer, R. A., & Parrish, J. T. (1990). Late Cretaceousearly Tertiary paleoclimates of northern high latitudes: A quantitative view. *Journal of the Geological Society, 147*, 329–341.

Spielhagen, R. F. (1991). Die Eisdrift in der Framstrasse whärend der letzten 200.000 Jahre. *GEOMAR Report, 4*, 133pp.

Spielhagen, R. F. (2001). Enigmatic Arctic ice sheets. *Nature, 410*, 427–428.

Spielhagen, R. F., Baumann, K.-H., Erlenkeuser, H., Nowaczyk, N. R., Nørgaard-Pedersen, N., Vogt, C., & Weiel, D. (2004). Arctic Ocean deep-sea record of Northern Eurasian ice sheet history. *Quaternary Science Review, 23*(11–13), 1455–1483.

Spielhagen, R. F., Bonani, G., Eisenhauer, A., Frank, M., Frederichs, T., Kassens, H., Kubik, P. W., Mangini, A., Nörgaard-Pedersen, N., Nowaczyk, N. R., Schäper, S., Stein, R., Thiede, J., Tiedemann, R., & Wahsner, M. (1997). Arctic Ocean evidence for Late Quaternary initiation of northern Eurasian ice sheets. *Geology, 25*, 783–786.

Spielhagen, R. F., & Erlenkeuser, H. (1994). Stable oxygen and carbon isotopes in planktic foraminifers from Arctic Ocean surface sediments: Reflection of the low salinity surface water layer. *Marine Geology, 119*(3/4), 227–250.

Spielhagen, R. F., Erlenkeuser, H., & Siegert, C. (2005). History of freshwater runoff across the Laptev Sea (Arctic) during the last deglaciation. *Global and Planetary Change, 48*, 187–207.

Spreen, G., Kaleschke, L., & Heygster, G. (2005). Operational sea ice remote sensing with AMSR-E 89 GHz channels. *International Geoscience and Remote Sensing Symposium Proceedings, IEEE 6*, 4033–4036.

Srivastava, S. P., Arthur, M. A., & Clement, B., et al. (1987). *Proceedings of ODP, Init. Report*, 105, College Station, Texas (Ocean Drilling Program), doi:10.2973/odp.proc.ir.105.1987.

Srivastava, S. P., Arthur, M. A., & Clement, B., et al. (1989). *Proceedings of ODP, Science Results*, 105, College Station, TX (Ocean Drilling Program), doi:10.2973/odp.proc.sr.105.1989.

St. John, K. (2008). Cenozoic ice-rafting history of the Central Arctic Ocean: Terrigenous sands on the Lomonosov Ridge. *Paleoceanography*, 23, PA1S05, doi: 10.1029/2007PA001483.

St. John, K., & Krissek, L. A. (2002). The late Miocene to Pleistocene ice-rafting history of southeast Greenland. *Boreas, 31*, 28–35.

St. John, K., & Willard, D. (2006). Cenozoic (0–46 Ma) ice-rafting history of the Central Arctic: Terrigenous sands on the Lomonosov Ridge. *EOS Transactions American Geophysics Union, 87*(52), Fall Meet. Suppl., Abstract U24A-07.

Stabeno, P., & Overland, J. E. (2001). Bering Sea shifts toward an earlier spring transition. *EOS Transactions American Geophysics Union, 82*, 317, 32.

Stein, A. B., & Syvitski, J. P. M. (1997). Glacially-influenced debris flow deposits: East Greenland slope. In: T. A. Davies, T. Bell, A. K. Cooper, H. Josenhans, L. Polyak, A. Solheim, M. S. Stoker & J. A. Stravers (Eds), *Glaciated continental margins: An atlas of acoustic images* (pp. 134–135). London: Chapman & Hall.

Stein, R. (1985). The Post- Eocene sediment record of DSDP-Site 366: Implications for African climate and plate tectonic drift. *Geological Society of America Memorials, 163*, 305–315.

Stein, R. (1986). Surface-water paleo-productivity as inferred from sediments deposited in oxic and anoxic deep-water environments of the Mesozoic Atlantic Ocean. In: E. T. Degens, P. A. Meyers & S. C. Brassell (Eds), *Biochemistry of Black Shales* (Vol. 60, pp. 55–70). Mitt. Geol. Paläont. Inst. University, Hamburg.

Stein, R. (1990). Organic carbon content/sedimentation rate relationship and its paleoenvironmental significance for marine sediments. *Geo-Marine Letters, 10*, 37–44.

Stein, R. (1991a). Accumulation of organic carbon in marine sediments. *Lecture Notes Earth Science*, Springer-Verlag, Heidelberg, 34, 217pp.

Stein, R. (1991b). Organic carbon accumulation in Baffin Bay and paleoenvironment in High-Northern Latitudes during the past 20 m.y. *Geology, 19*, 356–359.

Stein, R. (Ed). (1998). Arctic Paleo-River discharge (APARD) — A new research programme of the Arctic Ocean Science Board (AOSB). *Report Polar Research, 279*, 127pp.

Stein, R. (Ed). (2000). Circum Arctic Paleo-River Discharge and Its Geological Record. *International Journal of Earth Science, 89*, 447–616.

Stein, R. (2001). Lithostratigraphy of gravity cores and correlation with sediment echograph profiles ("Akademik Boris Petrov" Kara Sea expeditions 1999 and 2000). *Report on Polar Marine Research, 393*, 120–140.

Stein, R. (Ed.). (2005). Scientific cruise report of the Arctic Expedition ARK-XX/3 of RV "Polarstern" in 2004: Fram Strait, Yermak Plateau and East Greenland continental margin. *Reports on Polar Marine Research, 517*, 188pp.

Stein, R. (2007). Upper Cretaceous/Lower Tertiary black shales near the North Pole: Organic-carbon origin and source-rock potential. *Marine And Petroleum Geology, 24*, 67–73.

Stein, R., Backman, J., & Moran, K. (2007). The Arctic coring expedition: A break-through in Arctic Ocean geoscientific research. *Touch briefing: Exploration and production — oil and gas review 2007*, 47–49.

Stein, R., Boucsein, B., Fahl, K., Garcia de Oteyza, T., Knies, J., & Niessen, F. (2001). Accumulation of particulate organic carbon at the Eurasian continental margin during late Quaternary times: Controlling mechanisms and paleoenvironmental significance. *Global and Planetary Change, 31/1–4*, 87–102.

Stein, R., Boucsein, B., & Meyer, H. (2006a). Anoxia and high primary production in the Paleogene central Arctic Ocean: First detailed records from Lomonosov Ridge. *Geophysical Research Letters, 33*, L18606. doi: 10.1029/2006GL026776.

Stein, R., Diepenbroek, M., Evans, J., Grobe, H., Matthiessen, J., & Vogt, C. (1995). Sediment description and lithostratigraphy — east Greenland fjord systems and continental margin. *Reports on Polar Research, 174*, 32–47.

Stein, R., Dittmers, K., Fahl, K., Kraus, M., Matthiessen, J., Niessen, F., Pirrung, M., Polyakova, Ye., Schoster, F., Steinke, T., & Fütterer, D. K. (2004a). Arctic (Palaeo) river discharge and environmental change: Evidence from Holocene Kara Sea sedimentary records. *Quaternary Science Review, 23*, 1485–1511.

Stein, R., & Fahl, K. (2000). Holocene accumulation of organic carbon at the Laptev Sea Continental Margin (Arctic Ocean): Sources, pathways, and sinks. *Geo-Marine Letter, 20*, 27–36.

Stein, R., & Fahl, K. (2004a). The Kara Sea: Distribution, sources, variability and burial of organic carbon. In: R. Stein & R. W. Macdonald (Eds), *The organic carbon cycle in the Arctic Ocean* (pp. 237–266). Heidelberg: Springer-Verlag.

Stein, R., & Fahl, K. (2004b). The Laptev Sea: Distribution, sources, variability and burial of organic carbon. In: R. Stein & R. W. Macdonald (Eds), *The organic carbon cycle in the Arctic Ocean* (pp. 213–237). Heidelberg: Springer-Verlag.

Stein, R., Fahl, K., Dittmers, K., Niessen, F., & Stepanets, O. (2003a). Holocene siliciclastic and organic carbon fluxes in the Ob and Yenisei estuaries and the adjacent inner Kara Sea: Quantification, variability, and paleoenvironmental implications. In: R. Stein, K. Fahl, D. K. Fütterer, E. M. Galimov & O. V. Stepanets (Eds), *Siberian river run-off in the Kara Sea: Characterisation, quantification, variability, and environmental significance, proceedings in marine sciences* (Vol. 6, pp. 401–434). Amsterdam: Elsevier.

Stein, R., Fahl, K., Fütterer, D. K. & Galimov, E. M. (Eds). (2003b). *Siberian river run-off in the Kara Sea: Characterisation, quantification, variability, and environmental significance, proceedings in marine sciences* (Vol. 6, p. 484). Amsterdam: Elsevier.

Stein, R., Fahl, K., Niessen, F., & Siebold, M. (1999a). Late quaternary organic carbon and biomarker records from the Laptev Sea Continental margin: Implications for organic carbon flux and composition. In: H. Kassens, H. Bauch, I. Dmitrenko, H. Eicken, H.-W. Hubberten, M. Melles, J. Thiede & L. Timokhov (Eds), *Land-ocean systems in the Siberian Arctic: Dynamics and history* (pp. 635–655). Heidelberg: Springer-Verlag.

Stein, R., Grobe, H., Hubberten, H., Marienfeld, P., & Nam, S. (1993). Latest Pleistocene to Holocene changes in glaciomarine sedimentation in Scoresby Sund and along the adjacent East Greenland Continental Margin: Preliminary results. *Geo-Marine Letters, 13*, 9–16.

Stein, R., Grobe, H., & Wahsner, M. (1994c). Organic carbon, carbonate, and clay mineral distributions in eastern central Arctic Ocean surface sediments. *Marine Geology, 119*, 269–285.

Stein, R., Kanamatsu, T., Alvarez-Zarikian, C., Higgins, S. M., Channell, J. E. T., & The IODP Expedition 306 scientists. (2006b). Ocean drilling expedition explores North Atlantic Paleoceanography. *EOS, 87*, pp. 129, 133.

Stein, R., & Korolev, S. (1994). Shelf-to-basin sediment transport in the eastern Arctic Ocean. *Report on Polar Research, 144*, 87–100.

Stein, R., Kukina, N., Matthiessen, J., Müller, C., Nørgaard-Pedersen, N., & Usbeck, R. (1999b). Lithostratigraphy of Alpha Ridge sediments cores. In: W. Jokat (Ed.), *Arctic'98: The expedition ARKXIV/ 1a of RV Polarstern in 1998, Report on Polar Research, 308*, 60–75.

Stein, R., & Littke, R. (1990). Organic-carbon-rich sediments and paleoenvironment: Results from Baffin Bay (ODP-Leg 105) and the upwelling area off Northwest Africa (ODP-Leg 108). In: A. Huc (Ed), *Deposition of organic facies* (Vol. 30, pp. 41–56), *AAPG Studies in Geology.*

Stein, R., Littke, R., Stax, R., & Welte, D. H. (1989). Quantity, provenance, and maturity of organic matter at ODP-Sites 645, 646, and 647: Implications for reconstruction of paleoenvironments in Baffin Bay and Labrador Sea during Tertiary and Quaternary time. In: S. P. Srivastava, M. A. Arthur, B. Clement, et al. (Eds), *Proceeding of ODP, Science Results* (Vol. 105, pp. 185–208), College Station, Texas (Ocean Drilling Program).

Stein, R., & Macdonald, R. W. (Eds). (2004a). *The organic carbon cycle in the Arctic Ocean*. (p. 363). Heidelberg: Springer-Verlag.

Stein, R., & Macdonald, R. W. (2004b). Organic carbon budget: Arctiv Ocean vs. Global Ocean. In: R. Stein & R. W. Macdonald (Eds), *The organic carbon cycle in the Arctic Ocean* (pp. 315–322). Heidelberg: Springer-Verlag.

Stein, R., & Macdonald, R. W. (2004c). Geochemical proxies used for organic carbon source identification in Arctic Ocean sediments. In: R. Stein & R. W. Macdonald (Eds), *The organic carbon cycle in the Arctic Ocean* (pp. 24–32). Heidelberg: Springer-Verlag.

Stein, R., & Macdonald, R. W. (2004d). Organic carbon in Arctic Ocean sediments: A general introduction. In: R. Stein & R. W. Macdonald (Eds), *The organic carbon cycle in the Arctic Ocean* (pp. 169–177). Heidelberg: Springer-Verlag.

Stein, R., Nam, S.-I., Grobe, H., & Hubberten, H. (1996). Late quaternary glacial history and short-term ice-rafted debris fluctuations along the East Greenland continental margin. In: J. Andrews, W. E. N. Austin, H. Bergsten, & A. E. Jennings (Eds), *Late Quaternary Paleoceanography of the North Atlantic Margins* (pp. 135–151), Geological Society of London (Special Publication 111).

Stein, R., Nam, S.-I., Schubert, C., Vogt, C., Fütterer, D., & Heinemeier, J. (1994a). The last deglaciation event in the eastern central Arctic Ocean. *Science, 264*, 692–696.

Stein, R., Niessen, F., Dittmers, K., Levitan, M., Schoster, F., Simstich, J., Steinke, T., & Stepanets, O. V. (2002). Siberian river run-off and late Quaternary glaciation in the Southern Kara Sea: Preliminary results. *Polar Research, 21*, 315–322.

Stein, R., & Robert, C. (1985). Siliciclastic sediments at Sites 588, 590, and 591: Neogene and Paleogene evolution in the Southwest Pacific and Australian climate. In: J. P. Kennett, C. van der Borch, et al. (Eds), *Init. Reps. DSDP.* 90, Washington (U.S. Govt. Printing Office) (pp. 1437–1455).

Stein, R., Rullkötter, J., & Welte, D. H. (1986). Accumulation of organic-carbon-rich sediments in the Late Jurassic and Cretaceous Atlantic Ocean — A synthesis. *Chemical Geology, 56*, 1–32.

Stein, R., Schubert, C. J., Grobe, H., & Fütterer, D. (1994d). Late Quaternary changes in sediment composition in the central Arctic Ocean: Preliminary results of the ARCTIC '91 Expedition. In: D. K. Thurston, & K. Fujita (Eds), *Proceedings of the 1992 International Conference on Arctic Margins* (pp. 363–368), Anchorage, Alaska.

Stein, R., Schubert, C. J., Macdonald, R. W., Fahl, K., Harvey, H. R., & Weiel, D. (2004b). The central Arctic Ocean: Distribution, sources, variability and burial of organic carbon. In: R. Stein & R. W. Macdonald (Eds), *The organic carbon cycle in the Arctic Ocean* (pp. 295–314). Heidelberg: Springer-Verlag.

Stein, R., Schubert, C. J., Vogt, C., & Fütterer, D. (1994b). Stable isotope stratigraphy, sedimentation rates, and salinity changes in the Latest Pleistocene to Holocene eastern central Arctic Ocean. *Marine Geology, 119*, 333–355.

Stein, R., & Stax, R. (1991). Late Quaternary organic carbon cycles and paleoproductivity in the Labrador Sea. *Geo-Marine Letter, 11*, 90–95.

Stein, R., & Stepanets, O. V. (Eds). (2001). The German-Russian Project on Siberian River Run-off (SIRRO): Scientific Cruise Report of the Kara Sea Expedition "SIRRO 2000" of RV "Akademik Boris Petrov" and First Results. *Report on Polar Marine Research, 393*, 287pp.

Stein, R., & Stepanets, O. V. (Eds). (2002). Scientific Cruise Report of the Kara Sea Expedition "SIRRO 2001" of RV "Akademik Boris Petrov". *Report on Polar Marine Research, 419*, 278pp.

Steiner, N., Holloway, G., Gerdes, R., Häkkinen, S., Holland, D., Karcher, M., Kauker, F., Maslowski, W., Proshutinsky, A., Steele, M., & Zhang, J. (2004). Comparing modeled streamfunction, heat and freshwater content in the Arctic Ocean. *Ocean Model, 6*, 265–284.

Steinhauer, M. S., & Boehm, P. D. (1992). The composition and distribution of saturated and aromatic hydrocarbons in nearshore sediments, river sediments, and coastal peat of the Alaskan Beaufort Sea: Implications for detecting anghropogenic hydrocarbon inputs. *Marine Environmental Research, 33*, 223–253.

Steinsund, P. I., & Hald, M. (1994). Recent calcium carbonate dissolution in the Barents Sea: Paleoceanographic applications. *Marine Geology, 117*, 303–316.
Steuerwald, B. A., Clark, D. L., & Andrew, J. A. (1968). Magnetic stratigraphy and faunal patterns in Arctic Ocean sediments. *Earth and Planetary Science Letters, 5*, 79–85.
Stevenson, F. J., & Cheng, C. N. (1972). Organic geochemistry of the Argentine Basin sediments: Carbon-nitrogen relationships and Quaternary correlations. *Geochimica et Cosmochimica Acta, 36*, 653–671.
Stickley, C., Koç N., Brumsack H.-J., Jordan R., & Suto, I. (2008). A siliceous microfossil view of Middle Eocene Arctic palaeoenvironments: A window of biosilica production and preservation. *Paleoceanography, 23*, PA1S14, doi: 10.1029/2007PA001485.
Stocker, T., & Wright, D. (1991). Rapid transitions if the ocean's deep circulation induced by changes in surface water fluxes. *Nature, 351*, 729–732.
Stocker, T. F. (2003). Global change: South dials north. *Nature, 424*, 496–499.
Stoffyn-Egli, P. (1987). Iron and manganese micro-precipitates with a Cretaceous biosiliceous ooze from the Arctic Ocean: Possible hydrothermal source. *Geo-Marine Letter, 7*, 223–231.
Stoll, H. M., & Schrag, D. P. (1996). Evidence for glacial control of rapid sea-level changes in the Early Cretaceous. *Science, 272*, 1771–1774.
Stoner, J. S., Channell, J. E. T., & Hillaire-Marcel, C. (1996). The magnetic signature of rapidly deposited detrital layers from the deep Labrador Sea: Relationship to North Atlantic Heinrich layers. *Paleoceanography, 11*, 309–325.
Stoner, J. S., Channell, J. E. T., & Hillaire-Marcel, C. (1998). A 200 ka geomagnetic chronostratigraphy for the Labrador Sea: Indirect correlation of the sediment record to SPECMAP. *Earth and Planetary Science Letters, 159*, 165–181.
Stoner, J. S., Channell, J. E. T., Hillaire-Marcel, C., & Kissel, C. (2000). Geomagnetic paleointensity and environmental record from Labrador Sea core MD95-2024: Global marine sediment and ice core chronostratigraphy for the last 110 kyr. *Earth and Planetary Science Letters, 183*, 161–177.
Stoner, J. S., Channell, J. E. T., Hodell, D., & Charles, C. (2003). A 580 kyr paleomagnetic record from the sub-Antarctic South Atlantic (ODP Site 1089). *Journal of Geophysical Research, 108*, 2244. doi: 10.1029/2001JB001390.
Stow, D. A. V., & Shanmugam, G. (1980). Sequence of structures in fine-grained turbidites: Comparison of recent deep-sea and ancient flysch sediments. *Sedimentary Geology, 25*, 23–42.
Stroeve, J., Holland, M. M., Meier, W., Scambos, T., & Serreze, M. (2007). Arctic sea ice decline: Faster than forecast. *Geophysical Research Letters, 34*, L09501. doi: 10.1029/2007GL029703.
Stroh, A. (1988). *Quantitative röntgenographische Phasenanalyse von Gesteinen und Mineralgemischen*. Ph.D. dissertation, Justus-Liebig-Universität Gießen, 227pp.
Struck, U. (1997). Paleoecology of benthic foraminifera in the Norwegian-Greenland Sea during the past 500 ka. In: H. C. Hass, & M. A. Kaminski (Eds), *Contribution to the micropaleontology and paleoceanography of the Northern North Atlantic* (pp. 51–82), Grzybowsk i Found (Special Publication 5).
Stuiver, M., Grootes, P. M., & Braziunas, T. F. (1995). The GISP2 $\delta^{18}O$ climate record of the past 16,500 years and the role of the Sun, ocean, and volcanoes. *Quaternary Research, 44*, 341–354.
Subba Rao, D. V., & Platt, T. (1984). Primary production of Arctic waters. *Polar Biology, 3*, 191–201.
Sudgen, D. (1982). *Arctic and Antarctic — A modern geographical synthesis*. Oxford: Blackwell Publication, 472pp.
Suess, E. (1980). Particulate organic carbon flux in the oceans: Surface productivity and oxygen utilization. *Nature, 288*, 260–263.
Sundvor, E., & Austegard, A. (1990). The evolution of the Svalbard margins: Synthesis and new results. In: Bleil, U., & Thiede, J. (Eds), *Geological History of the Polar Oceans: Arctic versus Antarctic. NATO ASI Series C, 308*, pp. 77–94.
Svendsen, J. I., Alexanderson, H., Astakhov, V. I., Demidov, I., Dowdeswell, J. A., Funder, S., Gataullin, V., Henriksen, M., Hjort, C., Houmark-Nielsen, M., Hubberten, H.-W., Ingólfsson, O., Jakobsson, M., Kjær, K. H., Larsen, E., Lokrantz, H., Lunkka, J. P., Lyså, A., Mangerud, J., Matioushkov, A., Murray, A., Möller, P., Niessen, F., Nikolskaya, O., Polyak, L., Saarnisto, M.,

Siegert, R., Siegert, M. J., Spielhagen, R. F., & Stein, R. (2004). Late Quaternary ice sheet history of Northern Eurasia. *Quaternary Science Review, 23*, 1229–1272.

Svendsen, J. I., Astakhov, V. I., Bolshiyanov, D. Y., Demidov, I., Dowdeswell, J. A., Gataullin, V., Hjort, C., Hubberten, H. W., Larsen, E., Mangerud, J., Melles, M., Möller, P., Saarnisto, M., & Siegert, M. J. (1999). Maximum extent of the Eurasian ice sheets in the Barents and Kara Sea region during the Weichselian. *Boreas, 28*, 234–242.

Svendsen, O. (1997). *En geofysisk tolkning av sedimentasjonshistorien i Amundsenbassenget*. Unpublished Cand. Scient. Thesis, University of Tromsø.

Svindland, K. T., & Vorren, T. O. (2002). Late Cenozoic sedimentary environments in the Amundsen Basin, Arctic Ocean. *Marine Geology, 186*, 541–556.

Swanberg, N. R., & Eide, L. K. (1992). The radiolarian fauna at the ice edge in the Greenland Sea during summer, 1988. *Journal of Marine Research, 50*, 297–320.

Syvitski, J. P. M., Farrow, G. E., Atkinson, R. J. A., Moore, P. G., & Andrews, J. T. (1989). Baffin Island fjord macrobenthos: Bottom communities and environmental significance. *Arctic, 42*(3), 232–247.

Takahashi, K., Fujitani, N., & Yanada, M. (2002). Long term monitoring of particle fluxes in the Bering Sea and the central subarctic Pacific Ocean, 1990–2000. *Progress in Oceanography, 55*, 95–112.

Takahashi, K., Fujitani, N., Yanada, M., & Maita, Y. (2000). Longterm biogenic particle fluxes in the Bering Sea and the central subarctic Pacific Ocean, 1990–1995. *Deep-Sea Research. Part I, 47*, 1723–1759.

Takahashi, K., Ravelo, A. C., & Alvarez-Zarikian, C. A. (2007). Pliocene–Pleistocene paleoceanography and climate history of the Bering Sea. *IODP Scientific Prospectus, 318*, doi: 10.2204/iodp.sp.318.2007.

Taldenkova, E., Bauch, H. A., Stepanova, A., Dem'yankov, S., & Ovsepyan, A. (2005). Last postglacial environmental evolution of the Laptev Sea shelf as reflected in molluscan, ostracodal, and foraminiferal faunas. *Global and Planetary Change, 48*, 223–251.

Talwani, M., Udintsev, G., et al. (Eds). (1976). Initial Reports DSDP, 38, Washington (US Government Printing Office), 1256pp.

Tanaka, S., & Takahashi, K. (2005). Late Quaternary paleoceanographic changes in the Bering Sea and the western subarctic Pacific based on radiolarian assemblages. *Deep-Sea Research, 52*, 2131–2149.

Tarasov, G. A., Matishov, G. G., Samoilovich, Yu. G., & Kukina, N. A. (1999). The history of Russian geological investigation of bottom sediments in the Barents and Kara seas (with special emphasis on MMBI studies). In: R. Stein, K. Fahl, G. I. Ivanov, M. A. Levitan, & Tarasov, G. (Eds), *Modern and Late Quaternary Depositional Environment of the St. Anna Trough Area, Northern Kara Sea. Report on Polar Research*, 342, pp. 10–14.

Tarasov, L., & Peltier, W. R. (2005). Arctic freshwater forcing of the Younger Dryas cold reversal. *Nature, 435*, 662–665.

Tarduno, J. A., Brinkman, D. B., Renne, P. R., Cottrell, R. D., Scsher, H., & Castillo, P. (1998). Evidence for extreme climatic warmth from Late Cretaceous Arctic vertebrates. *Science, 282*, 2241–2244.

Tauxe, L. (1993). Sedimentary records of relative paleointensity of the geomagnetic field: Theory and practice. *Reviews of Geophysics, 31*, 319–354.

Taylor, G. H., Teichmüller, M., Davis, A., Dissel, C. F. K., Littke, R., & Robert, P. (1998). *Organic petrology* (pp. 0–704). Stuttgart: Schweitzerbart.

Taylor, J., Dowdeswell, J. A., & Kenyon, N. H. (2003). Canyons and channel systems in the Lofoten Basin, Norwegian margin. In: J. Mienert & P. Weaver (Eds), *European margin sediment dynamics* (pp. 93–97). Heidelberg: Springer Verlag.

Taylor, J., Dowdeswell, J. A., Kenyon, N. H., & Ó Cofaigh, C. (2002). Late Quaternary architecture of trough-mouth fans: Debris flows and suspended sediments on the Norwegian margin. In: J. A. Dowdeswell, & Ó Cofaigh, C. (Eds), *Glacier-influenced sedimentation on High-Latitude Continental Margins* (pp. 55–71), Geological Society of London (Special Publication 203).

Taylor, P. T., Kovacs, L. C., Vogt, P. R., & Johnson, G. L. (1981). Detailed aeromagnetic investigations of the Arctic Basin. *Journal of Geophysical Research, 86*, 6323–6333.

Teeri, J. A., & Stowe, L. G. (1976). Climatic pattern and the distribution of C4 grasses in North America. *Oceologica, 23*, 1–12.

Telang, S. A., Pocklington, R., Naidu, A. S., Romankevich, E. A., Gitelson, I. I., & Glachyshev, M. I. (1991). Carbon and mineral transport in major North American, Russian Arctic, and Siberian rivers: The St. Lawrence, the Mackenzie, the Yukon, the Arctic Alaskan rivers, the Arctic basin rivers in the Soviet Union, and the Yenisei. In: E. T. Degens, S. Kempe & J. E. Richey (Eds), *Biogeochemistry of major world rivers. SCOPE 42* (pp. 75–104). Chichester: Wiley.

Teller, J. T., Leverington, D. W., & Mann, J. D. (2002). Freshwater outbursts to the oceans from glacial Lake Agassiz and their role in climate change during the last deglaciation. *Quaternary Science Review, 21*, 879–887.

Thiebault, F., Cremer, M., Debrabant, P., Foulon, J., Nielsen, O. B., & Zimmerman, H. (1989). Analysis of sedimentary facies, clay mineralogy, and geochemistry of the Neogene-Quaternary sediments in ODP-Site 645, Baffin Bay. In: S. P. Srivastava, M. A. Arthur, B. Clement, et al. (Eds), *Proceedings of ODP, Science Results*, 105, College Station, Texas (Ocean Drilling Program), pp. 83–100.

Thiede, J. (Ed). (1988). Scientific cruise report of Arctic Expedition ARK-IV/3. *Report on Polar Research*, 43, 237pp.

Thiede, J. (Ed). (1996). Arctic paleoceanography — Quo vadis? In: R. Stein, G. Ivanov, M. Levitan, & K. Fahl (Eds), *Surface-sediment composition and sedimentary processes in the central Arctic Ocean and adjacent Eurasian continental margin. Report on Polar Research, 212*, 19–35.

Thiede, J. (Ed.). (2002). Polarstern ARKTIS XVII/2 cruise report: AMORE 2001 (Arctic Mid-Ocean Ridge Expedition). *Report on Polar and Marine Research, 421*, 397pp.

Thiede, J. (Ed). (2004). Quaternary environments of the Eurasian North (QUEEN). *Quaternary Science Review, 23*, pp. 1225–1511.

Thiede, J., Bauch, H. A., Hjort, C., & Mangerud, J. (Eds). (2001). The Quaternary stratigrapohy and environments of northern Eurasia and the adjacent Arctic seas – new contributions from QUEEN. *Global and Planetary Change, 31*, pp. 1–474.

Thiede, J., Clark, D. L., & Hermann, Y. (1990). Late Mesozoic and Cenozoic paleoceanography of the northern polar oceans. *The Geology of North America*, Vol. L, The Arctic Ocean Region, pp. 427–458.

Thiede, J., & Egerton, P. (2004). AURORA BOREALIS: A long-term European Science perspective for Deep Arctic Ocean Research 2006–2016. European Science Foundation (ESF), IREG Strasbourg — Dépôt légal: juillet 2004 — N° d'ordre 042478, 80pp.

Thiede, J., & Myhre, A. M. (1996). The paleoceanographic history of the North Atlantic-Arctic gateways: Synthesis of the Leg 151 drilling results. *Proceedings of Ocean Drilling Program Scientific Results, 151*, 645–658.

Thiede, J., Myhre, A. M., Firth, J. V., Johnson, G. L., & Ruddiman, W. F. (Eds). (1996). *Proceedings of ODP, Science Results*, 151. College Station, Texas (Ocean Drilling Program), doi:10.2973/odp.proc.sr.151.1996.

Thiede, J., & NAD Science Committee. (1992). The Arctic Ocean Record: Key to global change (Initial Science Plan). Polarforschung 61/1, 102pp.

Thiede, J., Winkler, A., Wolf-Welling, T., Eldholm, O., Myhre, A., Baumann, K.-H., Henrich, R., & Stein, R. (1998). Late Cenozoic History of the Polar North Atlantic: Results from Ocean drilling. In: A. Elverhøi, J. Dowdeswell, S. Funder, J. Mangerud, & R. Stein (Eds), *Glacial and Oceanic History of the Polar North Atlantic Margins. Quaternary Science Review, 17*, pp. 185–208.

Thiel, V. (1993). Untersuchungen von Biomarkern als Faziesindikatoren für limnische Ablagerungsräume. Hamburg University (Germany), Master thesis, 137pp.

Thiel, V., Jenisch, A., Landmann, G., Reimer, A., & Michaelis, W. (1997). Unusual distribution of long-chain alkenones and terahymanol from the highly alkaline Lake Va, Turkey. *Geochimica et Cosmochimica Acta, 61*(10), 2053–2064.

Thierstein, H. R., Geitzenauer, K. R., Molfino, B., & Shackleton, N. J. (1977). Global synchroneity of late quaternary coccolith datum levels: Validation by oxygen isotopes. *Geology, 5*, 400–404.

Thompson, D. W. J., & Wallace, J. M. (1998). The Arctic oscillation signature in the wintertime geopotential height and temperature fields. *Geophysical Research Letters, 25*(9), 1297–1300.

Thompson, R., & Oldfield, F. (1986). *Environmental magnetism.* London: Allen and Unwin, p. 205.

Thompson, W. G., & Goldstein, S. L. (2006). A radiometric calibration of the SPECMAP timescale. *Quaternary Science Reviews, 25*, 3207–3215.

Thomson, J., Nixon, S., Croudace, I. W., Pedersen, T. F., Brown, L., Cook, G. T., et al. (2001). Redox-sensitive element uptake in north-east Atlantic Ocean sediments (Benthic Boundary layer experiment sites). *Earth and Planetary Science Letters, 184*, 535–547.

Thorndike, A. S. (1986). Kinematics of sea ice. In: N. Untersteiner (Ed.), *The Geophysics of Sea Ice, Nato ASI Series B Physics, 146*, 489–549.

Thorndike, A. S., & Colony, R. (1982). Sea ice motion in response to geostrophic winds. *Journal of Geophysical Research, 87*, 5845–5852.

Tibbs, J. F. (1967). On some planktonic Protozoa taken from the track of Drift Station Arlis I, 1960–1961. *Journal of Arctic Institute of North America, 20*(4), 247–254.

Tiedemann, R., Sarnthein, M., & Shackleton, N. J. (1994). Astronomic timescale for the Pliocene Atlantic $\delta^{18}O$ and dust flux records of Ocean drilling program site 659. *Paleoceanography, 9*(4), 619–638. doi: 10.1029/94PA00208.

Timokhov, L. A. (1994). Regional characteristics of the laptev and East Siberian seas: Climate, topography, ice phases, thermohaline regime, and circulation. *Report on Polar Research, 144*, 15–31.

Tissot, B. P., & Welte, D. H. (1984). *Petroleum formation and occurrence.* Heidelberg: Springer Verlag, 699pp.

Tremblay, J.-E., Gratton, Y., Fauchot, J., & Price, N. M. (2002). Climatic and oceanic forcing of new, net, and diatom production in the North Water. *Deep-Sea Research Part II, 49*, 4927–4946.

Tremblay, L. B., Mysak, L. A., & Dyke, A. S. (1997). Evidence from driftwood records for century-to-millennium scale variations of the high latitude atmospheric circulation during the Holocene. *Geophysical Research Letters, 24*(16), 2027–2030.

Trettin, H. P. (Ed.). (1991). Geology of the Innitian Orogen and Arctic platform of Canada and Greenland. Geology of Canada, Ottawa (Geological Survey of Canada) 3, 569pp.

Trimble, S. M., & Baskaran, M. (2005). The role of suspended particulate matter in ^{234}Th scavenging and ^{234}Th-derived export fluxes of POC in the Canada Basin of the Arctic Ocean. *Marine Chemistry, 96*, 1–19.

Tripati, A., Backman, J., Elderfield, H., & Ferretti, P. (2005). Eocene bipolar glaciation associated with global carbon cycle changes. *Nature, 436*, 341–346.

Tripati, A., & Elderfield, H. (2005). Deep-sea temperature and circulation changes at the Paleocene-Eocene thermal maximum. *Science, 308*, 1894–1898.

Tripati, A., Zachos, J., Marincovich, L., Jr., & Bice, K. (2001). Late Paleocene Arctic coastal climate inferred from molluscan stable and radiogenic isotope ratios. *Palaeogeography, Palaeoclimatology, Palaeoecology, 170*, 101–113.

Tripati, A. K., Eagle, R. A., Morton, A., Dowdeswell, J. A., Atkinson, K. L., Bahé, Y., Dawber, C. F., Khadun, E., Shaw, R. M. H., Shorttle, O., & Thanabalasundaram, L. (2008). Evidence for glaciation in the Northern Hemisphere back to 44 Ma from ice-rafted debris in the Greenland Sea. *Earth and Planetary Science Letters, 265*, 112–122.

Tütken, T., Eisenhauer, A., Wiegand, B., & Hansen, B. T. (2002). Glacial-interglacial cycles in Sr and Nd isotopic composition of Arctic marine sediments triggered by the Svalbard/Barents Sea ice sheet. *Marine Geology, 182*(3–4), 351–372.

Turich, C., Freeman, K. H., Bruns, M. A., Conte, M., Jones, A. D., & Wakeham, S. G. (2007). Lipids of marine Archaea: Patterns and provenance in the water-column and sediments. *Geochimica et Cosmochimica Acta, 71*, 3272–3291.

Twichell, S. C., Meyers, P. A., & Diester-Haass, L. (2002). Significance of high C/N ratios in organic-carbon rich Neogene sediments under the Benguela Current upwelling system. *Organic Geochemistry, 33*, 715–722.

Tyson, R. V. (1995). *Sedimentary organic matter: Organic facies and palynofacies.* London: Chapman Hall.

U.S. Arctic Research Commission Permafrost Task Force. (2003). *Climate change, permafrost, and impacts on civil infrastructure.* Special Report 01-03, U.S. Arctic Research Commission, Arlington, Virginia, 62pp.

Uenzelmann-Neben, G., Jokat, W., & Vanneste, K. (1991). Quaternary sediments in Scoresby Sund, East Greenland: Their distribution as revealed by reflection seismic data. *Lundqua Report, 33,* 139–148.

Untersteiner, N., & Carmack, E. (1990). Arctic sea ice, freshwater, and global ocean climate: some possible connections. Appendix, WMO Working Group on Sea Ice and Climate, Report of the 4th Session.

USGS World Petroleum Assessment (2000). http://certmapper.cr.usgs.gov/rooms/we/index.jsp.

Vagnes, E. (1996). Cenozoic deposition in the Nansen Basin, a first-order estimate based on present-day bathymetry. *Global and Planetary Change, 12,* 149–156.

Vagnes, E., Faleide, J. I., & Gudlaugsson, S. T. (1992). Glacial erosion and tectonic uplift in the Barents Sea. *Nor. Geol. Tidsskr., 72.*

Van Andel, T. H. (1975). Mesozoic/Cenozoic calcite compensation depth and the global distribution of calcareous sediments. *Earth and Planetary Science Letters, 26,* 187–194.

Vanneste, M., Mienert, J., & Bünz, S. (2006). The Hinlopen Slide: A giant, submarine slope failure on the northern Svalbard Margin, Arctic Ocean. *Earth and Planetary Science Letters, 245,* 373–388.

Vasiliev, A., Kanevskiy, M., Cherkashov, G., & Vanshtein, B. (2005). Coastal dynamics at the Barents and Kara Sea key sites. *Geo-Marine Letters, 25,* 110–120.

Velichko, A. A., Andreev, A. A., & Klimanov, V. A. (1997b). Climate and vegetation dynamics in the tundra and forest zone during the late glacial and Holocene. *Quaternary International, 41/42,* 71–96.

Velichko, A. A., Catto, N., Drenova, A. N., Klimanov, V. A., Kremenetski, K. V., & Nechaev, V. P. (2002). Climate changes in East Europe and Siberia at the late glacial-Holocene transition. *Quaternary International, 91,* 75–99.

Velichko, A. A., Kononov, Y. M., & Faustova, M. A. (1997a). The last glaciation of Earth: Size and volume of ice-sheets. *Quaternary International, 41/42,* 43–51.

Velitchko, A. A., Dolukhanov, P. M., Rutter, N. W., & Catto, N. R. (Eds). (1997c). Quaternary of northern Eurasia: Late Pleistocene and Holocene landscapes, stratigraphy and environments. *Quaternary International, 41/42,* 1991pp.

Venegas, S. A., & Mysak, L. A. (2000). Are there natural time scales of climate. Variability in the Arctic? *Journal of Climate, 13,* 3412–3434.

Venkatesan, M. I., Kaplan, I. R., & Ruth, F. (1983). Hydrocarbon geochemistry in surface sediments of Alaskan outer continental shelf: Part 1: C_{15+} hydrocarbons. *Am. Ass. Petrol. Geol. Bull., 67,* 831–840.

Venkatesan, M. I., Ruth, E., Steinberg, S., & Kaplan, I. R. (1987). Organic geochemistry of sediments from the continental margin off southern New England, U.S.A. — Part II. *Lipids in Marine Chemistry, 21,* 267–299.

Vetrov, A., & Romankevich, E. A. (2004). The Barents Sea: Distribution, sources, variability, and burial of organic carbon. In: R. Stein & R. W. Macdonald (Eds), *The organic carbon cycle in the Arctic Ocean* (pp. 266–278). Heidelberg: Springer-Verlag.

Vilks, G. (1969). Recent foraminifera in the Canadian Arctic. *Micropaleontology, 15,* 35–60.

Vilks, G. (1989). Ecology of recent foraminifera on the Canadian Continental shelf of the Arctic Ocean. In: Y. Herman (Ed.), *The Arctic Seas, climatology, oceanography, geology and biology* (pp. 497–569). New York: Van Nostrand Reinhold.

Villanueva, J., Calvo, E., Pelejero, C., Grimalt, J. O., Boelaert, A., & Labeyrie, L. (2001). A latitudinal productivity band in the central North Atlantic over the last 270 kyr: An alkenone perspective. *Paleoceanography, 16,* 1–10.

Villanueva, J., Grimalt, J. O., Cortijo, E., Vidal, L., & Labeyrie, L. D. (1997). A biomarker approach to the organic matter deposited in the North Atlantic during the last climatic cycle. *Geochimica et Cosmochimica Acta, 61,* 4633–4646.

Vincent, E., & Berger, W. H. (1981). Planktonic foraminifera and their use in paleoceanography. In: Emiliani, C. (Ed), *The Oceanic Lithosphere: The Sea.* New York, pp. 1025–1119.

Vincent, E., & Berger, W. H. (1985). Carbon dioxide and polar cooling in Miocene: The Monterey hypothesis. In: E. T. Sundquist, & W. S. Broecker (Eds), *The carbon cycle and atmospheric CO_2: Natural variations Archean to Present, AGU, Geophysical Monographs Series*, Washington DC, pp. 455–468.

Vinje, T. (1987). Dynamics and morphology of the Barents Sea ice fields. *Procedings of 9th International Conferenceof POAC*, pp. 263–268.

Vinje et al. (1998). *Ice fluxes through Fram Strait*, http://www.gfi.uib.no/~svein/esop2tv.html

Vinnikov, K. Ya., Robock, A., Stouffer, R. J., Walsh, J. E., Parkinson, C. L., Cavalieri, D. J., Mitchell, J. F. B., Garrett, D., & Zakharov, V. F. (1999). Global warming and Northern Hemisphere sea ice extent. *Science, 286*, 1934–1936.

Viscosi-Shirley, C., Mammone, K., Pisias, N., & Dymond, J. (2003a). Clay mineralogy and multi-element chemistry of surface sediments on the Siberian-Arctic shelf: Implications for sediment provenance and grain size sorting. *Continental Shelf Research, 23*(11–23), 1175–1200.

Viscosi-Shirley, C., Pisias, N., & Mammone, K. (2003b). Sediment source strength, transport pathways and accumulation patterns on the Siberian Arctic's Chukchi and Laptev shelves. *Continental Shelf Research, 23*(11-23), 1201–1225.

Voelker, A. H. L. (1999). *Zur Deutung der Dansgaard-Oeschger Ereignisse in ultra-hochauflösenden Sedimentprofilen aus dem Europäischen Nordmeer.* Ph.D. thesis Thesis, Kiel University, 271pp.

Vogelsang, E. (1990). Paläo-Ozeanographie des Europäischen Nordmeeres an Hand stabiler Kohlenstoff-und Sauerstoffisotope. Ber. Sonderforschungsbereich 313, Universität Kiel, 23, 136pp.

Vogt, C. (1997). Regional and temporal variations of mineral assemblages in Arctic Ocean sediments as climatic indicator during glacial/interglacial changes. *Reps. Pol. Res., 251*, 309.

Vogt, C. (2004). Mineralogy of sediment core PS2185-6. PANGAEA, doi: 10.1594/PANGAEA.138270.

Vogt, C., Knies, J., Spielhagen, R. F., & Stein, R. (2001). Detailed mineralogical evidence for two nearly identical glacial/deglacial cycles and Atlantic water advection to the Arctic Ocean during the last 90,000 years. *Global and Planetary Change, 31/1-4*, 23–44.

Vogt, P. R., Crane, K., & Sundvor, E. (1994). Deep Pleistocene iceberg plowmarks on the Yermak Plateau: Sidescan and 3.5 kHz evidence for thick calving ice fronts and a possible marine ice sheet in the Arctic Ocean. *Geology, 22*, 403–406.

Vogt, P. R., Taylor, P. T., Kovacs, L. C., & Johnson, G. L. (1979). Detailed aeromagnetic investigation of the Arctic Basin. *Journal of Geophysical Research, 84*, 1071–1089.

Vogt, P. R., Taylor, P. T., Kovacs, L. C., & Johnson, G. L. (1982). The Canada basin; Aeromagnetic constraints on structure and evolution. *Tectonophysics, 89*, 295–336.

Volkman, J. K. (1986). A review of sterol markers for marine and terrigenous organic matter. *Organic Geochemistry, 9*, 83–99.

Volkman, J. K., Barrett, S. M., Blackburn, S. I., Mansour, M. P., Sikes, E. L., & Gelin, F. (1998). Microalgal biomarkers: A review of recent research developments. *Organic Geochemistry, 29*, 1163–1179.

Volkman, J. K., Barrett, S. M., Blackburrn, S. I., & Sikes, E. L. (1995). Alkenones in Gephyrocapsa oceanica: Implications for studies of paleoclimate. *Geochimica et Cosmochimica Acta, 59*(3), 513–520.

Volkman, J. K., Barrett, S. M., & Dunstan, G. A. (1994). C25 and C30 highly branched isoprenoid alkenes in laboratory cultures of two marine diatoms. *Organic Geochemistry, 21*, 407–414.

Volkman, J. K., Barrett, S. M., Dunstan, G. A., & Jeffrey, S. W. (1993). Geochemical significance of the occurrence of dinosterol and other 4-methyl sterols in a marine diatom. *Organic Geochemistry, 20*, 7–15.

Volkman, J. K., Eglinton, G., Corner, E. D. S., & Forsberg, T. E. V. (1980). Long-chain alkenes and alkenones in the marine coccolithophorid *Emiliania huxleyi*. *Phytochemistry, 19*, 2619–2622.

Volkmann, R. (2000a). Planktic foraminifer ecology and stable isotope geochemistry in the Arctic Ocean: Implications from water column and sediment surface studies for quantitative reconstructions of oceanic parameters. *Report on Polar Research, 361*, 100.

Volkmann, R. (2000b). Planktic foraminifers in the outer Laptev Sea and the Fram Strait — modern distribution and ecology. *Journal of Foraminiferal Research, 30*(3), 157–176.

Volkmann, R., & Mensch, M. (2001). Stable isotope composition ($\delta^{18}O$, $\delta^{13}C$) of living planktic foraminifers in the outer Laptev Sea and the Fram Strait. *Marine Micropaleontology, 42*, 163–188.

Vorobyeva, G. A. (1994). Paleoclimates around Lake Baikal in Pleistocene and the Holocene. In: Baikal as a nature laboratory for global change (Vol. 2, pp. 54–55). Irkutsk: Lisna Publishers.

Voronina, E., Polyak, L., Vernal, A. D., & Peyron, O. (2001). Holocene variations of sea-surface conditions in the southeastern Barents Sea, reconstructed from dinoflagellate cyst assemblages. *Journal of Quaternary Science, 16*(7), 717–726.

Vorren, T. O., Hald, M., & Lebesbye, E. (1988). Late Cenozoic environment in the Barents Sea. *Paleoceanography, 3*, 601–612.

Vorren, T. O., & Laberg, J. S. (1996). Late glacial air temparature, oceanographic and ice sheet interactions in the southern Barents Sea region. In: J. Andrews, W. E. N. Austin, H. Bergsten & A. E. Jennings (Eds), *Late Quaternary Paleoceanography of the North Atlantic Margins*. Geological Society of London (Special Publication 111), pp. 303–321.

Vorren, T. O., & Laberg, J. S. (1997). Trough mouth fans-palaeoclimate and ice-sheet monitors. *Quaternary Science Review, 16*, 865–881.

Vorren, T. O., Laberg, J. S., Blaume, F., Dowdeswell, J. A., Kenyon, N. H., Mienert, J., Rumohr, J., & Werner, F. (1998). The Norwegian-Greenland Sea continental margins: Morphology and late quaternary sedimentary processes and environment. *Quaternary Science Review, 17*, 273–302.

Vorren, T. O., Lebesbye, E., Andreassen, K., & Larsen, K.-B. (1989). Glacigenic debris flows on a passive continental margin as exemplified by the Barents Sea. *Marine Geology, 85*, 251–272.

Vorren, T. O., Richardsen, G., Knutsen, S. M., & Henriksen, E. (1991). Cenozoic erosion and sedimentation in the western Barents Sea. *Marine And Petroleum Geology, 8*, 317–340.

Voskresenskii, A. I., & Petrov, L. S. (1985). Climate peculiarities. In: *Arctic and Southern Oceans* (pp. 45–64). Leningrad: Nauka.

Vyssotski, A. V., Vyssotski, V. N., & Nezhdanov, A. A. (2006). Evolution of the West Siberian Basin. *Marine and Petroleum Geology, 23*, 93–126.

Wada, E., Minagawa, M., Mizutani, H., Tsuji, T., Imaizumi, R., & Karasawa, K. (1987). Biogeochemical studies on the transport of organic matter along the Otsuchi river watershed, Japan. *Estuarine, Coastal and Shelf Science, 25*, 321–336.

Waddell, L. A., & Moore, T. C. (2008). Salinity of the Eocnen Arctic Ocean from oxygen isotope analysis of fish bone carbonate. *Paleoceanography*, 23, PA1S12, doi:10.1029/2007PA001451.

Wadhams, P. (1997). Variability of Arctic sea ice thickness-statistical significance and its relationship to heat flux. In: J. H. Stel, H. W. A. Behrens, J. C. Borst, L. J. Droppert & J. v.d. Meulen (Eds), *Operational oceanography. The challenge for European co-operation* (pp. 368–384). Elsevier Science.

Waelbroeck, C., Duplessy, J.-C., Michel, E., Labeyrie, L., Paillard, D., & Duprat, J. (2001). The timing of the last deglaciation in North Atlantic climate records. *Nature, 412*, 724–726.

Waelbroeck, C., Labeyrie, L., Michel, E., Duplessy, J. C., McManus, J. F., Lambeck, K., Balbon, E., & Labracherie, M. (2002). Sea-level and deep water temperature changes derived from benthic foraminifera isotopic records. *Quaternary Science Review, 21*, 295–305.

Wagner, T. (1993). Organisches material in pelagischen Sedimenten: Glaziale-Interglaziale Variationen im Europaeischen Nordmeer. Ber. Sonderforschungsbereich 313, Universität Kiel, 42, 138pp.

Wagner, T. (1999). Petrology of sedimentary organic matter in modern and late quaternary deposits of the Equatorial Atlantic: Climatic and oceanographic links. *International Journal of Coal Geology, 39*(1-3), 155–184.

Wagner, T. (2002). Organic sedimentation in the Equatorial Atlantic: Evolution from Cretaceous to early quaternary depositional environments. *Palaeogeography, Palaeoclimatology, Palaeoecology, 179*, 113–147.

Wagner, T., & Dupont, L. M. (1999). Terrestrial organic matter in marine sediments: Analytical approaches and eolian-marine records in the central Equatorial Atlantic. In: G. Fischer & G. Wefer (Eds), *Use Proxies in Paleoceanography: Examples from the South Atlantic* (pp. 547–574). Heidelberg: Springer-Verlag.

Wagner, T., & Henrich, R. (1994). Organo-and lithofacies of glacial-interglacial deposits in the Norwegian Greenland sea: Responses to paleoceanographic and paleoclimatic changes. *Marine Geology, 120*, 335–364.

Wagner, T., & Hölemann, J. A. (1995). Deposition of organic matter in the Norwegian-Greenland sea during the last 2.7 million years. *Quaternary Research, 44*, 355–366.

Wagner, T., Kuhnt, W., & Sinninghe Damsté, J. S. (2003). Klimakapriolen der kreidezeit. *Spektrum der Wissenschaft, 12*, 48–56.

Wahsner, M., Müller, C., Stein, R., Ivanov, G., Levitan, M., Shelekova, E., & Tarasov, G. (1999). Clay mineral distributions in surface sediments from the Central Arctic Ocean and the Eurasian continental margin as indicator for source areas and transport pathways: A synthesis. *Boreas, 28*, 215–233.

Wahsner, M., & Shelekova, E. S. (1994). Clay mineral distribution in arctic deep sea and shelf surface sediments. *Greifswalder Geologische Beiträge A, 8*, 234 (abstract).

Wahsner, M., Tarasov, G., & Ivanov, G. (1996). Marine geological investigations of surface sediments in the Franz-Josef-Land area and the St. Anna Trough. In: R. Stein, G. Ivanov, M. Levitan & K. Fahl (Eds), *Surface-sediment Composition and Sedimentary Processes in the Central Arctic Ocean and along the Eurasian Continental Margin*. Reps. Pol. Res. 212, pp. 172–184.

Walsh, J. E. (1991). The arctic as a bellwether. *Nature, 352*, 19–20.

Walsh, J. J., & Dieterle, D. A. (1994). CO_2 cycling in the coastal ocean. I. A numerical analysis of the southeastern bering sea, with applications to the chukchi sea and the northern Gulf of Mexico. *Progress in Oceanography, 34*, 335–392.

Walter, H., & Breckle, S.-W. (1994). Spezielle Ökologie der gemäßigten und arktischen Zonen Euro-Nordasiens. Ökologie der Erde, Bd. 3. Gustav Fischer Verlag, Stuttgart, 726pp. (in German).

Warning, B., & Brumsack, H.-J. (2000). Trace metal signatures of mediterranean sapropels. *Palaeogeography, Palaeoclimatology, Palaeoecology, 158*, 293–309.

Warnke, D. A., & Hansen, M. E. (1977). Sediments of glacial origins in the area of DSDP Leg 38 (Norwegian-Greenland seas): Preliminary results from sites 336 and 344. *Naturforsch. Ges. Freib. Breisgau Ber., 67*, 371–392.

Washburn, A. L. (1980). *Geocryology: A survey of periglacial processes and environments*. New York, Seiten: Halsted Press.

Wassmann, P., Bauerfeind, E., Fortier, M., Fukuchi, M., Hargrave, B., Moran, B., Noji, T., Nöthig, E. M., Olli, K., Peinert, R., Sasaki, H., & Shevchenko, V. P. (2004). Particulate organic carbon fluy to the Arctic Ocean seafloor. In: R. Stein & R. W. Macdonald (Eds), *The organic carbon cycle in the Arctic Ocean* (pp. 101–138). Heidelberg: Springer Verlag.

Weaver, P. P. E., Chapman, M. R., Eglington, G., Zhao, M., Rutledge, D., & Read, G. (1999). Combined coccolith, foraminiferal, and biomarker reconstruction of paleoceanographic conditions over the past 120 kyr in the northern North Atlantic (59°N, 23°W). *Paleoceanography, 14*(3), 336–349.

Weaver, P. P. E., & Schultheiss, P. J. (1990). Current methods for obtaining, logging and splitting marine sediment cores. *Marine Geophysical Research, 12*, 85–100.

Weber, J. R., & Roots, E. F. (1990). Historical background; Exploration, concepts, and observations. In: A. Grantz, G. L. Johnson & J. F. Sweeney, (Eds), *The Arctic Ocean region* (pp. 5–36). The Geology of North America DNAG L. Boulder: Geological Society of America.

Weber, M. E., Niessen, F., Kuhn, G., & Wiedicke, M. (1997). Calibration and application of marine sedimentary physical properties using a multi-sensor core logger. *Marine Geology, 136*, 151–172.

Wedepohl, K. H. (1971). Environmental influences on the chemical composition of shales and clays. In: L. H. Ahrens, F. Press, S. K. Runcorn & H. C. Urey (Eds), *Physics and chemistry of the earth* (pp. 305–333). New York: Pergamon.

Wedepohl, K. H. (1991). The composition of the upper Earth's crust and the natural cycles of selected metals. Metals in natural raw materials. Natural resources. In: E. Merian (Ed.), *Metals and their compounds in the environment* (pp. 3–17). Weinheim: VCH.

Wehausen, R., & Brumsack, H. -J. (1998). The formation of Pliocene Mediterranean sapropels: Constraints from high-resolution major and minor element studies. In: A. H. F. Robertson, K.-C.

Emeis, C. Richter & A. Camerlenghi (Eds), *Proceedings of ODP Science Results* (Vol. 160, pp. 207–217). College Station, Texas.

Weiel, D. (1997). *Paläozeanographische Untersuchungen in der Vilkitsky Straße und östlich von Severnaya Zemlya mit sedimentologischen und geophysikalischen Methoden.* Unpubl. diploma thesis, University of Köln, 138pp.

Weigelt, E. (1998). Die Krustenstruktur und Sedimentdecke des Eurasischen Beckens, Arktischer Ozean: Resultate aus seismischen und gravimetrischen Untersuchungen. *Report on Polar Research, 261.*

Weijers, J. W. H., Schouten, S., Spaargaren, O. C., & Sinninghe Damsté, J. S. (2006). Occurrence and distribution of tetraether membrane lipids in soils: Implications for the use of the TEX86 proxy and the BIT index. *Organic Geochemistry, 37*, 1680–1693.

Weinelt, M., Kuhnt, W., Sarnthein, M., Altenbach, A., Costello, O., Erlenkeuser, H., Pflaumann, U., Simstich, J., Struck, U., Thies, A., Trauth, M. H., & Vogelsang, E. (2001). Paleoceanographic proxies in the North Atlantic. In: P. Schäfer, W. Ritzrau, M. Schlüter & J. Thiede (Eds), *The northern North Atlantic; a changing environment* (pp. 319–352). Heidelberg: Springer-Verlag.

Weis, D., Demaie, D., Souchez, R., Gow, A. J., & Meese, D. A. (1997). Ice sheet development in Central Greenland: Implications from the Nd, Sr, Pb isotopic compositions of basal material. *Earth and Planetary Science Letters, 150*, 161–169.

Weissert, H., & Lini, A. (1991). Ice age interludes during the time of Cretaceous greenhouse climate? In: D. W. Mueller, J. A. Mckenzie & H. Weissert (Eds), *Controversies in Modern Geology* (pp. 173–191). London: Academic Press.

Weller, P., & Stein, R. (2008). Paleogene biomarker records from the central Arctic Ocean (IODP Expedition 302): Organic-carbon sources, anoxia, and sea-surface temperature. *Paleoceanography, 23*, PA1S17, doi: 10.1029/2007PA001472.

Wheeler, P.A. (Ed). (1997). 1994 Arctic Ocean section. *Deep-Sea Res II, 44*(8), 1483–1757.

Wilkin, R. T., Arthur, M. A., & Dean, W. E. (1997). History of water-column anoxia in the black sea indicated by pyrite framboid size distributions. *Earth and Planetary Science Letters, 148*, 517–525.

Wilkin, R. T., Barnes, H. L., & Brantley, S. L. (1996). The size distribution of framboidal pyrite in modern sediments: An indicator of redox conditions. *Geochimica et Cosmochimica Acta, 60*, 3897–3912.

Wilken, M., & Mienert, J. (2006). Submarine glacigenic debris flows, deep-sea channels and past ice-stream behaviour of the East Greenland continental margin. *Quaternary Science Review, 25*, 784–810.

Wille, P. C. (2005). *Sound images of the ocean in research and monitoring.* Heidelberg: Springer-Verlag, 452pp.

Wilson, P. A., Norris, R. D., & Cooper, M. J. (2002). Testing the Cretaceous greenhouse hypothesis using glassy foraminiferal calcite from the core of the turonian tropics on demerara rise. *Geology, 30*, 607–610.

Wilson, P. A., & Opdyke, B. N. (1996). Equatorial sea-surface temperatures for the maastrichtian revealed through remarkable preservation of metastable carbonate. *Geology, 24*, 555–558.

Winkelmann, D. (2007). *Sediment dynamics of megaslides along the Svalbard continental magin and the relation to paleoenvironmental changes and climate history.* Unpublished Ph.D. thesis, Bremen University, 180pp.

Winkelmann, D., Geissler, W., Schneider, J., Stein, R., & Schenke, H.-W. (2008a). Dynamics and timing of the Hinlopen/Yermak Megaslide north of Spitsbergen, Arctic Ocean. *Marine Geology, 250*, 34–50.

Winkelmann, D., Jokat, W., Niessen, F., Stein, R., & Winkler, A. (2005). The Yermak slide - New constraints on extent and age. *22nd International Polar Meeting*, German Society of Polar Research, 18–24 September 2005, Jena, Germany, Abstract.

Winkelmann, D., Jokat, W., Niessen, F., Stein, R., & Winkler, A. (2006). Age and extent of the yermak slide north of Spitsbergen. *Geochemistry Geophysics Geosystems*, 7, Q06007. doi: 10.1029/2005GC001130.

Winkelmann, D., & Knies, J. (2005). Recent distribution and accumulation of organic carbon on the continental margin west off Spitsbergen. *Geochemistry Geophysics Geosystems, 6*(9), doi: 10.1029/2005GC000916.

Winkelmann, D., & Stein, R. (2007). Triggering of the Hinlopen/Yermak megaslide in relation to paleoceanography and climate history of the continental margin north of Spitsbergen. *Geochemistry Geophysics Geosystems, 1*, Q06018. doi: 10.1029/2006GC001485.

Winkelmann, D., Stein, R., Schäfer, C., & Mackensen, A. (2008b). Terrigenous events and climate history within the Sophia Basin, Arctic Ocean. *Geochemistry Geophysics Geosystems*, in press.

Winkler, A., Wolf-Welling, T. C. W., Stattegger, K., & Thiede, J. (2002). Clay mineral sedimentation in high northern latitude deep-sea basins since the middle miocene (ODP Leg 151, NAAG). *International Journal of Earth Science, 91*(1), 133–148.

Winter, B. L., Johnson, C. L., & Clark, D. L. (1997). Strontium, neodymium and lead isotope variations of authigenic and silicate sediment components from the Late Cenozoic Arctic Ocean: Implications for sediment provenance and source of trace metals in sea water. *Geochimica et Cosmochimica Acta, 61*, 4181–4200.

Wise, S. W., Breza, J. R., Harwood, D. M., & Wei, W. (1991). Palaeogene glacial history of Antarctica. In: D. W. Müller, J. A. McKenzie & H. Weissert (Eds), *Controversies in modern geology: Evolution of geological theories in sedimentology, earth history and tectonics* (pp. 133–171). London: Academic Press.

Wolf, T. C. W., & Thiede, J. (1991). History of terrigenous sedimentation during the past 10 m.y. in the North Atlantic (ODP Legs 104, 105, and DSDP 81). *Marine Geology, 101*, 83–102.

Wolfe, J. A. (1994). Tertiary climatic changes at middle latitudes of western North America. *Palaeogeography, Palaeoclimatology, Palaeoecology, 108*, 195–205.

Wolf-Welling, T. C. W., Cremer, M., O'Connell, S., Winkler, A., & Thiede, J. (1996). Cenozoic Arctic gateway paleoclimate variability: Indicatiosn from changes in coarse-fraction compositions (ODP Leg 151). In: J. Thiede, A. M. Myhre, J. Firth, G. L. Johnson, & W. F. Ruddiman (Eds), *Proceedings of ODP, Science Results 151* (pp. 515–525). College Station, Texas (Ocean Drilling Program).

Wollenburg, I. (1993). Sedimenttransport durch das arktische meereis: Die rezente lithogene und biogene materialfracht. *Report on Polar Research, 127*, 159.

Wollenburg, J. (1995). Benthische foraminiferenfaunen als wassermassen-, produktions- und Eisdriftanzeiger im Arktischen Ozean. *Report on Polar Research, 179*, 227.

Wollenburg, J. E., Knies, J., & Mackensen, A. (2004). High-resolution paleoproductivity fluctuations during the past 24 kyr as indicated by benthic foraminifera in the marginal Arctic Ocean. *Palaeogeography, Palaeoclimatology, Palaeoecology, 204*, 209–238.

Wollenburg, J. E., & Kuhnt, W. (2000). The response of benthic foraminifers to carbon flux and primary production in the Arctic Ocean. *Marine Micropaleontology, 40*, 189–231.

Wollenburg, J. E., Kuhnt, W., & Mackensen, A. (2001). Changes in Arctic Ocean palaeoproductivity and hydrography during the last 145 kyr: The benthic foraminiferal record. *Paleoceanography, 16*, 65–77.

Wollenburg, J. E., & Mackensen, A. (1998). Living benthic foraminifers from the central Arctic Ocean: Faunal composition, standing stock and diversity. *Marine Micropaleontology, 34*, 153–185.

Woodgate, R., Schauer, U., & Fahrbach, E. (1998). Moored current meters in Fram Strait at 79 °N: Preliminary results with special emphasis on the West Spitsbergen current. AWI Berichte aus dem Fachbereich Physik, 28pp.

Woodgate, R. A., Fahrbach, E., & Rohard, G. (1999). Structure and transports of the East Greenland Current at 75°N from moored current meters. *Journal of Geophysical Research, 104*, 18059–18072.

Woodworth-Lynas, C. M. T., Simms, A., & Rendell, C. M. (1985). Iceberg grounding and scouring on the labrador continental shelf. *Cold Regions Science and Technology, 10*, 163–186.

Worsley, D., Johansen, R., & Kristensen, S. E. (1988). The Mesozoic and Cenozoic succession of Tromsøflaket. In: A. Dalland, D. Worsley & K. Ofstad (Eds), *A lithostratigraphic scheme for the Mesozoic and Cenozoic succession offshore mid-and northern Norway. Norwegian Petrol* (Vol. 4, pp. 42–65). Directorate Bulletin.

Wraige, E. J., Belt, S. T., Lewis, C. A., Cooke, D. A., Robert, J.-M., Massé, G., & Rowland, S. J. (1997). Variations in structures and distributions of C_{25} highly branched isoprenoid (HBI) alkenes in cultures of the diatom, *Haslea ostrearia* (Simonsen). *Organic Geochemistry, 27*, 497–505.

Wright, J. D., & Miller, K. G. (1996). Control of North Atlantic deep water circulation by the Greenland-Scotland Ridge. *Paleoceanography, 11*, 157–170.

Wright, J. D., Miller, K. G., & Fairbanks, R. G. (1992). Early and middle miocene stable isotopes: Implications for deep water circulation and climate. *Paleoceanography, 7*, 357–389.

Wuchter, C., Schouten, S., Coolen, M. J. L., & Sinninghe Damste, J. S. (2004). Temperature dependent variation in the distribution of tetraether membrane lipids of marine Crenarchaeota: Implications for TEX86 paleothermometry. *Paleoceanography, 19*. doi: 10.1029/2004PA001041, PA4028.

Wuchter, C., Schouten, S., Wakeham, S. G., & Sinninghe Damsté, J. S. (2005). Temporal and spatial variation in tetraether membrane lipids of marine crenarchaeota in particulate organic matter. Implications for TEX86 paleothermometry. *Paleoceanography, 20*, PA3013. doi: 10.1029/2004PA001110.

Wüst, G., & Brögmus, W. (1955). Ozeanographische Ergebnisse einer Untersuchungsfahrt mit Forschungskutter "Südfall" durch die Ostsee Juni/Juli 1954 (anlässlich der totalen Sonnenfinsternis auf Öland). *Kieler Meeresf., 11*, 3–21.

Yamamoto, M., Ficken, K., Baas, M., Bosch, H.-J., & de Leuw, J. W. (1996). Molecular paleontology of the earliest danian at geulhemmerberg (the Nertherlands). *Geol. Mijnbouw, 75*, 255–267.

Yokoyama, Y., Lambeck, K., de Dekhar, P., Johnston, P., & Fifield, L. K. (2000). Timing of the last glacial maximum from observed sealevel minima. *Nature, 406*, 713–716 Correction 2001. Nature 412, 99.

Yunker, M. B., Belicka, L. L., Harvey, H. R., & Macdonald, R. W. (2005). Tracing the inputs and fate of marine and terrigenous organic matter in Arctic Ocean sediments: A multivariate analysis of lipid biomarkers. *Deep-sea Research Part II (Topical Studies in Oceanography), 52*, 3478–3508.

Yunker, M. B., & Macdonald, R. W. (1995). Composition and origins of polycyclic aromatic hydrocarbons in the Mackenzie River and on the Beaufort Sea shelf. *Arctic, 48*, 118–129.

Yunker, M. B., Macdonald, R. W., Cretney, W. J., Fowler, B. R., & McLaughlin, F. A. (1993). Alkane, terpene, and polycyclic aromatic hydrocarbon geochemistry of the Mackenzie River and Mackenzie Shelf: Riverine contributions to Beaufort Sea coastal sediment. *Geochimica et Cosmochimica Acta, 57*, 3041–3061.

Yunker, M. B., Macdonald, R. W., Fowler, B. R., Cretney, W. J., Dallimore, S. R., & McLaughlin, F. A. (1991). Geochemistry and fluxes of hydrocarbons to the Beaufort Sea shelf: A multivariate comparison of fluvial inputs and coastal erosion of peat using principal components analysis. *Geochimica et Cosmochimica Acta, 55*, 255–273.

Yunker, M. B., Macdonald, R. W., Veltkamp, D. J., & Cretney, W. J. (1995). Terrestrial and marine biomarkers in a seasonally ice-covered arctic estuary -integration of multivariate and biomarker approaches. *Marine Chemistry, 49*, 1–50.

Zachos, J. C., Dickens, G. R., & Zeebe, R. E. (2008). An early Cenozoic perspective on greenhouse warming and carbon-cycle dynamics. *Nature, 451*, 279–283.

Zachos, J., Pagani, M., Sloan, L., Thomas, E., & Billups, K. (2001). Trends, rhythms, and aberrations in global climate 65 Ma to present. *Science, 292*, 868–693.

Zachos, J. C., et al. (2003). A transient rise in tropical sea surface temperature during the Paleocene-Eocene thermal maximum. *Science, 302*, 1151–1154.

Zachos, J. C., Breza, J., & Wise, S. W. (1992). Earliest oligocene icesheet expansion on East Antarctica: Stable isotope and sedimentological data from Kerguelen Plateau. *Geology, 20*, 569–573.

Zachos, J. C., Stott, L. D., & Lohmann, K. C. (1994). Evolution of early cenozoic temperatures. *Paleoceanography, 9*, 353–387.

Zahn, R., Markussen, B., & Thiede, J. (1985). Stable isotope data and depositional environments in the Late Quaternary Arctic Ocean. *Nature, 314*, 433–435.

Zaripov, O. G. (2007). Heavy mineral composition of 0.100–0.050 mm size fraction fraction of sediments from Hole Pomorsk-1 drilled in the Barents Sea. All-Russian Research Institute for Geology and Mineral Resources of the World Ocean, PANGAEA, doi: 10.1594/PANGAEA.619104.

Zauderer, K. (1982). *Analysis of heavy minerals in Arenaceous Lutites from the Northern Canada Basin, Arctic Ocean*. Unpublished Master thesis, Old Dominion University, Norfolk, 83pp.

Zdanowicz, C. M., Zielinski, G. A., & Wake, C. P. (1998). Characteristics of modern atmospheric dust deposition in snow on the Penny Ice Cap, Baffin Island, Arctic Canada. *Tellus, 50,* 506–520.

Zhang, T., Barry, K., Knowles, K., Heginbottom, J. A., & Brown, J. (1999). Statistics and characteristics of permafrost and ground-ice distribution in the Northern Hemisphere. *Polar Geography, 23,* 132–154.

Zhang, T., Barry, R. G., Knowles, K., Ling, F., & Armstrong, R. L. (2003). Distribution of seasonally and perennially frozen ground in the Northern Hemisphere. In: M. Phillips, S. M. Springman & L. U. Arenson (Eds), *Proceedind of the Eighth International Conference on Permafrost* (pp. 1289–1294). Lisse: A.A. Balkema.

Ziegler, P. A. (Ed). (1988). Evolution of the Arctic-North Atlantic and the Western Tethys. *American Association of Petrology Geology, AAPG Mem.* 43, 198pp.

Zonenshain, L. P., & Natapov, L. M. (1989). Tectonic history of the Arctic region from the ordovician through the Cretaceous. In: Y. Herman (Ed.), *The Arctic Seas: Climatology, oceanography, geology and biology* (pp. 829–862). New York: van Nostrand Reinhold Company.

Zweck, C., & Huybrechts, P. (2003). Modeling the marine extent of northern hemisphere ice sheets during the last glacial cycle. *Annual Glaciology, 37,* 173–180.

Zweck, C., & Huybrechts, P. (2005). Modeling of the northern hemisphere ice sheets during the last glacial cycle and glaciological sensitivity. *Journal of Geophysical Research, 110,* D07103. doi: 10.1029/2004JD005489.

Subject Index

MIX
Papier | Fördert gute Waldnutzung
FSC® C083411

Kolibri 360 GmbH
Ferdinand-Jühlke-Straße 7
99095 Erfurt, Deutschland
produktsicherheit@kolibri360.de